IET ENERGY ENGINEERING SERIES 241

Power Electronic Converters and Systems, 2nd Edition

Other volumes in this series:

Volume 1 **Power Circuit Breaker Theory and Design** C.H. Flurscheim (Editor)
Volume 4 **Industrial Microwave Heating** A.C. Metaxas and R.J. Meredith
Volume 7 **Insulators for High Voltages** J.S.T. Looms
Volume 8 **Variable Frequency AC Motor Drive Systems** D. Finney
Volume 10 **SF_6 Switchgear** H.M. Ryan and G.R. Jones
Volume 11 **Conduction and Induction Heating** E.J. Davies
Volume 13 **Statistical Techniques for High Voltage Engineering** W. Hauschild and W. Mosch
Volume 14 **Uninterruptible Power Supplies** J. Platts and J.D. St Aubyn (Editors)
Volume 15 **Digital Protection for Power Systems** A.T. Johns and S.K. Salman
Volume 16 **Electricity Economics and Planning** T.W. Berrie
Volume 18 **Vacuum Switchgear** A. Greenwood
Volume 19 **Electrical Safety: A guide to causes and prevention of hazards** J. Maxwell Adams
Volume 21 **Electricity Distribution Network Design, 2nd Edition** E. Lakervi and E.J. Holmes
Volume 22 **Artificial Intelligence Techniques in Power Systems** K. Warwick, A.O. Ekwue and R. Aggarwal (Editors)
Volume 24 **Power System Commissioning and Maintenance Practice** K. Harker
Volume 25 **Engineers' Handbook of Industrial Microwave Heating** R.J. Meredith
Volume 26 **Small Electric Motors** H. Moczala *et al.*
Volume 27 **AC–DC Power System Analysis** J. Arrillaga and B.C. Smith
Volume 29 **High Voltage Direct Current Transmission, 2nd Edition** J. Arrillaga
Volume 30 **Flexible AC Transmission Systems (FACTS)** Y.-H. Song (Editor)
Volume 31 **Embedded Generation** N. Jenkins *et al.*
Volume 32 **High Voltage Engineering and Testing, 2nd Edition** H.M. Ryan (Editor)
Volume 33 **Overvoltage Protection of Low-Voltage Systems, Revised Edition** P. Hasse
Volume 36 **Voltage Quality in Electrical Power Systems** J. Schlabbach *et al.*
Volume 37 **Electrical Steels for Rotating Machines** P. Beckley
Volume 38 **The Electric Car: Development and future of battery, hybrid and fuel-cell cars** M. Westbrook
Volume 39 **Power Systems Electromagnetic Transients Simulation** J. Arrillaga and N. Watson
Volume 40 **Advances in High Voltage Engineering** M. Haddad and D. Warne
Volume 41 **Electrical Operation of Electrostatic Precipitators** K. Parker
Volume 43 **Thermal Power Plant Simulation and Control** D. Flynn
Volume 44 **Economic Evaluation of Projects in the Electricity Supply Industry** H. Khatib
Volume 45 **Propulsion Systems for Hybrid Vehicles** J. Miller
Volume 46 **Distribution Switchgear** S. Stewart
Volume 47 **Protection of Electricity Distribution Networks, 2nd Edition** J. Gers and E. Holmes
Volume 48 **Wood Pole Overhead Lines** B. Wareing
Volume 49 **Electric Fuses, 3rd Edition** A. Wright and G. Newbery
Volume 50 **Wind Power Integration: Connection and system operational aspects** B. Fox *et al.*
Volume 51 **Short Circuit Currents** J. Schlabbach
Volume 52 **Nuclear Power** J. Wood
Volume 53 **Condition Assessment of High Voltage Insulation in Power System Equipment** R.E. James and Q. Su
Volume 55 **Local Energy: Distributed generation of heat and power** J. Wood
Volume 56 **Condition Monitoring of Rotating Electrical Machines** P. Tavner, L. Ran, J. Penman and H. Sedding
Volume 57 **The Control Techniques Drives and Controls Handbook, 2nd Edition** B. Drury
Volume 58 **Lightning Protection** V. Cooray (Editor)
Volume 59 **Ultracapacitor Applications** J.M. Miller

Volume 62	**Lightning Electromagnetics** V. Cooray
Volume 63	**Energy Storage for Power Systems, 2nd Edition** A. Ter-Gazarian
Volume 65	**Protection of Electricity Distribution Networks, 3rd Edition** J. Gers
Volume 66	**High Voltage Engineering Testing, 3rd Edition** H. Ryan (Editor)
Volume 67	**Multicore Simulation of Power System Transients** F.M. Uriate
Volume 68	**Distribution System Analysis and Automation** J. Gers
Volume 69	**The Lightening Flash, 2nd Edition** V. Cooray (Editor)
Volume 70	**Economic Evaluation of Projects in the Electricity Supply Industry, 3rd Edition** H. Khatib
Volume 72	**Control Circuits in Power Electronics: Practical issues in design and implementation** M. Castilla (Editor)
Volume 73	**Wide Area Monitoring, Protection and Control Systems: The enabler for Smarter Grids** A. Vaccaro and A. Zobaa (Editors)
Volume 74	**Power Electronic Converters and Systems: Frontiers and applications** A.M. Trzynadlowski (Editor)
Volume 75	**Power Distribution Automation** B. Das (Editor)
Volume 76	**Power System Stability: Modelling, analysis and control** A.A. Sallam and B. Om P. Malik
Volume 78	**Numerical Analysis of Power System Transients and Dynamics** A. Ametani (Editor)
Volume 79	**Vehicle-to-Grid: Linking electric vehicles to the smart grid** J. Lu and J. Hossain (Editors)
Volume 81	**Cyber-Physical-Social Systems and Constructs in Electric Power Engineering** S. Suryanarayanan, R. Roche and T.M. Hansen (Editors)
Volume 82	**Periodic Control of Power Electronic Converters** F. Blaabjerg, K.Zhou, D. Wang and Y. Yang
Volume 86	**Advances in Power System Modelling, Control and Stability Analysis** F. Milano (Editor)
Volume 87	**Cogeneration: Technologies, optimisation and implementation** C.A. Frangopoulos (Editor)
Volume 88	**Smarter Energy: From smart metering to the smart grid** H. Sun, N. Hatziargyriou, H.V. Poor, L. Carpanini and M.A. Sánchez Fornié (Editors)
Volume 89	**Hydrogen Production, Separation and Purification for Energy** A. Basile, F. Dalena, J. Tong and T.N.Veziroğlu (Editors)
Volume 90	**Clean Energy Microgrids** S. Obara and J. Morel (Editors)
Volume 91	**Fuzzy Logic Control in Energy Systems With Design Applications in MATLAB®/Simulink®** İ.H. Altaş
Volume 92	**Power Quality in Future Electrical Power Systems** A.F. Zobaa and S.H.E.A. Aleem (Editors)
Volume 93	**Cogeneration and District Energy Systems: Modelling, analysis and optimization** M.A. Rosen and S. Koohi-Fayegh
Volume 94	**Introduction to the Smart Grid: Concepts, technologies and evolution** S.K. Salman
Volume 95	**Communication, Control and Security Challenges for the Smart Grid** S.M. Muyeen and S. Rahman (Editors)
Volume 96	**Industrial Power Systems with Distributed and Embedded Generation** R Belu
Volume 97	**Synchronized Phasor Measurements for Smart Grids** M.J.B. Reddy and D.K. Mohanta (Editors)
Volume 98	**Large Scale Grid Integration of Renewable Energy Sources** A. Moreno-Munoz (Editor)
Volume 100	**Modeling and Dynamic Behaviour of Hydropower Plants** N. Kishor and J. Fraile-Ardanuy (Editors)
Volume 101	**Methane and Hydrogen for Energy Storage** R. Carriveau and D.S.-K. Ting
Volume 104	**Power Transformer Condition Monitoring and Diagnosis** A. Abu-Siada (Editor)
Volume 106	**Surface Passivation of Industrial Crystalline Silicon Solar Cells** J. John (Editor)

Volume 151 **SiC Power Module Design: Performance, robustness and reliability** A. Castellazzi and A. Irace (Editors)
Volume 152 **Power Electronic Devices: Applications, failure mechanisms and reliability** F Iannuzzo (Editor)
Volume 153 **Signal Processing for Fault Detection and Diagnosis in Electric Machines and Systems** M. Benbouzid (Editor)
Volume 155 **Energy Generation and Efficiency Technologies for Green Residential Buildings** D. Ting and R. Carriveau (Editors)
Volume 156 **Lithium-ion Batteries Enabled by Silicon Anodes** C. Ban and K. Xu (Editors)
Volume 157 **Electrical Steels, 2 Volumes** A. Moses, K. Jenkins, Philip Anderson and H. Stanbury
Volume 158 **Advanced Dielectric Materials for Electrostatic Capacitors** Q. Li (Editor)
Volume 159 **Transforming the Grid Towards Fully Renewable Energy** O. Probst, S. Castellanos and R. Palacios (Editors)
Volume 160 **Microgrids for Rural Areas: Research and case studies** R.K. Chauhan, K. Chauhan and S.N. Singh (Editors)
Volume 161 **Artificial Intelligence for Smarter Power Systems: Fuzzy logic and neural networks** M.G. Simoes
Volume 165 **Digital Protection for Power Systems, 2nd Edition** Salman K. Salman
Volume 166 **Advanced Characterization of Thin Film Solar Cells** N. Haegel and M. Al-Jassim (Editors)
Volume 167 **Power Grids With Renewable Energy: Storage, integration and digitalization** A.A. Sallam and B. Om P. Malik
Volume 169 **Small Wind and Hydrokinetic Turbines** P. Clausen, J. Whale and D. Wood (Editors)
Volume 170 **Reliability of Power Electronics Converters for Solar Photovoltaic Applications** F. Blaabjerg, A.l Haque, H. Wang, Z. Abdin Jaffery and Y. Yang (Editors)
Volume 171 **Utility-scale Wind Turbines and Wind Farms** A. Vasel-Be-Hagh and D.S.-K. Ting
Volume 172 **Lightning Interaction With Power Systems, 2 Volumes** A. Piantini (Editor)
Volume 174 **Silicon Solar Cell Metallization and Module Technology** T. Dullweber (Editor)
Volume 175 **n-Type Crystalline Silicon Photovoltaics: Technology, applications and economics** D. Munoz and R. Kopecek (Editors)
Volume 178 **Integrated Motor Drives** Xu Deng and B. Mecrow (Editors)
Volume 180 **Protection of Electricity Distribution Networks, 4th Edition** J. Gers and E. Holmes
Volume 182 **Surge Protection for Low Voltage Systems** A. Rousseau (Editor)
Volume 184 **Compressed Air Energy Storage: Types, systems and applications** D. Ting and J. Stagner
Volume 186 **Synchronous Reluctance Machines: Analysis, optimization and applications** N. Bianchi, C. Babetto and G. Bacco
Volume 190 **Synchrophasor Technology: Real-time operation of power networks** N. Kishor and S.R. Mohanty (Editors)
Volume 191 **Electric Fuses: Fundamentals and new applications, 4th Edition** N. Nurse, A. Wright and P.G. Newbery
Volume 193 **Overhead Electric Power Lines: Theory and practice** S. Chattopadhyay and A. Das
Volume 194 **Offshore Wind Power: Reliability, availability and maintenance, 2nd Edition** P. Tavner
Volume 196 **Cyber Security for Microgrids** S. Sahoo, F. Blaabjerg and T. Dragicevic
Volume 198 **Battery Management Systems and Inductive Balancing** A. Van den Bossche and A. Farzan Moghaddam
Volume 199 **Model Predictive Control for Microgrids: From power electronic converters to energy management** J. Hu, J.M. Guerrero and S. Islam
Volume 201 **Design, Control and Monitoring of Tidal Stream Turbine Systems** M. Benbouzid (Editor)

Volume 204 **Electromagnetic Transients in Large HV Cable Networks: Modeling and calculations** Ametani, Xue, Ohno and Khalilnezhad
Volume 208 **Nanogrids and Picogrids and Their Integration with Electric Vehicles** S. Chattopadhyay
Volume 210 **Superconducting Magnetic Energy Storage in Power Grids** M.H. Ali
Volume 211 **Blockchain Technology for Smart Grids: Implementation, management and security** H.L. Gururaj, K.V. Ravi, F. Flammini, H. Lin, B. Goutham, K.B.R. Sunil and C. Sivapragash
Volume 212 **Battery State Estimation: Methods and models** S. Wang
Volume 215 **Industrial Demand Response: Methods, best practices, case studies, and applications** H.H. Alhelou, A. Moreno-Muñoz and P. Siano (Editors)
Volume 213 **Wide Area Monitoring of Interconnected Power Systems, 2nd Edition** A.R. Messina
Volume 214 **Distributed Energy Storage in Urban Smart Grids** P.F. Ribeiro and R.S. Salles (Editors)
Volume 217 **Advances in Power System Modelling, Control and Stability Analysis, 2nd Edition** F. Milano (Editor)
Volume 225 **Fusion–Fission Hybrid Nuclear Reactors: For enhanced nuclear fuel utilization and radioactive waste reduction** W.M. Stacey
Volume 227 **Processing and Manufacturing of Electrodes for Lithium-Ion Batteries** J. Li and C. Jin (Editors)
Volume 228 **Digital Technologies for Solar Photovoltaic Systems: From general to rural and remote installations** S. Motahhir (Editor)
Volume 235 **Clean Energy Past and Future** P. Tavner
Volume 236 **Green HV Switching Technologies for Modern Power Networks** K. Niayesh (Editor)
Volume 237 **High Voltage Power Transformers: State of the art and technological innovations** M. Sforna, S. Zunino and F. Ferrari
Volume 238 **AI for Status Monitoring of Utility Scale Batteries** S. Wang, K. Liu, Y. Wang, D. Stroe, C. Fernandez and J.M. Guerrero
Volume 239 **Intelligent Control of Medium and High Power Converters** M. Bendaoud, Y. Maleh and S. Padmanaban (Editors)
Volume 247 **Power System Strength Evaluation: Methods, best practice, case studies, and applications** H. Alhelou, N. Hosseinzadeh and B. Bahrani (Editors)
Volume 250 **Interactions of Wind Turbines With Aviation Radio and Radar Systems** Alan Collinson
Volume 256 **Wireless Power Transfer Technologies: Theory and technologies, 2nd Edition** N. Shinohara (Editor)
Volume 905 **Power System Protection, 4 Volumes**

Power Electronic Converters and Systems, 2nd Edition

Volume 1: Converters and machine drives

Edited by
Marcelo Godoy Simões and Tiago Davi Curi Busarello

The Institution of Engineering and Technology

Published by The Institution of Engineering and Technology, London, United Kingdom

The Institution of Engineering and Technology is registered as a Charity in England & Wales (no. 211014) and Scotland (no. SC038698).

First published 2015, second edition 2024

The Institution of Engineering and Technology
Futures Place
Kings Way, Stevenage
Hertfordshire, SG1 2UA, United Kingdom

www.theiet.org

British Library Cataloguing in Publication Data
A catalogue record for this product is available from the British Library

ISBN 978-1-83953-767-7 Vol 1 (hardback)

ISBN 978-1-83953-768-4 Vol 1 (PDF)

ISBN 978-1-83953-769-1 Vol 2 (hardback)

ISBN 978-1-83953-770-7 Vol 2 (PDF)

Typeset in India by MPS Limited

Mastery is breaking down theory and experiments, then building up what makes a job well done, why we have such a journey, from novice to virtuoso, is our companion, our path is such a joy who will share with all - then life, there is a realization of why we moved along the path of learning, to a possible understanding of why we did it; till Who we motivated to follow our ideas or dreams, befits to master What, Why, and When to continue what we started – Thank You To All: family, friends, students, colleagues, professionals, acquaintances, anyone with whom I learned something, or someone who I taught something. This book is a dedication to so many in my whole life.

— Marcelo Godoy Simões, Spring of 2024

I dedicate this book to my entire family and to the professionals who inspire me to be an increasingly better professor and researcher.

— Tiago Davi Curi Busarello.

Contents

About the editors

Marcelo Godoy Simões is a professor in Flexible and Smart Power Systems at the University of Vaasa, Finland. Prior positions include 11 years at the University of São Paulo (Brazil), 21 years at the Colorado School of Mines (USA), visiting professorships at Georgia Tech, L'École Normale Supérieure de Cachan and Université de Technologie Belfort-Montbéliard in France, Petroleum Institute in Abu Dhabi (UAE), and Fulbright Scholar at Aalborg University. He developed several AI-based power electronics for renewable energy and smart-grid integration, with several advanced applications, such as wind energy conversion, and early automation-control-based models of fuel cells. His current research interests include smart grid-based inverters for renewable energy, power electronics and power systems, power quality, renewable energy, and artificial intelligence. He is a Fellow of the IEEE with the citation "for applications of artificial intelligence in control of power electronics systems."

Tiago Davi Curi Busarello has been a professor since 2016 at the Federal University of Santa Catarina (UFSC)—Campus Blumenau. He holds a PhD and a Master's degree in Electrical Engineering from the State University of Campinas (UNICAMP), Brazil, obtaining the degrees in 2015 and 2013, respectively. He is an electrical engineer (2010) from the State University of Santa Catarina (UDESC). He worked as a postdoctoral researcher as well as a visiting professor from 2022 to 2023 at the University of Vaasa, Finland. He was a visiting researcher in 2014 at the Colorado School of Mines, United States. He is a senior member of the IEEE (Institute of Electrical and Electronics Engineers) and a member of the IEEE Power Electronics Society. He serves as a reviewer for IEEE journal articles. He is a member of GEPPEL-Group of Power Electronics and Electrical Energy Processing at UFSC Blumenau. His areas of interest encompass digital control for power electronics, digital twins, and smart grids.

First edition edited by Andrzej M. Trzynadlowski.

Andrzej M. Trzynadlowski is a professor emeritus of electrical engineering, at the University of Nevada, Reno, USA.

Foreword

Power Electronic Converters and Systems second edition is an extended and comprehensive edition of the book edited by Dr. Marcelo Godoy Simões and Tiago Davi Curi Busarello. For this edition, the Editors invited several well-known researchers, academics, and electrical engineering professionals to include new developments in the area of power electronics and applications. There are two volumes, the first one on the operation of power devices, power converters, switching strategies, and so on and the second volume on the power electronics applications.

The book reads very pleasantly, and it has excellent scope and breadth to support many courses at undergraduate and graduate levels by proper selection of the chapters. For example, based on the materials presented in this edition, a computer-based design course on power electronics and alternative energy could be introduced, another possibility could be a course on grid integration of renewable energy sources. The book has applicable content for both senior undergraduate and graduate course levels. Students and practicing engineers with basic knowledge of transient analysis of electric circuits and energy conversion can also benefit from self-study of the book.

The chapters are kept independent, but cross-references are provided whenever possible to align the chapters. At the beginning of each chapter, there is a brief introduction giving an idea about the contents of the chapter. Each chapter has state-of-the-art and novel ideas with comprehensive analysis and developments that make each presented chapter a self-study for independent learners.

The editors have extensive experience in modeling, simulation, and integration of power electronics in very diverse applications, from circuits to power systems, from machines to generators and turbines, from renewable energy based on wind, PV, fuel cells, batteries, hardware-in-the-loop, digital twin to smart-grid applications. They have worked very closely with the chapter authors bringing their individual expertise and contributions from fundamentals to advanced topics. Both volumes of the book will give the reader a plethora of knowledge of electric circuits, control of dc/dc converters and inverters, energy conversion and power electronics, electric vehicles, motor drives, power systems, microgrids, smart-grids, and distributed generation. This book will prepare students and practicing engineers to apply numerical and analytical methods for multi-domain simulation of energy systems and power electronics engineering problems.

I believe that this second edition is very innovative and a lasting contribution to the education of Electrical Engineering in this 21st century. This book is well advanced, and it could be adopted by instructors who would like to teach from fundamental to advanced concepts with power electronics applications.

Dr. Kaushik Rajashekara, FIEEE
Distinguished Professor of Engineering
Department of Electrical and Computer Engineering
University of Houston, Houston, Texas
Fall 2023

Introduction

Marcelo Godoy Simões[1,2] *and Tiago Davi Curi Busarello*[3]

The availability and cost of electricity is probably one of the most effective ways to reduce poverty in our world. Electricity has been conquered by humankind since Volta invented the first cell and Faraday discovered how electricity would flow in a circuit. In the past 150 years, there has been tremendous development in the electrotechnology field, from the time that electrical generators and the distribution of power were still being designed, through so many inventions and discoveries during the 20th century. Transistors, thyristors, and computers started to shape our electronics, power electronics, and information, making our life as modern as we have today.

It is clear that this 21st century brought the reality that resources are not everlasting, and now electrification is also for empowering individuals plus allowing industry and society with the most effective renewable generation technologies to supply a complex network and ever-evolving dynamic demand.

We now have a very dynamic and complex paradigm based on distributed computing, cloud-based services, ultrafast real-time execution of control with digital twins, incorporation of highly sophisticated mathematical models, and the maturity of artificial intelligence.

Upon receiving an invitation from IET to collaborate on a second edition of this book, the editors—resolute in their understanding that a mere straightforward organization would not suffice—decided to present the current version in two volumes. Decisions were made to invite previous authors from the 1st edition to enhance and update their chapters, as well as to invite new contributors for a creative development that became *Power Electronic Converters and Systems, 2nd Edition, Volume 1: Converters and Machine Drives, and Volume 2: Applications*. We hope to encourage the next few years of students, engineers, and professionals, who see not only their profession as a job but also as a path toward better and improved technical, social, economic, and environmental changes for our future generation. In the next paragraphs, we will try to comment a little on what each reader would be able to study and learn. Thank you for the decision to study this book.

[1]Professor in Flexible and Smart Power Systems, Department of Electrical Engineering, University of Vaasa, Finland
[2]Professor Emeritus, Colorado School of Mines, USA
[3]Department of Control, Automation and Computing Engineering, Federal University of Santa Catarina, Brazil

I.1 The contents of this two-volume book

In *Chapter 1* the power semiconductor devices, which are the key components of power converters, are described. They can convert from AC to AC, AC to DC, DC to AC, and DC to DC. Diode, MOSFET, IGBT, and SCR are typically used power semiconductor devices. For over 50 years, silicon (Si) has served as the primary material for power semiconductors, offering mature fabrication processes, high reliability, and low cost. However, the increasing demand for high power conversion efficiency, power density, and reduced system costs has driven exploration into alternative materials such as Silicon Carbide (SiC) and Gallium Nitride (GaN). These materials, known as wide band-gap materials, enable the development of lower resistance unipolar devices, which exhibit considerably lower switching losses compared to bipolar devices like IGBT and SCR due to the absence of stored carriers. The chapter describes the properties and physics of Si, SiC, and GaN discussing their bandgap energy and the higher critical electric field that enables them to realize considerably higher voltage and lower leakage current devices. The thermal conductivity of SiC is considerably higher relative to GaN and Si, which is important for the effective dissipation of the heat generated from the conduction and the switching losses.

Chapter 2 covers multilevel converters; they present advantages over classical converters, especially for medium and high-power applications. Common voltage source converters used for DC/AC unidirectional or bidirectional could be typically implemented as a two-level inverter, where output voltage would have two different levels, and it is common also to have three-level ones, where the output voltage could have three steps of voltage, positive, negative, and zero, or maybe 0, VDC, and 2VDC. Multilevel converters approach the DC/AC conversion with a topology that could have various DC sources or a multi-pulse transformer with multiple windings. They have the possibility of direct connection to the medium voltage network without the use of transformers with a considerable reduction of Electromagnetic Interference (EMI) levels due to the lower output voltage variation rate and the possibility of obtaining higher power without the need for series and/or parallel associations of switches. Very often the output waveform is generated with a staircase voltage synthesized of multilevel steps that are turned ON and OFF through the arrangement of semiconductors. There are CHB converters for their modularity and the reduced number of components compared to the NPC and FCC converters. The chapter describes the circuit topologies, the modulation schemes, and a very interesting comparison as to the number of components that are required to obtain a multilevel converter using transistors and diodes with the same nominal blocking voltage and capacitors with the same nominal voltage. The Modular Multilevel Converter has several characteristics and challenges, and they are an important topic of learning and understanding for further high-voltage applications of power electronics to utility-rated systems.

Chapter 3 approaches multi-input converters, they are implemented in applications where for example renewable sources, such as photovoltaic sources and wind turbines, are complementary in energy and a hybrid system may benefit from multiple energy sources in an integrated power system or a micro-grid. Designing a multi-input converter allows energy storage elements, such as batteries or fly-wheels in bidirectional ports to the system. In addition, electric and hybrid-electric vehicles, being mobile systems, extremely benefit from increased power density through the adoption of multi-input converters, further enhancing battery-powered vehicles, grid-connected automobiles, all-electric ships and airplanes, and satellites need multi-input vehicular systems for land, sea, air, and space applications. The topologies for multi-input converters are synthesized considering the power conditioning requirements of the sources and the loads, the chapter describes how to synthesize them with either a dc link or an ac link configuration with examples of recent multi-input and multi-port power converters in several applications. This chapter provides an overview of proper control techniques, and design procedures with recommendations from researchers, developers, and entrepreneurs.

Chapter 4 explains the modular multilevel cascade converters, starting from understanding voltage source converters and their applications in various power electronics such as power conditioning systems (PCSs) for renewable energy sources, battery energy storage systems (BESSs) for power leveling, active power line conditioners (APLCs) for harmonic minimization and reactive power compensation, as well as variable speed drives (VSDs), and high-voltage direct current (HVDC) systems. The fundamental characteristic of multilevel cascaded converters is combining identical low- or medium-power subsystems to achieve a system with higher power ratings. A high degree of modularity ensures high output energy quality, reliability, efficiency, and ease of maintenance, reduction in cost, weight, and volume, and is recognized for its exceptional performance in high-power applications with features in producing quasi-sinusoidal output waveform at low harmonic content. Furthermore, the modular solution emphasizes the idea of failure management, maintenance, and consistent redundancy. This chapter provides an in-depth overview of several MMCC topologies, such as double-star cascade converter, double-star bridge converter, single-star bridge converter, and triple-star bridge converter with their distinct characteristics and application domains, as well as procedures for the floating DC capacitors startup from basic resistor-based techniques to sophisticated closed-loop control systems.

Matrix converters (MC) are described in *Chapter 5*, starting initially with the discussion of how to interface ac sources with ac loads, and having a VSI or a CSI back-to-back for comparison, the foundations of how an MC enables ac–ac conversion without any intermediate energy storage element are discussed, with ample understanding of circuits and controls for direct matrix converters (DMC), which are single stage converters that connect an m-phase voltage source to an n-phase output load through an $m \times n$ array of bidirectional switches. Further coverage is given for indirect matrix converter (IMC) requiring separate stages for the voltage and current conversion. In this chapter, the most popular MC topologies along with their control and modulation strategies are presented. A brief discussion on the technological and practical issues facing MCs, and a comparative assessment of their performance with the voltage back-to-back converters, is given, plus traditional, and predictive control techniques. Due to intensive research on matrix converters during the past decades, these converters have found a wide variety of applications in the power converter industry. To date, two drive manufacturers, i.e., Yaskawa and Fuji Electric Systems, offer matrix converter products. Features such as a lack of large and fragile dc-link components make these converters suitable for deep sea remotely operated vehicle (ROV) applications where extreme undersea pressure is present. Furthermore, due to the higher power density and the elevated temperature capability, matrix converters are employed in the aircraft industry. The direct and indirect matrix converters have also been applied to doubly fed induction generator-based wind energy systems with the advantage of controlling a high-power generator with a relatively low-power, four-quadrant power converter. Matrix converters are finding applications in the power supply generation area, as well, where fixed voltage and frequency power supplies are implemented from variable frequency diesel generators.

Soft-switching converters are discussed in *Chapter 6* which starts from understanding the need for high-quality electrical energy conversion, with very high efficiency, decreased weight and volume, often requiring galvanic insulation through the use of high-frequency transformers. The increase in the switching frequency of the converters allowed to reduce the volume and consequently the weight of the power converters. But with increased frequency there is also increased electromagnetic interference and high switching losses in semi-conductors, motivating the research of soft-switched converters. In this chapter, the main techniques and topologies of soft-switched converters are reviewed, the operation of resonant converters is described, and switching strategies aim to reduce commutation losses associated with power semiconductors. The main feature of soft switching is to decrease the losses typically associated with hard switching, when several circuits and auxiliary electronics are described to make switches turn on and off under non-zero voltage and current, i.e., Zero Current Switching (ZCS)—occurs when the current in the switch is zero at the switching instant whereas Zero Voltage Switching (ZVS)—occurs when the voltage across the switch is zero at the switching instant. This chapter presents also the design of high-frequency transformers for applications with resonant converters, studies of losses in the main resonant topologies, industrial applications, among others, resonance concepts, soft switching, resonant and quasi-resonant converters, as well as their main topologies, design methodology of resonant elements, and mathematical formulation.

Z-source converters are covered in *Chapter 7*. In the last few years several impedance Z-sourced power converter topologies have been reported to overcome the limitations and problems of the traditional voltage source, current source as well as various classical buck-boost, unidirectional, and bidirectional power converter topologies. This circuit approaches impedance Z-sourced power converter covers the entire spectrum of electric power conversion applications (dc–dc, dc–ac, ac–dc, ac–ac) controlled and modulated by different modulation strategies to generate the desired dc/ac-voltage and current at the output. This chapter provides a comprehensive overview of impedance-source networks for electric-power conversion from the perspective of the development of different topologies, their modeling, control, and modulation techniques. A basic principle underlying the concept of shoot-through across the dc-link and corresponding voltage boost is described in detail with governing equations to comprehend the subject matter to the new readers. A Z-source network shows non-minimum phase behavior due to the presence of zero in the right half-plane which could impose a limitation on the controller design. To implement a good control strategy, it is imperative to have a good dynamic model of the converter. Various small-signal analyses and mathematical models are presented to study the dynamic behavior of the system, which then can be implemented in different closed-loop control strategies with different complexities based on various applications supporting a global picture and their pros and cons. A broad classification of the topologies and modulation techniques with further sub-classification aims to provide an easy selection of control and modulation techniques for appropriate topologies for particular applications.

Chapter 8 covers a subject that has always been important, the modeling and design of switching power supplies, which can be implemented with hard-switching pulse width modulated (PWM) topologies as well as with resonant and soft-switching converters. The chapter focuses on PWM converters, they have a low component count, high efficiency, constant frequency operation, relatively simple control and commercial availability of integrated circuit controllers, and the ability to achieve high conversion ratios for both step-down and step-up applications. PWM dc–dc converters may operate in three conduction modes: continuous conduction mode (CCM), critical conduction mode, and discontinuous conduction mode (DCM). In DCM, the inductor current drops to zero, remaining null for some time, and starts increasing only at the beginning of the next switching cycle. The benefits of DCM include simplified control, as the current in the inductor resets to zero at the beginning of each cycle, and reduced size requirements for the converter's inductor. Most of the high-power supplies will operate under CCM, and this chapter shows all details for non-isolated CCM structures, describing the seven basic dc–dc converter topologies that were analyzed: buck, boost, buck–boost, integrated buck and boost, SEPIC, Cuk, and Zeta, then also Quadratic Buck and Quadratic Boost converters. Modeling and the effects of parasitic on output voltage ripple, efficiency, and voltage transfer function of converters were discussed. For applications where galvanic isolation is needed, isolated converters such as flyback, forward, half-bridge, and full bridge were analyzed, especially topologies such as bidirectional power converters and interleaved converters are also covered. For closed loop operation, basic control principles of PWM dc–dc converters are presented, and two main control schemes, voltage-mode control and current-mode control include the application of the state-space average modeling technique.

Chapter 9 describes how power electronics building blocks started around 1994 and became a whole area called smart power electronic modules. The Office of Naval Research (ONR) sponsored and led a series of electronics, materials, and manufacturing techniques and also a program called "Ship Hull, Mechanical, and Electrical Systems Science and Technology Division" in their Advanced Electrical Power Systems (AEPS). Both PEBB and AEPS programs included academia and industry, with notable contributions from CPES at Virginia Tech and SatCon Technology Corporation, setting standardization in power electronics converters of medium power (tens to hundred kW). Similar programs were conducted within Oak Ridge National Laboratory (ORNL) with major power converter manufacturers. ABB Corporation has extended the PEBB program into the multi-megawatt range. Building on a series of principles, efforts have been made to address the modularity demands of production lines and after-market service expectations. This has led to advancements in hybrid integrated circuits, leveraging increasing technological capacity. Additionally, progress in computer-aided design enables the creation of new very-large-scale integrated circuits, achieving impressive performance in signal processing and contributing to size reduction in electronics equipment. Integrated circuits spread into various fields of application including power electronics conversion. This chapter analyzes these works, and discusses and compares several modules, for example, to drive motor with six power modules IPM/SPM/IRAMS setup. The power density is competitive as the entire power stage (with straight-pin mounting, six individual heat-sinks, passive LC filtering, and power connectors) accounts for (2 in $\times$ 2 in $\times$ 7.5 in =) 0.49 l for 2 kW delivered power (i.e., 4.1 kW/l). The advent of power semiconductor technology allowed the integration of semiconductor power devices and their control integrated circuits within the same hybrid device. This chapter has briefly reviewed the advantages of this integration, and the technology options, and provided a historical perspective on the topic. The conventional usage of SPM has been extended with a new class of multi-module power converters based on a network of switches implemented with multiple SPMs and ready to provide even better performance.

Converters for the power grid are explained in *Chapter 10*, since the electrical power system forms the essential backbone of the energy system that enables modern livelihood and it is composed of generation, transmission, distribution, and utilization systems, it is very important to understand how electrical power converters play in a holistic approach. The power grid is operated as a three-phase ac system, therefore, most of the discussion in this chapter is focused on three-phase power converters. The focus is mainly placed on the circuit configuration of the constituent components, and not on their control, modulation, switching strategy, or design aspects restricted to the functional details. While there is a large variety of power converters applied to the power grid there are some largely applied, even on SCR-based technologies. Advances in DC-link VSC-based technologies are mostly viable, and several can be possibly applied to FACTS applications. The recent introduction of MMCs for realizing high-power converters using gate-turn-off switch technologies appears to be a game changer in displacing SCR-based converter technologies. However, the technology of PWM ac power flow control and solid-state transformers is still in the realm of academic research. Emerging advances in power semiconductors such as silicon carbide and gallium nitride materials may tip the economies of scale to make such newer technologies more attractive in the future. This chapter provides a concise outline of different power converters used in the power grid, along with selected application examples, chosen to illustrate the diversity of approaches in use. The discussion is brief, the readers can refer to other chapters in this book for additional details. The references provided are drawn from the state of art literature instead of classical publications, and hopefully, the chapter can motivate students to learn and study.

Chapter 11 discusses permanent magnet synchronous motor drives. The control of electrical drives is a paradigm where a system must have dynamic stability under rather a variability of parameters, innovations are pushing the boundaries, with some technological constraints, and market requirements for cost and demand. PMSMs are used in industry for better energy efficiency and high performance, also in electric vehicles, due to the required peak power rating of typically 0 kW in mild hybrid configurations to 175 kW and above in high-performance strong hybrid and battery-electric drive trains. The range of applications of PM motors is also enhanced by the recent introduction of new constructive topologies, which are essentially the migration, at different degrees, of isotropic PM motors to pure synchronous reluctance (SynR) motors. The addition of a correct number of PMs to the synchronous reluctance rotor yields a higher power factor and improved efficiency, with limited back-EMF even in the flux-weakening speed range and without a necessary change in the stator design. The chapter also describes sensorless control highlighting the connections between theoretical algorithm and their implementation in the power electronic converter direct torque control (DTC) and the model predictive control (MPC) is included with examples of multi-object, energy-efficient control techniques application for PMSM drives. New hybrid (reluctance and PM) motor topologies have been studied to provide motors with improved efficiency and self-sensing features. The use of these motors complicates the control algorithms and makes the availability of a complete and reliable magnetic model unavoidable. The emerging energy awareness is going to draw new paradigms, based on multi-variable control, for the contemporary optimization of different power electronics and drive aspects, particularly with the applications of MPC with online compatibility will make this area even stronger and established for industry applications.

Modern adjustable induction motor speed drive control systems are presented in a clear mathematical and engineering perspective in *Chapter 12*. Induction machines are very important to the modern age, the chapter discusses how variable speed induction motors are important for energy-saving purposes. Most industrial applications would have centrifugal load torque applications, where the change is proportional to the square of the speed, and consequently power changes in proportion with the cube of speed with energy varying at different speeds. Variable frequency drives manage energy consumption by changing the speed, and continuous speed control is desirable for a variety of applications such as moving air, fluids, heating and ventilation, lifting water, moving belts, and loads in container systems, providing energy efficiency, reliable dynamics, and cost efficiency. After Nikola Tesla invented the induction machines and applied them initially for hydropower generation, the 20th century established mostly on large power plants powered by synchronous machines, then induction machine control had a breakthrough in 1969 when Hasse introduced the direct vector-control (DVC), a decoupled system achieved with flux measurements in the air-gap of the induction machine, and the indirect vector-control (IVC) was introduced by Blaschke in 1972. The analysis of the rotor d-q currents in the rotating reference frame made possible the vector control with the analysis of the synchronous reference frame and stationary reference frame dynamic models of the machine. This chapter shows the theoretical foundations of abc three-phase and d-q stationary and reference frame modeling and analysis to support the implementation of scalar control, indirect vector control based on rotor flux alignment, and direct vector control based on stator flux alignment. Scalar control is a V/f approach that may have slip frequency feedforward torque control. Field-oriented control can achieve excellent performance in transient and steady-state conditions. Two types of field-oriented control are defined based on the position of the rotor flux: indirect and direct. By adding the slip position to the measured rotor position, the flux position in indirect field-oriented control is obtained. The flux position in direct field-oriented control is calculated based on the terminal variables and rotor speed. Since field-oriented control stems from a frame transformation that requires rotor speed, the knowledge of rotor position needs to be acquired accurately to perform such transformation. Vector control relies on accurate machine parameter characterization, along with robust sensor monitoring and feedback of essential machine variables. This involves utilizing a powerful microcontroller or a processor based on digital signal processors (DSP) for rapid real-time computation employing efficient PWM methodology. Additionally, it requires effective protection mechanisms against over-currents or over-voltages. These elements collectively enable an inverter to control an induction machine, providing current-control capabilities within a closed-loop speed control or closed-loop torque control system.

The four-quadrant operation of the switched reluctance machines is rendered with a comprehensive review and discussion of control techniques and practical applications in *Chapter 13*. The switched reluctance machine stands out for its simplicity, robustness, and energy efficiency. It is capable of operating in motoring, generating, or braking modes and is widely applied in the industry, including automotive, aerospace, marine, wind, and other segments of the industry, rural, commercial, and residential uses. Such an electromechanical device operates based on the magnetic reluctance principle. Differently from induction and synchronous machines, the SRM does not have a magnetic field in the rotor, what happens is that as an electric current passes through the stator windings, such a magnetic field interacts with the rotor material, inducing torque via reluctance interaction, the chapter describes in all theoretical and practical topics, and initially the readers will understand direct average torque control, current profiling through torque sharing functions, and direct instantaneous torque control covered in the section of the operating principles. The four-quadrant model and ideal driving angles for SRM drive are presented in the third and fourth sections. The fifth and sixth section focuses on speed control and current control. The seventh section covers modern torque control methods used to operate the SRM efficiently. The eighth section presents the simulation results of the four-quadrant operation. The final section concludes the chapter with a summary of key findings, future research directions, and recommendations for practical applications. Switched Reluctance Motor has several advantages that render it suitable for various applications. These include a simple construction with fewer parts compared to other electric machines, resulting in reduced complexity and increased reliability. They are also cost-effective due to their uncomplicated design and absence of permanent magnets. Moreover, Switched Reluctance Motors exhibit high tolerance to mechanical and thermal stress, making them well-suited for harsh environments and high-temperature applications. Additionally, their high torque density makes them a suitable choice for electric vehicles and industrial drives. The switched reluctance machine and the induction machine are the ones that displace the mining of rare earth materials, making them a more environmentally friendly, sustainable option, and less dependent on economic fluctuation. It also has simple maintenance and repair. Acoustic noise drawback must be considered with several techniques presented in this chapter. The four-quadrant operation of SRMs presents many opportunities and challenges for research and development, while still being indicated for some industrial applications. Future work should focus on improving the accuracy and efficiency of control methods, as well as developing new techniques to overcome the limitations of SRMs, such as torque ripple and acoustic noise, fault detection and operation, and sensorless control.

Chapter 14 encompasses reliability in power electronics, a feature that may not be optimized for the average consumer's wants and needs but designing electronic systems with longer lifetimes and more accurate and reliable components is typically important for mission-driven projects, or longer lifetime industrial processes. Early failures may have negatively impacted the industry when occurring before or after the warranty period, but why reliability in power electronics should be taken as improving the performance or prolonging its operation time before failure are important for situations where catastrophic events, such as overcurrent or overvoltage, would cause component failure and the ability of an item to perform a required function under stated conditions for a stated period may be considered at the design phase. Devices may experience elevated leakage current, maximum thermal resistance variation, or increased voltage drop. Reliability engineering in power electronics centers on investigating the root cause of failures, understanding how failure mechanisms behave under specific stresses, and examining product design aspects that influence component lifetime. The comprehension of reliability metrics helps the engineers to consider the majority of the parameters and events that would wear out the component over time. This chapter aims to provide insights about the reliability of power electronic systems and readers will be introduced to the metrics of reliability, new modern approaches such as physics of failure and design for reliability, and also detailed step-by-step procedures to calculate the lifetime of power semiconductors and capacitors; it is then possible to find the system reliability, depending on the association of the components in the reliability diagram block. Several online condition monitoring methods can be applied to predict the real-time degradation of a certain failure mechanism. However, the application of these methods also inserts complexity into the system as one of the main challenges. Artificial intelligence is a newer tool that can be applied to develop automatic optimizing processes, improve monitoring methods, and support a Design for Reliability decision-making procedure. Fast reliability evaluation tools and advanced condition monitoring methods will certainly be implemented in the near future for power converters for the power grid and industry applications, and students, engineers, and professionals must be prepared and up-to-date to work together in multiphysics domains and improve the knowledge about reliability. This chapter allows us to understand better the failure mechanisms, propose new physics-based lifetime models, develop new techniques to manufacture new products with fewer parameter variances, investigate new packaging technologies, and improve methods and optimization.

Hardware-in-the-Loop (HIL) technology is defined as a setup with a Real-time Simulator (RTS) plus an external physical device, the theory and applications in power electronics are covered in Chapter 15. The external hardware is usually known as Hardware under Test (HUT) or Device Under Test (DUT). A control with a HIL setup has an RTS and external control unit, usually a microcontroller, a digital signal controller, or a digital signal processor where the connection of the RTS and the control unit may be direct or through an interface board, depending on the voltage compatibility of their input and output terminals. In C-HIL, the RTS simulates a system in a real-time frame and generates waveforms that describe the response and behavior of the modeled system. The control unit receives these waveforms, processes its control algorithm, and sends a control command signal back to the RTS. In this way, the modeled system and the control unit interact with each other in a loop. This chapter presents the concepts of hardware-in-the-loop technology and applications in power electronics. The definition of C-HIL and P-HIL setups are explained in detail. The principles of operation of these setups are addressed, together with a discussion about where to use them in power electronic applications-related research. The chapter shows three examples of power electronic applications.

Chapter 16 is a very comprehensive discussion of wind energy systems, starting with their technological history and then making comparisons of how the wind energy industry has advanced so much from 1980 to now. It also shows how there has been a greater incorporation of power electronics and advanced digital control to enhance the transfer of energy from the turbine, passing through the generator, and then reaching the electrical energy transmission and distribution networks. This chapter is an improved version of the first edition of this book. This chapter not only discusses traditional horizontal and vertical axis wind turbines of various geometries and electromechanics but also presents a very good explanation for the Magnus WT radial rotary axes. These units generate electricity even under very strong wind conditions when most other turbines should be closed for safety reasons. This chapter discusses the most well-known details and concepts of power electronic converters according to the sizes of small, medium, large, and very large turbines, leading to more precise details for manufacturers. The text shows how wound rotor synchronous generators and wound rotor induction generators (DFIG) are still very useful for power generation applications. This chapter also shows details of the permanent magnet (PM) synchronous generators. There is a comprehensive discussion on power converters and a very interesting combination of multilevel technology and specific technology for three-phase four-switch inverters, called the B4 inverter. A multilevel approach is also included, which considerably improves the performance of the B4 inverter by ensuring that the DC bus voltage is always higher than that of the conventional three-phase Var Static Compensator (VSC). The main advantage of this topology is the existing power semiconductor components. The entire PEC could be implemented in a single power electronic module with a bridge using commercially available three-phase neutral point clamped (NPC). Another promising technology addressing PMSG-based wind turbines is the quasi-Z source inverter (qZSI). It overcomes the limitations of VSCs and CSCs by enabling a buck-boost operation. The qZSI can handle all control functions on the grid side, allowing the three-phase diode bridge rectifier to be simpler on the generator side. Multi-MW power levels are still highly challenging because most high-power wind turbines are low voltage. To handle high currents, two-level back-to-back VSCs are generally connected in parallel using the same three-phase winding or even a single three-phase winding per converter. This approach could be implemented using redundancies since the reliability of the system is considerably lower than that of a single converter. Many control methods can be applied to different types of Maximum Power Point Tracking (MPPT), two basic ones are the most common. The first type requires prior knowledge of the turbine parameters to calculate the operating point and the second type iteratively searches for the optimal control point by power and rotational speed increments. Regular MPPT is described in Chapter 16 based on two well-known general methods to obtain the power characteristic of any wind system. In the first method, the power curve can be stored in a pre-programmed 2D look-up table, and in the second, by an adaptive approximation curve using a third-order polynomial and a fuzzy control and search logic. Students and engineers will learn a lot in this chapter.

Chapter 17 is about photovoltaic energy systems. Photovoltaic (PV) generation initially started in the aerospace industry, satellites, outer space sensors, monitors, and powering up the International Space Station. In the past few years, PV power plants have been growing in size for stationary use, the cost of PV panels has been dropping, and there is an estimation that in the next 10 years the contribution of PV plants to the total number of renewable energy power plants will grow significantly. The chapter discusses the technology, materials, and sizing, it goes through stand-alone and grid-tied PV systems composed of power converters with PV panel arrangement connected in series to provide the input voltage range to the inverter and in parallel for increased current availability. The inverter system may include a voltage booster stage. A transformer, which may or may not be included can be used for the final connection to the grid. Booster stages are common in residential and commercial inverters. Residential and commercial inverters have been called string inverters due to the way they are connected to the solar panel. Many of the larger commercial inverters (typically greater than 50 kilowatts) eliminate the booster stage. The same holds for utility-scale or centralized inverters. For residential applications, the use of microinverters is becoming popular, a microinverter is coupled directly to the DC generated from individual solar panels to convert it to AC for the power system. The microinverter has isolation transformers that operate at a high frequency, leading to a very good energy density ratio with isolation of the DC and the ability to perform optimization. The modular structure of the microinverter allows for aggregation on the AC side. The low power, low voltage of the microinverter is a good application field for GaN semiconductors and new advanced silicon power MOSFETs. For higher power applications multilevel inverters such as NPC (also known as Type I) and the modified NPC, generally known in the industry as MNPC, NPC2, or Type T. NPC topology requires components that withstand half of the applied voltage, but the MNPC arrangement could deliver a more efficient inverter topology. With the improvement of power semiconductors, the MNPC topology has been increasingly implemented in commercial solar inverters. In both the MNPC and the NPC, the operation can be explained as two buck converters, one each for the positive and negative parts of the sinusoidal. The two buck converters disconnect alternatively to avoid a short circuit between each other. Using the NPC structure, a three-level inverter can be described as two buck converters. The chapter discusses regulations and standards and gives the revisions made to IEEE 1547 in 2018. These revisions emphasized the safe, reliable, and robust interconnection of distributed energy resources (DERs) with electric power systems. The update addressed technical improvements and the continuous growth of distributed and interconnected renewable resources, such as solar energy, wind turbines, and energy storage. Photovoltaic systems have come a long way from specialized space applications to now being an integral and substantial part of the modern power system, alongside wind energy, stabilized battery storage, and traditional centralized energy generation.

Chapter 18 is titled Advanced Charging and Battery Management Systems for e-mobility. Electric vehicles (EVs) and hybrid vehicles are growing to occupy roads worldwide. The major components of an EV comprise a battery, battery charger, power train, DC–DC converter, and other components. The battery serves as the energy source or a reservoir of the vehicle, they are rechargeable and can be charged and discharged several times, it seems Lithium-ion batteries have been the best suited for vehicular applications, they have good specific energy and power density, but they need an intelligent way to manage their state of charge and become well optimized with power electronics and smart electronic controllers. This chapter covers Lithium-ion batteries (LIB) and the theory of LIB to be used in EVs in particular. Requirements of EV/PHEV batteries, protection strategies, lithium-ion battery management system, and battery state estimations are discussed. Finally, charging standards, protocols, wireless charging technology, and battery-swapping technologies are included. The electric vehicle market will continue to grow at an effective speed, and the push from governments across the globe for the adoption of EVs is encouraging. Despite the rapid growth and technological advancements, issues addressing range anxiety, standardization of charging connectors, and increasing the grid capacity still exist. Upcoming battery technologies and advanced charging solutions are very promising. In conclusion, this chapter discusses electric vehicle batteries including the fundamental definitions like battery voltage, capacity, energy density, and C-rate. The basics of the battery management system, battery states including state of charge, state of health, state of energy, and their estimation methods are discussed. Accurate estimates of battery life will assist in better utilization of battery capacity and increasing the vehicle range. Using advanced battery health estimation using EIS also benefits the prognosis of battery faults including battery thermal runaway. The important concept of battery thermal management is addressed. EV charging technology and infrastructure is rapidly growing. This chapter shows both AC and DC plugged charging standards. AC charging is also known as slow charging as the C-rate used to charge the batteries is considerably low. Standard connectors including SAE J1772 (Type-1) and IEC 61851-1 (Type-2) are discussed. Fast DC charging standards including CHAdeMo, CCS, and GB/T standards are discussed. With DC fast charging, the batteries can be charged in less than 10 min to 1 h, this will reduce the issue of long charging time. Wireless charging standards are also discussed. Worldwide, initiatives are being taken by governments and other private organizations to develop fast charging and ultra-fast charging infrastructure. Work is in progress to build roads embedded with wireless charging technology. Further research and development with more infrastructure in the near future will enable long-distance travel in EVs and also make them practical and convenient to use without any anxiety or fear of range.

Chapter 19 is about the design and control of DC–DC switched capacitor converters. The work in this chapter has been motivated by the need for fuel cell vehicles (FCV) with zero-emission, no pollution, and high efficiency. A typical system configuration of FCV includes a supercapacitor, fuel cells, a motor, DC–DC converters, and an inverter. The low output voltage of the fuel cell makes it difficult to meet the demand for DC bus voltage to feed the inverter, and a fuel cell has a fold-away output voltage characteristic, i.e., the output voltage drops fast with the increase of the output current. An isolated DC–DC converter can easily achieve high gain to compensate for this soft behavior of the fuel cell input by proper transformer turns ratio. However, due to the leakage inductance of the transformer, the circuit will produce a peak voltage, also the leakage inductance reduces the efficiency of the converter and causes electromagnetic interference problems. Considering the size, cost, and efficiency, the non-isolated DC–DC converter is more suitable for fuel cell vehicles, and typical control methods for DC–DC converters include proportional-integral (PI) control, sliding mode control, adaptive control, robust control, fuzzy control, and novel integrated control strategies. This chapter shows the categorization and derivation procedure of topologies of DC–DC converters, a new DC–DC converter topology based on switched-capacitors and switched-inductor is presented, with a control-oriented modeling process and then a closed-loop PID controller. To improve the robustness of the controller, robust PID and ADRC controller design methods are separately applied to the DC–DC converter. A simulation case study shows that the controller has good robustness and fast performance, the converter can achieve a wide range of operations, and the effectiveness of the design of the proposed converter is suitable for fuel cell vehicles.

A stationary-grid approach for batteries as an energy source compared to mobile applications is described in *Chapter 20*, where a comprehensive overview of battery integration and control is provided. Electrical power generation from renewable energy systems is intermittent and has low inertia, resulting in significant voltage and frequency instabilities in the distribution systems compared to traditional centralized power systems. The improvements to balance power supply can be accomplished by incorporating flexible energy sources (FESs), such as battery energy storage systems (BESSs). In addition, the demand for battery electric vehicles (BEVs) has been rising, but batteries constitute 25%–50% of the EV price, with market estimates that a cost parity between electric vehicles and traditional gasoline-powered vehicles will happen soon. Any BESS technology when used for stationary smart grid applications is connected through an inverter to the grid with active and reactive power and further services for the distribution grid. BESSs are described in this chapter in detail, where their capability of supporting various stationary applications will be described with their modeling, control, and integration methods in power systems, based on their time scales of utilization. The design methodology of the Lithium-ion BESS controllers includes their performance characteristics and battery aging. For electric vehicles, the right choice of battery type and chemistry can be such as Li-ion, $Na\text{-}NiCl_2$, Li-S, and Ni-MH. These distinct characteristics and performance in terms of charge–discharge efficiency, cost, reliability, and battery life are vital for an efficient battery management system to control and check the charging and discharging of batteries for reliable operations. The proposed battery management system operates by sensing the power level that appears on the DC bus of the EV charging station. The proposed charging setup utilizes maximum power from a solar PV system and prevents the battery from overcharging and undercharging. Inverter-based BESSs can be seen as ideal FERs due to their potential and versatility in providing multiple active and reactive power-related flexibility services for different DSO and TSO needs such as voltage and frequency control, black start, islanding, load leveling, and peak shaving. Thus, BESSs are useful for bridging the gap between nonconcurrent RES-based power generation and demand at different voltage levels of the power system. This chapter presents all issues necessary to understand technical characteristics that are highly nonlinear and are affected by various factors such as their ambient operational temperature, SOC, DOD, C-rate of charge and discharge, and most particularly, their aging patterns. The chapter also goes into the current and future prospects of the EV market and explores various battery charging methods for EVs. The integration of EVs into the power system is of paramount importance, covering energy required for EVs as relevant to their battery management power control. The chapter makes clear how to alleviate strain on the power grid and maximize the utilization of power from solar PV systems.

Chapter 21 brings in-depth discussions of shipboard power systems, fuel costs constitute a large portion of a ship's life-cycle costs, and environmental regulations in marine engineering have been steering the use of more power electronics-based drives and inverters for all the systems in the ship propulsion, auxiliary equipment, essentially making a ship a microgrid with very sophisticated energy management and control. Inverters with the capabilities to fulfill multiple different roles in addition to traditional AC drives are used to realize complex power electronics-based systems, ranging from shaft generators all the way to ship-wide DC distribution systems. The power electronics will have programmability and internal decision-making capabilities, making the system robust and functional in comparison to standard stationary power systems. The chapter has two case studies, the Aurora Botnia is a cruise ferry sailing between Vaasa (Finland) and Umea (Sweden) and started operation in August 2021. Aurora Botnia incorporates electric propulsion and an AC–DC hybrid distribution grid, where the four main generators are connected to the AC grid, and inverters running the bow thruster and main propulsion motors as well as ESS are divided into two separate DC-links. The inverter systems that consist of the DC-links are identical systems, as seen in Figure 21.15, and are divided electrically and physically onboard to different parts of the ship for safety reasons. Each inverter system includes a 6.0 MW azimuth thruster as main propulsion, a 1.5 MW bow tunnel thruster for maneuvering, and roughly 1.0 MWh ESS in the form of batteries. Each inverter system is then interfaced with the AC grid by two 3.0 MW grid inverters. The other case study is for M/F Grotte, an all-electric ferry sailing between Esjeberg and Fanø in Denmark, which started operation in 2021. Grotte has a shipwide DC distribution network that utilizes battery ESS as the main power source with charging capability at both ports and a traditional generator set on board to provide a traveling range extension and a secondary charging source. Pumps and fans that traditionally are located in the hotel grid on such ships are on-board Grotte located in the DC load sections. This allows a relatively large amount of loads, that would anyway be run by AC drives, to be fed directly from the main DC switchboards. This reduces the size of the hotel grid generating grid inverters considerably and removes the harmonic distortions from the hotel grid. It discusses shore connection integration, to keep the ship's electrical power on without running generator sets on-board.

Chapter 22 composes a formal understanding of distributed generation and microgrids. This chapter has been enriched since the 1st edition, delving into the evolution of the electricity and power system industry, particularly focusing on the progression of deregulation and the implementation of Distributed Generators (DG) since the mid-1970s. DGs are located along the electrical power system and mostly use renewable or alternative energy sources for their prime mover. Some examples are wind, solar, biofuels, hydrogen, geothermal, wave, and tidal technologies. Fuel cells may not be considered totally renewable, but they allow a mix of hydrogen that may have been made of electrolysis of surplus solar or wind, as well as from side production on excess of large nuclear power plants, or from cracking of fossil fuels; those are paths toward a future sustainable scenario. With power electronics as enabling technology, not only smart functionalities become a reality, but even the utilities could have beneficial services from prosumers in increasing the hosting capacity, decreasing losses, and furthering plasticity with the new paradigm of Electric Vehicles as a flexible resource, makes possible to have clusters of EVs contributing to such services as well. This chapter presents an overview of DG and microgrids. In section 2 the types of DGs are described with their mathematical models, their technical impacts on the power system, and some constraints imposed by standardization. In section 3 microgrids are presented and their features are briefly introduced. Throughout this chapter, a case study is conducted to clarify the presented concepts. There are sections on centralized control when the whole distribution system has some specific goals versus distributed control where agents may consider real-time decisions. Droop control is beyond the most popular control techniques that have a long history of use for synchronous generator control in power systems. Recently, it has been used for parallel inverter control, especially in the case of inverter-dominated microgrids. However, the future reality will be based on phasor measurement units (PMU) based on synchrophasor measurements or point-on-wave (POW) measurements to accurately reflect non-sinusoidal signals. The conventional power droop framework is still utilized, even in power electronic inverters mimicking electrical generators and imposing some artificial inertia on their response, when frequency falls, the output power of the generating unit is allowed to increase. A falling frequency indicates an increase in loading and a requirement for more active power. Multiple parallel units with the same droop characteristic can respond to the fall in frequency by increasing their output active powers simultaneously. However, local controls may be involved with one or more DG. They may have goals to improve local performance in a single feeder. The chapter shows a case study of feeders including PV, wind, and storage-based DGs as well as multifunctional inverter-based operation, with new enhancements of IEEE 1547 standards where services and functions can be performed simultaneously with power injection. Even though almost all system operators do not presently allow DG to perform these ancillary functions, these issues will certainly be under implementation in the coming years, primarily due to the simplicity of employing additional functions.

Uninterruptible power supply (UPS) systems have been used to supply and protect critical loads as described in *Chapter 23*. Needs for UPS systems have increased due to the advancements in information technology, with more sensitive electronic equipment, and mission-critical systems. Historically UPS approaches moved from rotary to off-line, then to on-line, and line-interactive systems. The chapter shows how it has been further evolving to multi-layer, multi-bus systems supporting complex infrastructure such as data centers. It discusses various UPS system topologies, control methods are explained, and some applications are discussed. How the battery bank is connected, managed, operated, and controlled is one of the main subjects, also how a phase-locked loop (PLL) is used to keep the output voltage in phase with the input waveform to ensure a smooth transition to the electric power system and the transition from energy storage mode to normal mode is explained with discussions on the common arrangements, the transfer time from normal mode to energy storage mode, and the tolerance for input voltage and frequency variations. There are configurations based on transformers to attain galvanic isolation for the load, two and three-winding transformer constructions, with also four-quadrant converter and controls. The chapter aims to present the main aspects related to the topic of UPS. This is a topic that attracts great interest from both academia and industry. Different topological arrangements are analyzed with their main operational characteristics as well as their advantages and disadvantages. A brief analysis of control strategies with recent advances aimed at improving energy storage systems, especially in the context of smart grids, microgrids, and distributed energy resources makes reading this chapter very important for anyone.

To electrify the transportation sector by promoting electric vehicles, it is key to read and study *Chapter 24* which covers wireless charging for electric vehicles. There are three ways for EV battery charging; battery swapping stations, conductive charging, and wireless charging (stationary and in-motion), as far as back the 18th century Hertz tried to spread the electromagnetic waves in space with a spark gap, then in 1890, Nicola Tesla examined radio waves to transport electric power wirelessly. Fast forwarding to more recent events, from 2007 to 2013, a group of researchers from the Massachusetts Institute of Technology (MIT) represented experiments and research of Tesla based on magnetic coupling between two resonators to transport 60 W through an airgap of 2 m with 40% transmission efficiency using 0.6-m coil diameter. In a short period various applications, including EVs, cell phones, laptops, home appliances, and medical devices, started to have wireless power transfer (WPT) electronics embedded in them. In the near-field-WPT technology, the power is transmitted wirelessly by electromagnetic induction, in which fields stay within a small region around the transmitter coil. Within this region, the fields can be divided into two types, electric fields that use capacitors (capacitive power transfer) or magnetic fields using coils (inductive power transfer [IPT]). This chapter focuses on IPT technology that applies to EV charging. Considering aspects related to power transfer capability, transfer distance, efficiency, and practicality, IPT shows promising results for EV applications. The chapter explores IPT technology including principles of operation, system architecture, R&D activities, challenges, and technology gaps. The working principle and components are presented for stationary and dynamic inductive charging. Various architectures of the transmitter and receiver pads are presented, compared, and discussed. The different compensation networks and converter topologies that are used in IPT and their characteristics have been compared as well as further activities related to stationary and in-motion IPT with the associated challenges and opportunities concerning the technology. This chapter provides a comprehensive guideline for inductive charging that will help researchers and engineers approach and understand the technology.

Advanced control of power electronic systems is examined in *Chapter 25* where the authors based their team experience and their articles and papers about PES control advancements in light of some non-linear control approaches. Power electronics are nonlinear hybrid dynamical systems, in addition to being a time-variant structure given to switching and discontinuities, there are dynamics at slow and fast scales. Historically, there has been work from varied perspectives and focus, which fall broadly into the following categories: (i) Sliding-mode control (SMC), (ii) Passivity-based control (PBC), (iii) Back-stepping control (BSC), (iv) Feedback linearizing control (FLC), and (v) Model-predictive control (MPC). Conventional analyses of PESs and their subsystems are based on average models, which ignore the fast-scale instability and analyze the stability on a reduced-order manifold, the validity of those models would need analyzing the stability of standalone and networked PESs based on linearized averaged (small-signal) models that require a smooth averaged model. Systems for smooth averaged models would have small-signal analyses of the nominal solution/orbit, there are three important characteristics: region of attraction of the nominal solution, dependence of the converter dynamics on the initial conditions of the states, and the post-instability dynamics. The chapter presents the unique approach taken by the researchers in the author's team defined as SBC—Switching-Sequence Based Control with further comparison with MPC—Model Predictive Control. The exploration of these two advanced control techniques has revealed their transformative potential in enhancing the reliable operation and control of power electronic systems. Both of these control approaches represent critical advancements in the field of power electronics control, holding the promise to revolutionize energy conversion and management across diverse applications, from renewable energy systems to electrified transportation and beyond.

Chapter 26 is about active power filter control methods for power quality improvement in more electric aircraft applications. Increasing demand for pollution-free and dependable transportation has accelerated the pace of adoption of more electric aircraft (MEA). Their power system has been discussed in detail, particularly various frequency-based and time-based harmonic current extraction algorithms have been developed, including capacitor voltage control, load current filtering, synchronous detection, neural networks, PQ theory, conservative power theory (CPT), Kalman filter-based phase-locked loops, and other techniques. There is a specialized approach to power electronic converter's role in MEA power systems, emphasizing the need for enhancing power quality. Conventional converters for power quality improvements delve into the utilization of traditional converters, and various configurations for electric aircraft are discussed. The chapter focuses on the control of active power filters in MEA applications and evaluating their performance. The chapter presents MATLAB simulation results supporting the effectiveness of three methods (PQ, DC voltage, and SRF control methods) in enhancing power quality for the MEA power system. In addition, this chapter offers a comprehensive view of the challenges and solutions related to power quality enhancement in more electric aircraft applications. This work may be useful for further research on improving the power quality by reducing the %THD of electric aircraft power systems. Our society is very dependent on aircraft and aviation applications to keep up with the economy and globalization, so there is a high expectation that in the near future, more electric aircraft could even become totally electric aircraft with some sustainable solution on the energy conversion path toward such a bright horizon.

Chapter 27 is an overview of fault ride-through strategies for grid-connected photovoltaic systems. The energy market has moved to renewable sources of energy, i.e., biomass energy, wind energy, hydropower, solar energy, etc. as conventional energy sources like petroleum, natural gas, and fossil fuels are not environmentally friendly. The chapter presents new principles and guidelines in the shape of grid codes for connecting PV systems to the grids, such as Fault Ride Through (FRT) ability and power generator performance during and in post-fault circumstances. There are many scientific faults aspects and strategies for High Voltage Ride Through (HVRT) and Low Voltage Ride Through (LVRT) with discussions in several scenarios. The FRT methods are divided based on improved controller-based and external device methods. In practice, FRT strategies are compared from various angles, including complexity, technical feasibility, and economic viability, contributing to the security, stability, power quality, and reliability of the power system. Most countries adapt and adopt their grid code regularly to provide a robust and efficient electricity infrastructure. Traditional Crowbar, Bridge Type Fault Current Limiter (BFCL), and Switch Type Fault Current Limiter (STFCL) with a properly tuned PI controller for a three-phase grid-connected system were effectively examined and verified under unbalanced fault situations, when compared to the traditional Crowbar method, the performance of BFCL and STFCL is resilient, smooth, stable, and ripple-free, reducing stress on semiconductor devices during and after the failure. There is a case study that explains the proper design and implementation of conventional Crowbar, STFCL, and BFCL as FRT approaches for a hundred-kilowatt three-phase PV system connected with a grid is presented in MATLAB/ Simulink. Performance evaluations such as IAE, ITAE, and ISE validate the improvement of power, grid voltage, DC link voltage, and grid current during asymmetrical failures. The active and reactive power parameters of modern grid codes should be thoroughly evaluated in future FRT approaches. A popular subject is the use of high-performance controllers and hybrid FRT techniques to improve overall efficiency and stability. This chapter will be useful to scholars, engineers, manufacturers, developers, and operators who work in the field of FRT techniques for evaluating and implementing grid-connected solar energy.

The support functions and grid-forming control on grid-connected inverters are addressed in *Chapter 28* because power electronics-based renewable energy resources are connected to the electricity grid through an inverter, and the theoretical and practical discussions are presented in this chapter to enhance the knowledge of students and engineers. Ancillary services, essential for regulating the voltage and frequency of the grid in high penetration scenarios, require a control technique to enable active and reactive power control in grid-connected (GC) mode and smooth transition between GC and islanded mode, among other features. The support functions will comply with standards, such as IEEE 1547 and UL 1741. The chapter explains how active and reactive power support functions, such as Volt-Watt, Volt-VAR, and Freq-Watt, allow the DER to work in cooperation with the grid, aiding voltage and frequency stability, as well as attenuating power mismatch between generation and loads. The GCIs may provide support functions while operating in grid-following (GFL) or grid-forming modes. Inverters operating in GFL act as current sources, while GFM inverters operate as voltage sources. They can provide frequency damping and inertial response to the system. These converters are also able to black start a system, and for these advantages, they are an appealing choice for enabling a healthy grid of the future. This chapter proposes a roadmap for designing power-based controllers to enable DER support functions and for designing grid-forming control systems considering DER operating points. Experimental results are shown to validate the proposed design methodologies. This chapter provides a guideline for designing grid-forming converters able to provide support functions to the grid. The basis for implementing and understanding the role of various support functions was given. These functions provide active and reactive power references to the power loop, responsible for generating voltage references to the inner loop. By using the concept of grid-forming inverters, the power loop was implemented by an active power controller, also responsible for the synchronization with the grid, and a reactive power loop, responsible for generating the reference of the voltage magnitudes to the inner controller. This controller regulates, with fast dynamics, the voltage at the output filter of the grid-forming converter and shows fast-tracking capability for real-time control variables magnitude and phase to regulate reactive and active power. The feasibility of the design procedure presented in this chapter has been validated experimentally and it is suitable for designing the next generation of grid-forming inverters.

I.2 This book in your hands, or maybe in your screen

This book has been a very pleasant editorial project, in bringing together so many outstanding professionals, researchers, and professors, and in the interaction with them, also with the support of IET, we editors also became fulfilled with experience. It is clear how advances in physics, applied math, and modern analysis and design in electrical engineering utterly transformed our society in the past 100+ years. The birth of power electronics occurred when GE invented the thyristor. Since then, advancements in electronics, information technology, and computing have led to incredible innovations in energy, making it a focal point in international affairs. We are now at a pivotal moment in the advancement of semiconductor physics. Whether progress has stalled or continues, silicon and new wide-band-gap materials are poised to drive the development of future transistors. Additionally, the emergence of quantum computing and quantum engineering promises to revolutionize our computational capabilities. The editors tried to put together in these two volumes traditional as well as contemporary topics and applications that show the present and the near future, yet we are at the crossover of another technological revolution in which artificial intelligence will transform how we live and work. There is no 21st-century Electrical Engineering without knowing the foundations, which hopefully are in your hands, maybe as an e-book on your screen, wishing for you to have a great learning experience!

Chapter 1

Semiconductor power devices

Arun Kadavelugu[1], Samir Hazra[2], Sachin Madhusoodhanan[3], Awneesh Tripathi[4], Kasunaidu Vechalapu[5], Dhaval Patel[6], Kamalesh Hatua[7] and Subhashish Bhattacharya[8]

1.1 Introduction

Power semiconductor devices are the fundamental components of power converters, which process over 70% of the total global electricity [1]. Power converters are used in a wide range of applications including power generation, transmission and distribution systems, household appliances and medical equipment, and industrial, transportation and consumer electronics. They can convert from AC to AC, AC to DC, DC to AC, and DC to DC. Diode, MOSFET, IGBT, and SCR are some of the typically used power semiconductor devices. These power devices are optimized for conduction and switching properties to achieve high power conversion efficiency and power density. For over half a century, silicon (Si) has been the material of choice for power semiconductors, resulting in an extremely matured fabrication process, high reliability, and low cost.

The ever-increasing demand for high power conversion efficiency, high power density, and lower system costs led to the exploration of alternative materials like silicon carbide (SiC) and gallium nitride (GaN), referred to as wide band-gap materials. The SiC and GaN materials enable lower resistance unipolar devices, which have considerably lower switching losses due to the absence of stored carriers as in bipolar devices like IGBT and SCR. The extremely low on-state resistance, as predicted by Baliga's figure of merit [2], made it viable to develop SiC MOSFETs rated up to 15 kV [3]. The SiC MOSFETs find their applications in automotive inverters, battery chargers, renewable energy converters, etc. The

[1]OmniOn Power, USA
[2]Department of Electrical Engineering, Indian Institute of Science, Bangalore, India
[3]Google, USA
[4]Mercury Marine, USA
[5]Infineon, USA
[6]Onsemi, USA
[7]Department of Electrical Engineering, Indian Institute of Technology Madras, India
[8]Department of Electrical and Computer Engineering, North Carolina State University, USA

commercially available GaN devices are high electron mobility transistor (HEMT) structures based on 2DEG [4]. The GaN HEMT enables considerably higher switching frequencies, up to several MHz, and typically finds its application in high power density, high-efficiency power supplies, adapters, etc. The lateral GaN HEMT devices are commercially available up to 650 V with some suppliers providing 900 V and 1,200 V ratings [5]. The GaN-on-GaN vertical device technology [6] enables further higher voltage-rated devices.

The properties of Si, SiC, and GaN materials are compared in Figure 1.1 [7]. Both SiC and GaN have three times the bandgap energy of Si, and the higher critical electric field enables them to realize considerably higher voltage and lower leakage current devices. The thermal conductivity of SiC is considerably higher relative to GaN and Si, which is important for the effective dissipation of the heat generated from the conduction and the switching losses. The drift layer-specific resistance versus voltage rating of the device is compared for the three materials in Figure 1.2 [8]. The benefit of SiC and GaN in realizing considerably lower resistance devices is shown in the figure. The data shown corresponds to the drift layer resistance, however, the practical devices have parasitic components of resistance like channel resistance, substrate resistance, etc., leading to higher overall specific on-state resistance than the values shown in Figure 1.2.

Similar to the case with silicon, bipolar devices such as IGBT and thyristors are more favorable as the voltage is scaled up to reduce the conduction losses of the unipolar devices. The injection of a large number of carriers to reduce the on-state resistance, known as conductivity modulation, enables lower loss of higher voltage bipolar devices. Owing to this advantage, SiC IGBT rated for 15 kV has been

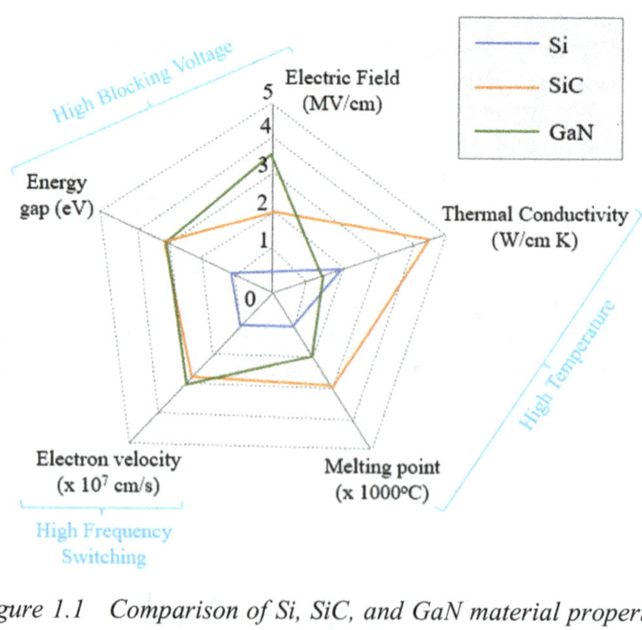

Figure 1.1 Comparison of Si, SiC, and GaN material properties

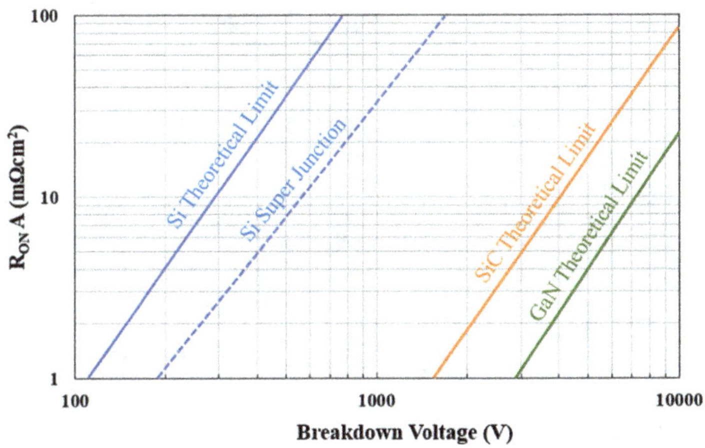

Figure 1.2 Comparison of ideal drift region resistance for Silicon, SiC, and GaN devices

developed and demonstrated in a solid-state transformer application [9,10]. The highest voltage single semiconductor device reported so far is a 27.5 kV SiC IGBT [11].

Several other materials with higher bandgap than SiC or GaN, known as ultra-wide bandgap (UWBG) materials, are being explored to push further the boundaries of the power semiconductor technology. There are reports of transistor development using gallium oxide (Ga_2O_3) [12], diamond [13], and aluminum nitride (AlN) [14].

1.2 Silicon power devices

The semiconductor industry has seen various stages of evolution of silicon power devices over the decades. These devices such as diode, MOSFET, thyristor, and IGBT are used in a variety of power conversion systems. The applications range from very low-voltage computer CPU, GPU, SOC, and FPGA power supply applications to medium-voltage motor drive applications to high-voltage direct current (HVDC) transmission systems. These applications also cover a broad range of power levels and switching frequencies. Basic details and theory of operation of a few of these silicon devices are provided in this section.

1.2.1 Diode

A diode consists of two terminals: positive (anode) and negative (cathode), and a connection between the two, known as the p–n junction. The two differently charged terminals are a result of different elements being added to the silicon. The anode is made from p-type silicon and is doped with boron or gallium. Adding phosphorous or arsenic creates the negative cathode, made from n-type silicon, in a similar manner [15,16]. The circuit symbol of the diode is shown in Figure 1.3.

Figure 1.3 Circuit symbol of a diode

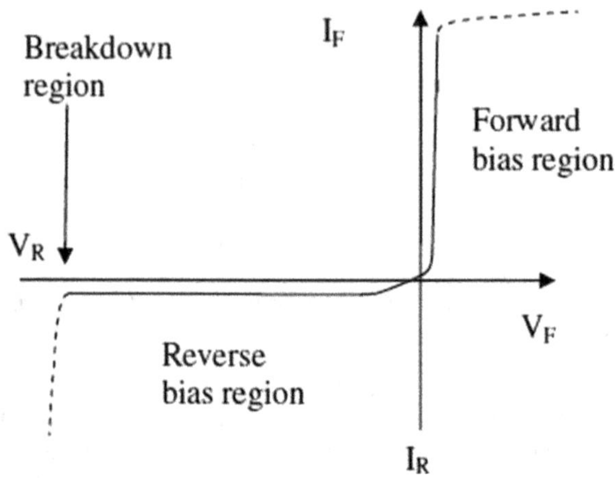

Figure 1.4 Forward and reverse V–I characteristics of a diode

Silicon power diode has a typical forward bias voltage of 0.7 V. Once this forward bias voltage is exceeded, it will conduct current across its p–n junction. It will also block the current from flowing in the reverse direction. Every silicon power diode has a maximum voltage that can be applied to it in the reverse direction, known as the breakdown voltage. It has a low (ideally zero) resistance in one direction, and high (ideally infinite) resistance in the other direction. The silicon power diode characteristics in the forward and reverse biased conditions are shown in Figure 1.4. Because of its unidirectional property, it is often used as a rectifier, which is a device that converts AC to DC with current flowing in one direction. Silicon power diodes are ubiquitous in power converters as rectifiers or co-packed with IGBT for power supplies and motor drives.

1.2.2 MOSFET

Metal oxide silicon field effect transistor (MOSFET) is a power electronic device used to switch or amplify voltages in power circuits. It is a voltage-controlled device and consists of three terminals. MOSFETs are classified into two categories: an enhancement mode device which is normally off at zero gate voltage; and a depletion mode device that is normally on at zero gate voltage. The circuit symbol of MOSFET is shown in Figure 1.5 [17].

In the depletion mode MOSFETs, when there is no voltage across the gate-source terminals, the channel has maximum conductance. When the voltage across

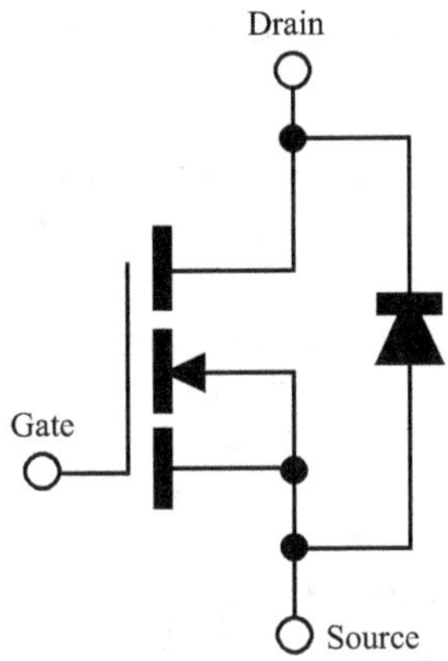

Figure 1.5 Circuit symbol of MOSFET

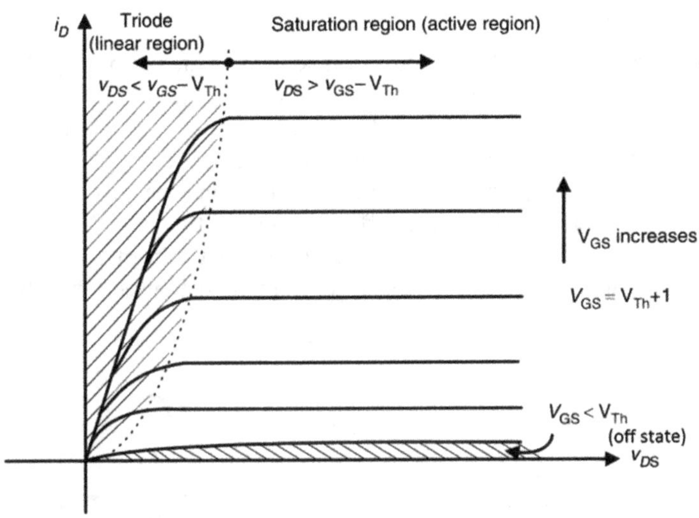

Figure 1.6 Forward characteristics of MOSFET

the gate-source terminals is increased, then the channel conductivity decreases. While in the enhancement mode MOSFETs, when there is no voltage across the gate-source terminals, the device does not conduct. When there is voltage across the gate-source terminals, the device has enhanced conductivity. The enhancement mode is more commonly used in practical applications to avoid shoot-through conditions during start-up or loss of auxiliary control supply.

A Silicon Power MOSFET has three operating regions in the forward direction, as shown in Figure 1.6, for an N-channel MOSFET. In the cut-off region (off-state), there will be no conduction and as a result, the MOSFET will be off. In this condition, the MOSFET behaves like an open switch. This happens when the applied gate-source voltage (V_{gs}) is less than the threshold voltage (V_{th}). The positive drain-source voltage (V_{ds}) applied to the MOSFET under this condition is called the forward-blocking voltage. Similarly, the negative V_{ds} applied to the MOSFET is called the reverse-blocking voltage. In power MOSFETs, due to the presence of a reverse polarity body diode, reverse blocking is not possible and it becomes a current bidirectional, voltage unidirectional two-quadrant switch. Moreover, the MOSFET channel conducts current in the reverse direction if the gate voltage is applied [2]. This mode of reverse conduction is typically used in power converters as a synchronous rectification to improve efficiency.

Silicon power MOSFETs are used in numerous applications such as DC–DC converters, AC–DC rectifiers, motor controllers, computer motherboards, server boards, etc. These devices below 300 V rating are more commonly used, as the higher voltage MOSFETs have higher conduction losses which can be addressed with superjunction or IGBT devices discussed later in this section.

1.2.3 Thyristor

A thyristor is a four-layered, three-junction semiconductor switching device with three terminals [2,15]. A thyristor is also a current unidirectional device like a diode. The Thyrisor circuit symbol and its three PN junctions and four layers are shown in Figure 1.7. Unlike a diode which is uncontrollable, the thyristor has a control or gate terminal to trigger the thyristor turn-on when a positive anode to cathode voltage is applied. There are different kinds of thyristors like SCR (silicon controlled rectifier), GTO (gate turn off), ETO (emitter turn off), etc. But SCR is the most widely used device, so thyristor is synonymously used for SCR.

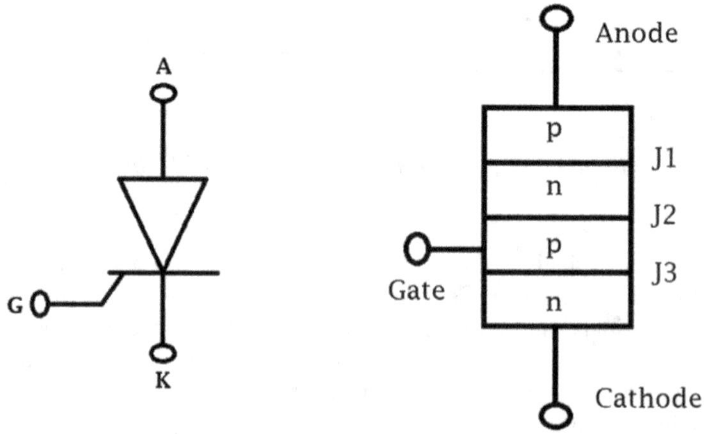

Figure 1.7 Circuit symbol and P–N junction layers of thyristor

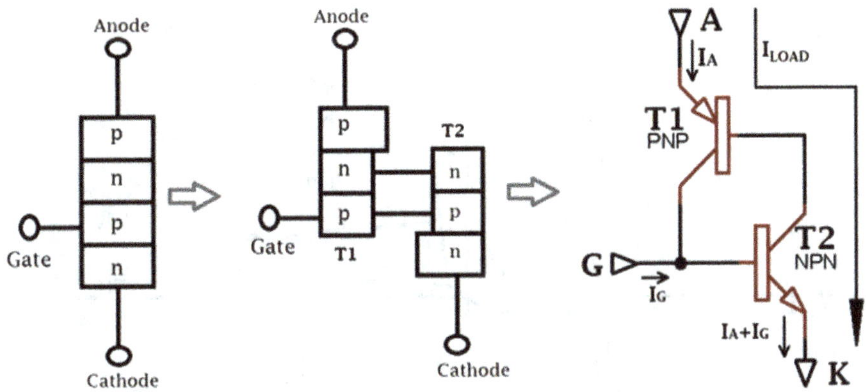

Figure 1.8 SCR P–N junction layers and equivalent circuit of two transistor analogy

The operation of a silicon thyristor could be understood by using a two-transistor analogy as shown in Figure 1.8. The equivalent circuit consists of two bipolar junction transistors (BJTs); one PNP transistor T1 and one NPN transistor T2. The base of PNP transistor T1 is fed by the collector current of NPN transistor T2, while the collector current of PNP transistor T1 feeds the base of NPN transistor T2. Applying a voltage at the gate of the thyristor triggers its turn-on state by the resulting gate current (I_g). The thyristor remains in on-state just by triggering it once. It cannot be turned off by removing the gate voltage, since both the transistors get their base continuously supplied by the collector current of the other [2]. The V–I characteristics of SCR are shown in Figure 1.9.

The SCR can be turned off by reducing the anode current (forward current) below the holding current (I_h). Once an SCR is in this forward conducting mode, different turn-off methods, known as commutation methods, are used to bring it again into the forward blocking mode. In an AC supply, the anode current goes to zero naturally at the end of the positive half cycle. Immediately reversing the voltage across the SCR turns it off naturally. This type of commutation is possible in applications such as controlled rectifiers, cyclo-converters, and AC voltage regulators. However, the current cannot go naturally to zero in the DC applications. Hence, in the case of DC circuits, the anode current is forced to zero. An external circuit known as the "forced commutation circuit" is used in such cases. Unlike the SCR, for other thyristors such as GTOs and ETOs, the thyristor design inherently takes care of the turn-off process [18,19]. The thyristors can also block reverse voltage when the cathode is made positive with respect to the anode and gate voltage is not applied.

The thyristors are mainly used where high currents and voltages are involved such as high-power motor drives, HVDC power transmission, relay circuits, oscillator circuits, level detector circuits, chopper circuits, light-dimming circuits, phase control circuits, etc.

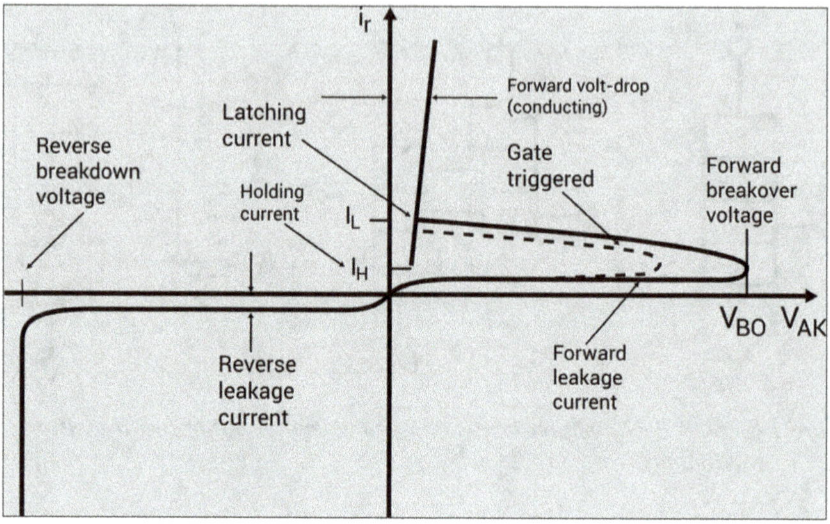

Figure 1.9 Forward and reverse V–I characteristics of SCR

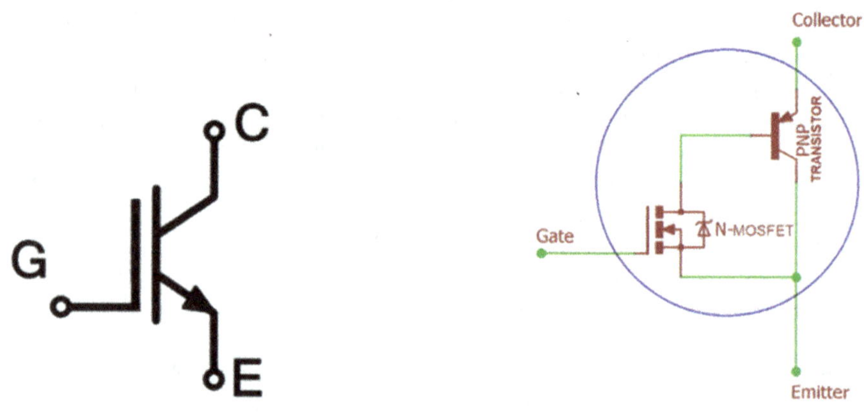

Figure 1.10 IGBT circuit symbol and equivalent representation using MOSFET and BJT

1.2.4 IGBT

The IGBT or insulated gate bipolar transistor can be equivalently represented as a combination of BJT and MOSFET [2,20]. Insulated gate refers to the input part of MOSFET having very high input impedance. Bipolar refers to the output part having a bipolar nature where the collector current flow is due to both types of charge carriers. It allows the IGBT to handle very large currents and voltages using voltage voltage-controlled gate. The circuit symbol and equivalent circuit representation of IGBT are shown in Figure 1.10.

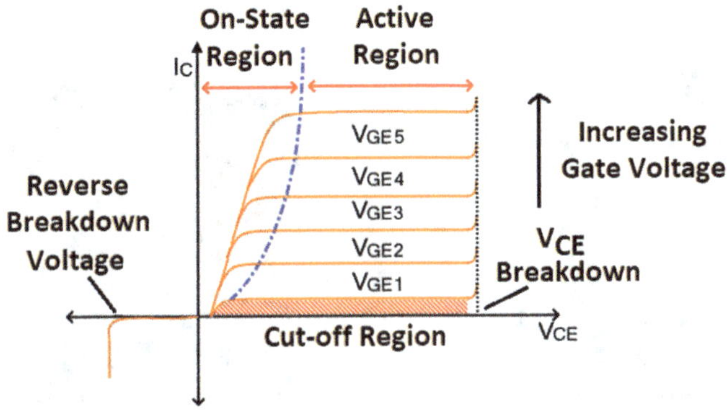

Figure 1.11 Forward and reverse V–I characteristics of IGBT

The IGBT forward characteristics are divided into three regions of operation—cut-off region, on-state region, and active region, as shown in Figure 1.11. When the gate voltage is removed, the IGBT enters the forward-blocking mode. When reverse polarity voltage is applied across the collector and emitter with gate voltage off, IGBT enters the reverse blocking mode [2,20]. IGBT is a voltage bidirectional, current unidirectional two-quadrant switch. In high-power applications, the IGBT is typically co-packaged with an anti-parallel diode to make it a current bidirectional and voltage unidirectional. In such co-packed IGBTs, the reverse blocking capability is considerably reduced by punch-through design to improve the forward conduction and switching properties. The IGBTs switch slower than MOSFET due to the tail current during turn-off [21]. Silicon IGBTs are more suited for high voltage, high current applications due to lower conduction loss [2]. In general, high voltage, high current, and lower switching frequency applications favor the IGBT while low voltage, medium current, and high switching frequency applications favor MOSFET.

The IGBTs are available with a voltage rating ranging from 400 V to 6.5 kV and a current rating ranging from a few amperes to thousands of amperes. They are widely used for industrial applications such as inverter systems, medium voltage high power industrial motor drives, uninterruptible power supplies (UPS), consumer applications such as air conditioners and induction cookers, and automotive applications such as electric vehicle (EV) motor controllers. IGBTs are also used for railway, HVDC transmission, and other such high-power applications.

1.2.5 Superjunction MOSFETs

Superjunction (SJ) MOSFETs are a special class of silicon power MOSFETs with relatively smaller on-state resistance compared to conventional MOSFETs, for the same blocking voltage [22]. The concept of superjunction was introduced by T. Fujihira in 1997. The first realization of power MOSFET based on the SJ concept, was the CoolMOS fabricated in 1998, with $R_{ds(on)}$ of 35 mΩ·cm^2 for a 600 V device [22].

The basic differences in the design of conventional power MOSFET and SJ MOSFET are highlighted in Figure 1.12. Instead of having the electrons flow

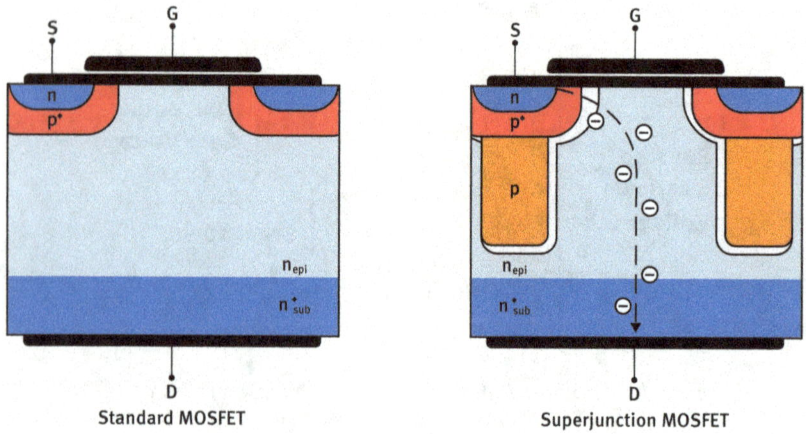

Figure 1.12 Cross-section representation of standard MOSFET and superjunction MOSFET

through a relatively high resistance n-drift area, in the SJ MOSFETs, the electrons flow in a very rich doped n-area. This creates a very low on-state resistance in the SJ MOSFETs. The main idea is to carefully balance the additional n-charge by adjacently positioned deep p-columns. These p-columns are realized all the way straight through the device close to the back-side n+ contact. The charges within the device are compensated precisely [22].

In an SJ MOSFET, both the output and reverse capacitances show a strong non-linearity. This is because, at very low voltages compared to the rated voltage, the p- and n-columns deplete each other leading to a fast expansion of the space charge layer throughout the structure. This means, for example, at a voltage beyond 50 V for 500 V rated devices, both output and reverse capacitances reach minimum values of only a few pF. This results in a dv/dt of more than 100 V/ns and di/dt of several thousand A/μs [22].

Because of this capacitance reduction, the turn-on and turn-off losses for the SJ MOSFET are about half in comparison to a standard MOSFET. Thus, the SJ MOSFET also has lower switching loss compared to a conventional planar power MOSFET. The reduction of capacitances also results in a reduced gate charge, which gives the benefit of reduced driver losses, enabling the use of a lower-cost driver [22]. The SJ MOSFETs are used in a number of applications, such as electric vehicle charging stations, power supplies, OLED television screens, etc. Efficiency in these applications is of the utmost importance and therefore, SJ MOSFETs are preferred compared to regular planar MOSFETs.

1.3 SiC power devices

SiC power devices have been explored for over 30 years for low resistance, high voltage unipolar devices as predicted by Baliga's figure of merit [2]. The high critical electrical field due to wide bandgap is the key enabler to realize low resistivity semiconductor devices. Even though MOSFET is an ideal unipolar device (normally-off device with fast switching), the initial explorations have faced challenges related to low channel mobility, gate oxide, and body diode reliability [23]. Alternative devices like SiC JFET [7] and SiC BJT [24] have been explored, where there are no gate channels or oxide or body diode. The SiC JFET can achieve lower resistance than MOSFETs but is a normally-on device, and requires a cascode configuration for normally-off operation. The SiC BJT is a normally-off device that requires relatively higher average gate power, despite considerably higher current gain than their silicon counterparts. The consistent efforts over the decades led to reliable SiC MOSFETs which have gained considerable market adoption over the last decade. Therefore, in the market, today, all these devices—SiC MOSFET, JFET, and BJT are available and tailored to meet the needs of different application requirements. This section gives an overview of some of the common SiC devices.

1.3.1 SiC diode

A major limitation of the higher voltage (1.2 kV and above) silicon diodes is the reverse recovery switching loss due to stored carriers, which can be eliminated with unipolar SiC Schottky diodes. Also, the SiC bipolar diodes have relatively much lower reverse recovery loss due to less stored carriers in comparison to their silicon counterparts. Therefore, even though SiC JBS diodes are most used today in low-voltage commercial applications, the PiN and MPS (mixed PiN and Schottky) diodes are attractive for medium-voltage applications.

The half-cell cross-sectional view of these diodes is shown in Figure 1.13. In the JBS diode, the P+/N- junction does not turn-on during on-state operation

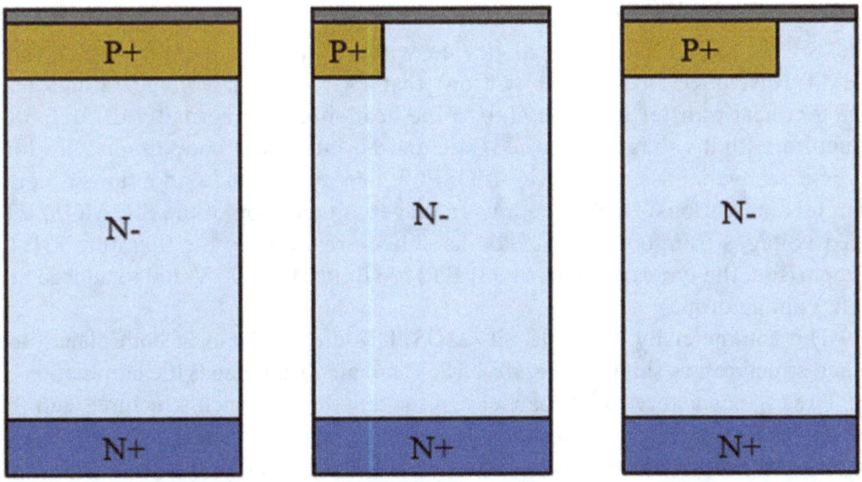

Figure 1.13 Half-cell cross-section representation of SiC PiN, JBS, and MPS diodes (from left to right)

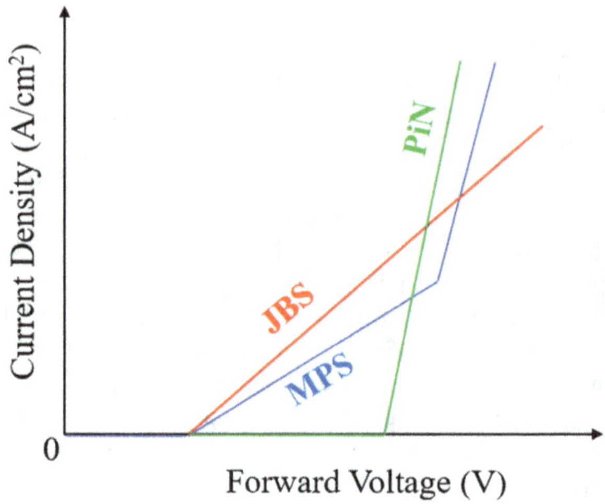

Figure 1.14 Piecewise representation of forward voltage drop of SiC JBS, MPS, and PiN diodes

leading to a high voltage drop. In the PiN diode, the minority carriers are injected as the P+/N- junction is turned on leading to a lower on-state voltage drop. The MPS diode is effectively a parallel connection of both the JBS and PiN diodes, where the junction is turned only at higher currents leading to lower on-state voltage as shown in Figure 1.14. Consequently, the JBS diode has no stored carriers, whereas the PiN diode has a larger number of stored carriers than the MPS diode contributing to the reverse recovery switching loss.

1.3.2 SiC MOSFET

The SiC MOSFET is a vertical device with no interfacial layers as in GaN HEMT (discussed in the next section) that enables high voltage ratings and efficient heat transfer from the chip to the heat sink. The normally-off unipolar structure with a voltage-controlled gate and an integrated body diode with low reverse recovery loss of the SiC MOSFET has gained wide adoption in commercial applications. Due to the low on-stage voltage drop of the SiC MOSFET, rated voltages from 600 V to 15 kV have been reported in the literature [3]. In comparison, the typical silicon MOSFETs are limited to 300 V due to a large on-state voltage drop.

The commercially available SiC MOSFETs are available in both planar and trench structures, as shown in Figure 1.15. The main difference is the elimination of JFET resistance and reduced channel resistance with the trench structure, with the planar being a simpler structure to fabricate. Both structures have an inherent bipolar body diode for reverse conduction. The body diode of the MOSFET eliminates the need for an extra device to be connected for the two-quadrant operation needed for the typical voltage source converters. However, it must be noted that the

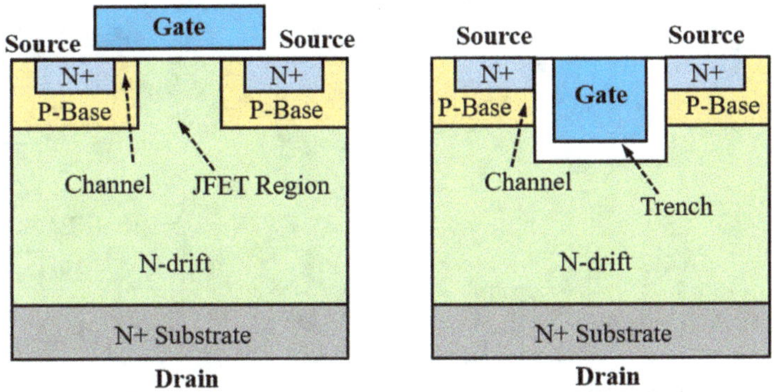

Figure 1.15 (Left) Simplified planar MOSFET structure, (right) simplified trench MOSFET structure

body diode of SiC MOSFET has a considerably larger voltage drop. So, its operation is restricted to dead-time periods to minimize thermal stress. The channel of the MOSFET is utilized for conduction in the reverse direction by the application of the gate voltage, a typical operation referred to as synchronous rectification. The large voltage drop of the SiC body diode also enables the addition of an antiparallel JBS diode to eliminate the reverse recovery loss. The antiparallel diodes also helped to overcome the body diode reliability issues with the older generation MOSFETs. With the advances in SiC technology, the newer generation devices are reported to have no body diode reliability issues.

1.3.3 SiC JFET

The JFET is a normally-on device, a favorable property for circuit breaker applications. However, for voltage source converters, this can possibly cause shoot-through during the start-up. JFET has promising characteristics like the absence of the gate oxide and considerably lower specific on-state resistance relative to SiC MOSFETs. This enabled them to extend their use for voltage source converter applications by realizing a normally-off device with cascode configuration shown in Figure 1.16. The cascode structure has a silicon MOSFET connected in series with high voltage SiC JFET, where the gate is controlled on the silicon MOSFET. A major benefit of this structure is that the gate driver of a typical silicon device can be used. A limitation of this arrangement is the lack of control of the switching speed of the JFET. It must be noted that the increase in the on-state resistance of cascode devices due to silicon MOSFET in series is a small fraction because of its low voltage rating.

Another advantage of cascode configuration is that more JFETs can be connected in series with the silicon MOSFET to realize higher voltage normally-off device, as shown in Figure 1.16. This configuration, referred to as super cascode [25], eliminates the need for high-voltage SiC devices which can be considerably more expensive. The multiple devices in the super cascode also help with thermal

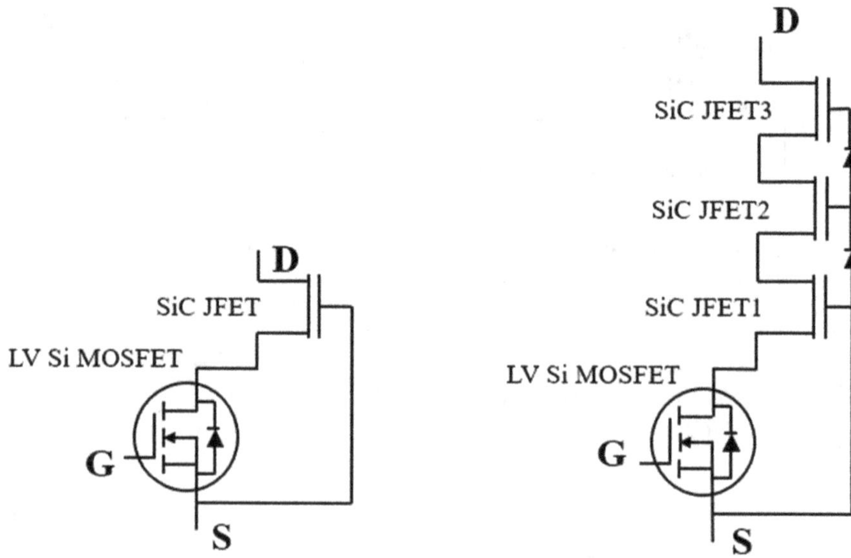

Figure 1.16 Cascode and super-cascode device configurations

dissipation as the heat is spread out. To ensure uniform voltage sharing among the series of connected devices during switching instants, passive balancing circuitry may be needed.

1.3.4 SiC IGBT

The unipolar devices like MOSFET have low switching loss, but as the voltage is scaled up their $R_{ds(on)}$ increases considerably which can offset their low switching loss. Therefore, at higher voltages, bipolar devices like IGBT are more attractive because of low conduction losses due to the injected carriers despite the increased switching losses. In silicon, the transition point from MOSFET to IGBT is around 600 V, whereas in SiC, it is around 9 kV [26]. Such considerably high transition voltage in SiC is due to the fact that the unipolar resistance is significantly lower due to a high critical electric field resulting from a wide bandgap.

The injection of minority carriers makes SiC IGBT promising in terms of conduction losses, at 10 kV and above. The switching losses are also relatively low, which enables them to switch at multiple kHz. The 15 kV N-IGBTs have been extensively evaluated for a medium voltage converter demonstration in [9,27]. There have also been reports showcasing the SiC IGBT blocking capability up to 27 kV [11]. A simplified cross section of 15 kV N-IGBT and a co-pack module using semiconductor dies from Cree and packaged by Powerex, are shown in Figure 1.17.

The IGBT switching versus conduction losses are adjusted by varying the field buffer layer thickness and doping of the punch-through structure. Figure 1.18 shows the turn-off switching loss of 15 kV SiC IGBT with 2 μm and 5 μm thick

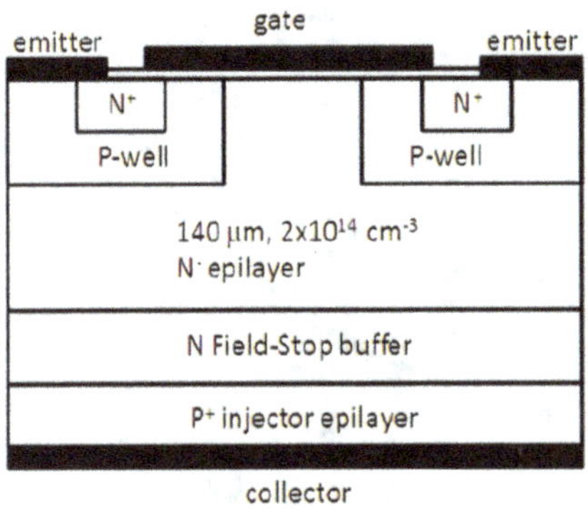

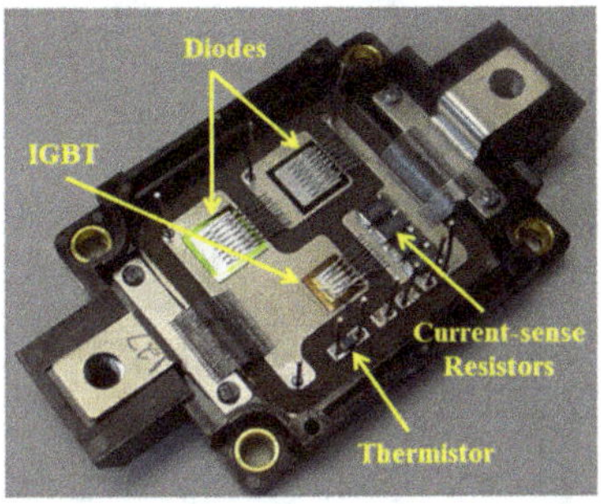

Figure 1.17 *(Left) Simplified cross-section of 15 kV N-IGBT; (right) 15 kV, 20 A, N-IGBT co-pack module*

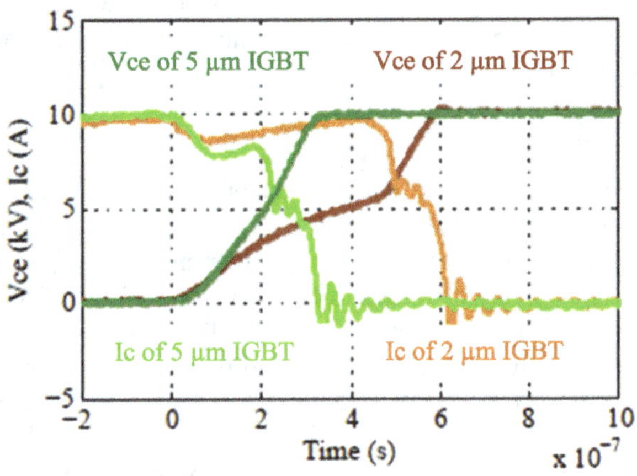

Figure 1.18 *Turn-off transitions of 15 kV, 2 μm and 5 μm thick buffer layer, SiC IGBT at 10 kV, 10 A, and 25°C*

buffer layers. The 5 μm buffer layer device has a faster transition, and thus lower switching loss. This is due to the fewer injected carriers from the P+ injector due to the thicker buffer layer. It must be noted that both SiC IGBTs have switching transition times well below 1 μs. Figure 1.19 compares the forward drop of both the IGBTs, where the 5 μm device has about 20% higher drop due to less concentration of the injected carriers.

The SiC IGBTs drastically simplify medium voltage power converter topologies due to their high voltage blocking capability. There are several reports highlighting the potential benefits of the SiC IGBTs for solid-state transformers (SST), renewable energy grid interfaces, and motor drive applications. An H-bridge converter is showcased with a 10 kV DC bus, that can be interfaced with a 7.2 kV AC single-phase grid using just four semiconductor devices. The 15 kV IGBT has been

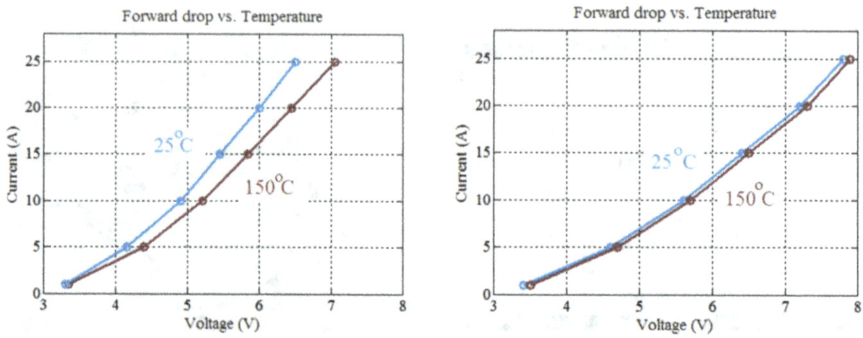

Figure 1.19 Forward voltage drop of 15 kV SiC N-IGBTs: (left) 2 μm buffer device, (right) 5 μm buffer device

paper demonstrated at 5 kHz switching capability, while operating at 10 kV DC bus [9]. The high switching frequency capability enables high power density by reducing the filtering requirement.

1.3.5 Summary

The low voltage ($\leq$ 1,200 V) SiC semiconductor devices have gained significant market adoption over the last decade. They are widely used in electric vehicle drive trains, battery charging, renewable energy converters, power supplies, auxiliary power supplies, etc. The main benefits of SiC devices are high efficiency and high power density enabled by low switching losses. The higher switching frequency reduces filter requirements which saves cost and improves power density. The SiC devices are also promising for soft switching topologies due to lower dead-time requirements.

The high-voltage devices have shown the capability of the SiC technology to enable simpler topologies with high efficiency and high power density. Several demonstrators at medium voltage showcased these benefits, for example, solid-state power substation (SSPS) [28] based on 10 kV SiC MOSFETs, transformerless intelligent power substation (TIPS) [29] based on 15 kV SiC IGBTs and medium voltage DC circuit breaker using 15 kV SiC thyristors [30]. The superjunction MOSFETs with 6.5 kV blocking voltage have been developed [31] to showcase the possibility of extending the unipolar devices SiC technology.

1.4 GaN power devices

Gallium Nitride (GaN) is a promising wide bandgap semiconductor material that enables high efficiency, high-frequency power conversion. Currently, GaN lateral devices are commercially available with voltage ratings from 15 V [32] up to 1,200 V [33], even though the typical offering is limited to 650 V. The lateral GaN devices complement the SiC devices which are available at higher voltages from 600 V and above, for developing high-performance power converters. As vertical GaN-on-GaN devices become available, the GaN technology can offer a broad range of devices from 30 V to several kilovolt ratings.

1.4.1 Basic GaN HEMT

A unique feature of GaN is that it can be grown on silicon substrates, thus significantly reducing cost. A simplified structure of GaN HEMT is shown in Figure 1.20. An important feature of this device structure is the AlGaN/GaN heterojunction. At the interface between these two layers, a layer of high-mobility electrons called "two-dimensional electron gas" (2DEG) is formed. The 2DEG is a highly conductive native channel between the source and drain of the device [34]. Because of the native 2DEG channel, the GaN HEMT is inherently a normally-on device. As power converter designers prefer normally-off devices to prevent a potential shoot-through during startup or loss of control power, several methods have been explored to develop normally-off GaN HEMTs.

1.4.2 GaN cascode device

A popular method of converting the HEMT into a normally-off device is by using the cascode structure shown in Figure 1.21. The cascode configuration is a co-package of the depletion-mode HEMT with a low-voltage enhancement mode MOSFET, typically a silicon device in commercial implementations. The two dies are connected in such that the output (drain–source) voltage of the MOSFET determines the input (gate–source) voltage of the HEMT. Essentially, the gate pulses applied to the low voltage silicon MOSFET control the HEMT switching transitions, while the cascode configuration behaves as normally-off switch. An attractive feature of the cascode device is that a typical gate driver used for silicon MOSFETs can be employed. The switching performance of the cascode device is influenced by the parasitic inductances in the package, especially between the two dies, and also by how well the junction capacitances of the two are matched. If the inductances are too high, or the capacitances are not matched well, the switching losses can increase significantly [34].

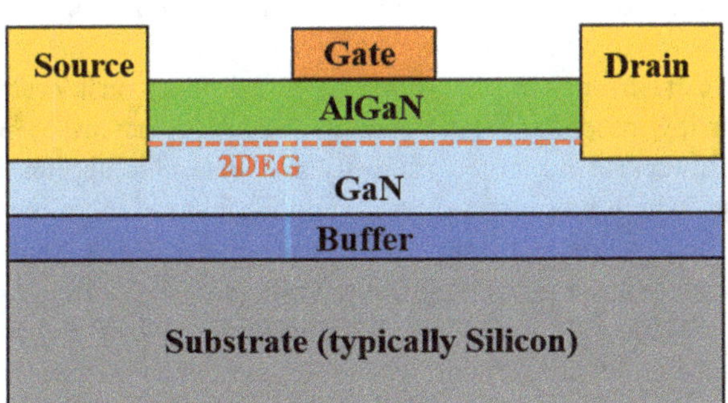

Figure 1.20 Simplified cross-section of GaN HEMT

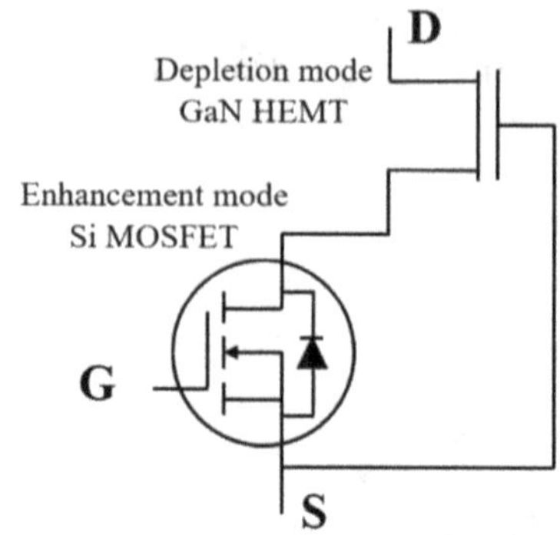

Figure 1.21 Cascode configuration of normally-off GaN HEMT

1.4.3 E-mode GaN device

The GaN HEMT can also be converted to a normally-off enhancement mode (e-mode) device by modifying the gate to shift the threshold voltage to a positive value. This avoids the need for a second device to be connected in series, as in the cascode configuration, and therefore adopted by multiple GaN semiconductor manufacturers. One key difference from the cascode configuration is that the e-mode GaN device has a lower gate-source voltage requirement, and cannot be driven with a standard gate driver used for silicon devices. However, the lateral GaN devices have significantly lower gate-drain capacitance (C_{gd}) and gate-source capacitance (C_{gs}) compared to Si devices. The lower gate charge, due to lower capacitances, enables significantly faster and low-loss switching transients making the GaN devices suitable for very high frequency applications. The lower gate charge also results in significantly lower losses on the gate driver and therefore requires lower power isolation supplies to drive these devices.

Unlike in a MOSFET structure, there is no inherent body-diode in the lateral GaN devices. However, there are reverse conduction mechanisms in both cascode and e-mode GaN devices. In the cascode structure, the body-diode (of silicon MOSFET) starts conducting the reverse current, which will then turn-on the channel of the GaN device. The reverse recovery charge during the switching events will come from only the Si MOSFET and it is a considerably small component due to its low blocking. There is no reverse recovery loss component in the GaN device, as there are no stored minority carriers. This applies to the e-mode GaN device also, which has zero reverse recovery loss, as there is no low voltage MOSFET like in cascode configuration. The e-mode GaN device has reverse conduction behavior like a diode with knee voltage of V_{th}, when the gate voltage with respect to the drain (V_{gd}) exceeds V_{th}, and then exhibits a channel resistance like in the forward conduction mode due to the symmetry of the device. The forward and reverse conduction characteristics of an e-mode GaN device are shown in Figure 1.22. If negative gate voltage is used, the knee voltage will further increase (red dashed line in the figure) by the magnitude of the negative V_{gs}. Therefore, it is

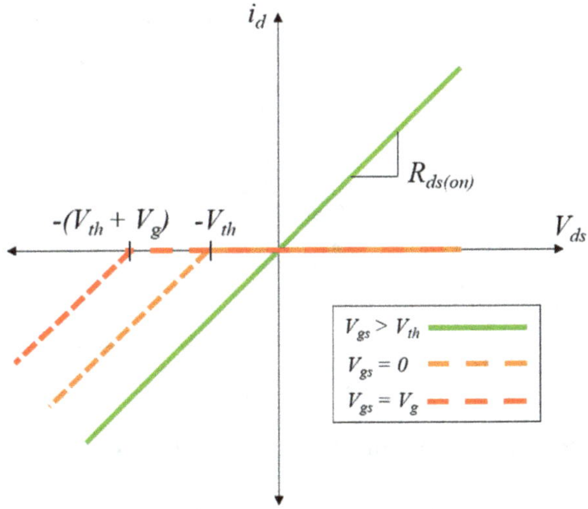

Figure 1.22 Forward and reverse conduction characteristics of e-mode GaN device

recommended to turn-on the gate, like in a synchronous rectifier, to obtain similar conduction properties as in forward conduction, after the dead time to avoid shoot-through.

1.4.4 Summary

GaN devices complement the wide bandgap space at lower voltages (<650 V), as SiC devices are available at 650 V and above, for high efficiency and high power-density applications. At 650 V, both GaN and SiC devices are available, with GaN preferred for higher switching frequency applications. The use of multi-level converter topologies can extend the benefit of the GaN devices in higher voltage applications. Vertical GaN devices [35] are also being explored for higher voltage-rated GaN devices. This development is enabled by the availability of higher quality, bulk GaN substrates [36].

An important feature of the lateral GaN devices is using silicon substrates which are cost-effective and enable monolithic integration of driver or control circuitry. The integrated driver circuit minimizes gate voltage oscillations, improves power density, and potentially reduces cost. The lack of reverse recovery losses is another important property of lateral GaN devices that enable high-frequency switching. Also, the GaN devices have significantly lower device capacitances as shown in Table 1.1, where a 650 V GaN device is compared with a 650 V SiC device. The lower capacitances lead to significantly faster switching transients with GaN and also the lower gate charge reduces the isolated power supply requirements for the gate driver circuit. It is to be noted that the higher voltage lateral GaN devices have a considerable increase in $R_{ds(on)}$ with respect to temperature, unlike the SiC device of similar voltage ratings as compared in Table 1.1.

Table 1.1 Comparison of 650 V normally-off GaN and SiC devices

Parameter	GaN (650 V) GS66508B	SiC (650 V) C3M0060065D
$R_{ds(on)}$ at 25°C	50 mΩ	60 mΩ
$R_{ds(on)}$ at 150°C	2.6 x 50 mΩ	1.3 x 60 mΩ
C_{iss} (at 400 V)	242 pF	1,000 pF
C_{oss} (at 400 V)	65 pF	80 pF
C_{rss} (at 400 V)	1.5 pF	9 pF
Q_G	6.1 nC	46 nC
Q_{rr}	0 nC	181 nC

1.5 Characterization of power semiconductor devices

In a power electronic converter, a semiconductor power device operates either in saturation or cut-off region on I–V plane which is defined as on and off states respectively. The power loss in the device is zero when the device is in an off state. During the on state, due to the small saturation voltage drop, a power loss occurs which is termed as conduction loss. The power loss due to the switching of the states between on and off is called switching loss.

The static characteristics of devices are used to estimate the total conduction loss of the converter. The forward characteristics can be obtained from an I–V curve tracer at different junction temperatures. The datasheet of a power device typically provides conduction characteristics at different temperatures, gate voltages, and currents. This data is useful in calculating the conduction loss of a power converter. However, the switching losses are more dependent on the converter setup and can vary significantly with the stray inductance of the commutation loop. Therefore, it is important to characterize the switching losses in the converter setup.

1.5.1 Switching characterization of a power device

A clamped inductive double pulse circuit shown in Figure 1.23 is widely used for the characterization of turn-on and turn-off transients by applying two switching pulses. When the device under test (DUT) is turned on by the first pulse, the current in the inductor builds from zero as shown in Figure 1.24. When it is turned off the inductor current freewheels through an anti-parallel diode FWD. At the end of the first pulse, the DUT turn-off transient is measured. While the inductor current is freewheeling, a second pulse is applied to DUT to test the turn-on transient. The width of the first pulse is decided by the current at which the device is to be tested for turn-off transient. The free-wheeling time is kept short to minimize the decay of the current so that the device can be tested for turn-on transient at the same level of current. The second pulse is also short to restrict unnecessary current buildup in the inductor. For device characterization device, voltage V_{ds} and current I_d are measured with high bandwidth voltage and current probes.

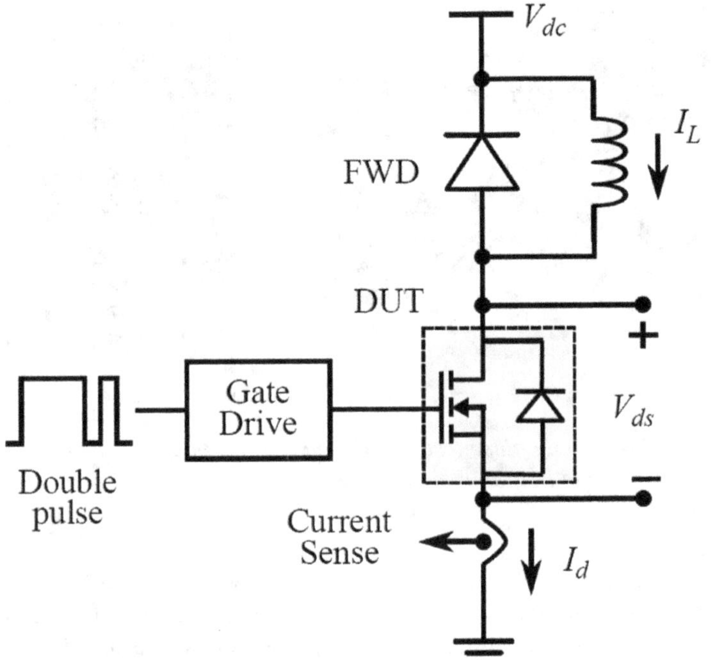

Figure 1.23 Double pulse device characterization circuit

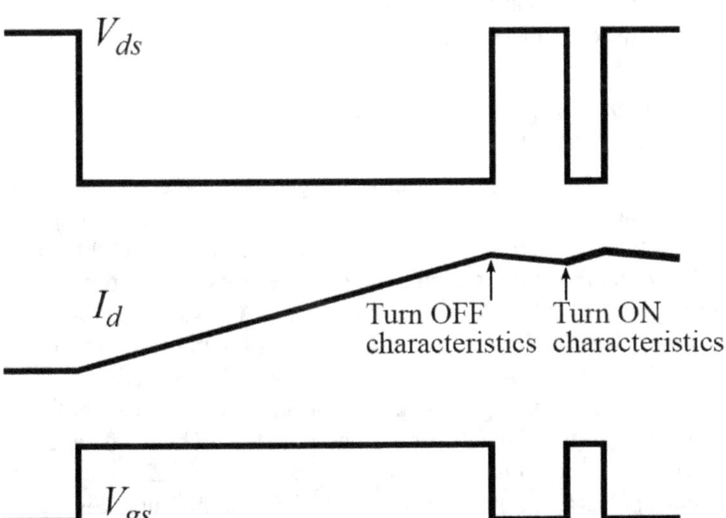

Figure 1.24 Device voltage and current waveform for double pulse test

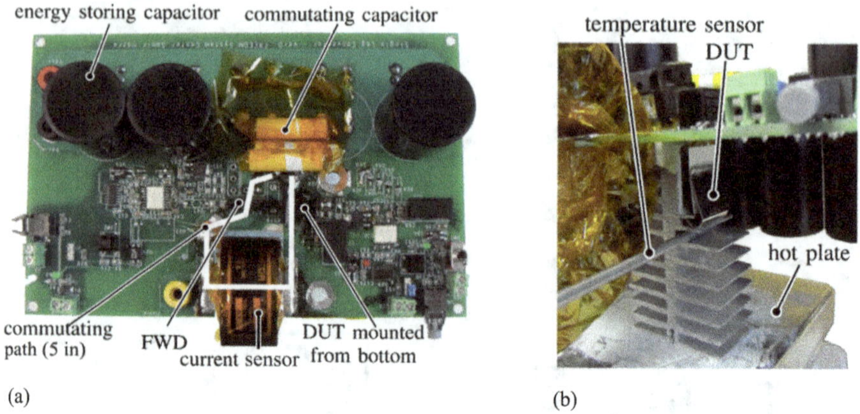

Figure 1.25 (a) Double pulse test circuit setup. (b) Temperature measurement example in a double pulse setup

An example of a double-pulse test setup is shown in Figure 1.25(a). A high-bandwidth Pearson current probe is used to measure the device current as shown in the figure. The Rogowski coil-based current probes and shunts are also commonly used for measuring switching current. It is important to not introduce a significant inductance in the commutation path while adding the current probe for measurements. The higher inductance will lead to V_{ds} overvoltage spike and also distort the switching losses of the DUT. An arrangement is shown in Figure 1.25(b) to control the temperature of the setup to be able to test the device at a particular junction temperature. The use of hotplates to mount the devices is also a common method for high-temperature characterization.

1.5.2 Difference in switching transients with a complementary device

The DUT turn-on characteristics vary based on the complementary device as shown in Figure 1.26 due to the higher output capacitance. A practical example of this is presented with the characterization of a 1,200 V, 100 A SiC MOSFET half-bridge module [37]. In the first case, the high-side device is bypassed from the circuit by not connecting it to the DC bus. Instead, a separate freewheeling diode is used as per the double pulse circuit in Figure 1.26(a). Turn-on and turn-off characteristics at 800 V and 100 A are shown in Figures 1.27 and 1.28 respectively.

In the second case, the high-side device of the module is used in the circuit as shown in Figure 1.26(b). The anti-parallel diode of the high-side device freewheels inductor current. From the turn-on characteristics in Figure 1.29, it can be seen that the device current has a large turn-on spike which is not present in Figure 1.27. This is due to the high drain to source capacitance (C_{ds}) of the top device which has both MOSFET and diode co-packed.

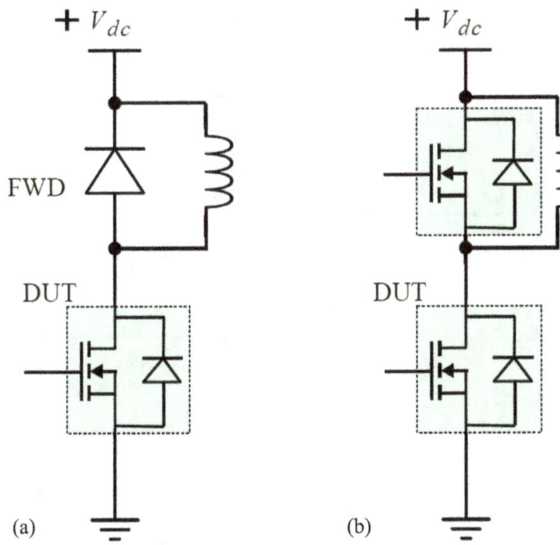

Figure 1.26 Device characterization with (a) free-wheeling diode (b) MOSFET for free-wheeling

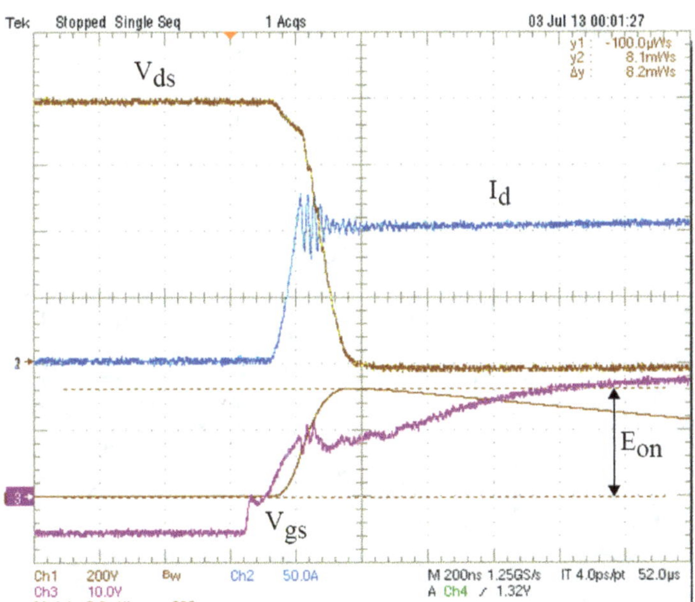

Figure 1.27 Turn-on characteristics with measured E_{on} = 8.2 mJ (scale: V_{ds} ⇒ 200 V/div, I_d ⇒ 50 A/div, V_{gs} ⇒ 10 V/div, Energy ⇒ 5 mJ/div, Time ⇒ 200 ns/div)

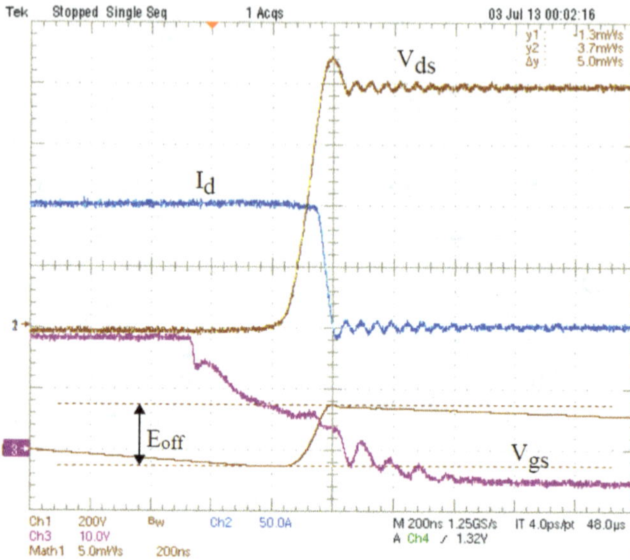

Figure 1.28 Turn-off characteristics with measured E_{off} = 5.0 mJ (scale: $V_{ds} \Rightarrow$ 800 V/div, $I_d \Rightarrow$ 50 A/div, $V_{gs} \Rightarrow$ 10 V/div, Energy $\Rightarrow$ 5 mJ/div, Time $\Rightarrow$ 200 ns/div)

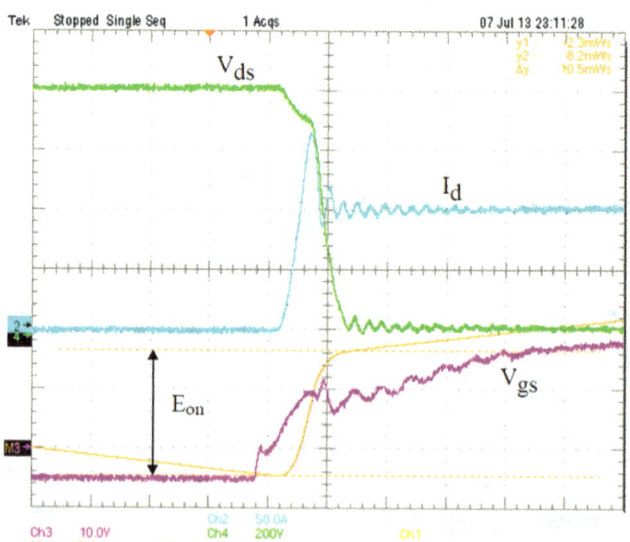

Figure 1.29 Module turn-on characteristics with measured E_{on} = 10.5 mJ (scale: $V_{ds} \Rightarrow$ 200 V/div, $I_d \Rightarrow$ 50 A/div, $V_{gs} \Rightarrow$ 10 V/div, Energy $\Rightarrow$ 5 mJ/div, Time $\Rightarrow$ 200 ns/div)

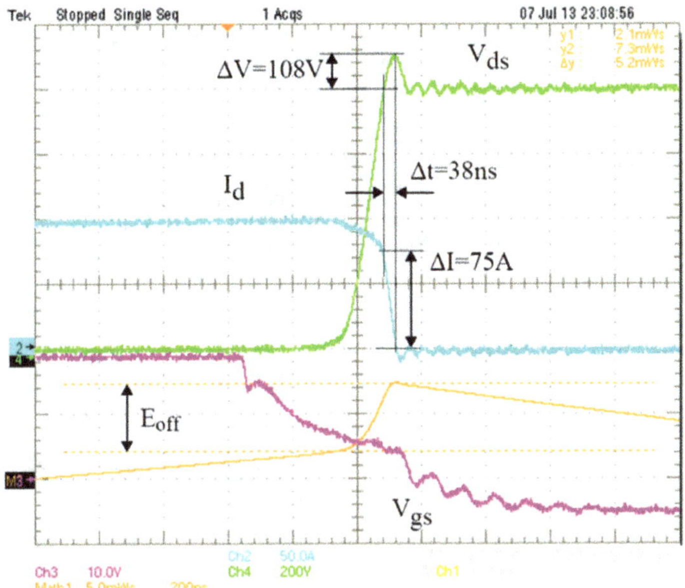

Figure 1.30 Module turn-off characteristics with measured E_{off} = 5.2 mJ (scale: $V_{ds} \Rightarrow$ 200 V/div, $I_d \Rightarrow$ 50 A/div, $V_{gs} \Rightarrow$ 10 V/div, Energy $\Rightarrow$ 5 mJ/div, Time $\Rightarrow$ 200 ns/div)

Table 1.2 Loss variation of a SiC device with different freewheeling devices

Test configuration	Turn-on loss (mJ)	Turn-off loss (mJ)
Separate FWD	8.2	5.0
Anti-parallel diode of a high-side device of the module	10.5	5.2

The current spike causes additional turn-on loss. However, the turn-off characteristics of the module remain the same as shown in Figure 1.30. This switching loss variation between the two cases is compared in Table 1.2.

The variation of gate resistance has an impact on the switching loss. With smaller gate resistance, device switching loss reduces, but the dv/dt increases. A practical example of the effect of the gate resistance variation on the turn-on and turn-off behavior of a 1,200 V, 100 A SiC MOSFET half-bridge module is shown in Figures 1.31 and 1.32, with the drain-source voltages, drain currents, and gate voltages [37].

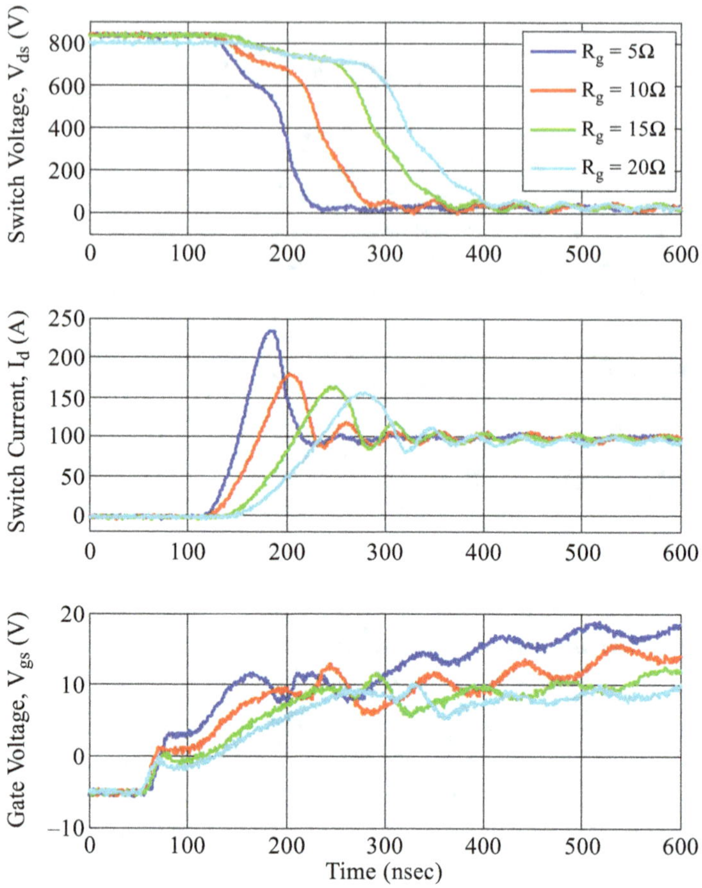

Figure 1.31 Example waveforms of turn-on transient behavior with different gate resistances, R_g.

The double-pulse test circuit shown in Figure 1.23 can be used for measuring V_{ds} overshoot at overload or other corner cases of the converter operation to ensure that the voltage is within the device ratings. Also, it can be used for diode reverse recovery loss characterization and cross-conduction (or Miller turn-on) phenomenon. It is a simple circuit that requires just two PWM pulses that can be generated using a benchtop signal generator. The evaluation of switching losses, V_{ds} overshoot, and ensuring no cross-conduction phenomenon are critical for a converter design.

It is to be noted that the double pulse circuit shown in Figure 1.23 is for characterizing typical hard-switched conditions. However, for soft-switching characterization, there are modified circuits to emulate zero voltage switching (ZVS) [38] and zero current switching (ZCS) [39] behavior of power semiconductor devices.

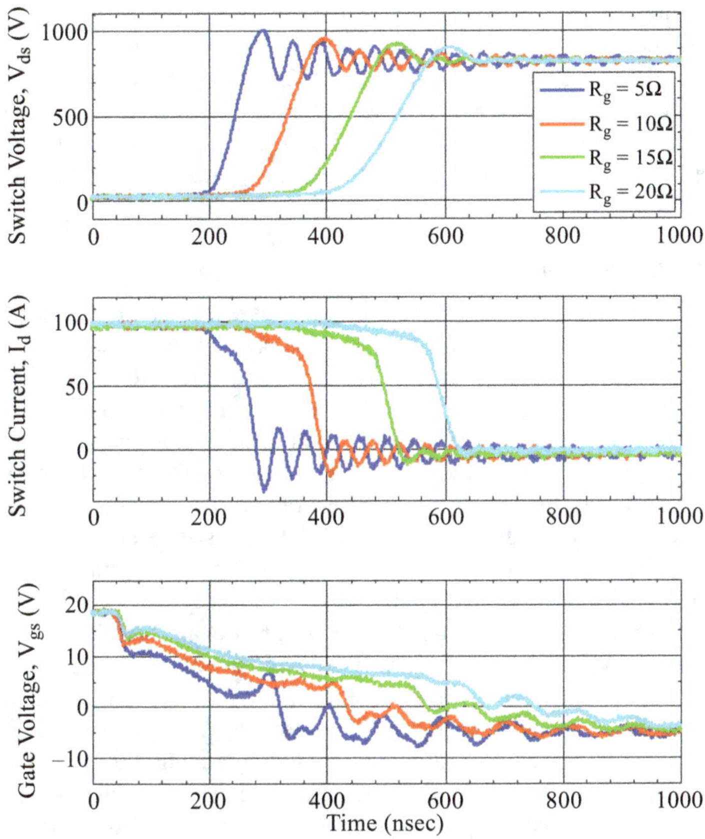

Figure 1.32 Example waveforms of turn-off transient behavior with different gate resistances, R_g.

1.6 Parallel connection of devices

The semiconductor chip size is limited by the wafer yields and the thermo-mechanical considerations due to temperature cycling. Therefore, high-current semiconductors are realized by paralleling multiple chips in a semiconductor module, or by paralleling discrete semiconductor devices, or by paralleling modules for higher current. Ensuring uniform current sharing between parallel devices is pivotal to avoiding non-uniform thermal stress and device failures over time.

There are several factors that contribute to static and dynamic current sharing. The static current sharing depends on variables that are resistive in nature; for example, the ohmic value of the parallel power loops, the applied on-state gate voltage (V_{gs}), and the on-state resistance or voltage drop. On the other hand, the dynamic switching characteristics are primarily governed by the stray inductance on the power loops of each parallel branch (bus structure and connections); the common source/emitter stray inductances—the stray inductances that are common to both, the power loop and gate loop paths; differences on device's threshold voltage V_{th}; and device's capacitances C_{iss} and C_{rss}. If separate gate drivers are used for each device, the dynamic behavior is also influenced by their propagation delays and their output stage impedance. The current sharing is also influenced by differences in device turn-on/turn-off delays, gate loop resistances, and inductances.

1.6.1 Static current sharing

The basic requirement for static current sharing is ensuring the resistance of the current flowing paths is identical for each of the parallel semiconductors. This can be achieved by a symmetric layout or busbar structure. However, the parallel semiconductor devices can have different on-state resistance values due to process variations during the fabrication. This can lead to non-uniform current sharing in spite of the symmetric circuit layout. A simple approach to address this issue is by adjusting the V_{gs}, as higher V_{gs} reduce the on-state resistance and vice versa. However, typical semiconductors like MOSFET (a resistive device with positive temperature coefficient) and IGBT (typically positive temperature coefficient of on-state voltage drop) can have a natural current balance due to negative feedback from their temperature versus on-state voltage drop characteristics. On the other hand, a device like SCR has a negative temperature coefficient of the on-state voltage drop and also is not gate-controllable. By prescreening the devices and paralleling SCRs with similar on-state voltage drops can potentially avoid the thermal runaway and ensure uniform current sharing among the parallel devices.

A practical example of the current sharing among the paralleled high-current SiC modules is reported in [40]. The double-pulse waveforms are shown in Figure 1.33, where the current ramps of the three parallel SiC modules are different

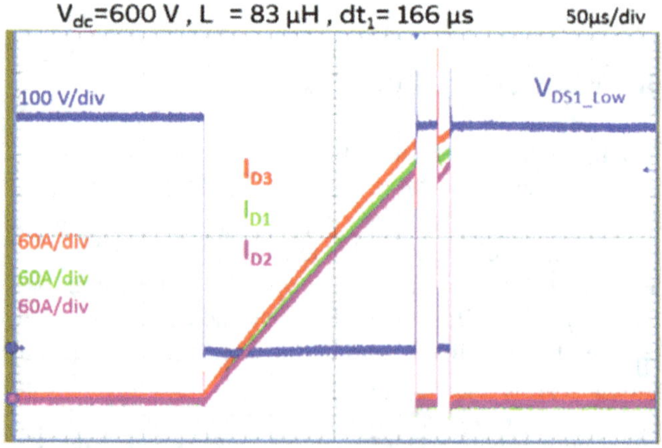

Figure 1.33 Double-pulse waveforms of three paralleled 1,200 V, 400 A SiC modules [39]

due to the difference in their on-state resistances. The larger value of I_{D3} (drain current of third SiC MOSFET) implies lower on-state resistance and the lower I_{D2} implies higher on-state resistance. In most power electronic applications, the semiconductors switch between on-state to off-state and vice versa and ensuring uniform current sharing during those switching events is also critical, as discussed below.

1.6.2 Dynamic current sharing

The basic requirement for paralleling multiple semiconductors is to ensure a symmetric layout design with no mismatches in inductance in both power commutation and gate loops. However, this may not result in uniform current sharing between the parallel semiconductors if, for example, there is a mismatch in the V_{th} or gate driver delays. These mismatches can lead to circulating currents in the gate driver loop, as shown by the red dashed line in Figure 1.34.

The circulation current in the gate driver loop can be minimized by a simple passive technique of adding an impedance [41], Z_S (usually resistor) in the source loop of the MOSFET as shown in Figure 1.34. An alternative approach is to use common mode chokes as shown in the right-side figure, to balance the gate voltages. However, care should be taken not to introduce excessive differential inductance from the use of common mode chokes, to avoid oscillation or overvoltage spikes in the gate voltage.

The use of SiC and GaN semiconductors has gained wide acceptance due to improved power conversion efficiencies, high power densities, and potentially lower system costs. However, due to their faster slew rates, the converter circuit designer needs to carefully consider the aforementioned points while paralleling discrete devices or modules to ensure uniform current sharing. In addition to the passive techniques for paralleling semiconductors, numerous active gating techniques have also been reported in the literature. The active gating techniques are based on analog circuitry or digital logic devices like CPLD or FPGA, where the

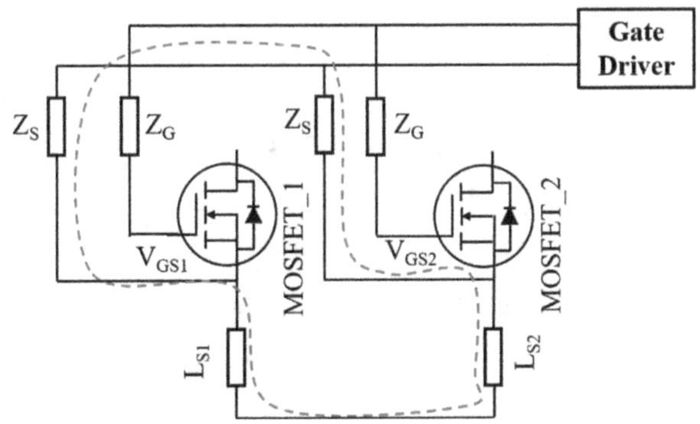

Figure 1.34 Simplified schematics of paralleling semiconductors using a common gate driver

gate signals are adjusted to match the switching instants by utilizing the data (current, voltage, etc.) of the parallel devices. A simplified schematic of an active gate driving approach is illustrated in Figure 1.35.

For reliable operation of power converters, it is important to ensure uniform current sharing among the parallel devices. The uniform distribution of the conduction loss for commonly used devices like MOSFET and IGBT can be ensured by the positive temperature coefficient of their on-state resistance or voltage drop. However, the switching losses have a negative temperature coefficient—the hotter device has a lower V_{th}, as illustrated in Figure 1.36, which leads to even more switching loss. Essentially, the device with lower V_{th} will be the first to switch on among the parallel set leading to higher turn-on loss; and similarly, it will be the last to switch off leading to higher turn-off losses also. This must be taken into consideration if the part-to-part V_{th} variation is expected to be significant, or if the switching losses contribution is large, e.g., in a very high frequency hard-switched application.

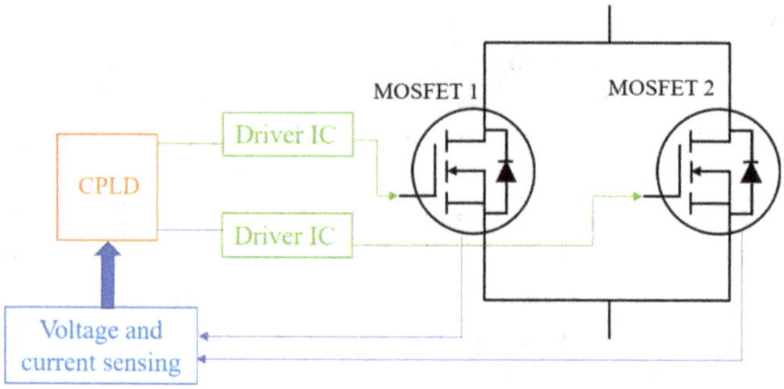

Figure 1.35 Simplified representation of an active gate driver using CPLD

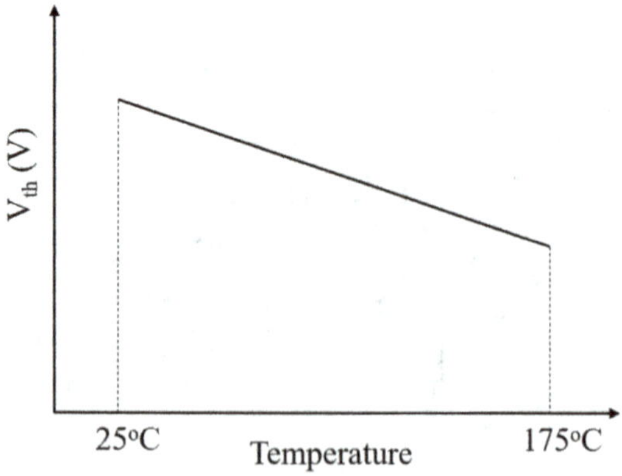

Figure 1.36 Representation of monotonic decrease of V_{th} with temperature of a MOSFET or an IGBT

1.7 Application example

The demonstration results of a three-phase solid-state transformer (SST) using 15 kV SiC IGBTs and 1.2 kV SiC MOSFETs are presented in this section to showcase the potential of the state-of-the-art power semiconductor devices. The front end of the SST is interfaced to a medium voltage grid, and it is followed by an MV to LV dual active bridge (DAB) converter and the LV DC is converted to 480 V three-phase output using the inverter. The all-SiC SST topology demonstrated at FREEDM Systems Center at North Carolina State University is referred to as a transformerless intelligent power substation (TIPS) which interfaces the three-phase, 13.8 kV and 480 V distribution grids [42,43].

TIPS has three power conversion stages as shown in Figure 1.37. The active front-end converter (AFEC) is tied to a medium voltage (MV) distribution grid with a three-level neutral point clamped (3L-NPC) converter built using the 15 kV SiC IGBTs. It is a PWM boost rectifier with a 22 kV DC bus on the output and is operated with hard-switched PWM at 3–5 kHz. The AFEC handles many functions like reactive power compensation to improve the grid voltage profile on the MV grid, power quality improvement, harmonic compensation, power factor control, stability, and control of the 22 kV DC bus voltage [41]. The AFEC DC bus voltage (22 kV) is converted to low voltage (LV) side DC bus voltage (800 V) using a high-frequency DC link, enabled by a three-phase dual active bridge converter (DABC). The DABC is responsible for isolation and voltage conversion in the TIPS system. The DABC is a ZVS soft-switched converter operated at 10 kHz. The MV side of DABC is also a 3L-NPC converter built using 15 kV SiC IGBTs [43]. The LV side of DABC has a 1.2 kV SiC MOSFET half-bridge module-based two-level converter, for each star- and delta-connected secondary windings. The output LV inverter stage is developed using 1.2 kV SiC MOSFET modules with three interleaved 17–20 kHz, 35 kVA converters [37]. The experimental demonstrator of the TIPS is shown in Figure 1.38, and the design of the MV power converter using SiC IGBTs is described in [9]. The objective of this demonstrator is to showcase the advantage of the 15 kV SiC IGBT module in simplifying converter topologies for MV applications.

Figure 1.39 and Figure 1.40 show the test waveforms of the AFEC and DABC under an integrated mode of operation of the complete TIPS system [27,44]. The test results are with the 3-phase, 4.16 kV, 60 Hz grid at unity power factor on the input side with the MV DC bus at 8 kV, LV DC bus at 480 V, and transferring 10 kW to the load. Figure 1.39 shows the AFEC line currents and the pole voltage under this condition. The switching frequency is 5 kHz.

Figure 1.40 shows the DABC waveforms under the integrated mode when DABC is switching at 10 kHz. The ringing in the DABC currents is contributed by the parasitic coupling capacitances of the high-frequency transformer. Along with the SiC devices, improvements in the magnetic design have to be made to harness the benefits of SiC devices. Two-level, three-phase converters for the LV side of the DABC and the inverter stage for the 480 V grid interface are built with 1.2 kV,

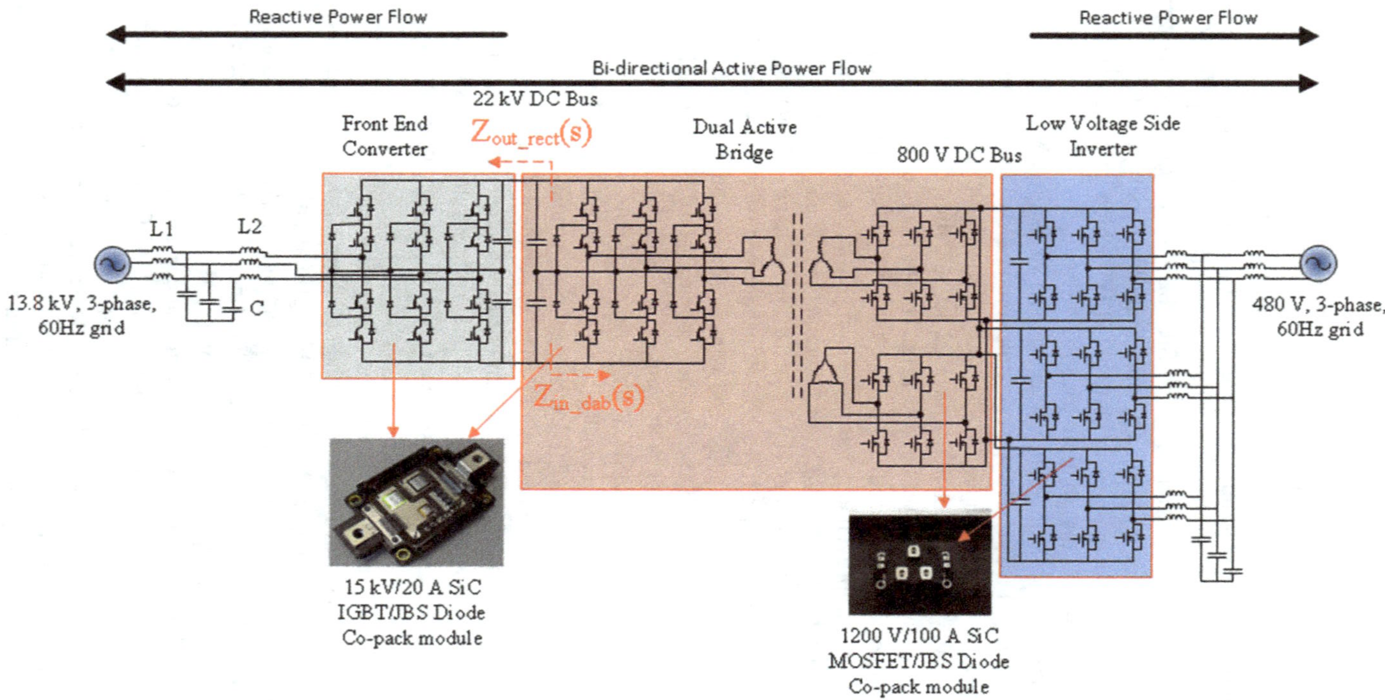

Figure 1.37 Schematic of the TIPS—a 13.8 kV to 480 V grid interfaced three-phase SST

Figure 1.38 *TIPS system experimental setup at NCSU FREEDM Systems Center laboratory*

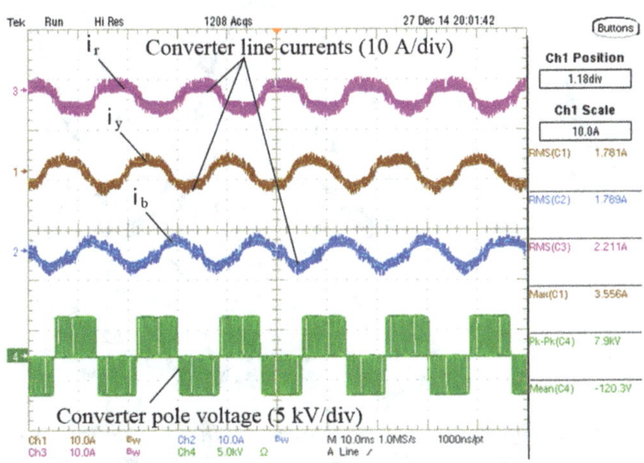

Figure 1.39 *AFEC line currents and R-phase pole voltage at 4.16 kV AC voltage, 8 kV DC bus voltage, 5 kHz switching frequency, and 10 kW power*

100 A SiC MOSFET half-bridge modules packaged by Powerex with dies from Cree. Two of these converters are tested in parallel up to 50 kVA and the corresponding waveforms are shown in Figure 1.41.

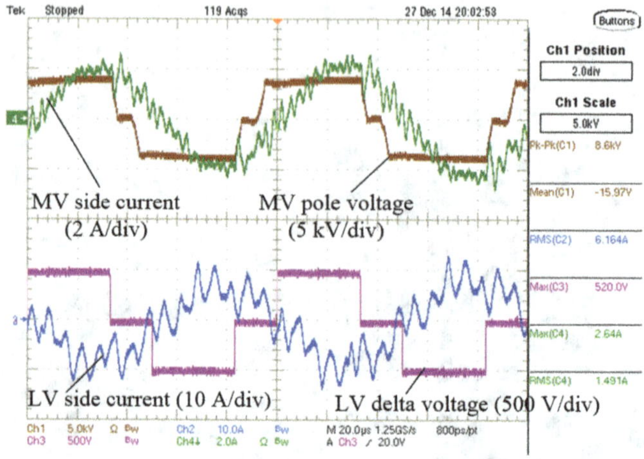

Figure 1.40 DABC waveforms at 8 kV MV DC bus voltage, 10 kHz switching frequency, and 10 kW power

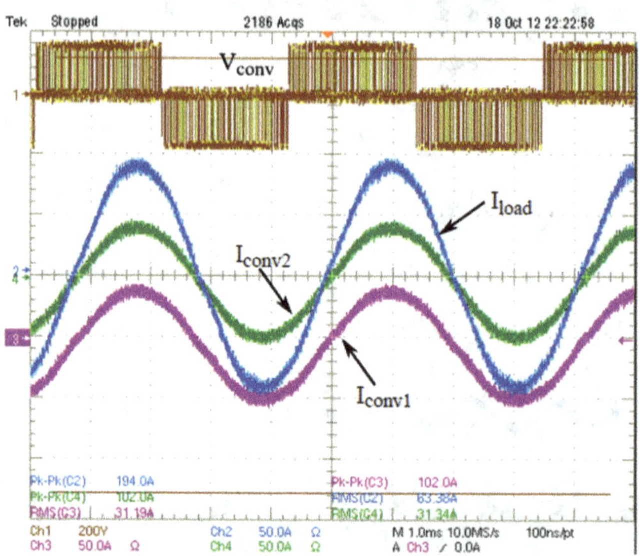

Figure 1.41 Parallel operation of two converters in an open loop supplying a total load of 50 kVA (scale: currents—50 A/div, voltage—1 kV/div, time—1 ms/div)

1.8 Conclusions

Power semiconductor devices are the fundamental components of power conversion products ubiquitous in energy conversion applications. Advances in power electronics devices would impact all these applications and open avenues for more applications that were previously not possible. This chapter presented an overview of conventional silicon devices and their real-world applications. Due to the ever-increasing demand for higher efficiency and power density, wide bandgap materials like SiC and GaN-based devices have gained significant adoption in recent years. Various SiC and GaN devices and their basic properties have been provided. The chapter included several examples of real semiconductor characterization waveforms to illustrate the practical behavior of the semiconductors for power converter designers. Also, several considerations on measuring switching losses and paralleling semiconductor devices have been provided. There are ongoing efforts to develop even more efficient power devices using ultra-wide bandgap materials like gallium oxide, aluminum nitride, and diamond that can enable even higher efficiencies and power densities in future power conversion systems.

Acknowledgment

This work is supported by the US Government through the DOE ARPA-E program under contract no. DE-AR0000110. This work made use of FREEDM ERC shared facilities supported by NSF under award no. EEC- 0812121. The authors also thank Wolfspeed, Powerex, and DOE PowerAmerica for their support.

Disclaimer

References

[1] "Power electronics: Revolutionizing the world's future energy systems," Hitachi ABB Power Grids, August 2021.

[2] B. Jayant Baliga, *Fundamentals of Power Semiconductor Devices,* Springer, New York, 2016.

[3] J. W. Palmour, L. Cheng, V. Pala, *et al.*, "Silicon carbide power MOSFETs: Breakthrough performance from 900 V up to 15 kV," *2014 IEEE 26th International Symposium on Power Semiconductor Devices & IC's (ISPSD)*, 2014.

[4] N. Keshmiri, D. Wang, B. Agarwal, R. Hou, and A. Emadi, "Current status and future trends of GaN HEMTs in electrified transportation," *IEEE Access*, Vol. 8, 2020.

[5] 1200 V GaN device datasheet: https://iganpower.com/wp-content/uploads/2021/10/GPIHV3QDFN_v3.pdf

[6] D. Ramanathan, C. Coles, and W. Meier, "Not all GaN transistors are built equal: The benefits of vertical GaN-on-GaN," *PCIM Europe*, 2020.

[7] J. Millan, P. Godignon, X. Perpina, *et al.*, "A survey of wide bandgap power semiconductor devices," *IEEE Transaction on Power Electronics*, Vol. 29, 2014.

[8] A. Imbruglia, F. Gennaro, and G. Di Marco, "Silicon and wide bandgap technologies in automotive power electronics and their applications," *2021 AEIT International Conference*, 2021.

[9] A. Kadavelugu, K. Mainali, D. Patel, *et al.*, "Medium voltage power converter design and demonstration using 15 kV SiC N-IGBTs," *IEEE Applied Power Electronics Conference and Exposition (APEC)*, 2015.

[10] K. Mainali, A. Tripathi, S. Madhusoodhanan, *et al.*, "A transformerless intelligent power substation: A three-phase SST enabled by a 15-kV SiC IGBT," *IEEE Power Electronics Magazine*, Vol. 2, 2015.

[11] J. W. Palmour, "Silicon carbide power device development for industrial markets," *IEEE International Electron Devices Meeting*, 2014.

[12] H. Zhang, L. Yuan, X. Tang, *et al.*, "Progress of ultra-wide bandgap Ga_2O_3 semiconductor materials in power MOSFETs," *IEEE Transactions on Power Electronics*, Vol. 35, 2020.

[13] H. Umezawa, T. Matsumoto, and S. Shikata, "Diamond metal–semiconductor field-effect transistor with breakdown voltage over 1.5 kV," *IEEE Electron Device Letters*, Vol. 35, 2014.

[14] M. Hiroki, Y. Taniyasu, and K. Kumakura, "High-temperature performance of AlN MESFETs with epitaxially grown n-type AlN channel layers," *IEEE Electron Device Letters*, Vol. 43, 2022.

[15] S.M. Sze, *Physics of Semiconductor Devices*, 2nd Edition, Wiley, New York, 1981.

[16] S.K. Ghandhi, *Semiconductor Power Devices*, Wiley, New York, 1977.

[17] D.A. Grant and J. Gowar, *Power MOSFETs: Theory and Applications*, Wiley, New York, 1989.

[18] B.J. Baliga and H.R. Chang, "Gate turn-off capability of depletion-mode thyristors," *IEEE Electron Device Letters*, Vol. 10, no.10, 1989.

[19] Y. Li, A. Huang, and F.C. Lee, "Introducing the emitter turn-off thyristor (ETO)", *Conference Record of 1998 IEEE Industry Applications Conference. Thirty-Third IAS Annual Meeting*, 1998.

[20] B.J. Baliga, M. S. Adler, R. P. Love, *et al.*, "The insulated gate transistor: A new three-terminal MOS-controlled bipolar power device," *IEEE Transactions on Electron Devices*, Vol. 31, no.6, 1984.

[21] P. Ranstad and H.P. Nee, "On dynamic effects influencing IGBT losses in soft-switching converters," *IEEE Transactions on Power Electronics*, Vol. 26, no.1, 2011.

[22] Infineon Application Note: AN 2012-04, "500 V CoolMOS™ CE, 500 V Superjunction MOSFET for consumer and lighting applications."

[23] Z. Chbili, K. P. Cheung, P. Campbell, *et al.*, "Unusual bias temperature instability in SiC DMOSFET," *IEEE International Integrated Reliability Workshop Final Report*, 2013.

[24] S. Sundaresan, S. Jeliazkov, H. Issa, *et al.*, "Static and switching characteristics of 1200 V SiC junction transistors with on-chip integrated Schottky rectifiers," *IEEE International Symposium on Power Semiconductor Devices & IC's (ISPSD)*, 2014.

[25] Z. Li and A. Bhalla, "USCi SiC JFET cascode and super cascode technologies," PCIM Asia, 2018

[26] R. J. Callanan, A. Agarwal, A. Burk, *et al.*, "Recent progress in SiC DMOSFETs and JBS diodes at Cree," *IEEE Industrial Electronics Conference*, 2008.

[27] S. Madhusoodhanan, A. Tripathi, K. Mainali *et al.*, "Three-phase 4.16 kV medium voltage grid tied AC-DC converter based on 15 kV/40 a SiC IGBTs," *IEEE Energy Conversion Congress and Exposition (ECCE)*, 2015.

[28] M. K. Das, C. Capell, D. E. Grider, *et al.*, "10 kV, 120 A SiC half H-bridge power MOSFET modules suitable for high frequency, medium voltage applications," *IEEE Energy Conversion Congress and Exposition (ECCE)*, 2011.

[29] K. Hatua, S. Dutta, A. Tripathi, *et al.*, "Transformerless intelligent power substation design with 15kV SiC IGBT for grid interconnection," *IEEE Energy Conversion Congress and Exposition (ECCE)*, 2011.

[30] X. Song, C. Peng, and A.Q. Huang, "A medium-voltage hybrid DC circuit breaker, part I: Solid-state main breaker based on 15 kV SiC emitter turn-OFF Thyristor," *IEEE Journal of Emerging and Selected Topics in Power Electronics*, 2017.

[31] R. Kosugi, S. Ji, K. Mochizuki, *et al.*, "Breaking the theoretical limit of 6. 5 kV-Class 4H-SiC super-junction (SJ) MOSFETs by Trench-filling epitaxial growth," *IEEE International Symposium on Power Semiconductor Devices and IC's (ISPSD)*, 2019.

[32] 15 V GaN device datasheet: https://epc-co.com/epc/Portals/0/epc/documents/datasheets/EPC2040_datasheet.pdf

[33] 1200 V GaN device datasheet: https://iganpower.com/wp-content/uploads/2021/10/GPIHV30DFN_v3.pdf

[34] E. A. Jones, F. Wang, and D. Costinett, "Review of commercial GaN power devices and GaN-based converter design challenges," *IEEE Journal of Emerging and Selected Topics in Power Electronics*, Vol. 4, 2016.

[35] D. Ramanathan, C. Coles, and W. Meier, "Not all GaN transistors are built equal: The benefits of vertical GaN-on-GaN," PCIM Europe, 2020.

[36] J. Suda, "Current status and future prospects of GaN-on-GaN vertical power devices," *International Electron Devices Meeting (IEDM)*, 2022.

[37] S. Hazra, S. Madhusoodhanan, S. Bhattacharya, *et al.*, "Design considerations and performance evaluation of 1200 V, 100 A SiC MOSFET based

converter for high power density application," *IEEE Energy Conversion Congress and Exposition*, 2013.

[38] A. Kurnia, O.H. Stielau, G. Venkataramanan, *et al.*, "Loss mechanisms in IGBTs under zero voltage switching," *Power Electronics Specialists Conference*, 1992. PESC '92 Record., 23rd Annual IEEE, Vol. 2, pp. 1011–1017.

[39] A. Mertens, H.C. Skudelny, P. Caldeira, *et al.*, "Characterization of GTOs under different modes of zero current switching," *22nd Annual IEEE Power Electronics Specialists Conference*, 1991.

[40] E. Aeloiza, A. Kadavelugu, R. Rodrigues, *et al.*, "Skin modules paper evaluation and characterization of parallel connected ultra-low-inductance 400A SiC MOSFET modules," *IEEE Energy Conversion Congress and Exposition*, 2019.

[41] Paralleling of IGBT modules, Application note 5SYA 2098, Hitachi.

[42] S. Madhusoodhanan, A. Tripathi, D. Patel, *et al.*, "Solid state transformer and MV grid tie applications enabled by 15 kV SiC IGBTs and 10 kV SiC MOSFETs based multilevel converters," *IEEE Transactions on Industry Applications*, Vol. 51, no. 4, pp. 3343–3360, 2015.

[43] A. Tripathi, K. Mainali, D. Patel, *et al.*, "Design considerations of a 15 kV SiC IGBT enabled high frequency isolated dc-dc converter," *IEEE Transactions on Industry Applications*, Vol. 51, no. 4, pp. 3284–3294, 2015.

[44] A. Tripathi, K. Mainali, S. Madhusoodhanan, *et al.*, "MV 3-phase DAB for DC Micro-grids with experimental validation at 8kV," *Proceedings of the 2015 IEEE Energy Conversion Congress and Exposition, Montreal, Canada*, 2015.

Chapter 2
Multilevel converters

João I.Y. Ota[1], Tiago D.C. Busarello[2] and José A. Pomilio[1]

2.1 Introduction

This chapter introduces the concept and application of power converters using voltage sources in multilevel converters to increase the quality of output voltage and better utilize the power rating of semiconductor devices. They could operate either as converters or inverters and their attributes are made to improve efficiency in industrial processes and expand applications in new areas such as renewable energy conversion, power transmission and distribution, and transportation.

Multilevel converters present advantages over converters seen so far, especially for medium and high-power applications. Among them, can be listed (i) the possibility of direct connection to the medium voltage network without the use of transformers; (ii) considerable reduction of electromagnetic interference (EMI) levels due to the lower output voltage variation rate; (iii) the possibility of obtaining higher power without the need for series and/or parallel associations of switches, (iv) among others.

Most of the converters used today work with two or three levels of voltage and contribute to an important part of total harmonic distortion (THD). The reduction of THD strongly depends on the switching frequency of the power semiconductors and is limited to low and medium power conversion. These reasons have generated much interest in the topic of multilevel converters [1–5]. Multilevel converters make use of a series connection of switches, which allows the use of switches with reduced voltage ratings. Lower voltage rating switches have lower conduction losses and can operate at a relatively higher frequency. Higher switching frequencies with many voltage levels result in higher-quality voltage waveforms.

There are also some disadvantages that should be considered when choosing a multilevel topology as a DC/AC converter. The modulation strategies are more complex than for conventional converters, and microcontrollers (μC) and digital

[1]School of Electrical Engineering, University of Campinas, Brazil
[2]Department of Control, Automation and Computing Engineering, Federal University of Santa Catarina, Brazil

signal processors (DSP) may not have adequate hardware to perform the modulation strategies for large numbers of levels or submodules, making their implementation difficult. This disadvantage can be overcome by means of suitable algorithms and/or with the inclusion of external analog and digital circuits, which must be developed especially for modulation purposes. There are also control challenges that depend on the application, such as the balancing of the voltages in the various capacitors in the cascaded multilevel converters, as well as the balancing of the neutral voltage in the diode-clamped converter.

2.2 Basic understanding of multilevel voltage-source converters

Multilevel converters comprise specific topologies that produce "cleaner" voltage and current waveforms, i.e., voltage synthesis that reduces the harmonics significantly. Also, the special topologies that can be implemented allow application in high-power conversion, which was very difficult to achieve with conventional two-level converter topologies. Usually, "multilevel" topologies consist of DC/AC conversion from more than one power source, and one of the first commercially available converters of this type is the three-level neutral point clamped (NPC) converter. Subsequently, other multilevel topologies, such as cascaded H-bridge (CHB) and flying clamped capacitors (FCCs), have been developed. CHBs can reach high output voltage and power levels and high reliability due to their modular topology. Also, H-bridge converters can be combined with NPC topologies to improve energy conversion quality and reach higher levels of power.

2.2.1 One-branch converter

Typically, the output waveform is generated with a staircase voltage synthesized of multilevel steps that are turned ON and OFF through the arrangement of semiconductors with specific DC voltage sources in the overall topology. Figure 2.1 shows the schematic diagram of voltage source inverters with different numbers of levels, which utilize only one branch of switches. As it is well known, a two-level inverter, like the one shown in Figure 2.1(a), generates an output voltage with two different levels: 0 and V_{DC}, while a three-level module, like the one shown in Figure 2.1(b), generates three different output voltages: 0, V_{DC}, and $2V_{DC}$. The different positions of the ideal switches are implemented with a number of semiconductors that are in direct relation with the number of levels.

It can be observed that the larger the number of DC sources, the greater the number of steps that can be generated and the number of sources is directly related with the number of levels through the relationship in (2.1):

$$s = n - 1, \tag{2.1}$$

where s is the number of DC supplies connected in series and n is the number of voltage levels generated.

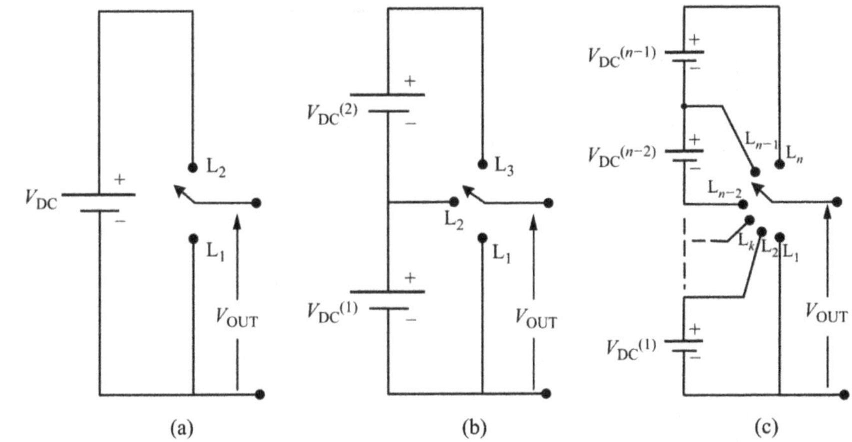

Figure 2.1 Basic multilevel converter topologies: (a) two levels, (b) three levels, and (c) n levels.

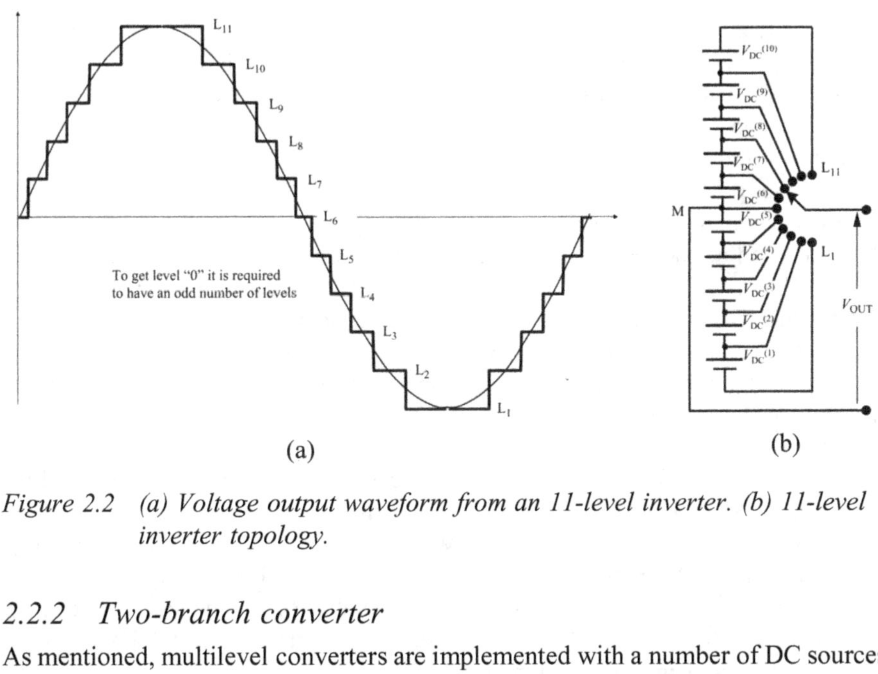

Figure 2.2 (a) Voltage output waveform from an 11-level inverter. (b) 11-level inverter topology.

2.2.2 Two-branch converter

As mentioned, multilevel converters are implemented with a number of DC sources to form a staircase AC waveform, which follows a given reference template. For example, with ten DC sources, an 11-level waveform can be obtained (five positives, five negatives, and zero with respect to the middle point between the ten sources). If the template is a sinusoidal waveform like the one shown in Figure 2.2, it is possible to obtain a waveform with very low THD when compared to square-wave waveforms.

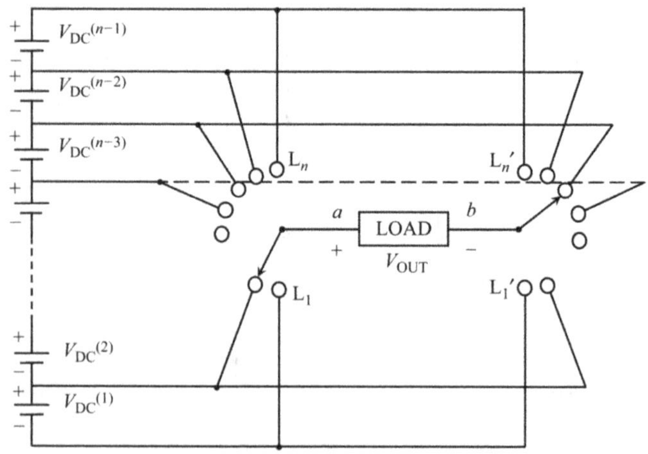

Figure 2.3 "H-bridge" inverter topology

To increase the number of levels, two branches of the previous topologies are combined to get the well-known "H-bridge" (HB) converter. It consists of connecting two of the previously discussed topologies using a common set of DC sources, which almost doubles the number of levels generated. This topology is displayed in Figure 2.3.

In this topology, the concept of "redundant levels" appears. The "H-bridge" topology has many redundant combinations to produce the same voltage level. As an example, the level "zero" can be generated with the load connected between L_1 and L_1', or in general, between L_k and L_k'. In a similar way, the level that produces $V_{OUT} = 2V_{DC}$ can be generated connecting $L_3 - L_1'$ or $L_{k+2} - L_k'$. Negative levels are generated in reverse way (e.g., $V_{OUT} = -2V_{DC}$ is obtained between positions $L_k - L_{k+2}'$). In general, all levels generated with inner switches of any of the two branches are redundant, which means, from L_2' to L_{n-1}' (or L_2 to L_{n-1}).

Figure 2.4 shows, with full dots, all the positions (or switches) that could be removed, because they only generate redundant levels, and this removal can be done in whichever of the two branches. When these switches are eliminated, only one redundant level remains, which is the state $L_1 - L_1'$ because it generates the same level as $L_n - L_n'$ ($V_{OUT} = 0$). The total number of redundant levels can be easily obtained from Figure 2.4, because each $n-2$ removed switch produces n_R redundant levels, then the total number is

$$n_R = n(n-2) + 1, \tag{2.2}$$

where n_R is the number of redundant levels and n is the number of voltage levels in one branch. Equation (2.2) includes the redundant state $L_1 - L_1'$ (or its pair $L_n - L_n'$).

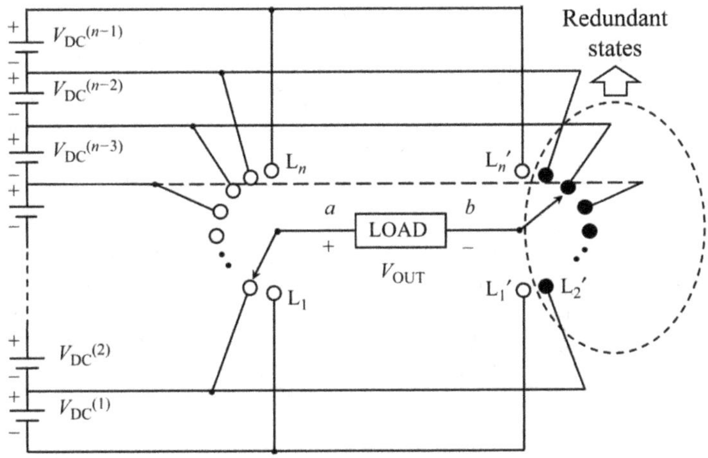

Figure 2.4 Redundant switches in an "H-bridge" converter

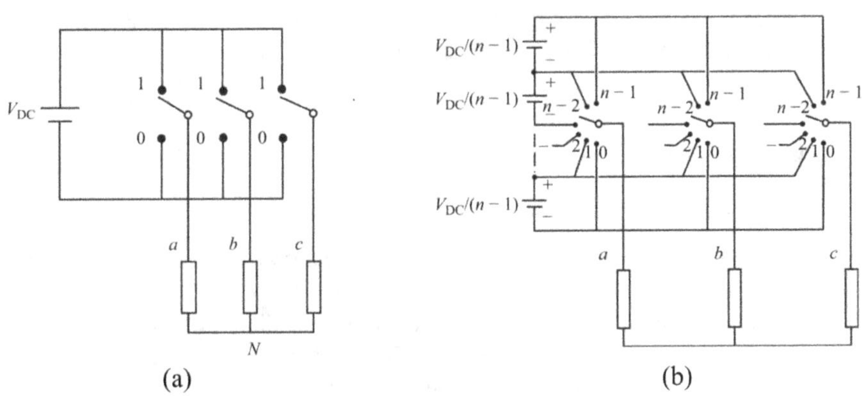

Figure 2.5 Three-phase converters. (a) Two-level and (b) n-level converters.

2.2.3 Voltage levels and space vector representation

Generally speaking, three-phase converters have a load connected in wye with floating neutral. This connection, shown in Figure 2.5, generates more voltage levels at the load because it corresponds to the phase-to-neutral voltage. As the phase-to-phase voltage has the same number of levels as an H-bridge, phase-to-neutral voltage can be obtained from the following relation:

$$n_{f-f} = n_{ab} = 2n - 1, \tag{2.3}$$

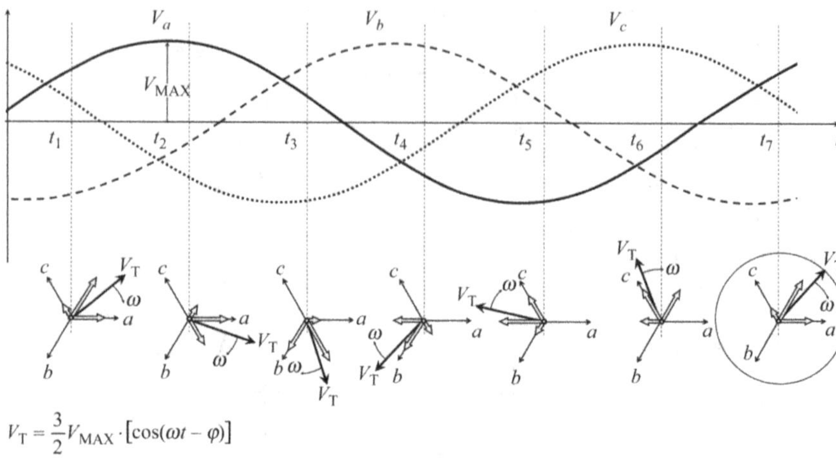

Figure 2.6 Vector diagram of a pure sinusoidal power source

where n_{f-f} is the phase-to-phase voltage and n is the number of levels in one branch of the inverter. Similarly, the number of levels in the phase-to-neutral connection with respect to phase-to-phase voltage is:

$$n_{f-N} = 2n_{ab} - 1. \tag{2.4}$$

Replacing (2.4) in (2.3) it yields:

$$n_{f-N} = 2(2n - 1) - 1 = 4n - 3, \tag{2.5}$$

where n_{f-N} is the number of levels in the phase-to-neutral voltage in whichever of the three phases, V_{aN}, V_{bN}, or V_{cN}.

Three-phase systems can be analyzed using vector diagrams, which are especially useful for motor drive applications. The vector diagram shows, in a two-dimensional way, the instantaneous magnitude of each phase-to-neutral voltage at the load. Figure 2.6 shows the vector diagram of an ideal sinusoidal source and Figure 2.7 shows the vector diagram of a two-level converter. The composition of the three phases generates a rotating voltage vector with an angular speed of ω rad/s. The trajectory describes a perfect circle, whose magnitude is $V_T = \frac{3}{2} V_{MAX}$, where V_{MAX} is the peak value of phase-to-neutral voltage.

Assuming that:

$$\begin{aligned} v_a &= V_{\text{MAX}} \cos \omega t, \; v_b = V_{\text{MAX}} \cos(\omega t - 120°), \\ v_c &= V_{\text{MAX}} \cos(\omega t - 240°), \end{aligned} \tag{2.6}$$

where v_a, v_b, and v_c are the instantaneous voltages of phases a, b, and c, respectively, and ω is the angular frequency of the main supply.

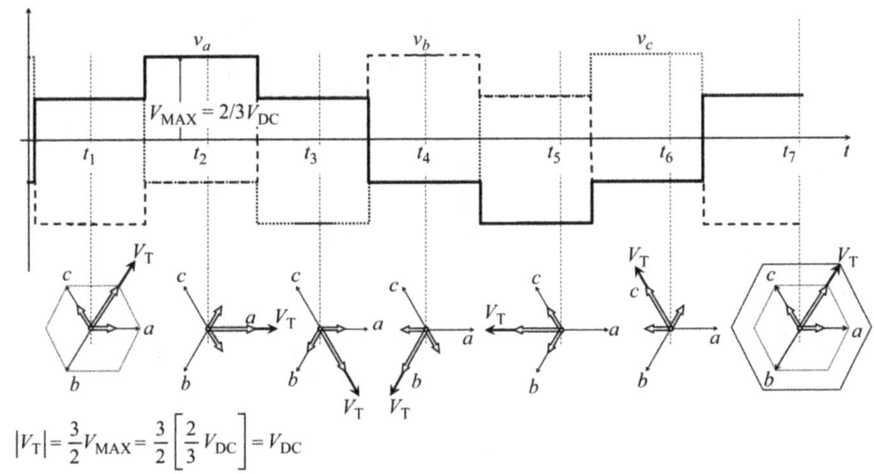

Figure 2.7 Phase-to-neutral voltages and vector diagram of a two-level inverter

Assuming that the vector V_T is measured at a specific angle ϕ with respect to phase a:

$$V_{\mathrm{T}} = v_a \cos\phi + v_b \cos(\phi - 120^\circ) + v_c \cos(\phi - 240^\circ). \tag{2.7}$$

Replacing (2.6) in (2.7):

$$V_{\mathrm{T}} = V_{\mathrm{MAX}}\omega t \cos\phi + \cos(\omega t - 120^\circ)\cos(\phi - 120^\circ) + \cos(\omega t - 240^\circ)\cos(\phi - 240^\circ) \tag{2.8}$$

Equation (2.8) yields:

$$V_{\mathrm{T}} = \frac{1}{2}V_{\mathrm{MAX}}\,[\cos(\omega t + \phi) + \cos(\omega t - \phi) + \cos(\omega t + \phi - 240^\circ)\cos(\omega t - \phi) + \cos(\omega t - \phi) + \cos(\omega t + \phi - 120^\circ) + \cos(\omega t - \phi)\,]. \tag{2.9}$$

However,

$$\cos(\omega t + \phi) + \cos(\omega t + \phi - 240^\circ) + \cos(\omega t + \phi - 120^\circ) = 0, \tag{2.10}$$

and finally,

$$V_T = \frac{3}{2}V_{\mathrm{MAX}}\cos(\omega t - \phi). \tag{2.11}$$

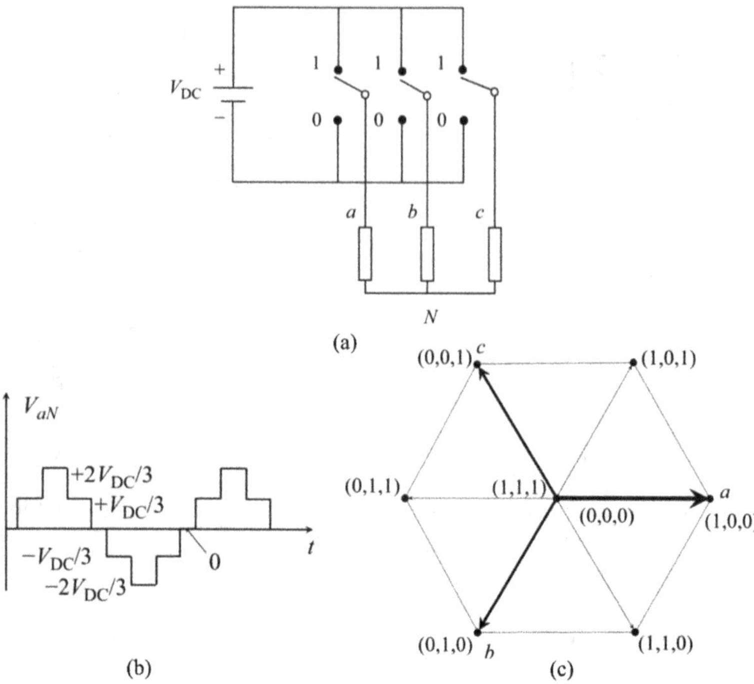

Figure 2.8 (a) Two-level inverter. (b) Phase-to-neutral voltage. (c) Space vector diagram.

Equation (2.10) represents a rotating vector that travels at ω rad/s and has a constant magnitude of $\frac{3}{2}V_{MAX}$. In the case of a two-level converter, the vector V_T can be located in six non-zero positions only, because each phase can be at two different locations (two levels): "0" and "V_{DC}," where V_{DC} is the voltage of the power source of the converter. Besides, there are two redundant vectors of magnitude "0," which are generated with switches in position (0,0,0) or (1,1,1).

Figure 2.8 shows the phase-to-neutral voltage waveform and vector diagram of a two-level converter, which can be associated with a six-sided polygon, i.e., hexagon. In this hexagon, each vertex represents one of the six vectors and at the center are located vectors (0,0,0) and (1,1,1). In this case, the magnitude of the vector $\boldsymbol{V}_T$ equals V_{DC}, which is the voltage source that feeds the converter:

$$|\boldsymbol{V}_{\mathrm{T}}| = \frac{3}{2}V_{\mathrm{MAX}} = \frac{3}{2}\left[\frac{2}{3}V_{\mathrm{DC}}\right] = V_{\mathrm{DC}}. \tag{2.12}$$

When the number of levels is increased, more vectors with different magnitudes and angles appear. Figures 2.9 and 2.10 show similar diagrams for two-level and four-level converters, respectively. No matter the number of levels, the peak values of phase-to-neutral voltages are always $\pm\frac{2}{3V_{\mathrm{DC}}}$ (assuming that $V_{\mathrm{DC}} = \Sigma_i V_{\mathrm{DC}}/i$).

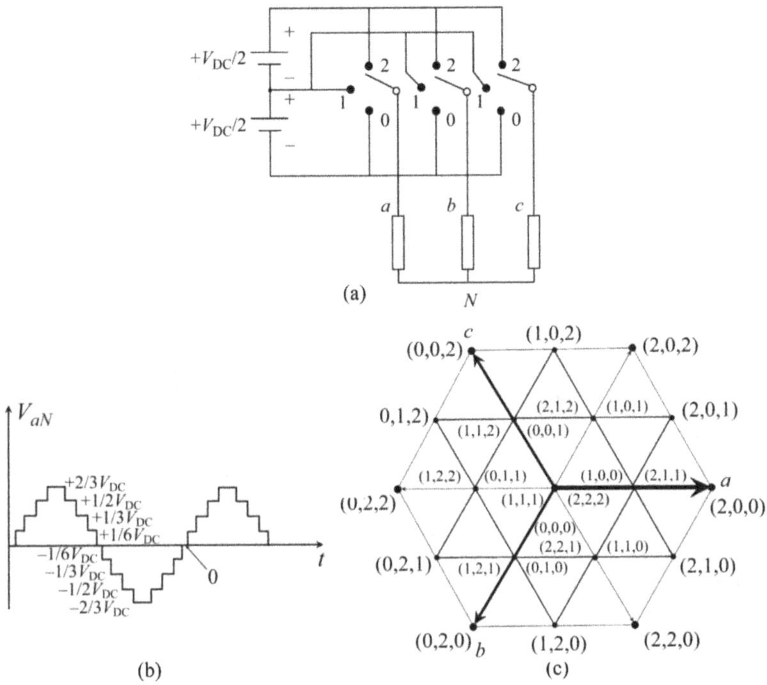

Figure 2.9 (a) Three-level inverter. (b) Phase-to-neutral voltage. (c) Space vector diagram.

Each dot represents a new voltage vector that starts from the center of the hexagon. The number of vectors with magnitude "zero" is equal to the number of levels and hence, an n-level converter has n vectors at the origin: $(0, 0, \ldots, 0)$, $(1, 1, \ldots, 1), \ldots, (n-1, n-1, \ldots, n-1)$. There are no redundant levels at the periphery of the hexagon, but they increase by one unit when going to the center of the polygon. At the center, there are $(n-1)$ redundant levels.

Figure 2.11 shows the space vector diagram of an n-level inverter. The total number of non-redundant vectors in an n-level inverter can easily be found from the space vector diagram. The hexagon can be decomposed in six sectors (six triangles in Figure 2.11), where each triangle has a number of vectors v_Δ given by:

$$v_\Delta = \frac{n(n-1)}{2}, \tag{2.13}$$

where n is the number of levels of the three-phase converter. Hence, the total number of non-redundant vectors is:

$$v_{NR} = 6\frac{n(n-1)}{2} + 1 = 3n(n-1) + 1. \tag{2.14}$$

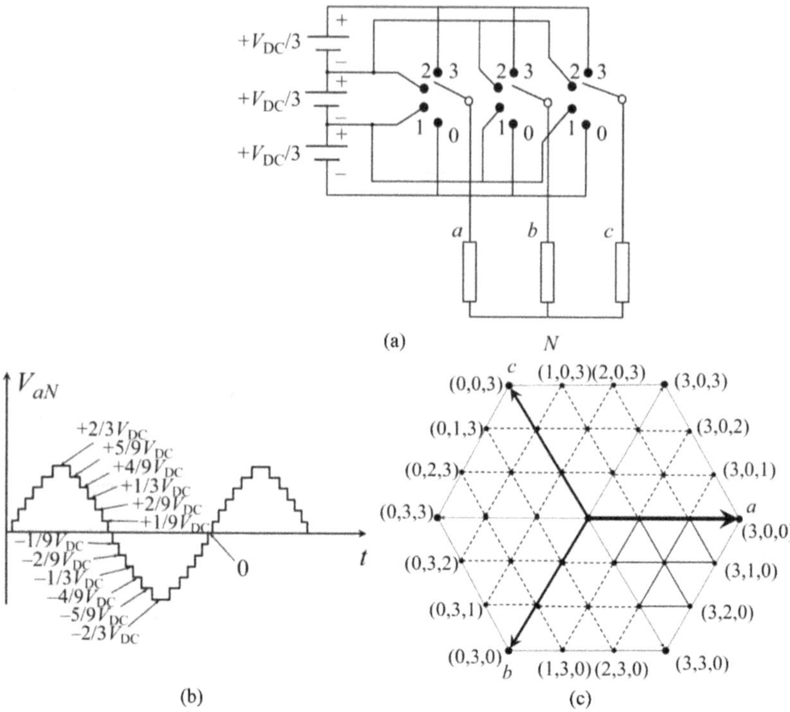

Figure 2.10 (a) Four-level inverter. (b) Phase-to-neutral voltage. (c) Space vector diagram.

The total number of vectors (redundant and non-redundant) may also be evaluated from Figure 2.11, and they correspond to all combinations of switches of the three-phase inverter. The switches of phase *a* may be in *n* different positions for each one of the *n* positions of the phase *b* and also for each one of the n positions of phase *c*. As a result, the total number of vectors v_T is:

$$v_T = n^3. \tag{2.15}$$

The total number of redundant vectors, v_R is obtained from (2.14) and (2.15):

$$v_R = v_T - v_{NR} = (n-1)^3. \tag{2.16}$$

Figure 2.12 resumes the results obtained in this section.

The multilevel converters described so far need electronic switches to operate, like metal oxide semiconductor field effect transistors, insulated gate bipolar transistors, or gate turn-off thyristors. These devices cannot replace the ideal switches shown in previous sections, because they have some limitations. They are unidirectional, are limited in switching speed, and produce power losses. Besides, none of them can commutate from one power source to another; it requires the use

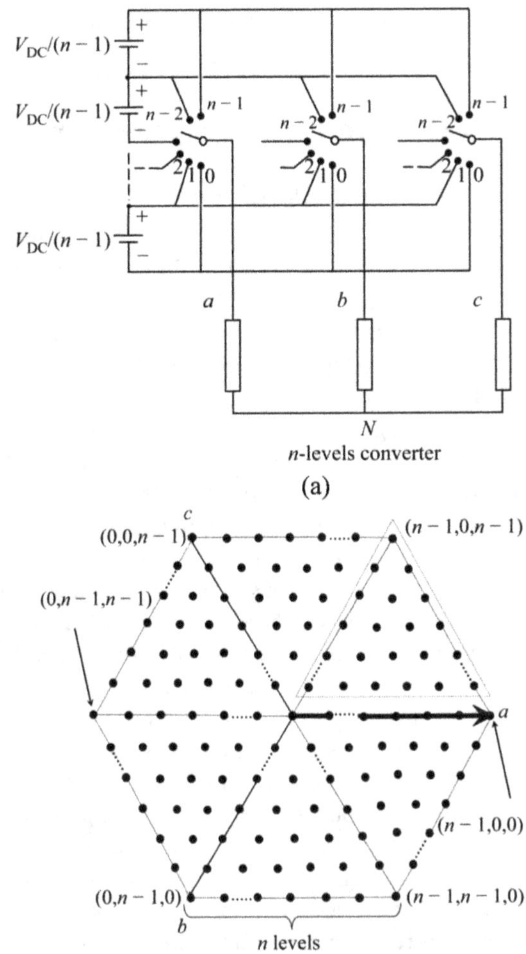

Figure 2.11 (a) n-level inverter. (b) Space vector diagram.

Space vector for *n*-level converter

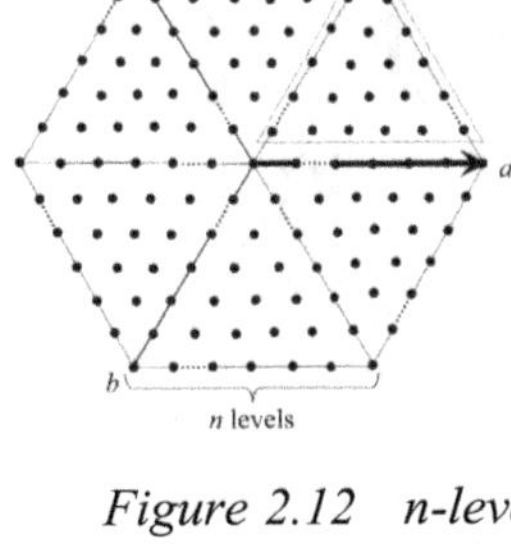

Number of levels	: n
Number of levels V_{f-f}	: $n_{f-f} = 2n - 1$
Number of levels V_{f-N}	: $n_{f-N} = 2n_{f-f} - 1 = 4n - 3$
Non-redundant vectors	: $v_{NR} = 6\frac{(n-1)n}{2} + 1 = 3n(n-1) + 1$
Redundant vectors	: $v_R = n_S - n_V = (n-1)^3$
Total number of vectors	: $v_T = n^3$

Figure 2.12 n-level space vector diagram and related equations

of special topologies for multilevel implementation, which will be described in the next sections.

2.3 Neutral-point clamped multilevel converter

2.3.1 Three-phase three-level NPC converter

The circuit for the three-phase neutral-point clamped converter (NPC) [1], also known as diode clamped converter, is shown in Figure 2.13(a). The structure of Figure 2.13(a) is generally called "3-level." The output voltage between a phase and neutral of the inverter is a PWM waveform with three levels ($+V_{DC}/2$, 0, and $-V_{DC}/2$), although the output voltage between two phases has five levels ($+V_{DC}$, $+V_{DC}/2$, 0, $-V_{DC}/2$, and $-V_{DC}/2$). The spectral content (mainly for harmonics) is quite reduced. The NPC topology is regarded as an invention by Akira Nabae *et al.* [1] at Nagaoka University of Technology in 1980. Currently, the NPC topology is the one most commonly used in industry applications.

Besides the advantages listed, one of the main characteristics of the NPC is to enable, for the same semiconductor switch technology, an inverter with twice the power compared to a traditional three-phase inverter circuit. The reason is due to the fact that the voltage over each switch becomes half of the total voltage of the DC bus. This is one important reason that popularized the use of the clamped neutral inverter mainly in motor applications, such as, for example, the inverter that performs the control of electric motors in high-speed trains [2].

A disadvantage of the NPC is its non-uniform distribution of losses. This occurs because the path through which the current flows between the positive (or negative) terminal and the AC output is different from the path between the neutral point and the AC output [3]. Also, according to the voltage and current references at the AC output, the number of components that are effectively switching changes. The losses become non-uniform among the semiconductor components, which can cause either issues for long-term operation and reliability due to the non-uniform heating or more complex and bulky heat sinking for the converter.

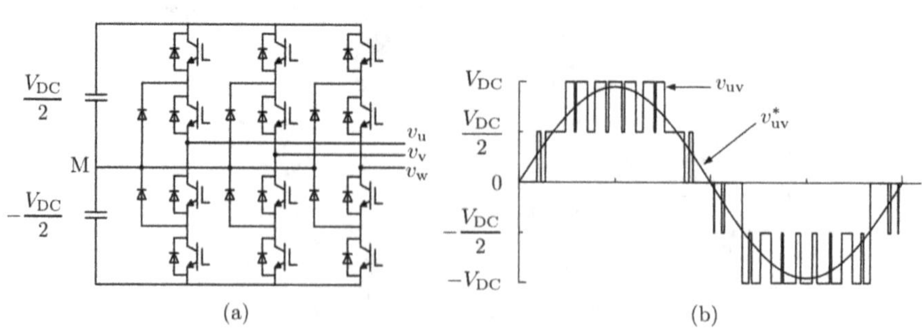

Figure 2.13 (a) Circuit of a three-phase three-level NPC inverter. (b) Line-to-line voltage on AC terminals.

An alternative to balance such losses is the active clamped neutral multi-level converter (*Active NPC*) [3]. This structure consists of replacing each clipping diode with a semiconductor switch. Thus, one can "balance" the losses between all the components of the converter. However, it increases the complexity of the control, and in the case of a feedback strategy, monitoring of the junction temperatures of the components would be required.

2.3.2 Neutral-point voltage balancing

Analyses of the conduction modes of the NPC converter show that the incident current at the neutral point varies with switching due to the phase angle of the AC current (power factor) and other factors. Thus, the voltages on the DC capacitors tend to oscillate, which may impact their sizing or the reliability of the system, since an excessively large voltage may be applied on some of their switches. Compensation methods should be applied, as well as capacitor and DC bus voltage sizing should be carefully studied.

There is plenty of available literature, but in this chapter, just points will be presented regarding the case of the 3-level three-phase NPC Converter for a reactive compensation application with PWM modulation [4]. Figure 2.14 shows simulation results of the neutral current and its compensation for a 3-level NPC converter application with a current phase angle at 90°. It is known that part of the variation of the neutral voltage occurs at low frequency (i.e., well below the switching frequency), and is directly linked to the phase angle. It can be seen that, in this case, the neutral current has an average value of zero, dependent on the modulation level, and oscillates at a frequency three times greater than the AC output frequency. Therefore, a compensation method is proposed through the synthesis of a neutral current, considering that this is a 3-wire application. It is noted that the compensation is no longer effective for high modulation levels, although it is effective until the modulation index is below 60%. In this case, if such

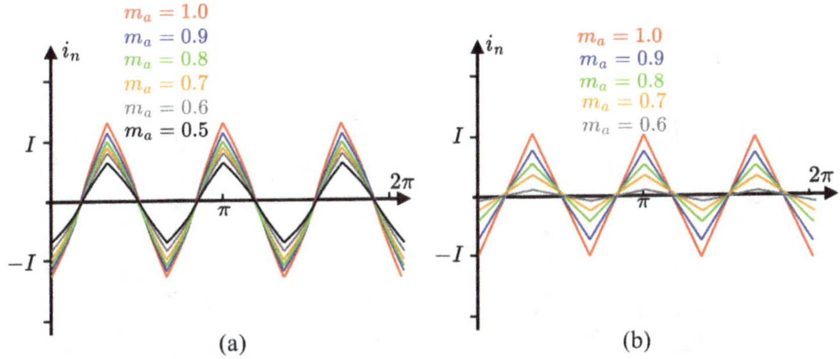

Figure 2.14 Simulated neutral currents of a 3-level NPC converter applied to a STATCOM (a) when there is no compensation and (b) when compensation is applied.

oscillations are not acceptable for the application, larger capacitors and a higher DC bus voltage should be considered in the design of the 3-level NPC converter.

These results are corroborated by other references [5,6]. Although small differences in the methodology for calculating and obtaining the low-frequency oscillations are dependent on the modulation strategy chosen (e.g., *space vector modulation*), it is thus understood that capacitors and DC bus voltage should be carefully studied according to the final application.

2.3.3 Expansion and generalization to more levels

The structure shown in Figure 2.13 could be expanded to more levels. Figure 2.15 shows a specific example using a seven-level NPC converter. To further exemplify the switching strategy, Figure 2.15 shows four of the seven possible outputs: $V_{\mathrm{OUT}} = 0$, $V_{\mathrm{OUT}} = 2V_{\mathrm{DC}}$, $V_{\mathrm{OUT}} = 3V_{\mathrm{DC}}$, and $V_{\mathrm{OUT}} = 5V_{\mathrm{DC}}$. As can be seen, a chain of six transistors is moved together to obtain the desired level.

Generally speaking, if the required voltage at the output is kV_{DC}, the entire chain of transistors located between the diodes $D_{(k)}$ and $D_{(k)}'$ should be turned on. This allows the current to flow bidirectionally. In this way, all the DC sources located at the bottom of diodes $D_{(k)}$ and $D_{(k)}'$ synthesizes the output voltage V_{OUT}. For example, if a $3V_{\mathrm{DC}}$ voltage is required at the output, then all transistors located between diodes $D_{(3)}$ and $D_{(k)}'$ must be turned on:

$$V_{\mathrm{OUT}} = V_{\mathrm{DC}}^{(1)} + V_{\mathrm{DC}}^{(2)} + V_{\mathrm{DC}}^{(3)} = 3V_{\mathrm{DC}}. \tag{2.17}$$

The main advantage of the NPC converter, when compared to other multilevel structures described in the next sections is that only one DC source is used. This is a striking difference mainly from the CHB converter structure. However, when a

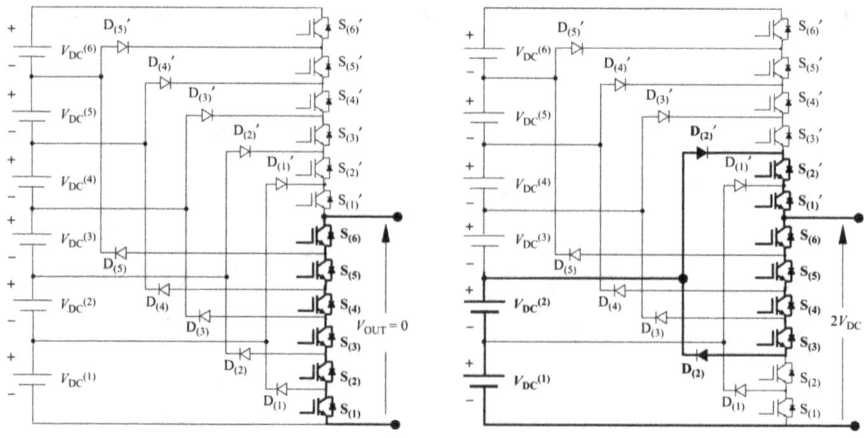

Figure 2.15 Seven-level NPC topology. (a) $V_{\mathrm{OUT}} = 0$. (b) $V_{\mathrm{OUT}} = 2V_{\mathrm{DC}}$. (c) $V_{\mathrm{OUT}} = 3V_{\mathrm{DC}}$. (d) $V_{\mathrm{OUT}} = 5V_{\mathrm{DC}}$.

larger number of levels is desired, NPC converter needs a larger number of components in relation to the CHB topology. Other aspect is the lack of modularity. Note, in Figure 2.15, that only one diode is represented in each "level" conduction path. However, the real diode requires a blocking voltage equivalent to the DC voltage applied over them, which could be up to four times the blocking voltage of a diode of a 3-level NPC converter.

The NPC converter with more than three levels has the drawback of DC capacitor voltage balancing connected to the DC bus. As presented in the previous subsection, the issues involving unbalanced conditions due to the neutral current of the 3-level NPC converter are also seen, but with more intermediate points between the DC bus capacitors. Depending on the application, this balancing needs to be implemented using external circuits, which makes the application of such a converter less attractive [7].

2.3.4 T-Type NPC converter

Figure 2.16 presents the clamped neutral T-type converter (NPC T-type), an alternative to the NPC that has become popular mainly in photovoltaic converter applications in recent years [8]. Despite the recent popularization, such a possibility had already been observed in the first NPC proposal [1], which is presented in

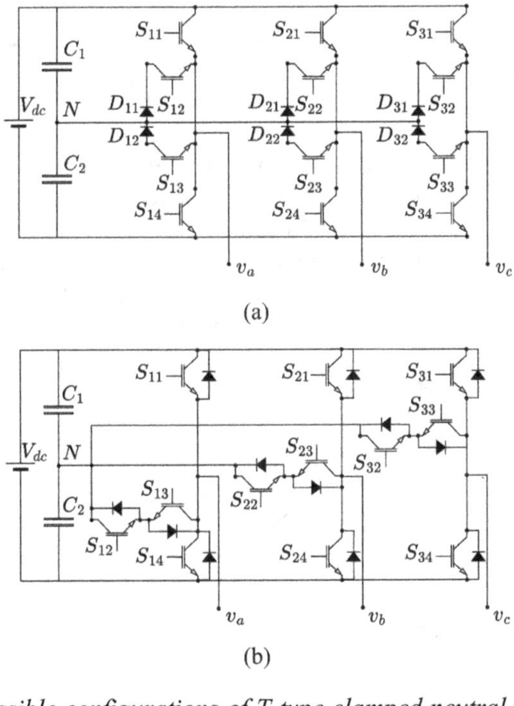

Figure 2.16 Possible configurations of T-type clamped neutral converter, (a) as presented in 1980 and (b) modern applications.

Figure 2.16(a). Figure 2.16(b) shows a typical circuit configuration of modern applications. The creation of three voltage levels occurs through the current path made possible by the combination of the switches between the neutral point and the AC output, which allows bidirectional current and voltage reverse voltage drop. This topology, although presenting the advantages of the multilevel inverters listed at the beginning of the section, does not allow the realization of an inverter with doubled power for the same semiconductor switch technology compared to a standard three-phase inverter as shown in Figure 2.13(a). Thus, its use is justified by the reduced harmonic content and in applications with low DC bus voltages [9].

2.4 Flying clamped capacitor multilevel converter

2.4.1 Three-phase three-level FCC converter

Figure 2.17 presents the *flying clamped capacitor* (FCC) converter, also known as clamped capacitor multilevel converter, which is an alternative to the NPC converter and was proposed by Meynard and Foch [10]. It presents the same feature of voltage clamping from the NPC converter but performed by capacitors or intermediate DC sources. Therefore, each semiconductor switch actually supports only part of the applied DC bus voltage during operation. The voltage applied in each switch in the case of the 3-level FCC converter is half the DC bus voltage.

The FCC converter was later classified as a potential candidate for applications in electrical grids due to its similar advantages to the NPC converter and the CHB converter [11]. However, the number of capacitors for a larger number of voltage levels is excessive. Another disadvantage is that the voltage of the intermediate capacitors tends to be very close to the main DC bus voltage, which makes the use of the FCC converter less attractive for high-voltage applications. Practical applications with the FCC converter can be found in electric drives and some applications in electrical networks, all aimed towards medium voltage applications [12].

Figure 2.18 shows how the voltage levels of a four-level FCC converter are generated. Level 1 ($V_{OUT} = 0$) is generated by turning on the transistors $S_{(1)}$, $S_{(2)}$, and $S_{(3)}$. Level 2 ($V_{\mathrm{OUT}} = V_{\mathrm{DC}}$) is obtained by turning on transistors $S_{(2)}$, $S_{(3)}$, and

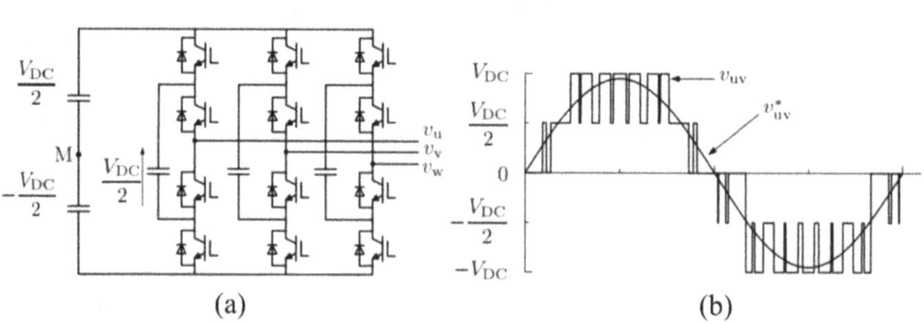

Figure 2.17 (a) Three-phase three-level FCC converter and (b) waveforms at the AC terminals.

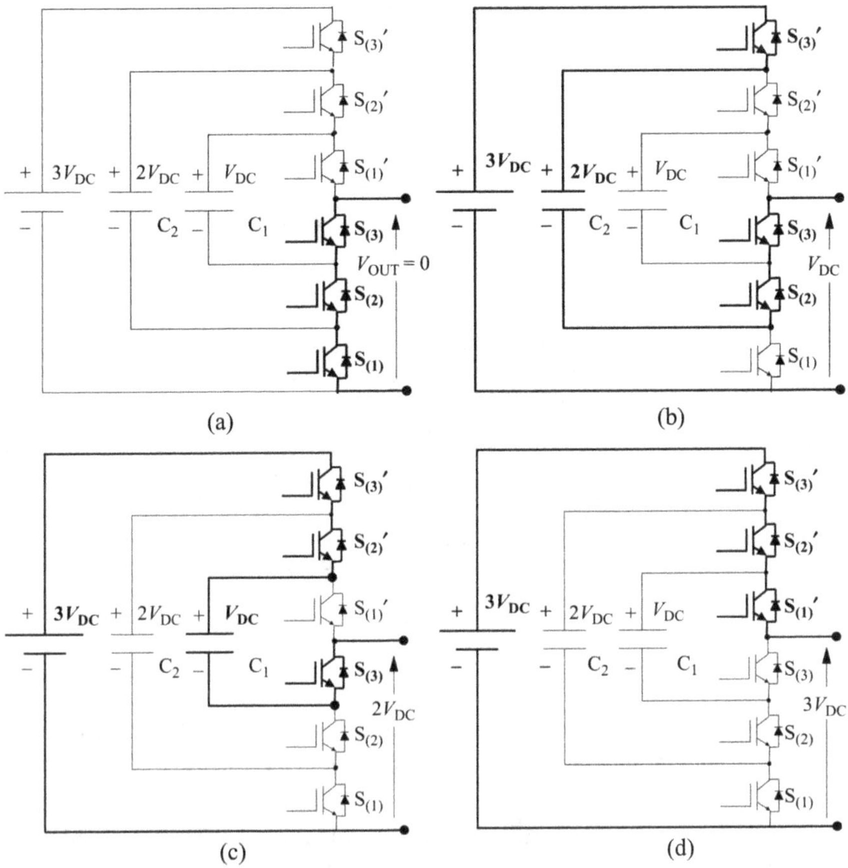

Figure 2.18 Voltage synthesis in a 4-level FCC converter. (a) $V_{\mathrm{OUT}} = 0$. (b) $V_{\mathrm{OUT}} = V_{\mathrm{DC}}$. (c) $V_{\mathrm{OUT}} = 2V_{\mathrm{DC}}$. (d) $V_{\mathrm{OUT}} = 3V_{\mathrm{DC}}$.

$S_{(3)}'$. Level 3 ($V_{\mathrm{OUT}} = 2V_{\mathrm{DC}}$) is obtained by turning on $S_{(3)}$, $S_{(2)}'$, and $S_{(3)}'$. Finally, Level 4 ($V_{OUT} = 3V_{DC}$) is obtained by turning on $S_{(1)}'$, $S_{(2)}'$, and $S_{(3)}'$. Note that $V_{OUT} = V_{DC}$ is not generated from the voltage of the capacitor C_1, but from the DC source and the voltage of the capacitor C_2, i.e., $3V_{DC} - V_{C2}$. In a similar way, $V_{\mathrm{OUT}} = 2V_{\mathrm{DC}}$ is obtained.

2.4.2 Capacitor voltage self-balancing

In the first studies of the FCC Converter, it was observed that capacitors in the intermediate DC voltages could be "self-regulated," i.e., converge to the equilibrium voltage without the need for an active feedback control for voltage balancing. This phenomenon can be observed when the phase shift PWM method described in Figure 2.19 (see more about multicarrier PWM in Section 2.7.3) is

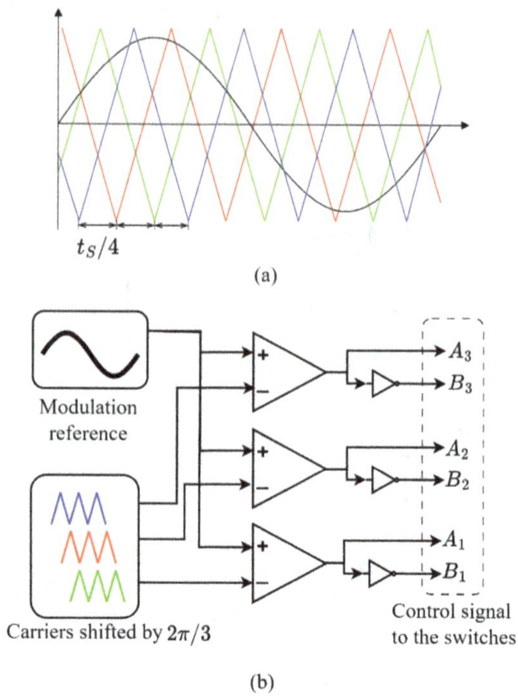

Figure 2.19 Phase-shift PWM for an FCC converter. (a) Modulation carriers and modulation signal. (b) PWM schematic.

applied [12]. For a converter with p basic cells (see Section 2.4.3), p modulation carriers are used, such that the phase shift between the carriers is $\frac{2\pi}{p}$.

To demonstrate such characteristics, simulation results are presented. Figures 2.20–2.22 show waveforms for FCC inverters with 3, 4, and 5 levels, respectively. The voltage on the DC bus is 200 V, and the load is a 1-Ω resistor. The AC voltage reference is set to be sinusoidal, with a frequency of 60 Hz, and a modulation index of 0.95. An inductor is used to filter the voltage at the AC output. No feedback control is used, i.e., the voltage is synthesized through an open-loop control. Note that the intermediate voltages reach, after an initial transient, the voltages according to the "basic cells" voltages (see Section 2.4.3). It is also possible to observe the formation of the multilevel voltage waveform at the AC terminals of the FCC converter.

Further studies show, however, that the natural balancing of capacitors depends on the components present in the AC terminals, i.e., the load in the case of an inverter application or the grid impedance in the case of grid-connected applications. It has been shown that for a purely reactive load, natural balancing is not achievable [13]. Another aspect is the accommodation time for the voltage in the intermediate stages to reach the steady state, which is also dependent on the passive components in the AC side and of the flying clamped capacitors. For the balancing

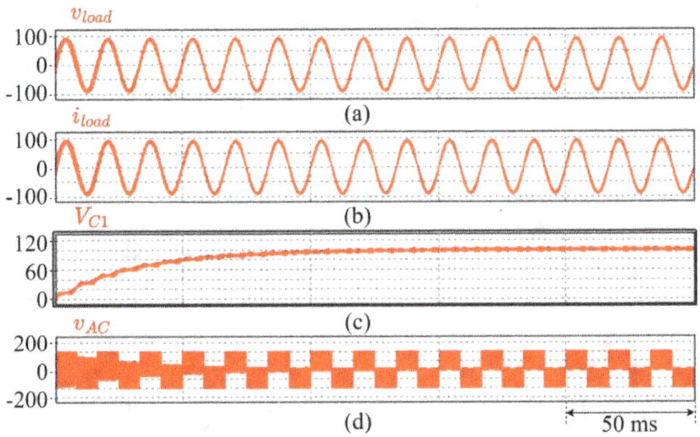

Figure 2.20 Three-level FCC converter. (a) Voltage on resistive load. (b) AC current. (c) DC voltage at intermediate stages. (d) Voltage at AC terminals.

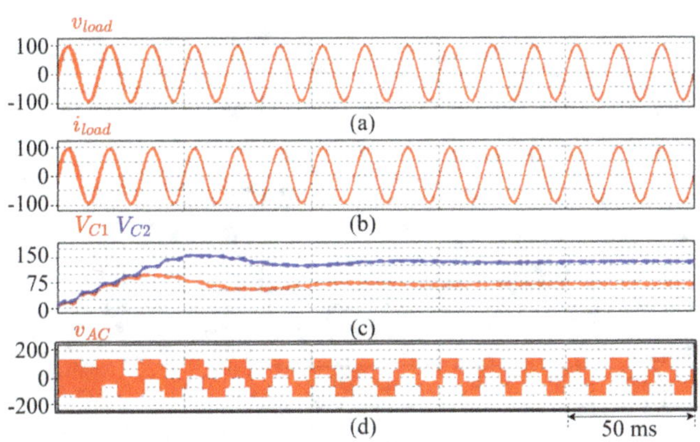

Figure 2.21 Four-level FCC converter. (a) Voltage on resistive load. (b) AC current. (c) DC voltage at intermediate stages. (d) Voltage at AC terminals.

dynamics to be accelerated, the use of coupled filters in the AC output called "*balance boosters*" has been proposed in the literature [14,15].

2.4.3 Expansion and generalization to more levels

The structure shown in Figure 2.17 can be expanded to more levels. Figure 2.23(a) shows a single-phase 3-level FCC converter and Figure 2.23(b) shows a generic *n*-level FCC converter. As the number of levels increases, however, one can understand the FCC converter as an inverter with several intermediate DC voltage

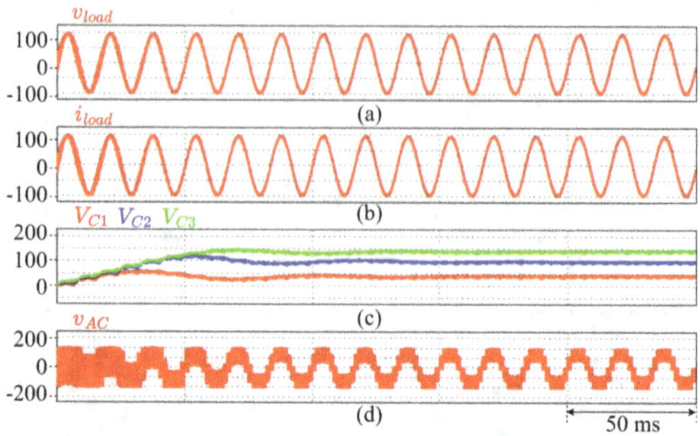

Figure 2.22 Five-level FCC converter. (a) Voltage on resistive load. (b) AC current. (c) DC voltage at intermediate stages. (d) Voltage at AC terminals.

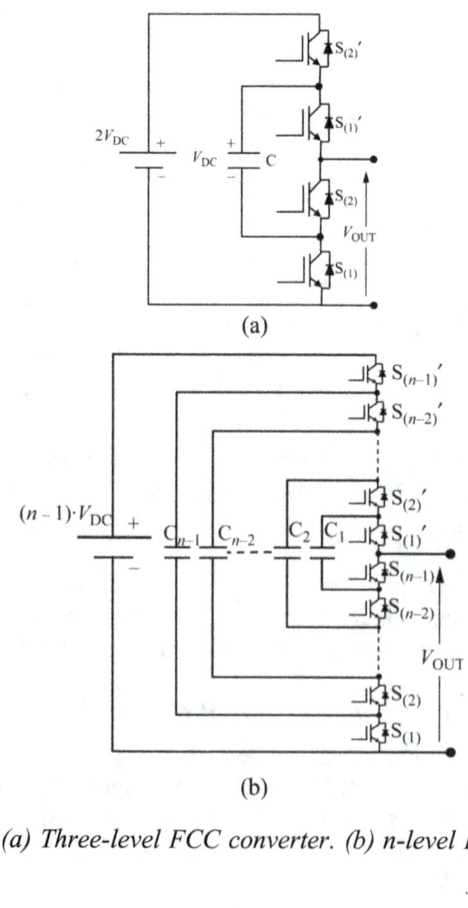

Figure 2.23 (a) Three-level FCC converter. (b) n-level FCC converter.

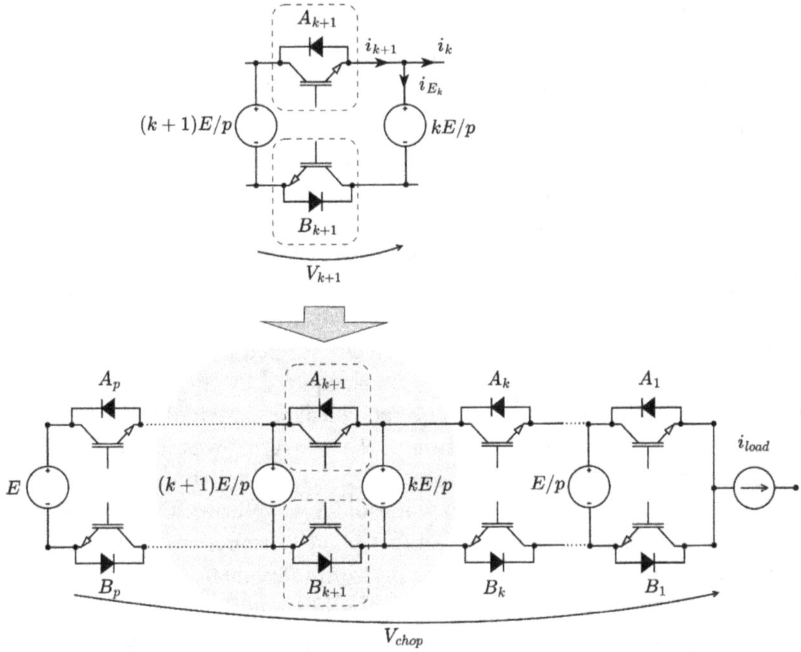

Figure 2.24 Basic cell of an FCC converter and a single-phase (p+1)-level FCC converter

stages [16], which are obtained with the clamped (or floating—also known as "flying") capacitors in "basic cells."

Figure 2.24 presents the construction of the basic cell as well as its structure with p elementary cells. The number of voltage levels achieved is $m = p + 1$, and k denotes the k-th fundamental cell in the FCC converter. Let k equal to 1 from the cell interfacing with the AC side and let it increment when a new cell is added to the DC side. Therefore, the following relations can be applied for the DC voltage at each intermediate stage and for the voltage drop at each semiconductor device, as shown by (2.18) and (2.19), respectively [12]:

$$V_{Ck} = k\frac{E}{p}, \tag{2.18}$$

$$V_{A_k} = V_{B_k} = \frac{V_{chop}}{p} = (k-1)\frac{E}{p} - k\frac{E}{p} = \frac{E}{p}. \tag{2.19}$$

As an example, a 5-level FCC Converter with a DC bus voltage E consists of four basic cells, intermediate voltages $\frac{E}{4}$, $\frac{E}{2}$, and $\frac{3E}{4}$, and a voltage drop $\frac{E}{4}$ on each semiconductor switch.

2.5 Cascaded H-bridge multilevel converter

2.5.1 Three-phase three-level CHB converter

The cascade H-bridge (CHB) converter can be considered the most successful multilevel converter regarding the number of existing medium and high power applications both in industry and in medium voltage grids. The main advantage is the modularity of the basic cell (or submodule), which allows (a) feasible expansion to a higher number of voltage levels in spite of the NPC and FCC converters, (b) choice of AC voltage terminals which allows connection without a transformer, (c) easier maintenance, and (d) increased reliability. One of the first industrial applications of the CHB was an electric drive presented by Hammond and manufactured by ROBICON [17], which was later acquired by Siemens AG. A special medium voltage multi-winding transformer consisting of multiple phase-shifted low voltage secondary terminals and the CHB converter was applied for voltage synthesis aiming medium voltage motors (>2.4 kV) with nominal power up to 1,000 hp (approx. 750 kW), with quasi-sinusoidal waveforms. Concomitant to this proposal, Lai and Peng presented a reactive power compensation converter composed of several full-bridge circuits feeding from individual DC sources [18].

A schematic diagram of the symmetrical CHB multilevel converter (DC voltages are equal), with the H-complete bridge being the basic module, is shown in Figure 2.25(a). PWM modulation also allows the multilevel inverter to further reduce the distortion of the AC voltage and minimize the filtering requirement. Such a waveform is shown in Figure 2.25(b). The cutoff frequency of AC filters to obtain a sinusoidal voltage can be designed considering the resulting spectral components in multiples of $2n + 1$, where n is the number of levels of the voltage output [19]. The filter attenuation is, as a result, not large, since the amplitude of the harmonics becomes relatively small.

Another important aspect of the CHB converter is the regulation of the DC voltages of its submodules, especially in applications where DC sources are not used and the voltage must be maintained by capacitors, which, in general, must be

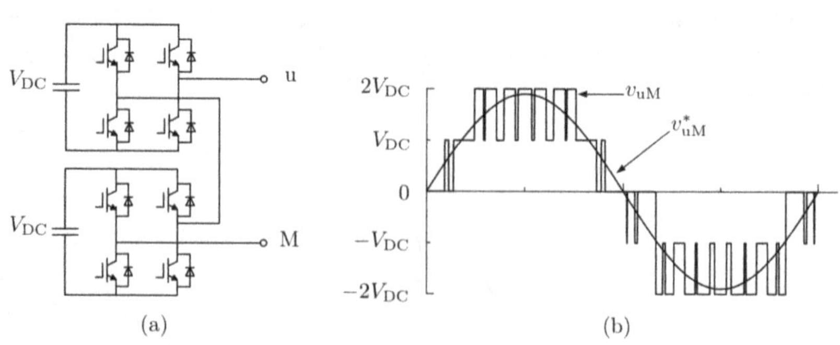

Figure 2.25 (a) Single-leg of a three-phase five-level CHB converter. (b) Line-to-neutral voltage on AC terminals.

sized to maintain a constant DC voltage during operation. Both the final application of the CHB converter (reactive power compensation, electric drives, active filter, etc.) and the modulation strategies used exert influence on the way the load on the DC capacitors is discharged and charged, which can change the voltage value on duty. In general, hierarchical DC voltage control strategies can be applied [20].

2.5.2 Delta- and star-connected three-phase CHB converters

When applied to three-phase systems, the CHB converter presents, in principle, two configuration possibilities: star (wye) or delta. Figure 2.26 shows the circuit of both configurations. The main differences are in the voltage applied to each phase cluster (considering the same AC line voltage), which changes the blocking voltage requirement of the switches. Another feature is the power exchange that takes place between the phases, which is important when the balancing of the capacitor voltages of each submodule has to be assured [20].

The wye connection of the three-phase CHB star-type converter provides a common voltage among the three clusters. For a 3-wire system, this voltage can fluctuate while the phase-to-phase voltages at the AC terminals of the converter are constant. Such characteristic, which can be considered a degree of freedom of this application [21], allows both the design of redundant modules (aiming for higher reliability) and the energy exchange between phases, which is essential to assure the submodule capacitor voltage balancing.

The three-phase CHB delta-type converter has no common neutral voltage, since the AC line voltage results from the sum of the output voltages of the

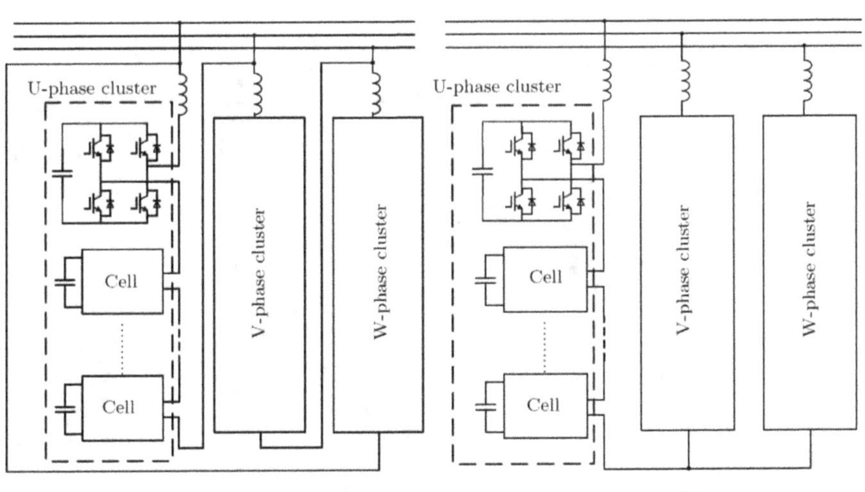

Figure 2.26 (a) Three-phase delta connected CHB converter and (b) three-phase star connected CHB converter.

submodules that form a phase cluster. However, the presence of coupling inductors among the clusters allows the circulation of an internal current in the delta connection without affecting the AC current. The circulating current, which can be considered as a degree of freedom of this application, can be used for energy exchange between phase clusters. It is also shown that such current allows the delta CHB converter to synthesize asymmetrical currents in its AC terminals with acceptable unbalance of the submodule capacitor voltages. As a result, one of the main applications of this configuration is STATCOMs aiming for the compensation of negative sequence currents [22].

2.5.3 Asymmetrical and symmetrical CHB converters

The CHB inverter uses a combination of many H-bridges in cascade, usually of the two-level type. There are two kinds of CHB converters: symmetric CHB (SCHB) and asymmetric CHB (ACHB). In the SCHB, each H-bridge uses power sources of the same value ($V_{DCj} = V_{DCk}$), while in the ACHB the power sources have different values ($V_{DCj} \neq V_{DCk}$). One of the advantages of the ACHB is that with the same number of H-bridges and power sources, more voltage levels are obtained. Its drawback is that the H-bridges are not interchangeable.

Figure 2.27(a) shows an SCHB inverter with three H-bridges, and Figure 2.27(b) shows an ACHB of the same type but with the power sources scaled in power of 3. In the first case, only seven levels are obtained, while the second topology is able to generate 27 levels of voltage.

In general, ACHB converters produce more levels because they use a redundant combination of switches from the SCHB converters. ACHB converters scaled in the power of 3 optimize the number of levels because a two-level H-bridge can

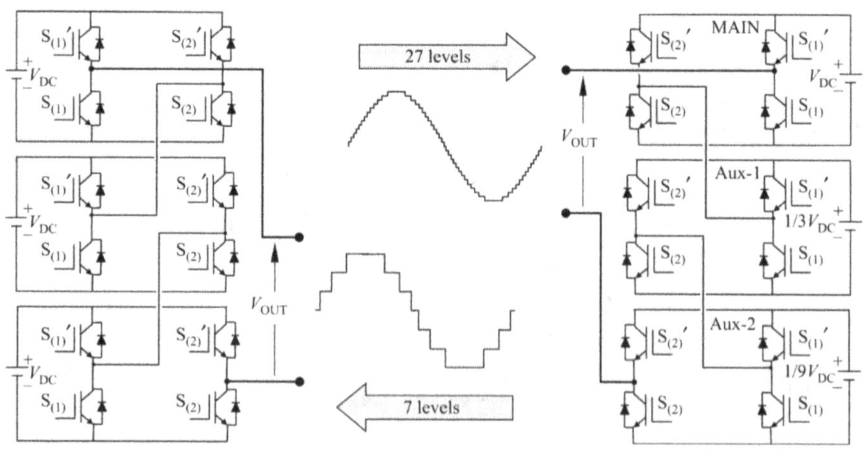

Figure 2.27 (a) Symmetrical CHB converter with three cells. (b) Asymmetrical CHB converter with three cells.

generate three different levels: "1," "0," and "1." The total number of levels that an ACHB converter with m bridges scaled in power of 3 is 3^m.

2.5.4 Comparison with NPC and FCC converters

One advantage of the CHB converter is its modularity and the reduced number of components compared to the NPC and FCC Converters. This is a feature that justifies the use of the CHB converters over the NPC and FCC converters for certain applications. Figure 2.28 shows a comparison of the number of components that are required to obtain a multilevel converter using transistors and diodes with the same nominal blocking voltage and capacitors with the same nominal voltage.

From the comparison, the following observations are made:

- NPC converter: it presents an excessive number of clipping diodes for a high number of levels in the output, which makes it unfeasible for configurations with more than three or four levels.

Neutral Point Clamped Converter	Flying Capacitor Converter	Cascaded H-Bridge Converter
Device count for a m-level converter (m ≥ 3)		
• Switches: $2(m-1)$ • Capacitors: $m-1$ • Clamp diodes: $(m-1)(m-2)$	• Switches: $2(m-1)$ • Capacitors: $(m-1)+$ $+\frac{1}{2}(m-1)(m-2)$	• Switches: $2(m-1)$ • Capacitors: $m-1$ • Clamp diodes: $(m-1)(m-2)$
Device count for a 5-level converter		
• 8 switches • 4 capacitors • 12 clamp diodes	• 8 switches • 10 capacitors	• 8 switches • 2 capacitors
Circuit of a single-phase 5-level converter		

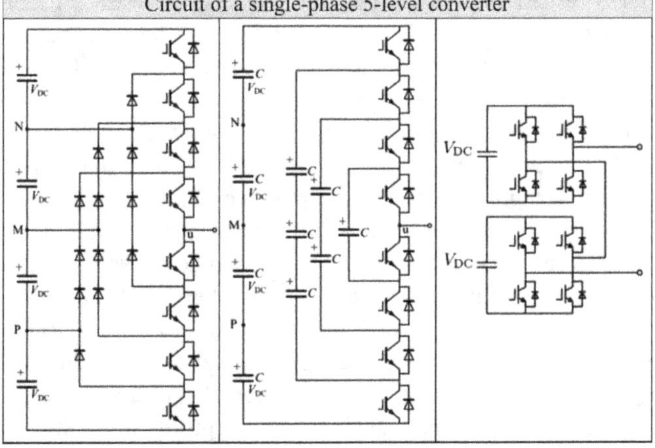

Figure 2.28 Comparison between multilevel converters

- FCC converter: it presents the need for voltages close to the DC bus in the intermediate capacitors for a high number of levels on the AC side, which results in a large count of capacitors with the same higher nominal voltage or near the DC voltage source rating.
- CHB converter: it presents a linear relationship between the number of levels in the output voltage and the components required. The CHB converter also has another advantage, which is modularity. The possibility of using several modules with a single basic structure facilitates the design, manufacturing, and maintenance.

2.6 Modular and hybrid multilevel converters

2.6.1 Modular multilevel converter

The modular multilevel converter has several characteristics and challenges similar to the CHB converter, but its structure allows to use only one DC source in the DC-to-AC conversion or vice versa. From the various research conducted for the CHB converters, it was noted that the DC–AC conversion, which characterizes the inverter, should be individualized for each submodule (or basic cell), while each submodule could be modeled as an individual voltage source. This characteristic is striking in the electric drive presented by ROBICON [17], where the special multi-winding phase-shifted transformer with multiple secondary terminals connected to individual diode rectifiers enabled a sufficient number of DC sources from the AC mains supply.

Based on these observations, Marquadt *et al.* proposed the configurations shown in Figure 2.29 [23–25]. Nowadays it is widely known as modular multilevel converter, or MMC, although the acronym used initially was *M2LC*. The MMC has as its main advantage the possibility of using a single DC source for the DC–AC conversion. This made it a suitable candidate for applications in HVDC converter stations as an alternative to two-level converters, besides presenting similar advantages to the CHB converter. The basic module that forms the converter can be composed of a full bridge submodule, in a similar manner to the CHB converter, or a *chopper,* also known as *a* half-bridge submodule.

Note that the choice between full-bridge or half-bridge modules is important to evaluate the MMC application. As shown in Figure 2.29(a), the half-bridge MMC has the submodule string connected at the positive DC terminal synthesizing voltages from 0 to $+nV_C$, while the submodule string at the negative terminal synthesizes voltages from 0 to $-nV_C$, where n *is* the number of submodules in a string. The full-bridge MMC in Figure 2.29(b), on the other hand, has both submodule strings between the AC side terminals and the positive DC terminal and between the AC side terminal and the negative DC terminal synthesizing voltages from $-nV_C$ to $+nV_C$. One could argue that the number of levels synthesized at the AC terminals by each configuration is an important feature of the MMC. However, practical applications, mainly in HVDC, have presented circuits with tens or hundreds of modules, making the evaluation by the number of levels innocuous.

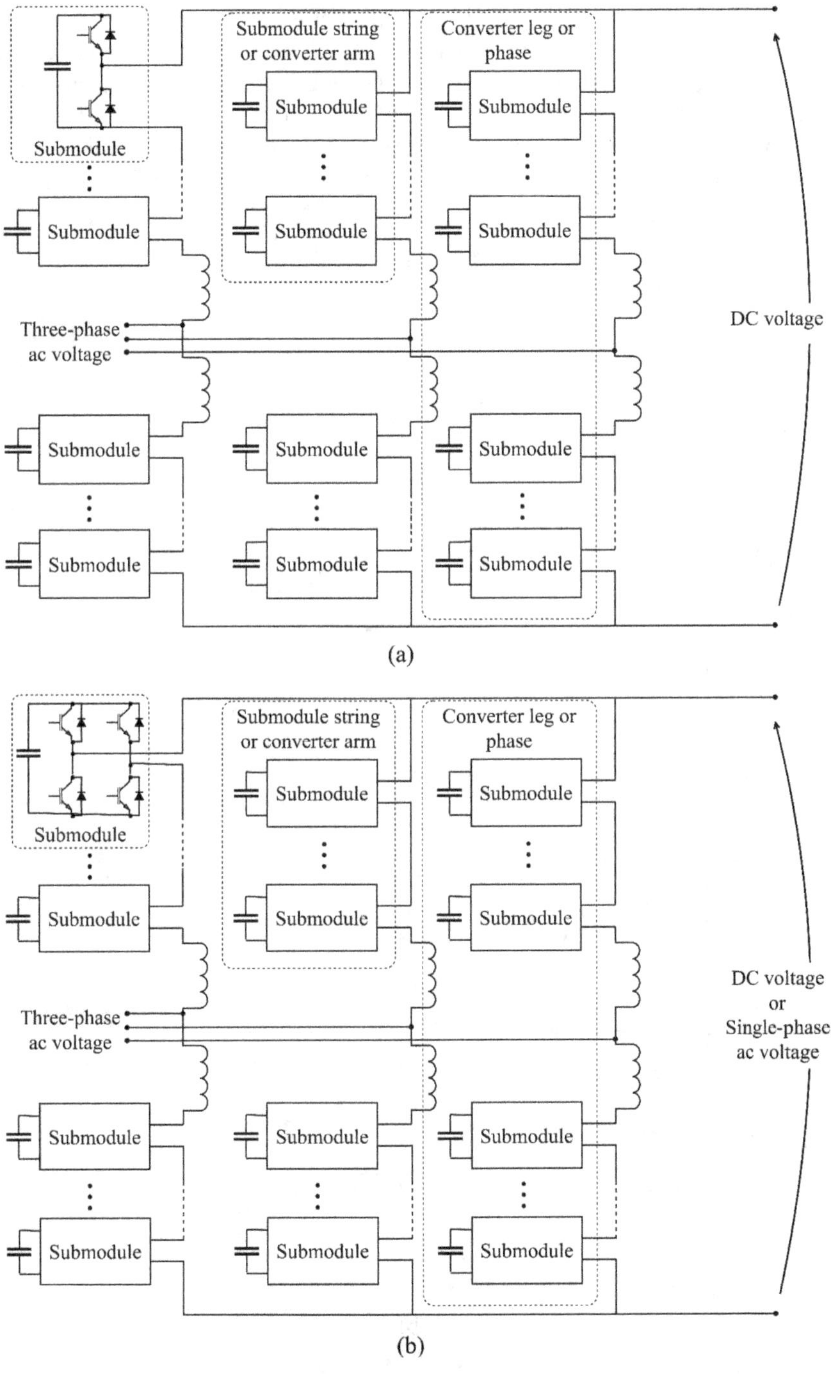

Figure 2.29 MMC converter circuit. (a) Half-bridge submodules. (b) Full-bridge submodules.

Figure 2.30 shows the voltage waveforms that each cascaded combination of submodules is capable of synthesizing. For the half-bridge submodules, the resulting voltage is restricted to one polarity and a DC voltage level appears on the submodule string. For the full-bridge submodules, a voltage waveform with no DC bias is presented in Figure 2.30. Nevertheless, the full-bridge submodules are capable of AC-only component synthesis and combined AC and DC component

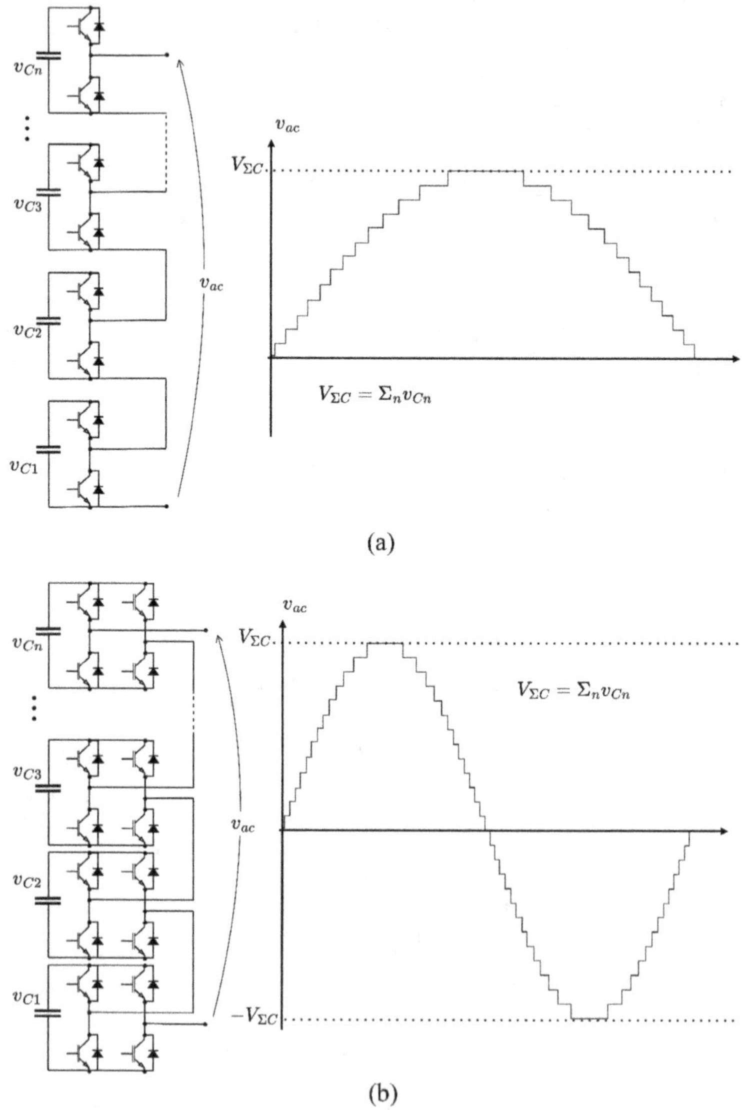

Figure 2.30 (a) Voltage synthesized by a half-bridge submodule string. (b) Voltage synthesized by a full-bridge submodule string.

synthesis. The flexibility of the full-bridge MMC allows the voltage between the terminals on the DC side to vary, and it is possible for this voltage to be either a DC voltage or a controlled AC voltage. In the case of half-bridge MMC, the voltage between the terminals on the DC side is restricted by $V_{DC} \geq \left(2\sqrt{2}/\sqrt{3}\right)V_{RMS}$, where V_{RMS} is the AC grid rms voltage. Such a restriction results in a DC voltage always existing between the DC terminals as the voltage is synthesized in the AC terminals. As it has been used extensively in HVDC voltage-source applications or DC–AC applications with a single DC source, the MMC half-bridge with half-bridge submodules is the most common configuration found in the field, so the term "MMC" is used in industry to refer to the half-bridge MMC directly.

The modulation strategies typically applied to MMC are the staircase modulation [26] and the PWM modulation with phase shift [27] or level shift [28]. The space vector modulation, although proposed together with the MMC initially [24], is less used due to its complexity. Other existing modulations are non-linear or heuristic, like nearest level modulation [29], which are justified in applications with many modules per phase.

The aspects and challenges involving the regulation of the DC capacitor voltages of the MMC modules are also relevant and objects of several research and studies. Due to the intense energy exchange and the high current circulating through the modules, the capacitance is relatively high, in order to keep the voltage of the capacitors constant. This implies a module with a larger volume. Some observations of commercial modules show that the capacitors fill up to $\frac{3}{4}$ of the total volume of a submodule.

2.6.2 Circulating current in MMC

The MMC has an intrinsic characteristic which is the presence of currents circulating through the modules and the DC bus. Figure 2.31(a) illustrates a single-phase circuit of a half-bridge MMC to illustrate and model the circulating currents within the circuit.

At first sight, two circulating currents are observed, i_p and i_n, which circulate among the submodules, the AC terminals (in this case, load), and the DC bus. Note that the analysis considers the "instantaneous" values of the electrical quantities. An important feature is the relationship between the circulating currents and i_L, which is the current at the AC terminals:

$$i_L = i_p - i_n. \tag{2.20}$$

This relationship could be used for the control of the MMC. However, both currents i_p and i_n contain DC and AC components, even assuming that i_L is purely sinusoidal. Furthermore, both i_p and i_n are dependent on each other directly. Therefore, using (2.20) as a starting point for the controller is not a straightforward option. However, another approach is to define a current i_z, which circulates between the submodules and the DC bus:

$$i_z = \frac{1}{2}\left(i_p + i_n\right). \tag{2.21}$$

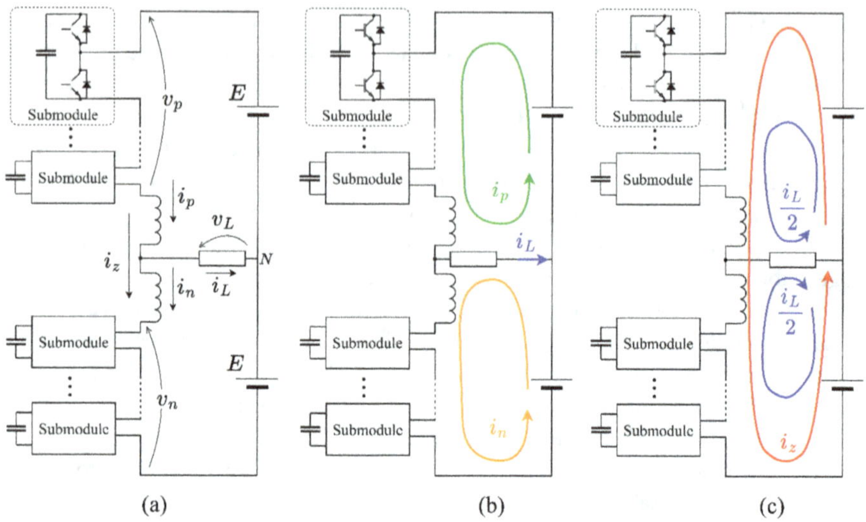

Figure 2.31 (a) Currents and voltages at the MMC. (b) Relation among i_L, i_n, and i_p. (c) Circulating current and load current.

Thus, the currents i_p and i_n are redefined as follows by Kirchhoff's Law:

$$i_p = i_z - \frac{1}{2} i_L \tag{2.22}$$

$$i_n = i_z + \frac{1}{2} i_L \tag{2.23}$$

From Equations (2.21)–(2.23) it can be seen that the circulating current i_z has only DC components. Also, note that the decoupling between the AC and DC current is visible clearly in (2.22) and (2.23) [30]. Figure 2.31(b) and (c) illustrates this decoupling [20].

Therefore, it is possible to determine the current i_z by calculating the instantaneous powers of the AC terminals and the DC bus, assuming that the circuit is operating in a steady state and without losses, i.e., the average power in the half-bridge modules is zero. The circulating current i_z can be derived in Figure 2.31(c) by:

$$i_z = \frac{1}{3} \frac{VI \cos \theta}{2E}, \tag{2.24}$$

where θ is the phase angle of the AC current, V and I are the rms values of the AC voltage (phase-neutral) and AC current, respectively.

Figure 2.32 shows waveforms for the circuit of Figure 2.31, without switching components. The significance of defining the "circulating current" i_z, which is exclusively composed of DC components, was pivotal in the advancement of the MMC. This became evident when the ability to separate the current components due to DC

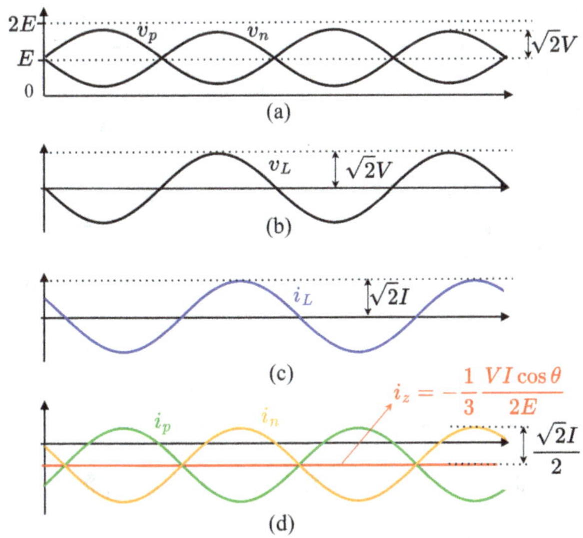

Figure 2.32 MMC simulated waveforms. (a) Submodule string voltages. (b) Load voltage. (c) Load current. (d) Circulating currents.

bus and the AC side, potentially the AC mains, was demonstrated as extremely useful to model the MMC. Note that, for the three-phase MMC, the AC components of the currents i_p and i_n cancel each other when added at the DC terminals, so only the contribution of the DC component of the circulating currents appears at the DC bus, which is i_z basically. From this principle, the MMC is applied to HVDC voltage source converter stations, where the control of the power transmitted by the high-voltage DC lines is directly done by regulating the active and reactive powers at the AC terminals.

2.6.3 Hybrid multilevel converters

Other topologies can be obtained using combination of different converters. Figure 2.33 shows a simple two-level H-bridge that uses sub-cells to generate as many levels as the number of sub-cells. As each sub-cell can generate two different levels (level "0" and level "1"), the number of levels generated by the sub-cells is optimized when their power sources are scaled in power of 2. In this case, redundant levels do not exist, and with m sub-cells, $2m$ positive levels are generated. The H-bridge is in charge of producing positive and negative half cycles, and then the total number of levels is given by:

$$n_{ab} = 2m^2 - 1. \tag{2.25}$$

Figure 2.34 shows a combined topology for a single-phase H-bridge, using one branch with a three-level converter and one branch with a two-level converter. Section 2.7.5 presents other hybrid topologies alongside hybrid modulation methods.

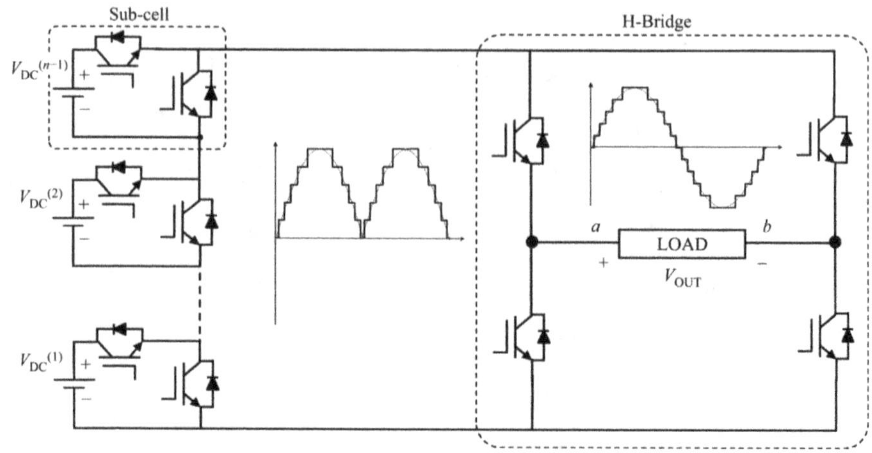

Figure 2.33 Multilevel inverter using subcells and a H-bridge

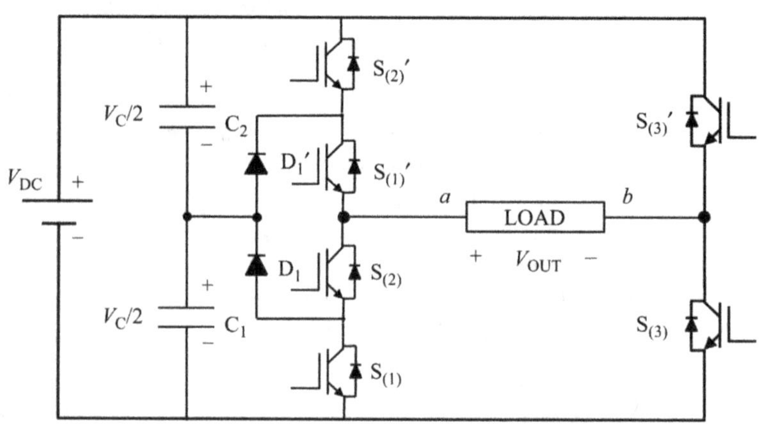

Figure 2.34 Combined H-bridge with NPC three-level and two-level converters

2.7 Modulation strategies of multilevel converters

2.7.1 About the classification of modulation strategies of multilevel converters

With the development of multilevel inverters and topologies came the challenge to extend traditional modulation methods for two-level converters to the multilevel case. On one side, there is the inherent additional complexity of having more powerful electronic devices to control, and on the other hand, the possibility to take advantage of the extra degrees of freedom provided by the redundant switching states generated by these topologies. As a consequence, a large number of different

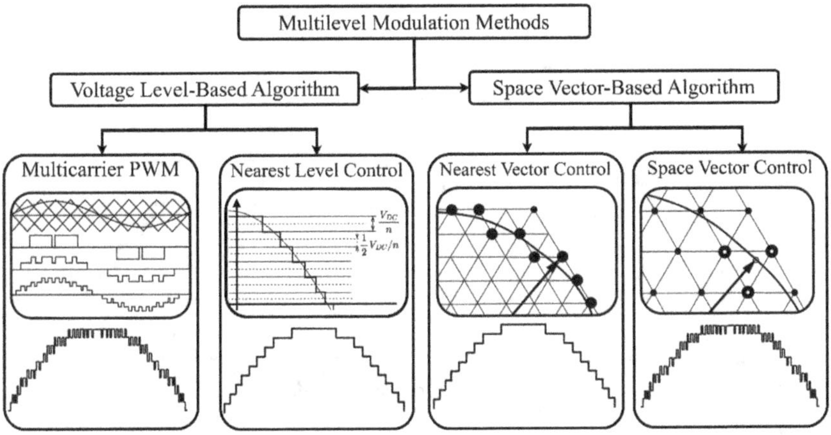

Figure 2.35 Classification of modulation methods for multilevel converters based on algorithm paradigm

modulation algorithms have been developed or adapted, each one with unique features and drawbacks, which are significantly weighted on the final application.

Nevertheless, the literature presents diverse classifications of modulation methods for multilevel converter. We briefly discuss two of them. Figure 2.35 shows a classification based on the time domain, i.e., the modulation method considers the synthesis of voltage levels over a time period, or in the space vector domain, i.e., the operating principle is based on the voltage vector generation [31].

On the other hand, Figure 2.36 shows the classification of modulation methods based on how the resulting switching frequency is set [32]. This approach highlights a point of view considered mainly for high-power applications, where the limitations of the switching devices and losses are important in the design of multilevel converters. Note that even space vector control, hybrid modulation, or multicarrier PWM, especially level shifted PWM, can be still put into different classifications, which are not discussed in this chapter.

2.7.2 Staircase modulation

The *staircase modulation* is exemplified in Figure 2.37 for a CHB converter [18]. This strategy allows switching at low frequency (usually at the fundamental frequency), which is advantageous for the first medium- and high-voltage semiconductor components, in addition to a reduced harmonic distortion. The so-called staircase modulation was the first implementation for the multilevel converters, but it is usually presented alongside techniques aiming at the selective elimination of harmonics. Such an approach is also known as "selective harmonic elimination," or SHE. Nabae *et al.* [1] have actually presented an angle firing technique in order to reduce the harmonic content of the NPC inverter.

Generally speaking, staircase modulation techniques impose a switching frequency for the semiconductor switch equal to the fundamental frequency with a

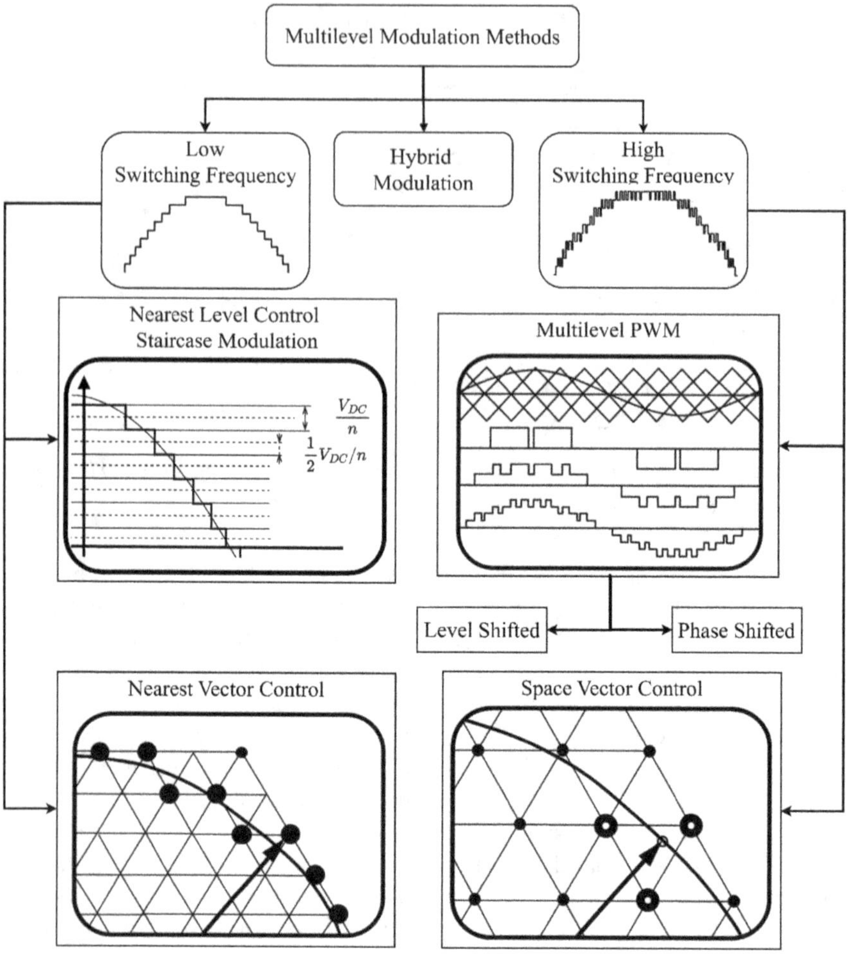

Figure 2.36 Classification of modulation methods for multilevel converters based on switching frequency paradigm

quasi-square voltage waveform. Voltage pulses are generated according to the desired firing angle, which should be precisely tracked through the voltage. For example, zero-crossing techniques were widely used. The final level count, on the other hand, raises the equivalent switching frequency, and the final voltage waveform also changes. SHE methods increase the switching frequency as extra pulses are added to promote the desired harmonic elimination.

2.7.3 Multicarrier pulse-width modulation

The Multicarrier PWM is, probably, the most well-known modulation method aimed at multilevel converters. In the first years of the multilevel converters,

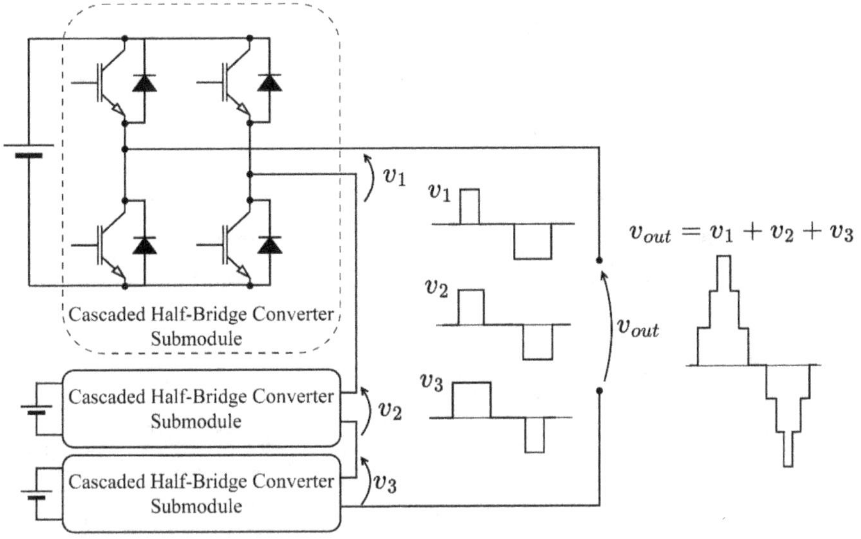

Figure 2.37 Staircase modulation for a symmetrical CHB converter

applying PWM methods was prohibitive due to the semiconductor technology available and switching losses. However, technological development led to its popularization, besides multicarrier implementation being a fair simple method, both for analog and digital implementation. Ready-to-use PWM modules and blocks in modern digital control hardware are widely available nowadays as well.

Generally speaking, multicarrier PWM produces the pulses by comparing the modulation signal to multiple modulation carriers, which imposes the switching frequency for the individual switch. Each modulation carrier is assigned to a specific set of switches, which depends on the chosen topology according to the description made in Sections 2.3, 2.4, 2.5, and 2.6. Note that although the switching occurs within the equivalent time period, the actual switching action changes according to the comparator output. This is the reason for the frequency sidebands when the frequency spectra of the voltage waveform are analyzed.

The use of multicarrier PWM in multilevel converters can be divided mainly into two classes when considering voltage levels in a time frame: the phase-shifted method, also known as "*PWM phase-shifted*," and the level-shifted method, also known as "*PWM level-shifted*." A great feature of PWM modulation is the further reduction of harmonic distortion.

Figure 2.38 shows an example of the phase-shift PWM for a 12-level CHB converter. The optimized harmonic reduction is obtained by shifting the modulation between each of the n modules composing the CHB by π/n [33]. PS-PWM assumes that DC voltage in submodules is equal to provide the optimal harmonic cancellation, otherwise harmonic cancellation may lose effectiveness.

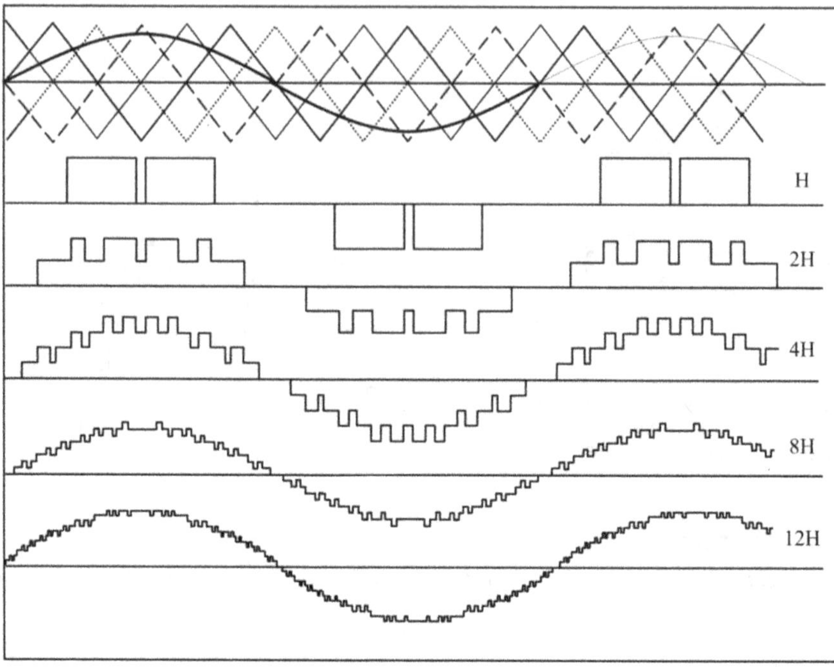

Figure 2.38 Modulation carriers, modulation signal, and voltage output for a 25-level CHB converter with PS-PWM

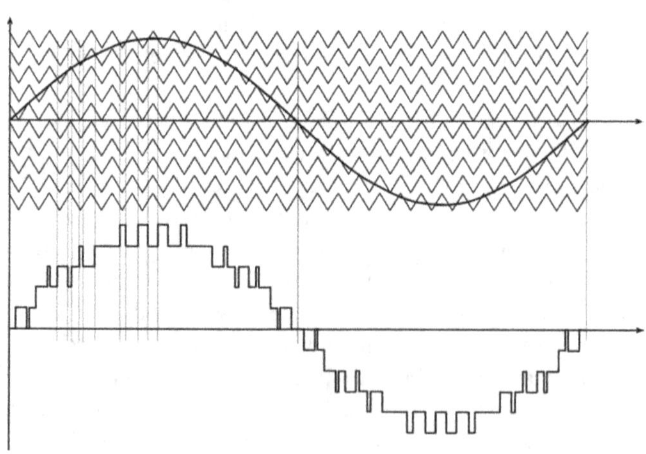

Figure 2.39 Modulation carriers, modulation signal, and voltage output for an 11-level converter with LS-PWM

Figure 2.39 shows an example of the level-shift PWM for an 11-level multilevel converter. This method consists of arranging voltage level shifts, setting each carrier between two voltage levels. Since each carrier is associated with two levels,

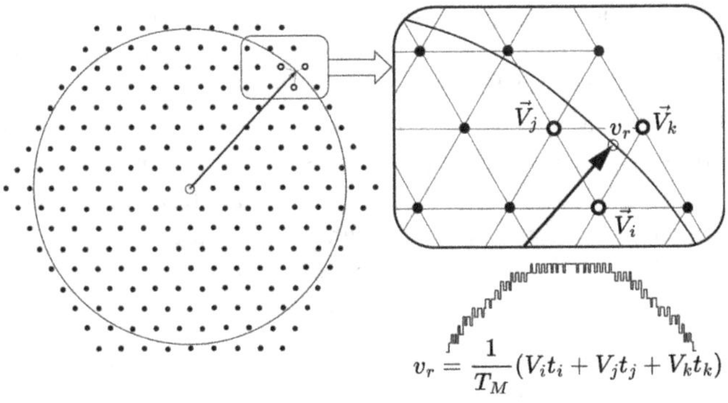

Figure 2.40 Vector diagram and space vector modulation (SVM) for a 9-level converter

the voltage output due to the carriers can cover the full amplitude range that can be generated by the converter. Certain components in the lower (and more significant) frequencies may be not effectively canceled in LS-PWM [34]. Moreover, its effectiveness may be reduced when the modulation index is low, which is the case for certain motor drive applications.

When applying the space vector approach, the alternative is to use the *space vector modulation*. The SVM method has the advantage of being a fully digitally implementable method, but it suffers from inherent complexity when dealing with several levels. Figure 2.40 shows the space vector diagram and possible voltage vectors for a 9-level multilevel converter.

The main idea behind SVM is to synthesize the rotating reference vector by a time-averaged combination of selected fixed vectors generated by the inverter. For a given reference vector magnitude and position (angle), the three closest vectors are selected to generate the voltage over the modulation period according to its reference. This strategy reduces distortion in voltage by reducing the instances of dv/dt_s. Consider, for example, the reference vector v_r illustrated in Figure 2.40. The mean value of the inverter voltage can be expressed as:

$$v_r = \frac{1}{T_M}\left(V_i t_i + V_j t_j + V_k t_k\right), \tag{2.26}$$

where t_i, t_j, and t_k are the times in which the fixed vectors V_i, V_j, and V_k are applied to the load, with the restriction:

$$T_M = t_i + t_j + t_k, \tag{2.27}$$

where T_M is the modulation period, which is related to the designed switching frequency.

2.7.4 Nearest level control

Currently widely used, heuristic modulation techniques such as nearest level control (NLC) produce pretty similar voltage waveforms to the staircase modulation.

Therefore, much of the same advantages also applies. While NLC uses a comparison of instantaneous voltage reference in a designed time frame to select the voltage level to synthesize, NVC tracks the reference through the rotating space vector. Modern digital control and hardware allow simplified implementation when compared to previous firing angle control based on analog electronics.

The NLC is also known as the rounding modulation method. It consists of selecting the nearest voltage level that can be generated by the inverter to the desired output voltage reference. It can be expressed as:

$$v_{\mathrm{OUT}} = V_{\mathrm{DC}}\,\mathrm{Round}\left(\frac{v_{ref}}{V_{\mathrm{DC}}}\right), \tag{2.28}$$

where V_{DC} is the voltage of each power source and v_{ref} is the voltage reference, which is a sinusoidal waveform usually. Figure 2.41 shows the waveform generation for an n-level inverter using the NLC strategy. As can be seen, the NLC method only produces one commutation between two voltage levels, which minimizes the switching frequency. However, the THD is large unless the number of levels is larger than 20 levels because the THD of voltage becomes less than 4% [35].

The equivalent space vector approach is the nearest vector control (NVC), first introduced by [36], as an alternative to selected harmonic elimination or staircase modulation, to provide a low-switching frequency modulation method, without the offline requirements and poor dynamic performance. The basic idea is to take advantage of the high number of voltage vectors generated by a multilevel inverter, by simply approximating the reference to the closest voltage vector that can be generated. This principle is illustrated in Figure 2.42, where a nine-level space vector complex plane is illustrated. Each circle is one of the possible voltage vectors generated by the inverter. They are surrounded by hexagons that represent the boundary of the area closest to the corresponding vector. The circle is the voltage space vector reference trajectory through the plane. When the reference falls into a certain hexagon, the closest vector V_T will be generated by the inverter.

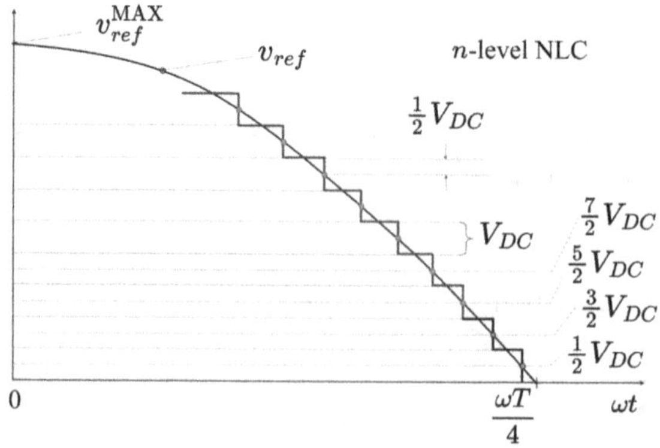

Figure 2.41 AC waveform synthesis for a n-level multilevel converter by NLC

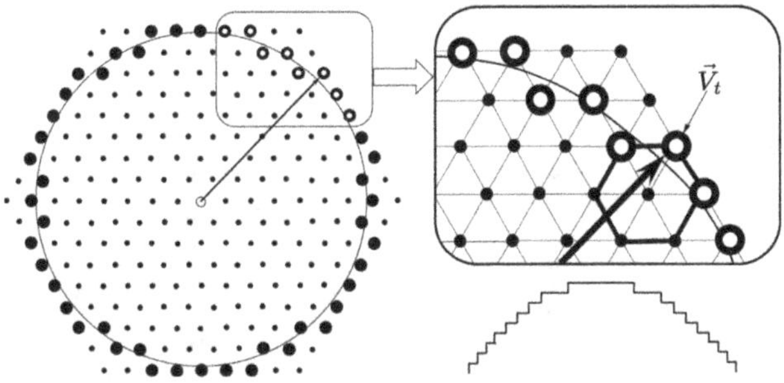

Figure 2.42 Nearest vector control (NVC)

2.7.5 Hybrid or combined modulation strategies

Although space vector-based algorithms are a technology useful for any application, they are especially applicable to electrical machines. Vector control permits a dynamic behavior not reachable with other technologies because they allow changing vector position in a matter of milliseconds. Besides, different topologies of multilevel inverters can be combined to increase the size and number of vectors. Section 2.6.3 presented the possibility of hybrid multilevel topologies.

Figure 2.43 shows a hybrid multilevel converter consisting of a two-level converter combined with a three-level converter using H-bridge topologies. As these converters are fed with independent power supplies, their voltages can be selected using different magnitudes. It is interesting to mention that the H-bridges can operate with capacitors instead of power sources, using an adequate control strategy. Figure 2.43 also shows how the space vector diagram changes when the relative amplitudes of V_{DC1} and V_{DC2} change. Figure 2.43(c) shows the hexagons when V_{DC1} is kept constant and Figure 2.43(d) shows the hexagons when V_{DC2} is kept constant [37].

Figure 2.44 shows a combined topology using three different converter schemes. Figure 2.44(a)–(c) shows a two three-level NPC, a three-level NPC with one three-level HB, and two three-level HBs, respectively. As in Figure 2.43, the space vector diagram can be modified by changing the voltage relation between the first and the second converter. The number of levels increases until the smaller hexagons do not interact with the larger one. Figure 2.44(d) shows that up to a limit of nine levels can be reached so all vectors are within the overall hexagon area. If the ratio between V_{DC1} and V_{DC2} is extended, up to 13 levels are obtainable, but some vectors are not achievable (i.e., it is not possible to synthesize them), which is observable from the non-equidistant distance between the vectors in the diagrams in Figure 2.44(e) [37].

It is possible to get other combined topologies using more than two converters and each one with different values of voltage supplies. Besides, as was mentioned before, with an adequate strategy some converters can work with capacitors instead of power sources. In this case, these converters do not supply any power but they

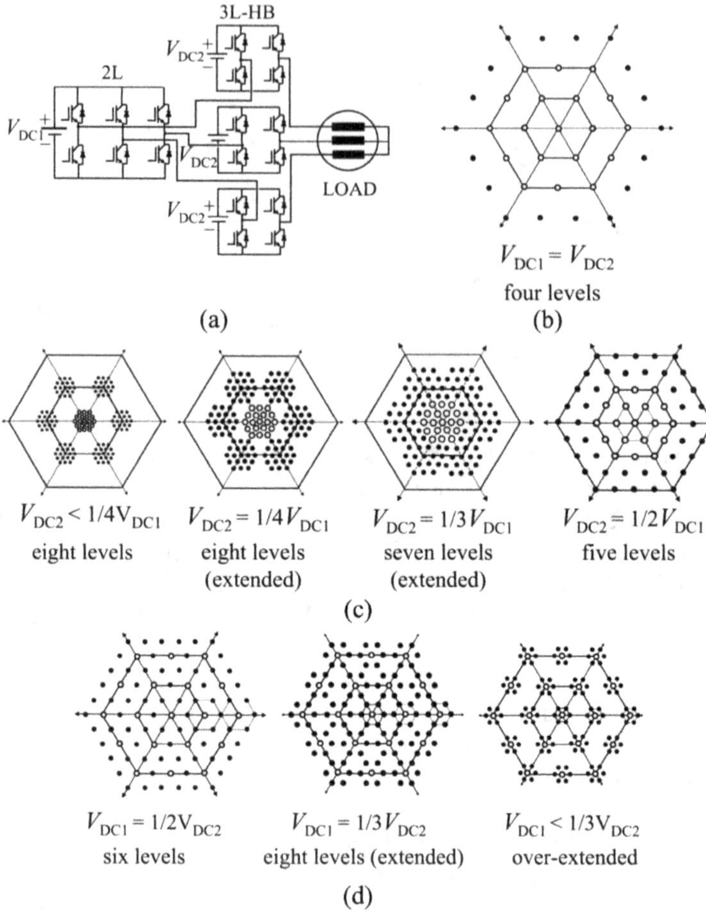

Figure 2.43 (a)Two-level inverter with three-level H-Bridges. (b) Space vector for $V_{DC1} = V_{DC2}$. (c) Space vector for $V_{DC1} < V_{DC2}$. (d) Space Vector for $V_{DC1} > V_{DC2}$.

increase the number of vectors improving the quality of the output voltage. When some converters work using capacitors, they work as a series of active power filters.

Hybrid modulation for multilevel converter can be achieved through the "conventional" topologies as well. Figure 2.45 presents the modulation of an asymmetric CHB converter composed of three submodules [38]. In this case, the DC voltages are different among each submodule, which allows a larger amount of voltage levels compared to an equivalent symmetrical CHB converter. In such applications, the optimized DC voltage ratio between the modules can be obtained by $V_n = 2\Sigma_{k=1}^{n-1} V_k$, where V_n is the DC voltage of the n-th module as a function of the voltage V_1. In this case, this ratio is V_{DC}, $2V_{DC}$, and $6V_{DC}$, and

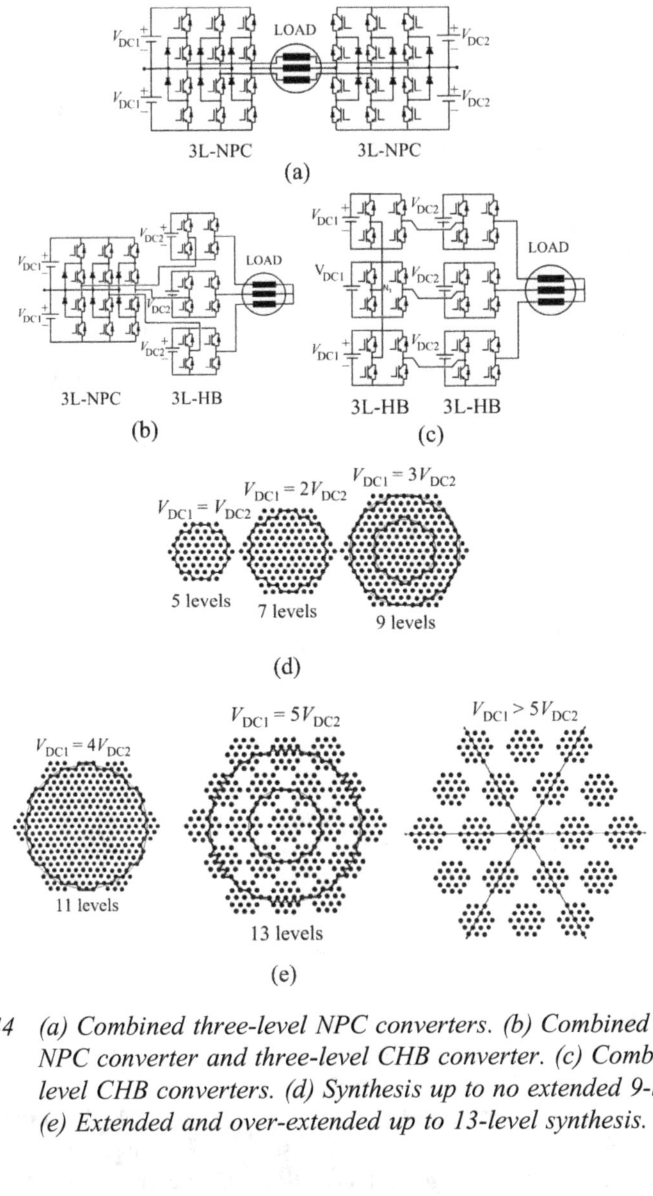

Figure 2.44 *(a) Combined three-level NPC converters. (b) Combined three-level NPC converter and three-level CHB converter. (c) Combined three-level CHB converters. (d) Synthesis up to no extended 9-levels. (e) Extended and over-extended up to 13-level synthesis.*

$V_1 = V_{DC}$. Note that the staircase modulation is applied, but with a "reduced" modulation signal for the submodules with lower voltages. Such reduction results in a quasi-PWM voltage waveform for such submodules. Recent applications in research studies presented a mix of semiconductor switches, such as Si IGBT and SiC MOSFETs [39], in which hybrid modulation methods could optimize the overall performance.

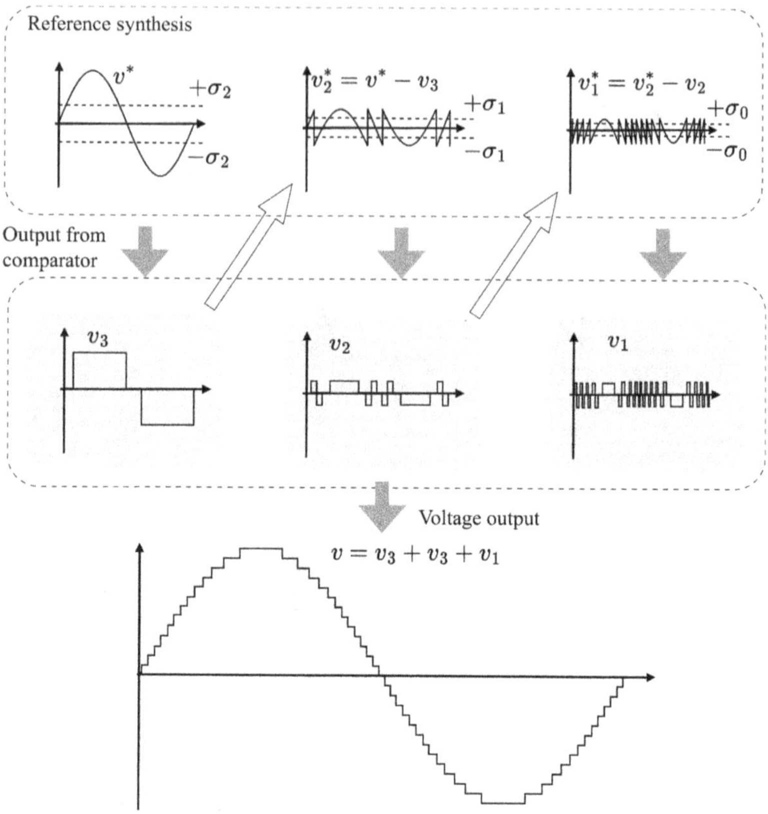

Figure 2.45 Voltage waveform synthesis of an asymmetric CHB converter with alternative staircase modulation

References

[1] A. Nabae, I. Takahashi and H. Akagi, "A new neutral-point-clamped PWM inverter," in *IEEE Transactions on Industry Applications*, vol. IA-17, no. 5, pp. 518–523, 1981.

[2] R. Sommer, A. Mertens, M. Griggs, H. Conraths, M. Bruckmann and T. Greif, "New medium voltage drive systems using three-level neutral point clamped inverter with high voltage IGBT," Conference Record of the 1999 IEEE Industry Applications Conference. Thirty-Forth IAS Annual Meeting (Cat. No.99CH36370), Phoenix, AZ, USA, vol. 3, pp. 1513–1519, 1999.

[3] T. Bruckner, S. Bernet and H. Guldner, "The active NPC converter and its loss-balancing control," in *IEEE Transactions on Industrial Electronics*, vol. 52, no. 3, pp. 855–868, 2005.

[4] S. Ogasawara and H. Akagi, "Analysis of variation of neutral point potential in neutral-point-clamped voltage source PWM inverters," Conference

Record of the 1993 IEEE Industry Applications Conference Twenty-Eighth IAS Annual Meeting, Toronto, Ontario, Canada, vol. 2, pp. 965–970, 1993.

[5] N. Celanovic and D. Boroyevich, "A comprehensive study of neutral-point voltage balancing problem in three-level neutral-point-clamped voltage source PWM inverters," *IEEE Transactions on Power Electronics*, vol. 15, no. 2, pp. 242–249, 2000.

[6] J. Pou, R. Pindado, D. Boroyevich and P. Rodriguez, "Evaluation of the low-frequency neutral-point voltage oscillations in the three-level inverter," in *IEEE Transactions on Industrial Electronics*, vol. 52, no. 6, pp. 1582–1588, 2005.

[7] H. Akagi, H. Fujita, S. Yonetani and Y. Kondo, "A 6.6-kV transformerless STATCOM based on a five-level diode-clamped PWM converter: System design and experimentation of a 200-V 10-kVA laboratory model," in *IEEE Transactions on Industry Applications*, vol. 44, no. 2, pp. 672–680, 2008.

[8] A. Anthon, Z. Zhang, M. A. E. Andersen, D. G. Holmes, B. McGrath and C. A. Teixeira, "The benefits of SiC mosfets in a T-type inverter for grid-tie applications," in *IEEE Transactions on Power Electronics*, vol. 32, no. 4, pp. 2808–2821, 2017.

[9] Ingo Staudt, "Application Note AN-11001," Semikron, 2015.

[10] T. A. Meynard and H. Foch, "Multi-level conversion: High voltage choppers and voltage-source inverters," PESC '92 Record. 23rd Annual IEEE Power Electronics Specialists Conference, Toledo, Spain, vol. 1, pp. 397–403, 1992.

[11] Jih-Sheng Lai and Fang Zheng Peng, "Multilevel converters-a new breed of power converters," in *IEEE Transactions on Industry Applications*, vol. 32, no. 3, pp. 509–517, 1996.

[12] T. A. Meynard, H. Foch, P. Thomas, J. Courault, R. Jakob and M. Nahrstaedt, "Multicell converters: Basic concepts and industry applications," in *IEEE Transactions on Industrial Electronics*, vol. 49, no. 5, pp. 955–964, 2002.

[13] Xiaoming Yuang, H. Stemmler and I. Barbi, "Self-balancing of the clamping-capacitor-voltages in the multilevel capacitor-clamping-inverter under sub-harmonic PWM modulation," in *IEEE Transactions on Power Electronics*, vol. 16, no. 2, pp. 256–263, 2001.

[14] R. H. Wilkinson, T. A. Meynard and H. du Toit Mouton, "Natural balance of multicell converters: The general case," in *IEEE Transactions on Power Electronics*, vol. 21, no. 6, pp. 1658–1666, 2006.

[15] T. A. Meynard, M. Fadel and N. Aouda, "Modeling of multilevel converters," in *IEEE Transactions on Industrial Electronics*, vol. 44, no. 3, pp. 356–364, 1997.

[16] T.A. Meynard and H. Foch, "Multi-level choppers for high voltage applications," in *EPE Journal*, vol. 2, no. 1, pp.45–50, 1992.

[17] P. W. Hammond, "A new approach to enhance power quality for medium voltage AC drives," in *IEEE Transactions on Industry Applications*, vol. 33, no. 1, pp. 202–208, 1997.

[18] Jih-Sheng Lai and Fang Zheng Peng, “Multilevel converters-a new breed of power converters,” in *IEEE Transactions on Industry Applications*, vol. 32, no. 3, pp. 509–517, 1996.
[19] L. de A. Silva, S. P. Pimentel and J. A. Pomilio: “Active filtering system with nineteen-level cascaded asymmetrical multilevel inverter and DC bus voltage control,” *Power Electronics*, vol. 11, no.1, pp. 17–24, 2006.
[20] H. Akagi, “What led to success in academic research on the family of modular multilevel cascade converters?,” 2018 International Power Electronics Conference (IPEC-Niigata 2018 – ECCE Asia), Niigata, Japan, pp. 2352–2359, 2018.
[21] R. E. Betz, T. Summers and T. Furney, “Symmetry compensation using an H-Bridge multilevel STATCOM with zero sequence injection,” Conference Record of the 2006 IEEE Industry Applications Conference Forty-First IAS Annual Meeting, Tampa, FL, USA, pp. 1724–1731, 2006.
[22] M. Hagiwara, R. Maeda and H. Akagi, “Negative-sequence reactive-power control by a PWM STATCOM based on a modular multilevel cascade converter (MMCC- SDBC),” in *IEEE Transactions on Industry Applications*, vol. 48, no. 2, pp. 720–729, 2012.
[23] R. Marquardt and A. Lesnicar, “A new modular voltage source inverter topology,” in *Proceedings of EPE*, 2003.
[24] A. Lesnicar and R. Marquardt, “An innovative modular multilevel converter topology suitable for a wide power range,” in *Proceedings of IEEE Bologna Power Tech*, vol. 3, pp. 23–26, 2003.
[25] M. Glinka and R. Marquardt, “A new ac/ac-multilevel converter family applied to a single-phase converter,” in *Proceedings of IEEE-PEDS*, vol. 1, pp. 16–23, 2003.
[26] K. Ilves, A. Antonopoulos, S. Norrga and H. Nee, “A new modulation method for the modular multilevel converter allowing fundamental switching frequency,” in *IEEE Transactions on Power Electronics*, vol. 27, no. 8, pp. 3482–3494, 2012.
[27] K. Ilves, L. Harnefors, S. Norrga and H. Nee, “Analysis and operation of modular multilevel converters with phase-shifted carrier PWM,” in *IEEE Transactions on Power Electronics*, vol. 30, no. 1, pp. 268–283, 2015.
[28] S. Fan, K. Zhang, J. Xiong and Y. Xue, “An improved control system for modular multilevel converters with new modulation strategy and voltage balancing control,” in *IEEE Transactions on Power Electronics*, vol. 30, no. 1, pp. 358–371, 2015.
[29] P. M. Meshram and V. B. Borghate, “A simplified nearest level control (NLC) voltage balancing method for modular multilevel converter (MMC),” in *IEEE Transactions on Power Electronics*, vol. 30, no. 1, pp. 450–462, 2015.
[30] M. Hagiwara and H. Akagi, “Control and experiment of pulsewidth-modulated modular multilevel converters,” in *IEEE Transactions on Power Electronics*, vol. 24, no. 7, pp. 1737–1746, 2009.

[31] L. G. Franquelo, J. Rodriguez, J. I. Leon, S. Kouro, R. Portillo and M. A. M. Prats, "The age of multilevel converters arrives," in *IEEE Industrial Electronics Magazine*, vol. 2, no. 2, pp. 28–39, 2008.

[32] J. Rodriguez, L. G. Franquelo, S. Kouro *et al.*, "Multilevel converters: An enabling technology for high-power applications," in *Proceedings of the IEEE*, vol. 97, no. 11, pp. 1786–1817, 2009.

[33] D. G. Holmes and B. P. McGrath, "Opportunities for harmonic cancellation with carrier-based PWM for a two-level and multilevel cascaded inverters," in *IEEE Transactions on Industry Applications*, vol. 37, no. 2, pp. 574–582, 2001.

[34] B. P. McGrath and D. G. Holmes, "Multicarrier PWM strategies for multilevel inverters," in *IEEE Transactions on Industrial Electronics*, vol. 49, no. 4, pp. 858–867, 2002.

[35] J. Pereda and J. Dixon, "23-level inverter for electric vehicles using a single battery pack and series active filters," in *IEEE Transactions on Vehicular Technology*, vol. 61, no. 3, pp. 1043–1051, 2012.

[36] J. Rodriguez, L. Moran, P. Correa and C. Silva, "A vector control technique for medium-voltage multilevel inverters," in *IEEE Transactions on Industrial Electronics*, vol. 49, no. 4, pp. 882–888, 2002.

[37] J. Pereda and J. Dixon, "Cascaded multilevel converters: Optimal asymmetries and floating capacitor control," in *IEEE Transactions on Industrial Electronics*, vol. 60, no. 11, pp. 4784–4793, 2013.

[38] L. A. Silva, S. P. Pimentel and J. A. Pomilio, "Nineteen-level active filter system using asymmetrical cascaded converter with DC voltages control," 2005 IEEE 36th Power Electronics Specialists Conference, Dresden, Germany, 2005.

[39] Y. Koyama, Y. Nakazawa, H. Mochikawa, A. Kuzumaki, K. Sano and N. Okada, "A transformerless 6.6-kV STATCOM based on a hybrid cascade multilevel converter using SiC devices," in *IEEE Transactions on Power Electronics*, vol. 33, no. 9, pp. 7411–7423, 2018.

Chapter 3
Multi-input Converters

Babak Farhangi[1] *and Hamid A. Toliyat*[2]

3.1 Introduction

Unidirectional power electronic converters mostly have a single input and a single output, although it is possible to have multiple outputs. In order to perform analysis and design, it is often considered that such converters are time-variant, nonlinear, single-input, single-output electric systems [1–3]. On the other hand, hybrid systems or integrated power systems may have more than one source supplying the load. It is possible to configure such systems through a set of single-input, single-output power converters; where one converter is employed from each source to the load. In this case, the converters may be coupled to the load in a parallel scheme through a capacitive filter or in a series scheme through an inductive filter [4,5]. If the separated power conversion paths combine into one power converter, they form a multi-input power converter. This approach can potentially reduce the component count. Moreover, the multi-input converters may be designed to improve the functionality and reliability of the system.

In multiple-input, single-output converters, each source is conditioned through unidirectional power converter stages [5–21], with the consideration that sources are not able to exchange energy between them. When bidirectional sources or energy storage elements exist in such a system, bidirectional power conditioning stages are needed to condition each input. Then, sources may freely exchange power among them. Such systems are called multi-port systems as each port can be either a load or a source to the system [22–30]. In this situation, a group of sources supply a group of loads in a multi-input, multi-output system. Multiple-input, multiple-output systems can also interconnect a group of unidirectional sources to a group of active and passive loads [31,32]. The multi-input converters may be DC–DC, DC–AC, AC–DC or hybrid DC and AC systems, according to the nature of the loads and the sources. The galvanic isolation between the sources and loads depends on the application requirements.

[1]Principal Power Electronics Engineer, GAF Energy, Research and Development, San Jose, CA, USA
[2]Texas A&M University, Department of Electrical and Computer Engineering, College Station, TX, USA

Emerging applications make use of multi-input converters, especially when integrating renewable sources like photovoltaic panels and wind turbines. This approach proves beneficial as these sources complement each other in energy generation, addressing the potential issue of neither being constantly available. A hybrid renewable system may benefit from multiple energy sources in an integrated power system or a micro-grid. This will improve the versatility of the system. Moreover, introducing energy storage elements, such as batteries or flywheels, introduce bidirectional ports to the system. Integrated power systems and micro-grids can be implemented in a small physical space; such applications benefit from multi-input converters [7,12,19,20,24,27,33]. The application of the electric and hybrid-electric vehicles, being mobile systems, extremely benefit from increased power density through the adoption of multi-input converters. Battery-powered vehicles, grid-connected automobiles, all electric ships and airplanes, and satellites need multi-input vehicular systems for land, sea, air, and space applications [4,20,22,23,34–40].

This chapter is dedicated to recent advances in multi-input converters. The novel applications for multi-input converters need customized power converter topologies. These topologies need to be synthesized considering the power conditioning requirements of the sources, loads, and energy storage elements. The techniques proposed for synthesizing multi-input power converter topologies will be introduced in Section 3.2. Section 3.3 is dedicated to multi-port converters with an emphasis on methods for enabling bi-directional power flow between power converter terminals. Furthermore, coupling multiple power conversion stages through DC and AC energy links is discussed in this section. In Section 3.4, major applications of the multiport converters are explored: renewable energy systems, microgrids, and integrated and vehicular power systems. Eventually, this chapter is summarized in Section 3.5

3.2 Realizing multi-input converter topologies

Multi-input converters have applications in emerging systems, so it is important to develop topologies customized to each application, becoming a novel structure. The fundamental concepts for developing single-input, single-output power converters can be extended to the synthesis of multi-input converters. Fundamentally, power converters consist of switching networks and passive elements between the sources and loads [1,2]. Mathematically it means that between each source and load, there is a graph that contains time-variant components, such as switches and diodes, and passive elements (inductors and capacitors). By cascading switching stages and passive networks, a multi-stage converter is formed. While there are different definitions for power conversion stages, the one used in this chapter is related to the number of switching networks separated by passive elements between sources and loads, defining the count of stages. According to this definition, basic buck and boost converters are single-stage converters, as well as the differential synchronous buck converter, and the h-bridge converter. In contrast, the z-source converter is not a single-stage converter because a diode is separated from the other switches by the z-source network [34,35,41]. Extending the multi-stage switching power conversion networks, it is possible to connect several sources through switching networks and passive elements to form a multi-input power converter. If each converter is directly connected to the load, it is a separate solution. In contrast, a multi-input power converter is realized if there is a common power conversion path between the multiple sources and loads.

Assuming that each source is either a voltage source or a current source, it is preferred to connect each source through a switching stage to the power converter network. If a number of voltage sources are in series or a number of current sources are in parallel, they will add up according to Kirchhoff's Voltage and Current laws (respectively known as KVL and KCL) and represent one unified source to the power converter network. It is not possible to condition these sources individually, which does not satisfy the goal of developing multi-input converters. The fundamentals of realizing multi-input power converters depend on the conceptual theory presented in this section

In [1], the fundamental theory of power converters as repetitive switching networks is applied to the analytical synthesis of single-input, single-output switching power converters; moreover, the properties of different switching power converters are discussed. The buck and the boost converters are two basic converters, and several converters can be derived by a cascade or differential connection of these two basic converters as shown in Figure 3.1. As an example, the buck-boost converter is derived by cascading buck and boost converters and simplifying the passive elements in between. The H-bridge converter can be derived by differential connection of two buck converters. The boost converter is actually a dual of the buck converter [42]; it can also be realized by the inversion of the source and the load in the buck converter. Not all power converters can be derived similarly from buck and boost converters. The single ended primary inductor converter (SEPIC) and Watkins–Johnson converters are two examples shown in [1]. These two topologies are shown in Figure 3.2. In both converters, the passive network between the switches is more complicated than a simple inductor or a single capacitor; a transformer, which is a network of inductors and capacitors, is used within the sources and switching networks. A generalized approach is proposed to realize single-input, single-output power converters using the state

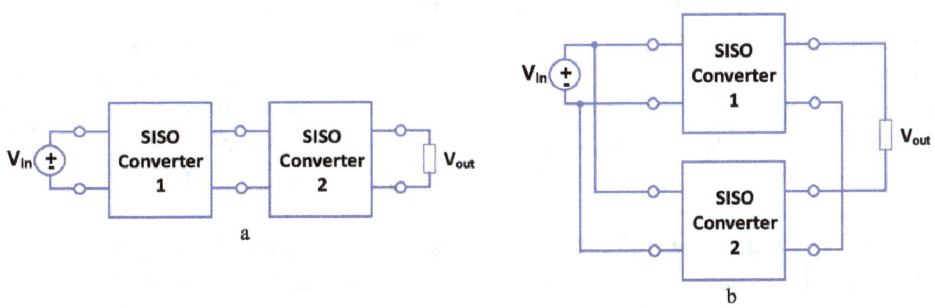

Figure 3.1 *Converters can be derived from the existing converters. (a) The cascade connection of the converters. (b) The differential connection of the converters.*

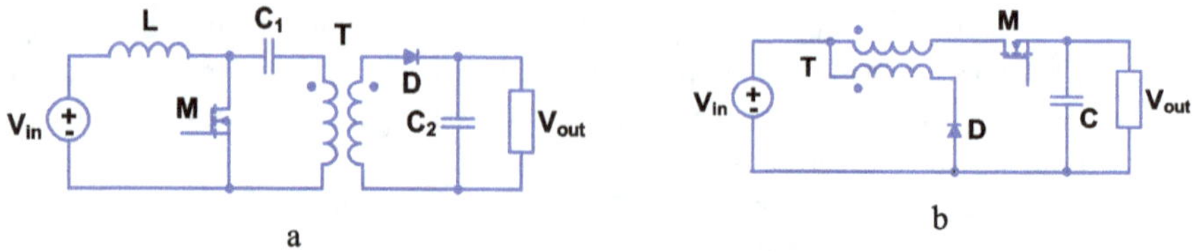

Figure 3.2 SEPIC and Watkins–Johnson topologies cannot be derived from buck and boost converters. (a) SEPIC. (b) Watkins–Johnson Converter.

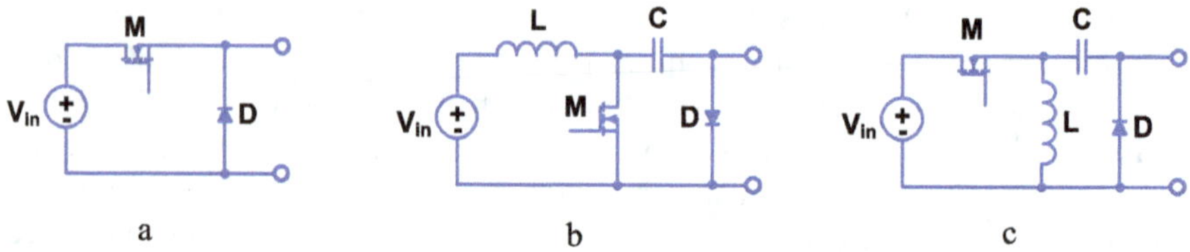

Figure 3.3 Networks suggested by [6] to introduce new voltage sources into existing converters of Figure 3.2. The configurations are inspired from: (a) Buck converter. (b) Cuk converter. (c) Zeta converter. These networks are connected in series with an inductor of the existing converters.

space equations in [1]. This method uses connection matrices to describe the power converters' graphs.

In [3], the single-input, single-output DC–DC power converters are realized by defining various converter cells. Each converter cell is a three-terminal graph of switches, inductors, and capacitors. Rotating the cell between the input and output generates different converters with the same converter cell. A conceptual method extending the existing single-input, single-output power converters to multi-input single-output converters is proposed in [6]. This method is limited to multiple-input, single-output topologies and only allows unidirectional power flow out of the sources. Only non-isolated DC–DC converters can be derived using this method. Three networks consisting of a switch, a diode, and reactive elements are considered for the connection of the voltage sources to the existing power converters as shown in Figure 3.3. These connection schemes are inspired from existing power converter topologies, specifically, buck, Cuk, and Zeta converters. The source connection schemes shown in Figure 3.3 are called pulsating voltage source cells (PVSC) in [14].

The conceptual idea of [14] is synthesizing new multi-input converters by connecting the voltage source cells in series with a current buffer or a current sink (an inductor) in one of six existing power converters, specifically, the Buck, Boost, Buck-boost, Cuk, Zeta, and SEPIC converters. Similarly, pulsating current source cells (PCSCs) are inspired from boost, buck-boost, and SEPIC converters as shown in Figure 3.4. These networks have to be connected in parallel with a voltage buffer

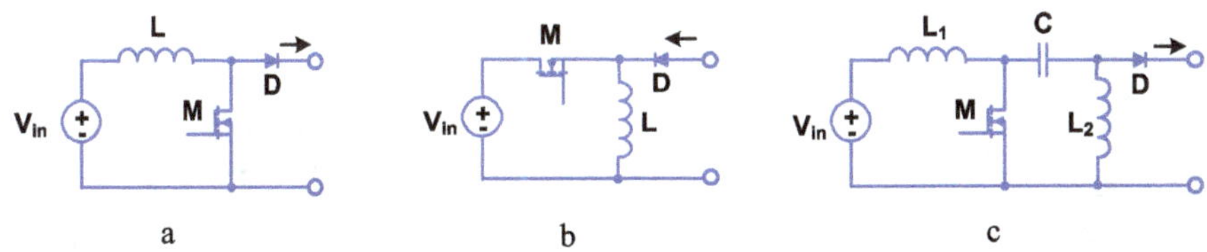

Figure 3.4 Networks suggested by [6] to introduce new voltage sources into existing converters of Figure 3.2. The configurations are inspired from (a) Boost converter, (b) Buck-Boost converter, (c) SEPIC. These networks are connected in series with an inductor of the existing converters.

or a voltage sink (a capacitor) in one of the six existing converters. This method is addressed as the nested approach in this chapter. The nested approach for realizing multi-input converters is adding a new source through a power conversion stage to an existing power converter topology. The method can be repetitively applied to realize a more complicated multi-input topology.

Acronyms, such as PVSC and PCSC, are repetitively used in [6,14]. In this chapter, PVSCs are called voltage input cells, and PCSCs are called current input cells. Alternatively, the network that connects a source to the power converter circuit is called the source's power conditioning stage. This is based on the general theory of power converters as receptive networks of switching and passive elements, which was introduced at the beginning of this chapter.

The source and its power conditioning network are combined into one input cell according to [6,14]; hence, the cell has only one free terminal. This will limit the realized topologies, as the power conditioning stage cannot be altered. In contrast, in [3] a three-terminal network of switches and passive elements was defined as a power converter cell. The power converter cells defined in [3] do not include any source or load (input or output). The three-terminal power converter cells can rotate between the input and output and form alternative power converter topologies; whereas, the input cells that are defined in [6,14] are fixed and cannot rotate.

The voltage input cells need to be connected within a mesh containing an output sink. Alternatively, the current input cells need to form a mesh with an output sink. This will assure all the inputs can transfer power to the load separately; otherwise, the sources have to work together to supply the load simultaneously. Such configurations are called quasi-multi-input converters in [6].

An example of the nested extension of the existing DC–DC converters is illustrated for a dual-input, single-output DC–DC converter for renewable energy applications in [19]. A voltage input cell is introduced to the buck-boost converter as shown in Figure 3.5. This topology saves an inductor and an output capacitor when compared with two individual buck or buck-boost converters. The multi-input converter of Figure 3.5 saves one inductor compared with two separate DC–DC converters supplying a common load.

The steady-state transfer function of this converter is formulated in (3.1), where d_1 and d_2 are the duty cycles of M_1 and M_2, respectively. If d_2 is zero, V_{s1}

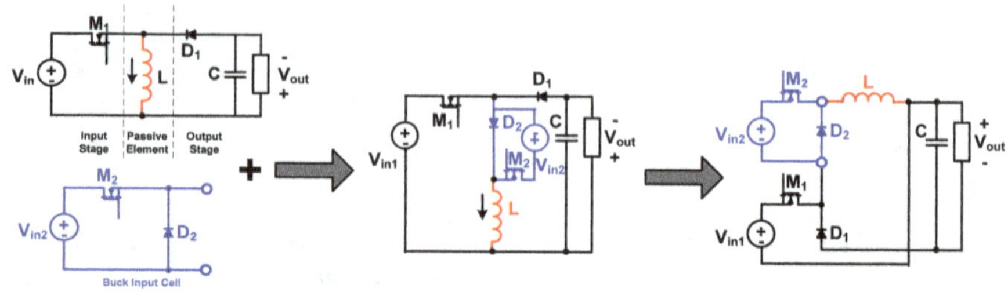

Figure 3.5 Extending a buck-boost converter to a dual-input single-output converter through a buck-type voltage input cell in [19]

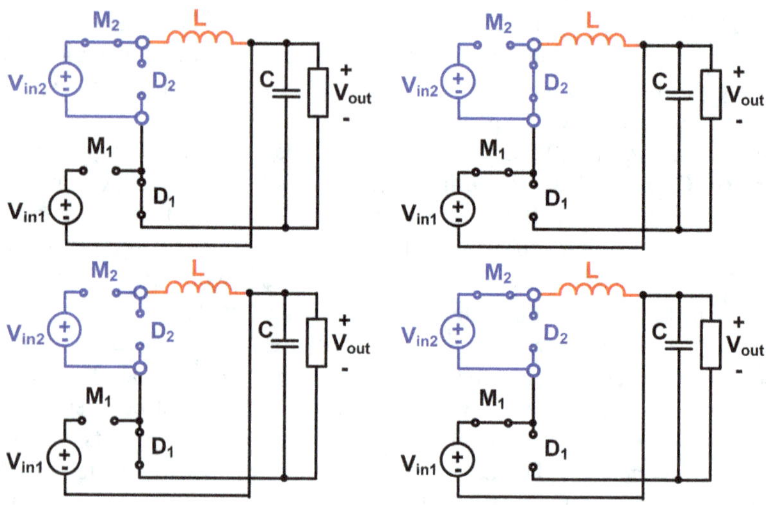

Figure 3.6 Switching states of the dual-input converter shown in Figure 3.5

supplies the load through a buck converter with a duty cycle of d_1. If d_2 is not zero and d_1 is zero, V_{s2} supplies the load through a buck-boost converter. Finally, both voltages can supply the load according to (3.1), when both d_1 and d_2 are not zero. There are two switches and four switching states for this converter, as shown in Figure 3.6.

$$V_o = \frac{d_1}{1-d_1} V_{in1} + \frac{d_2}{1-d_1} V_{in2} \tag{3.1}$$

The idea of building multi-input converters by introducing input cells is extended in [14] by introducing isolation transformers to the input cells. Examples of the isolated input cells are shown in Figure 3.7. The approach proposed in [14] is a modular approach. Filter cells are defined, in addition to source cells. The multi-input converters are formed by series and parallel connections of the input cells and

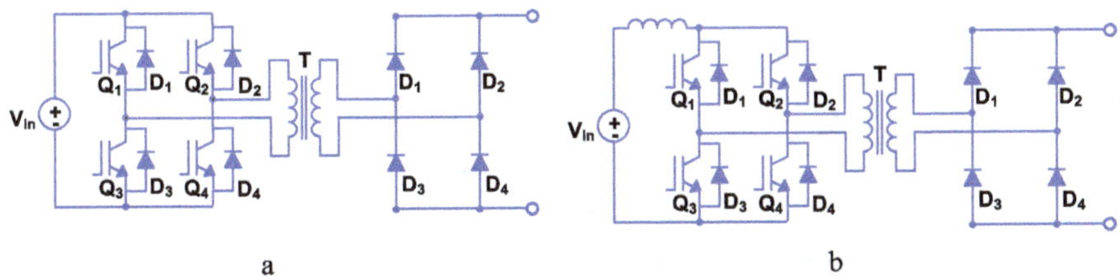

Figure 3.7 Examples of the isolated input cells. (a) Full bridge-type network for a voltage input cell. (b) Current source full bridge network as a current input cell.

filter cells to the load. The voltage input cells are connected in series in a loop containing an LC filter cell and the source. The current input cells are connected in parallel with a capacitive filter to the source. The parallel connection of the voltage input cells and the series connection of the current input cells are discussed, but only the parallel connection of the voltage input cells is recommended in [14]. Compared to the nested approach discussed earlier, this approach can realize fewer power conversion graphs. Similarly to the nested approach, realized multi-input converters will be unidirectional and single-output.

Looking at Figure 3.7, the whole isolated full bridge converter is selected as the input voltage cell. When a number of similar cells are connected to an LC filter, only an output inductor is saved compared to the case in which separate power converters were used for different sources. This criticism is partially addressed in the same reference by employing common rectifiers or transformers in the multi-input converter as shown in Figure 3.8. Initially, an example of cascaded full bridge converters connected to the single load through the LC filter is shown. In this realization, each input has a dedicated high-frequency inverter, an isolation transformer, and a rectifier. The outputs are cascaded. Next, the rectifier stage is shared between both sources through the secondary of the isolation transformer. Inputs excite the primaries of the isolation transformer, and the secondary supplies a common rectifier. A winding, a transformer core, and four diodes are reduced compared to the cascaded full bridges. The third alternative of this topology cascades the outputs of H-bridges and eliminates one more winding. In this case, the input voltages need to be isolated. Both sources share the same turn-ratio to the output side winding when they share the same primary winding. This simplification is not seen in the fundamental synthesizing rules suggested by [14]. The described shortcoming, presented in Figure 3.8, suggests the power conditioning stages include too many components and the suggested connection schemes are oversimplified.

The hybrid input cells by combining known power input cells according to the rules suggested for the nested approach are discussed in [14]. It will make it possible to realize more complicated power conversion graphs; whereas, the input cells will be even larger. An additional observation from the example shown in Figure 3.8 is that, the more features that are introduced into the power converter,

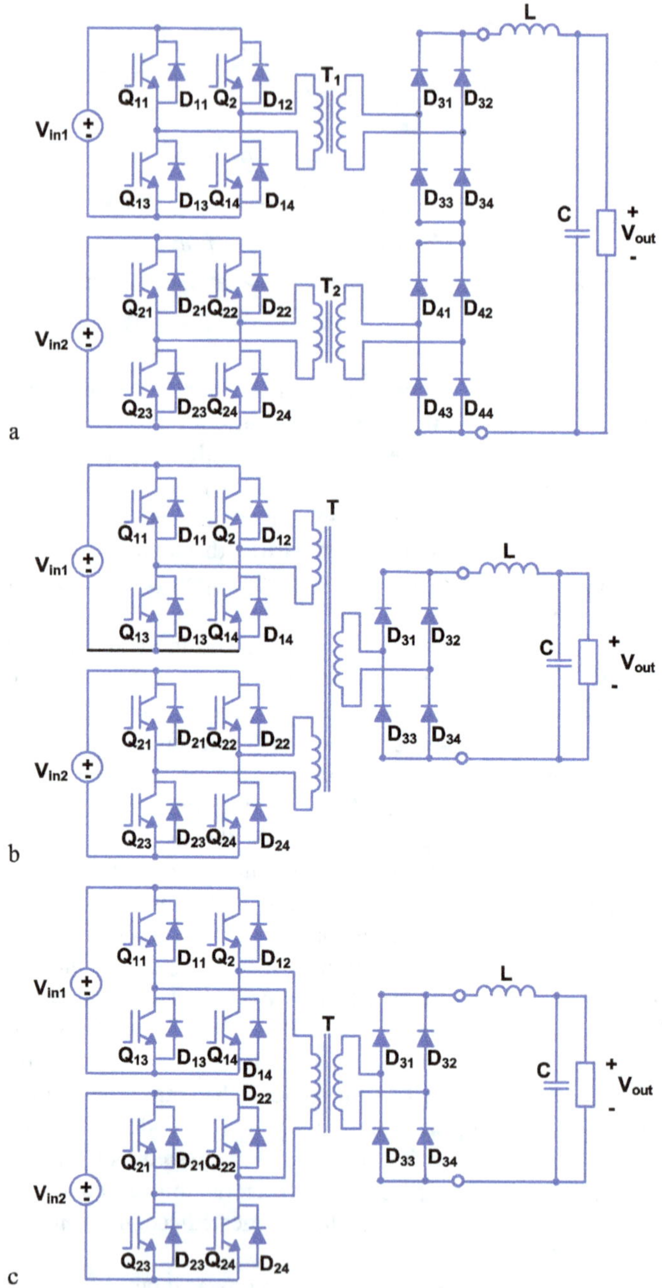

Figure 3.8 *Simplifying isolated multi-input converters by employing common transformer or rectifier stages: (a) two cascaded full bridge converters supplying a load. (b) Separated primary windings supply the common rectifier. (c) The cascaded AC voltage is applied to the common primary winding and rectified by a common rectifying stage.*

the more effective a multi-input topology becomes. Adding an isolation transformer and rectifier stages provided more flexibility for combining power converter stages between the multiple sources and the loads.

In [14], the filter cells were defined to condition the load. A second look at the example shown in Figure 3.8 suggests more common elements can be gathered into the load-side cells in multi-input converters. This is the approach suggested in [5] for synthesizing multi-input power converters by extending single-input, and single-output power converters. The common output cell is defined, in addition to the input cells. The multiple-input, single-output converter is divided into input cells and the common output cell. Multiple input cells will be employed for conditioning additional inputs in multi-input converters. The common output cell is the common portion of the power converter network between all the input cells and the load. In order to achieve the minimum component count in the multi-input converter, the maximum possible number of components should be moved into the common output cell. At the same time, each input needs to be controlled individually. The following rules are developed to achieve the described multi-input power converter:

Rule 1: The input cell must contain at least one forward conducting and bidirectional blocking switch.

Rule 2: Only one switch terminal can be connected to the input cell terminals. This assures the input cell will not have a redundant switch between several input cells.

Rule 3: If the input cell has a center capacitor, this capacitor's average voltage should be independent of the source voltage.

Rule 4: The input source cannot directly be connected to both of the input cell terminals.

The various converter cells listed in [3] are evaluated as potential candidates according to the mentioned rules. This results in four input cells; buck, buck-boost, G1, and G6 satisfying these rules as feasible input cells. G1 and G6 topologies are introduced in [3]. These topologies are shown in Figure 3.9. If these input cells draw pulsating current from the source, a source conditioning filter can be employed as shown in Figure 3.10. More feasible input cells can qualify if the third rule is relaxed by moving the middle capacitor to the input cells from the common stage; accordingly, additional cells, which are shown in Figure 3.11, will be feasible. Non-minimal multi-input power converters are realized from the input cells shown in Figure 3.11.

Two examples of multi-input converters realized by following the rules mentioned in [5] are shown in Figures 3.12 and 3.13. Figure 3.12 shows a multi-input converter extended from the G6 topology shown in Figure 3.9. It has the minimum possible components in the input cell. The multi-input topology shown in Figure 3.13 is based on the SEPIC topology. The middle capacitor filter is distributed among all the input cells. Although the approach proposed in [5] systematically leads to multi-input converters with minimal components, the derived topologies are limited to those with identical input cells. It does not discuss if the common output stages can be shared among several different input cells.

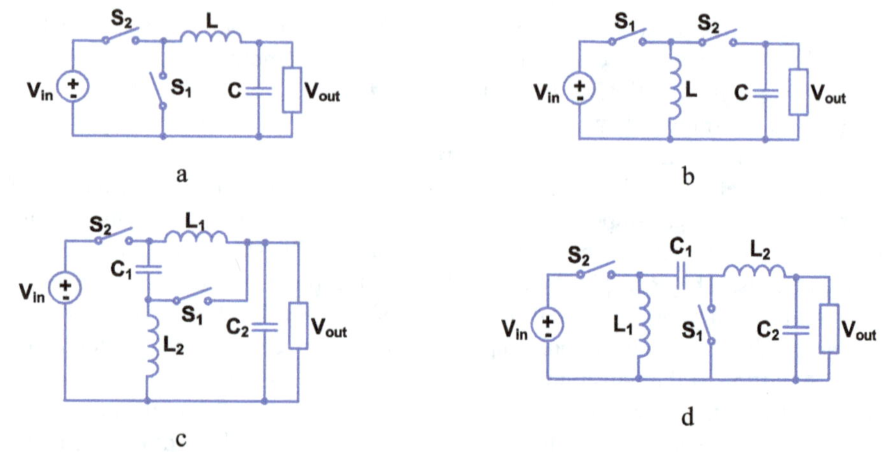

Figure 3.9 Feasible single-input, single-output topologies to be extended into multi-input, single-output topologies according to [5]. (a) Buck. (b) Buck-boost. (c) G1. (d) G6.

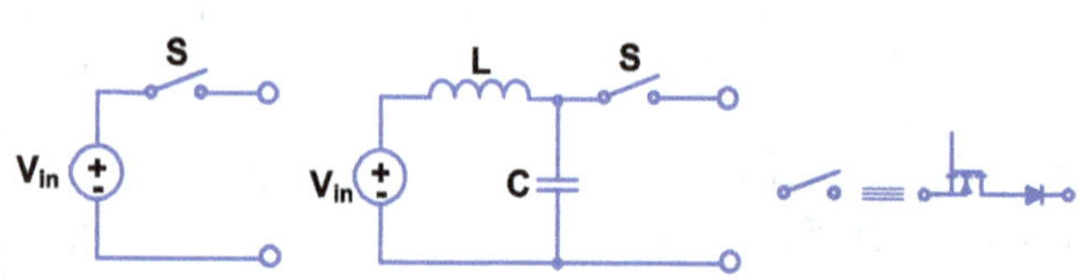

Figure 3.10 Employing current source conditioning filter (L and C) into the buck type input cell

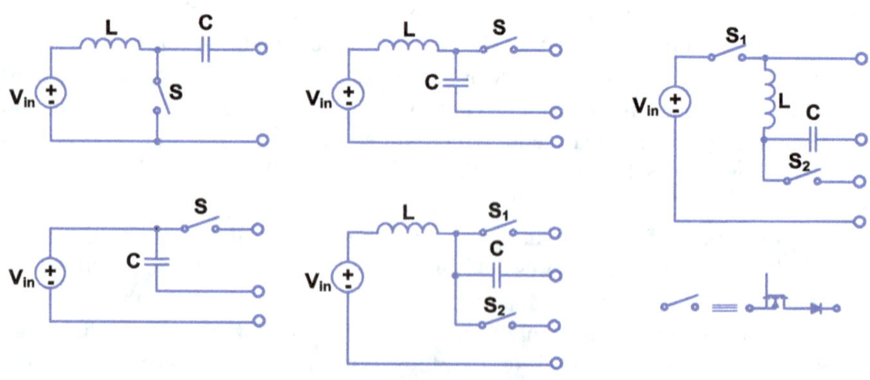

Figure 3.11 More feasible input cells if the middle capacitor is distributed into the non-minimal input cell

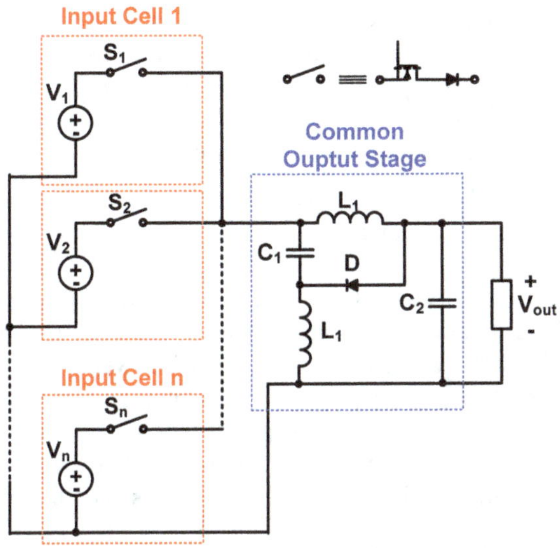

Figure 3.12 A multiple-input topology extending the G6 converter is shown in Figure 3.9

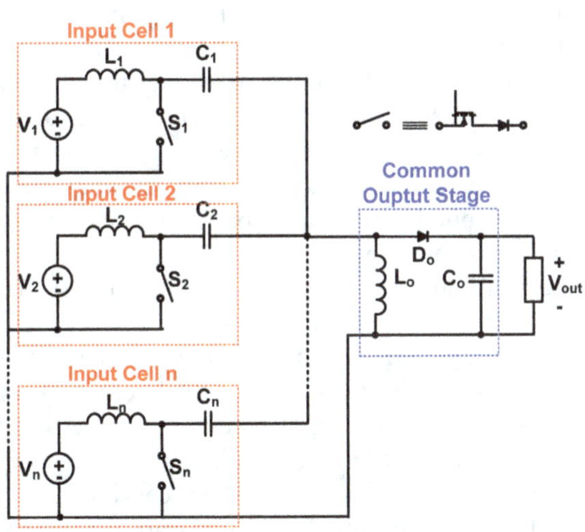

Figure 3.13 A multiple-input topology extending SEPIC

This section provided an overview of the proposed methods for realizing multi-input converters. The fundamental procedure suggested by each reference was briefly presented and the limitations of each method were discussed. All the methods were able to creatively realize new multi-input topologies; however, none of the discussed methods were able to address all the requirements of the practical applications. On the contrary, the trivial rules of extending the power converters through DC and AC links remain as two flexible methods that can be applied to several multi-input, multi-output power converter systems. The multi-input converters formed around a link will be discussed in the next section which is dedicated to the realization of the multi-port converters.

3.3 Multi-port converters

Multi-input converters only enable unidirectional power conversion paths from sources to the load. The AC grid and batteries are examples of sources and energy storage elements that accept bidirectional power flow. As discussed in the introduction, multi-port converters can provide bidirectional power flow for the bidirectional components. Multi-port converters can be synthesized similarly to multi-input converters. Multi-port converters may also be realized through DC link and AC link configurations. All these methods will be presented in this section.

3.3.1 Synthesis of multi-port converters by extending multi-input topologies

In Section 3.2, extending multi-input, single-output converters from the single-input, single-output converters was discussed. In [25], an approach for extending double-input, single-output converters to three-port converters is proposed. These three-port converters have one bidirectional port, one unidirectional source, and one unidirectional load. An example of this system is a renewable system with an energy storage element. When the source and the energy storage element supply the load, the converter is a double-input single-output converter. Alternatively, the converter is a single-input double-output converter when the source supplies the load and the energy storage element.

In [25], adding a unidirectional power conversion path from the bidirectional port (energy storage element) to the input (source) in order to realize a three-port converter is suggested. The procedure is applied to a boost dual-input converter, which was realized in [6], as shown in Figure 3.14. Initially, a boost path is added; then, the converter is simplified in a few steps to the final topology. Only one inductor, three active switches, and three diodes exist in the final power converter topology. The simplification process is called topology optimization in [25]. Although conceptual directions are given for topology optimization, this procedure needs to be manually customized for each power converter topology.

The relationship between the outputs and the input in the dual output mode is written in (3.2) for steady-state conditions. The outputs are not decoupled and need to be controlled together through a central controller. The output is related to the inputs according to (3.3) during the dual-input mode. The steady-state relationship between the inputs and outputs in (3.2) and (3.3) shows the control variables are coupled. In [27,29,30], a decoupling method is suggested for a converter with two

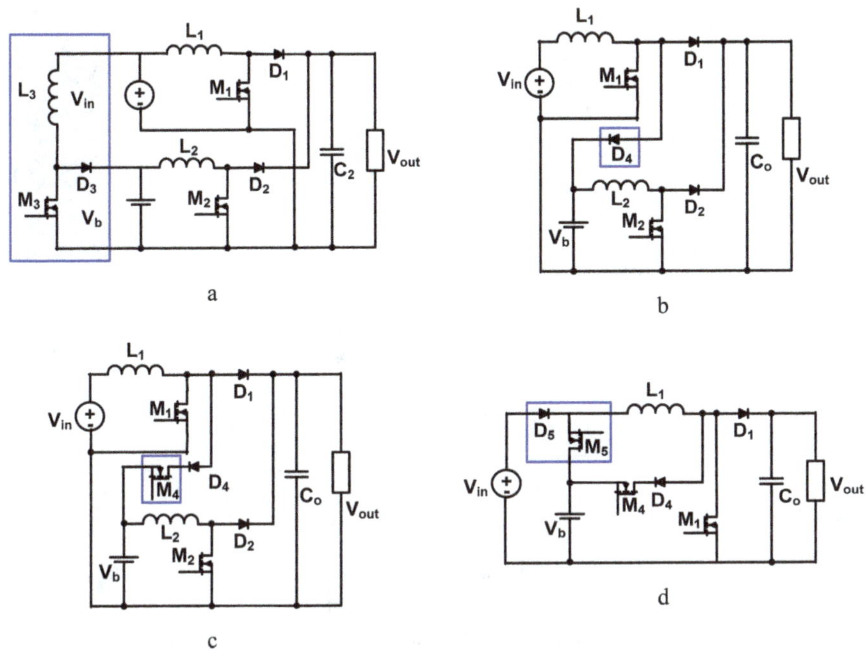

Figure 3.14 Extension of a three-port converter from the dual-input boost converter as described in [25], (a) A power conversion path is added from the source to the battery. (b) L1-M1 and L3-M3 paths are merged. (c) Adding M4 enables independent power flow from V_{in} to V_b and V_{out}. (d) Two boost paths are paralleled in the previous sub-figure (L1-M1-D1 and L2-M2-D2), these two paths are merged and M5-D5 are placed in order to regulate V_{in} output and block reverse current into V_{in}.

inputs, one bidirectional port, and one DC load.

$$V_{in} = V_o(1 - D_1) - V_b D_2 \tag{3.2}$$

$$V_o = \frac{V_{in}(1 - D_3) + D_3 V_b}{1 - D_1} \tag{3.3}$$

If a three-port converter has two bidirectional ports, following a similar concept will result in four unidirectional power conversion paths or two unidirectional paths and one bidirectional path. If all the components of a three-port converter need bidirectional power flow, then six unidirectional paths or three bidirectional paths are needed in the converter. In a four-port converter, up to six bidirectional paths or 12 unidirectional paths are required. This is illustrated in Figure 3.15. Generally, in an n-port converter, the required number of power conversion paths

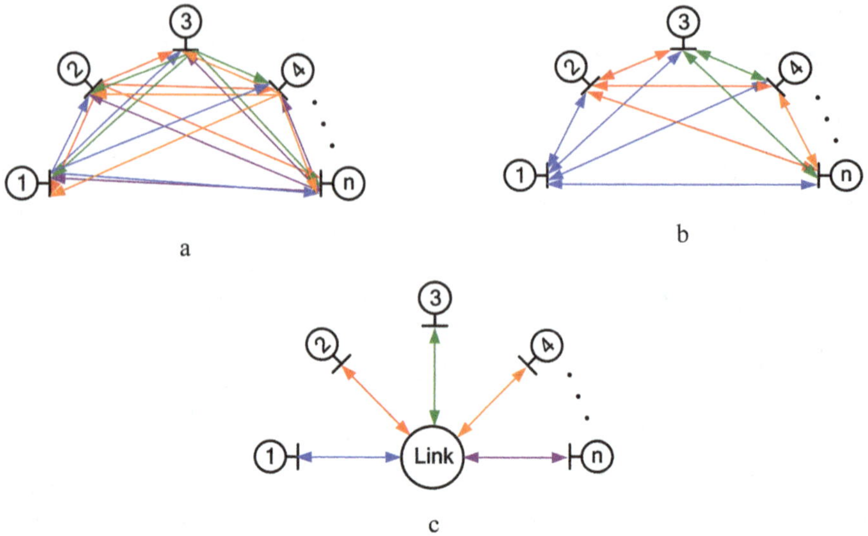

Figure 3.15 Power conversion paths in multi-port converters; the number of paths for an n-port converter is: (a) n(n – 1) unidirectional paths. (b) n(n – 1)/2 bidirectional paths. (c) 2n unidirectional paths or n bidirectional paths to the link.

can be calculated as k_n in (3.4).

$$k_{n(\max)} = \begin{cases} \dfrac{n(n-1)}{2} \ \textit{Bidirectional Paths} \\ n(n-1) \ \textit{Unidirectional Paths} \end{cases} \tag{3.4}$$

If each power conversion path is implemented with a diode and an active switch, then the n-port converter needs $2n(n-1)$ switches; whereas, the number of switches is $2n$ if each power conversion path is connected to a DC link. This analysis suggests the benefit of developing a multi-port converter around a DC or AC link. This benefit becomes more significant as the number of ports grows. Multi-port converters formed around a DC link will be discussed in the next subsection.

3.3.2 Multiport converters with DC link

Developing multi-port power converters around a DC link is a trivial approach. The capacitive DC link integrates multiple voltage source power converters into one multi-port power converter. All the loads, sources, and energy storage elements are connected to the DC link through a power conditioning stage. The capacitive DC link is a parallel DC link as all the power conditioning stages will be in parallel with the link. The DC link decouples the controller function of each stage from the rest of the power converter according to KVL.

In [28], the implementation and control of a five-port power converter around a DC link is considered. The sources include a fuel cell, a solar panel, and a low-power wind turbine. Three unidirectional boost converters have been coupled to the DC link. The wind turbine generator has an AC output and is conditioned through a rectifier prior to connecting to the DC–DC stage. A battery is connected through a bidirectional boost (buck) converter. A bidirectional H-bridge inverter (rectifier) is the grid's power conditioning stage. This topology includes six legs in parallel with the DC link. The general topology of a similar multi-port system formed around a DC link is shown in Figure 3.16.

In the topology of Figure 3.16, each source is conditioned through an inductor. The inductor currents and the DC link voltage are the state variables of the system. Thus the system described in [28] is a fifth-order system. The average and small signal models of the system are developed; moreover, the control system is developed in [28]. The system is partitioned into the sources and loads. The main control goal is regulating the DC link voltage. One converter may regulate the DC link voltage, and the other converters will exchange power accordingly. Similarly, to the single-input, single-output converters with a DC link, the sources' and the loads' power flows are synchronized if the DC link voltage is regulated.

In contrast to the capacitive DC link, the inductive DC link can be employed in a power converter system. In this case, all the power converters are in series with an inductor. An example is shown in [31]. The general topology of this multi-input, multi-output power converter is shown in Figure 3.17. Each source is connected

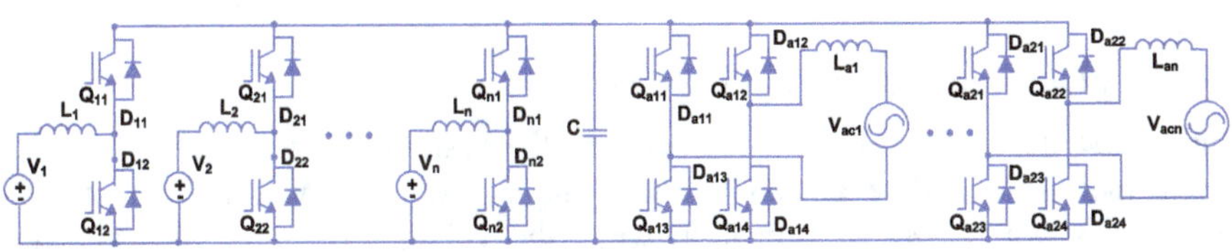

Figure 3.16 Multi-port power converter formed around the capacitive DC link

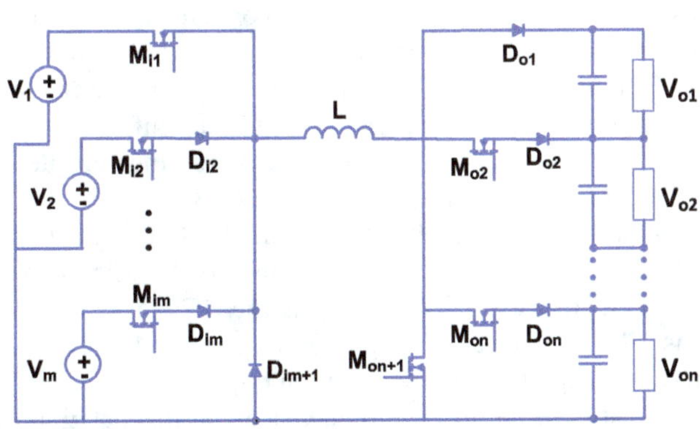

Figure 3.17 Multiple-input multiple-output converter formed around the inductive DC link

through a forward-conducting bidirectional blocking switch similar to the switches used in [5]. The described switches are realized with a MOSFET in series with a diode, in this paper. The diode is needed to avoid paralleling different voltage sources when the switches are "ON" simultaneously. The small signal model of the system is derived by the state space averaging method for a two-input, two-output system. The controller for the same system is designed. This system is evaluated by implementing a three-input, two-output prototype. Although the inductor L in Figure 3.17 serves as an inductive DC link, the converter is essentially an extension of single-input, single-output converters to multi-input, multi-output converters.

This topology is a limited multi-input, multi-output topology formed around an inductive DC link. When several sources are connected to the DC link inductor, only the one with the highest voltage transfers energy. The sources with lower voltages will be blocked by the series diodes. Only sequential power transfer is possible with this topology. The accommodation of bidirectional components such as batteries is not discussed in [31,32]. One solution can be connecting the battery to the switching cells on both sides of the series DC inductor. In the proposed switching scheme, all the input switches are initially turned "ON". The switches will be turned off from the highest input source to the lowest input source sequentially. Similarly, the output switches are turned off sequentially. The source and load power conditioning stages are over-simplified in [32]. Looking at the source cells, only one switch is used. This will limit the controllability of the sources. The flexibility of the system is lost in exchange for simplicity when compared to Figure 3.16.

3.3.3 AC link multi-port power converters

When galvanic isolation is required between the system components, an isolation transformer needs to be placed between the ports. Moreover, the transformer improves power converter utilization when the voltages of either side are very different. In order to reduce the transformer's volume and weight, the switching frequency needs to be increased. This is especially important for portable applications, such as vehicular applications. AC link converters are ideal topologies for accommodating isolation transformers. Besides galvanic isolation, AC link converters provide benefits such as soft switching, resonant power conversion, and multiple-quadrant power conversion. The AC link current and voltage are alternating and have zero crossings. Switching at zero crossings enables zero voltage switching (ZVS) and (or) zero current switching (ZCS).

Similarly, to the DC link implementation, the AC link is implemented by one or more passive components and an isolation transformer, when galvanic isolation is required. In contrast to the DC link approach, the DC bias is avoided for the AC link components. This is mandatory for proper transformer utilization. The AC link components may be integrated into the isolation transformer. The electric model of the transformer consists of the leakage inductance and magnetizing inductance. If the leakage inductance is utilized as the AC link, this is a series inductive AC link. If the magnetizing inductance is utilized as the AC link, this is a parallel AC link.

The Dual Active Bridge (DAB) converter is an example of a single-input, single-output, series AC link topology. The DAB topology was introduced in the early 1990s [43]. The two-quadrant DAB is shown in Figure 3.18. A four-quadrant DAB is investigated in [44] as shown in Figure 3.19. The equivalent circuit of a DAB is shown in Figure 3.20. The equivalent circuit includes two equivalent AC sources which are connected through an inductor, L, in series. The AC inductor, L, shown in Figure 3.20, is the key component for transferring power between the AC sources. The power flow is controlled by controlling the phase shift between the primary side bridge's AC voltage and the secondary side bridge's AC voltage. This modulation method is called phase shift modulation [43]. Phase shift modulation enables maximum power transfer between the bridges.

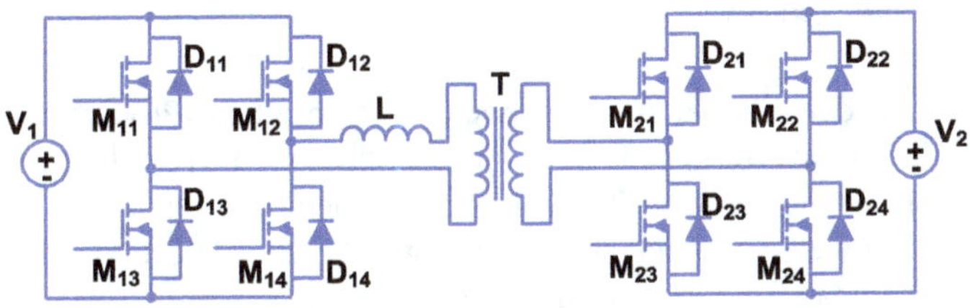

Figure 3.18 A two-quadrant DAB is a series AC link DC–DC converter

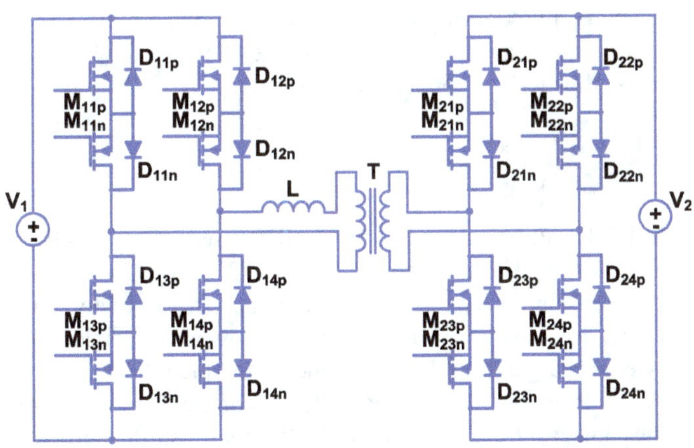

Figure 3.19 A four-quadrant DAB is implemented with four-quadrant switches

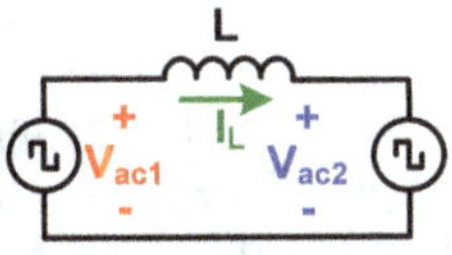

Figure 3.20 DAB equivalent circuit

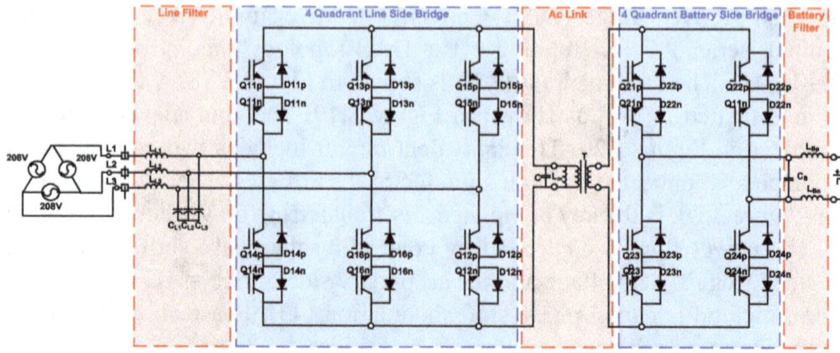

Figure 3.21 Single-input single-output AC–DC converter based on a parallel AC link converter

The flyback converter and the converter proposed in [26] are examples of parallel AC link converters. A single-input, single-output, parallel AC link converter is shown in Figure 3.21. The modulation method proposed in [45] can be applied to the power converter shown in Figure 3.21. This modulation method results in a triangular AC link current waveform.

In order to compare the AC link topologies, the AC link Utilization Factor (UF) is defined in (3.5) [39]. The AC link UF is the ratio of the average power to the maximum power transferred through the AC link. The AC link is cascaded with the power path; thus, this definition determines the power utilization of the converter and; furthermore, evaluates the power density. In (3.5), i_{link} is the link current and v_{sec} is the transformer's secondary side voltage.

$$UF = \frac{\frac{1}{T_s}\int_{T_s}(i_{\text{link}}(t).v_{\text{sec}}(t))dt}{i_{\text{link(max)}} \cdot v_{\text{sec(max)}}} \tag{3.5}$$

The AC link UF will be calculated for the discussed topologies by considering the AC link waveforms. The diagrams of Figures 3.22 and 3.23 are the AC link waveforms of parallel and series AC link topologies, respectively. In the parallel AC link topology, the link is parallel with the power conversion path. All the transferred power is charged into the link inductor and then discharged to the output port alternatively during each switching cycle. In contrast, the series AC link is in series with the power conversion path. With phase shift modulation, square waveform voltages are applied to each side of the link, with a phase shift. Only the difference in primary and secondary sides' voltages will store energy in the link. Thus, the AC link inductance causes less circulating power for the series configuration with the phase shift modulation.

The proposed AC link UF criterion supports the previous statement about the series AC link configuration's excellence. Following (3.5) suggests the maximum UF for a square waveform, which is 1. The phase shift modulation with low phase shift is the closest to a unity UF. On the other hand, UF is less than 0.5 for a triangular link current. Since only half of the switching cycle is dedicated to each side

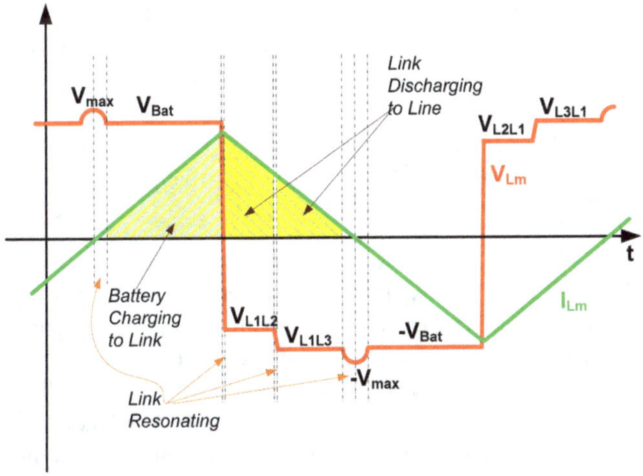

Figure 3.22 AC link waveforms of the parallel AC link converter are shown in Figure 3.21

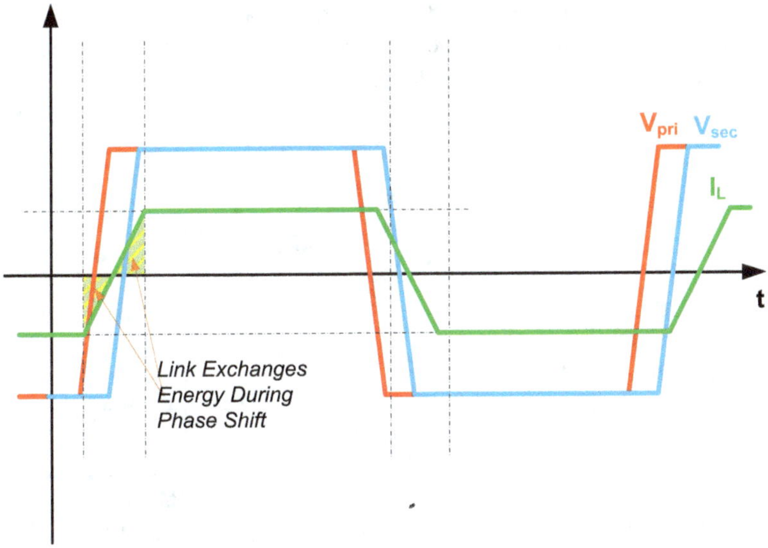

Figure 3.23 AC link waveforms of the phase shift modulated DAB as the series AC link converter

in a parallel series AC link converter, the UF is less than 0.25 for this configuration. The UF for the discussed topologies is approximately calculated as follows. The parallel AC link converter's UF is estimated as (3.6). If the resonating cycles are neglected, the UF is 0.25 for this topology. Similarly, the series AC link converter UF is calculated in (3.7). In (3.7), PS is the phase shift between the bridges in radians. The phase shift is always below 180°; hence, the UF is always more than 0.5. In practice, the phase shift is controlled below 90° in order to avoid excessive reactive power. Thus, the UF for series AC link topology would be more than 0.75.

$$UF_{Par} \approx \frac{\frac{1}{Ts} \cdot V_{Bat} \cdot I_{L\ \max} \cdot \frac{T_s}{4}}{V_{Bat} \cdot I_{L\ \max}} = \frac{1}{4} \tag{3.6}$$

$$UF_{Ser} \approx 1 - \frac{PS}{2\pi} \tag{3.7}$$

When galvanic isolation is required, the magnetizing inductance will be the AC link. In other words, the energy needs to be stored in the magnetizing inductance, and the magnetizing inductance acts as an AC inductor. The magnetizing inductance easily saturates; hence an air gap needs to be integrated into the core. This reduces the magnetic coupling and increases the leakage flux. The leakage inductance is not utilized in this topology and leads to low efficiency for the magnetic components.

Both the series and parallel AC link topologies can be extended to multi-port configurations. This can be achieved by introducing more windings around the transformer's core. In [26], an example of a multiport parallel AC link converter is presented. In [46–50], examples of multi-active bridge converters are suggested. Figure 3.24 shows a triple-active bridge converter as an example of a multi-active bridge topology. Alternatively, Figure 3.25 shows a system based on the three-port, parallel, high-frequency AC link topology.

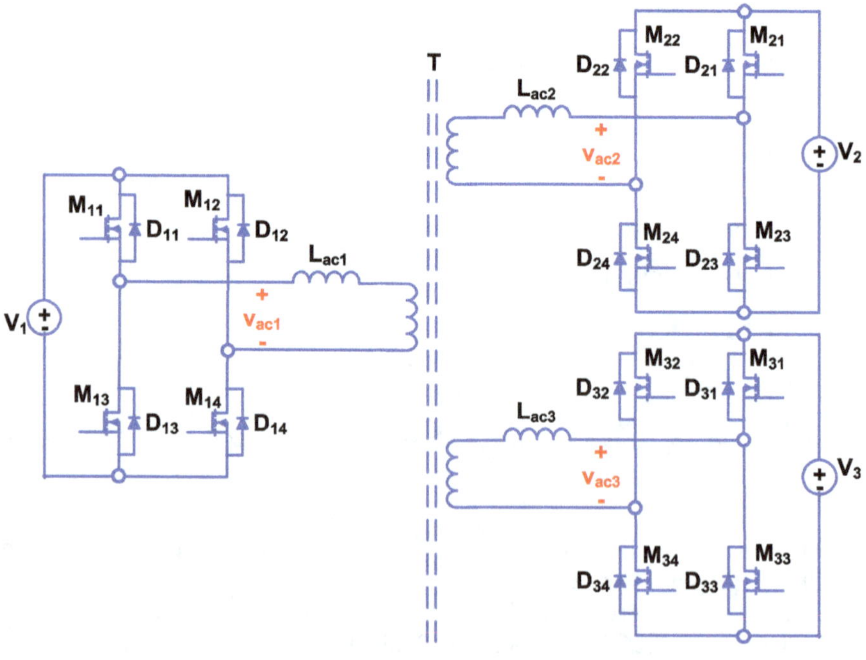

Figure 3.24 MOSFET realization of the triple active bridge

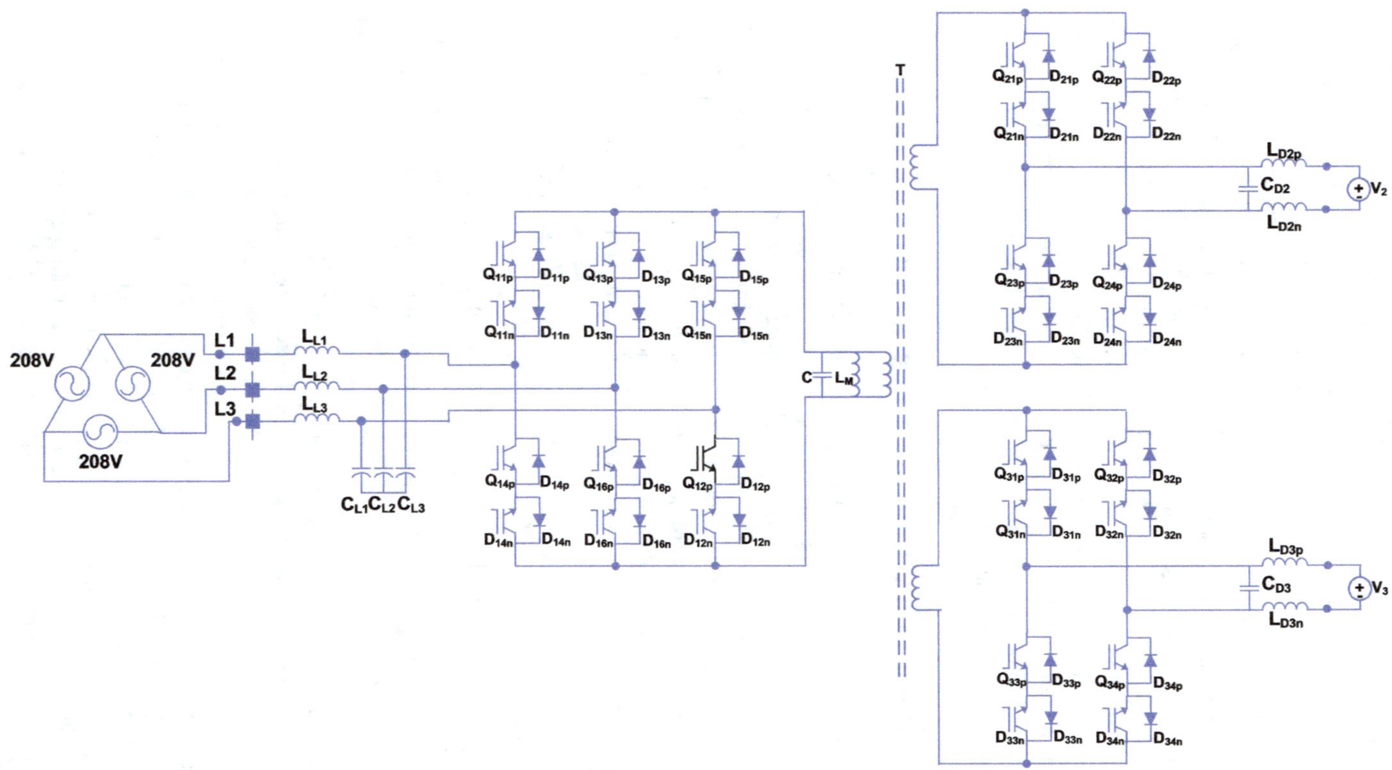

Figure 3.25 Three-port parallel high-frequency AC link converter

The multiport variations of these two topologies inherit the characteristics that were discussed earlier in this subsection. Additionally, the multi-active bridge is more flexible for power flow control between the ports. All the ports may simultaneously increase or decrease the magnetic flux in the magnetic circuit. On the other hand, in the parallel AC link multi-port converter, the link is charged and discharged sequentially by the ports, and it is not possible for several ports to charge or discharge the link at the same time. As a result, the UF can potentially approach the unity for the multi-active bridge topology, but the UF will be less than 0.25 for any of the sources that are attached to the parallel AC link converter. Indeed, the more sources that are added, the lower UF each has from the AC link. This translates to higher power density and more flexibility for the series AC link configuration that utilizes the phase shift modulation. In conclusion, the fundamental comparison between the isolated multi-port topologies suggested that the multi-active bridge topology is the superior topology to interconnect all isolated sources.

3.4 Applications of multi-port power converters

Different methods for realizing multi-port power converter topologies were introduced in the previous sections. In this section, the major applications of multi-port power converters will be reviewed. Renewable energy systems, micro-grids, and vehicular power systems are among the major applications for multi-port power converters. These applications have been the subject of interest in recent publications. The following subsections are dedicated to the multi-port converters proposed for the applications mentioned above. In each subsection, a brief introduction to the application will be provided and a few examples of recent works will be reviewed.

3.4.1 Multi-port power converters for renewable energy systems

Distributed generation is possible due to renewable energy resources. Renewable energy resources, including photovoltaic panels and wind turbines, are programmable by the load. A combination of multiple renewable sources and energy storage elements increases the availability of renewable power in grid-connected and stand-alone systems [7,12,19,20,24,27,33].

A hybrid renewable energy system or renewable backup system includes a renewable source, an energy storage system, and loads. Renewable backup systems are being commercialized. A photovoltaic backup system includes a photovoltaic plant, a battery pack, and interconnections to the AC grid and local AC and DC loads. A solution practiced in industry includes separate commercial converters that are all connected to the low-frequency AC grid [33]. In contrast, a multi-port converter can integrate all these sources and loads into one power conditioner [7,19,24,25].

Figure 3.26 shows a multi-port power converter proposed in [7] for renewable energy systems. The core of this topology is a non-isolated multi-port converter, which is realized with two switching legs (half-bridges). The battery can be conditioned by both switching legs. The AC load is supplied through the differential connection of the

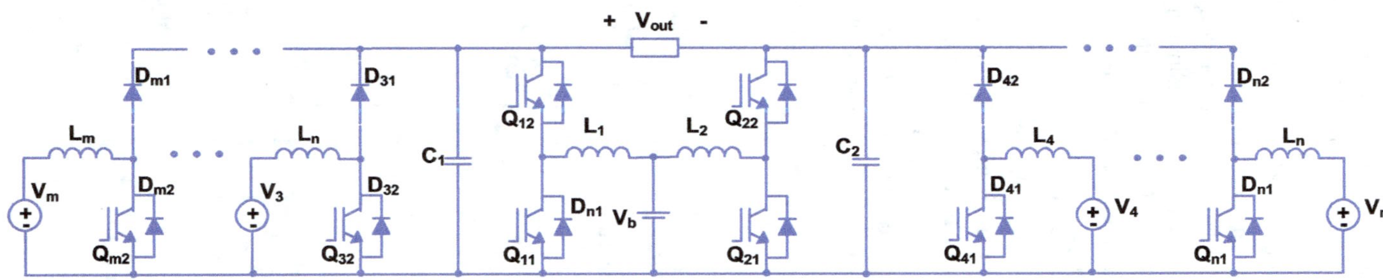

Figure 3.26 A multi-port power converter proposed in [7] for renewable energy systems

switching legs. Each leg is supplied by a number of loads connected through a DC link. This solution is a multi-port power converter extended through two DC links. The authors discuss different operating modes and suggest the state-feedback control method for the converter. The battery does not have a dedicated power conditioning cell in this system. It balances the supply and demand in this system. If the source provides more power than the load, the battery is charged. Alternatively, the battery is discharged to supply the load if the source power is less than the demanded load power. In Figure 3.26, the battery does not have a dedicated power conditioning stage. The switches are reduced at the cost of less flexibility in the power conditioning of the battery.

The solution used in [33] is a bidirectional, high-frequency, isolated inverter that connects the battery to the low-frequency (50 Hz/60 Hz) AC grid. This solution aims to form energy storage systems from blocks that are all connected to the terrestrial AC grid. In contrast, a more integrated photovoltaic backup system is developed in [24]. In this system, both the photovoltaic array and the batteries are connected to the DC link. The photovoltaic array is conditioned through a non-isolated boost converter. The battery is conditioned through a bidirectional push-pull forward isolated DC–DC converter. The DC link is also connected to the grid through a bidirectional AC–DC converter. These photovoltaic backup systems are compared in Figure 3.27.

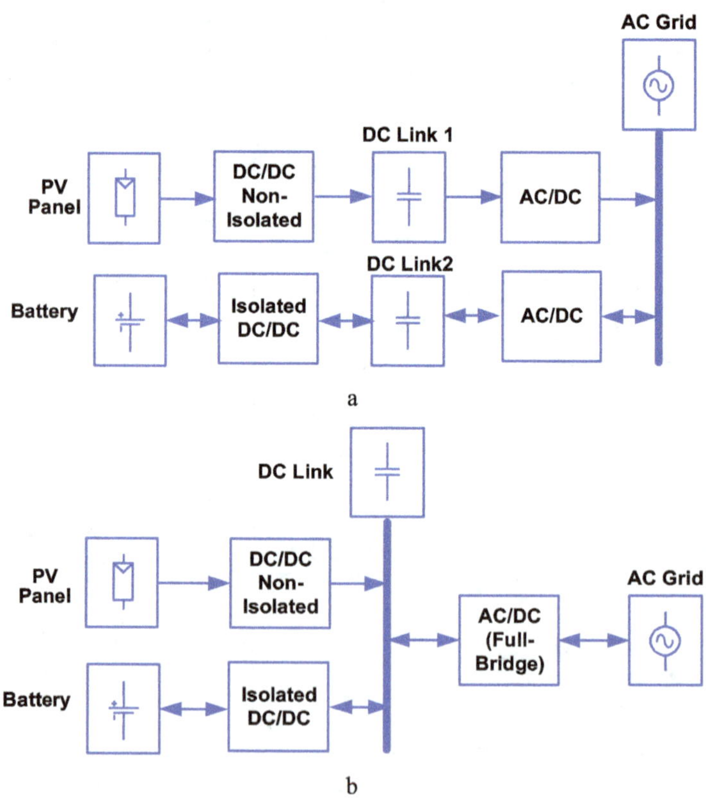

Figure 3.27 Photovoltaic backup systems. (a) The system is formed around the AC grid. (b) The sources are integrated through the DC link.

It is possible for more than one source to need to be isolated from the rest of the system. In this case, the sources may use a multi-input converter. In [12], multiple renewable sources are conditioned by the source cells shown in Figures 3.3 and 3.4, which were suggested by [6]. The output of the resulting converter is cascaded with an isolated DC–DC converter. This approach includes a non-isolated DC–DC stage and an isolated DC–DC stage, which is illustrated in Figure 3.28. It has one more stage when compared with the approach proposed in [14], which is shown in Figure 3.8. However, fewer switches are used to condition each DC source.

The idea of a multi-port converter cascaded with an isolated DC–DC converter is also illustrated in [29]. A four-port converter is proposed, as shown in Figure 3.29. Two unidirectional inputs, which are solar and wind sources, are

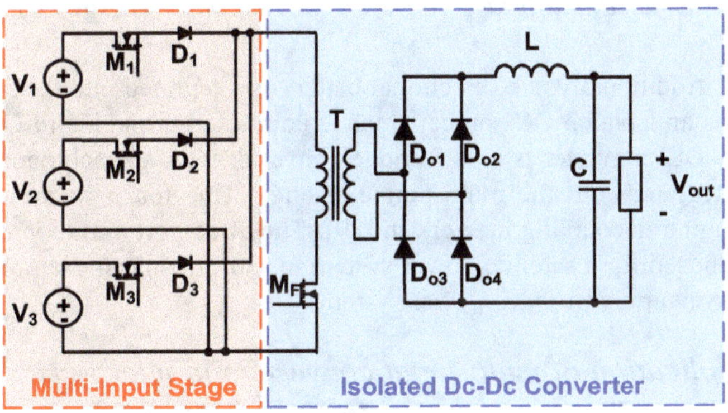

a

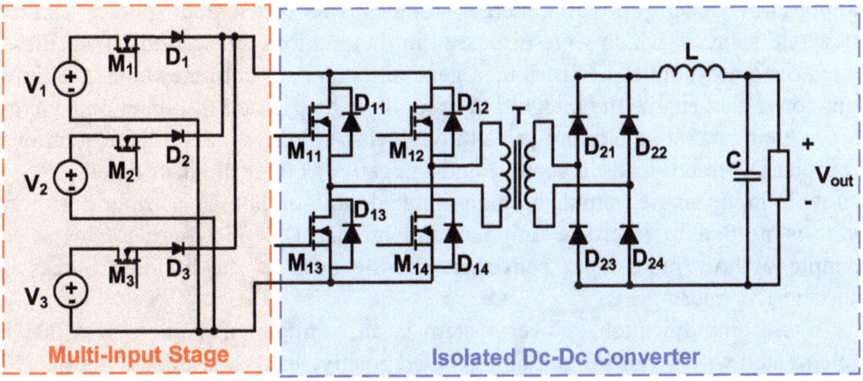

b

Figure 3.28 Multi-input converters cascaded with an isolated DC–DC converter. (a) Three buck input cells in parallel are cascaded with the Flyback converter. (b) Three buck input cells in parallel are cascaded with the Full-Bridge converter.

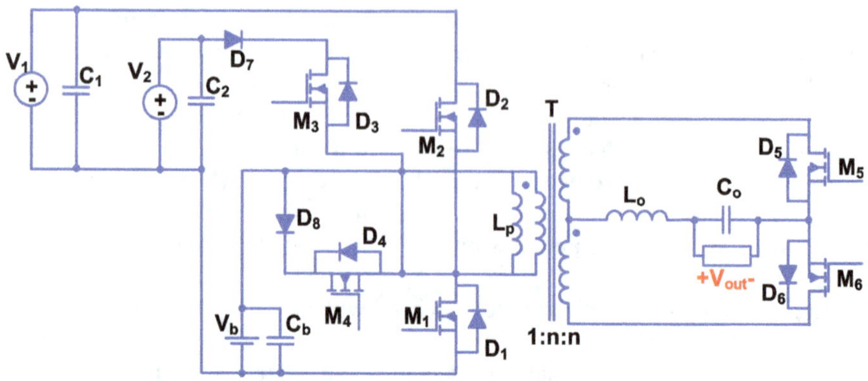

Figure 3.29 A four-port converter built with a multi-port converter and an isolated half-bridge converter for renewable energy applications

conditioned. Additionally, a bidirectional battery is integrated into the system. The converter has an isolated DC source which is connected through an LC filter. The isolated DC–DC converter is a half-bridge converter with a synchronous rectifier, which is integrated into the multi-port converter. This four-port system is controlled through a decoupling network in [29]. The three-port variation of this converter is proposed for a satellite power system in [30]; this is an example of multi-port power converters in space power systems.

3.4.2 Application of multi-input converters in micro-grids

Micro-grids are areas in a grid that may either work independently or in conjunction with the main grid. Each micro-grid needs to be supplied by distributed generation and backed up with energy storage. The distributed sources can be renewable sources, which were discussed in the previous subsection. Small diesel generators are examples of distributed generators that are not renewable. Multiple-input converters enable the integration of multiple distributed resources and energy storage elements into the micro-grids through cost-effective and efficient solutions. In bipolar DC micro-grids, a positive and a negative voltage are distributed [16,51]. A double-input single-output, a single-input double-output, or a triple port converter is needed to interface unipolar and bipolar DC micro-grids. This is an example of how multi-input converters can be used as the building blocks of futuristic DC micro-grids.

A telecommunications power system is an example of a micro-grid that is implemented with a multi-input, non-isolated positive buck-boost converter in [15]. This converter is formed around an inductive DC link similar to the converter shown in Figure 3.17; in contrast, the multiple-input positive buck-boost converter does not condition multiple outputs. Alternatively, bidirectional power flow through the DC link inductor is allowed [17]. The switches are unidirectional conducting and bidirectional blocking in [15]. Bidirectional switches are utilized in [17] to condition several current sources.

3.4.3 Multi-port converters for vehicular power systems

Hybrid electric and fully electric vehicles include several electric components gathered in a mobile electric power system [52]. If all the sources and loads are interconnected within a vehicle, a vehicular-integrated power system is formed [23]. The integrated power system concept was specifically investigated for the shipboard power systems [4]. An integrated shipboard power system requires less reserve power and offers more reliability than a segregated shipboard power system.

Multi-port power converters are attractive topologies for vehicular power systems because multi-port power converters can reduce the component count and increase the power density of the converter. Vehicular power systems are mobile, and power density matters for such applications. Multi-port power converters can be designed for interconnecting several components in a vehicular system. This can improve the reliability and reduce the required power capacity of the system as used in integrated shipboard power systems. This subsection provides a few examples of multi-port power converters that have been proposed for battery-powered vehicles.

A conceptual block diagram of a battery-powered vehicle is shown in Figure 3.30 [23]. This system includes two DC buses, a high-voltage DC bus and a low-voltage DC bus. The voltage of the high-voltage bus is determined by the high-voltage battery of the vehicle. The traction drive is connected to this battery. This battery is charged from the AC grid. Additionally, the battery's power can be transferred to the AC grid for vehicle-to-grid operation [38,39,53,54]. The low-voltage bus is determined by the low-voltage battery in the vehicle. The control

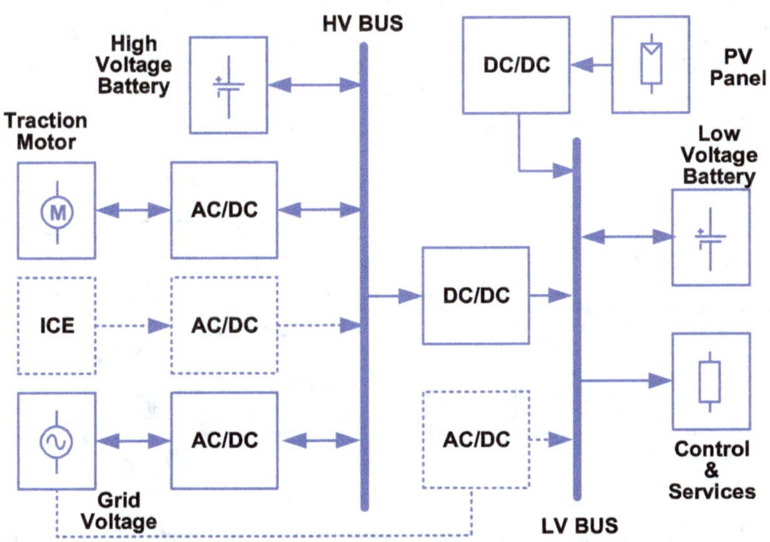

Figure 3.30 Conceptual single-line block diagram of a conventional vehicular power system

system and the cabin loads are connected to this low-voltage bus. The low-voltage battery is charged from the high-voltage bus. Renewable sources may be connected to the low voltage or the high voltage DC bus. The low-voltage components are grounded to the chassis. The high-voltage components are referenced to the chassis through a high-impedance grounding network. The high-voltage components need to be isolated from the low-voltage components due to safety standards [38].

In [36], a hybrid battery and ultra-capacitor energy storage system are investigated for battery-powered vehicles. The authors indicate that the multi-input converter is expensive for this particular application. They have placed the ultra-capacitor as the DC link and interconnected the battery and the ultra-capacitor through a DC–DC converter. The power flow between the energy storage elements is controlled by adjusting the ultra-capacitor's voltage. In addition to the battery and ultra-capacitor, a fuel cell is integrated into the vehicle's propulsion system in [20]. A multiple-input topology based on the DC link connection is utilized. Each source is conditioned through a synchronous buck-boost converter. All the DC–DC converters are connected to the DC link that supplies the drive's inverter in [20].

A charger that is able to charge both the low-voltage and the high-voltage batteries in the vehicle is an attractive product. In [55], a unidirectional auxiliary power conditioning pass is added to a two-stage non-isolated vehicular utility interface. The shortcomings are that the auxiliary mode requires the power converter to be reconfigured and there are no means to transfer energy from the low-voltage battery to the high-voltage battery. In [56], an isolated three-stage charger connects to the high-voltage DC bus, and an auxiliary unidirectional DC–DC converter interconnects the low-voltage battery bus to the high-voltage DC bus. In order to avoid the mentioned shortcomings, a multi-port series AC link converter is proposed for interconnecting the low-voltage battery and the grid-side converter in [23].

The system shown in Figure 3.31 interconnects all the sources and loads of a vehicle into a vehicular integrated power system [23]. The core of this

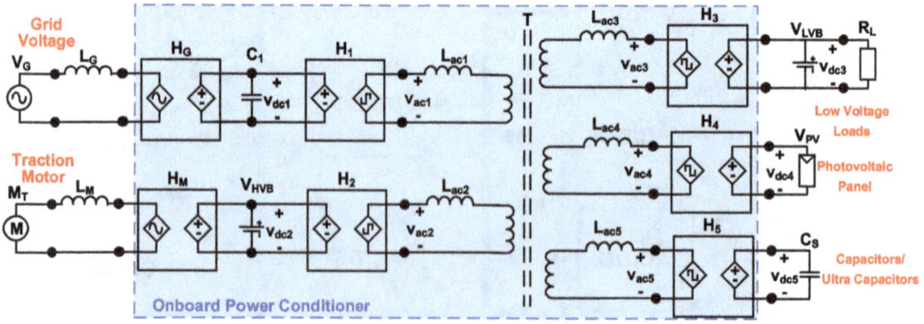

Figure 3.31 A vehicular integrated power system realized with AC–DC power conditioning stages connected through a multi-port series AC link converter

vehicular integrated power system is a multi-port isolated converter. The isolated components are conditioned through a series AC link, which offers high power density power conversion. Several DC links extend the system where needed. The design procedure for the magnetic components of this system is presented in [22,37].

3.5 Conclusions

Recent advances in multi-input and multi-port converters were reviewed in this chapter. Multi-input converters bring new features to emerging applications such as renewable energy systems, micro-grids, and vehicular systems. Synthesis of multi-input converters by extending existing single-input, single-output converters was presented. Additionally, realizing multi-port systems through DC link and AC link converters was discussed. The examples of recent multi-input and multi-port power converters in several applications were presented. Multi-port converters can be developed in order to reduce the component count. Alternatively, multi-port converters can offer more flexibility to the power conversion system. Both aspects of realizing multi-port converters were discussed. This chapter provided an overview of realizing the proper multiple-input topologies for several applications. Furthermore, the proper control techniques and design procedures were referenced from the latest works. Finally, potential applications of the multi-input converters were evaluated by the researchers, developers, and entrepreneurs.

References

[1] R. W. Erickson, "Synthesis of switched-mode converters," in *IEEE Power Electronics Specialists Conference*, pp. 9–22, 1983.

[2] P. Wood, *Switching Power Converters*. New York: Van Nostrand Reinhold Co., 1981.

[3] R. Tymerski and V. Vorperian, "Generation and classification of PWM DC-to-DC converters," *Aerospace and Electronic Systems, IEEE Transactions on*, vol. 24, pp. 743–754, 1988.

[4] B. Farhangi and K. Butler-Purry, "Transient study of DC zonal electrical distribution system in next generation shipboard integrated power systems using PSCAD™," in *North American Power Symposium (NAPS)*, 2009, pp. 1–8, 2009.

[5] A. Kwasinski, "Identification of feasible topologies for multiple-input DC–DC converters," *Power Electronics, IEEE Transactions on*, vol. 24, pp. 856–861, 2009.

[6] L. Yuan-Chuan and C. Yaow-Ming, "A systematic approach to synthesizing multi-input DC-DC converters," *Power Electronics, IEEE Transactions on*, vol. 24, pp. 116–127, 2009.

[7] S. Danyali, S. H. Hosseini and G. B. Gharehpetian, "New extendable single-stage multi-input DC-DC/AC boost converter," *Power Electronics, IEEE Transactions on*, vol. 29, pp. 775–788, 2014.

[8] A. Shoaei, K. Abbaszadeh and H. Allahyari, "A single-inductor multi-input multilevel high step-up DC–DC converter based on switched-diode-capacitor cells for PV applications," *IEEE Journal of Emerging and Selected Topics in Industrial Electronics*, vol. 4, no. 1, pp. 18–27, 2023.
[9] M. Veerachary and A. Trivedi, "Linear matrix inequality-based multi-variable controller design for boost cascaded charge-pump-based double-input DC–DC converter," *IEEE Transactions on Industry Applications*, vol. 58, no. 6, pp. 7515–7528, 2022.
[10] X. L. Li, C. K. Tse and D. D. -C. Lu, "Synthesis of reconfigurable and scalable single-inductor multiport converters with no cross regulation," *IEEE Transactions on Power Electronics*, vol. 37, no. 9, pp. 10889–10902, 2022.
[11] L. Mo, J. Huang, G. Chen, X. Qing and Y. Hu, "Computer-aided systematic topology derivation of single-inductor multi-input multi-output converters from working principle," *IEEE Transactions on Circuits and Systems I: Regular Papers*, vol. 69, no. 6, pp. 2637–2649, 2022.
[12] W. Qin, Z. Jie, R. Xinbo and J. Ke, "Isolated single primary winding multiple-input converters," *Power Electronics, IEEE Transactions on*, vol. 26, pp. 3435–3442, 2011.
[13] D. Chen, T. Zhao, L. Han and Z. Feng, "Single-stage multi-input buck type high-frequency link's inverters with series and simultaneous power supply," *IEEE Transactions on Power Electronics*, vol. 37, no. 6, pp. 7411–7421, 2022.
[14] L. Yan, R. Xinbo, Y. Dongsheng, L. Fuxin and C. K. Tse, "Synthesis of multiple-input DC/DC converters," *Power Electronics, IEEE Transactions on*, vol. 25, pp. 2372–2385, 2010.
[15] O. C. Onar, O. H. A. Shirazi and A. Khaligh, "Grid interaction operation of a telecommunications power system with a novel topology for multiple-input buck-boost converter," *Power Delivery, IEEE Transactions on*, vol. 25, pp. 2633–2645, 2010.
[16] H. Kakigano, Y. Miura and T. Ise, "Low-voltage bipolar-type DC microgrid for super high quality distribution," *Power Electronics, IEEE Transactions on*, vol. 25, pp. 3066–3075, 2010.
[17] A. Khaligh, J. Cao and L. Young-Joo, "A multiple-input DC-DC converter topology," *Power Electronics, IEEE Transactions on*, vol. 24, pp. 862–868, 2009.
[18] A. Khaligh, "A multiple-input dc-dc positive buck-boost converter topology," in *Applied Power Electronics Conference and Exposition, 2008. APEC 2008. Twenty-Third Annual IEEE*, pp. 1522–1526, 2008.
[19] C. Yaow-Ming, L. Yuan-Chuan and L. Sheng-Hsien, "Double-input PWM DC/DC converter for high-/low-voltage sources," *Industrial Electronics, IEEE Transactions on*, vol. 53, pp. 1538–1545, 2006.
[20] L. Solero, A. Lidozzi and J. A. Pomilio, "Design of multiple-input power converter for hybrid vehicles," *Power Electronics, IEEE Transactions on*, vol. 20, pp. 1007–1016, 2005.
[21] G. R. Walker and P. C. Sernia, "Cascaded DC–DC converter connection of photovoltaic modules," *Power Electronics, IEEE Transactions on*, vol. 19, pp. 1130–1139, 2004.

[22] B. Farhangi and H. A. Toliyat, "Modeling and analyzing multi-port isolation transformer capacitive components for onboard vehicular power conditioners," *Industrial Electronics, IEEE Transactions on*, vol. 62, pp. 3134–3142, 2015.

[23] B. Farhangi and H. Toliyat, "A novel vehicular integrated power system realized with multi-port series AC link converter," in *Applied Power Electronics Conference and Exposition, 2015. APEC 2015. 30th Annual IEEE*, pp. 1–7, 2015.

[24] M. Moosavi, S. Farhangi, H. Iman-Eini and A. Haddadi, "An LCL-based interface connecting photovoltaic back-up inverter to load and grid," in *Power Electronics, Drive Systems and Technologies Conference (PEDSTC)*, 2013 4th, pp. 465–470, 2013.

[25] W. Hongfei, S. Kai, D. Shun and X. Yan, "Topology derivation of non-isolated three-port DC-DC converters from DIC and DOC," *Power Electronics, IEEE Transactions on*, vol. 28, pp. 3297–3307, 2013.

[26] M. Amirabadi, H. A. Toliyat and W. C. Alexander, "A multiport AC link PV inverter with reduced size and weight for stand-alone application," *Industry Applications, IEEE Transactions on*, vol. 49, pp. 2217–2228, 2013.

[27] F. Nejabatkhah, S. Danyali, S. H. Hosseini, M. Sabahi and S. M. Niapour, "Modeling and control of a new three-input DC-DC boost converter for hybrid PV/FC/battery power system," *Power Electronics, IEEE Transactions on*, vol. 27, pp. 2309–2324, 2012.

[28] J. Wei and B. Fahimi, "Multiport power electronic interface-concept, modeling, and design," *Power Electronics, IEEE Transactions on*, vol. 26, pp. 1890–1900, 2011.

[29] Q. Zhijun, O. Abdel-Rahman and I. Batarseh, "An integrated four-port DC/DC converter for renewable energy applications," *Power Electronics, IEEE Transactions on*, vol. 25, pp. 1877–1887, 2010.

[30] Q. Zhijun, O. Abdel-Rahman, H. Al-Atrash and I. Batarseh, "Modeling and control of three-port DC/DC converter interface for satellite applications," *Power Electronics, IEEE Transactions on*, vol. 25, pp. 637–649, 2010.

[31] H. Behjati and A. Davoudi, "A multiple-input multiple-output DC-DC converter," *Industry Applications, IEEE Transactions on*, vol. 49, pp. 1464–1479, 2013.

[32] G. Chen, Y. Liu, X. Qing, M. Ma and Z. Lin, "Principle and topology derivation of single-inductor multi-input multi-output DC–DC converters," *IEEE Transactions on Industrial Electronics*, vol. 68, no. 1, pp. 25–36, 2021.

[33] A. Falk, "Bidirectional battery power inverter," ed: Google Patents, 2014.

[34] B. Farhangi and S. Farhangi, "Application of Z-source converter in photovoltaic grid-connected transformer-less inverter," *Electrical Power Quality and Utilisation, Journal*, vol. 12, pp. 41–45, 2006.

[35] B. Farhangi and S. Farhangi, "Comparison of z-source and boost-buck inverter topologies as a single phase transformer-less photovoltaic grid-connected power conditioner," in *Power Electronics Specialists Conference*, 2006. PESC'06. 37th *IEEE*, pp. 74–79, 2006.

[36] J. Cao and A. Emadi, "A new battery/ultracapacitor hybrid energy storage system for electric, hybrid, and plug-in hybrid electric vehicles," *Power Electronics, IEEE Transactions on*, vol. 27, pp. 122–132, 2012.

[37] B. Farhangi and H. A. Toliyat, "Modeling isolation transformer capacitive components in a dual active bridge power conditioner," in *Energy Conversion Congress and Exposition (ECCE), 2013 IEEE*, pp. 5476–5480, 2013.

[38] B. Farhangi, H. A. Toliyat and A. Balaster, "High impedance grounding for onboard plug-in hybrid electric vehicle chargers," in *Power Engineering, Energy and Electrical Drives (POWERENG), 2013 Fourth International Conference on*, pp. 609–613, 2013.

[39] B. Farhangi, "Power conditioning for plug-in hybrid electric vehicles," Ph.D. Dissertation, Texas A&M University, College Station, TX, 2014.

[40] B. Farhangi and H. A. Toliyat, "Piecewise linear modeling of snubberless dual active bridge commutation," in *Energy Conversion Congress and Exposition (ECCE), 2014 IEEE*, pp. 2065–2071, 2014.

[41] P. Fang Zheng, "Z-source inverter," *Industry Applications, IEEE Transactions on*, vol. 39, pp. 504–510, 2003.

[42] R. P. Severns and G. Bloom, *Modern DC-to-DC Switchmode Power Converter Circuits*. New York: Van Nostrand Reinhold Co., 1985.

[43] R. W. A. A. De Doncker, D. M. Divan, and M. H. Kheraluwala, "A three-phase soft-switched high-power-density DC/DC converter for high-power applications," *Industry Applications, IEEE Transactions on*, vol. 27, pp. 63–73, 1991.

[44] Q. Hengsi and J. W. Kimball, "Solid-state transformer architecture using AC–AC dual-active-bridge converter," *Industrial Electronics, IEEE Transactions on*, vol. 60, pp. 3720–3730, 2013.

[45] M. Amirabadi, A. Balakrishnan, H. A. Toliyat and W. C. Alexander, "High-frequency AC-link PV inverter," *Industrial Electronics, IEEE Transactions on*, vol. 61, pp. 281–291, 2014.

[46] T. Haimin, A. Kotsopoulos, J. L. Duarte and M. A. M. Hendrix, "Transformer-coupled multiport ZVS bidirectional DC–DC converter with wide input range," *Power Electronics, IEEE Transactions on*, vol. 23, pp. 771–781, 2008.

[47] J. L. Duarte, M. Hendrix and M. G. Simões, "Three-port bidirectional converter for hybrid fuel cell systems," *Power Electronics, IEEE Transactions on*, vol. 22, no. 2, 480–487, 2007.

[48] S. Dey, A. Mallik and A. Akturk, "Investigation of ZVS criteria and optimization of switching loss in a triple active bridge converter using penta-phase-shift modulation," *IEEE Journal of Emerging and Selected Topics in Power Electronics*, vol. 10, no. 6, pp. 7014–7028, 2022.

[49] S. Chakraborty, MD Weiss and MG Simoes, "Distributed intelligent energy management system for a single-phase high-frequency AC microgrid," *IEEE Transactions on Industrial Electronics*, vol. 54, no. 1, pp. 97–10, 2007.

[50] F. Jauch and J. Biela, "An innovative bidirectional isolated multi-port converter with multi-phase AC ports and DC ports," in *Power Electronics and Applications (EPE), 2013 15th European Conference on*, pp. 1–7, 2013.

[51] A. Kwasinski, "Quantitative evaluation of DC microgrids availability: effects of system architecture and converter topology design choices," *Power Electronics, IEEE Transactions on*, vol. 26, pp. 835–851, 2011.

[52] A. Emadi, S. S. Williamson and A. Khaligh, "Power electronics intensive solutions for advanced electric, hybrid electric, and fuel cell vehicular power systems," *Power Electronics, IEEE Transactions on*, vol. 21, pp. 567–577, 2006.

[53] W. Kempton and J. Tomić, "Vehicle-to-grid power implementation: from stabilizing the grid to supporting large-scale renewable energy," *Journal of Power Sources*, vol. 144, pp. 280–294, 2005.

[54] B. Farhangi, "A novel modified deadbeat controller for vehicle to grid application," in *Power Electronic & Drive Systems & Technologies Conference (PEDSTC), 2015 6th*, pp. 1–6, 2015.

[55] J. G. Pinto, V. Monteiro, H. Goncalves and J. L. Afonso, "On-board reconfigurable battery charger for electric vehicles with traction-to-auxiliary mode," *Vehicular Technology, IEEE Transactions on*, vol. PP, pp. 1–1, 2013.

[56] M. Pahlevaninezhad, D. Hamza and P. K. Jain, "An improved layout strategy for common-mode EMI suppression applicable to high-frequency planar transformers in high-power DC/DC converters used for electric vehicles," *Power Electronics, IEEE Transactions on*, vol. 29, pp. 1211–1228, 2014.

Chapter 4

Modular multilevel cascade converters

Bruno Emmanuel de O. B. Luna[1], Italo Roger F. M. P. da Silva[2], Cursino B. Jacobina[3] and Alexandre C. Oliveira[3]

4.1 Introduction

Voltage-source converter (VSC) topologies have been very popular, they are based on modular structures that have been widely adopted in various power electronics applications, including power conditioning systems (PCSs) for renewable energy sources, battery energy storage systems (BESSs) for power leveling, active power line conditioners (APLCs) for harmonic minimization and reactive power compensation, as well as variable speed drives (VSDs) and high-voltage direct current (HVDC) systems. Modular converters have a fundamental characteristic of combining identical low- or medium-power subsystems in order to achieve a system with higher power ratings. Such a feature is achieved through the series connection of submodules (SMs), which are identical in rated power and can be viewed as power cells with similar circuit topology, all controlled by the same modulation and control schemes. The high degree of modularity ensures high output energy quality, reliability, efficiency, ease of. maintenance, and a reduction in cost, weight, and volume. In high-power systems, where large-scale modular structures typically contain more than ten power cells per arm, designing submodules using standard low- or medium-voltage technology devices can significantly improve the converter reliability. Furthermore, designing large-scale modular structures with a high number of voltage levels and a significant decrease in the average switching frequency can optimize the performance and efficiency of voltage source converters, while preserving power quality.

The cascaded H-bridge (CHB) topology has been the subject of extensive research since the mid-1990s due to its modular feature [1,2]. Inherent to its design, the CHB topology is composed of multiple single-phase converters, which are simple and established building blocks in power electronics systems. As a result, the term "modular converters" has become increasingly popular, although it led to a

[1]Department of Engineering and Technology, Federal Rural University of the Semi-Arid Region, Brazil
[2]Department of Renewable Energy Engineering, Federal University of Paraíba, Brazil
[3]Department of Electrical Engineering, Federal University of Campina Grande, Brazil

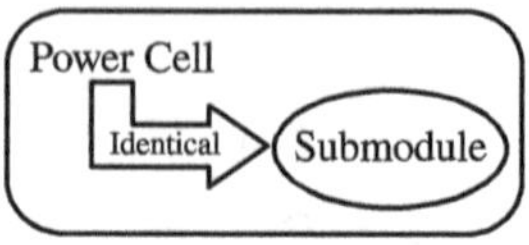

Figure 4.1 Submodule definition

vast group of topologies, whose structure is based on the connection of submodules or power cells. Therefore, it is essential to define the concept of a submodule and power cell in order to provide a reasonable classification of modular converters. In this context, both the submodule and power cell are arrangements of switching devices, DC capacitors, or sources based on the same electrical circuit. A submodule must be designed with the same electrical and thermal characteristics, as well as the same power semiconductor technology, making it identical and interchangeable with other submodules. This design approach ensures modularity and redundant operation because a defective submodule might be replaced by another identical one or bypassed, without interrupting the energy transfer in the converter. From Figure 4.1 it can be observed that all submodules are power cells, but not all power cells are submodules.

A few years after the introduction of the CHB topology, an innovative alternative based on single-phase converters was proposed by Marquardt *et al.* [3]. This new topology, initially referred to as the modular multilevel converter (M^2LC) [4], consists of half-bridge (HB) converters connected in series, creating a common DC link without direct capacitor connections, making it highly suitable for back-to-back (BTB) topologies. The modular structure allows the use of different types of voltage-source converter topologies as submodules, including the conventional half-bridge and full-bridge (FB) configurations, as well as advanced ones such as the neutral point-clamped (NPC), flying capacitor (FC), and neutral point-piloted (NPP) [5]. The use of advanced submodule topologies can result in high output energy quality with fewer submodules and improved fault tolerance, or even a minimized count of submodule components, extending to unidirectional topologies. However, this can also lead to a loss of reliability or redundant-based faulty modes, as fewer submodules are needed to produce the desired output waveform. It is important to note that, based on the duality principle, a current source circuit can be derived from a voltage source circuit. For high-voltage direct current systems, Nami *et al.* [6] investigated different modular topologies of current sources, which are very important, for example, as described in [7] and [8]. However, the focus of this chapter is on the main modular topologies with voltage source-based submodules given their widespread adoption and utilization.

The recent focus of research on modular multilevel converters has been directed towards a family of multilevel topologies known as modular multilevel cascade converters (MMCC) [6,9,10]. This family of converters is primarily based on half-bridge submodule (HB-SM) and full-bridge submodule (FB-SM) topologies. The term MMCC reflects the arrangement of arm clusters, which are groups of submodules per arm m, and the submodule circuit configuration. The multilevel

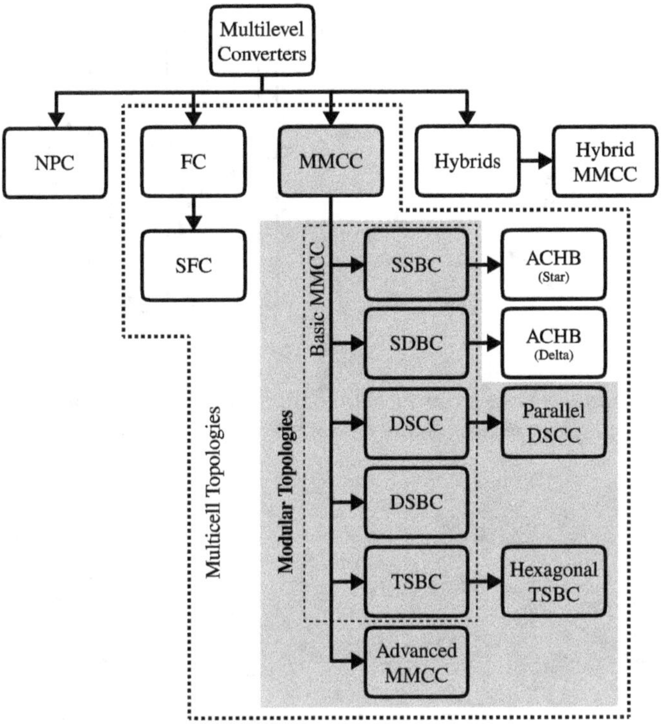

Figure 4.2 Multilevel converter classification

converters can be classified based on the diagram illustrated in Figure 4.2, following the classification proposed in [11]. The star- and delta-connected cascade H-bridge converters are classified as single-star bridge-cells (SSBC) and single-delta bridge cells (SDBC), respectively. On the other hand, the modular multilevel converter is an alternative name for the double-star chopper-cells (DSCC). Both the modular multilevel cascade converter and flying capacitor topologies are multicell topologies, as they are composed of a set of power cells. However, for a topology to be classified as modular, the series connection of submodules should be capable of scaling to different voltage and power levels exclusively by varying the number of submodules, without requiring any additional central components. Figure 4.3(a) shows one phase-leg of a flying capacitor (FC) converter, and Figure 4.3(b) illustrates one phase-leg of a double-star chopper-cells converter. The DSCC has twice the number of power cells per phase M compared to the flying capacitor topology, as it consists of two arm clusters per phase-leg.

In order to illustrate the proposed classification, a comparison between multicell and modular topologies is made using both the flying capacitor and double-star chopper-cell converter topologies. Figure 4.4 shows the parameter M as a function of voltage and current levels. As shown in Figure 4.4(a), the voltage level in each power cell of the flying capacitor converter is unequal and increases

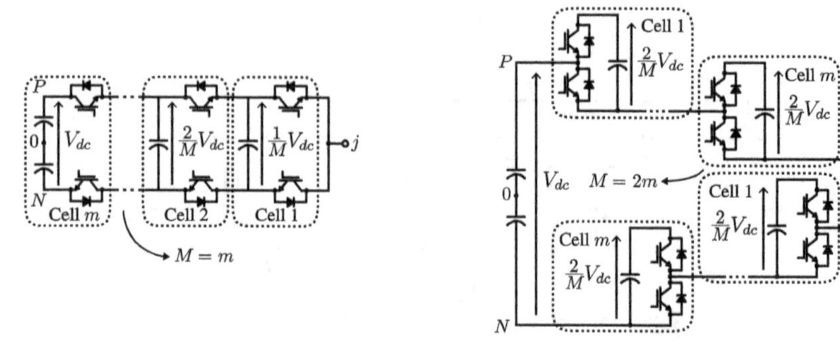

(a) (b)

Figure 4.3 Multilevel converter circuit configuration. (a) One phase-leg of FC converter. (b) One phase-leg of DSCC converter.

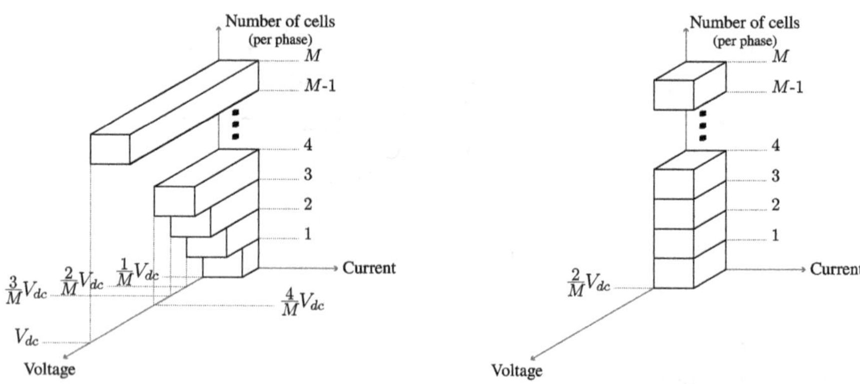

(a) (b)

Figure 4.4 Number of cells as a function of voltage and current levels. (a) Multicell FC topology. (b) Modular DSCC topology.

proportionally with the expansion of M. In contrast, as shown in Figure 4.4(b), the DSCC maintains an equal voltage distribution even with an increase in M. Therefore, the flying capacitor converter and its derivatives, such as the stacked flying capacitor (SFC) converter, are unable to meet the requirements of a modular structure. Alternative topologies have been proposed by combining different converter technologies, referred to as hybrid multilevel cascade converter. These topologies exhibit some degree of modularity and may improve certain aspects of multilevel topologies. However, their design is not entirely modular, potentially reducing modularity and converter reliability [6,12]. Therefore, these hybrid topologies are not considered part of the modular family and are not included in the classification diagram shown in Figure 4.2. Another topology that cannot be classified as modular is the asymmetric cascaded H-bridge (ACHB) converter, due to the need for different switching device technologies for medium- and high-power applications, which leads to different thermal designs for each power cell, thereby eliminating the modularity benefit. The performance and operation of converters were evaluated in [13].

The modular topologies are represented by the darkened boxes in the classification diagram, highlighting their distinct modularity advantage. This chapter presents the basic modular topologies based on the proposed classification of multilevel converters. It covers the fundamental features of the modular multilevel cascade converter family, including circuit configurations, control strategies, modulation techniques, operational issues, and main applications. This chapter covers both basic and advanced modular topologies, with a main focus on the basic topologies, which have gained popularity in academia and industry due to their unique benefits and the increasing demand for sustainable energy solutions [14]. The following sections of the chapter are structured as follows: Section 4.2 discusses the circuit configuration and provides a comprehensive review of the different modular arrangements and submodule topologies that can be utilized in converter design. Section 4.3 presents the key contributions to voltage-balancing and circulating current control strategies. Modulation techniques are then discussed in Section 4.4. Finally, Section 4.5 examines the recent developments in operational issues, including fault-tolerant operation and pre-charging procedures of floating DC capacitors.

4.2 Modular converter topologies and description

Multilevel converters are widely recognized for their exceptional performance in high-power applications [15,16]. Their unique ability to produce a quasi-sinusoidal output waveform with low harmonic content significantly reduces or eliminates the need for passive filters, resulting in lower voltage stress on semiconductor devices and the requirement for smaller cooling systems. However, as the number of voltage levels increases to expand the power capacity of multilevel converters, it leads to design complexity, capacitor voltage unbalanced conditions, and greater computational efforts. When the number of voltage levels is high, topologies such as neutral point clamped and flying capacitors face challenges due to an excessive number of clamping diodes and bulk capacitors, respectively. To overcome these design complexities, modular topologies can be employed as an alternative approach to stepped output waveform synthesis. Figure 4.5(a) and (b) illustrates the conventional multilevel solution based on the neutral point clamped topology and the modular solution based on the modular converter topology, respectively. In both cases, stepped voltages are selected from a single DC-link voltage V_{dc}, resulting in the output phase voltage v_j, where j is the converter phase, as shown in Figure 4.5(c). By employing modular topologies, a different approach can be taken to achieve the stepped output waveform, avoiding the aforementioned design complexities.

The primary difference between the conventional multilevel and modular solutions lies in the proposed arrangement of the capacitive DC-link. In the

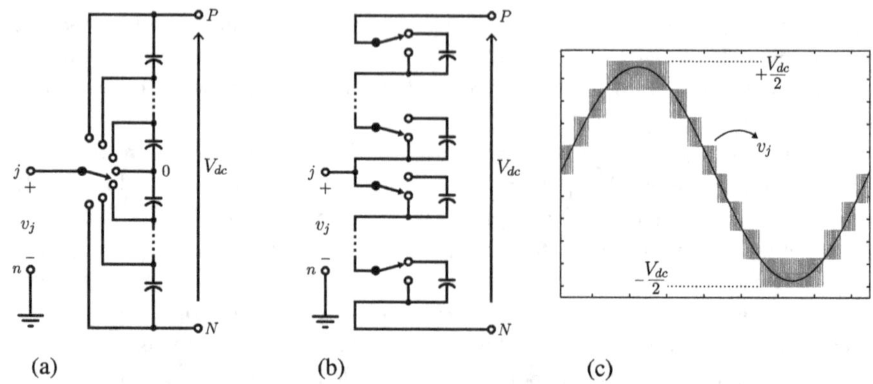

Figure 4.5 Multilevel converter topologies and waveform. (a) NPC topology. (b) DSCC topology. (c) Nine-level waveform.

conventional multilevel solution shown in Figure 4.5(a), a central DC-link capacitor is necessary for DC energy storage, whereas the modular solution offers distributed DC energy storage through floating DC capacitors. This eliminates the need for a common DC-link capacitor between the terminals (*P*, *N*), as shown in Figure 4.5(b). Such an arrangement increases the overall converter reliability against high surge currents and potential damage in case of short circuits occurring at the DC terminals [17]. Furthermore, the modular solution emphasizes the idea of failure management, maintenance, and consistent redundancy. Although the solution is based on the modular multilevel cascade converter topology, the fundamental concept of separated DC-link capacitors and their advantages can be applied to other modular topologies.

4.2.1 Modular multilevel cascade converter arrangements and submodule topologies

The MMCC topologies are composed of series-connected submodules that form an arm-cluster. Different arm-clusters can be combined to create groups of modular phase-leg arrangements: single-star (SS), single-delta (SD), double-star (DS), and triple-star (TS). The initial letter indicates the number of sets of three arm-clusters (single, double, or triple), and the subsequent letter indicates the disposition (star or delta) of these arm-clusters. Figure 4.6(a) and (b) illustrates MMCC topologies based on star arrangements. The single-star arrangement can be classified into positive (*P-SS*) and negative (*N-SS*) subconverters, depending on the connections of the arm-clusters in relation to the common terminals (*P*, *N*). The combination of subconverters creates the double-star arrangement, as shown in Figure 4.6(c). Furthermore, the parallel connection of three subconverters results in the triple-star arrangement as shown in Figure 4.7.

The MMCC family consists of both basic and advanced topologies that utilize submodules based on circuits other than half-bridge and full-bridge topologies.

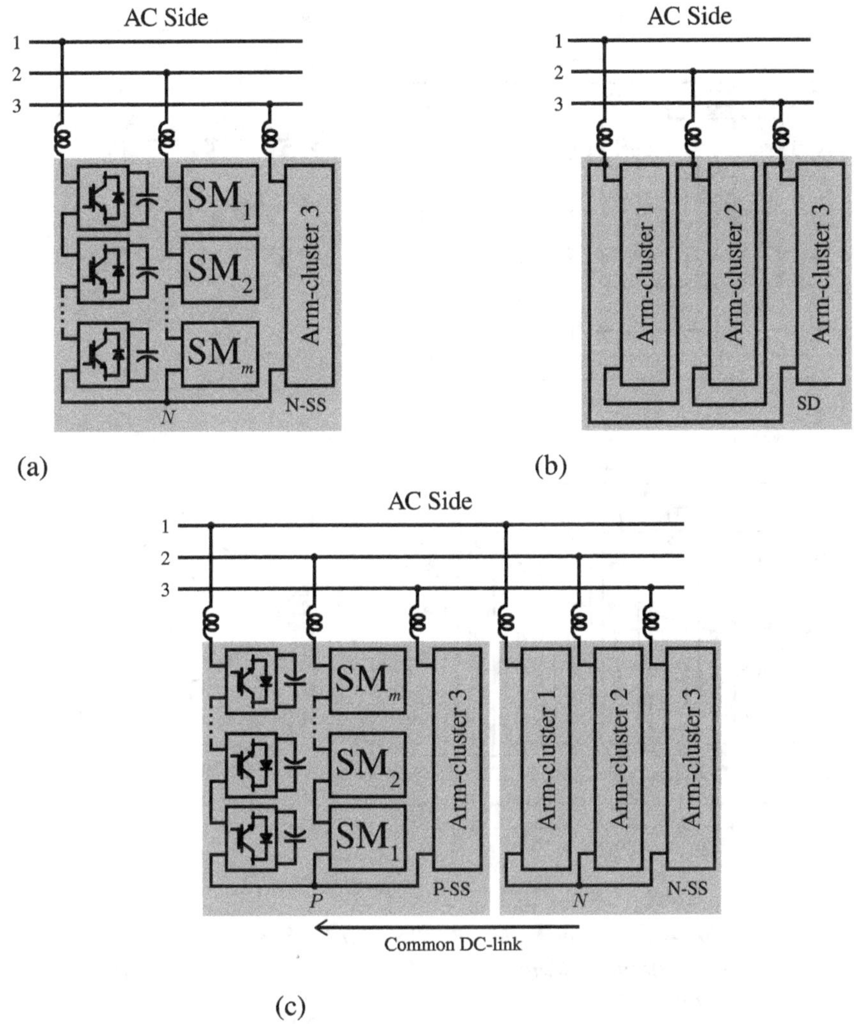

Figure 4.6 Modular multilevel cascade converter arrangements. (a) Single-star arrangement. (b) Single-delta arrangement. (c) Double-star arrangement.

Figure 4.8 shows the submodule circuit configurations proposed over the past two decades, and Table 4.1 describes the output voltage of each submodule (v_m) in relation to the floating DC capacitor voltage (v_c). The full-bridge submodule shown in Figure 4.8(a) can generate a symmetrical three-level bipolar voltage [10], whereas the half-bridge submodule in Figure 4.8(b) can only produce two unipolar voltage levels [3,4]. In MMCC topologies with DC link availability, in the event of a subsequent DC short-circuit, the full-bridge submodule offers superior DC fault handling capability compared to the half-bridge submodule. Bipolar submodules

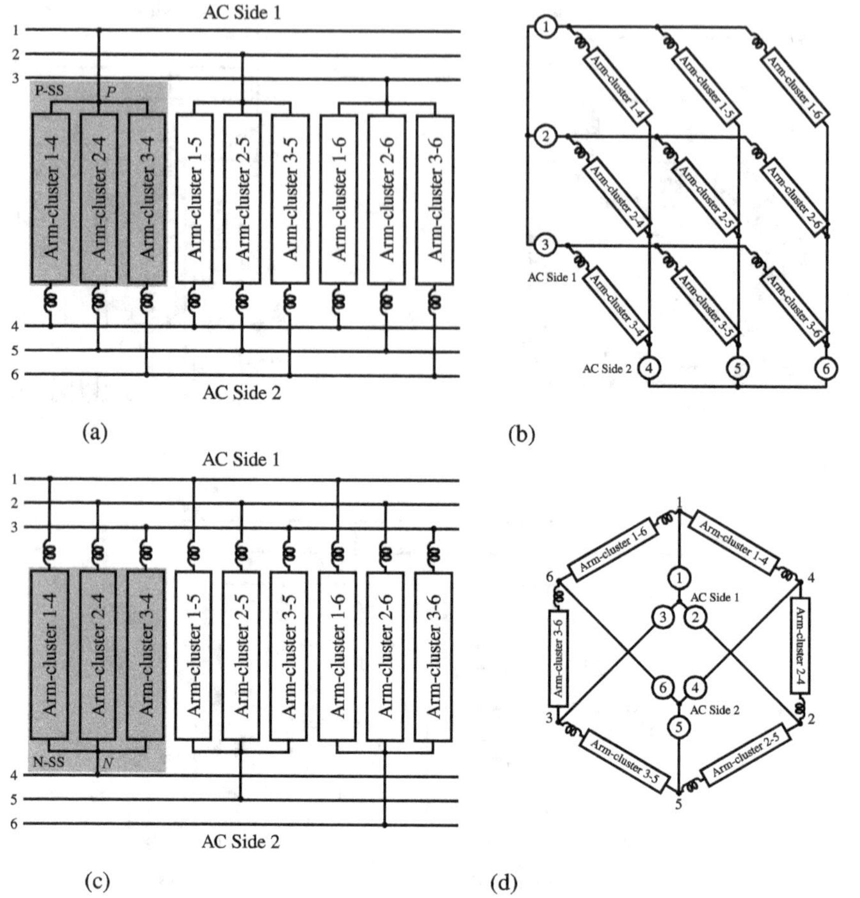

Figure 4.7 Triple-star arrangements. (a) TS with P-SS subconverters. (b) Typical representation of the TS. (c) TS with N-SS subconverter. (d) Hexagonal representation of the TS.

can interrupt arm currents in any direction by applying the appropriate polarity of terminal voltages in the arm [18]. When all switches of the full-bridge submodule are turned off, a reverse voltage can be generated from the capacitor voltage to block AC side currents, potentially reducing the fault current to zero [5]. However, since the half-bridge submodule cannot provide DC fault-handling capability, thyristors must be installed in parallel with the diode to bypass the fault current.

Submodules based on multilevel topologies offer the advantage of reducing the overall size of the converter. However, this advantage comes at the cost of increased complexity in submodule design and operation. Common multilevel topologies, such as neutral point-clamped and flying capacitor topologies, have been applied as submodules [19], as illustrated in Figure 4.8(c) and (d), respectively. Both multilevel submodule topologies can generate three-level unipolar

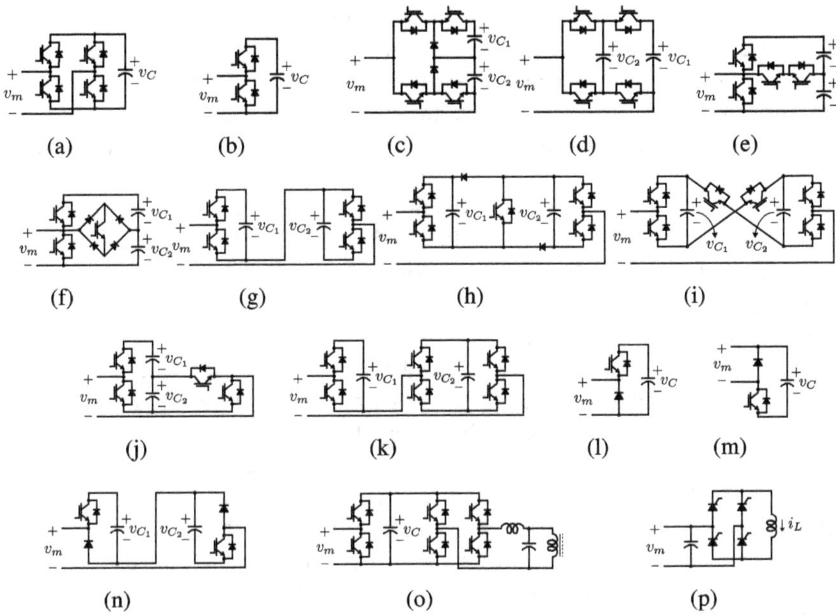

Figure 4.8 Submodules topologies. (a) FB. (b) HB. (c) NPC. (d) FC. (e) NPP1. (f) NPP2. (g) TC. (h) CD. (i) CCC. (j) ACD. (k) MC. (l) RC1. (m) RC2. (n) RC3. (o) IPT (p) CS.

Table 4.1 Voltage output of submodules in voltage-source-based topologies

Submodule topology	**Number of voltage levels**	**Output voltage of submodule (v_m)**
FB	3	$-v_C, 0, v_C$
HB	2	$0,\ v_C$
NPC	3	$0,\ v_{C1}, v_{C2}, (v_{C1}+v_{C2})$
FC1	3	$0,\ v_{C2}, (v_{C1}-v_{C2}), v_{C1}$
NPP1	3	$0,\ v_{C1}, (v_{C1}+v_{C2})$
NPP2	3	$0,\ v_{C1}, (v_{C1}+v_{C2})$
TC	3	$0,\ v_{C1}, v_{C2}, (v_{C1}+v_{C2})$
CD	4	$-v_C, 0, v_C, v_{C1}, v_{C2}, (v_{C1}+v_{C2})$
CCC	5	$(-v_{C1}-v_{C2}), -v_{C1}, -v_{C2}, 0, v_{C1}, v_{C2}, (v_{C1}+v_{C2})$
ACD	4	$-v_{C2}, 0, v_{C1}, v_{C2}, (v_{C1}+v_{C2})$
MC	4	$-v_{C2}, (v_{C1}-v_{C2}), 0, v_{C1}, v_{C2}, (v_{C1}+v_{C2})$
RC1	2	$0,\ v_C$
RC2	2	$0,\ v_C$
RC3	3	$0,\ v_{C1}, (v_{C1}+v_{C2})$

voltage. However, combining two submodules in parallel to form a bridge allows the generation of symmetrical bipolar voltages as well [20]. Despite this benefit, these solutions result in reduced modularity and inherit the inherent issues associated with both topologies. To address these limitations, Sahoo *et al.* [21] introduced two novel multilevel submodule topologies based on the neutral point-piloted (NPP) topology, as shown in Figure 4.8(e) and (f). The first topology, named NPP-Type 1 (NPP1), follows the conventional circuit of the converter. The second topology, named NPP-Type 2 (NPP2), offers an alternative solution to Type 1, operating in a similar way. Both submodule topologies are capable of synthesizing three-level unipolar voltage.

Several submodule topologies have been proposed to improve power density, fault-handling capacity, and increase the number of output voltage levels. Combining two half-bridge submodules connected by the terminals of the floating DC capacitors can result in four different submodule concepts: twin-cells (TC) [22], clamp-double (CD) [17], cross-connected cells (CCC) [23], and asymmetric clamp-double (ACD) [6], as shown in Figure 4.8(g)–(j), respectively. Furthermore, the combination of half-bridge and full-bridge submodules originates a mixed cell (MC) [6], as shown in Figure 4.8(k). During normal operation, both the clamp-double and twin-cell submodules function as equivalent units of two half-bridges, operating similarly to synthesize three-level unipolar voltage. However, in the event of a DC fault, the clamp-double submodule transforms into a full-bridge submodule by opening the intermediate switch between the two half-bridges. This configuration results in energy absorption or voltage clamping, where the clamp-double submodule can provide reverse voltage to clear a DC fault current and minimize over-voltage [17]. An alternative approach to doubling the commutation cells and achieving reverse voltage is presented by the asymmetric clamp-double submodule. On the other hand, mixed cell submodules can generate asymmetrical four-level bipolar voltage, while the cross-connected cells submodule can produce symmetrical five-level bipolar voltage [6].

Unidirectional topologies can offer a simpler and cost-effective alternative for energy conversion systems that do not require bidirectional power flow. In comparison to full-controlled structures, unidirectional topologies reduce the number of switches and simplify designs, making them an attractive choice. This concept of unidirectional topologies can be expanded to the MMCC converter by utilizing submodules based on reduced cells (RCs). The half-bridge submodule can generate two distinct unidirectional submodule topologies, known as RC-Type 1 (RC1) and RC-Type 2 (RC2), as illustrated in Figure 4.8(l) and (m), respectively. Additionally, the twin-cells submodule can produce RC-Type 3 (RC3), as shown in Figure 4.8(n). In these configurations, the output voltage can be switched only when there is a positive current flow through the submodule. Conversely, the output voltage remains constant when there is a negative current flow. Both Type 1 and Type 2 submodules are two-level unipolar submodules, whereas Type 3 can achieve an additional voltage level. Similar to neutral point-clamped and flying capacitor submodules, the bridge-based reduced cells submodule can synthesize symmetrical bipolar voltage.

The use of parallel subsystems for each submodule can offer additional advantages to the characteristics of submodule topologies. Inductive power transfer (IPT) technology can be utilized to maintain the floating DC capacitors in the half-bridge submodule balanced within tight bounds [24]. An example of this is shown in Figure 4.8(o), where an IPT-based converter was parallel-connected to a half-bridge submodule for energy exchange among the floating DC capacitors, effectively regulating them to their nominal values. Additionally, to improve fault-tolerant operation and achieve better DC voltage regulation, boost rectifiers have been incorporated in parallel with full-bridge submodules [25]. Furthermore, the application of the duality principle can enhance fault-handling capabilities in submodule topologies based on current sources [6]. Figure 4.8(p) illustrates a dual version of the full-bridge submodule, as presented in [26]. It is important to note that while current source submodules can transmit higher power compared to standard voltage source submodules, they also come with increased losses and necessitate larger footprints for the DC inductors [6]. Careful consideration of trade-offs and thorough evaluation of each topology and subsystem are vital. The choice of submodule topology and configuration depends on various factors, including the specific application, performance requirements, cost constraints, and reliability considerations. Thus, optimization is essential to achieve the best balance between system performance, cost-effectiveness, and complexity.

4.2.2 Basic modular multilevel cascade converters

The modular design inherent in MMCC topologies allows the theoretical expansion of cascading submodules to an unlimited number without increasing circuit complexity, making them highly scalable to different power and voltage levels. These topologies can produce incremental low-voltage steps, resulting in improved output waveform quality, reduced harmonic content, and minimal electromagnetic interference emissions. Consequently, they demand smaller or even no passive filters, leading to space-efficient and cost-effective systems. Furthermore, MMCC topologies employ well-established industrial low-voltage components, making them suitable for a wide range of high-power applications.

In basic MMCC topologies, inductors are connected in series with each arm cluster to regulate arm currents and limit fault currents. In double-star topologies, a single coupled inductor can be used in each phase-leg, reducing both size and weight compared to employing two non-coupled inductors. In other basic MMCC topologies, except for the single-star bridge cells, interconnected arm clusters form inner current loops, with the arm currents carrying a component known as circulating current. Its primary function is to exchange active power between the DC-link and the arm clusters or among the arm clusters themselves. However, the fundamental AC component that flows through each submodule may cause voltage fluctuations in the floating DC capacitors, resulting in circulating current harmonics. Controlling circulating currents is essential to improve performance, reduce losses and energy fluctuations, and achieve operational stability in conditions that were previously challenging. The main purpose of controlling circulating currents is to achieve an average voltage balance among the arm clusters. To synthesize the correct voltage level, it is necessary to regulate the voltage of each floating DC capacitor. The strategies for voltage balancing and circulating current control, as well as modulation techniques, will be discussed in Sections 4.3 and 4.4, respectively.

The selection of one of the basic MMCC topologies depends mainly on its application, semiconductor ratings, and energy storage requirements. Each topology has its own characteristics in terms of performance and viability [11]. This chapter focuses on the five basic MMCC topologies: single-star bridge-cells, single-delta bridge cells, double-star chopper-cell converters, double-star bridge-cells, and triple-star bridge-cells. These topologies use both full-bridge and half-bridge submodules as the default power cells. The main characteristics of each basic modular topology will be compared, taking into account physical and technical constraints, as well as suitability for specific application segments.

4.2.2.1 Single-star or -delta bridge-cells topologies

The SSBC and SDBC topologies were initially presented in [1,2]. These converters are arranged with one arm-cluster per phase-leg and do not share a common DC-link. To achieve direct DC to AC power inversion in these topologies, isolated DC sources for each submodule are necessary, especially in variable speed drive applications. Isolated DC sources are typically provided by rectifiers powered by phase-shifting transformers, such as multipulse transformers [16]. These converters are suitable for grid-connected applications, such as reactive power and harmonic compensation, which improve power quality [27,28]. However, they are not commonly used for regenerative applications due to the need for active front-end rectifiers to improve input power quality.

In these topologies, floating DC capacitors are used instead of isolated DC sources [10]. The SSBC topology can effectively control positive-sequence active-power [11]. However, compensating for negative-sequence reactive-power requires an increase in converter voltage ratings under unbalanced conditions. This is because the converter must handle both active and reactive power exchange with the grid [29]. In practice, the initial exchange between the grid and the converter results in the charge and discharge of the floating DC capacitors, which is inconsistent with the operation of a static synchronous compensator (STATCOM) [30]. Furthermore, the SSBC lacks a circulating current, making it unsuitable for the exchanging of both active and reactive power with the grid [11].

Despite the challenges associated with compensating for negative-sequence, several studies have developed control strategies relying on zero-sequence voltage injection to achieve the desired operating range for practical unbalanced conditions. When utilizing the SSBC, it is essential to consider the power rating and the degree of unbalance [31]. This is critical because an increase in the maximum achievable output voltage is associated with higher converter costs [31,32]. Hagiwara *et al.* [33] demonstrated that the SDBC can control both positive- and negative-sequence reactive power, assuming symmetrical three-phase grid voltages. However, in the case of serious unbalanced conditions, such as an unbalanced load under an asymmetrical grid fault, a large circulating current is required to balance the floating DC capacitors. Consequently, the converter cannot compensate for negative-sequence reactive power. Additionally, active power control can also be applied to regulate the state-of-charge (SOC) in SSBC battery energy storage systems [34,35] and to control low-frequency active power in SDBC for flicker compensation [33].

4.2.2.2 Double-star chopper-cells topology

The DSCC topology was originally developed for use for application in high-voltage direct current transmission systems. The modular design provides high scalability, and the double-star arrangement provides a common DC-link while effectively limiting parasitic and fault currents using internal arm inductors. However, the DSCC has also found applications in medium-voltage high-power variable-speed drives by enabling AC-to-AC indirect power conversion and improving power quality in grid-connected systems.

The DSCC uses arm-clusters composed of HB-SMs to generate unipolar voltages with a 180° phase difference between the arm-clusters of the same phase-leg. In numerous applications, the common DC-link capacitor can be removed because the energy storage in the DSCC is distributed among the floating DC capacitors, such as in back-to-back grid-connected systems. However, the ripple in the floating DC capacitor voltage is a significant issue. It is inversely proportional to the frequency of the load current and directly proportional to the magnitude of the load current [36]. This imposes limitations on operating motors at low frequencies, particularly for constant-torque loads, where the ripple magnitude directly impacts the frequency. This issue is less pronounced when driving quadratic-torque loads.

An appropriate control strategy, based on common-mode voltage and AC circulating current injection, can enable the DSCC to operate effectively over a wide range of frequencies. This includes start-up and low-speed operation for constant-torque motor drives [37–39]. The DSCC can exchange negative-sequence reactive power with the grid due to the presence of circulating currents [11]. Unlike the SDBC, the DSCC can fully compensate for negative-sequence reactive power with a small amount of DC circulating current. It can also handle unbalanced nonlinear loads, even under asymmetrical grid fault conditions [40]. To meet high-current requirements, a parallel configuration of the DSCC, known as the extended MMC (EMMC), was proposed by Pirouz and Bina [41]. Initially, the EMMC or parallel DSCC had separate DC-links, but a version sharing the same DC-link was presented in [42].

4.2.2.3 Double-star bridge-cells topology

The double-star bridge-cell (DSBC) topology represents a variation of the DSCC, offering similar control and modulation techniques but with increased practical flexibility. By substituting half-bridge submodules with full-bridge submodules, the DSBC can synthesize bipolar voltages, providing an additional capability to decrease or increase the DC-link voltage during power rectification or inversion [11]. Consequently, the DSBC can produce AC voltage independently of the DC-link voltage, enabling it to perform AC-to-AC power conversion when connected to a single-phase AC voltage source [43–45]. Furthermore, depending on the utilization of the DC-link voltage, the number of voltage levels can vary [46]. Therefore, the DSBC can operate not only as a DC-to-AC converter [46], similar to the DSCC but also as a single-phase to three-phase and single-phase to single-phase AC-to-AC converter. However, when operating as a single-phase to three-phase AC-to-AC converter, the input and output frequencies are not the same for proper operation [45]. This topology is especially well-suited for distributed power conditioning systems for renewable energy resources, where the DC-link voltage can exhibit significant variations [11].

4.2.2.4 Triple-star bridge-cells topology

The triple-star bridge-cell (TSBC) topology, also known as the modular multilevel matrix converter (M3C), consists of three SSBC subconverters. This topology was originally proposed by Erickson and Al-Naseem [47] and is particularly useful for medium-voltage high-power motor drives at low frequencies and regenerative braking, enabling AC-to-AC power conversion [47–51]. However, similar to the DSBC, the TSBC was not recommended for synchronous frequency operation due to an increase in ripple in the floating DC capacitors. Without proper control of circulating currents, voltage fluctuations in the floating DC capacitors increase as the output frequency is reduced below one-third of the input frequency [48,49]. To address this issue and achieve a broad frequency range of operation, a control method was proposed in [50]. While this method enables lower frequency operation with input and output frequencies close or equal, it increases the ripple in the floating DC capacitors compared to previous methods proposed in [48,49]. Therefore, this control method for synchronous operation is primarily intended for lower frequencies. However, since some energy variations may still occur at lower frequencies, it is necessary to select the semiconductor ratings that can support such operations [51].

An alternative topology to the TSBC is the Hexverter, based on a hexagonal configuration, as presented in [52]. The Hexverter consists of a reduced version of the TSBC, with one arm cluster in each subconverter, resulting in a total of six arm clusters for the entire topology, as shown in Figure 4.7(d). The Hexverter delivers similar performance to the conventional TSBC, with the advantage of having fewer submodules. Comparisons were conducted between DSCC and TSBC topologies, and both were found to be suitable for three-phase AC-to-AC conversion across a broad frequency range, exhibiting improved performance and stable operation with appropriate control techniques [51]. The DSCC-based back-to-back system is the most suitable solution for synchronous frequency operation at higher frequencies, while the TSBC topologies can provide superior performance and efficiency at low-frequency applications.

4.3 Control strategies

4.3.1 Voltage-balancing control

In practical scenarios, voltage imbalances between the submodules are caused by losses and non-ideal conditions, which can result in non-zero active power flow into the phase legs in MMCC topologies. Because the capacitors in these topologies can be floating, it becomes essential to implement voltage-balancing control strategies to maintain a constant voltage level. Circulating current control strategies have been developed to improve the operational range and assist in the regulation of the floating DC capacitors. Typically, the voltage-balancing and circulating current control strategies are based on the overall control schemes illustrated in Figures 4.9 and 4.10, which describe the individual regulation schemes for the floating DC capacitors.

To maintain energy balance between the converter legs, voltage-balancing control is generally achieved by regulating the average voltage per phase and the voltage of each submodule. Averaging voltage control, also known as cluster-balancing control, is a closed-loop strategy that ensures the average value of the sum of all floating DC capacitors is regulated, resulting in a balanced energy distribution in a phase-leg [53–56]. Two individual regulation methods are employed: closed-loop individual-balancing voltage control, where the capacitor voltage of

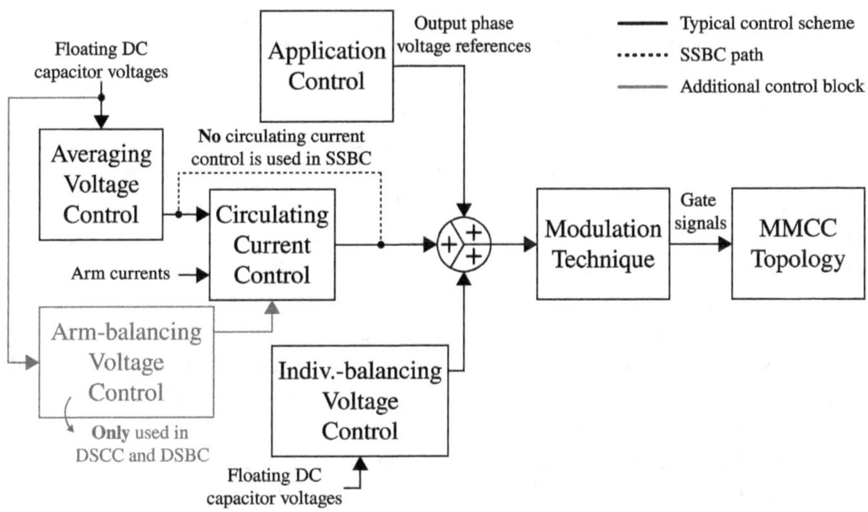

Figure 4.9 Block diagram of the overall voltage control system

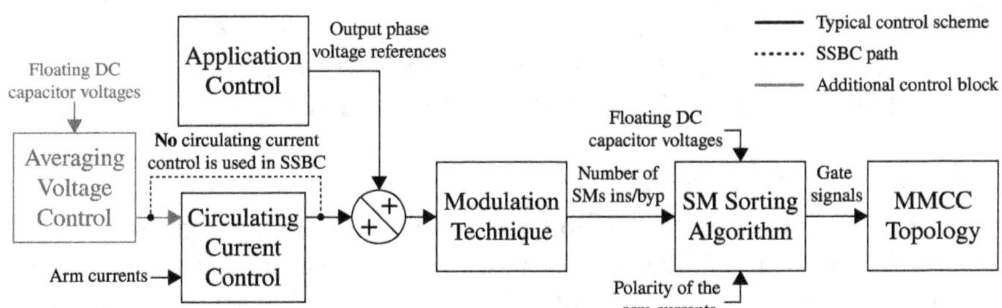

Figure 4.10 Block diagram of the submodule voltage sorting algorithm

each submodule is fed back in a closed-loop control, and an algorithm-based sorting method, which determines which submodules require bypassing or insertion to achieve balanced voltage [53–55].

Modifications to the MMCC topology can be implemented to achieve voltage balance, depending on the specific arrangement of the converter. For example, adding a zero-sequence voltage or slightly changing the switching pattern can help compensate for converter losses and maintain submodule balance [11,56–59]. Since the SSBC lacks circulating current, such modifications can be sufficient to achieve voltage balance. This control scheme is illustrated in Figures 4.9 and 4.10, and it excludes the circulating current control block, indicated by a dashed line path. In contrast, the SDBC, DSCC, DSBC, and TSBC, which involve circulating currents, necessitate both averaging voltage control and inner circulating current control. Further details on circulating current control strategies will be discussed in Section 4.3.4.

4.3.2 Individual-balancing voltage control

Peng *et al.* [1] proposed a method to achieve voltage balancing in SxBC topologies by combining the selective harmonic elimination (SHE) technique with a closed-loop controller. The SHE switching pattern was slightly shifted to achieve voltage balancing. This approach was further optimized in [27] by interchanging the pulse patterns among the submodules, reducing the need to measure all floating DC capacitors. Fujii *et al.* [60] implemented average voltage control with submodules using unipolar sinusoidal pulse-width modulation with phase-shifted carriers (PSC-PWM) while individual-balancing voltage control was implemented through individual voltage feedback. Akagi *et al.* [54] presented a voltage-balancing strategy for the SSBC topology, incorporating voltage feedback from all floating DC capacitors and introducing an inner decoupled current loop for independent active and reactive power control within the average voltage control [34]. The delay time of the modulation pattern in voltage-balancing control should also be considered when expanding the number of submodules [55]. Hagiwara *et al.* [53,61] introduced arm-balancing voltage control for double-star-based topologies, which have two arms per phase-leg, aiming to balance the voltage between the converter arms and maintain voltage-balancing control stability during four-quadrant operation, as proposed in [62]. In addition to these methods, several studies have investigated PSC-PWM voltage-balancing control, including average and individual-balancing voltage control approaches, for all fundamental MMCC topologies such as SSBC [54,55], SDBC [33], DSCC [62], DSBC [45,46], and TSBC [50].

4.3.3 Voltage sorting algorithm

The submodule voltage sorting algorithms were originally presented for the DSCC topology [56–58]. These algorithms achieve the required voltage level by selectively inserting or bypassing submodules. This decision is based on the voltage magnitude of each floating DC capacitor, sorted either in ascending or descending order, and the direction of the arm currents. The fundamental principle of the submodule voltage sorting algorithms was described in [4,43]. Angquist *et al.* [63] developed an open-loop control strategy that relies on estimating stored energy. This method incorporates the principles of submodule sorting algorithms to mitigate the effects of delays in the measurement and communication systems of the converter. It proves particularly useful for large-scale MMCC topologies [64]. Although the open-loop scheme does not use voltage feedback controllers to balance the floating DC capacitors, the voltage of each submodule capacitor is measured and then compared directly in the sorting process [63].

Alternatively, Ilves *et al.* [65] introduced a submodule sorting algorithm based on a round-robin schedule, eliminating the necessity of measuring the voltage of floating DC capacitors to achieve voltage balance. To accomplish this, the converter must be controlled to ensure that the average energy transfer is zero. Ilves *et al.* [65] also introduced a predictive submodule voltage sorting algorithm that minimizes capacitor voltage ripple under low switching frequency modulation. Traditional voltage-balancing strategies have a trade-off between switching frequency and the capacitor voltage ripple, so the proposed algorithm aims to minimize this trade-off. Submodule voltage sorting algorithms maintain all floating DC capacitor voltages within a narrow voltage range [57], determined by the selection of inserted or bypassed submodules. This selection depends on the sampling frequency [64,66].

Initially designed for the DSCC topology, most submodule voltage sorting algorithms have now been adapted for various other MMCC topologies, including basic [44,52,67] and advanced topologies [21,68]. Furthermore, several studies have proposed combining these algorithms with different modulation techniques to enhance their efficiency [57,64–66,69–73]. These techniques aim to improve switching frequency efficiency, output waveform quality [74,75], reduce the need for sensors [65,69,76], and simplify algorithm complexity [66,76].

4.3.4 Circulating current control

There is significant research interest in the control of circulating currents in the MMCC topologies, with a particular emphasis on the DSCC. While circulating currents also exist in other MMCC topologies, there has been limited exploration of circulating current control in these topologies, as evidenced in [33]. As a result, most research on circulating current control has focused on DSCC. Several studies have conducted an in-depth analysis of the mathematical model of the DSCC, examining both its dynamic behavior and steady-state characteristics to develop active and passive solutions for regulating circulating currents [77–82].

Under ideal conditions, circulating currents in the DSCC topology are composed of a fraction of the DC-link current or the current itself in the case of a single phase-leg. This allows the transfer of power energy between the DC-link and arm-clusters. In the DSBC topology, there is an additional fundamental frequency component present to exchange active power between the upper and lower arm clusters in the same phase-leg. In contrast, in the SDBC and TSBC topologies, circulating currents are solely composed of the fundamental frequency component, as neither topology includes a DC-link. Circulating currents are fundamental for exchanging active power among the arm clusters of MMCC topologies and do not affect the AC-side signals, as presented in [11]. However, under practical conditions, oscillating components emerge within the circulating currents, which degrade the arm currents. These oscillating components are referred to as circulating harmonics. In balanced systems, these components are primarily composed of low-order harmonics at negative-sequence double-line frequency. However, in unbalanced voltage conditions, it is also necessary to consider zero- and positive-sequence components [53]. The modulation and control schemes, as well as the converter operating point, have a significant influence on the harmonics in circulating currents [57]. Additionally, the interaction between the arm inductances and submodule capacitances can contribute to the amplification of these harmonics, leading to resonance effects [83,84].

The presence of circulating harmonics directly impacts the ripple of the floating DC capacitors by distorting the arm currents. If left uncontrolled, they can compromise the stability of the converter, negatively impact the rating of capacitors and semiconductor devices, and increase total power losses. Therefore, suppressing undesired circulating harmonics is essential for achieving high performance and efficiency of the converter, as only active power should be exchanged within the converter. While selective harmonic injection into circulating currents can address specific operational issues, such as reducing voltage fluctuations in the floating DC capacitors, it is essential to ensure that it does not compromise the overall performance and stability of the converter.

The primary aim of circulating current control strategies is to maintain energy balance among the floating DC capacitors during dynamic state and to regulate them in steady-state operation. In a DSCC operating under ideal conditions, where the circulating current is purely DC, a proportional-integral (PI) controller would be sufficient for achieving a satisfactory dynamic response [62]. However, due to the limited action of this controller in harmonic frequencies, it cannot eliminate circulating harmonics. As a result, several studies have aimed to identify effective solutions for eliminating or suppressing these harmonics to an acceptable level. Various control schemes for controlling or suppressing circulating harmonics can be found in the literature, including closed-loop and open-loop strategies, commonly referred to as active methods. In contrast, passive methods involve physical modifications to minimize circulating harmonics. For instance, Ilves *et al.* [78] have demonstrated that the values of the arm inductances and submodule capacitances can be calculated to avoid a specific resonant harmonic frequency. Modifications to the arm inductors have been evaluated for DSCC converters' storage requirements and the maximum permissible fluctuation magnitude [77,78]. Tu *et al.* [77] developed a theoretical analysis that resulted in arm inductances as a function of the second-order circulating harmonic and the fault current rise rate. While increasing the arm inductances can minimize this harmonic order to a certain level, this solution increases the overall size and weight of the converter and can affect performance due to high voltage across the arm inductors. Using coupled inductors is a simple and attractive alternative for suppressing circulating harmonic currents and reducing the size, weight, and cost of the magnetic core [62]. Another passive solution proposed in [78] involves the insertion of a tuned LC filter to eliminate the second-order circulating harmonic. However, passive methods have limitations in suppressing circulating harmonics, as they can only mitigate low-order circulating harmonics to a certain level. An effective approach is to combine these methods with active methods.

Regarding active suppression methods based on closed-loop controllers for the DSCC topology, Hagiwara and Akagi [62] proposed a control strategy that had certain limitations related to operating points. These limitations were addressed in [53,61] by introducing arm-balancing voltage control into the voltage-balancing strategy. This modification transformed the circulating current controller into a proportional (P) controller, resulting in a reduction of circulating harmonics. However, it is worth noting that the primary focus of these studies was not solely on eliminating circulating harmonics but rather on achieving stable four-quadrant operation. In contrast, Tu *et al.* [70] used PI controllers to suppress second-order circulating harmonics in a negative-sequence dq coordinate. While this approach effectively suppresses circulating harmonics, it is limited to three-phase systems and only applicable under balanced conditions. Additionally, other even-order harmonics can also be present in circulating currents under real conditions [84]. To address this, proportional-resonant (PR) controllers were introduced to suppress these additional even-order circulating harmonics, and they are suitable for both single-phase and multi-phase systems [79]. As described in [84], circulating harmonics can be considered a series of even-order harmonics, and resonant transfer functions are fine-tuned to target specific harmonic orders for suppression. However, implementing PR controllers can be complex when dealing with multiple harmonics that require elimination.

As described in [80], the utilization of multi-resonant controllers to minimize circulating harmonics above the tenth order has proven to be ineffective. To address this limitation, a plug-in strategy combining repetitive control with a PI controller was proposed in [80,81]. The repetitive control scheme, based on the internal mode principle, has demonstrated robust steady-state performance in suppressing circulating harmonics due to its ability to offer higher gains at the fundamental frequency and its multiples. However, this approach does have some drawbacks, including the requirement for precise control parameters such as sampling frequency and the number of samples in a cycle, as well as susceptibility to noise at intermediate frequencies. As mentioned in Section 4.3.1, open-loop control strategies have also been explored with the aim of reducing the number of closed-loop controllers. While these methods have displayed effective dynamic performance and circulating harmonics suppression, they demand a precise analytical model and parameter values. Harnefors *et al.* [82] integrated circulating current control into the open-loop system proposed in [63], making slight adjustments derived from a continuous-variable dynamic model. In this circulating current control, a P controller, referred to as active resistance, is employed to regulate the internal arm voltage, alongside a feedforward path for the circulating current reference. Additionally, circulating currents can be managed to mitigate voltage fluctuations in the floating DC capacitors by introducing

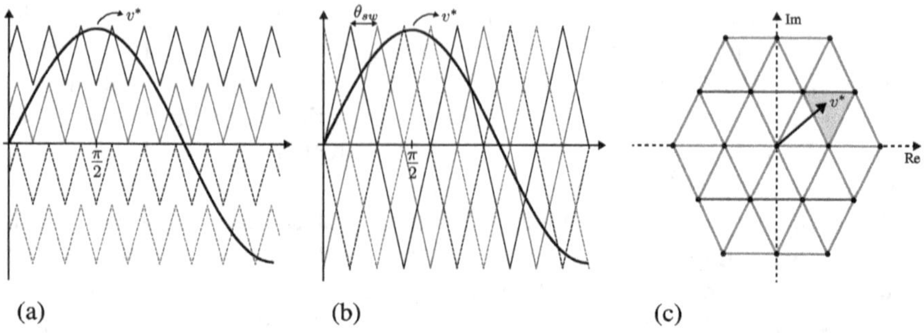

Figure 4.11 High-frequency modulation techniques. (a) LSC-PWM technique. (b) PSC-PWM technique. (c) SV-PWM technique.

zero-sequence components [37–39,48–51]. As investigated in [39], a trade-off exists between reducing voltage fluctuations and peak current. Nevertheless, the effectiveness of these control strategies hinges on their ability to operate effectively across a broad frequency range for both the DSCC and the TSBC topologies.

4.4 Modulation techniques

High-frequency switching techniques can be classified into two main categories: carrier-based and space-vector (SV) techniques, as shown in Figure 4.11. Among the MMCC family, level-shifted carrier (LSC) and phase-shifted carrier (PSC) techniques are the most widely used carrier-based PWM techniques [21,85]. These techniques are popular due to their ease of implementation. Both LSC- and PSC-PWM techniques are based on multiple carrier arrangements.

4.4.1 High switching frequency techniques

The LSC-PWM technique involves stacking m carriers on top of each other, with each voltage level represented by one carrier, as shown in Figure 4.11(a). However, this technique leads to an uneven power distribution across the submodule terminals. To address this issue, several modifications have been proposed, including alternating the level between contiguous carriers at each switching frequency [86], signal rotation [85,87], carrier rotation at each fundamental frequency [85,88], and continuous carrier rotation at the end of each carrier cycle [89]. For the DSCC topology, LSC-PWM produces $m+1$ voltage levels [58], and the number of voltage levels can be increased up to $2m+1$ by controlling the number of inserted submodules Non or the number of on-state submodules per arm [57,72,90], and by doubling the number of carriers compared to the conventional method [90].

The conventional PSC-PWM technique, as shown in Figure 4.11(b), is based on the phase shifting of carriers and effectively reduces low-order harmonics in multilevel converters [70]. The PSC-PWM technique ensures equal power distribution among submodules by commanding all submodules with the same voltage reference v_{ref}, and phase-shifting carriers such that they are equally spaced in a single cycle [21]. To synthesize $2m+1$ pole voltage levels, m carriers are required for SxBC and TSBC topologies and $2m$ carriers are required for DSCC topology, with the phase-shifted angle θ_{sw} is defined by $180°/m$ [86,87]. However, by modifying the conventional technique and controlling the number of inserted submodules, the number of carriers can be reduced by half for the DSCC topology [91]. In the case of the DSBC topology, $2m$ carriers are still required for the implementation of PSC-PWM, but the number of voltage levels may vary based on the utilization of the DC-link voltage [46].

The LSC-PWM technique improves voltage harmonic performance and lowers semiconductor losses compared to the conventional PSC-PWM technique, despite having the same number of switching transitions [5,85,92]. Previous studies have investigated the combination of both techniques [21,93]. To maintain energy balance among the converter phases and submodules in MMCC topologies, feedback voltage-balancing control strategies are typically applied when using the conventional PSC-PWM. However, the implementation of the LSC-PWM technique is slightly more complex than the PSC-PWM, as it requires modifications to ensure equal power distribution among all submodules.

In addition to carrier-based techniques, the Space Vector Pulse Width Modulation (SV-PWM) technique is also widely used in MMCC topologies [94–96]. This technique is particularly useful when the converter does not require a large number of voltage levels. In comparison to carrier-based techniques, SV-PWM offers higher utilization of the DC-link voltage [95] and more flexibility in optimizing the switching pattern design [96], as additional switching states can be selected to reduce output voltage distortion and avoid unnecessary switching transitions [94]. The SV-PWM technique can produce a large number of switching states in multilevel converters, represented by vectors, which are obtained from the voltage reference mapping, as shown in Figure 4.11(c). However, as the level number increases, a large number of redundant state sequences are generated, making the implementation of the SV-PWM technique computationally complex. Nevertheless, research has been conducted to simplify the implementation of the SV-PWM technique [95,96].

High-frequency PWM techniques can effectively meet the total harmonic distortion (THD) level standards with low-order harmonic distortion in the output waveforms, even across the entire modulation range. The higher frequency of the carrier allows the harmonic sidebands to be shifted to higher frequencies, making it easier to filter out with passive filters [87]. However, the increased number of commutations per cycle results in significant switching losses. To mitigate this, the zero-sequence voltage injection technique has been implemented in high-frequency PWM techniques for three-phase systems, extending the linearity range of voltage utilization and improving the output waveform quality while reducing switching losses. Despite these benefits, high-frequency PWM techniques are only suitable for MMCC topologies with a low number of submodules per arm, typically less than ten, as THD is not the major issue for large-scale MMCC topologies.

Instead, the computational effort and algorithm complexity of these PWM techniques, as well as the voltage balancing strategies, must be considered. For this reason, low switching frequency techniques have been implemented to take advantage of the high modularity degree of MMCC topologies in synthesizing good quality output waveforms with minimal switching commutation, ultimately improving the overall efficiency and reliability of the converter.

4.4.2 Low switching frequency techniques

The implementation of low switching frequency techniques in large-scale MMCC topologies has proven to be an attractive approach for achieving low THD in output waveforms, which are associated with uniform sampling frequency, and semiconductor losses that are closer to those of thyristor technology [64]. This leads to an improvement in the overall efficiency and reliability of the converter. In high-power applications, low switching frequencies are defined as those below 1 kHz [15]. As a result, low-frequency switching techniques can be classified into two categories: those based on the fundamental frequency, such as nearest vector control (NVC) and nearest level control (NLC), and those with a frequency above the fundamental frequency but below 1 kHz, such as spatial harmonic elimination and its variations.

The selective harmonic elimination technique has been applied in traditional voltage-source converters to eliminate low-order harmonics by determining the switching angles (α_m) per quarter of the fundamental period [15], as shown in Figure 4.12(a). These angles are pre-calculated and stored in lookup tables for different modulation indices, which is why this technique is often referred to as the programmed technique. In MMCC topologies with full-bridge submodules, the output waveform synthesized by the SHE technique typically has a staircase shape, with each submodule assigned a switching angle at the fundamental frequency, resulting in $2m+1$ pole voltage levels [1,15]. To further improve the THD levels, including higher-order harmonics, the active harmonic elimination (AHE) technique was proposed in [97], in which new switching angles were added at high-order harmonic frequencies. For the DSCC topology, SHE techniques have also been

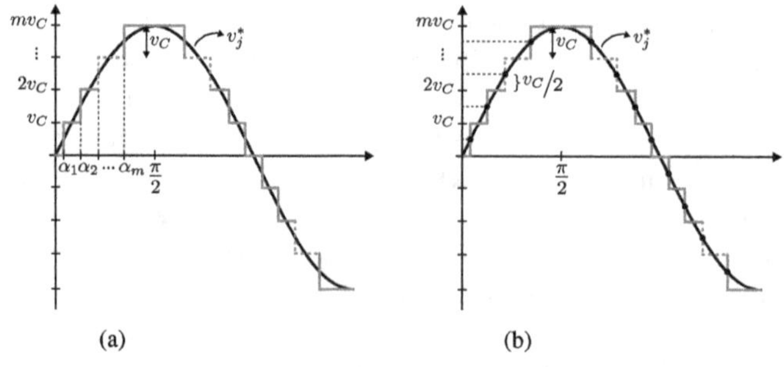

Figure 4.12 Low-frequency modulation techniques. (a) SHE technique. (b) NLC technique.

developed at the fundamental frequency [65]. Konstantinou *et al.* [75] proposed a SHE technique that takes into account a large number (up to 17) of switching angles per quarter-wave, resulting in high-quality output waveforms. The number of inserted submodules per arm can also affect the number of pole voltage levels achievable by the DSCC, with $m+1$ or $2m+1$ levels possible, leading to the presence or absence of high-frequency components in the DC-link voltage, respectively.

The selective harmonic elimination technique and its variants have been applied to reduce the THD levels in the output waveforms of MMCC topologies. However, the technique requires pre-calculation of a large number of non-linear equations, which increases the computational complexity and results in a slow dynamic response, making it unsuitable for practical implementation. The nearest vector control technique, a low-frequency version of the SV-PWM, has been proposed by Rodríguez *et al.* [94]. However, finding the nearest vector becomes more challenging in large-scale MMCC topologies [15]. As a result, the nearest level control has been considered the most efficient technique for large-scale MMCC topologies operating at the fundamental frequency [64,66], as shown in Figure 4.12(b). The NLC technique is based on a simple mathematical operation in the time domain, producing pulse patterns by finding the nearest voltage level. Due to its simplicity, several studies have implemented the NLC in combination with submodule voltage sorting algorithms to drive the DSCC [64,66,73,98]. This enables the NLC technique to determine the number of on-state submodules needed to generate the desired output voltage level. However, as the number of submodules increases, the sampling frequency of the digital implementation must be taken into consideration. If the sampling frequency is not high enough to ensure that all submodules can produce the required voltage level, the voltage step will increase, resulting in a degraded output voltage waveform quality [64]. Similarly to the SHE technique, the NLC technique can synthesize $m+1$ and $2m+1$ pole voltage levels with staircase-shaped waveforms for MMCC topologies based on full-bridge and half-bridge submodules.

4.5 Operational issues in modular multilevel cascade

4.5.1 Fault-tolerant operation

Fault tolerance is fundamental to increase system reliability. In MMCC topologies, faults can be classified into two types: external and internal. External faults can arise on either the AC or DC side, especially in converters with a shared DC-link. On the other hand, internal faults can be caused by faults in the submodules or short circuits in the floating DC capacitors.

4.5.1.1 External faults

In the event of AC side short circuits, several studies have explored MMCC topologies to maintain operation even during asymmetrical grid fault conditions. These topologies have demonstrated excellent fault ride-through (FRT) capability without incurring additional hardware costs [32,40,58,83]. The rapid interruption of DC fault current is crucial in the case of a DC-link short-circuit to avoid permanent damage to the system. Three commonly employed solutions for extinguishing DC fault current include opening the AC side circuit breakers, opening the DC side circuit breakers, or utilizing advanced submodule topologies. Opening the AC side circuit breakers is typically considered the least favorable solution due to the potential for delays that could result in component damage [5,99]. In the case of half-bridge submodules, the initial step in DC-link fault protection is to bypass all submodules once the fault is detected to prevent the converter from feeding the DC fault current. The basic solution for this procedure is to connect thyristors to protect the free-wheeling diodes from over-current. However, this setup establishes a one-way path for the DC fault current, making rapid interruption challenging as the DC arc cannot be extinguished until the AC side circuit breaker is opened. To address this issue, Li *et al.* [99] proposed the use of double-thyristor switches to provide a bidirectional path for the DC non-permanent fault current, eliminating the diode-bridge rectifier mode. This allows the natural extinguishing of the DC arc and allows the converter to be restarted after a few fundamental cycles. Although the double-thyristor strategy is a simple, reliable, and cost-effective protection scheme, the DC fault current cannot be extinguished faster than using DC side circuit breakers or a clamp-double submodule [5,99]. On the other hand, using a DC side circuit breaker can quickly extinguish the DC fault current, but this technology is still under development [18,99].

Advanced submodule topologies can provide high-performance fault-tolerant operation [18,23,68]. Topologies that generate bipolar voltages, such as full-bridge submodules and cross-connected cells submodules, can impose reverse voltage polarity to block AC side currents during a DC short-circuit. However, in applications like HVDC systems, reverse voltage has no practical use in normal operation and the increased number of semiconductor devices raises the design cost and losses (approximately 70% more than half-bridge submodules) [18]. On the other hand, the clamp-double submodule operates as a half-bridge submodule with an additional current path. However, during a DC short-circuit, it operates as a full-bridge submodule, generating reverse voltage polarity through two capacitors to clear the DC fault current. Compared to the half-bridge submodule, the clamp-double submodule topology has around 35% more losses due to the extra current path [18]. Nonetheless, the clamp-double submodule is considered one of the best submodule topologies for HVDC systems, as it provides high reliability and low losses. While the clamp-double submodule has more semiconductor devices and slightly higher conduction losses compared to the half-bridge submodule, it is important to consider the overall system cost, as using half-bridge submodules in current technology may require a bulky and expensive DC side circuit breaker to achieve the same level of performance in DC fault clearance as the clamp-double submodule.

4.5.1.2 Internal faults

To maintain operation in the presence of faulty submodules, the system can compensate using redundant submodules [71,72] or software modifications [88,91].

Redundancy is an effective method for maintaining the functionality of MMCC topologies without decreasing the output power quality. The basic approach to using redundant submodules involves designing the converter with additional submodules that are pre-charged and in standby mode. When a failure occurs, a highly reliable high-speed switch quickly bypasses the faulty submodule and replaces it with a redundant submodule. Son *et al.* [71] explored the use of spare submodules to ensure fail-safe operation and enhance the redundancy process. In their approach, spare submodules are not only bypassed like redundant ones; they are also disconnected from the system, allowing them to discharge. Spare submodules are connected to the system only when the number of faulty submodules exceeds the redundancy level, potentially causing voltage transients due to the capacitor charging process. Konstantinou *et al.* [100] proposed using redundant submodules that do not remain bypassed but actively participate in normal operation. In this approach, submodules in the same arm are swapped and selected by a voltage algorithm, resulting in fault-tolerant operation and reduced ripple in the floating DC capacitors. Another method to achieve fail-safe operation in case of submodule failure involves software modifications in the control and modulation strategies. Maharjan *et al.* [101] introduced a technique based on fundamental-frequency zero-sequence voltage injection, known as neutral shift. This technique adds the zero-sequence component to each healthy submodule to compensate for the operation of the faulty submodule and maintain balanced line voltages. Shen *et al.* [88] proposed modifying the LSC-PWM technique based on carrier rotation to sustain system operation even in the presence of submodule failure. While hardware-based redundant solutions enhance system reliability by offering backup, they also increase costs. On the other hand, software solutions may require voltages higher than the rated value and can degrade the quality of the phase voltage waveform.

4.5.2 Floating DC capacitor pre-charging procedures

Before initiating the operation of MMCC topologies, the floating DC capacitors are usually in a de-energized state. To prevent large inrush currents during startup, it is necessary to pre-charge the capacitors to their nominal or near-nominal values. The same consideration applies after a DC fault (for topologies with a DC-link) during the re-charging process. Various pre-charging or black-start solutions have been devised to ensure a rapid, smooth, and fail-safe startup [4,68,102,103]. A straightforward pre-charging procedure utilizes an AC side current-limiting resistor in series with a magnetic contactor and an AC circuit breaker. The floating DC capacitors are charged to a value close to their rated voltage through the supply currents that pass through the current-limiting resistor. Once the capacitors are pre-charged, the AC circuit breaker is activated, and the magnetic contactor bypasses the current-limiting resistor. Although this method is simple, it lacks control over the charging time, is inefficient, and can result in bulky resistors in high-power systems [9]. Another common solution involves the use of an auxiliary DC voltage source and a DC circuit breaker to charge the capacitors one by one in a staggered sequence, gradually reaching their rated values over a constant time period [4]. However, this solution can be cost-prohibitive due to the requirement for an expensive DC circuit breaker [104]. The start-up procedure based on current-limiting resistors has been expanded to include different controlled schemes [68,102,104]. Li *et al.* [104] proposed a closed-loop pre-charging control that maintains a constant charging current from either the DC-side or AC-side main voltage. This approach eliminates the need for an auxiliary power supply, reduces the charging time, and prevents inrush currents.

4.6 Conclusion

This chapter provides an in-depth overview of modular multilevel cascade converters, offering insights into their fundamental principles and addressing key challenges. Modular multilevel cascade converters represent a notable advancement in the field of power electronics, providing a versatile and scalable solution. These distinctive characteristics of MMCC topologies have garnered significant attention across all application domains, particularly those requiring high-quality output power. Notably, the star- and delta-connected cascade H-bridge topologies have found widespread utility as active power line conditioners in high-voltage transmission grids and distribution systems, often referred to as custom power devices. Among active power line conditioners, the STATCOM, based on these topologies, stands as the most widely adopted device due to its straightforward design and the significant advantage of not requiring isolated DC sources. The core components and operational principles of MMCCs were discussed, with a focus on the role of submodules and floating DC capacitors in producing multilevel voltage waveforms. Several MMCC topologies are introduced, including the double-star cascade converter, double-star bridge converter, single-star bridge converter, and Triple-Star Bridge Converter, each with its distinct characteristics and application domains. Furthermore, the operational nuances of MMCCs are discussed, addressing the challenges and presenting the solutions to enhance their performance. Circulating current control is examined in detail, underscoring its importance in regulating circulating currents and mitigating undesirable harmonics to improve converter efficiency and stability. High-frequency and low-frequency modulation techniques are elucidated, providing valuable insights into strategies for minimizing harmonic distortions and switching losses while meeting specific application requirements. Fault-tolerant operation, a critical aspect of power electronics, is thoroughly discussed. Both external and internal fault scenarios are explored, elucidating strategies for ensuring fault ride-through capability, rapidly interrupting DC fault currents, and maintaining system integrity even in the presence of submodule failures. These fault-tolerant mechanisms significantly contribute to the reliability and robustness of MMCC systems. Additionally, pre-charging procedures for the floating DC capacitors are scrutinized, a crucial step in the startup process of MMCC topologies. Various pre-charging methods, ranging from basic resistor-based techniques to sophisticated closed-loop control systems, are examined to assist in selecting the most suitable approach tailored to specific application requirements. As research and development in MMCC technology continue to progress, these converters are expected to play a pivotal role in addressing the challenges and opportunities presented by the dynamic energy landscape of the future.

References

[1] Peng FZ, Lai JS, McKeever JW, *et al.* A multilevel voltage-source inverter with separate DC sources for static var generation. *IEEE Transactions on Industry Applications*. 1996;32(5):1130–1138.

[2] Lai JS and Peng FZ. Multilevel converters-a new breed of power converters. *IEEE Transactions on Industry Applications*. 1996;32(3):509–517.

[3] Marquardt R, Lesnicar A, Hildinger J, *et al.* Modulares stromrichterkonzept für netzkupplungsanwendung bei hohen spannungen. ETG-Fachtagung, Bad Nauheim, Germany. 2002;114.

[4] Lesnicar A and Marquardt R. An innovative modular multilevel converter topology suitable for a wide power range. In: 2003 IEEE Bologna Power Tech Conference Proceedings, vol. 3. IEEE; 2003. p. 6.

[5] Debnath S, Qin J, Bahrani B, *et al.* Operation, control, and applications of the modular multilevel converter: A review. *IEEE Transactions on Power Electronics*. 2014;30(1):37–53.

[6] Nami A, Liang J, Dijkhuizen F, *et al.* Modular multilevel converters for HVDC applications: Review on converter cells and functionalities. *IEEE Transactions on Power Electronics*. 2014;30(1):18–36.

[7] Alskran F and Simões MG. Multilevel current source converter-based STATCOM suitable for medium-voltage applications. *IEEE Transactions on Power Delivery*. 2020;36(2):1222–1232.

[8] Alskran F and Simões MG. Current balancing algorithm for three-phase multilevel current source inverters. *Energies*. 2020;13(4):860.

[9] Perez MA, Bernet S, Rodriguez J, *et al.* Circuit topologies, modeling, control schemes, and applications of modular multilevel converters. *IEEE Transactions on Power Electronics*. 2014;30(1):4–17.

[10] Perez MA, Ceballos S, Konstantinou G, *et al.* Modular multilevel converters: Recent achievements and challenges. IEEE Open *Journal of the Industrial Electronics Society*. 2021;2:224–239.

[11] Akagi H. Classification, terminology, and application of the modular multilevel cascade converter (MMCC). *IEEE Transactions on Power Electronics*. 2011;26(11):3119–3130.

[12] Dekka A, Wu B, Fuentes RL, *et al.* Evolution of topologies, modeling, control schemes, and applications of modular multilevel converters. *IEEE Journal of Emerging and Selected Topics in Power Electronics*. 2017;5(4):1631–1656.

[13] Kurtoğlu M, Eroğlu F, Arslan AO, *et al.* Recent contributions and future prospects of the modular multilevel converters: A comprehensive review. *International Transactions on Electrical Energy Systems*. 2019;29(3):e2763.

[14] Luna BEdOB, Jacobina CB, and Oliveira AC. Internal Energy Balance of a Modular Multilevel Cascade Converter Based on Chopper-Cells with Distributed Energy Resources for Grid-Connected Photovoltaic Systems. *IEEE Transactions on Industry Applications*. 2022;59(2):1935–1943.

[15] Wang Y, Aksoz A, Geury T, *et al.* A review of modular multilevel converters for stationary applications. *Applied Sciences*. 2020;10(21):7719.

[16] Sau S and Fernandes BG. Modular multilevel converter based variable speed drive with reduced capacitor ripple voltage. *IEEE Transactions on Industrial Electronics*. 2018;66(5):3412–3421.

[17] Marquardt R. Modular Multilevel Converter: An universal concept for HVDC-Networks and extended DC-Bus-applications. In: The 2010 International Power Electronics Conference-ECCE ASIA-. IEEE; 2010. p. 502–507.

[18] Ahmed N, Norrga S, Nee HP, *et al.* HVDC SuperGrids with modular multilevel converters—The power transmission backbone of the future. In: International Multi-Conference on Systems, Signals & Devices. IEEE; 2012. p. 1–7.

[19] Solas E, Abad G, Barrena JA, *et al.* Modular multilevel converter with different submodule concepts—Part II: Experimental validation and comparison for HVDC application. *IEEE Transactions on Industrial Electronics*. 2012;60(10):4536–4545.

[20] Kurtoğlu M, Eroğlu F, Arslan AO, *et al.* Recent contributions and future prospects of the modular multilevel converters: A comprehensive review. *International Transactions on Electrical Energy Systems*. 2019;29(3):e2763.

[21] Sahoo AK, Otero-De-Leon R, Chandrasekaran V, *et al.* New 3-level submodules for a modular multilevel converter based HVDC system with advanced features. In: IECON 2013-39th Annual Conference of the IEEE Industrial Electronics Society. IEEE; 2013. p. 6269–6274.

[22] Rohner S, Hiller M, and Sommer R. A new highly modular medium voltage converter topology for industrial drive application. In: Power electronics and applications, EPE 13th European conference; 2009.

[23] Nami A, Wang L, Dijkhuizen F, *et al.* Five level cross connected cell for cascaded converters. In: 2013 15th European conference on power electronics and applications (EPE). IEEE; 2013. p. 1–9.

[24] Riar BS, Madawala UK, and Thrimawithana DJ. Analysis and control of a three-phase modular multi-level converter based on inductive power transfer technology (M2LC-IPT). In: 2013 IEEE International Conference on Industrial Technology (ICIT). IEEE; 2013. p. 475–480.

[25] Thitichaiworakorn N, Hagiwara M, and Akagi H. A medium-voltage large wind turbine generation system using an AC/AC modular multilevel cascade converter. *IEEE Journal of Emerging and Selected Topics in Power Electronics*. 2015;4(2):534–546.

[26] Perez MA, Lizana R, Azocar C, *et al.* Modular multilevel cascaded converter based on current source H-bridges cells. In: IECON 2012-38th Annual Conference on IEEE Industrial Electronics Society. IEEE; 2012. p. 3443–3448.

[27] Peng FZ, McKeever JW, and Adams DJ. A power line conditioner using cascade multilevel inverters for distribution systems. *IEEE Transactions on Industry Applications*. 1998;34(6):1293–1298.

[28] Junling C, Yaohua L, Ping W, *et al.* A closed-loop selective harmonic compensation with capacitor voltage balancing control of cascaded multilevel inverter for high-power active power filters. In: 2008 IEEE Power Electronics Specialists Conference. IEEE; 2008. p. 569–573.

[29] Peng FZ, Ott GW, and Adams DJ. Harmonic and reactive power compensation based on the generalized instantaneous reactive power theory for three-phase four-wire systems. *IEEE Transactions on Power Electronics*. 1998;13(6):1174–1181.

[30] Cupertino AF, Farias JVM, Pereira HA, *et al.* Comparison of DSCC and SDBC modular multilevel converters for STATCOM application during negative sequence compensation. *IEEE Transactions on Industrial Electronics*. 2018;66(3):2302–2312.

[31] Song Q and Liu W. Control of a cascade STATCOM with star configuration under unbalanced conditions. *IEEE Transactions on Power electronics*. 2009;24(1):45–58.

[32] Ota JIY, Shibano Y, and Akagi H. Low-voltage-ride-through (LVRT) capability of a phase-shifted-PWM STATCOM using the modular multilevel cascade converter based on single-star bridge-cells (MMCC-SSBC). In: *2013 IEEE Energy Conversion Congress and Exposition*. IEEE; 2013. p. 3062–3069.

[33] Hagiwara M, Maeda R, and Akagi H. Negative-sequence reactive-power control by a PWM STATCOM based on a modular multilevel cascade converter (MMCC-SDBC). *IEEE Transactions on Industry Applications*. 2011;48(2):720–729.

[34] Konidaris N, Karunathilake D, Vilathgamuwa M, *et al.* Modular multilevel series parallel converter prototype design for li-ion battery management systems. In: 2021 *IEEE PES Innovative Smart Grid Technologies-Asia (ISGT Asia)*. IEEE; 2021. p. 1–5.

[35] Li N, Gao F, Hao T, *et al.* SOH balancing control method for the MMC battery energy storage system. *IEEE Transactions on Industrial Electronics*. 2017;65(8):6581–6591.

[36] Du S, Wu B, Tian K, *et al.* An active cross-connected modular multilevel converter (AC-MMC) for a medium-voltage motor drive. *IEEE Transactions on Industrial Electronics*. 2016;63(8):4707–4717.

[37] Antonopoulos A, Ängquist L, Norrga S, *et al.* Modular multilevel converter ac motor drives with constant torque from zero to nominal speed. *IEEE Transactions on Industry Applications*. 2013;50(3):1982–1993.

[38] Fang J, Blaabjerg F, Liu S, *et al.* A review of multilevel converters with parallel connectivity. *IEEE Transactions on Power Electronics*. 2021;36 (11):12468–12489.

[39] Kawamura W, Chiba Y, and Akagi H. A broad range of speed control of a permanent magnet synchronous motor driven by a modular multilevel

TSBC converter. *IEEE Transactions on Industry Applications*. 2017;53 (4):3821–3830.

[40] Ota JIY, Shibano Y, Niimura N, *et al.* A phase-shifted-PWM D-STATCOM using a modular multilevel cascade converter (SSBC)—Part I: Modeling, analysis, and design of current control. *IEEE Transactions on Industry Applications*. 2014;51(1):279–288.

[41] H. M. Pirouz and M. T. Bina, A transformerless medium-voltage STATCOM topology based on extended modular multilevel converters. *IEEE Transactions on Power Electronics*, 2011;26(5):1534–1545. doi: 10.1109/TPEL.2010.2085088.

[42] Gao F, Niu D, Tian H, *et al.* Control of parallel-connected modular multilevel converters. *IEEE Transactions on Power Electronics*. 2014;30 (1):372–386.

[43] Glinka M and Marquardt R. A new AC/AC multilevel converter family. *IEEE Transactions on Industrial Electronics*. 2005;52(3):662–669.

[44] Pan X, Zhang L, Li Y, *et al.* Modulated model predictive control with branch and band scheme for unbalanced load compensation by MMCC-STATCOM. *IEEE Transactions on Power Electronics*. 2022;37(8):8948–8962.

[45] Thitichaiworakorn N, Hagiwara M, and Akagi H. A single-phase to three-phase direct AC/AC modular multilevel cascade converter based on double-star bridge-cells (MMCC-DSBC). In: 2013 1st International Future Energy Electronics Conference (IFEEC). IEEE; 2013. p. 476–481.

[46] Thitichaiworakorn N, Hagiwara M, and Akagi H. Experimental verification of a modular multilevel cascade inverter based on double-star bridge cells. *IEEE Transactions on Industry Applications*. 2013;50(1):509–519.

[47] Erickson RW and Al-Naseem OA. A new family of matrix converters. In: IECON'01. 27th Annual Conference of the IEEE Industrial Electronics Society (Cat. No. 37243). vol. 2. IEEE; 2001. p. 1515–1520.

[48] Akagi H. Multilevel converters: Fundamental circuits and systems. *Proceedings of the IEEE*. 2017;105(11):2048–2065.

[49] Kawamura W, Hagiwara M, and Akagi H. Control and experiment of a modular multilevel cascade converter based on triple-star bridge cells. *IEEE Transactions on Industry Applications*. 2014;50(5):3536–3548.

[50] Kawamura W, Hagiwara M, and Akagi H. A broad range of frequency control for the modular multilevel cascade converter based on triple-star bridge-cells (MMCC-TSBC). In: *2013 IEEE Energy Conversion Congress and Exposition*. IEEE; 2013. p. 4014–4021.

[51] Ilves K, Bessegato L, and Norrga S. Comparison of cascaded multilevel converter topologies for AC/AC conversion. In: 2014 International Power Electronics Conference (IPEC-Hiroshima 2014-ECCE ASIA). IEEE; 2014. p. 1087–1094.

[52] Baruschka L and Mertens A. A new three-phase AC/AC modular multilevel converter with six branches in hexagonal configuration. *IEEE Transactions on Industry Applications*. 2013;49(3):1400–1410.

[53] Hagiwara M, Maeda R, and Akagi H. Control and analysis of the modular multilevel cascade converter based on double-star chopper-cells (MMCC-DSCC). *IEEE Transactions on Power Electronics*. 2010;26(6):1649–1658.

[54] Akagi H, Inoue S, and Yoshii T. Control and performance of a transformerless cascade PWM STATCOM with star configuration. *IEEE Transactions on Industry Applications*. 2007;43(4):1041–1049.

[55] Ota JIY, Shibano Y, Niimura N, *et al.* Current control of a phase-shifted-PWM STATCOM using the modular multilevel cascade converter based on single-star bridge-cells (MMCC-SSBC). In: *2013 IEEE Energy Conversion Congress and Exposition*. IEEE; 2013. p. 420–427.

[56] Li Z, Lizana R, Sha S, *et al.* Module implementation and modulation strategy for sensorless balancing in modular multilevel converters. *IEEE Transactions on Power Electronics*. 2018;34(9):8405–8416.

[57] Rohner S, Bernet S, Hiller M, *et al.* Modulation, losses, and semiconductor requirements of modular multilevel converters. *IEEE Transactions on Industrial Electronics*. 2009;57(8):2633–2642.

[58] Patro SK and Shukla A. Modular directed series multilevel converter for HVDC applications. *IEEE Transactions on Industry Applications*. 2020;56 (2):1618–1630.

[59] Jin Y, Xiao Q, Dong C, *et al.* A novel submodule voltage balancing scheme for modular multilevel cascade converter—Double-star chopper-cells (MMCC-DSCC) based STATCOM. *IEEE Access*. 2019;7:83058–83073.

[60] Fujii K and De Doncker RW. A novel DC-link voltage control of PWM-switched cascade cell multi-level inverter applied to STATCOM. In: Fortieth IAS Annual Meeting. Conference Record of the 2005 Industry Applications Conference, 2005. vol. 2. IEEE; 2005. p. 961–967.

[61] Hagiwara M, Maeda R, and Akagi H. Negative-sequence reactive-power control by the modular multilevel cascade converter based on double-star chopper-cells (MMCC-DSCC). In: *2010 IEEE Energy Conversion Congress and Exposition*. IEEE; 2010. p. 3949–3954.

[62] Hagiwara M and Akagi H. Control and experiment of pulsewidth-modulated modular multilevel converters. *IEEE Transactions on Power Electronics*. 2009;24(7):1737–1746.

[63] Angquist L, Antonopoulos A, Siemaszko D, *et al.* Open-loop control of modular multilevel converters using estimation of stored energy. *IEEE Transactions on Industry Applications*. 2011;47(6):2516–2524.

[64] Yang S, Shao Z, Zheng W, *et al.* Mitigation of background harmonics effect on MMC controller based on a novel coordinate transformation technique. *IEEE Access*. 2019;7:167113–167126.

[65] Ilves K, Antonopoulos A, Norrga S, *et al.* A new modulation method for the modular multilevel converter allowing fundamental switching frequency. *IEEE Transactions on Power Electronics*. 2012;27(8):3482–3494.

[66] Samajdar D, Bhattacharya T, and Dey S. A reduced switching frequency sorting algorithm for modular multilevel converter with circulating current

suppression feature. *IEEE Transactions on Power Electronics*. 2019;34 (11):10480–10491.

[67] Jia G, Chen M, Tang S, *et al.* A modular multilevel converter with active power filter for submodule capacitor voltage ripples and power losses reduction. *IEEE Transactions on Power Electronics*. 2020;35(11):11401–11417.

[68] Xue Y, Xu Z, and Tang G. Self-start control with grouping sequentially precharge for the C-MMC-based HVDC system. *IEEE Transactions on Power Delivery*. 2013;29(1):187–198.

[69] Ilves K, Harnefors L, Norrga S, *et al.* Predictive sorting algorithm for modular multilevel converters minimizing the spread in the submodule capacitor voltages. *IEEE Transactions on Power Electronics*. 2014;30(1):440–449.

[70] Tu Q, Xu Z, and Xu L. Reduced switching-frequency modulation and circulating current suppression for modular multilevel converters. *IEEE Transactions on Power Delivery*. 2011;26(3):2009–2017.

[71] Son GT, Lee HJ, Nam TS, *et al.* Design and control of a modular multilevel HVDC converter with redundant power modules for noninterruptible energy transfer. *IEEE Transactions on Power Delivery*. 2012;27(3):1611–1619.

[72] Sochor P, Tan NML, and Akagi H. Low-voltage-ride-through control of a modular multilevel single-delta bridge-cell (SDBC) inverter for utility-scale photovoltaic systems. *IEEE Transactions on Industry Applications*. 2018;54 (5):4739–4751.

[73] Du S, Liu J, and Liu T. Modulation and closed-loop-based DC capacitor voltage control for MMC with fundamental switching frequency. *IEEE Transactions on Power Electronics*. 2014;30(1):327–338.

[74] Li Z, Wang P, Zhu H, *et al.* An improved pulse width modulation method for chopper-cell-based modular multilevel converters. *IEEE Transactions on Power Electronics*. 2012;27(8):3472–3481.

[75] Konstantinou G, Ciobotaru M, and Agelidis V. Selective harmonic elimination pulse-width modulation of modular multilevel converters. *IET Power Electronics*. 2013;6(1):96–107.

[76] Deng F and Chen Z. A control method for voltage balancing in modular multilevel converters. *IEEE Transactions on Power Electronics*. 2013;29(1):66–76.

[77] Tu Q, Xu Z, Huang H, *et al.* Parameter design principle of the arm inductor in modular multilevel converter based HVDC. In: 2010 International Conference on Power System Technology. IEEE; 2010. p. 1–6.

[78] Ilves K, Norrga S, Harnefors L, *et al.* Analysis of arm current harmonics in modular multilevel converters with main-circuit filters. In: International Multi-Conference on Systems, Signals & Devices. IEEE; 2012. p. 1–6.

[79] Li Z, Wang P, Chu Z, *et al.* An inner current suppressing method for modular multilevel converters. *IEEE Transactions on Power Electronics*. 2013;28(11):4873–4879.

[80] Zhang M, Huang L, Yao W, *et al.* Circulating harmonic current elimination of a CPS-PWM-based modular multilevel converter with a plug-in repetitive controller. *IEEE Transactions on Power Electronics*. 2013;29(4):2083–2097.

[81] He L, Zhang K, Xiong J, *et al.* A repetitive control scheme for harmonic suppression of circulating current in modular multilevel converters. *IEEE Transactions on Power Electronics*. 2014;30(1):471–481.

[82] Harnefors L, Antonopoulos A, Norrga S, *et al.* Dynamic analysis of modular multilevel converters. *IEEE Transactions on Industrial Electronics*. 2012;60(7):2526–2537.

[83] Isik S, Alharbi M, and Bhattacharya S. An optimized circulating current control method based on PR and PI controller for MMC applications. *IEEE Transactions on Industry Applications*. 2021;57(5):5074–5085.

[84] Lyu J, Zhang X, Cai X, *et al.* Harmonic state-space based small-signal impedance modeling of a modular multilevel converter with consideration of internal harmonic dynamics. *IEEE Transactions on Power Electronics*. 2018;34(3):2134–2148.

[85] Sedghi S, Dastfan A, and Ahmadyfard A. A new multilevel carrier based pulse width modulation method for modular multilevel inverter. In: 8th International Conference on Power Electronics-ECCE Asia. IEEE; 2011. p. 1432–1439.

[86] Angulo M, Lezana P, Kouro S, *et al.* Level-shifted PWM for cascaded multilevel inverters with even power distribution. In: 2007 IEEE Power Electronics Specialists Conference. IEEE; 2007. p. 2373–2378.

[87] Konstantinou GS and Agelidis VG. Performance evaluation of half-bridge cascaded multilevel converters operated with multicarrier sinusoidal PWM techniques. In: 2009 4th IEEE Conference on Industrial Electronics and Applications. IEEE; 2009. p. 3399–3404.

[88] Shen K, Xiao B, Mei J, *et al.* A modulation reconfiguration based fault-tolerant control scheme for modular multilevel converters. In: 2013 Twenty-Eighth Annual IEEE Applied Power Electronics Conference and Exposition (APEC). IEEE; 2013. p. 3251–3255.

[89] Sreenivasarao D, Agarwal P, and Das B. Performance evaluation of carrier rotation strategy in level-shifted pulse-width modulation technique. *IET Power Electronics*. 2014;7(3):667–680.

[90] Amankwah EK, Clare JC, Wheeler PW, *et al.* Multi carrier PWM of the modular multilevel VSC for medium voltage applications. In: 2012 Twenty-Seventh Annual IEEE Applied Power Electronics Conference and Exposition (APEC). IEEE; 2012. p. 2398–2406.

[91] Pinheiro da Silva IRFM, Jacobina CB, Oliveira AC, *et al.* Single-phase AC–AC double-star chopper cells (DSCC) converter without common DC-link capacitor. *IEEE Transactions on Industry Applications*. 2015;51 (6):4642–4652.

[92] Shi X, Wang Z, Tolbert LM, *et al.* A comparison of phase disposition and phase shift PWM strategies for modular multilevel converters. In: *2013 IEEE Energy Conversion Congress and Exposition*. IEEE; 2013. p. 4089–4096.

[93] Naderi R and Rahmati A. Phase-shifted carrier PWM technique for general cascaded inverters. *IEEE Transactions on Power Electronics*. 2008;23 (3):1257–1269.

[94] Rodrguez J, Morán L, Correa P, *et al.* A vector control technique for medium-voltage multilevel inverters. *IEEE Transactions on Industrial Electronics*. 2002;49(4):882–888.

[95] Deng Y, Wang Y, Teo KH, *et al.* Optimized control of the modular multilevel converter based on space vector modulation. *IEEE Transactions on Power Electronics*. 2017;33(7):5697–5711.

[96] Ahmed I and Borghate VB. Simplified space vector modulation technique for seven-level cascaded H-bridge inverter. *IET Power Electronics*. 2014;7(3):604–613.

[97] Jiao N, Wang S, Ma J, *et al.* Influencing factors of harmonic coupling in modular multilevel converter. *IEEE Transactions on Applied Superconductivity*. 2021;31(8):1–4.

[98] Meshram PM and Borghate VB. A simplified nearest level control (NLC) voltage balancing method for modular multilevel converter (MMC). *IEEE Transactions on Power Electronics*. 2014;30(1):450–462.

[99] Li X, Song Q, Liu W, *et al.* Protection of nonpermanent faults on DC overhead lines in MMC-based HVDC systems. *IEEE Transactions on Power Delivery*. 2012;28(1):483–490.

[100] Konstantinou G, Pou J, Ceballos S, *et al.* Active redundant submodule configuration in modular multilevel converters. *IEEE Transactions on Power Delivery*. 2013;28(4):2333–2341.

[101] Maharjan L, Yamagishi T, Akagi H, *et al.* Fault-tolerant operation of a battery-energy-storage system based on a multilevel cascade PWM converter with star configuration. *IEEE Transactions on Power Electronics*. 2010;25(9):2386–2396.

[102] Dekka A, Wu B, and Zargari NR. Start-up operation of a modular multilevel converter with flying capacitor submodules. *IEEE Transactions on Power Electronics*. 2017;32(8):5873–5877.

[103] Ali S, Ling Z, Tian K, *et al.* Recent advancements in submodule topologies and applications of MMC. *IEEE Journal of Emerging and Selected Topics in Power Electronics*. 2020;9(3):3407–3435.

[104] Li B, Xu D, Zhang Y, *et al.* Closed-loop precharge control of modular multilevel converters during start-up processes. *IEEE Transactions on Power Electronics*. 2014;30(2):524–531.

Chapter 5

Matrix converters

Mehdi Farasat[1]

5.1 Introduction

Voltage and/or current back-to-back converters are traditionally used to interface an AC source with an AC load. An energy storage element is used to couple the DC-link of the front-end AC–DC rectifier to the back-end DC–AC inverter. A matrix converter (MC), however, enables AC–AC conversion without any intermediate energy storage element. Conventional matrix converters, known as direct matrix converters (DMC), are single-stage converters that connect an m-phase voltage source to an n-phase output load through a $m \times n$ array of bidirectional switches. On the other hand, an indirect matrix converter (IMC) requires separate stages for the voltage and current conversion. In this chapter, the most popular MC topologies along with their control and modulation strategies are presented. A brief discussion on the technological and practical issues facing MCs, and a comparative assessment of their performance with the voltage back-to-back converters is given.

5.2 Direct matrix converter

5.2.1 Circuit topology

Figure 5.1 depicts the circuit topology of a three-phase to three-phase DMC. The circuit consists of nine bidirectional switches which enable direct connection of a three-phase input source to a three-phase load. The L-C filter is employed to compensate the input current harmonics so that sinusoidal currents are drawn from the source. Furthermore, the L-C filter mitigates the overvoltages caused by the input current commutation.

5.2.2 Modulation techniques

The modulation techniques can be categorized into scalar and pulse width modulation (PWM) techniques. The PWM techniques in turn are divided into

[1]Division of Electrical and Computer Engineering, Louisiana State University, Baton Rouge, USA

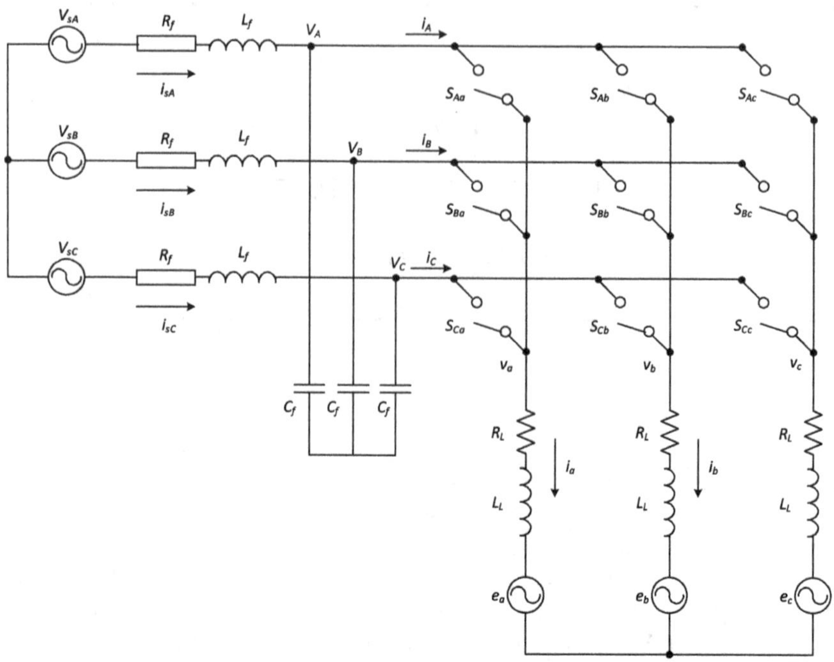

Figure 5.1 Three-phase direct matrix converter topology

carrier-based PWM and space vector modulation (SVM) methods. A detailed description of the scalar and SVM techniques is presented in the following subsections. Recently, a predictive control technique has been applied to the MCs. This technique is discussed, as well.

5.2.2.1 Scalar technique

In this method, a desired set of three-phase output voltages is synthesized from the three-phase input voltages by sequential piecewise sampling [1]

$$\begin{bmatrix} \bar{v}_a \\ \bar{v}_b \\ \bar{v}_c \end{bmatrix} = \frac{1}{T_s} \begin{bmatrix} t_{Aa} & t_{Ba} & t_{Ca} \\ t_{Ab} & t_{Bb} & t_{Cb} \\ t_{Ca} & t_{Cb} & t_{Cc} \end{bmatrix} \begin{bmatrix} V_A \\ V_B \\ V_C \end{bmatrix} \tag{5.1}$$

where $\bar{v}_j; j = \{a, b, c\}$ is the average value of the output phase voltage, T_s is the sampling period, and $t_{ij}; i \in \{A, B, C\}; j \in \{a, b, c\}$ is the time during which the switch S_{ij} is "ON" (see Figure 5.1). The following duty ratios can be defined assuming $T_s = t_{Aj} + t_{Bj} + t_{Cj}$

$$m_{Aj}(t) = \frac{t_{Aj}}{T_s} \quad m_{Bj}(t) = \frac{t_{Bj}}{T_s} \quad m_{Cj}(t) = \frac{t_{Cj}}{T_s} \tag{5.2}$$

Therefore, (5.1) can be rewritten as

$$\begin{bmatrix} \bar{v}_a \\ \bar{v}_b \\ \bar{v}_c \end{bmatrix} \bar{v}_{out}(t) = \begin{bmatrix} m_{Aa}(t) & m_{Ba}(t) & m_{Ca}(t) \\ m_{Ab}(t) & m_{Bb}(t) & m_{Cb}(t) \\ m_{Ca}(t) & m_{Cb}(t) & m_{Cc}(t) \end{bmatrix} M(t) \begin{bmatrix} V_A \\ V_B \\ V_C \end{bmatrix} v_{in}(t) \tag{5.3}$$

where $\bar{\boldsymbol{v}}_{out}(t)$ represents the average output voltage vector, $\boldsymbol{v}_{in}(t)$ is the input voltage vector, and $\boldsymbol{M}(t)$ is the transfer matrix.

Following the same procedure, the equation that describes the average input current vector, $\bar{\boldsymbol{i}}_{in}(t)$, in terms of the transfer matrix and output current vector, $i_{out}(t)$, can be derived as follows:

$$\bar{\boldsymbol{i}}_{in}(t) = \boldsymbol{M}^T(t)\boldsymbol{i}_{out}(t) \tag{5.4}$$

Here, $\boldsymbol{M}^T(t)$ is the transpose of the transfer matrix.

Consider the three-phase input voltages and the output currents as follows:

$$\boldsymbol{v}_{in}(t) = V_{in} \begin{bmatrix} \cos(\omega_{in}t) \\ \cos(\omega_{in}t - 2\pi/3) \\ \cos(\omega_{in}t - 4\pi/3) \end{bmatrix} \tag{5.5}$$

$$\boldsymbol{i}_{out}(t) = I_{out} \begin{bmatrix} \cos(\omega_{out}t + \phi_{out}) \\ \cos(\omega_{out}t + \phi_{out} - 2\pi/3) \\ \cos(\omega_{out}t + \phi_{out} - 4\pi/3) \end{bmatrix} \tag{5.6}$$

where ω_{in} and ω_{out} are the source and load frequencies, respectively, V_{in} and I_{out} are the input voltage and output current amplitudes, respectively, and ϕ_{out} is the load phase angle. The input currents can be written as

$$\boldsymbol{i}_{in}(t) = I_{in} \begin{bmatrix} \cos(\omega_{in}t + \phi_{in}) \\ \cos(\omega_{in}t + \phi_{in} - 2\pi/3) \\ \cos(\omega_{in}t + \phi_{in} - 4\pi/3) \end{bmatrix} \tag{5.7}$$

with I_{in} and ϕ_{in} denoting the input current amplitude and input displacement angle, respectively.

Assume that the output voltages are described by the following equation

$$\boldsymbol{v}_{out}(t) = qV_{in} \begin{bmatrix} \cos(\omega_{out}t) \\ \cos(\omega_{out}t - 2\pi/3) \\ \cos(\omega_{out}t - 4\pi/3) \end{bmatrix} \tag{5.8}$$

Here, q is the voltage gain of the MC.

Venturini proposed the transfer matrix that yields output voltages as (5.8) with the following elements [1]:

$$m_{ij}(t) = \frac{1}{3}\left[1 + \frac{2v_i\bar{v}_j}{V_{in}^2}\right] \tag{5.9}$$

This solution is of little practical significance since the voltage gain of the MC cannot exceed 50%. This is because the reference output voltage to be synthesized must remain within an envelope formed by the input voltages [2]. In order to enlarge the area within the input voltage envelope, common-mode voltages are added to the reference output voltages:

$$\boldsymbol{v}_{out}(t) = qV_{in}\begin{bmatrix} \cos(\omega_{out}t) - \frac{1}{6}\cos(3\omega_{out}t) + \frac{1}{2\sqrt{3}}\cos(3\omega_{in}t) \\ \cos(\omega_{out}t - 2\pi/3) - \frac{1}{6}\cos(3\omega_{out}t) + \frac{1}{2\sqrt{3}}\cos(3\omega_{in}t) \\ \cos(\omega_{out}t - 4\pi/3) - \frac{1}{6}\cos(3\omega_{out}t) + \frac{1}{2\sqrt{3}}\cos(3\omega_{in}t) \end{bmatrix} \tag{5.10}$$

The transfer matrix that yields output voltages as (5.10) is given as

$$m_{ij}(t) = \frac{1}{3}\left[1 + \frac{2v_i\bar{v}_j}{V_{in}^2} + \frac{4q}{3\sqrt{3}}\sin(\omega_{in}t + \beta_i)\sin(3\omega_{in}t)\right] \tag{5.11}$$

where $\beta_i = 0, 2\pi/3, 4\pi/3$ for $i = \{A, B, C\}$ [1]. This method yields the maximum voltage gain of $\sqrt{3}/2$. Typical waveforms of the output voltage and current are shown in Figure 5.2.

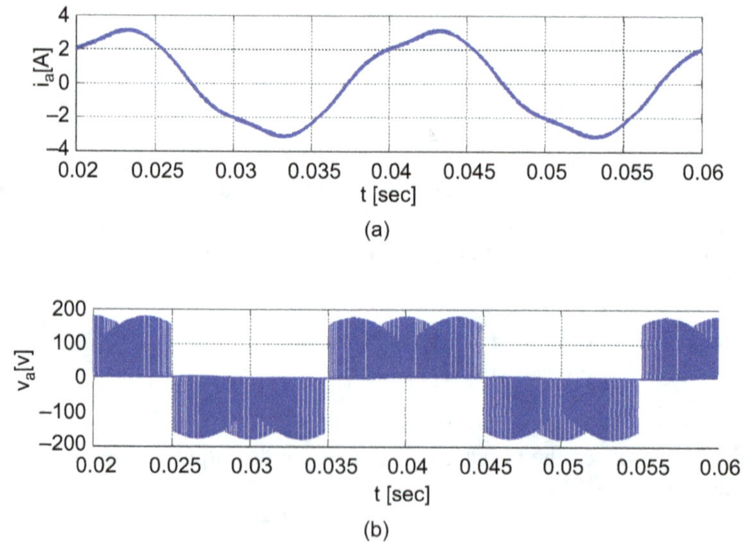

Figure 5.2 (a) Output phase current. (b) Output phase voltage of DMC with Venturini's method.

Another scalar method proposed by Roy yields the same voltage gain of $\sqrt{3}/2$. In this technique, the modulation duty ratios are expressed as follows [2]:

$$m_{ij}(t) = \frac{1}{3}\left[1 + \frac{2v_i\bar{v}_j}{V_{in}^2} + \frac{2}{3}\sin\left(\omega_{in}t + \beta_i\right)\sin\left(3\omega_{in}t\right)\right] \tag{5.12}$$

Comparing (5.11) and (5.12), it can be concluded that the only difference is that the term q is used in the Venturini method and is fixed at its maximum in Roy's method. The two methods result in the same output voltage quality except at low switching frequencies where the Venturini method is superior [2].

5.2.2.2 Space vector modulation

The MC is typically supplied from a voltage source. Therefore, the input terminals should not be short-circuited. On the other hand, the output current should never be interrupted as the load is mostly inductive. Based on these considerations, there are 27 switching states available as given in Table 5.1. The available switching states produce three groups of output vectors, as follows:

Group I: Output voltage vectors with constant amplitude and rotating at the input frequency. These vectors are generated by connecting each output phase to a different input phase. Hence, there are six vectors in this group.

Group II: Output stationary vectors with variable amplitude. This group can be classified into three subsets with each subset consisting of six vectors which are generated by connecting two output phases to the same input phase.

Group III: Zero output vectors. This group consists of three vectors which are generated by simultaneous connection of the three-phase output phases to the same input phase.

In the SVM, only the Group II and III vectors are used. The SVM is based on the indirect transfer function approach which emulates the back-to-back voltage source rectification and inversion, shown in Figure 5.3 [3]. V_{pn} represents a fictitious DC link voltage. In the following subsections, voltage source rectification (VSR) and voltage source inversion (VSI) are reviewed, and the steps to combine those methods for control of the MC are explained.

VSR

Based on the available switching states, there are six active and two zero current vectors, shown in Figure 5.4. The location of the reference vector within one of the 60° sectors defines the framing vectors I_λ and I_δ. For example, in Figure 5.4 the current reference angle is in the first sector.

The reference current vector is synthesized from the adjacent vectors and a zero vector. Denoting the local angle of the reference current by θ_R, the duty ratios of the active, d_λ, d_δ, and zero, d_{0R}, vectors are calculated as

$$d_\lambda = m_c \sin\left(\frac{\pi}{3} - \theta_R\right) \tag{5.13}$$

$$d_\delta = m_c \sin\left(\theta_R\right) \tag{5.14}$$

Table 5.1 Switching states of the DMC

	No.	a	b	c	S_{Aa}	S_{Ab}	S_{Ac}	S_{Ba}	S_{Bb}	S_{Bc}	S_{Ca}	S_{Cb}	S_{Cc}	v_{ab}	v_{bc}	v_{ca}	i_A	i_B	i_C
Group III	1	*A*	*A*	*A*	1	0	0	1	0	0	1	0	0	0	0	0	0	0	0
	2	*B*	*B*	*B*	0	1	0	0	1	0	0	1	0	0	0	0	0	0	0
	3	*C*	*C*	*C*	0	0	1	0	0	1	0	0	1	0	0	0	0	0	0
Group I	4	*A*	*C*	*C*	1	0	0	0	0	1	0	0	1	$-V_{CA}$	0	V_{CA}	i_a	0	$-i_a$
	5	*B*	*C*	*C*	0	1	0	0	0	1	0	0	1	V_{BC}	0	$-V_{BC}$	0	i_a	$-i_a$
	6	*B*	*A*	*A*	0	1	0	1	0	0	1	0	0	$-V_{AB}$	0	V_{AB}	$-i_a$	i_a	0
	7	*C*	*A*	*A*	0	0	1	1	0	0	1	0	0	V_{CA}	0	$-V_{CA}$	$-i_a$	0	i_a
	8	*C*	*B*	*B*	0	0	1	0	1	0	0	1	0	$-V_{BC}$	0	V_{BC}	0	$-i_a$	i_a
	9	*A*	*B*	*B*	1	0	0	0	1	0	0	1	0	V_{AB}	0	$-V_{AB}$	i_a	$-i_a$	0
Group II	10	*C*	*A*	*C*	0	0	1	1	0	0	0	0	1	V_{CA}	$-V_{CA}$	0	i_b	0	$-i_b$
	11	*C*	*B*	*C*	0	0	1	0	1	0	0	0	1	$-V_{BC}$	V_{BC}	0	0	i_b	$-i_b$
	12	*A*	*B*	*A*	1	0	0	0	1	0	0	0	1	V_{AB}	$-V_{AB}$	0	$-i_b$	i_b	0
	13	*A*	*C*	*A*	1	0	0	0	0	1	1	0	0	$-V_{CA}$	V_{CA}	0	$-i_b$	0	i_b
	14	*B*	*C*	*B*	0	1	0	0	0	1	0	1	0	V_{BC}	$-V_{BC}$	0	0	$-i_b$	i_b
	15	*B*	*A*	*B*	0	1	0	1	0	0	0	1	0	$-V_{AB}$	V_{AB}	0	i_b	$-i_b$	0
	16	*C*	*C*	*A*	0	0	1	0	0	1	1	0	0	0	V_{CA}	$-V_{CA}$	i_c	0	$-i_c$
	17	*C*	*C*	*B*	0	0	1	0	0	1	0	1	0	0	$-V_{BC}$	V_{BC}	0	i_c	$-i_c$
	18	*A*	*A*	*B*	1	0	0	1	0	0	0	1	0	0	V_{AB}	$-V_{AB}$	$-i_c$	i_c	0
	19	*A*	*A*	*C*	1	0	0	1	0	0	0	0	1	0	$-V_{CA}$	V_{CA}	$-i_c$	0	i_c
	20	*B*	*B*	*C*	0	1	0	0	1	0	0	0	1	0	V_{BC}	$-V_{BC}$	0	$-i_c$	i_c
	21	*B*	*B*	*A*	0	1	0	0	1	0	1	0	0	0	$-V_{AB}$	V_{AB}	i_c	$-i_c$	0
	22	*A*	*B*	*C*	1	0	0	0	1	0	0	0	1	V_{AB}	V_{BC}	V_{CA}	i_a	i_b	i_c
	23	*A*	*C*	*B*	1	0	0	0	0	1	0	1	0	$-V_{CA}$	$-V_{BC}$	$-V_{AB}$	i_a	i_c	i_b
	24	*B*	*A*	*C*	0	1	0	1	0	0	0	0	1	$-V_{AB}$	$-V_{CA}$	$-V_{BC}$	i_b	i_a	i_c
	25	*B*	*C*	*A*	0	1	0	0	0	1	1	0	0	V_{BC}	V_{CA}	V_{AB}	i_c	i_a	i_b
	26	*C*	*A*	*B*	0	0	1	1	0	0	0	1	0	V_{CA}	V_{AB}	V_{BC}	i_b	i_c	i_a
	27	*C*	*B*	*A*	0	0	1	0	0	1	1	0	0	$-V_{BC}$	$-V_{AB}$	$-V_{CA}$	i_c	i_b	i_a

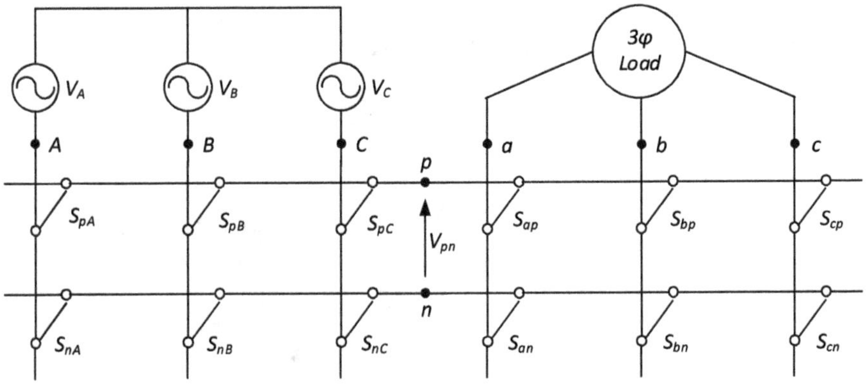

Figure 5.3 Emulation of VSR–VSI conversion

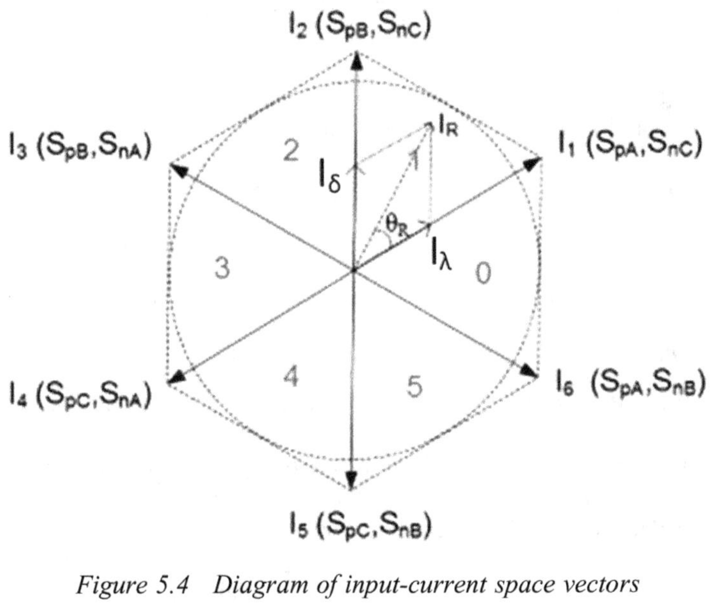

Figure 5.4 Diagram of input-current space vectors

$$d_{0R} = 1 - (d_\lambda + d_\delta) \tag{5.15}$$

where $0 \leq m_c \leq 1$ is the current modulation index [4].

VSI

Similar to VSR, the reference voltage vector is synthesized from six active and two zero voltage vectors, shown in Figure 5.5. The location of the reference vector within one of the 60° sectors defines the framing vectors V_α and V_β. For example, in Figure 5.5, the reference vector of output voltage is in the zero sector.

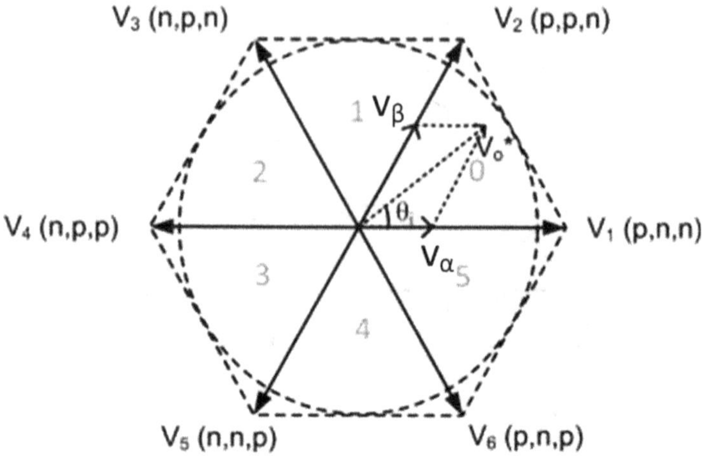

Figure 5.5 Diagram of output-voltage space vectors

Denoting the local angle of the reference voltage by θ_i, the duty ratios of the active, d_α, d_β, and zero, d_{0i}, vectors are calculated as

$$d_\alpha = m_v \sin\left(\frac{\pi}{3} - \theta_i\right) \tag{5.16}$$

$$d_\beta = m_v \sin(\theta_i) \tag{5.17}$$

$$d_{0i} = 1 - \left(d_\alpha + d_\beta\right) \tag{5.18}$$

where $0 \le m_v \le 1$ is the voltage modulation index [4].

Combined space vector rectification (SVR) and space vector inversion (SVI)

The time-averaged power outputs from the rectifying stage and the inversion stage are equal. This allows the combining of the two modulation strategies. Assuming the input displacement angle of zero, the combined duty ratios are given by

$$d_{\lambda\alpha} = d_\lambda * d_\alpha = m_c m_v \sin\left(\frac{\pi}{3} - \theta_R\right) \sin\left(\frac{\pi}{3} - \theta_i\right) = T_{\lambda\alpha}/T_s \tag{5.19}$$

$$d_{\lambda\beta} = d_\lambda * d_\beta = m_c m_v \sin\left(\frac{\pi}{3} - \theta_R\right) \sin(\theta_i) = T_{\lambda\beta}/T_s \tag{5.20}$$

$$d_{\delta\alpha} = d_\delta * d_\alpha = m_c m_v \sin(\theta_R) \sin\left(\frac{\pi}{3} - \theta_i\right) = T_{\delta\alpha}/T_s \tag{5.21}$$

$$d_{\delta\beta} = d_\delta * d_\beta = m_c m_v \sin(\theta_R) \sin(\theta_i) = T_{\delta\beta}/T_s \tag{5.22}$$

$$d_0 = 1 - \left(d_{\lambda\alpha} + d_{\lambda\beta} + d_{\delta\alpha} + d_{\delta\beta}\right) = T_0/T_s \tag{5.23}$$

The final step is to determine the switching sequence within one switching period. The selection of zero vectors in the sequence is important. Those zero vectors that require switches to change state only once during a switching period are normally selected. A commonly used sequence is $d_{\lambda\alpha}d_{\lambda\beta}d_{\delta\beta}d_{\delta\alpha}d_0$.

The relationship between the voltage modulation index and the fundamental component of the output voltage, $\widehat{V}_{out}$, can be expressed as follows:

$$m_v = \frac{\widehat{V}_{out}}{V_{pn}} \tag{5.24}$$

Neglecting the losses in the rectification stage, the input power to the MC, P_{in}, equals the fictitious dc link power, P_{dc}

$$P_{in} = P_{dc} \Rightarrow \frac{\sqrt{3}}{2}\widehat{V}_{in}\widehat{I}_{in}\cos\phi_{in} = V_{pn}I_p \tag{5.25}$$

where $\widehat{V}_{in}$ and $\widehat{I}_{in}$ represent the fundamental input line voltage and current components, respectively, and I_p is the DC link current. The current modulation index relates the fundamental input current to the DC link current as follows:

$$m_c = \frac{\widehat{I}_{in}}{I_p} \tag{5.26}$$

Using (5.25) and (5.26), it can be written

$$V_{pn} = \frac{\sqrt{3}}{2}\widehat{V}_{in}m_c\cos\phi_{in} \tag{5.27}$$

Substituting (5.27) into (5.24), the following equation can be derived to express the relation between the input and the average output voltages of the MC

$$\widehat{V}_{out} = \frac{\sqrt{3}}{2}m_v m_c \widehat{V}_{in}\cos\phi_{in} \tag{5.28}$$

Assuming a unity input power factor, i.e., $\cos\phi_{in} = 1$, the maximum voltage ratio at maximum voltage and current modulation is $\sqrt{3}/2$ [4]. Figure 5.6 depicts the typical output voltage and current waveforms.

5.2.2.3 Predictive control technique

The main characteristic of predictive control is the use of a model of the system for predicting the future behavior of the controlled variables. This information is used by the controller to obtain the optimal actuation, here the switching states of the converter, to minimize a cost function that describes the desired behavior of the system [5]. The most desired objectives that must be achieved are control of the input power factor and the load current.

The model of the MC for predictive current control considers (5.3) and (5.4), which relate the input and output currents and voltages. However, the load and input filter models should be derived, as well.

The load model gives the value of the load current in the next sampling time for each of the 27 switching states of the MC. The equation for the *R-L-E* load in

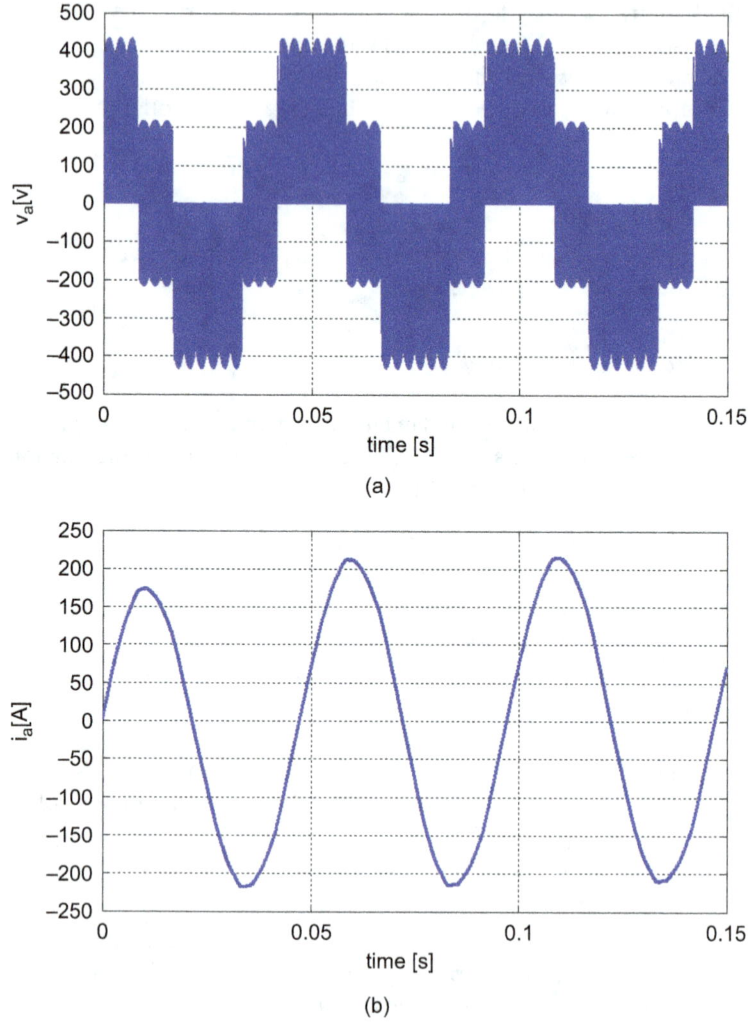

Figure 5.6 (a) Output phase voltage. (b) Output phase current of DMC with SVM.

Figure 5.1 is

$$L_L \frac{d\boldsymbol{i}_{out}(t)}{dt} = \boldsymbol{v}_{out}(t) - R_L \boldsymbol{i}_{out}(t) - \boldsymbol{e}(t) \tag{5.29}$$

Discretizing the above equation yields the value of the load current in the next sampling time, $\boldsymbol{i}_{out}(k+1)$, as follows:

$$\boldsymbol{i}^p_{out}(k+1) = \left(1 - \frac{R_L T_{sam}}{L_L}\right)\boldsymbol{i}_{out}(k) + \frac{T_{sam}}{L_L}(\boldsymbol{v}_{out}(k) - \widehat{\boldsymbol{e}}(k)) \tag{5.30}$$

where T_{sam} represents the sampling time.

The filter model can be described by the following equation in matrix form (see Figure 5.1)

$$\begin{bmatrix} \dot{v}_{in}(t) \\ \dot{i}_s(t) \end{bmatrix} \dot{x}(t) = \begin{bmatrix} 0 & 1/C_f \\ -1/L_f & -R_f/L_f \end{bmatrix} A \begin{bmatrix} v_{in}(t) \\ i_s(t) \end{bmatrix} x(t)$$

$$+ \begin{bmatrix} 0 & -1/C_f \\ 1/L_f & 0 \end{bmatrix} B \begin{bmatrix} v_s(t) \\ i_{in}(t) \end{bmatrix} u(t) \tag{5.31}$$

where

$$\boldsymbol{v}_s(t) = 2/3\left(V_{sA} + e^{j120^\circ} V_{sB} + e^{j240^\circ} V_{sC}\right) \tag{5.32}$$

$$\boldsymbol{i}_s(t) = 2/3\left(i_{sA} + e^{j120^\circ} i_{sB} + e^{j240^\circ} i_{sC}\right) \tag{5.33}$$

For the MC predictive control, the value of the source current in the next sampling time, $\boldsymbol{i}_s(k+1)$, is required, which can be calculated as follows:

$$\boldsymbol{i}_s(k+1) = B_d(2,1)\boldsymbol{v}_s(k) + A_d(2,1)\boldsymbol{v}_{in}(k) + A_d(2,2)\boldsymbol{i}_s(k) + B_d(2,2)\boldsymbol{i}_{in}(k) \tag{5.34}$$

where

$$A_d = e^{AT_{sam}} = \mathscr{L}^{-1}\left\{(s\boldsymbol{I} - A)^{-1}\right\}|_{t=T_{sam}} \tag{5.35}$$

$$B_d = \int_{\tau=0}^{T_{sam}} e^{A(T_{sam}-\tau)} B d\tau = A^{-1}(A - \boldsymbol{I})B \tag{5.36}$$

with $\boldsymbol{I}$ representing the identity matrix.

The most commonly used cost function, g, improves the input power factor by minimizing the instantaneous reactive power, and controls the load current to follow the reference with acceptable accuracy:

$$g = \left|i^*_{out_\alpha}(k+1) - i^p_{out_\alpha}(k+1)\right| + \left|i^*_{out_\beta}(k+1) - i^p_{out_\beta}(k+1)\right|$$

$$+ C\left|Q^* - Q^p(k+1)\right| \tag{5.37}$$

Here, $i^*_{out_\alpha}$ and $i^*_{out_\beta}$ are the $\alpha - \beta$ components of the reference load current, $i^p_{out_\alpha}$ and $i^p_{out_\beta}$ are the $\alpha - \beta$ components of the predicted load current calculated by

(5.30), and C is the weighting factor. The values of the instantaneous reactive power along with the predicted reactive power, $Q^p(k+1)$, can be calculated by the following equations

$$Q = \left(v_{s\beta} i_{s\alpha} - v_{s\alpha} i_{s\beta}\right) \tag{5.38}$$

$$Q^p(k+1) = v_{s\beta}(k+1) i_{s\alpha}(k+1) - v_{s\alpha}(k+1) i_{s_\beta}(k+1) \tag{5.39}$$

Subscripts α and β represent the $\alpha - \beta$ components of the corresponding vectors [2,5].

The weighting factor, C, is the only parameter that should be adjusted in the predictive control. However, there is no analytical method to determine the weighting factor for a system with certain operational constraints and specific structures of the cost function. Figure 5.7 depicts the block diagram of the predictive current control strategy with minimization of the input reactive power.

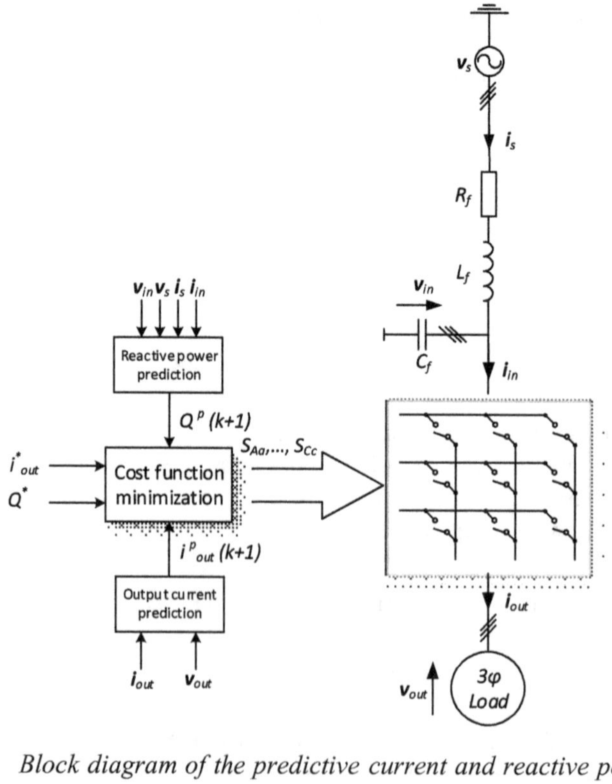

Figure 5.7 Block diagram of the predictive current and reactive power control

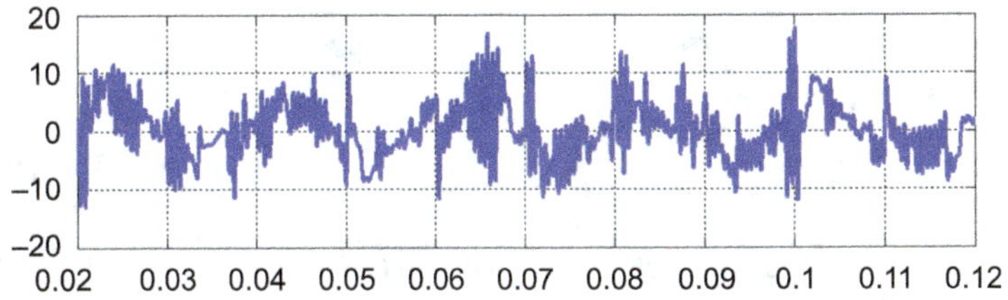

(a)

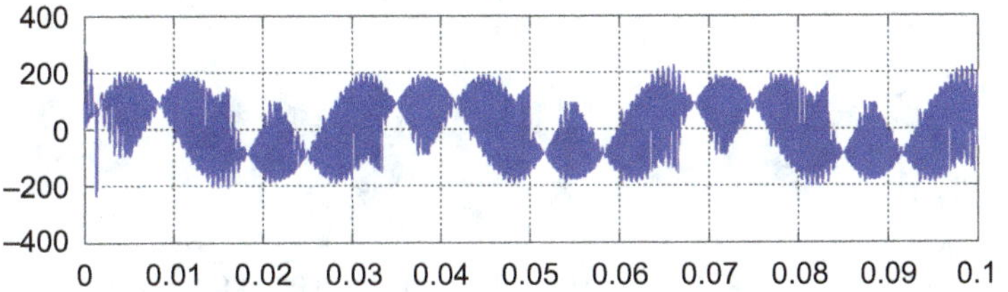

(b)

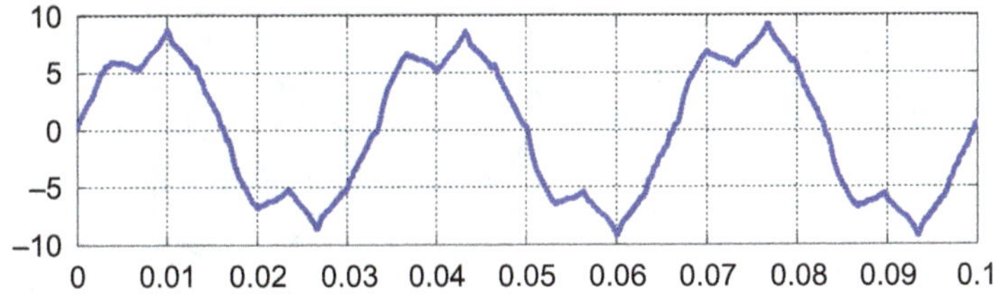

(c)

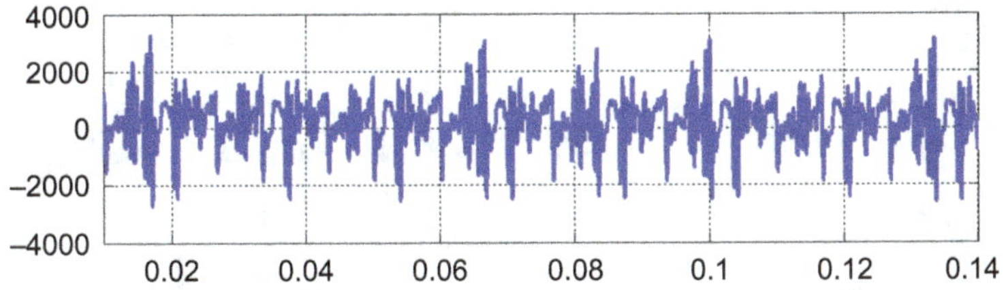

(d)

Figure 5.8 *Model predictive control of DMC without reactive power control. (a) Input phase current. (b) Output phase voltage. (c) Output phase current. (d) Input reactive power.*

Figure 5.8 illustrates the behavior of the model predictive control scheme for the DMC when the cost function has a value of $C = 0$ for the weighting factor. The output current is sinusoidal. However, the input current is highly distorted. In order to improve the quality of the input current as well as the input power factor, the weighting factor is increased as shown in Figure 5.9.

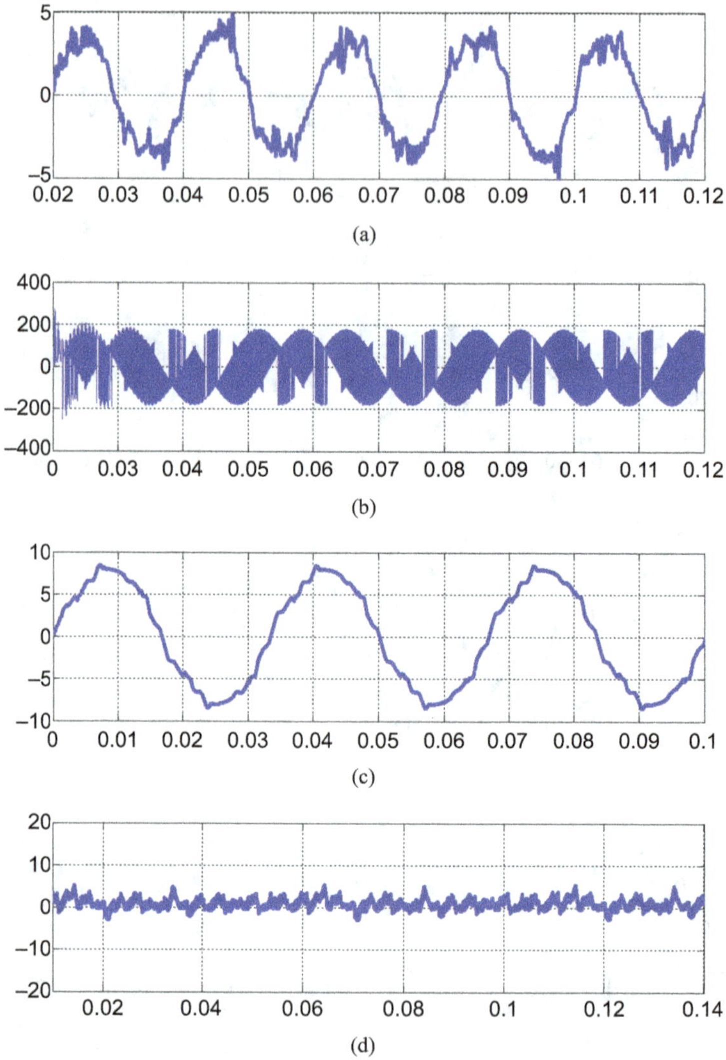

Figure 5.9 Model predictive control of DMC with reactive power control. (a) Input phase current. (b) Output phase voltage. (c) Output phase current. (d) Input reactive power.

5.3 Indirect matrix converter

The IMC, shown in Figure 5.10, is the physical implementation of the voltage source rectification and inversion idea which is used to control the DMC. In this type of MC, a conventional voltage-source-type inverter is supplied from a current-source-type rectifier which is able to operate with positive and negative DC

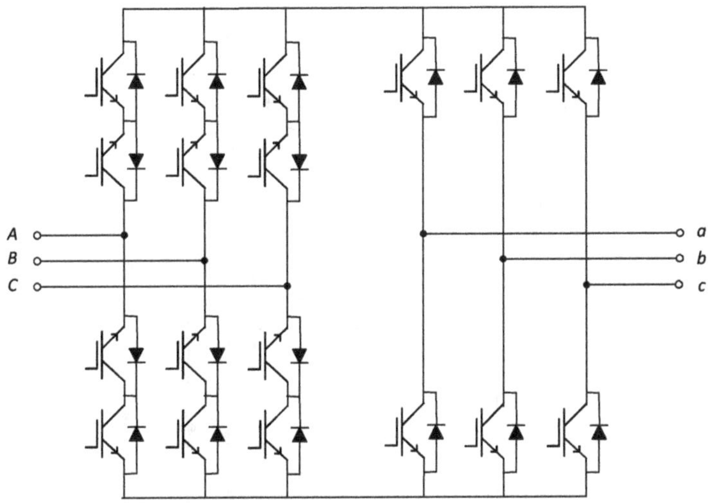

Figure 5.10 Circuit topology of the indirect matrix converter

currents and provides bipolar DC link voltage [6]. Similar to the DMC, the IMC employs 18 unipolar switches and 18 diodes. Sparse and ultra-sparse matrix converters are two types of IMCs with a reduced number of switching devices. In the following subsections, the circuit topology and modulation techniques of these converters are explained.

5.3.1 Circuit topology

The sparse matrix converter (SMC) is derived from the fact that the voltage-source-type inverter does not require a bipolar DC link voltage to operate. Figure 5.11(a) depicts one leg of the IMC. Assuming that S_{Ap} and S_{pA} are conducting, it is obvious that for a positive DC link polarity, $V_{pn} > 0$, S_{An} is blocking voltage, while for a negative DC link, S_{nA} is blocking voltage. By restricting the operation to $V_{pn} > 0$, blocking of S_{nA} within the conduction interval of S_{Ap} is not necessary. Therefore, the two transistors can be combined to form a single transistor which connects a to p or a to n, as shown in Figure 5.11(b). The resulting leg topology provides a bidirectional current flow for $V_{pn} > 0$. Therefore, the functionality of the SMC is realized by employing 15 switches and 18 diodes. It should be mentioned that despite the reduction in the number of switches, the controllability and the operating range of the SMC are not restricted compared to the DMC [6,7].

Further reduction in the number of switches can be achieved by omitting S_{pi} and S_{in} switches in the rectifier stage. However, the operating range of the resulting topology is restricted to unidirectional power flow $\left(V_{pn} > 0, i > 0\right)$. Furthermore, the controllability of the phase displacement of input voltage and input current fundamental is limited to $\pm\pi/6$. The same restriction exists for the phase

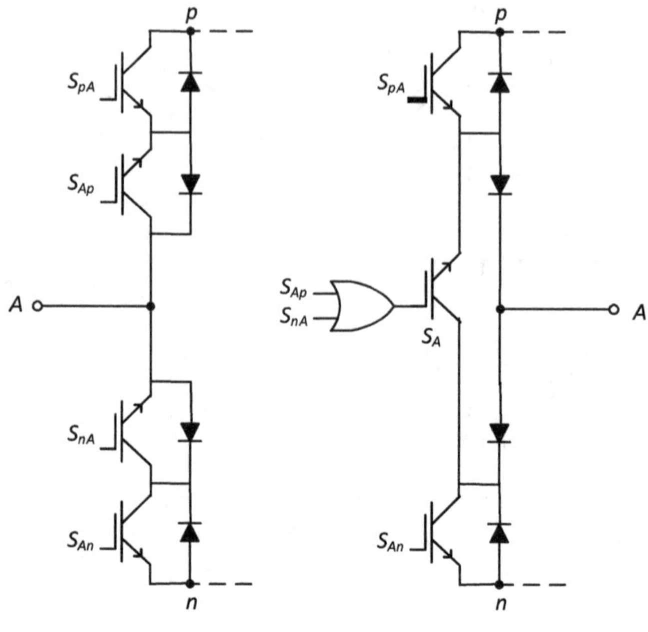

(a) (b)

Figure 5.11 Structure of a bride leg of the input rectifier stage. (a) IMC. (b) SMC.

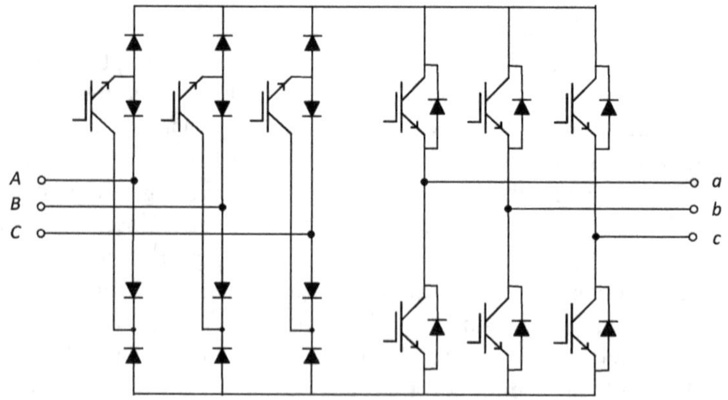

Figure 5.12 Circuit topology of the ultra-sparse matrix converter

displacement of load current and load voltage fundamental. This topology, shown in Figure 5.12, consists of nine switches and 18 diodes and is known as an ultra-sparse matrix converter (USMC) [6,7]. Due to a significant reduction in the number of switches, this converter is of high practical interest.

5.3.2 Space vector modulation

The modulation concept explained in this section is applicable to both SMC and USMC. The rectifier stage should provide a maximum voltage for the inverter stage in order to achieve a maximum voltage gain. Therefore, in every $\pi/3$-wide interval, a phase voltage with the highest absolute value is clamped to the positive or negative DC-link bus. Consequently, the required operating condition $V_{pn} > 0$ of the SMC and USMC is inherently satisfied. The following discussions are limited to the $0 < \omega_{in}t < \pi/6$ interval where phase a remains clamped to the positive DC bus p assuming the following input voltages:

$$\begin{aligned} V_A &= V_{in}\cos(\omega_{in}t) \\ V_B &= V_{in}\cos(\omega_{in}t - 2\pi/3) \\ V_C &= V_{in}\cos(\omega_{in}t - 4\pi/3) \end{aligned} \tag{5.40}$$

It is also assumed that the average value of the DC link current, I, remains constant in each rectifier switching state. In this interval, the DC link voltage, V_{pn}, is formed by segments of the input line-to-line voltages V_{AB} and V_{AC}. Therefore, the output voltage formed by the inverter shows two different levels within each pulse half period. In order to simplify the modulation scheme, the switching of the rectifier stage should occur during the free-wheeling interval of the inverter and the output voltage zero vector should be generated by operating only the inverter stage with free-wheeling intervals. At the rectifier input stage during $0 < \omega_{in}t < \pi/6$ interval, this can be achieved by

$$d_{ab} + d_{ac} = 1 \tag{5.41}$$

where d_{ab} and d_{ac} are the relative on-time of the switching states that will result in a DC link voltage of V_{AB} and V_{AC}, respectively. The average input phase currents in this interval are given as follows:

$$\bar{i}_a = (d_{ab} + d_{ac})I \quad \bar{i}_b = -d_{ab}I \quad \bar{i}_c = -d_{ac}I \tag{5.42}$$

In order to achieve a unity input power factor, the average value of an input phase current during a pulse period must be proportional to the average value of the corresponding input phase voltage:

$$d_{ab} = -\frac{\bar{i}_b}{\bar{i}_a} = -\frac{\bar{v}_b}{\bar{v}_a} d_{ac} = -\frac{\bar{i}_c}{\bar{i}_a} = -\frac{\bar{v}_c}{\bar{v}_a} \tag{5.43}$$

Equation (5.44) is derived assuming that $\bar{v}_a + \bar{v}_b + \bar{v}_c = 0$.

At the inverter output a voltage space vector v^*_{out} of given amplitude $|v^*_{out}|$ and phase $\phi^*_{out} = \omega^*_{out}t$ has to be formed over half a pulse period, $T_p/2$. The following discussions are limited to the $0 < \phi_{out} < \pi/6$ interval. In this interval, the output voltage is synthesized by active voltage space vectors (100) and (110) and zero voltage space vectors (111) or (000). In the time intervals $T_{ac} = d_{ac}T_p/2$ and $T_{ab} = d_{ab}T_p/2$ the DC link voltage is V_{AC} and V_{AB}, respectively. For calculating the on-times of the active switching states the local average value of the DC link voltage can be employed. The following equations express the on-times of the output voltage space vectors in the $0 < \phi_{out} < \pi/6$ interval:

$$T_{100(ac)} = -\frac{2}{3\sqrt{3}} T_p \frac{V^*_{out}}{V^2_{in}} V_c \cos\left(\phi^*_{out} + \frac{\pi}{6}\right) \tag{5.44}$$

$$T_{100(ab)} = -\frac{2}{3\sqrt{3}} T_p \frac{V^*_{out}}{V^2_{in}} V_b \cos\left(\phi^*_{out} + \frac{\pi}{6}\right) \tag{5.45}$$

$$T_{110(ac)} = -\frac{2}{3\sqrt{3}} T_p \frac{V^*_{out}}{V^2_{in}} V_c \sin\phi^*_{out} \tag{5.46}$$

$$T_{110(ab)} = -\frac{2}{3\sqrt{3}} T_p \frac{V^*_{out}}{V^2_{in}} V_b \sin\phi^*_{out} \tag{5.47}$$

The voltage transfer ratio of the SMC and UMC is given by

$$M = \frac{V^*_{out}}{V_{in}} \leq \frac{\sqrt{3}}{2} \tag{5.48}$$

The maximum voltage transfer ratio is only available for the unity input power factor, i.e., $\cos\phi_{in} = 1$ [6]. Figure 5.13 [6] depicts the formation of the DC-link voltage and DC-link current along with the switching states of the rectifier and inverter stages for $0 < \omega_{in}t < \pi/6$ and $0 < \omega_{out}t < \pi/6$. In this figure, S_a, S_b, and S_c are the switching states of the upper switches in each leg of the output inverter stage. Figure 5.14 shows the typical output voltage and current waveforms of a USMC.

5.3.3 Predictive control

The predictive control of SMC and USMC is conceptually very similar to that of the DMC. However, in addition to minimizing the line side current error and reactive power, the DC link voltage of the SMC and USMC must be maintained positive for proper operation. Therefore, the cost function should be modified as follows:

$$g = Q^{k+1} + C.\Delta i_1^{k+1} + h^{k+1} \tag{5.49}$$

where

$$\begin{aligned}
\Delta i_1^{k+1} &= \left|i^*_{out_\alpha}(k+1) - i^p_{out_\alpha}(k+1)\right| + \left|i^*_{out_\beta}(k+1) - i^p_{out_\beta}(k+1)\right| \\
Q^{k+1} &= |Q^* - Q^p(k+1)| \\
h^{k+1} &= \begin{cases} 0, & V^p_{pn}(k+1) > 0 \\ N, & V^p_{pn}(k+1) < 0 \end{cases}
\end{aligned} \tag{5.50}$$

Here, N is the maximum positive number that can be generated by the arithmetic unit of the controller [8].

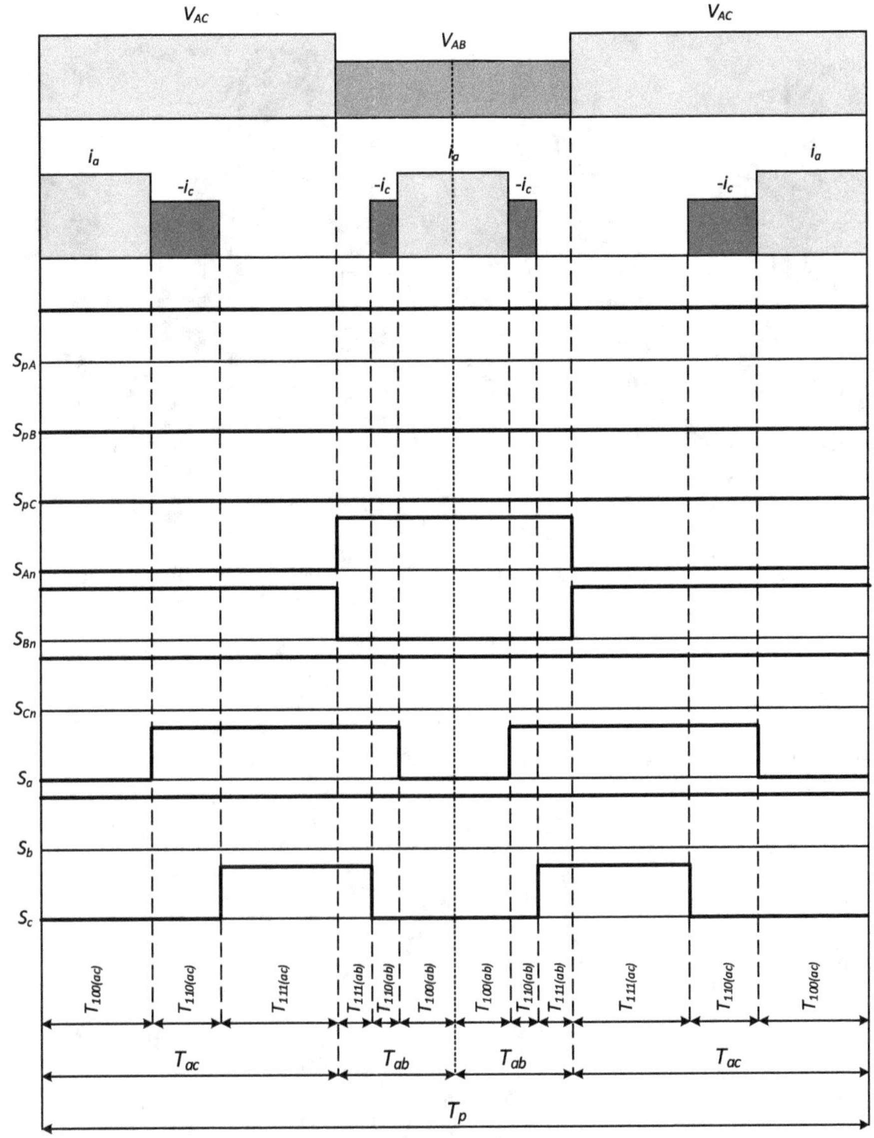

Figure 5.13 Top to bottom: DC-link voltage, DC-link current, switching states of the rectifier stage, switching states of the inverter stage

The weighting factor C can be adjusted empirically such that the output current has no noticeable deviations from the reference and in the meantime, the input currents are not highly distorted.

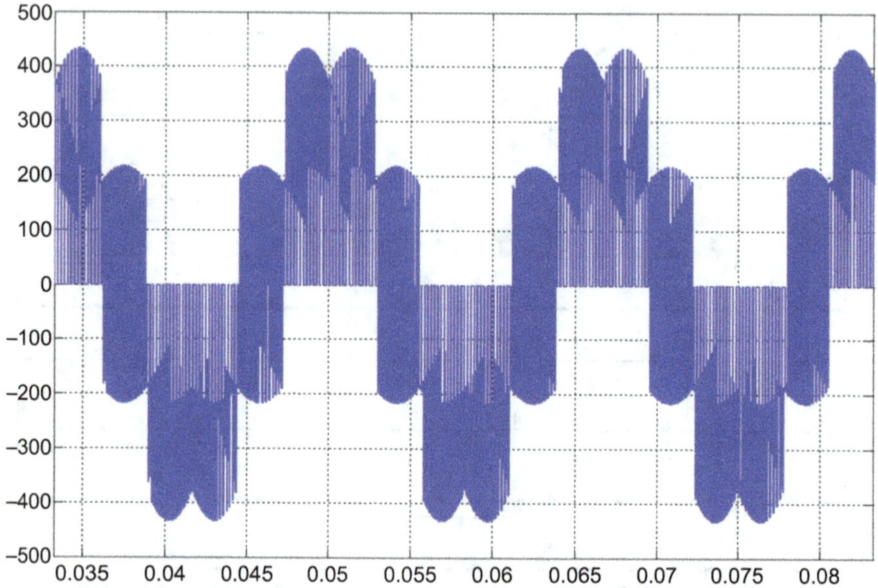

(a)

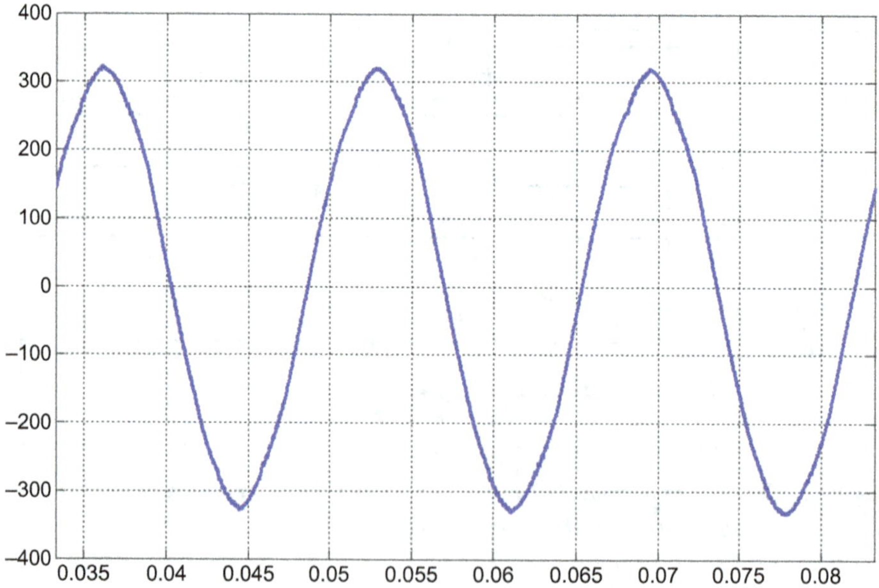

(b)

Figure 5.14 (a) Output phase voltage. (b) Output phase current of USMC with SVM.

5.4 Z-source matrix converters

Combining Z-source (ZS) networks with MC topologies is considered a viable solution for addressing the limited voltage transfer ratio (VTR) [9] issue since overmodulation strategies proposed for increasing VTR [10–12] result in low-quality input currents and output voltages. In [13] and [14], a comprehensive review of ZS networks employed with DMCs is provided. However, due to the advantages of IMC compared to DMC, Z-source IMC, in particular, Z-source USMC (ZSUSMC) is considered and studied in this chapter. In [15], the ZS network is suggested to be placed between the three-phase input voltage source and the IMC, and in [16] performance of Z-source indirect matrix converter (ZSIMC) with three ZS configurations is evaluated in terms of VTR, current and voltage ripple, and stress on the switches. This arrangement realizes high VTR with small ZS components, but it increases the cost and reduces the reliability of the converter due to additional switches in its structure. The second option is to place the ZS network between the rectifier and inverter stages of IMC. In this manner, no additional switches are required in the converter structure. The second alternative is considered in this chapter.

5.4.1 Topology

The topology of the ZSUSMC, where an X-shaped impedance network is inserted between the rectifier and inverter stages of a USMC, is depicted in Figure 5.15. Thanks to the Z-source network, the inverter stage can be operated at the shoot-through stage and boosting the input voltage can be achieved. It is noteworthy to mention that other derivations of the Z-source network, such as series, switched-inductor, quasi, etc. could be used, each offering a different boosting capability and size, number of, and voltage and current ratings of the passive components of the Z-source network.

5.4.2 Operation modes

Figure 5.16 illustrates the two operation modes of the converter. In the first mode, the inverter stage works in a non-shoot through state (active and zero states) and the

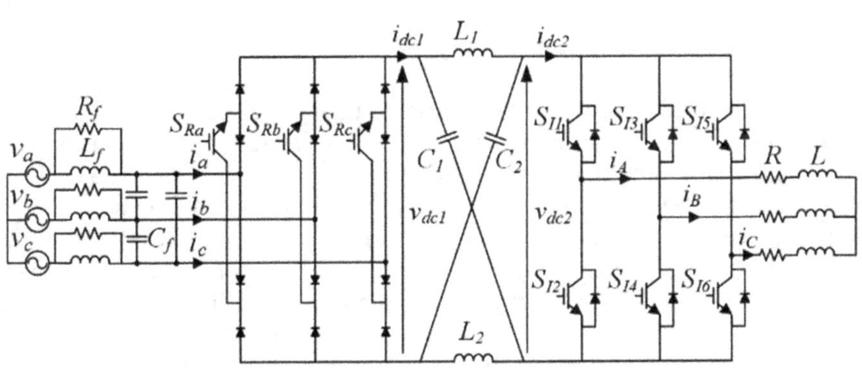

Figure 5.15 Topology of Z-source ultra-sparse matrix converter (ZSUSMC)

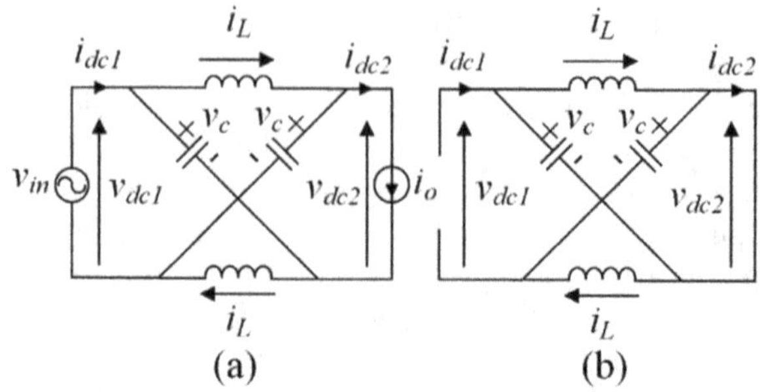

Figure 5.16 ZSUSMC equivalent circuit in (a) non-shoot-through state and (b) shoot-through state

output of the ZS network can equivalently be modeled as a current source. It is obvious that in the zero states, the current source is open-circuited, i.e., the value of the current is zero. In addition, during zero switching states of the rectifier stage, the input voltage is zero and it can be considered as short-circuited. In the second mode (shoot-through state), the switches on one, two, or all three legs of the inverter stage are turned on simultaneously and therefore, the output of the ZS network is short-circuited. Under this condition, the input voltage of the ZS network (v_{dc1}) becomes higher than the input voltage source. As a result, the series diodes in the rectifier stage are reverse-biased; hence, the ZS network is isolated from the input source.

5.4.3 Space vector modulation

SVM is achieved by a combination of SVR and SVI techniques. Consider the following as the three-phase input voltages

$$v_a(t) = V_{im}\cos(\theta_i),\ v_b(t) = V_{im}\cos\left(\theta_i - \frac{2\pi}{3}\right),\ v_c(t) = V_{im}\cos\left(\theta_i + \frac{2\pi}{3}\right) \tag{5.51}$$

where V_{im} is the amplitude of the input phase voltages and $\theta_i = \omega_i t$ is the angle of the phase a voltage.

The normalized line-to-line input voltages, instantaneous voltages at the output of the rectifier stage, v_{dc1} (see Figure 5.15), and its average, V_{dc1} are illustrated in Figure 5.17. In order to achieve maximum v_{dc1}, the input phase voltage with maximum absolute value within every $\pi/3$ interval is continuously connected to the positive or negative rail of the fictitious DC-link and the other two phases are switched based on the modulation requirements. In other words, the rectifier stage is controlled in such a way that the maximum and medium line-to-line input voltages in each $\pi/3$ interval appear at its output (see Figure 5.17). In this manner, one input voltage period can be divided into six equal intervals. In order to produce zero DC-link voltage, it is sufficient to turn on only one out of the three switches of the

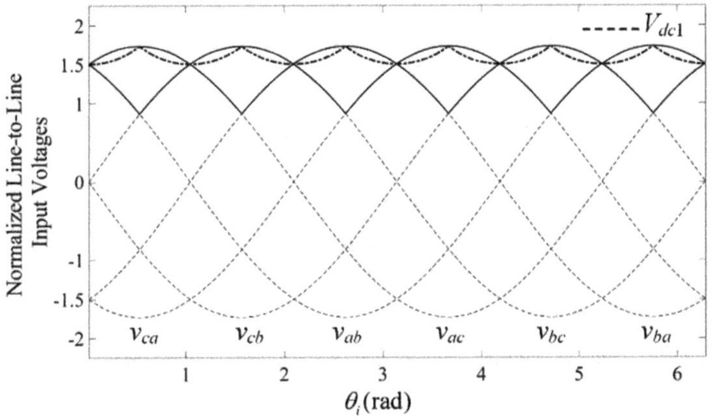

Figure 5.17 Normalized line-to-line input voltages and local average of rectifier output voltage

Table 5.2 Switching states and fictitious DC-link voltage

		On switch	**Modulated switch**		**DC-link voltage (v_{dc1})**	
Sector	θ_i	S_{on}	S_{i1}	S_{i2}	v_{i1}	v_{i2}
1	$(11\pi/6, \pi/6]$	S_{Ra}	S_{Rb}	S_{Rc}	v_{ab}	v_{ac}
2	$(\pi/6, \pi/2]$	S_{Rc}	S_{Ra}	S_{Rb}	v_{ac}	v_{bc}
3	$(\pi/2, 5\pi/6]$	S_{Rb}	S_{Rc}	S_{Ra}	v_{bc}	v_{ba}
4	$(5\pi/6, 7\pi/6]$	S_{Ra}	S_{Rb}	S_{Rc}	v_{ba}	v_{ca}
5	$(7\pi/6, 3\pi/2]$	S_{Rc}	S_{Ra}	S_{Rb}	v_{ca}	v_{cb}
6	$(3\pi/2, 11\pi/6]$	S_{Rb}	S_{Rc}	S_{Ra}	v_{cb}	v_{ab}

rectifier stage. Table 5.2 summarizes the rectifier switching states and the corresponding DC-link voltage over one input voltage period [17,18].

The input current space vector diagram is depicted in Figure 5.18(a). Duty cycles of the rectifier stage switches can be obtained as

$$d_{i1} = m_i \sin\left(\frac{\pi}{3} - \tilde{\beta}_i\right), d_{i2} = m_i \sin\left(\tilde{\beta}_i\right), d_{i0} = 1 - d_{i1} - d_{i2} \tag{5.52}$$

where $0 \le m_i \le 1$ is the modulation index of the rectifier stage, and

$$\tilde{\beta}_i = \beta_i - (K_i - 1)\frac{\pi}{3} + \frac{\pi}{6} \tag{5.53}$$

Here, β_i is the angle of reference input current vector and K_i is the sector in which it is located. In order to achieve unity power factor operation, $\beta_i = \theta_i$. This assumption is made throughout the rest of the chapter.

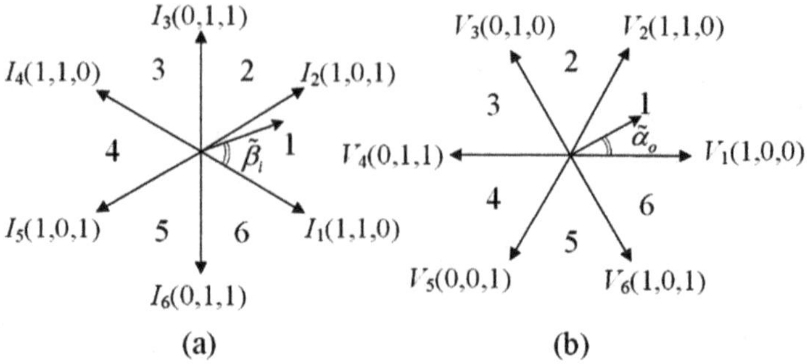

Figure 5.18 Space vector diagrams of (a) rectifier stage, (b) inverter stage

Table 5.3 Conventional switching pattern for Z-source ultra-sparse matrix converter over one switching period

d_{i1}				d_{i2}				d_{i0}			
d_{o1}	d_{o2}	d_{st}	d_{o0}	d_{o1}	d_{o2}	d_{st}	d_{o0}	d_{o1}	d_{o2}	d_{st}	d_{o0}

The same procedure can be applied to the inverter stage. Figure 5.18(b) shows the output voltage space vector diagram. Duty cycles of the inverter stage switches can be calculated as

$$d_{o1} = m_o\sin\left(\frac{\pi}{3} - \tilde{\alpha}_o\right), d_{o2} = m_o\sin(\tilde{\alpha}_o), d_{o0} = 1 - d_{o1} - d_{o2} \tag{5.54}$$

where $0 \leq m_o \leq 1$ is the modulation index of the inverter stage, and

$$\tilde{\alpha}_o = \alpha_o - (K_o - 1)\frac{\pi}{3} \tag{5.55}$$

Here, α_o and K_o are the angle of reference output voltage vector and the sector in which it is located, respectively.

Duty cycles obtained from SVR and SVI are combined to yield the duty cycles of the USMC. This is achieved by cross-multiplication of rectifier and inverter stages duty cycles. In ZSUSMC, the shoot-through state is also included in the inverter stage switching states to provide the desired boosting. Two different modulation schemes can be developed for ZSIMC depending on using or not using zero state in SVR. If zero state in SVR is not used, the number of sub-durations in each switching period will be reduced to eight. While, as shown in Table 5.3, this number is 12 with zero state. With this modulation scheme, the number of changes in the switching states is high and the duration of combined duty cycles is extremely short. Therefore, an field programmable gate arrays (FPGA) is usually employed besides a digital signal processor (DSP) for implementation.

5.4.3.1 Modulation scheme with zero state in SVR

Without loss of generality, the following discussions are based on the case where reference input current and output voltage space vectors are located in sector 1 (see Figure 5.18). In this condition, v_{dc1} will be equal to v_{ab}, v_{ac} or 0 V and i_{dc2} will be equal to i_A, $-i_C$ or 0 A. During the shoot-through state, output of the ZS network will be short-circuited and according to Figure 5.16(b), v_{dc1} will be equal to $2V_C$, where V_C is the average voltage of the ZS capacitor over a switching period. Since this voltage is higher than the magnitude of input voltages, diodes of the rectifier stage will be off and the ZS network will be isolated from the input voltage source.

The average of the rectifier stage output voltage over a switching period, V_{dc1}, can be calculated as follows:

$$V_{dc1} = (d_{i1o1} + d_{i1o2} + d_{i1o0})v_{ab} + (d_{i2o1} + d_{i2o2} + d_{i2o0})v_{ac} + (d_{i1st} + d_{i2st} + d_{i0st})2V_C \quad (5.56)$$

Since $d_{o1} + d_{o2} + d_{o0} + d_{st} = 1$, after some manipulations (5.56) can be rewritten as

$$V_{dc1} = (1 - d_{st})d_{i1}v_{ab} + (1 - d_{st})d_{i2}v_{ac} + 2d_{st}V_C = (1 - d_{st})\frac{3}{2}V_{im} + 2d_{st}V_c \quad (5.57)$$

Considering that the average voltage of the ZS network inductors over a switching period is zero, i.e., $V_L = 0$, *the* following equations can be derived

$$V_{dc1} = V_{dc2} = V_C = \frac{3V_{im}(1 - d_{st})}{2(1 - 2d_{st})} \quad (5.58)$$

$$V_{om} = \frac{V_{dc2}}{\sqrt{3}} = \frac{\sqrt{3}(1 - d_{st})V_{im}}{2(1 - 2d_{st})} \quad (5.59)$$

where V_{om} is the amplitude of output voltage. Since V_{om} is a constant, sinusoidal output voltage can be achieved through this modulation technique. The output to input VTR can be obtained as

$$B = \frac{V_{om}}{V_{im}} = \frac{\sqrt{3}(1 - d_{st})}{2(1 - 2d_{st})} \quad (5.60)$$

With $d_{st} = 0$ in (5.60), $B = \sqrt{3}/2$. This corresponds to the maximum VTR of a conventional matrix converter.

Based on Figure 5.16(b), one can observe that during the shoot-through state $i_{dc2} = 2I_L$. Following the same above procedure, DC current at the input of the inverter stage can be derived as follows:

$$\begin{aligned} I_{dc2} &= (d_{i1o1} + d_{i2o1} + d_{i0o1})i_A - (d_{i1o2} + d_{i2o2} + d_{i0o2})i_C \\ &\quad + (d_{i1st} + d_{i2st} + d_{i0st})2I_L \\ &= d_{o1}i_A - d_{o2}i_C + 2d_{st}I_L \end{aligned} \tag{5.61}$$

Here, I_L is the average of the Z-source network inductor current over a switching period. According to capacitor charge (amp-second) balance, the average of capacitor current, I_C, is equal to zero. Hence, the following equations can be written

$$I_L = I_{dc1} = I_{dc2} = \frac{\sqrt{3}(1 - d_{st})I_o\cos(\varphi_o)}{2(1 - 2d_{st})} \tag{5.62}$$

$$I_{im} = m_i I_{dc1} = \frac{\sqrt{3}(1 - d_{st})I_o\cos(\varphi_o)}{2(1 - 2d_{st})} \tag{5.63}$$

where φ_o is the phase displacement between the output voltage and current. As it is observed from (5.63), the amplitude of input currents is constant and sinusoidal input currents can be accomplished if SVM with sub-durations presented in Table 5.3 is employed. It is noteworthy to mention that the derived equations in this section are based on the assumption that the rectifier and inverter stage modulation indexes are set to their maximum values, i.e., $m_i = 1$ and $m_o = 1 - d_{st}$.

A ZSUSMC feeding an R-L load with B = 1.5 and 0.3 is modeled on a Hardware-in-the-loop (HIL) setup, consisting of an OP4510 real-time simulator from Opal-RT Technologies Inc. with a 4-core Intel Xeon 3.2 GHz CPU and a TI TMS320F28335 DSP. The entire system, including the source, ZSUSMC, and the load, is modeled on the real-time simulator, and the ZSUSMC modulation strategy is implemented on the DSP. Results are depicted in Figures 5.19 and 5.20, respectively. The input current is in phase with the input voltage and unity power factor operation is achieved. In addition, the input voltage of the ZS network, v_{dc1}, and its output voltage (the input voltage of the inverter stage), v_{dc2}, are shown. As it can be seen, the converter is capable of providing close to sinusoidal input and output currents for a wide range of VTR.

5.5 Technological issues of matrix converters

The MC topology is based on the bidirectional semiconductor switches to both conduct current in each direction and block voltage in both directions. However,

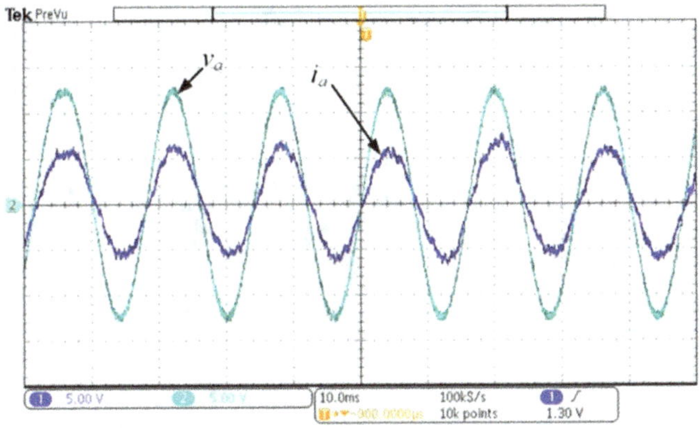

(a)

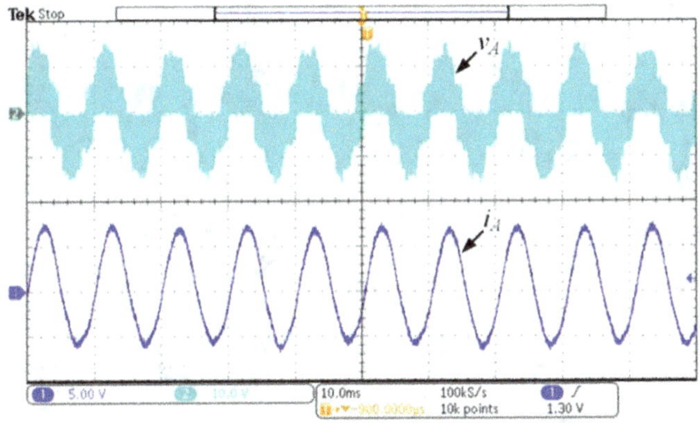

(b)

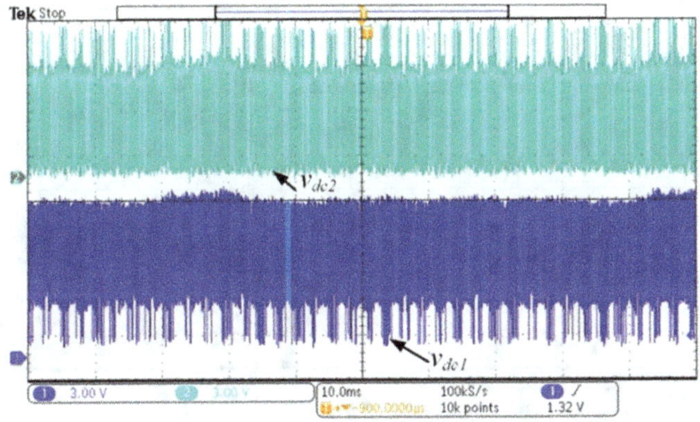

(c)

Figure 5.19 *HIL results for B = 1.5. (a) Input voltage [62.5 V per div] and input current [16.5 A per div], (b) output voltage [250 V per div] and output current [12.5 A per div], and (c)* v_{dc1} *[130 V per div] and* v_{dc2} *[130 V per div].*

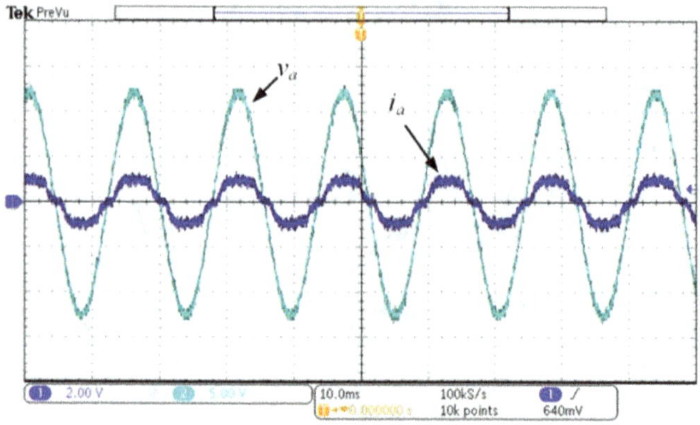

(a)

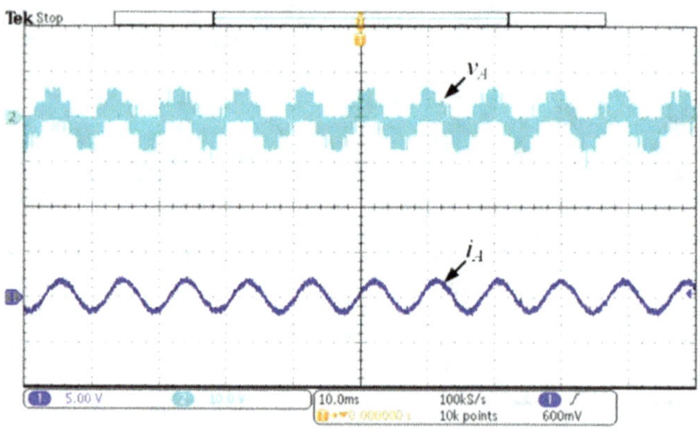

(b)

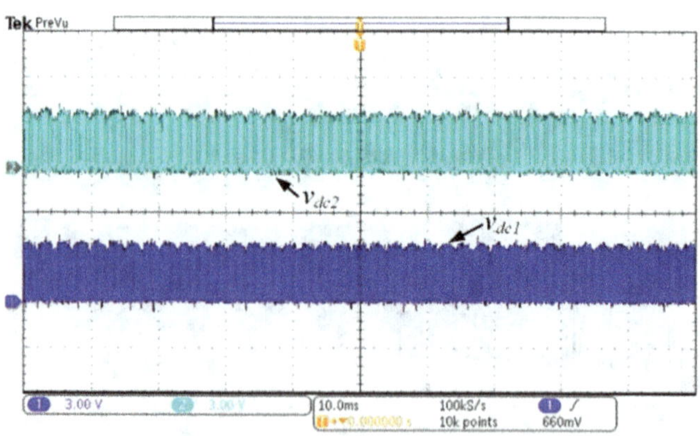

(c)

Figure 5.20 HIL results for B = 0.3. (a) Input voltage [62.5 V per div] and input current [2 A per div], (b) v_{dc1} *[130 V per div] and* v_{dc2} *[130 V per div], and (c) output voltage [250 V per div] and output current [12.5 A per div].*

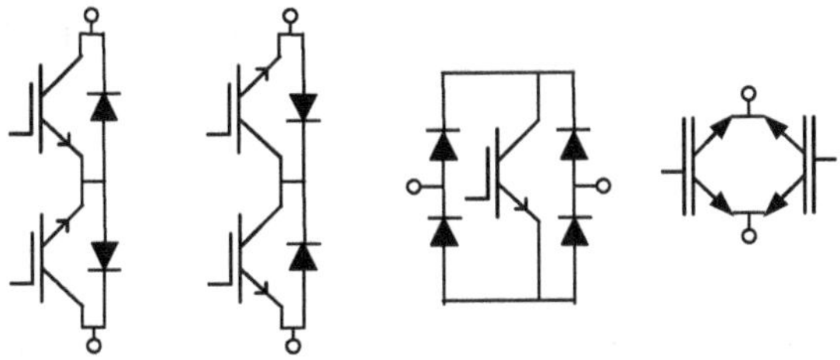

Figure 5.21 *Possible bidirectional switch arrangements. (a) Common emitter, (b) common collector, (c) diode bridge, and (d) reverse blocking IGBT.*

such a device is currently not commercially available. Therefore, insulated gate bipolar transistor (IGBT) devices and diodes are mostly used to create the power circuit. The reverse blocking IGBT has been reportedly used in the MC topology since antiparallel diodes can be eliminated from the converter. Other arrangements include the diode bridge arrangement which uses only one active device, and the common emitter and common collector arrangements which use two active devices and two diodes, as shown in Figure 5.21 [19].

Similar to any converter, a simple LC filter is often used at the input of the MC in order to reduce the switching harmonics of the input current. However, the LC circuit creates overvoltages during the transient operation, especially the power-up procedure of the MC. Some passive solutions such as connecting damping resistors in parallel with the input inductors have been proposed to reduce overvoltages. Overvoltages can also occur due to an open circuit of the load as there are no freewheeling paths within the MC circuit for the inductive load [19]. Several schemes, such as passive and active clamp circuits, have been proposed and successfully applied to protect the MCs against these overvoltages.

5.6 Matrix converter versus voltage back-to-back converter

Within a certain switching frequency range, the DMC and IMC enable higher maximum power density and maximum power-to-mass ratio at a higher efficiency compared to the voltage back-to-back converter. The back-to-back converter requires a significantly smaller semiconductor chip area compared to the DMC and IMC for switching frequencies below a specific switching frequency, whereas above that frequency, the back-to-back converter requires the largest chip area, followed by the DMC and IMC.

Alas, those advantages are outweighed by the low voltage transfer ratio, constrained input reactive power compensation, dependency of control of the input currents on the output currents, and restriction on the single-phase operation of MCs. Consequently, the MC is not suitable for applications where a bidirectional, low-voltage, and low-power AC–AC converter system is required. The voltage back-to-back converter clearly is the preferred choice for such requirements [20].

5.7 Conclusions

Due to intensive research on matrix converters during the past two decades, these converters are finding a wide variety of applications in the power converter industry, to date, two drive manufacturers, i.e., Yaskawa and Fuji Electric Systems, offer matrix converter products. Features such as lack of large and fragile DC-link components make these converters suitable for deep sea remotely operated vehicle (ROV) applications where extreme undersea pressure is present. Furthermore, due to the higher power density and the elevated temperature capability, matrix converters are employed in the aircraft industry. The direct and indirect matrix converters have also been applied to doubly fed induction generator-based wind energy systems with the advantage of controlling a high-power generator with a relatively low-power, four-quadrant power converter. Matrix converters are finding applications in the power supply generation area, as well, where a fixed voltage and frequency power supplies are implemented from variable frequency diesel generators.

References

[1] P. W. Wheeler, J. Rodriguez, J. C. Clare, L. Empringham, and A. Weinstein, "Matrix converters: a technology review," *IEEE Trans. Ind. Elec.*, vol. 49, no. 2, pp. 276–287, 2002.

[2] J. Rodriguea, M. Rivera, J. W. Kolar, and P. W. Wheeler, "A review of control and modulation methods for matrix converters," *IEEE Trans. Ind. Elec.*, vol. 59, no. 1, pp. 58–70, 2012.

[3] L. Huber, and D. Borojevic, "Space vector modulated three-phase to three-phase matrix converter with input power factor correction," *IEEE Trans. Ind. Appl.*, vol. 31, no. 6, pp. 1234–1246, 1995.

[4] E. Karaman, M. Farasat, and A. M. Trzynadlowski, "A comparative study of series and cascaded Z-Source matrix converters," *IEEE Trans. Ind. Elec.*, vol. 61, no. 10, pp. 5164–5173, 2014.

[5] J. Rodriguez, and P. Cortes, "*Predictive Control of Power Converters and Electric Drives,*" John Wiley, 2012.

[6] J. W. Kolar, F. Schafmeister, S. D. Round, and H. Ertl, "Novel three-phase AC-AC sparse matrix converters," *IEEE Trans. Power Elec.*, vol. 22, no. 5, pp. 1649–1661, 2007.

[7] J. W. Kolar, T. Friedli, J. Rodriguez, and P. W. Wheeler, "Review of three-phase PWM AC-AC converter topologies," *IEEE Trans. Ind. Elec.*, vol. 58, no. 11, pp. 4988–5006, 2011.

[8] P. Correa, J. Rodriguez, M. Rivera, J. R. Espinoza, and J. W. Kolar, "Predictive control of an indirect matrix converter," *IEEE Trans. Ind. Elec.*, vol. 56, no. 6, pp. 1847–1853, 2009.

[9] V. Siva, A. Singh, M. Raghuram, A. Kumar, and S. Singh, "Enhance voltage gain of USMC by modified quasi Z source impedance network," in *Proc. IEEE Ind. Appl. Ann. Mtg (IAS)*, pp. 16, 2022.

[10] A. M. Bozorgi, M. Monfared, and H. Mashhadi, "Two simple overmodulation algorithms for space modulated three-phase to three-phase matrix converter," *IET Power Electron.*, vol. 7, no. 7, pp. 1915–1924, 2014.

[11] Y. Xia, X. Zhang, M. Qiao, F. Yu, Y. Wei, and P. Zhu, "Research on a new indirect space-vector overmodulation strategy in matrix converter," *IEEE Trans. Ind. Electron.*, vol. 63, no. 2, pp. 1130–1141, 2016.

[12] S. Li, C. Xia, Y. Yan, and T. Shi, "Space-vector overmodulation strategy for ultrasparse matrix converter based on the maximum output voltage vector," *IEEE Trans. Power Electron.*, vol. 32, no. 7, pp. 5388–5397, 2017.

[13] O. Ellabban, H. Abu-Rub, and S. Bayhan, "Z-source matrix converter: an overview," *IEEE Trans. Power Electron.*, vol. PP, pp. 1–1, 2015.

[14] B. Ge, Q. Lei, W. Qian, and F. Z. Peng, "A family of Z-source matrix converters," *IEEE Trans. Ind. Electron.*, vol. 59, no. 1, pp. 35–46, 2012.

[15] S. Liu, B. Ge, H. Abu-Rub, and F. Z. Peng, "Modeling, analysis, and motor drive application of quasi-Z-source indirect matrix converter," *COMPEL: Int. J. Comput. Math. Elect. Electron. Eng.*, vol. 33, no. 1/2, pp. 28–28, 2013.

[16] S. Liu, B. Ge, X. Jiang, H. Abu-Rub, and F. Z. Peng, "Comparative evaluation of three Z-source/quasi-Z-source indirect matrix converters," *IEEE Trans. Ind. Electron.*, vol. 62, no. 2, pp. 692–701, 2015.

[17] A. Bozorgi, and M. Farasat, "An in-depth investigation of a Z-source ultrasparse matrix converter in buck and boost modes of operation," *IEEE Trans. Ind. Electon.*, vol. 65, no. 6, pp. 5177–5187, 2018.

[18] A. Bozorgi, and M. Farasat, "Improved design and space vector modulation of a Z-source ultra sparse matrix converter: Analysis, implementation, and performance evaluation," *IEEE Trans. Ind. Appl.*, vol. 54, no. 4, pp. 3737–3748, 2018.

[19] L. Empringham, J. W. Kolar, J. Rodriguez, P. W. Wheeler, and J. C. Clare, "Technological issues and industrial application of matrix converters: a review," *IEEE Trans. Ind. Elec.*, vol. 60, no. 10, pp. 4260–4271, 2013.

[20] T. Friedli, J. W. Kolar, J. Rodriguea, and P. W. Wheeler, "Comparative evaluation of three-phase AC-AC matrix converter and voltage DC-link back-to-back converter systems," *IEEE Trans. Ind. Elec.*, vol. 59, no. 12, pp. 4487–4510, 2012.

Chapter 6

Soft-switching converters

Paulo Peixoto Praça[1] and Luiz Carlos Gomes Freitas[2]

6.1 Introduction

Switched power supplies are popular and essential because they provide high-quality electrical energy conversion, possibly with very high efficiency, decreased weight and volume, and very often galvanic insulation through the use of high-frequency transformers. It is important to design and manufacture them for applications in residential, commercial, industrial, as well as rural, transportation, for key sectors of our society.

The increase in the switching frequency of the converters allowed to reduce the volume and consequently the weight of the power converters. However, with the frequency increase, other problems appeared, such as electromagnetic interference and high switching losses in semiconductors, thus motivating the research for topologies that could mitigate these effects. These new topologies are called soft-switched converters and started with the resonant converters.

The study of resonant converters emerged in the mid-1970s with the main objective of reducing the switching losses on thyristors, when forced commutation techniques would have too much losses on the external circuit to turn the device off [1]. For many years several topologies were introduced with transistor-based circuits, based on resonance or quasi-resonance operations. In recent decades, studies have advanced toward reducing losses in semiconductor switching. The first converter that was very popular was a series resonant converter. In this chapter, the main techniques and topologies of soft-switched converters will be reviewed.

6.2 What is resonance

Resonance in electrical circuits is a phenomenon that occurs between energy storage elements (inductors and capacitors), which can be in a series or parallel configuration. A circuit is said to be resonant when the voltage applied to an

[1]Department of Electrical Engineering, Federal University of Ceará, Brazil
[2]Faculty of Electrical Engineering, Federal University of Uberlandia, Brazil

electrical circuit composed of a resistor, a capacitor, and an inductor is in phase with the resulting current.

At resonance, the equivalent Z impedance of a circuit is purely resistive, as the voltage and current are in phase. Thus, the power factor of a resonant circuit is unity.

In electricity, there are two types of resonance: an association in which the circuit elements are in series and the input impedance is minimum, and another one, with elements in parallel, where the input impedance is maximum.

6.2.1 Series RLC resonant circuits

Figure 6.1 shows the classic configuration of a series RLC (resistor - R, inductor - L, and capacitor - C) circuit. The impedance seen from the source terminals V_i is given by:

$$Z(w) = R + j\left(w.L - \frac{1}{w.C}\right) = R + j.X(w) \tag{6.1}$$

The circuit is in resonance when the inductive and capacitive reactances cancel each other out. Thus, $Z(w = w_n) = R$ when:

$$X(w_n) = w_n.L - \frac{1}{w_n.C} = 0 \quad w_n = \frac{1}{\sqrt{L.C}} \tag{6.2}$$

where the angular frequency w_n is also called resonant frequency, given in rad/s. The same frequency, in hertz, is presented in (6.3).

$$f_n = \frac{1}{2.\pi\sqrt{L.C}} \tag{6.3}$$

Another important parameter to characterize resonant circuits is the quality factor of the circuit, Q. This factor can be defined as the ratio between the energy stored in the circuit and the energy lost by the circuit, per cycle, at resonance. The higher the quality factor of a circuit, the better its selectivity and the sharper its resonance curve, as can be seen in Figure 6.2. Equation (6.4) shows the quality factor of a series RLC circuit.

$$Q = \frac{w_n.L}{R} = \frac{1}{w_n.R.C} \tag{6.4}$$

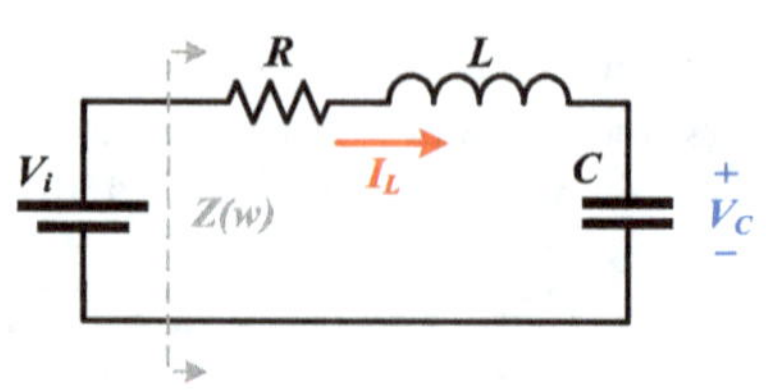

Figure 6.1 Series RLC circuit

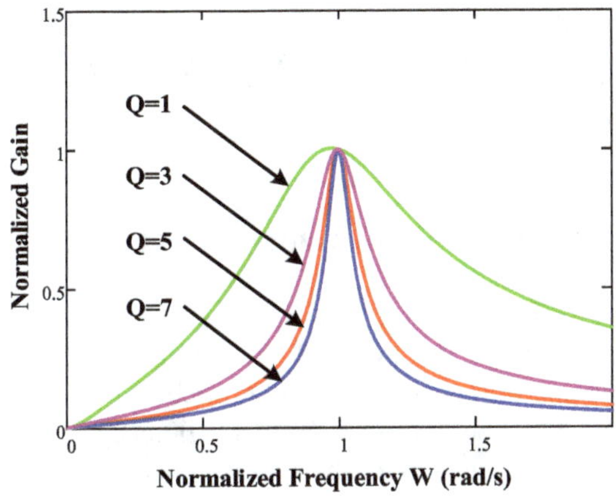

Figure 6.2 Effect of the quality factor Q on frequency response selectivity in a series RLC circuit

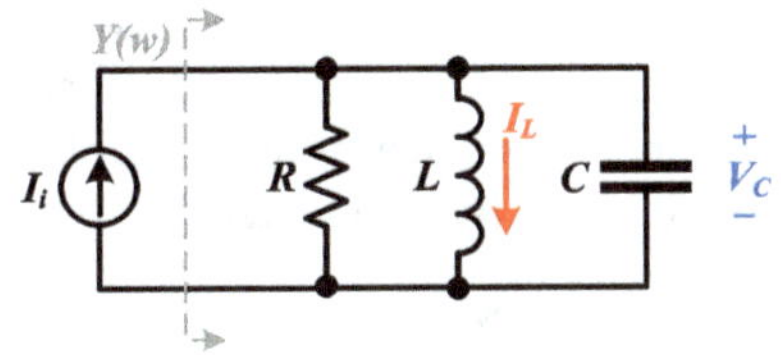

Figure 6.3 Parallel RLC circuit

However, at resonance, $w_n = 1/\sqrt{L.C}$. Thus:

$$Q = \frac{w_n.L}{R} = \frac{1}{\sqrt{L.C}}\cdot\left(\frac{L}{R}\right)Q = \frac{1}{R}\cdot\sqrt{\frac{L}{C}} \tag{6.5}$$

6.2.2 Parallel RLC resonant circuits

Figure 6.3 presents the classic configuration of a parallel RLC circuit. In this configuration, a current source feeds an RLC load, in which all elements of the circuit are in parallel. The admittance seen from the terminals of the current source I_i is given by:

$$Y(w) = \frac{1}{R} + j\left(w.C - \frac{1}{w.L}\right) = \frac{1}{R} + j.X(w) \tag{6.6}$$

The circuit is in resonance when the inductive admittance cancels with the capacitive admittance. Thus, $Y(w = w_n) = 1/R$ when:

$$Y(w_n) = w_n.C - \frac{1}{w_n.L} = 0 \quad w_n = \frac{1}{\sqrt{L.C}} \tag{6.7}$$

The resonant frequency given in hertz is the same as shown in (6.3).

The quality factor in a parallel RLC circuit is given by:

$$Q = w_n.R.C = \frac{R}{w_n.L} \tag{6.8}$$

6.3 Characteristics of soft switching converters

Resonant converters can be characterized as a set of topologies and switching strategies that aim to reduce commutation losses associated with power semiconductors, thus seeking to increase the overall circuit efficiency.

One of the main contributors to the loss of efficiency of converters is forced switching. This situation occurs when the semiconductor switches turn on and off under non-zero voltage and current, as in this transition the switching devices absorb power, as shown in Figure 6.4. The larger the highlighted area, the greater the switching losses and, consequently, the lower the converter efficiency.

Several switching strategies are used in the literature to avoid forced commutation and obtain soft switching [2–8]. Under this condition, the semiconductor switch turns on and off under zero voltage and/or current, thus avoiding simultaneous voltage and current transitions. Therefore, it enables a significant reduction of switching losses. There are basically two categories of soft switching:

- Zero current switching (ZCS)—occurs when the current in the switch is zero at the switching instant.
- Zero voltage switching (ZVS)—occurs when the voltage across the switch is zero at the switching instant.

Figure 6.5 shows the switching behavior for the two aforementioned cases.

Switching devices absorb power by turning on or off when passing through a transition if the current or voltage is not zero. With increasing switching frequency in converters, these transitions occur more often and, consequently, the average power lost on these devices rises. In resonant switching circuits, where the transition occurs with zero voltage or current, switching losses are eliminated. Among the types of resonant converters, the following stand out:

- Converters with resonant switch.
- Converters with resonant load.
- Converters with resonant DC link.

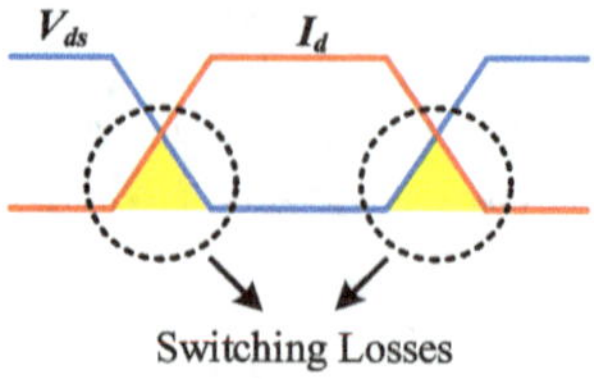

Figure 6.4 Voltage and current across a MOSFET with hard switching

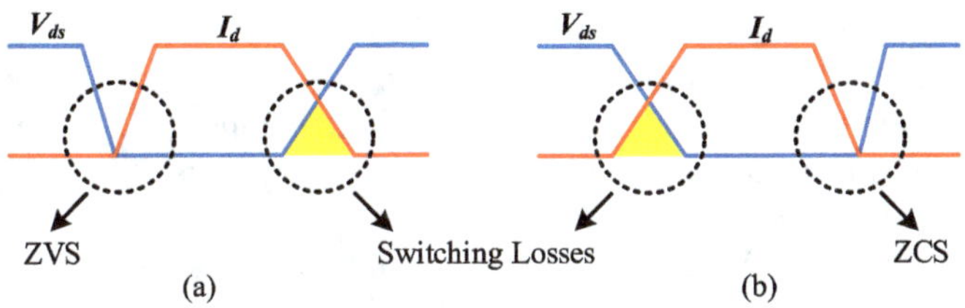

Figure 6.5 *Voltage and current across a soft-switched MOSFET. (a) Commutation in ZVS. (b) Commutation in ZCS.*

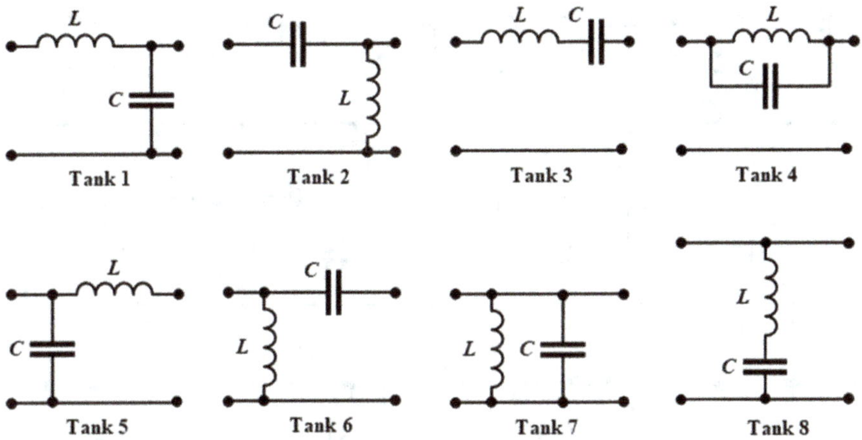

Figure 6.6 *Combinations in a two-element resonant tank*

In general, ZVS is preferred over ZCS for high frequencies. The reason is due to the intrinsic capacitances of semiconductor switches. When the switch is turned on under zero current (ZCS), but with voltage across its terminals, the charge stored in the internal capacitances is dissipated into the component. This phenomenon becomes more significant at very high frequencies. On the other hand, no losses occur in ZVS.

6.3.1 Energy storage elements for resonant converters

The use of energy storage elements (inductors and capacitors) is fundamental to obtaining soft-switching in power converters. There are several families of converters that use different configurations of these two elements, aiming to reduce switching losses in semiconductors. The most common configurations are those with arrangements of two, three, or four storage elements associated in series or parallel. Basically, there are eight possible combinations in a 2-element (inductor - L and capacitor - C) tank, as seen in Figure 6.6, 18 combinations with three elements, as presented in Figure 6.7, and a large amount with 4-element associations [9].

Among all the combinations illustrated in Figure 6.6, resonant tank circuits shown in Figure 6.6(a) and (c) have attracted more attention. These resonant tanks, which are called series load resonant tanks and parallel load resonant tanks respectively, form the traditional series resonant converters (SRC) and parallel resonant converters (PRC).

Both SRC and PRC have limitations that can affect the overall performance of the system. To overcome these shortcomings, hybrid resonant converters, in which the resonant tank is formed by three or more resonating elements, were proposed. There are numerous combinations to model a resonant tank network with three elements. Some of these combinations are shown in Figure 6.7.

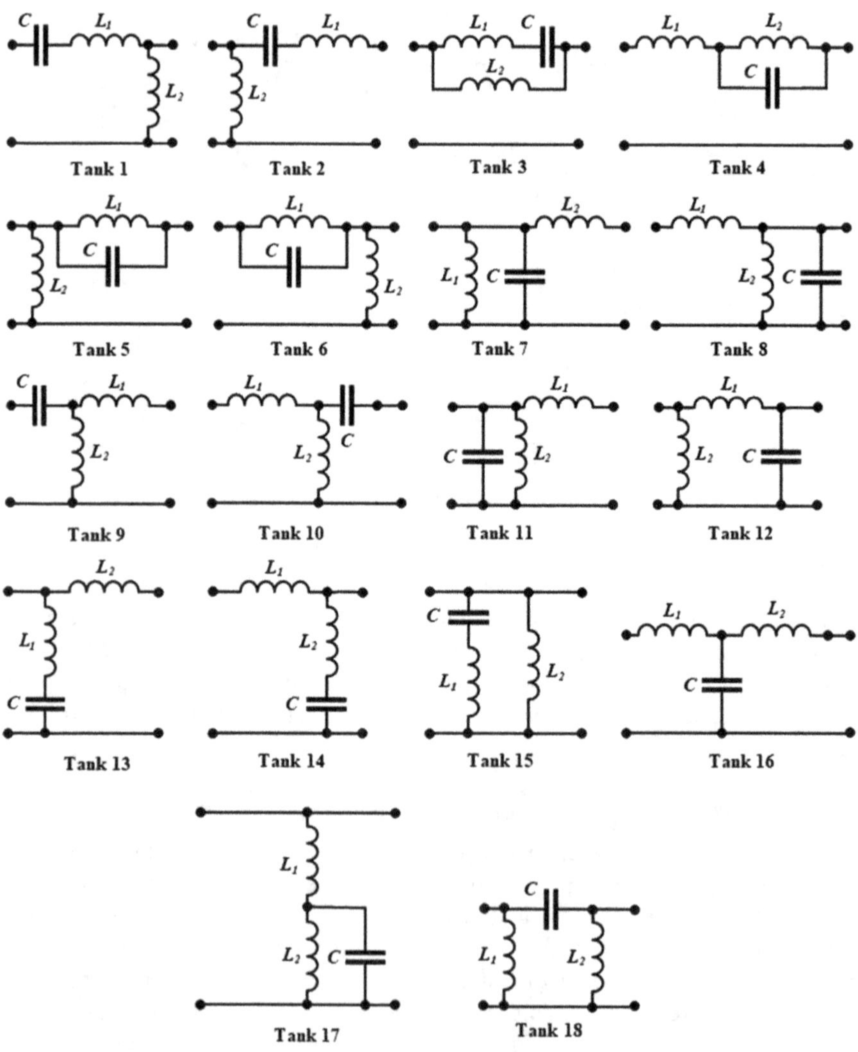

Figure 6.7 Combinations of a three-element resonant tank

6.4 Quasi-resonant converters

The quasi-resonant converters are those which, through the association of resonant elements (inductors and capacitors) with the semiconductor switches, make the commutation soft, with no energy dissipation during the power switches commutation. These types of converters aim to associate soft switching techniques present in resonant converters with different topologies, and may even be associated with classic topologies such as Buck, Boost, Buck-Boost, Cuk, and many others [10–18].

Resonant switches are represented by a subcircuit composed of an inductor L_r, a switch $S_{W,}$ and a capacitor C_r. The characteristic impedance of the resonant tank and the resonant frequency, in rad/s, are defined as:

$$Z_o = \sqrt{\frac{L_r}{C_r}} \tag{6.9}$$

$$w_o = \frac{1}{\sqrt{L_r.C_r}} \tag{6.10}$$

Figure 6.8 presents the basic structures of resonant switches, where it would be possible to obtain soft-switching by replacing the power switches for one of these auxiliary circuits [19].

If the ZCS configuration is adopted and the current that passes through the power switch is only in one direction, it is said that the configuration is in a half wave. Nevertheless, if the current through the switch can circulate in both directions, the configuration is said to be a full wave. Figure 6.9 shows the switch configurations in ZCS for a half-wave and full-wave arrangement, respectively [20–22].

As in the configuration previously presented in ZCS, there is also the ZVS configuration for the resonant switch, where the half-wave configuration occurs when the voltage across the power switch can only assume one polarity, and the full wave configuration when the voltage across the switch of power can assume both

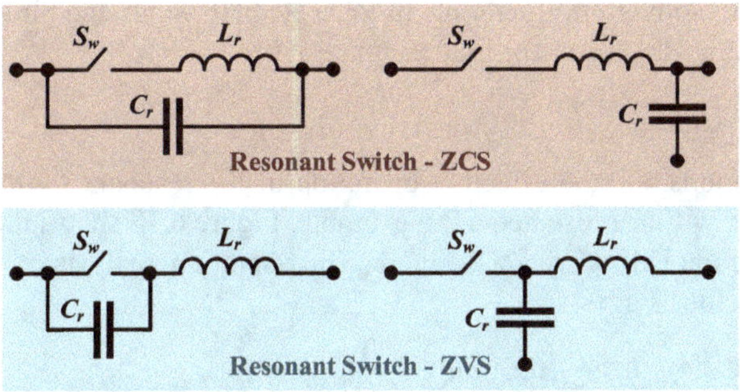

Figure 6.8 Resonant switches models

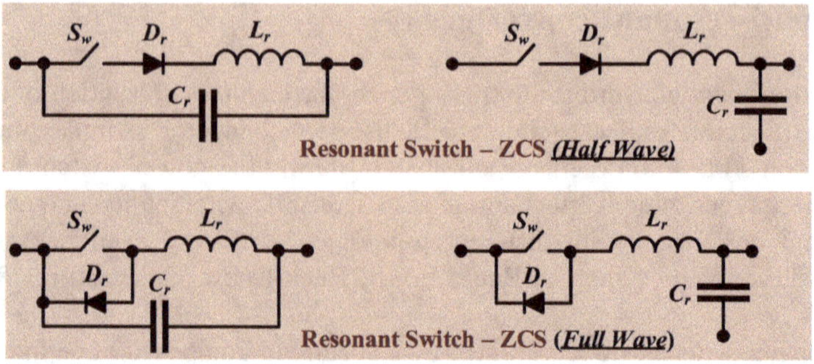

Figure 6.9 Half-wave and full-wave ZCS resonant switch models

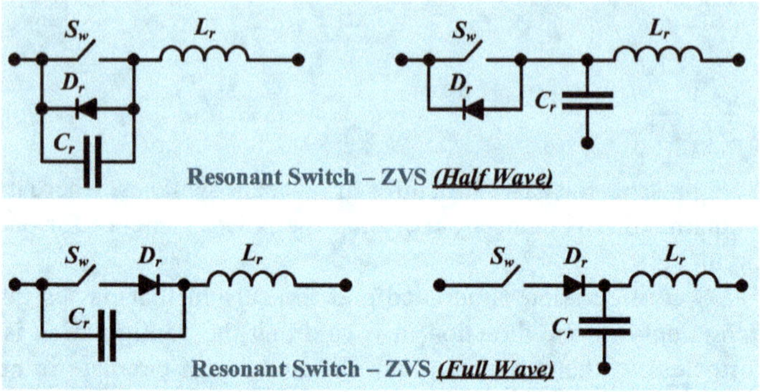

Figure 6.10 Half-wave and full-wave ZVS resonant switch models

polarities. Figure 6.10 shows the switch models in ZVS for a half-wave and full-wave arrangement.

Thus classic DC–DC converter topologies can be adapted to operate under ZCS and ZVS conditions, simply replacing the traditional power switch for the auxiliary circuits shown above. Figure 6.11 presents the configuration of the main classic DC–DC converters operating in ZCS in half wave and full wave, while Figure 6.12 shows the same configurations in ZVS.

6.4.1 Quasi-resonant ZCS boost converter

In order to demonstrate the effective operation of the resonant switches, the ZCS boost converter will be presented as an example. Figure 6.13 shows the circuit to be analyzed using the half-wave ZCS switch. The operation analysis of this converter is divided into four stages.

- *First stage* $[t_0, t_1]$*—*Δt_1 *(first linear stage):*

This stage begins when the switch S_w is turned on, and ends when the current through the resonant inductor i_{Lr} equals the input current I_{in}. During this step, the

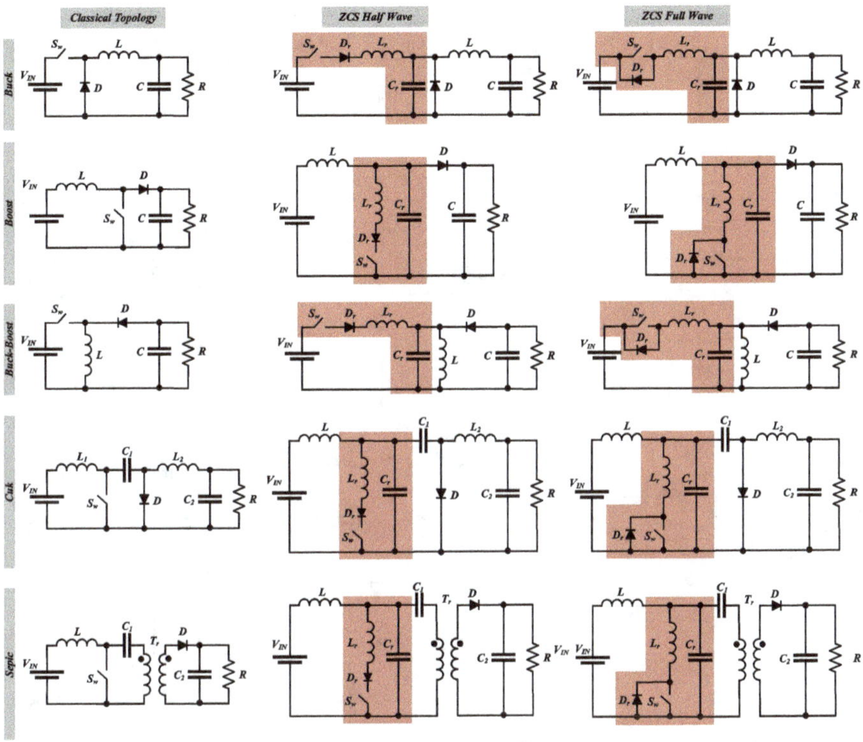

Figure 6.11 ZCS quasi-resonant classic converters

current in the resonant inductor increases linearly. Figure 6.14 shows the equivalent circuit for this stage.

The differential equation and the boundary condition are given by:

$$i_{Lr}(t) = \frac{V_o}{L_r}.t \tag{6.11}$$

$$\Delta t1 = \frac{\alpha}{w_o} \tag{6.12}$$

$$\alpha = \frac{I_{in}}{V_o}.\sqrt{\frac{L_r}{C_r}} \quad \alpha = \frac{I_{in}}{V_o}.Z_o \tag{6.13}$$

- Second stage $[t_1, t_2]$—Δt_2 (resonant stage):

This stage begins when the current in the resonant inductor i_{Lr} equals the input current I_{in}, and ends when the current in the resonant inductor i_{Lr} reaches zero. During this step, the resonant circuit oscillates. The current in the inductor L_r increases, passes through a maximum value, and decreases sinusoidally until it reaches zero. Meanwhile, the voltage across the resonant capacitor v_{Cr} decreases,

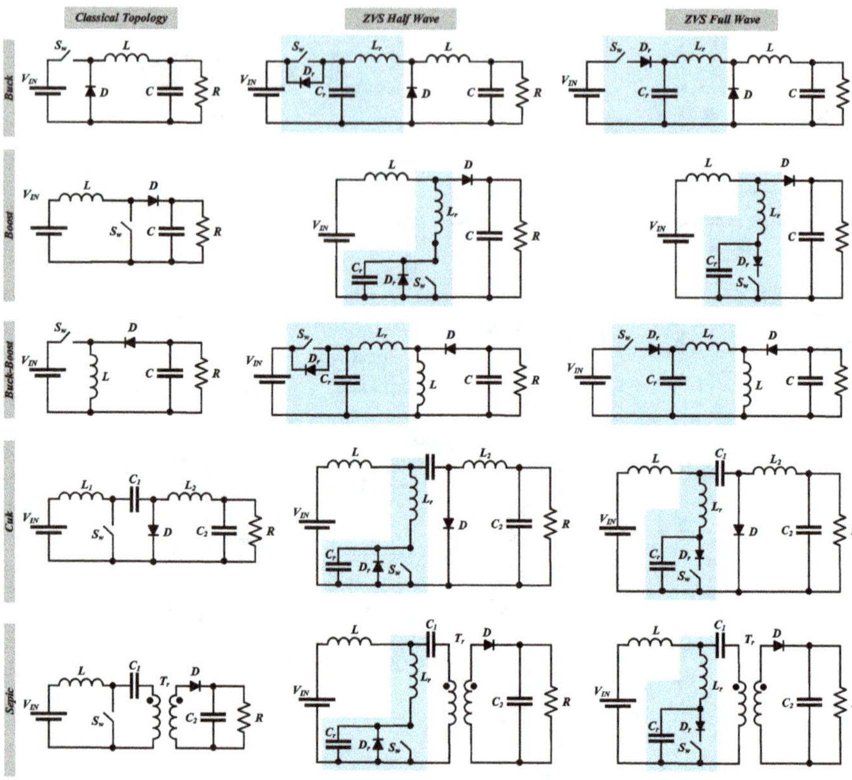

Figure 6.12 ZVS quasi-resonant classic converters

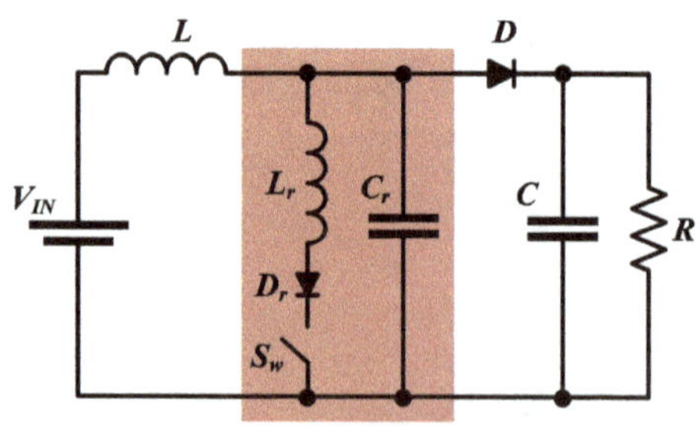

Figure 6.13 Quasi-resonant half-wave ZCS boost converter

reaches zero, continues decreasing, passes through a minimum value, and increases co-sinusoidally. In this way:

$$v_{Cr}(t) = V_o.\cos(w_o.t) \tag{6.14}$$

$$\sqrt{\frac{L_r}{C_r}}.i_{Lr}(t) = V_o.\sin(w_o.t) + \text{Iin}.\sqrt{\frac{L_r}{C_r}} \tag{6.15}$$

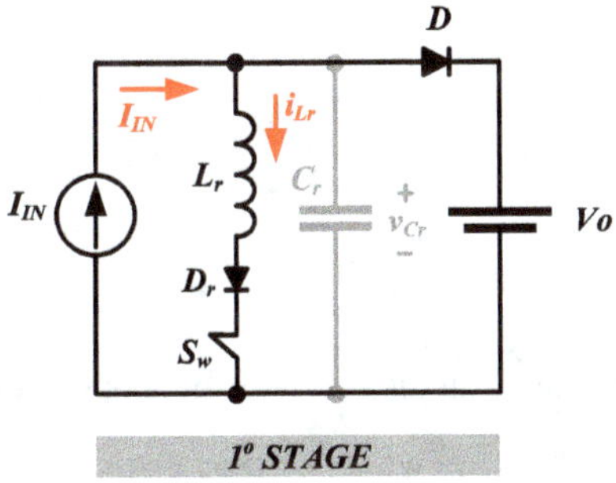

Figure 6.14 First stage of the half-wave ZCS quasi-resonant boost converter

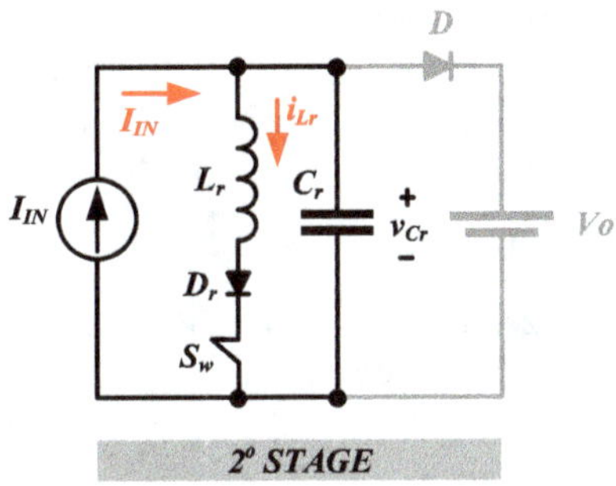

Figure 6.15 Second stage of the half-wave ZCS quasi-resonant boost converter

$$\Delta t2 = t2 - t1 \tag{6.16}$$

When $i_{Lr}\ (t) = 0$. Thus, the duration of this stage is given by:

$$\Delta t2 = \frac{1}{w_o}.\left(\pi + \sin^{-1}(\alpha)\right) \tag{6.17}$$

The following equation determines the voltage across the resonant capacitor at the end of this stage:

$$v'_{Cr} = -V_o.\sqrt{1-\alpha^2} \tag{6.18}$$

Figure 6.15 presents the equivalent circuit of this stage.

- Third stage $[t_2, t_3]$—Δt_3 (second linear stage):

This stage begins when the current through the resonant inductor i_{Lr} is zero, and ends when the voltage across the resonant capacitor v_{Cr} equals the output voltage V_o. During this step, the resonant capacitor charges linearly until it reaches the output voltage.

$$v_{Cr}(t) = v_{Cr} + \frac{\text{Iin}}{C_r}.t \tag{6.19}$$

$$\Delta t3 = t3 - t2 \tag{6.20}$$

When $v_{Cr}(t) = V_o$. Thus, the duration of this stage is given by:

$$\Delta t3 = \frac{1}{w_o} \cdot \left(\frac{1}{\alpha} - \sqrt{\left(\frac{1}{\alpha}\right)^2 - 1} \right) \tag{6.21}$$

Figure 6.16 presents the equivalent circuit of this stage.

- Fourth stage $[t_3, t_4]$—Δt_4 (freewheel stage):

This stage begins when the voltage across the resonant capacitor v_{Cr} equals the output voltage V_o, and ends when the switch S_w is turned on, initiating again the first stage. During this step, the current source I_{in} supplies power to the load (Figure 6.17).

Figure 6.18 presents the main waveforms of the half-wave ZCS quasi-resonant boost converter and their respective stages.

- Static gain and switch-off time

The equation that characterizes the static gain (V_o/V_{in}) of the quasi-resonant half-wave ZCS boost converter is presented below:

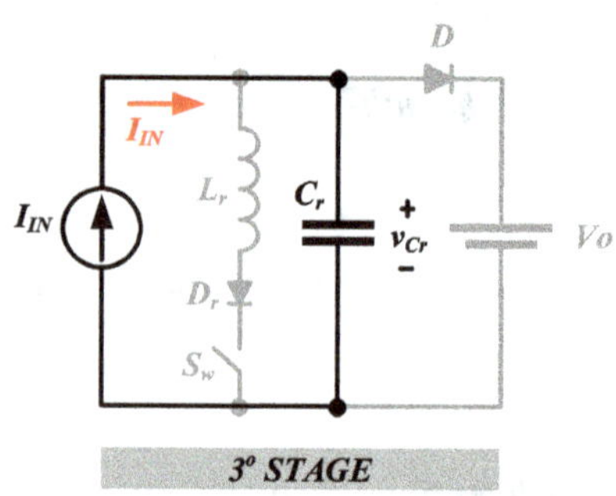

Figure 6.16 Third stage of the half-wave ZCS quasi-resonant boost converter

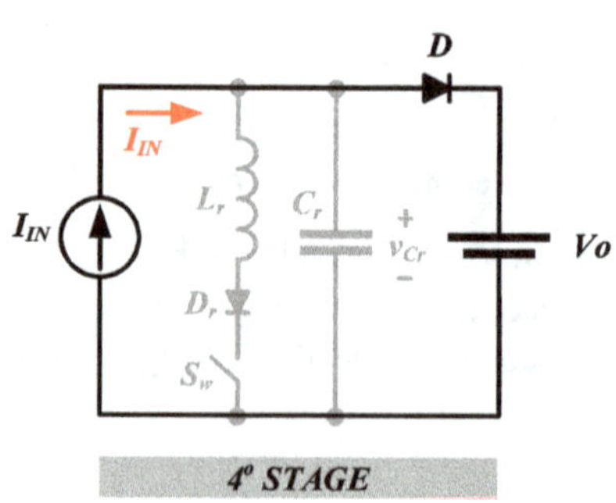

Figure 6.17 Fourth stage of the half-wave ZCS quasi-resonant boost converter

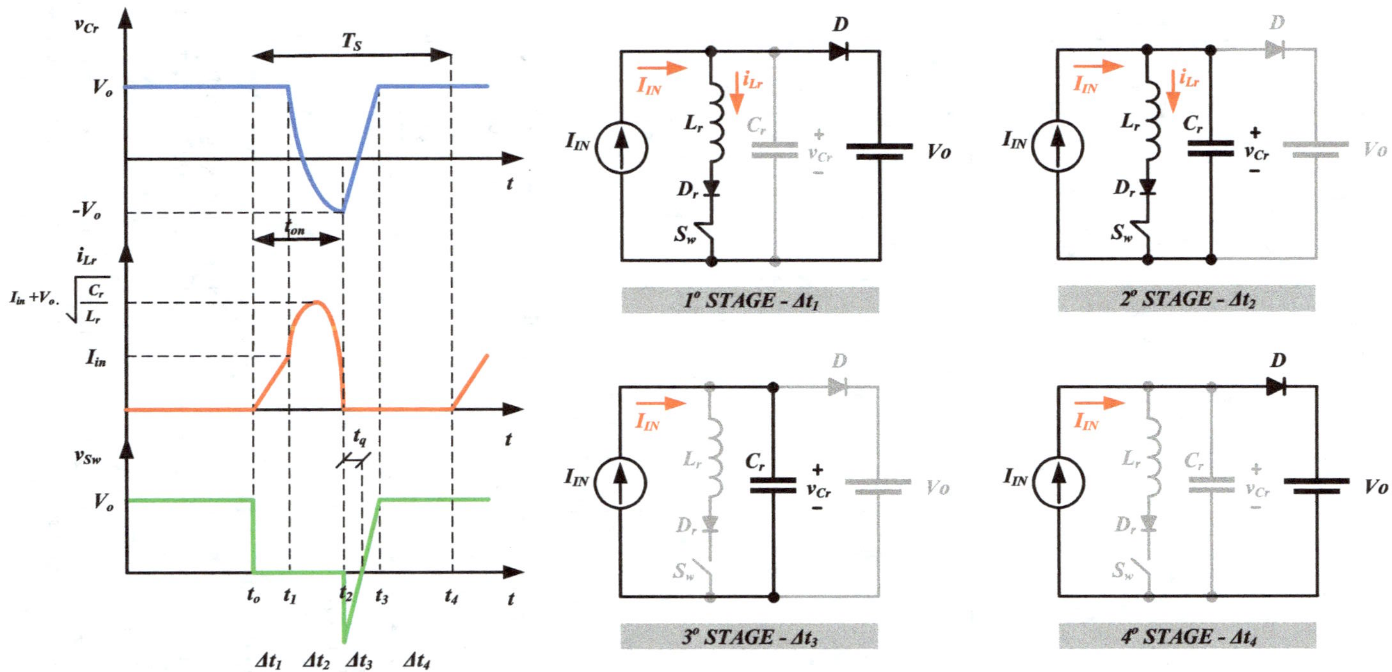

Figure 6.18   Main waveforms and operation stages

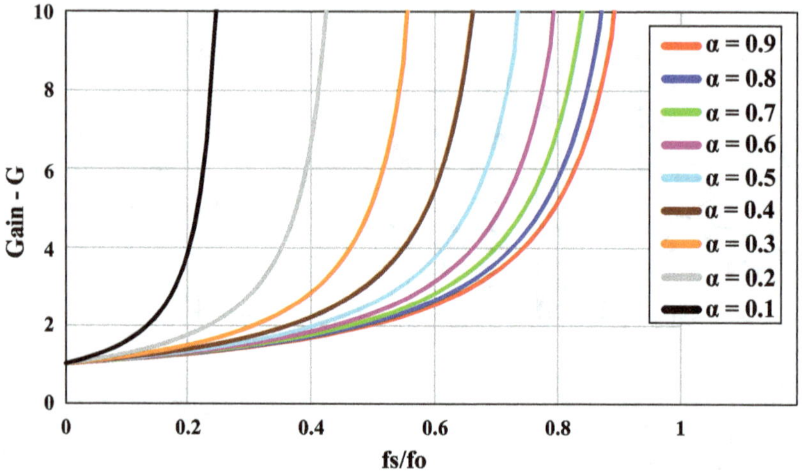

Figure 6.19 Static gain for α < 1

$$G = \frac{V_o}{V_{in}} = \frac{1}{1 - \frac{1}{2.\pi} \cdot \frac{fs}{fo} \cdot \left(\frac{\alpha}{2} + \frac{1}{\alpha} + \sqrt{\left(\frac{1}{\alpha}\right)^2 - 1} + \pi + \sin^{-1}(\alpha) \right)} \quad (6.22)$$

Where:

- f_s is the switching frequency, in Hz.
- f_o is the resonance frequency, in Hz.

Examining (6.22), it can be observed that the normal operation of the circuit is only obtained for ($\alpha \leq 1$). Figure 6.19 presents the relationship between the gain (G) of the converter and the parameterized frequency (f_s/f_o).

The following equation defines the available time for turning off switch S_w:

$$t_q = \frac{1}{w_o} \cdot \left(\sqrt{\left(\frac{1}{\alpha}\right)^2 - 1} \right) \quad (6.23)$$

- Step-by-step design example

Based on the circuit presented in Figure 6.13, determine the value of the resonant elements L_r and C_r, considering V_{in}=100 V, V_o = 400 V, f_s = 100 kHz, L = 1 mH, C = 3.75 μF, and P_o = 800 W.

Answer:

Bearing in mind that the converter gain is four times the input voltage and, based on Figure 6.19, it must be chosen a value of α less than 1 to determine the resonant frequency f_o, it was selected α = **0.8** for two reasons: first, for a value of α equal to 1, it would be taken into account that the converter is ideal, which does not happen in practice and could lead to problems in practice; second, is due to the fact that small values of α result in very high resonant peaks. For these reasons, it will be adopted $\alpha = 0.8$.

$$\alpha = \frac{I_{in}}{V_o}.Z_o \quad Z_o = \frac{V_o}{I_{in}}.\alpha \quad Z_o = \frac{400}{8}.(0.8) \quad \boldsymbol{Z_o = 40}$$

Based on Figure 6.19, it can be defined the ratio $f_s/f_o = 0.77$.

$$0.77 = \frac{f_s}{f_o} \quad f_o = \frac{100k}{0.77} \quad \boldsymbol{f_o = 129.9\ \mathrm{kHz}}$$

$$w_o = 2.\pi.f_o \quad \boldsymbol{w_o = 815.9\ \mathrm{krad/s}}$$

Isolating the value of C_r from (6.9):

$$C_r = \frac{L_r}{Z_o^2}$$

Substituting the value of C_r in (6.10):

$$L_r = \frac{Z_o}{w_o} \quad \boldsymbol{L_r = 49\ \mathrm{uH}}$$

In this way, it can be deduced that the value of C_r is:

$$C_r = \frac{L_r}{Z_o^2} \quad \boldsymbol{C_r = 30.6\ \mathrm{nF}}$$

The time periods of each stage are, respectively (Figure 6.20):

$$\Delta t1 = \frac{\alpha}{w_o} \quad \boldsymbol{\Delta t1 = 980.394\ \mathrm{ns}}$$

$$\Delta t2 = \frac{1}{w_o}.\left(\pi + \sin^{-1}(\alpha)\right) \quad \boldsymbol{\Delta t2 = 4.986\ \mathrm{us}}$$

$$\Delta t3 = \frac{1}{w_o}.\left(\frac{1}{\alpha} - \sqrt{\left(\frac{1}{\alpha}\right)^2 - 1}\right) \quad \boldsymbol{\Delta t3 = 2.451\ \mathrm{us}}$$

$$\Delta t4 = \frac{1}{f_s} - (\Delta t1 + \Delta t2 + \Delta t3) \quad \boldsymbol{\Delta t4 = 1.582\ \mathrm{us}}$$

$$t_{on} = \Delta t1 + \Delta t2 \quad \boldsymbol{t_{on} = 5.967\ \mathrm{us}}$$

6.4.2 Quasi-resonant ZVS boost converter

In order to demonstrate the effective operation of the resonant switches, the ZVS boost converter will be used as an example. Figure 6.21 presents the circuit to be analyzed using the half-wave ZVS switch. The auxiliary circuit used to obtain the

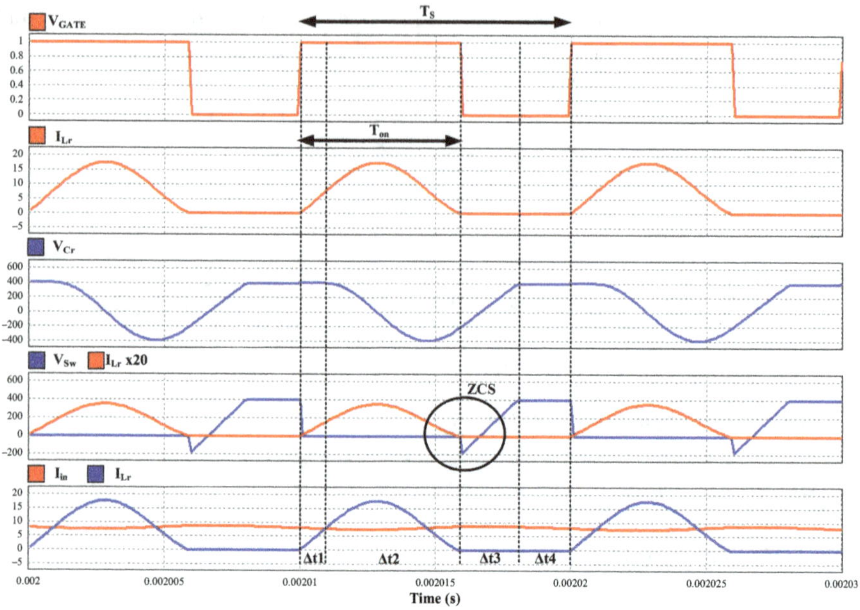

Figure 6.20 Main waveforms

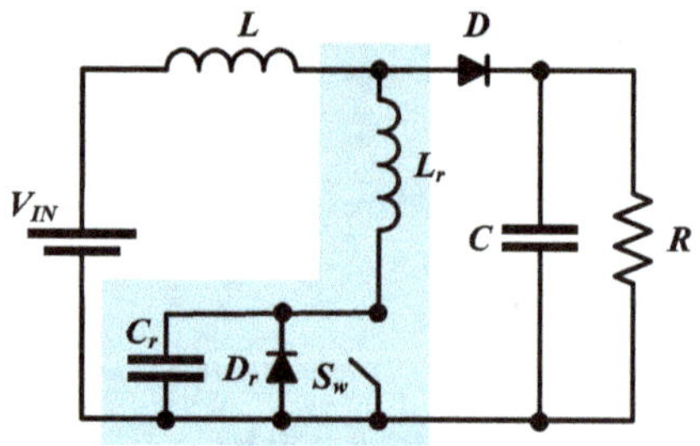

Figure 6.21 Quasi-resonant half-wave ZVS boost converter

soft-switching is basically composed of a diode D_r in anti-parallel with the switch and the capacitor C_r. The operation analysis of this converter is divided into four stages.

- First stage $[t_0, t_1]$—$\Delta t1$ (first linear stage):

This stage begins when the switch S_w is turned off, and ends when the voltage across the resonant capacitor v_{Cr} equals the output voltage V_o. During this step, the resonant capacitor charges linearly. Figure 6.22 shows the equivalent circuit of this stage.

The differential equation and boundary condition are given by:

$$v_{Cr}(t) = \frac{I_{in}}{C_r}.t \tag{6.24}$$

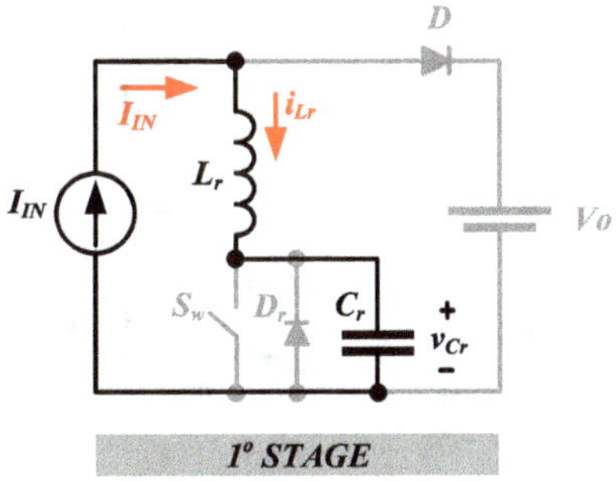

Figure 6.22 First stage of the half-wave ZVS quasi-resonant boost converter

$$\Delta t1 = \frac{1}{w_o.\alpha} \tag{6.25}$$

- <u>Second stage $[t_1, t_2]$—Δt_2 (resonant stage):</u>

This stage begins when the voltage across the resonant capacitor v_{Cr} equals the output voltage V_o, and ends when the voltage on the resonant capacitor v_{Cr} reaches zero. During this step, the resonant circuit oscillates. The voltage across the resonant capacitor C_r increases, passes through a maximum value, and decreases sinusoidally until it reaches zero. Meanwhile, the current in the resonant inductor i_{Lr} decreases, reaches zero, continues decreasing, passes through a minimum value, and increases co-sinusoidally. In this way:

$$v_{Cr}(t) = V_o + \text{Iin}.\sqrt{\frac{L_r}{C_r}}.\sin(w_o.t) \tag{6.26}$$

$$\sqrt{\frac{L_r}{C_r}}.i_{Lr}(t) = \text{Iin}.\sqrt{\frac{L_r}{C_r}} - \text{Iin}.\sqrt{\frac{L_r}{C_r}}.\cos(w_o.t) \tag{6.27}$$

$$\Delta t2 = t2 - t1 \tag{6.28}$$

When v_{Cr} $(t) = 0$. Thus, the duration of this stage is given by:

$$\Delta t2 = \frac{1}{w_o}.\left(\pi + \sin^{-1}\left(\frac{1}{\alpha}\right)\right) \tag{6.29}$$

The following equation determines the current through the resonant inductor at the end of this stage:

$$i_{Lr} = \text{Iin} + \text{Iin}.\sqrt{1 - \left(\frac{1}{\alpha}\right)^2} \tag{6.30}$$

Figure 6.23 presents the equivalent circuit of the second stage.

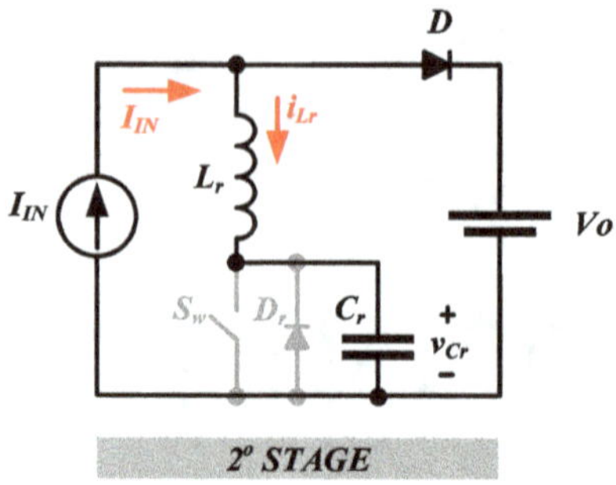

Figure 6.23 Second stage of the half-wave ZVS quasi-resonant boost converter

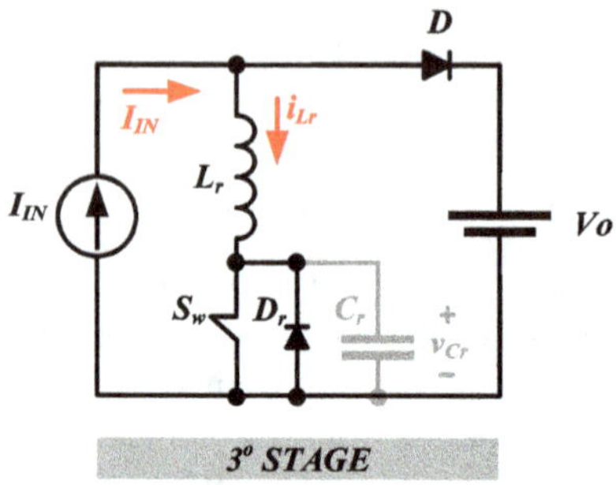

Figure 6.24 Third stage of the half-wave ZVS quasi-resonant boost converter

- Third stage $[t_2, t_3]$—Δt_3 (second linear stage):

This stage begins when the voltage across the resonant capacitor v_{Cr} is zero, and ends when the current through the resonant inductor i_{Lr} equals the input current I_{in}. During this step, the current in the resonant inductor i_{Lr} increases linearly until it reaches the input current.

$$i_{Lr}(t) = i_{Lr} + \frac{Vo}{L_r}.t \tag{6.31}$$

$$\Delta t3 = t3 - t2 \tag{6.32}$$

When i_{Lr} (t) = 0. Thus, the duration of this stage is given by:

$$\Delta t3 = \frac{1}{w_o}.\left(\alpha + \sqrt{\alpha^2 - 1}\right) \tag{6.33}$$

Figure 6.24 presents the equivalent circuit of the third stage.

- Fourth stage $[t_3, t_4]$—Δt_4 (freewheel stage):

This stage begins when the current in the resonant inductor i_{Lr} equals the input current I_{in}, and ends when the switch S_w is turned off, initiating again the first stage.

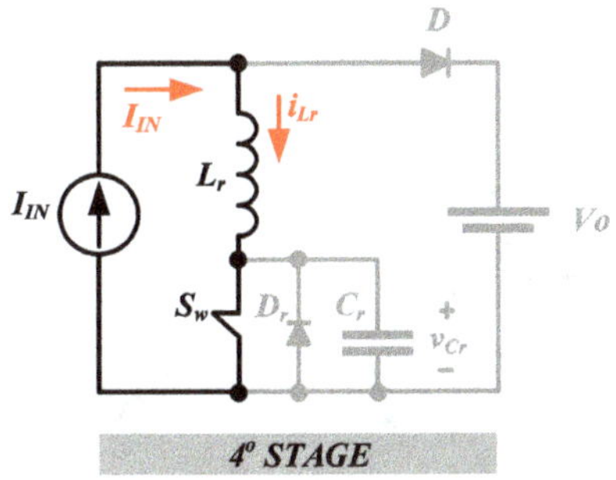

Figure 6.25 Fourth stage of the half-wave ZVS quasi-resonant boost converter

During this stage, the current source I_{in} is freewheeling through switch S_W. Figure 6.25 presents the equivalent circuit of this step.

Figure 6.26 presents the main waveforms of the half-wave ZVS quasi-resonant boost converter and their respective stages.

- Static gain and switch-off time:

The equation that characterizes the static gain (V_o/V_{in}) of the quasi-resonant half-wave ZVS boost converter is presented below:

$$G = \frac{V_o}{V_{in}} = \frac{1}{\frac{1}{2.\pi} \cdot \frac{fs}{fo} \cdot \left(\frac{1}{2.\alpha} + \alpha + \sqrt{\alpha^2 - 1} + \pi + \sin^{-1}\left(\frac{1}{\alpha}\right) \right)} \tag{6.34}$$

Where:

- f_s is the switching frequency, in Hz.
- f_o is the resonant frequency, in Hz.

Examining (6.34), it can be observed that the normal operation of the circuit is only obtained for ($\alpha \geq 1$). Figure 6.27 presents the relationship between the gain (G) of the converter and the parameterized frequency (f_s/f_o).

The following equation defines the available time for turning off switch S_W:

$$t_q = \frac{1}{w_o} \cdot \left(\sqrt{\alpha^2 - 1} \right) \tag{6.35}$$

- Step-by-step design example

Based on the circuit presented in Figure 6.21, determine the value of the resonant elements L_r and C_r, considering $V_{in} = 100$ V, $V_o = 400$ V, $f_s = 100$ kHz, $L = 1$ mH, $C = 3.75$ μF, and $P_o = 800$ W.

Answer:

Knowing that the converter gain is four times the input voltage and, based on Figure 6.27, it must be chosen a value of α greater than 1 to determine the resonant frequency f_o, it was selected α = **1.5** for a very important reason: values of α much

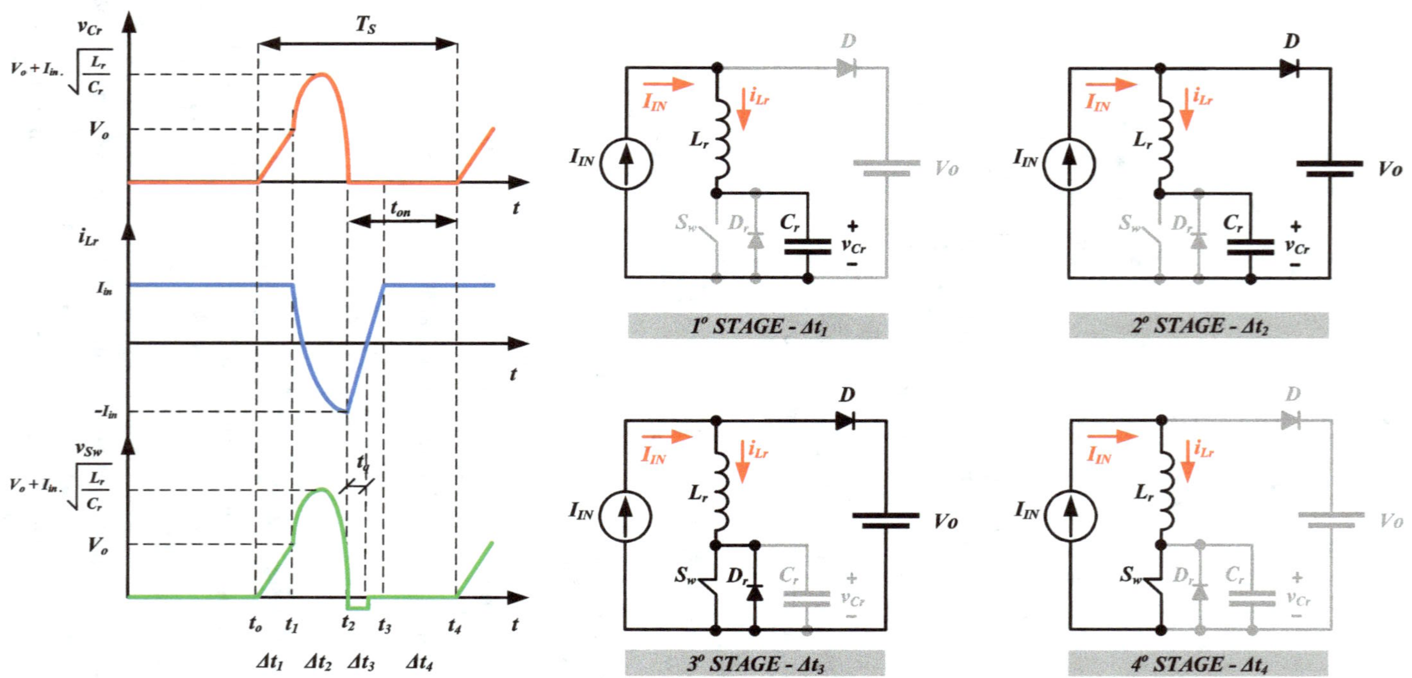

Figure 6.26 Main waveforms and operation stages

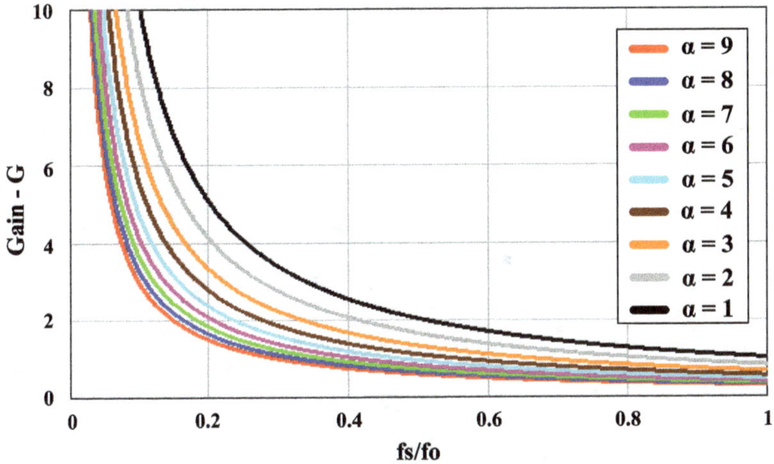

Figure 6.27 Static gain for α > 1

greater than 1 result in very small time periods for switching off S_w, despite reducing the resonant current peaks.

$$\alpha = \frac{I_{in}}{V_o}.Z_o \quad Z_o = \frac{V_o}{I_{in}}.\alpha \quad Z_o = \frac{400}{8}.(1.5)\ \boldsymbol{Z_o = 75}$$

Based on Figure 6.27, it can be defined the ratio $f_s/f_o = 0.23$.

$$0.23 = \frac{f_s}{f_o} \quad f_o = \frac{100k}{0.23} \quad \boldsymbol{f_o = 434.3\ \mathrm{kHz}}$$

$$w_o = 2.\pi.f_o \quad \boldsymbol{w_o = 2.729\ \mathrm{M\ rad/s}}$$

Isolating the value of C_r from (6.9):

$$C_r = \frac{L_r}{Z_o^2}$$

Substituting the value of C_r in (6.10):

$$L_r = \frac{Z_o}{w_o} \quad \boldsymbol{L_r = 27.4\ \mathrm{uH}}$$

In this way, it can be deduced that the value of C_r is:

$$C_r = \frac{L_r}{Z_o^2} \quad \boldsymbol{C_r = 4.88\ \mathrm{nF}}$$

The time periods of each stage are, respectively (Figure 6.28):

$$\Delta t1 = \frac{1}{w_o.\alpha} \quad \boldsymbol{\Delta t1 = 244.28\ \mathrm{ns}}$$

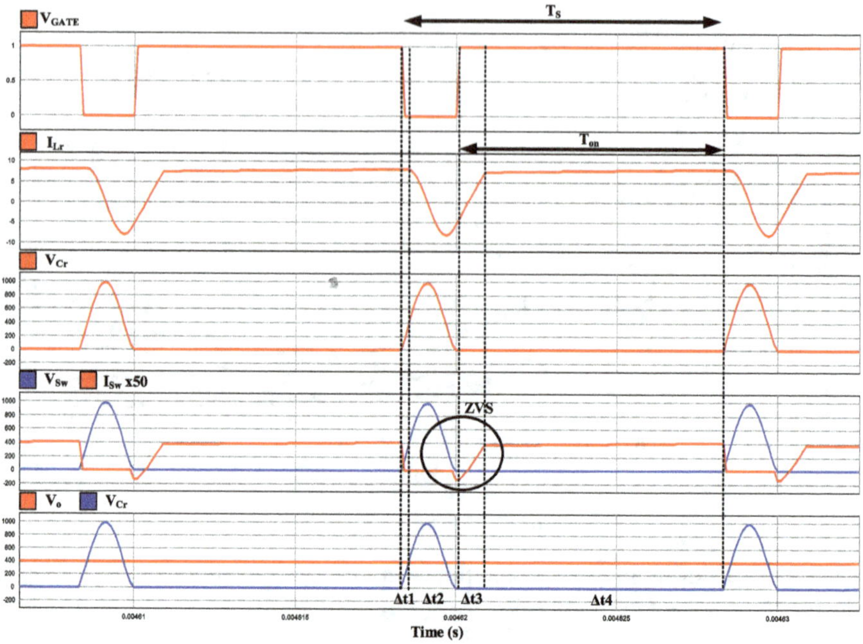

Figure 6.28 Main waveforms

$$\Delta t2 = \frac{1}{w_o}\cdot\left(\pi + \sin^{-1}\left(\frac{1}{\alpha}\right)\right) \qquad \mathbf{\Delta t2 = 1.419\ us}$$

$$\Delta t3 = \frac{1}{w_o}\cdot\left(\alpha + \sqrt{\alpha^2 - 1}\right) \qquad \mathbf{\Delta t3 = 959.312\ ns}$$

$$\Delta t4 = \frac{1}{f_s} - (\Delta t1 + \Delta t2 + \Delta t3) \qquad \mathbf{\Delta t4 = 7.378\ us}$$

$$t_{on} = \Delta t3 + \Delta t4 \qquad \mathbf{t_{on} = 8.33\ us}$$

6.4.3 Quasi-resonant ZVT converters

Quasi-resonant converters have another category to characterize soft-switching, called ZVT (zero-voltage transition). In this category of converters, other resonant elements are inserted so that switching losses can be further reduced, even if it is necessary to add auxiliary switches, diodes, inductors, and extra capacitors.

The ZVT technique is a very popular active soft switching method since it can effectively eliminate switching and capacitive turn-on losses, while the converter

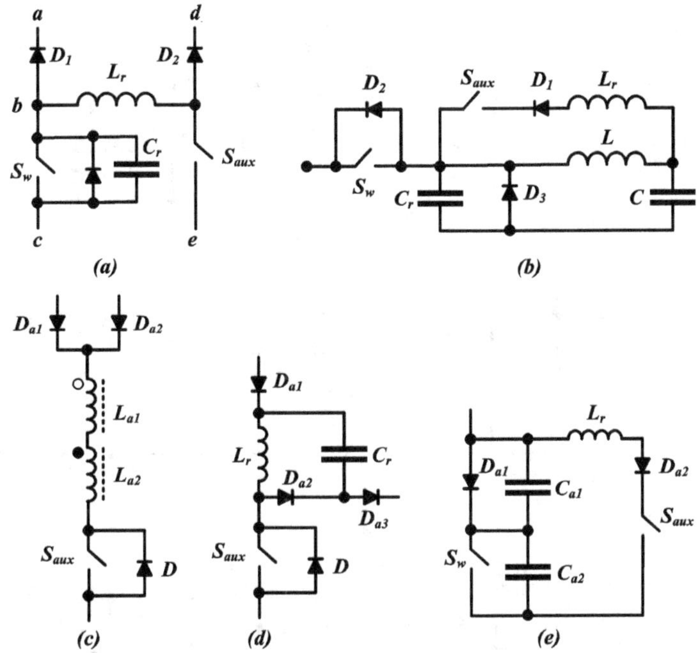

Figure 6.29 ZVT commutation cells based on: (a) F. C. Lee [31], (b) L. C Fretas [33], (c) B. Akhlaghi [26], (d) H. N. Tran [28], and (e) R. N. A. L. S. Aquino [34]

control keeps with pulse width modulation. Therefore, many researches have been carried out on ZVT converters [23–30].

Some ZVT converters require separate ZVT cells with active switches [31,32]. Figure 6.29 presents some soft-switching ZVT commutation cells.

The first ZVT switching cells were proposed by [31] and [33] in the early 90s, and, soon after, several other topologies appeared as can be seen in [35–40]. The article proposed by [31] presents the six classical converters with ZVT commutation which can be observed in Figure 6.30.

The converters presented in Figure 6.30 do not present additional voltage stresses compared to the basic PWM converters. The main switch will not present overcurrent, however, a small current stress may appear in the resonant inductor L_r and in the auxiliary switch S_{aux}. This current stress will not cause a significant reduction in efficiency, since the auxiliary switch conducts current for a small time interval, just so that the ZVS commutation takes place on the main switch. The auxiliary switch turn-on will occur in ZCS, while its turn-off will be dissipative.

This soft switching technique brings a significant reduction in switching losses and also considerably reduces the size and volume of the used heatsink when compared to the conventional PWM converter. The presence of auxiliary

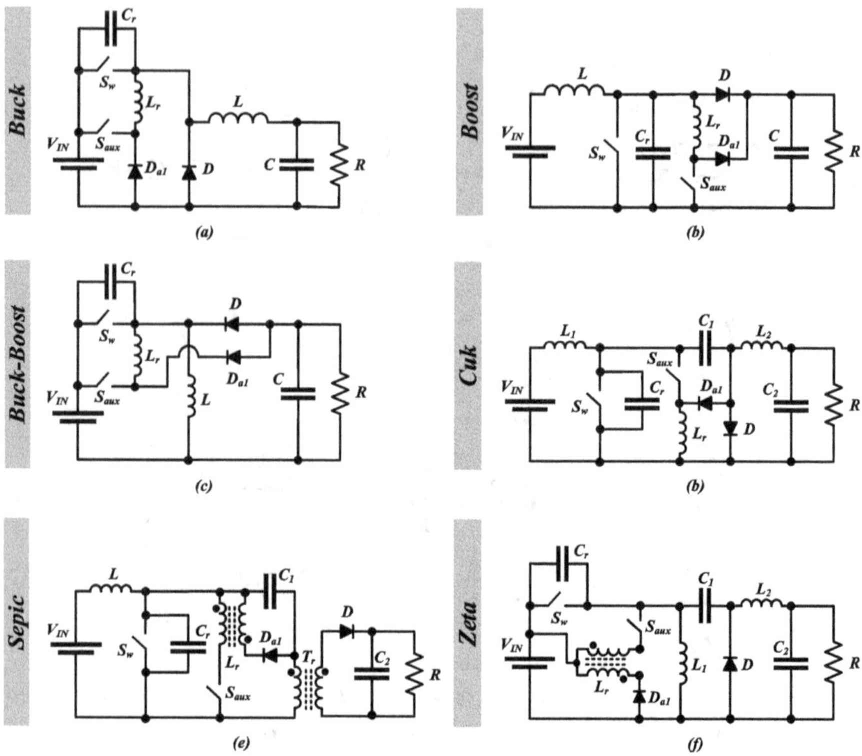

Figure 6.30 Classical topologies of the ZVT-PWM converters. (a) Buck, (b) Boost, (c) Buck-Boost, (d) Cuk, (e) Sepic, and (f) Zeta.

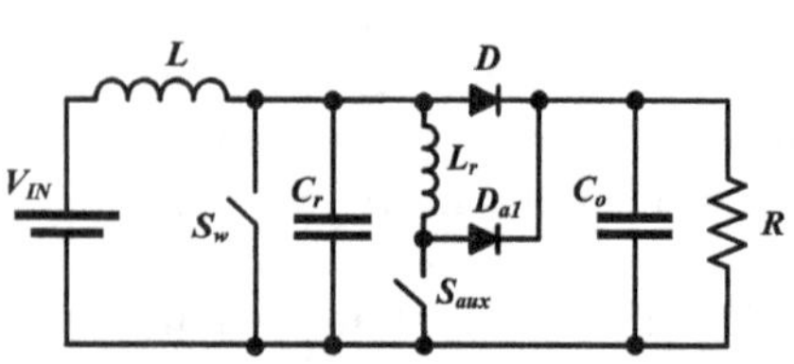

Switching Frequency:	fs = 300 KHz
Switching Period:	Ts = 33.33 us
Output Power:	Po = 600 W
Input Voltage:	Vin = 175 V
Input Current:	Iin = 3.42 A
Output Voltage:	Vo = 300 V
Resonant Inductor:	Lr = 10 uH
Resonant Capacitor:	Cr = 330 pF
Duty Cycle Sw:	Daux = 0.061
Duty Cycle Saux:	D = 0.4 (22.08° Delayed)
Boost Inductor:	L = 0.2 mH
Boost Capacitor:	Co = 10 uF

Figure 6.31 Circuit diagram of the ZVT boost converter

semiconductors will not contribute to the reduction in the total efficiency of the structure. Figure 6.31 presents the circuit and specifications of the ZVT boost converter proposed by [31], while Figure 6.32 shows its theoretical waveforms.

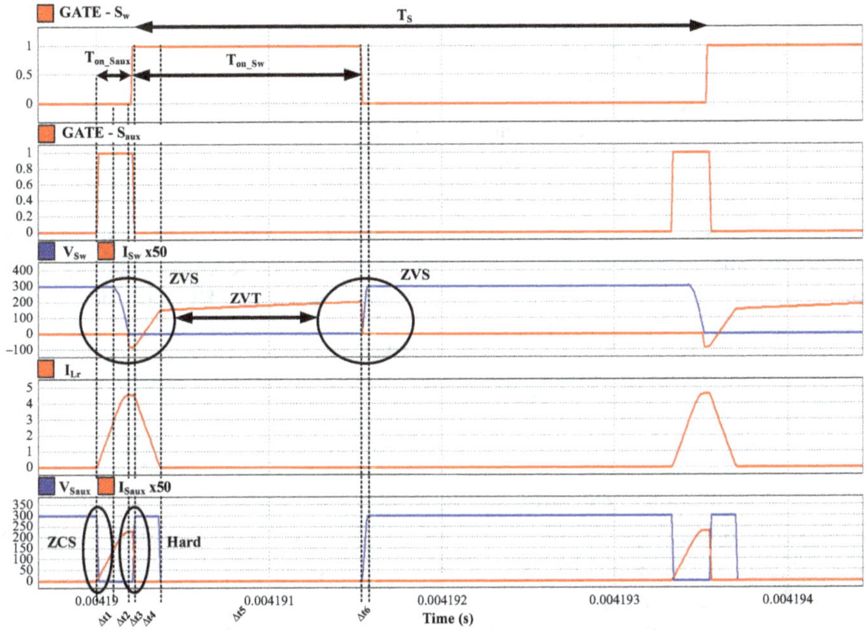

Figure 6.32 Circuit diagram waveforms

6.5 Resonant converters

The resonant converters are defined as the combination of converter topologies and switching strategies that result in zero-voltage and/or zero-current switching.

The dissipative commutation of converter switches is one of the main contributors to the reduction of efficiency and consequently of power loss in converters. Switched devices absorb power during their turn-on or -off transitions when voltage or current is non-zero. As the switching frequency increases, these transitions occur more often and the average power lost in these devices increases, resulting in high temperatures and shortening the lifespan of semiconductors. On the other hand, high switching frequencies decrease the size, weight, and volume of the magnetic elements of the converters, which is a very desirable feature for the overall design. Furthermore, high values of di/dt and dv/dt are potential causes of electromagnetic interference (EMI).

It will now analyze some classic topologies that make such non-dissipative switching possible. The load "seen" by the converter is formed by a resonant circuit and a source (voltage or current). The proper sizing of the L/C pair causes the current and/or voltage to invert, allowing switches to be switched in situations of zero current and/or voltage, eliminating commutation losses. One way to categorize resonant converters is:

- Quasi-resonant converters.
- Load resonant converters.
- Resonant DC link converters.

The analyses made on this topic were based on [41].

6.5.1 Load resonant converters

In these converter circuits, the power flow to the load is controlled by the resonant impedance tank, which, in turn, is controlled by the switching frequency f_s, in comparison to the resonant frequency f_o, of the tank. These DC/DC and DC/AC converters can be subclassified as follows:

- Series-load resonant (SLR) converters.
- Parallel-load resonant (PLR) converters.
- Hybrid-resonant converters.

6.5.1.1 Series-load resonant converters

Figure 6.33a presents the configuration of a half-bridge series resonant converter (SLR). The resonant tank consists of an inductor L_r and a capacitor C_r connected in series with the load through a full-wave rectifier bridge.

In this configuration, the output current low-pass filter is composed by the output capacitor C_o and the load R_o, and can be characterized by a ripple-free voltage source with a value equal to V_o. Switches S_1 and S_2 operate complementary and, together with their respective antiparallel diodes D_1 and D_2, define the switched voltage at the input of the resonant tank, v_s. Figure 6.33b shows the simplified circuit that will be used to describe the three conduction modes of this converter.

The classic voltage gain control is related to frequency modulation, with the duty cycle of the switches at 50% for continuous conduction modes, and less than 50% for the discontinuous conduction mode, while keeping a phase shift of 180° between switches. Some papers with this converter can be seen in [42–44].

A. Discontinuous conduction mode (DCM)—$w_s < 0.5\ w_o$

In this operating mode, the switches turn-on and -off are soft, in ZCS. In this way, there are virtually no switching losses. Furthermore, the peak voltage across the

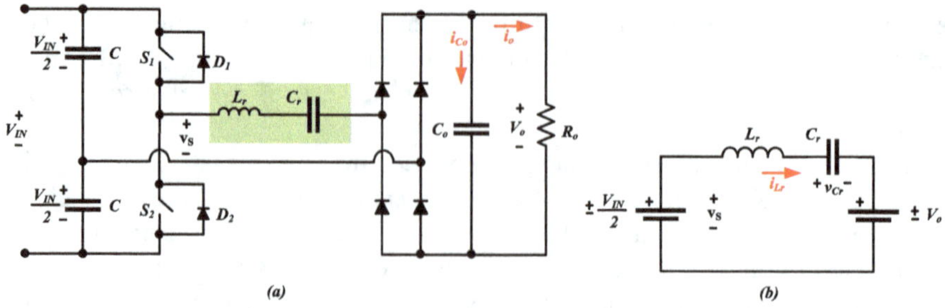

Figure 6.33 Series-load resonant converter. (a) Circuit. (b) Simplified model.

resonant capacitor is limited to the value of the input voltage V_{in}. However, the current peaks in the switches are high, increasing conduction losses.

In continuous conduction mode, the switching frequency varies between $0.5f_o \leq f_s \leq f_o$, but in discontinuous conduction mode, it varies from $0 \leq f_s \leq 0.5f_o$, and the limit between the two conduction modes is called critical conduction mode, and occurs at $f_s = 0.5f_o$.

The discontinuous conduction mode has six stages of operation, and the first three are analogous to the other three stages. Figure 6.34 shows the equivalent circuits of the first three stages of operation for the discontinuous mode.

In this operation mode it must be taken into account the initial voltage of the resonant capacitor $v_{Cr}(0)$, which is different from zero, and the initial current through the resonant inductor $i_{Lr}(0)$, which is equal to zero.

- First stage $[t_0, t_1]$—Δt_1:

In this stage, switch S_1 is turned on, the current through the resonant inductor is zero, and the voltage across the resonant capacitor is negative, i.e., $i_{Lr}(0) = 0$ and $v_{Cr}(0) < 0$. The current through the inductor increases sinusoidally, and the capacitor, which has a negative voltage, begins to discharge. During this stage, the source transfers energy to the load. Analyzing the voltage in the equivalent circuit of the first stage:

$$\frac{di_{Lr}(t)}{dt} = \frac{\frac{V_{IN}}{2} - v_{Cr}(t) - V_o}{L_r} \tag{6.36}$$

Using the current ratio in a capacitor, and being $i_{Lr}(t) = i_{Cr}(t)$, it can be stated that:

$$\frac{dv_{Cr}(t)}{dt} = \frac{i_{Lr}(t)}{C_r} \tag{6.37}$$

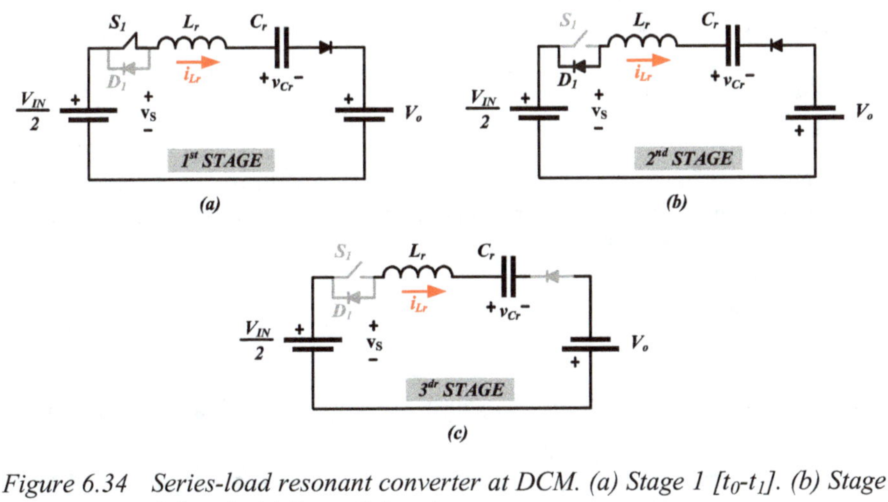

Figure 6.34 Series-load resonant converter at DCM. (a) Stage 1 [t_0-t_1]. (b) Stage 2 [t_1-t_2]. (c) Stage 3 [t_2-t_3].

By manipulating (6.36) and (6.37), a second-order equation is obtained. Using the initial conditions of $i_{Lr}(0)$ and $v_{Cr}(0)$, the following equation can be achieved:

$$i_{Lr}(t) = \frac{\frac{V_{IN}}{2} - v_{Cr}(0) - V_o}{Z_o}.\sin(w_o.t) \tag{6.38}$$

Thus, the voltage across capacitor C_r is:

$$v_{Cr}(t) = \frac{V_{IN}}{2} - V_o - \left(\frac{V_{IN}}{2} - v_{Cr}(0) - V_o\right).\cos(w_o.t) \tag{6.39}$$

This stage lasts until the current through L_r reaches zero and D_1 conducts, making it possible to softly turn-off switch S_1. Thus:

$$0 = \frac{\frac{V_{IN}}{2} - v_{Cr}(0) - V_o}{Z_o}.\sin(w_o.t_1) \tag{6.40}$$

It can be concluded that the duration of this stage is $\Delta t_1 = t_1\text{-}t_0$, where $t_0 = 0$ is:

$$\Delta t_1 = \frac{\pi}{w_o} \tag{6.41}$$

The voltage across the resonant capacitor C_r at t_1 reaches its peak and is given by:

$$v_{Cr}(t_1) = V_{IN} - 2.V_o - v_{Cr}(0) \tag{6.42}$$

- Second stage $[t_1, t_2]$—Δt_2:

In this stage, switch S_1 is blocked and diode D_1 is conducting. Because the current in the resonant inductor has a negative direction, the conduction of the rectifier bridge diodes reverses the polarity of the output voltage source connected to the resonant tank. Based on the voltage across the inductor L_r:

$$\frac{di_{Lr}(t)}{dt} = \frac{\frac{V_{IN}}{2} - v_{Cr}(t) - V_o}{L_r} \tag{6.43}$$

Using the current ratio in a capacitor, in conjunction with (6.43), a second-order equation is obtained. Considering the initial conditions $i_{Lr}(t_1) = 0$ and $v_{Cr}(t_1)$, the following equations are achieved:

$$i_{Lr}(t) = \frac{\frac{V_{IN}}{2} - v_{Cr}(0) + V_o}{Z_o}.\sin(w_o.[t_2 - t_1]) \tag{6.44}$$

Thus, the voltage across capacitor C_r can be calculated as:

$$v_{Cr}(t) = \frac{V_{IN}}{2} + V_o - \left(\frac{V_{IN}}{2} - v_{Cr}(0) + V_o\right).\cos(w_o.[t_2 - t_1]) \tag{6.45}$$

It can be concluded that the duration of this stage $\Delta t_2 = t_2\text{-}t_1$ is:

$$\Delta t_2 = \frac{\pi}{w_o} \tag{6.46}$$

This stage ends when the current through the resonant inductor reaches zero. The voltage across the resonant capacitor C_r at instant t_2 can be calculated as:

$$v_{Cr}(t_2) = 4.V_o + v_{Cr}(0) \tag{6.47}$$

By operation symmetry, the equality that $v_{Cr}(t_2) = -v_{Cr}(0)$ is valid. Therefore, the values of the voltages across the resonant capacitor at the instants $t = 0$ and $t = t_2$ are:

$$v_{Cr}(0) = -2.V_o \tag{6.48}$$

$$v_{Cr}(t_2) = 2.V_o \tag{6.49}$$

- Third stage $[t_2, t_3]$—Δt_3:

When the current through the resonant inductor i_{Lr} reaches zero, at t_2, diode D_1 is blocked. As there is no other switch conducting, there will be no current passing through the circuit, causing the voltage in C_r to remain constant and equal to $2V_o$, until the beginning of the next half-switching cycle, in $T_s/2$, when switch S_2 is turned on. In this stage, there is no current circulation, as can be seen in Figure 6.34c.

The voltage gain is done by controlling the duration of steps 3 and 6. Changing these times results in changing the switching frequency, given that the durations of the other stages are constant, and depend only on the values of the components of the resonant tank and the output voltage. Figure 6.35 shows the main elements waveforms for this conduction mode.

B. Continuous conduction mode (CCM)—$0.5w_o < w_s < w_o$

In this operation mode, the switches operate complementary, with a 50% duty cycle. In this condition, it must be taken into account the initial conditions of the resonant capacitor $v_{Cr}(0)$ and the initial current in the resonant inductor $i_{Lr}(0)$, as both differ from zero.

This operation mode has four stages. However, as in the discontinuous mode, only the first two stages of operation will be analyzed, i.e., until half of the switching period ($T_s/2$), since the other two stages are analogous. In this switching condition where $0.5w_o < w_s < w_o$, there can be observed a situation where there is no current discontinuity, so one of the commutations is dissipative. The switches turn-on is dissipative, while their turn-off is non-dissipative, in ZCS. In addition, the peak voltage across the resonant capacitor, which is a function of the switching frequency, can reach high values, even exceeding the input voltage of the source. Figure 6.36 shows the equivalent circuits of the first two stages of operation for the continuous conduction mode.

- First stage $[t_0, t_1]$—Δt_1:

In this stage, switch S_1 is turned on when the current i_{Lr} and the voltage v_{Cr} are not zero. The analysis of this stage starts when the current in the resonant inductor L_r is positive and the voltage on the resonant capacitor C_r is negative.

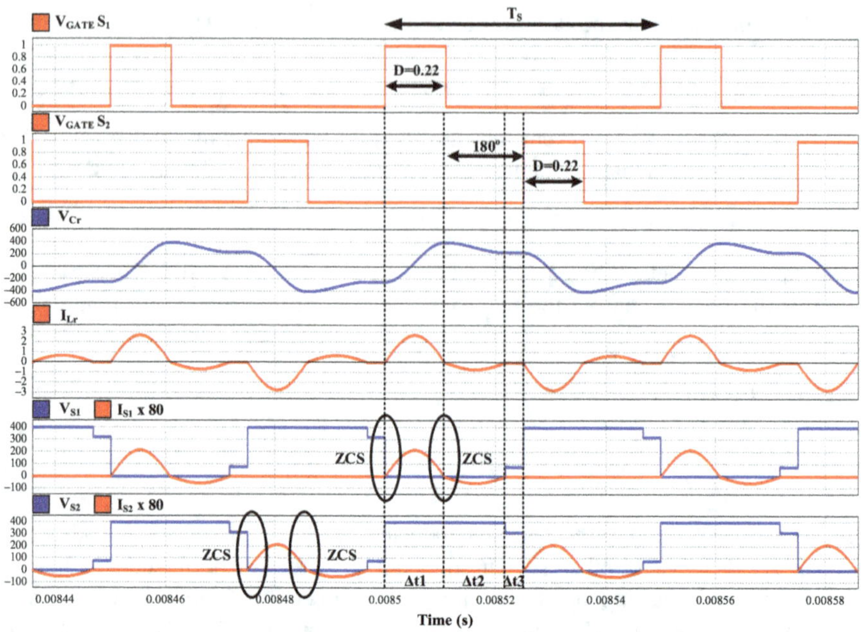

Figure 6.35 Series-load resonant converter at DCM waveforms

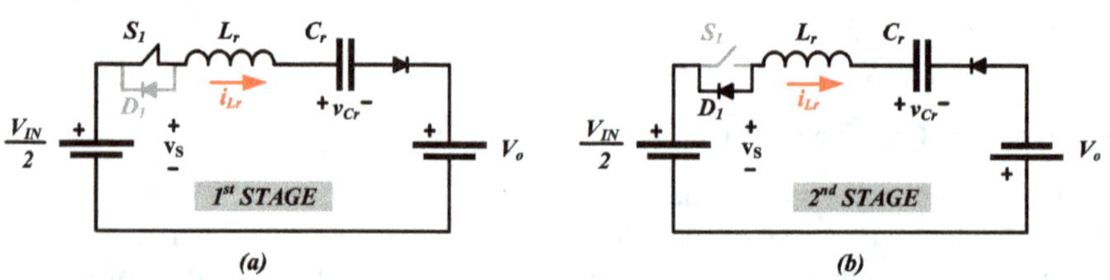

Figure 6.36 Series-load resonant converter at CCM. (a) Stage 1 [t_0-t_1]. (b) Stage 2 [t_1-t_2].

The analysis of the circuit results in:

$$\frac{di_{Lr}(t)}{dt} = \frac{\frac{V_{IN}}{2} - v_{Cr}(t) - V_o}{L_r} \tag{6.50}$$

The current in the resonant capacitor is given by the following equation:

$$\frac{dv_{Cr}(t)}{dt} = \frac{i_{Lr}(t)}{C_r} \tag{6.51}$$

The equation that defines the current in the resonant inductor during this stage is given by:

$$i_{Lr}(t) = \frac{\frac{V_{IN}}{2} - v_{Cr}(0) - V_o}{Z_o}.\sin(w_o.t) + i_{Lr}(0).\cos(w_o.t) \tag{6.52}$$

The voltage across the resonant capacitor C_r is given by the following equation:

$$v_{Cr}(t) = \frac{V_{IN}}{2} - V_o - \left(\frac{V_{IN}}{2} - v_{Cr}(0) - V_o\right).\cos(w_o.t) + Z_o.i_{Lr}(0).\sin(w_o.t) \tag{6.53}$$

This stage continues until the current in the resonant inductor reaches zero, which means that:

$$0 = \frac{\frac{V_{IN}}{2} - v_{Cr}(0) - V_o}{Z_o}.\sin(w_o.t_1) + i_{Lr}(0).\cos(w_o.t_1) \tag{6.54}$$

Manipulating the previous equation, we have:

$$\Delta t_1 = \frac{1}{w_o}.\tan^{-1}\left(-\frac{i_{Lr}(0).Z_o}{\frac{V_{IN}}{2} - v_{Cr}(0) - V_o}\right) \tag{6.55}$$

When the current in the resonant inductor is zero, the voltage across the resonant capacitor will reach its maximum value.
When $i_{Lr}(t_1) = 0$, the voltage across capacitor C_r, at time t_1, denoted as $v_{Cr}(t_1)$, will have its maximum value.

- Second stage $[t_1, t_2]$—Δt_2:

This stage begins when the current in the resonant inductor reaches zero, $i_{Lr} = 0$. At this point, diode D_1 conducts due to the reversal of the current direction in the resonant inductor, while switch S_1 can be turned off without switching losses, in ZCS. As current i_{Lr} has its direction reversed, the rectifier bridge connects the resonant tank to the output source with a negative value, as shown in Figure 6.36b. The analysis of the circuit results in:

$$\frac{di_{Lr}(t)}{dt} = \frac{\frac{V_{IN}}{2} - v_{Cr}(t) + V_o}{L_r} \tag{6.56}$$

The current in the resonant capacitor is given by the following equation:

$$\frac{dv_{Cr}(t)}{dt} = \frac{i_{Lr}(t)}{C_r} \tag{6.57}$$

The equation that defines the current in the resonant inductor during this stage is given by:

$$i_{Lr}(t) = \frac{\frac{V_{IN}}{2} - v_{Cr}(t_1) + V_o}{Z_o}.\sin(w_o.\Delta t_2) \tag{6.58}$$

The voltage across the resonant capacitor C_r is given by the following equation:

$$v_{Cr}(t) = \frac{V_{IN}}{2} + V_o - \left(\frac{V_{IN}}{2} - v_{Cr}(t_1) + V_o\right).\cos(w_o.\Delta t_2) \tag{6.59}$$

The end of this stage occurs when switch S_2 turns on at time $T_s/2$. Figure 6.37 shows the waveforms in the main elements for this mode of operation.

C. Continuous conduction mode (CCM)—$w_s \geq w_o$

In this operating mode, the switches operate in the same way as in topic B, i.e., with a duty cycle of 50% in both switches, operating complementarily. In general, the resonant behavior of the current i_{Lr} and the voltage v_{Cr} have general equations for temporal evaluation, which can be observed below:

$$i_{Lr}(t) = \frac{V - v_{Cr}(t_i)}{Z_o}.\sin(w_o.t - t_i) + i_{Lr}(t_i).\cos(w_o.t - t_i) \tag{6.60}$$

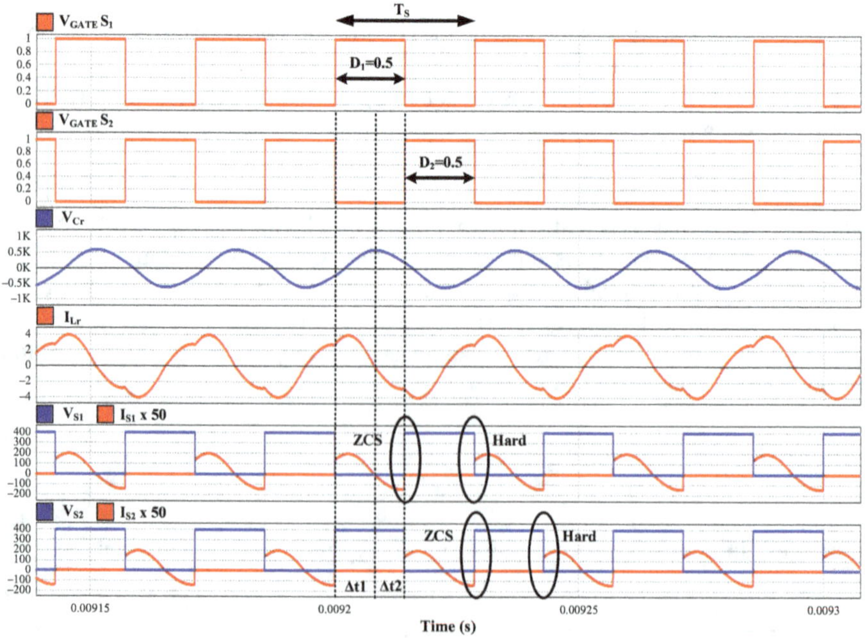

Figure 6.37 Series-load resonant converter at CCM ($0.5w_o < w_s < w_o$) waveforms

$$v_{Cr}(t) = V - (V - v_{Cr}(t_i)).\cos(w_o.t - t_i) + Z_o.i_{Lr}(t_i).\sin(w_o.t - t_i) \tag{6.61}$$

Where $v_{Cr}(t_i)$ and $i_{Lr}(t_i)$ are the initial conditions of the resonant capacitor voltage and resonant current, respectively; t_i is the initial instant of the stage evaluation; V is the resulting voltage from the addition or subtraction of the input voltage, $\pm V_{in}/2$, with the output voltage, $\pm V_o$, depending on the operation stage behavior.

Equations (6.60) and (6.61) can be used to obtain any of the equations already presented under resonance conditions, whether in continuous or discontinuous conduction mode. Therefore, the equations of this stage will not be developed in this chapter. However, the equivalent circuits necessary to develop such equations follow the same methodology as described before, and can be seen in Figure 6.38.

- First stage $[t_0, t_1]$—Δt_1:

This stage begins when switch S_2 is turned off and the current flows through diode D_1. The first stage ends when the current i_{Lr} reverses its direction. During this interval, switch S_1 must be turned on to achieve ZVS at turn-on transition.

- Second stage $[t_1, t_2]$—Δt_2:

After the inversion of current i_{Lr}, this direction being now positive, switch S_1 is already in conduction and remains turned on until the half of the switching period, where it is then turned off in a dissipative way.

Figure 6.39 shows the waveforms of the main elements for this conduction mode.

The voltage gain G of the converter can be defined as:

$$G = \frac{V_o}{\frac{V_{in}}{2}} = \frac{1}{\sqrt{1 + Q^2.\left(\frac{fo}{fs} - \frac{fs}{fo}\right)^2}} \tag{6.62}$$

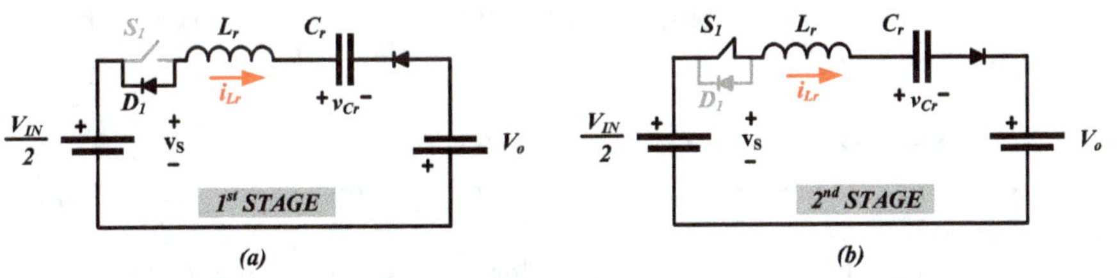

Figure 6.38 Series-load resonant converter at CCM. (a) Stage 1 [t_0-t_1]. (b) Stage 2 [t_1-t_2].

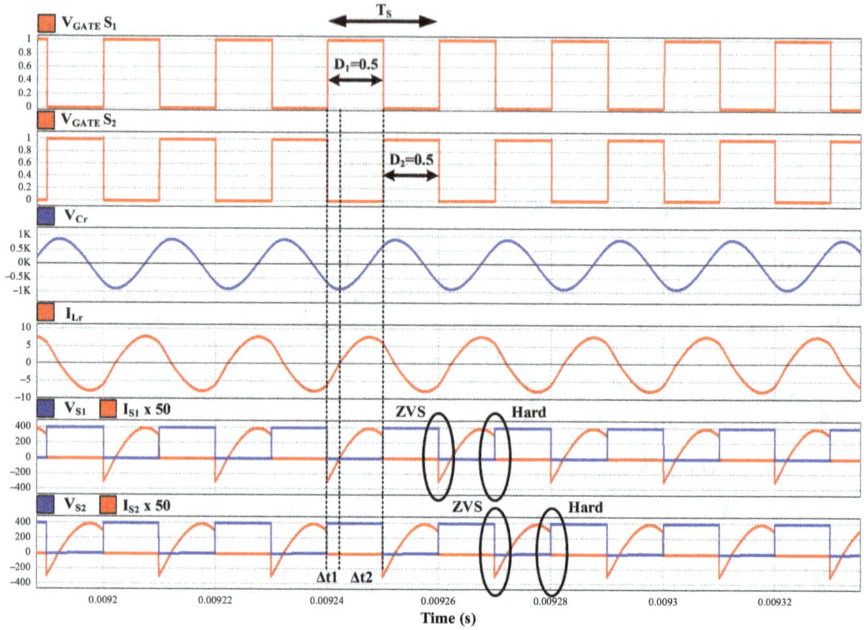

Figure 6.39 Series-load resonant converter at CCM ($w_s \geq w_o$) waveforms

Where:

- f_s is the switching frequency, in Hz.
- f_o is the resonant frequency, in Hz.
- Q is the quality factor.

$$Q = \frac{Z_o}{R_e} \tag{6.63}$$

$$R_e = \frac{8.R_o}{\pi^2} \tag{6.64}$$

Figure 6.40 shows the relation between the converter gain (G) and the normalized frequency (f_s/f_o).

6.5.1.2 Parallel-load resonant converters

Figure 6.41a presents the configuration of a parallel resonant converter (PRC) in a half-bridge configuration. The resonant tank is composed of an inductor L_r and a capacitor C_r connected in parallel with the load as a full-wave rectifier bridge.

In this configuration, the low-pass filter of the output voltage is composed of the output capacitor C_o, the output inductor L_o, and the load R_o, and can be characterized by a ripple-free current source with a value equal to I_o. Switches S_1 and S_2

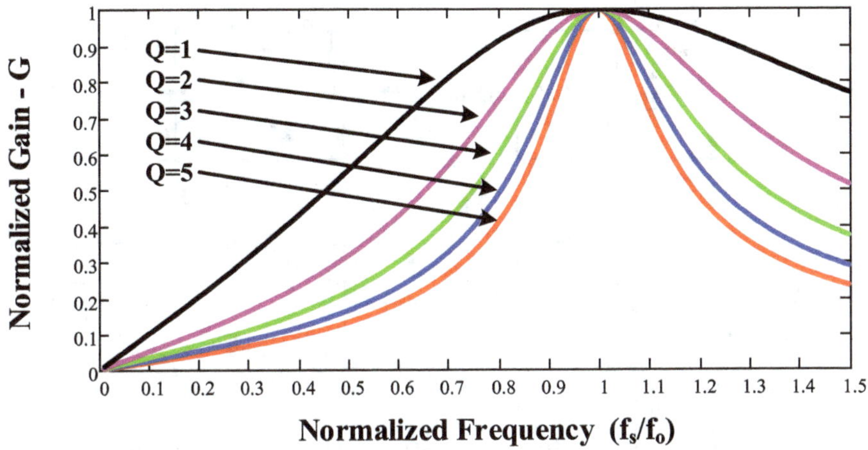

Figure 6.40 Series-load resonant converter: Gain × Normalized Frequency

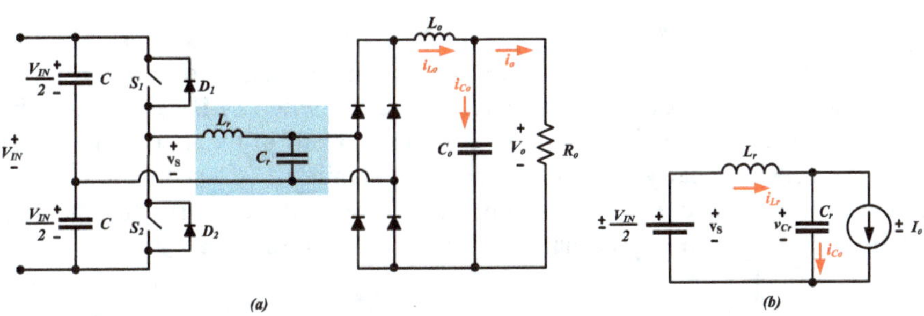

Figure 6.41 Parallel-load resonant converter. (a) Circuit. (b) Simplified model.

operate complementarily and, together with their respective anti-parallel diodes D_1 and D_2, define the switched voltage at the input of the resonant tank, v_S. Figure 6.41b shows the simplified circuit that will be used to describe the three conduction modes of this converter.

This converter operates as a voltage source, being able to operate without load and supporting a wide range of variations in the output current.

The traditional voltage gain control is related to frequency modulation, with a duty cycle of 50% for continuous conduction mode and less than 50% for discontinuous conduction mode, while keeping a phase shift of 180° between switches. Some papers with this converter can be seen in [45–47].

A. Discontinuous conduction mode (DCM)—$w_S < 0.5w_O$

The discontinuous conduction mode has eight stages of operation, with the last four being analogous to the first ones. Figure 6.42 shows the equivalent circuits of the first four operating stages for the discontinuous mode.

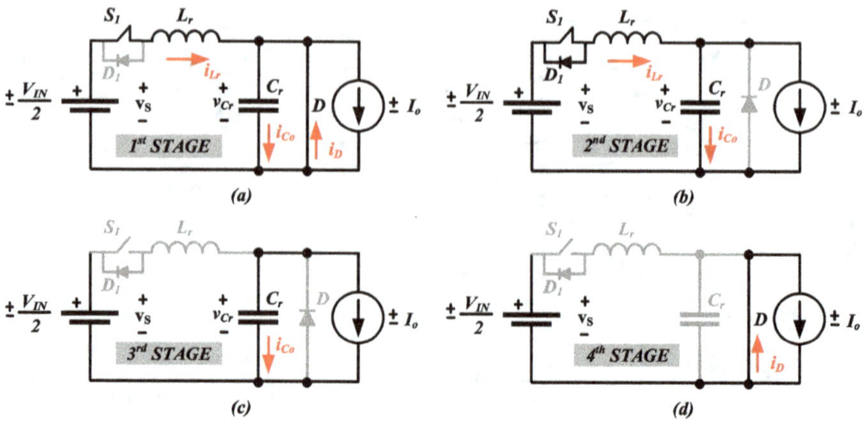

Figure 6.42 Parallel-load resonant converter at DCM. (a) Stage 1 [t_0-t_1]. (b) Stage 2 [t_1-t_3]. (c) Stage 3 [t_3-t_4]. (d) Stage 4 [t_4-t_5].

- First stage [t_0, t_1]—Δt_1:

In this operating mode, both the current through the resonant inductor i_{Lr} and the voltage across the resonant capacitor v_{Cr} are zero. Switch S_1 is turned on, initiating the stage. As the voltage across L_r is equal to half of the input voltage, the current increases linearly, as defined in (6.65). While i_{Lr} is less than the output current I_o, the output current flows through the diodes of the rectifier bridge, thus keeping the voltage in the resonant capacitor v_{Cr} equal to zero.

$$i_{Lr}(t) = \frac{\left(\frac{V_{IN}}{2}\right)}{L_r}.t \tag{6.65}$$

This stage ends when the current in the resonant inductor reaches the value of the output current I_o, at instant t_1. Therefore:

$$\Delta t_1 = \frac{I_o.L_r}{\frac{V_{in}}{2}} \tag{6.66}$$

- Second stage [t_1, t_3]—Δt_2:

When the current $i_{Lr} > I_o$, at t_1, the difference between these two currents ($i_{Lr} - I_o$) flows through the resonant capacitor C_r, increasing the voltage v_{Cr}. Due to the resonance between L_r and C_r, the current tends to oscillate. When the magnitude of i_{Lr} becomes smaller than I_o again, the capacitor Cr begins to discharge, providing the complementary current to the output. At t_2, the current is reversed and flows through D_1. To achieve commutation with zero voltage and current, the command of S1 must be turned off before t_3. The capacitor starts to supply alone the output current, causing its voltage to linearly decrease to zero.

Analyzing the equivalent circuit of this conduction stage:

$$\frac{dv_{Cr}(t)}{dt} = \frac{i_{Lr}(t) - I_o}{C_r} \tag{6.67}$$

$$\frac{di_{Lr}(t)}{dt} = \frac{\frac{V_{IN}}{2} - v_{Cr}(t)}{L_r} \tag{6.68}$$

Developing the second-order differential equations, whose initial conditions are $v_{Cr}(t) = 0$ and $i_{Lr}(t) = I_o$:

$$i_{Lr}(t) = I_o + \frac{\frac{V_{IN}}{2}}{Z_o}.\sin(w_o.\Delta t_2) \tag{6.69}$$

The equation for the voltage across the resonant capacitor is:

$$v_{Cr}(t) = \frac{V_{IN}}{2}.(1 - \cos(w_o.\Delta t_2)) \tag{6.70}$$

From (6.69), the time interval Δt_2 is calculated when the current $i_{Lr}(t_2) = 0$.

$$\Delta t_2 = \frac{1}{w_o}.\left[\pi + \sin^{-1}\left(\frac{I_o.Z_o}{\frac{V_{IN}}{2}}\right)\right] \tag{6.71}$$

Switch S_1 conducts until instant t_2 when the current I_{Lr} reaches zero. After t_2, the current in the resonant inductor has its direction reversed and is conducted by diode D_1. The switch S_1 must be turned off before t_3 in order to commute with zero voltage and current. The time interval between t_2 and t_3 is given by:

$$\Delta t_3 = \frac{1}{w_o}.\left[2\pi - \sin^{-1}\left(\frac{I_o.Z_o}{\frac{V_{IN}}{2}}\right)\right] \tag{6.72}$$

- Third stage $[t_3, t_4]$—Δt_4:

The switches S_1 and S_2 are blocked, as well as their respective antiparallel diodes. The current in capacitor C_r is negative and has a value of $-I_o$, causing the voltage across the capacitor to decrease linearly. The equation for the resonant capacitor voltage in this stage is given by:

$$v_{Cr}(t) = \frac{1}{C_r}\int_{t_3}^{t_4} -I_o.dt + v_{Cr}(t_3) = v_{Cr}(t_3) - \frac{I_o}{C_r}.\Delta t_4 \tag{6.73}$$

The equation for v_{Cr} is valid until the voltage across the capacitor reaches zero and the diodes of the rectifier bridge conduct simultaneously to start freewheeling. Assuming that the voltage on the resonant capacitor reaches zero at instant t_4:

$$\Delta t_4 = \frac{v_{Cr}(t_3).C_r}{I_o} = \frac{C_r.\frac{V_{IN}}{2}.(1 - \cos(w_o.\Delta t_2))}{I_o} \tag{6.74}$$

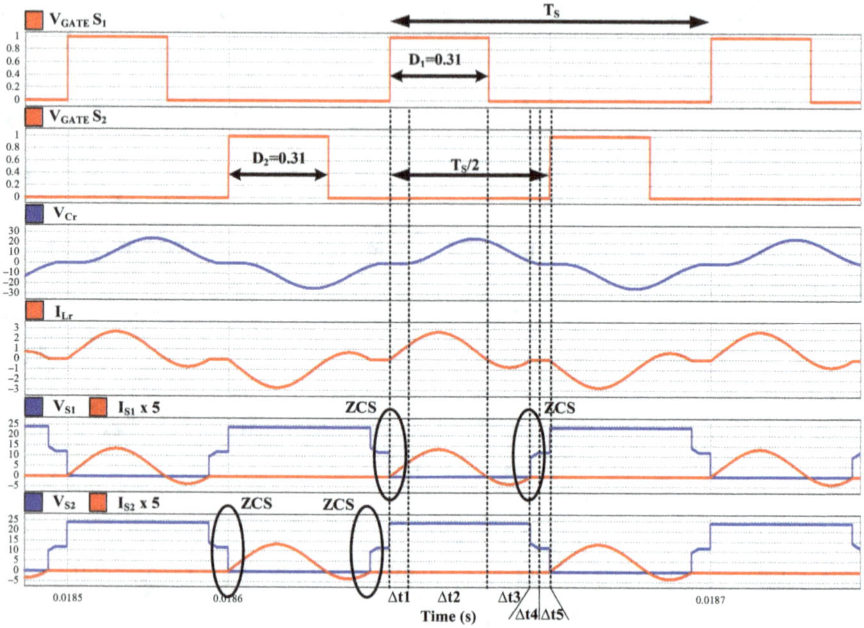

Figure 6.43 Parallel-load resonant converter at DCM waveforms

- Fourth stage $[t_4, t_5]$—Δt_5:

At this stage, both the current through the resonant inductor and the voltage across the resonant capacitor are zero. At half of the switching period, i.e., $T_s/2$, switch S_2 is commanded to conduct. This completes the first half-cycle of operation. The subsequent stages are analogous to the previously analyzed stages but with the directions of voltage and current inverted. Figure 6.43 presents the waveforms of the main components for this conduction mode.

B. Continuous conduction mode (CCM)—$0.5w_o < w_s < w_o$

In this operating mode, the output current source I_o is connected to the resonant tank throughout all the operation stages, without ever entering freewheeling mode. The switches operate with a complementary 50% duty cycle.

In this condition, it is essential to consider the initial conditions of the resonant capacitor $v_{Cr}(0)$ and the initial current in the resonant inductor $i_{Lr}(0)$, as both values differ from zero. This operating mode has four stages, however, similar to the discontinuous mode, it will only be analyzed in the first two stages, until half of the switching period ($T_s/2$), since the other two stages are analogous.

- First stage $[t_0, t_1]$—Δt_1:

In this operating stage, both the resonant inductor current i_{Lr} and resonant capacitor voltage v_{Cr} are zero. Switch S_1 is turned on, starting the operating stage. The resonant capacitor voltage is negative, and the resonant inductor current is positive.

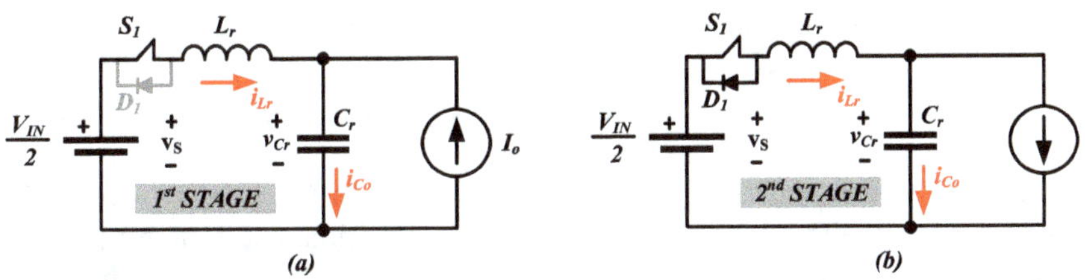

Figure 6.44 Parallel-load resonant converter CCM. (a) Stage 1 [t_0-t_1]. (b) Stage 2 [t_1-t_3].

Analyzing the circuit in Figure 6.44a:

$$\frac{V_{IN}}{2} = L_r\frac{\mathrm{d}i_{Lr}(t)}{\mathrm{dt}} + v_{Cr}(t) \tag{6.75}$$

Analyzing the currents, it can be stated:

$$\frac{\mathrm{d}v_{Cr}(t)}{\mathrm{dt}} = \frac{i_{Lr}(t) + I_o}{C_r} \tag{6.76}$$

Developing the second-order differential equations:

$$i_{Lr}(t) = (i_{Lr}(0) + I_o).\cos(w_o.t) + \frac{\frac{V_{IN}}{2} - v_{Cr}(0)}{Z_o}.\sin(w_o.t) - I_o \tag{6.77}$$

The equation for the voltage across the resonant capacitor is:

$$v_{Cr}(t) = \left(v_{Cr}(0) + \frac{V_{IN}}{2}\right).\cos(w_o.t) + Z_o.(i_{Lr}(0) + I_o).\sin(w_o.t) + \frac{V_{IN}}{2} \tag{6.78}$$

This stage lasts until the voltage across the resonant capacitor reaches zero.

- Second stage [t_1, t_2]—Δt_2:

The switch S_1 remains tuned-on, while the voltage across the resonant capacitor is positive, causing the output current source to conduct in the forward direction. The circuit analysis yields:

$$\frac{V_{IN}}{2} = L_r\frac{\mathrm{d}i_{Lr}(t)}{\mathrm{dt}} + v_{Cr}(t) \tag{6.79}$$

Analyzing the currents:

$$\frac{\mathrm{d}v_{Cr}(t)}{\mathrm{dt}} = \frac{i_{Lr}(t) + I_o}{C_r} \tag{6.80}$$

Developing the second-order differential equations:

$$i_{Lr}(t) = (i_{Lr}(t_1) - I_o).\cos(w_o.\Delta t_1) + \frac{\frac{V_{IN}}{2}}{Z_o}.\sin(w_o.\Delta t_1) + I_o \tag{6.81}$$

The equation for the voltage across the resonant capacitor is:

$$v_{Cr}(t) = -\frac{V_{IN}}{2}.\cos(w_o.\Delta t_1) + Z_o.(i_{Lr}(t_1) - I_o).\sin(w_o.\Delta t_1) + \frac{V_{IN}}{2} \quad (6.82)$$

After the resonant current in the inductor reaches zero at instant t_2, diode D_1 starts conducting due to the reversal of the current direction, and switch S_1 can be turned off without losses. At the instant $T_s/2$, switch S_2 begins conducting, initiating the second half-cycle operation of the converter, whose analysis is analogous to the previous stages. Figure 6.45 presents the main component waveforms for this conduction mode.

C. Continuous conduction mode (CCM)—$w_s \geq w_o$

This operating mode follows the same switching behavior described in topic B, with both switches operating complementary, with a 50% duty cycle. The resonant behavior of the current i_{Lr} and voltage v_{Cr} can be evaluated using general equations, which are presented below:

$$i_{Lr}(t) = (i_{Lr}(t_1) - I).\cos(w_o.[t - t_i]) + \frac{V - v_{Cr}(t_i)}{Z_o}.\sin(w_o.[t - t_i]) + I \quad (6.83)$$

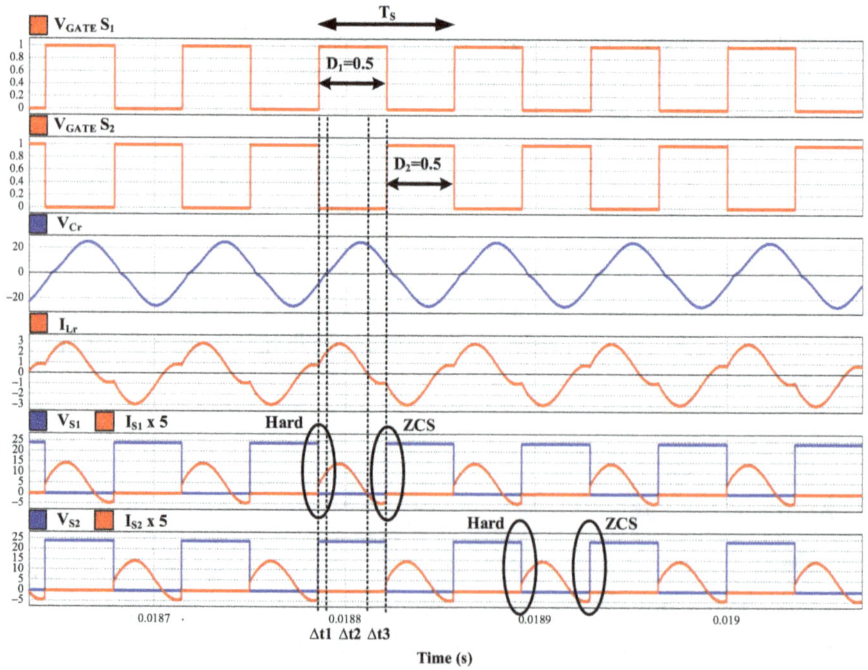

Figure 6.45 Parallel-load resonant converter at CCM ($0.5w_o < w_s < w_o$) waveforms

$$v_{Cr}(t) = -(V - v_{Cr}(t_i)).\cos(w_o.[t - t_i]) + Z_o.(i_{Lr}(t_1) - I).\sin(w_o.[t - t_i]) + V \tag{6.84}$$

Here, $v_{Cr}(t_i)$ and $i_{Lr}(t_i)$ are the initial conditions of the resonant capacitor voltage and resonant current respectively; t_i is the initial instant of the stage; V is the voltage applied to the input of the resonant tank, where $V = \pm V_{in}/2$ depending on the operation stage behavior; I is the output current source connected to the resonant tank, where $I = \pm Io$.

The equations presented can be used to obtain any of the equations previously introduced under resonant conditions, whether in continuous or discontinuous conduction mode. Therefore, the equations for this stage will not be further developed in this chapter. However, the equivalent circuits required to derive such equations follow the same methodology as the modes described earlier and can be observed in Figure 6.46.

- First stage $[t_0, t_2]$—Δt_1:

This stage starts when switch S_2 is blocked, and current flows through diode D_1. Switch S_1 can be turned on from this instant until the current reaches zero (at instant t_1) for soft-switching. The first stage ends when the voltage across capacitor C_r drops to zero.

- Second stage $[t_2, t_3]$—Δt_2:

In t_2, the voltage across the resonant capacitor reverses its polarity, causing the output current source to reverse its direction This stage ends at $T_s/2$, when switch S_2 is turned on and diode D_2 conducts.

Figure 6.47 shows the main element waveforms for this conduction mode.

The equation below defines the voltage gain G:

$$G = \frac{V_o}{\frac{V_{in}}{2}} = \frac{8}{\pi^2} \cdot \frac{1}{\sqrt{\left(1 - \left(\frac{fs}{fo}\right)^2\right)^2 + \left(\frac{1}{Q} \cdot \frac{fs}{fo}\right)^2}} \tag{6.85}$$

Where:

- f_s is the switching frequency, in Hz.
- f_o is the resonant frequency, in Hz.

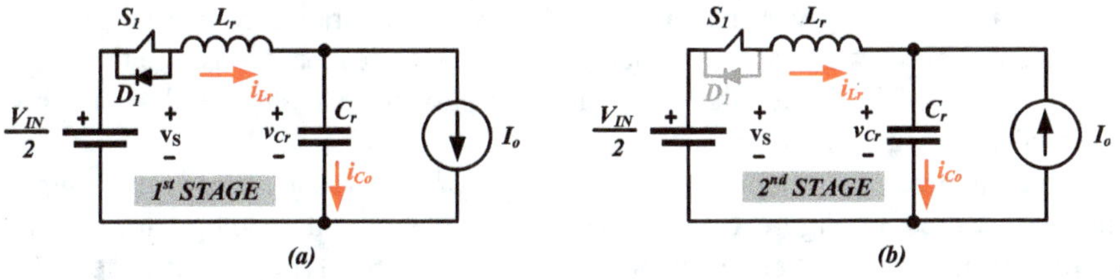

Figure 6.46 Parallel-load resonant converter at CCM. (a) Stage 1 $[t_0$-$t_1]$. (b) Stage 2 $[t_1$-$t_2]$.

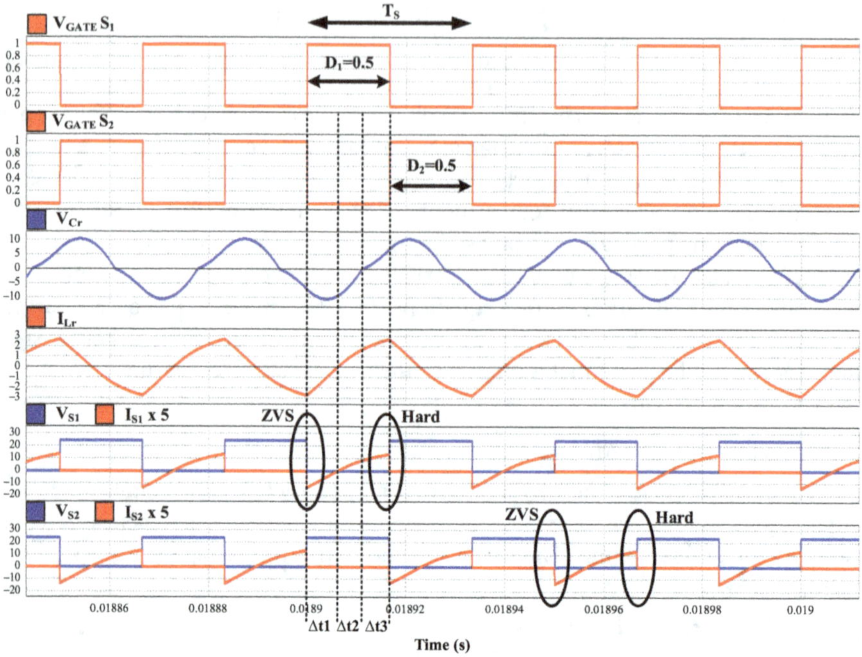

Figure 6.47 Parallel-load resonant converter at CCM ($w_s \geq w_o$) waveforms

- Q is the quality factor.

$$Q = \frac{R_e}{w_o} \tag{6.86}$$

$$R_e = \frac{\pi^2.R_o}{8} \tag{6.87}$$

Figure 6.48 presents the relation between the gain (G) and the parameterized frequency (f_s/f_o).

6.5.1.3 Series-parallel-load resonant converters

There are other configurations of resonant converters based on series and parallel topologies. Hybrid converters, also known as series-parallel converters, can be designed by using transformers connected between the resonant tank and the output rectifier bridge. They offer the advantages of galvanic isolation and a gain proportional to the turns ratio, $n = N_1/N_2$.

The (inductor - L, inductor - L and capacitor - C) resonant converter, which can be observed in Figure 6.49, is the most widely used topology for isolated resonant converters. When using a transformer in this type of converter, it is important to consider the two main inductances: the leakage inductance and the magnetizing inductance. Some papers with this converter can be seen in [48–50].

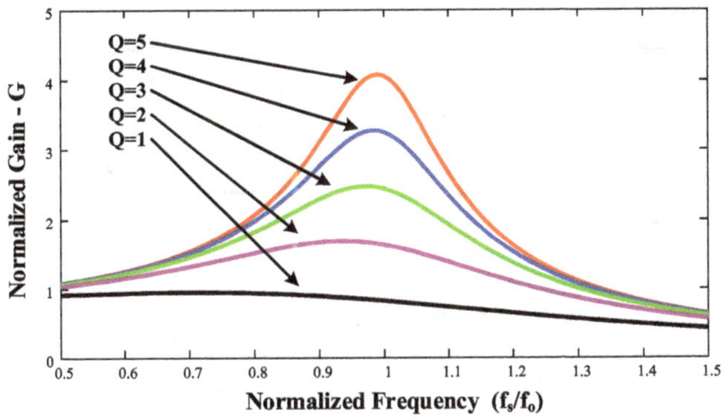

Figure 6.48 Parallel-load resonant converter: Gain × Normalized Frequency

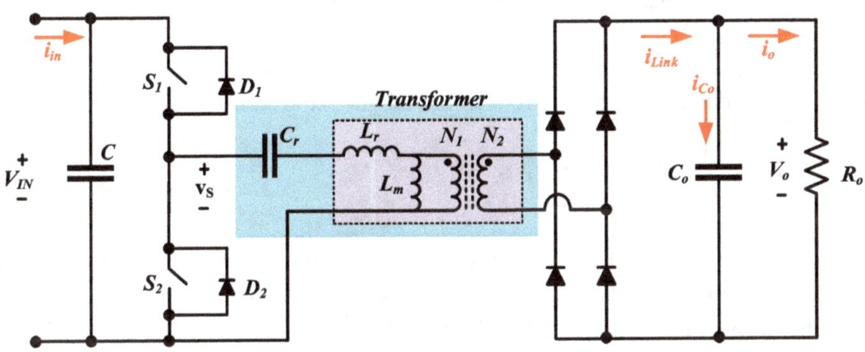

Figure 6.49 Series-parallel-load resonant converter

In the case of this converter, the design of the resonant tank can be adapted so that the resonant inductance can be utilized as part of the transformer design, thereby composing the leakage and magnetizing inductance. In this section, only continuous conduction mode will be analyzed for a switching frequency lower than the resonant frequency of the converter.

This converter has six operating stages, with the last three being analogous to the first ones. The equations presented are the final results of the development of differential equations obtained from the converter behavior. Figure 6.50 shows the equivalent circuit of each operating stage.

- First stage $[t_0, t_1]$—Δt_1:

In this stage, diode D_1 is conducting since the switch S_2 is blocked. The output voltage source V_o is reflected to the primary side and connected in parallel with the magnetizing inductance L_m, causing the current to increase linearly. Thus, the

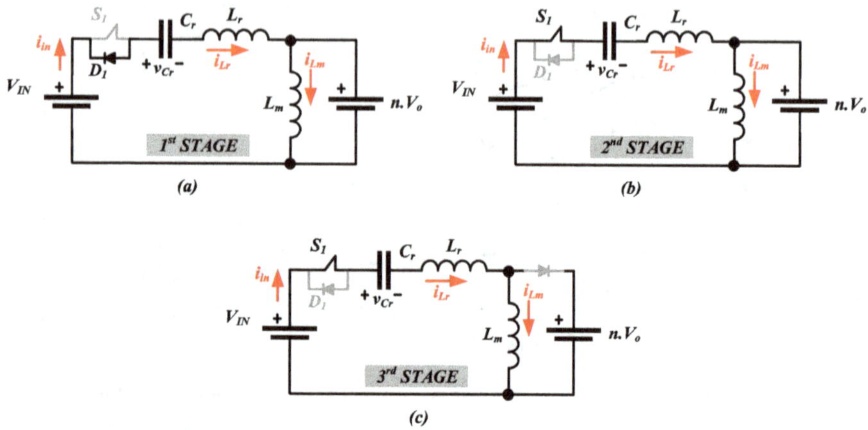

Figure 6.50 Series-parallel-load resonant converter at CCM. (a) Stage 1 [t_0-t_1]. (b) Stage 2 [t_1-t_3]. (c) Stage 3 [t_3-t_4].

equation that determines the current in the magnetizing inductor is given by:

$$i_{Lm}(t) = \frac{1}{L_m}.n.V_o.t + i_{Lm}(0) \tag{6.88}$$

Analyzing the circuit of this stage, it can be observed that components L_r and C_r are connected similarly to the series-resonant converter. Therefore, the current in the resonant inductor can be defined by the following equation:

$$i_{Lr}(t) = \frac{V_{IN} - V_o - v_{Cr}(0)}{Z_o}.\sin(w_o.t) + i_{Lr}(0).\cos(w_o.t) \tag{6.89}$$

- Second stage [t_1, t_2]—Δt_2:

At this stage, the current through the resonant inductor changes direction, becoming positive, and the switch S_1 starts conducting. The output voltage source maintains its polarity, and the current in the magnetizing inductor L_m continues to increase linearly and still does not participate in resonance. This stage ends when the current through the resonant inductor i_{Lr} equals the magnetizing current of the transformer i_{Lm}.

- Third stage [t_2, t_3]—Δt_3:

When the current i_{Lm} is equal to i_{Lr}, the output current in the rectifier bridge given by $i_{Link} = n.(i_{Lr} - i_{Lm})$ is equal to zero. For this case, the resulting voltage in the inductor L_m reflected to the secondary side, is lower than the output voltage V_o, resulting in the blocking of the rectifier bridge diodes. The current passing through L_m is the same as the one passing through L_r, causing the magnetizing inductor to

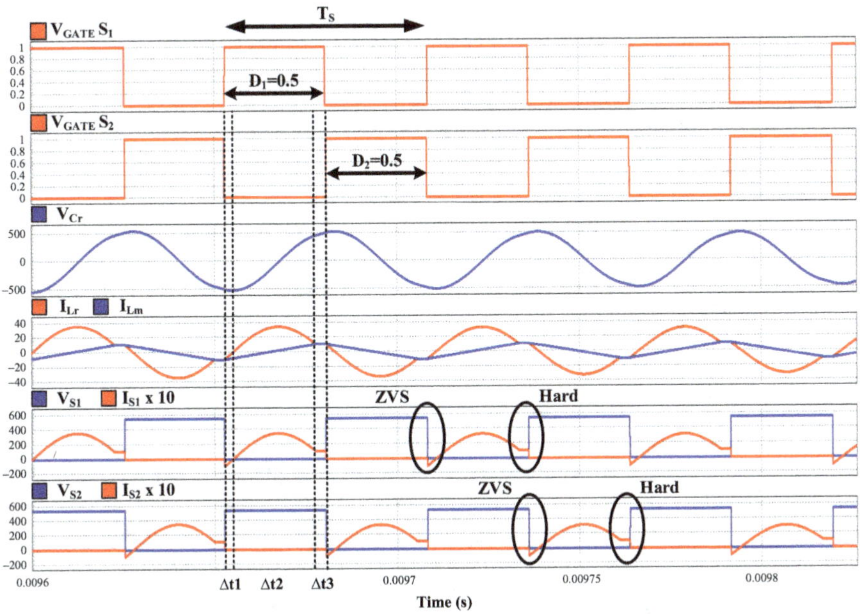

Figure 6.51 Series-parallel-load resonant converter at CCM ($w_s < w_o$) waveforms

participate in resonance since it is in series with the resonant tank L_r and C_r. The behavior of the current in the resonant inductor is given by (Figure 6.51):

$$i_{Lr}(t) = i_{Lm}(t) = \frac{v_{Cr}(t_2)}{Z_{o2}}.\sin(w_{o2}.\Delta t_3) + i_{Lr}(t_2).Z_{o2}.C_r.\cos(w_o.\Delta t_3) \qquad (6.90)$$

Where:

$$w_{o2} = \frac{1}{\sqrt{(L_r + L_m).C_r}} Z_{o2} = \sqrt{\frac{L_r + L_m}{C_r}}$$

The equation below defines the voltage gain G:

$$G = \frac{2.n.V_o}{V_{in}}$$

$$= L_n.\left(\frac{f_s}{f_o}\right)^2.\frac{1}{\sqrt{\left((L_n+1).\left(\frac{fs}{fo}\right)^2 - 1\right)^2 + \left(\left(\left(\frac{fs}{fo}\right)^2 - 1\right).\left(\frac{fs}{fo}.Q.L_n\right)\right)^2}} \qquad (6.91)$$

Where:

- f_s is the switching frequency, in Hz.
- f_o is the resonant frequency, in Hz.
- Q is the quality factor.
- L_n is the parameterized inductance.

$$Q = \frac{1}{R_e} \cdot \sqrt{\frac{L_r}{C_r}} \tag{6.92}$$

$$R_e = \frac{8.n^2{}_o}{\pi^2}.R_o \tag{6.93}$$

$$L_n = \frac{L_m}{L_r} \tag{6.94}$$

Figure 6.52 shows the relationship between the gain (G) and the parameterized frequency (f_s/f_o).

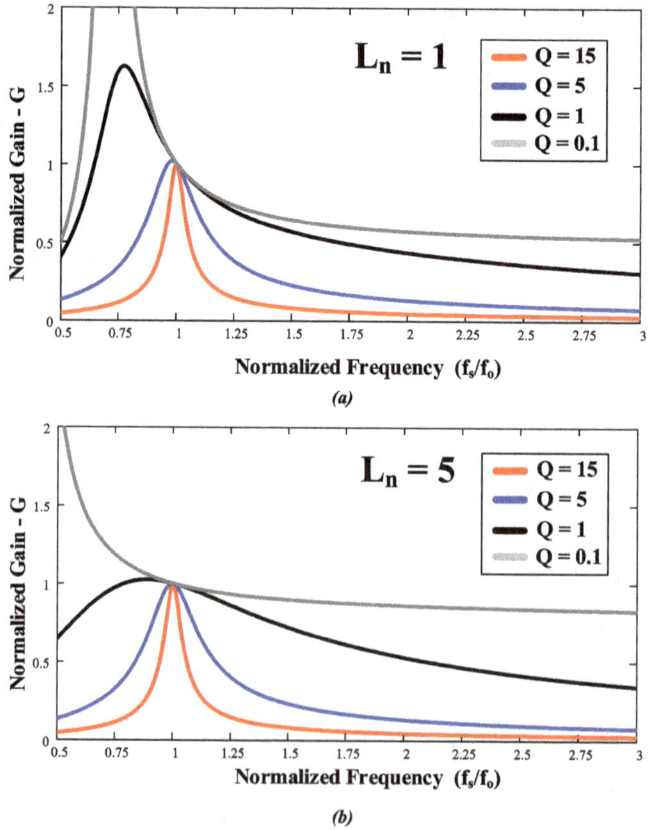

Figure 6.52 Series-parallel-load resonant converter: Gain × Normalized Frequency: (a) $L_n = 1$. (b) $L_n = 5$

6.6 Conclusion

This chapter aimed to present the main aspects related to the topic of resonant converters and their characteristics. This subject is extensive, and undoubtedly, there are other aspects to be explored within this theme, such as the design of high-frequency transformers for applications with resonant converters, the study of losses in the main resonant topologies, and industrial applications, among others. The key topics were addressed in this chapter, including resonance concepts, soft switching, resonant and quasi-resonant converters, as well as their main topologies, design methodology of resonant elements, and mathematical formulation.

References

[1] F. C. Schwarz, "An Improved Method of Resonant Current Pulse Modulation for Power Converters," in *IEEE Transactions on Industrial Electronics and Control Instrumentation*, vol. IECI-23, no. 2, pp. 133–141, 1976, doi: 10.1109/TIECI.1976.351365.

[2] R. H. Ashique and Z. Salam, "A Family of True Zero Voltage Zero Current Switching (ZVZCS) Nonisolated Bidirectional DC–DC Converter with Wide Soft Switching Range," in *IEEE Transactions on Industrial Electronics*, vol. 64, no. 7, pp. 5416–5427, 2017, doi: 10.1109/TIE.2017.2669884.

[3] S. Hamada, S. Moisseev, and M. Nakaoka, "Two-Switch Forward Soft-Switching PWM DC-DC Power Converter," in *Electronics Letters*, vol. 36, no. 25, p. 2055–2056, doi: 10.1049/el:20001474.

[4] J. Xiong, F. Zhou, and Q. Li, "Soft-Switching Modulation of the Multilevel Cascaded H-Bridge Inverter under DC Source Fault," in *IET Generation, Transmission & Distribution*, vol. 15, pp. 1337–1347, 2021, doi.org/10.1049/gtd2.12105.

[5] T. Yao, C. Nan, and R. Ayyanar, "A New Soft-Switching Topology for Switched Inductor High Gain Boost," in *IEEE Transactions on Industry Applications*, vol. 54, no. 3, pp. 2449–2458, 2018, doi: 10.1109/TIA.2018.2791972.

[6] D. -D. Tran, H. -N. Vu, S. Yu, and W. Choi, "A Novel Soft-Switching Full-Bridge Converter with a Combination of a Secondary Switch and a Non-dissipative Snubber," in *IEEE Transactions on Power Electronics*, vol. 33, no. 2, pp. 1440–1452, 2018, doi: 10.1109/TPEL.2017.2688580.

[7] R. C. M. Gomes, M. A. Vitorino, D. A. Acevedo-Bueno, and M. B. R. Corrêa, "Two Half-Bridge Resonant Inverters with Coupled Coils for AC/AC Induction Heating Application," *2018 IEEE Energy Conversion Congress and Exposition (ECCE)*, Portland, OR, USA, pp. 928–935, 2018, doi: 10.1109/ECCE.2018.8558476.

[8] S. A. Q. Mohammed and J. -W. Jung, "A State-of-the-Art Review on Soft-Switching Techniques for DC–DC, DC–AC, AC–DC, and AC–AC Power

Converters," in *IEEE Transactions on Industrial Informatics*, vol. 17, no. 10, pp. 6569–6582, 2021, doi:10.1109/TII.2021.3058218.

[9] Muhammad H. Rashid. *Power Electronica Handbook.* Butterworth-Heinemann, 2011.

[10] L. H. S. C. Barreto, E. A. A. Coelho, V. J. Farias, J. C. de Oliveira, L. C. de Freitas, and J. Joao Batista Vieira, "A Quasi-Resonant Quadratic Boost Converter Using a Single Resonant Network," in *IEEE Transactions on Industrial Electronics*, vol. 52, no. 2, pp. 552–557, 2005, doi: 10.1109/TIE.2005.844255.

[11] S. Y. R. Hui, E. S. Gogani, and Jian Zhang, "Analysis of a Quasi-Resonant Circuit for Soft-Switched Inverters," in *IEEE Transactions on Power Electronics*, vol. 11, no. 1, pp. 106–114, 1996, doi: 10.1109/63.484423.

[12] J. -M. Kwon, W. -Y. Choi, and B. -H. Kwon, "Single-Switch Quasi-Resonant Converter," in *IEEE Transactions on Industrial Electronics*, vol. 56, no. 4, pp. 1158–1163, 2009, doi: 10.1109/TIE.2008.2006239.

[13] J. Hu, A. D. Sagneri, J. M. Rivas, Y. Han, S. M. Davis, and D. J. Perreault, "High-Frequency Resonant SEPIC Converter with Wide Input and Output Voltage Ranges," in *IEEE Transactions on Power Electronics*, vol. 27, no. 1, pp. 189–200, 2012, doi: 10.1109/TPEL.2011.2149543.

[14] C. Wang, S. Xu, W. Shen, S. Lu, and W. Sun, "A Single-Switched High-Switching-Frequency Quasi-Resonant Flyback Converter," in *IEEE Transactions on Power Electronics*, vol. 34, no. 9, pp. 8775–8786, 2019, doi: 10.1109/TPEL.2018.2884937.

[15] I. Barbi, J. C. O. Bolacell, D. C. Martins, and F. B. Libano, "Buck Quasi-Resonant Converter Operating at Constant Frequency: Analysis, Design, and Experimentation," in *IEEE Transactions on Power Electronics*, vol. 5, no. 3, pp. 276–283, 1990, doi: 10.1109/63.56518.

[16] A. Esmaeili and H. Gholizade-Narm, "Robust Nonlinear Control of a Quasi-Resonant DC-DC Converter with Turn-On and Turn-Off Zero Current Switching," in *IET Power Electronics*, vol. 15, pp. 325–336, 2022, https://doi.org/10.1049/pel2.12233

[17] S. Sharifi and M. Jabbari, "Family of Single-Switch Quasi-Resonant Converters with Reduced Inductor Size," in *IET Power Electronics*, vol. 7, pp. 2544–2554, 2014, https://doi.org/10.1049/iet-pel.2013.0615.

[18] L. Schmitz, R.G.A. Cacau, T.P. Horn, D.C. Martins, and R.F. Coelho, "High Step-Up Non-Isolated ZVS-ZCS DC–DC Cúk-Based Converter," in *IET Power Electronics*, vol. 13, pp. 1343–1352, 2020, https://doi.org/10.1049/iet-pel.2019.1262.

[19] K. -H. Liu and F. C. Y. Lee, "Zero-Voltage Switching Technique in DC/DC Converters," in *IEEE Transactions on Power Electronics*, vol. 5, no. 3, pp. 293–304, 1990, doi:10.1109/63.56520.

[20] K. -H. Liu and F. C. Lee, "Resonant Switches – A Unified Approach to Improve Performances of Switching Converters," INTELEC '84 – International Telecommunications Energy Conference, New Orleans, LA, USA, pp. 344–351, 1984, doi: 10.1109/INTLEC.1984.4794149.

[21] K. Liu, R. Oruganti, and F. C. Lee, "Resonant Switches – Topologies and Characteristics," 1985 IEEE Power Electronics Specialists Conference, Toulouse, France, pp. 106–116, 1985, doi: 10.1109/PESC.1985.7070935.

[22] F. C. Lee, "High-Frequency Quasi-Resonant and Multi-Resonant Converter Technologies," Proceedings. 14 Annual Conference of Industrial Electronics Society, Singapore, pp. 509–521, 1988, doi: 10.1109/IECON.1988.665736.

[23] B. Akın, "An Improved ZVT–ZCT PWM DC–DC Boost Converter with Increased Efficiency," in *IEEE Transactions on Power Electronics*, vol. 29, no. 4, pp. 1919–1926, 2014, doi: 10.1109/TPEL.2013.2269172.

[24] R. Fani, E. Farshidi, E. Adib, and A. Kosarian, "Analysis, Design, and Implementation of a ZVT High Step-Up DC–DC Converter with Continuous Input Current," in *IEEE Transactions on Industrial Electronics*, vol. 67, no. 12, pp. 10455–10463, 2020, doi: 10.1109/TIE.2019.2960727.

[25] H. N. Tran and S. Choi, "A Family of ZVT DC–DC Converters with Low-Voltage Ringing," in *IEEE Transactions on Power Electronics*, vol. 35, no. 1, pp. 59–69, 2020, doi: 10.1109/TPEL.2019.2911040.

[26] B. Akhlaghi and H. Farzanehfard, "Family of ZVT Interleaved Converters with Low Number of Components," in *IEEE Transactions on Industrial Electronics*, vol. 65, no. 11, pp. 8565–8573, 2018, doi: 10.1109/TIE.2018.2808915.

[27] Ching-Jung Tseng and Chern-Lin Chen, "Novel ZVT-PWM Converters with Active Snubbers," in *IEEE Transactions on Power Electronics*, vol. 13, no. 5, pp. 861–869, 1998, doi: 10.1109/63.712292.

[28] H. Bodur and A. F. Bakan, "A New ZVT-PWM DC-DC Converter," in *IEEE Transactions on Power Electronics*, vol. 17, no. 1, pp. 40–47, 2002, doi: 10.1109/63.988668.

[29] B. Zhu, Q. Zeng, Y. Chen, Y. Zhao, and S. Liu, "A Dual-Input High Step-Up DC/DC Converter with ZVT Auxiliary Circuit," in *IEEE Transactions on Energy Conversion*, vol. 34, no. 1, pp. 161–169, 2019, doi:10.1109/TEC.2018.2876303.

[30] L. Schuch, C. Rech, H. L. Hey, H. A. Grundling, H. Pinheiro, and J. R. Pinheiro, "Analysis and Design of a New High-Efficiency Bidirectional Integrated ZVT PWM Converter for DC-Bus and Battery-Bank Interface," in *IEEE Transactions on Industry Applications*, vol. 42, no. 5, pp. 1321–1332, 2006, doi:10.1109/TIA.2006.880847.

[31] G. Hua, C. S. Leu, and F. C. Lee, "Novel Zero-Voltage-Transition PWM Converters," PESC '92 Record. 23rd Annual IEEE Power Electronics Specialists Conference, Toledo, Spain, pp. 55–61, vol.1, 1992, doi:10.1109/PESC.1992.254691.

[32] B. Akhlaghi and H. Farzanehfard, "Efficient ZVT Cell for Interleaved DC–DC Converters," in *IET Power Electronics*, vol. 13, pp. 1925–1933, 2020, https://doi.org/10.1049/iet-pel.2019.1102

[33] L. C. de Freitas, J. B. Vieira, V. J. Farias, H. L. Hey, P. S. Caparelli, and D. F. de Cruz, "An Optimum ZVS-PWM DC-to-DC Converter Family: Analysis, Simulation and Experimental Results," PESC '92 Record. 23rd Annual

IEEE Power Electronics Specialists Conference, Toledo, Spain, pp. 229–235, vol. 1, 1992, doi:10.1109/PESC.1992.254668.

[34] R.N.A.L. Silva Aquino, F.L. Tofoli, P.P. Praca, D.d.S. Jr Oliveira, and L.H. S.C. Barreto, "Soft Switching High-Voltage Gain DC–DC Interleaved Boost Converter," in *IET Power Electronics*, vol. 8, pp. 120–129, 2015, https://doi.org/10.1049/iet-pel.2014.0275.

[35] L. C. de Freitas and P. R. C. Gomes, "A High-Power High-Frequency ZCS-ZVS-PWM Buck Converter Using a Feedback Resonant Circuit," *Proceedings of IEEE Power Electronics Specialist Conference - PESC '93*, Seattle, WA, USA, pp. 330–336, 1993, doi: 10.1109/PESC.1993.471905.

[36] S. Dusmez, A. Khaligh, and A. Hasanzadeh, "A Zero-Voltage-Transition Bidirectional DC/DC Converter," in *IEEE Transactions on Industrial Electronics*, vol. 62, no. 5, pp. 3152–3162, 2015, doi: 10.1109/TIE.2015.2404825.

[37] N. Altintaş, A. F. Bakan, and İ. Aksoy, "A Novel ZVT-ZCT-PWM Boost Converter," in *IEEE Transactions on Power Electronics*, vol. 29, no. 1, pp. 256–265, 2014, doi: 10.1109/TPEL.2013.2252197.

[38] T. Bang and J. -W. Park, "Development of a ZVT-PWM Buck Cascaded Buck–Boost PFC Converter of 2 kW with the Widest Range of Input Voltage," in *IEEE Transactions on Industrial Electronics*, vol. 65, no. 3, pp. 2090–2099, 2018, doi: 10.1109/TIE.2017.2739703.

[39] M. Heidari, M. Esteki, S. A. Khajehoddin, and H. Farzanehfard, "A High Voltage Gain ZVT Quasi-Z-Source Converter with Reduced Voltage Stress," in *IEEE Transactions on Power Electronics*, vol. 37, no. 11, pp. 13696–13710, 2022, doi: 10.1109/TPEL.2022.3178140.

[40] W. Li, X. Xiang, C. Li, W. Li, and X. He, "Interleaved High Step-Up ZVT Converter with Built-In Transformer Voltage Doubler Cell for Distributed PV Generation System," in *IEEE Transactions on Power Electronics*, vol. 28, no. 1, pp. 300–313, 2013, doi: 10.1109/TPEL.2012.2199771.

[41] Montiê A. Vitorino. Eletrônica de Potência: Fundamentos, Conceitors e Aplicações. Appris, 2019.

[42] R. L. Steigerwald, "A Comparison of Half-Bridge Resonant Converter Topologies," in *IEEE Transactions on Power Electronics*, vol. 3, no. 2, pp. 174–182, 1988, doi: 10.1109/63.4347.

[43] S. D. Johnson, A. F. Witulski, and R. W. Erickson, "Comparison of Resonant Topologies in High-Voltage DC Applications," in *IEEE Transactions on Aerospace and Electronic Systems*, vol. 24, no. 3, pp. 263–274, 1988, doi: 10.1109/7.192094.

[44] F. Ibanez, J.M. Echeverria, and L. Fontan, "Novel Technique for Bidirectional Series-Resonant DC/DC Converter in Discontinuous Mode," in *IET Power Electronics*, vol. 6, pp. 1019–1028, 2013, https://doi.org/10.1049/iet-pel.2012.0540.

[45] A. K. S. Bhat and M. M. Swamy, "Loss Calculations in Transistorized Parallel Resonant Converters Operating above Resonance," in *IEEE*

Transactions on Power Electronics, vol. 4, no. 4, pp. 391–401, 1989, doi: 10.1109/63.41766.

[46] A. K. Rathore and V. R. Vakacharla, "A Simple Technique for Fundamental Harmonic Approximation Analysis in Parallel and Series–Parallel Resonant Converters," in *IEEE Transactions on Industrial Electronics*, vol. 67, no. 11, pp. 9963–9968, 2020, doi: 10.1109/TIE.2019.2952820.

[47] N. C. D. Pont, D. G. Bandeira, T. B. Lazzarin, and I. Barbi, "A ZVS APWM Half-Bridge Parallel Resonant DC–DC Converter with Capacitive Output," in *IEEE Transactions on Industrial Electronics*, vol. 66, no. 7, pp. 5231–5241, 2019, doi: 10.1109/TIE.2018.2868270.

[48] X. Tan and X. Ruan, "Equivalence Relations of Resonant Tanks: A New Perspective for Selection and Design of Resonant Converters," in *IEEE Transactions on Industrial Electronics*, vol. 63, no. 4, pp. 2111–2123, 2016, doi: 10.1109/TIE.2015.2506151.

[49] Siu-Chung Wong and A. D. Brown, "Analysis, Modeling, and Simulation of Series-Parallel Resonant Converter Circuits," in *IEEE Transactions on Power Electronics*, vol. 10, no. 5, pp. 605–614, 1995, doi: 10.1109/63.406849.

[50] M. K. Kazimierczuk, N. Thirunarayan, and S. Wang, "Analysis of Series-Parallel Resonant Converter," in *IEEE Transactions on Aerospace and Electronic Systems*, vol. 29, no. 1, pp. 88–99, 1993, doi: 10.1109/7.249115.

Chapter 7
Impedance (Z-) source power converter

Yam P. Siwakoti[1], *Frede Blaabjerg*[2] *and Poh Chiang Loh*[3]

7.1 Introduction

Over the past decades, various impedance Z-sourced power converter topologies have been reported to overcome the limitations and problems of the traditional voltage source, current source as well as various classical buck-boost, unidirectional, and bidirectional power converter topologies. This impedance Z-sourced power converter covers the entire spectrum of electric power conversion applications (dc–dc, dc–ac, ac–dc, ac–ac) controlled and modulated by different modulation strategies to generate the desired dc/ac-voltage and current at the output. Proper implementation of the impedance source network with appropriate switching configurations, modulation, and control methods reduces the number of power conversion stages in the power conversion chain, which may improve the reliability and performance of the power electronic system. However, the number of combinations is large and the different topologies are hard always to follow. This chapter aims to provide a comprehensive overview of impedance-source networks for electric-power conversion from the perspective of the development of different topologies, their modeling, control, and modulation techniques. A basic principle underlying the concept of shoot-through across the dc-link and corresponding voltage boost is described in detail with governing equations to comprehend the subject matter to the new readers. Further, a broad classification and sub-classification of the topologies and modulation techniques with comparisons provide a quick reference for researchers and practicing engineers in deciding which control and modulation method to consider for an application in a given topology at a certain power level, switching frequency, and demanded dynamic response.

[1]School of Electrical and Data Engineering, Faculty of Engineering and IT, University of Technology Sydney, Australia
[2]Department of Energy, Aalborg University, Denmark
[3]Department of Electronic Engineering, Chinese University of Hong Kong, China

7.1.1 General overview

Impedance-sourced networks provide an efficient means of power conversion between source and load in a wide range of electric power conversion applications (dc–dc, dc–ac, ac–dc, ac–ac). Various topologies and control methods using different impedance source networks have been presented in the literature, e.g., for adjustable-speed drives (ASD), uninterruptible power supply (UPS), distributed generation (fuel cell, PV, wind, etc.), battery or super-capacitor energy storage, electric vehicles, distributed dc power systems, avionics, flywheel energy storage systems, electronic loads, dc circuit breaker, and many more [1–13]. A variety of converter topologies with buck, boost, buck-boost, unidirectional, bidirectional, isolated as well as non-isolated converters are possible by proper implementation of the impedance-source network with various switching devices, topologies, and configurations. Figure 7.1 shows the general configuration of an impedance-source network for electric power conversion.

The impedance source network was originally invented to overcome the limitations of the voltage source inverter (VSI) and current source inverter (CSI) topologies, which are commonly used in electric power conversion [1]. The ac output voltage of the VSI is limited to be below the input voltage, i.e., the VSI is a buck-type inverter that cannot serve the need of distributed generation and ac drives without another buck-boost stage. It requires an additional dc–dc boost converter to obtain the desired ac output, which increases system cost and lowers the efficiency. In addition, the switching devices are vulnerable to electromagnetic interference as mitigating causes a short circuit across the inverter bridge and destroys the switching devices. The dead time introduced in such cases causes waveform distortion at the output. On the other hand, in the case of the CSI, the output voltage cannot be less than the input voltage. For applications where a wide voltage range is desirable, an additional dc–dc buck converter is needed. Also, the upper and the lower switches of the inverter have to be gated on and maintained at any time. Otherwise, an open circuit of the dc inductor would occur and destroy the devices.

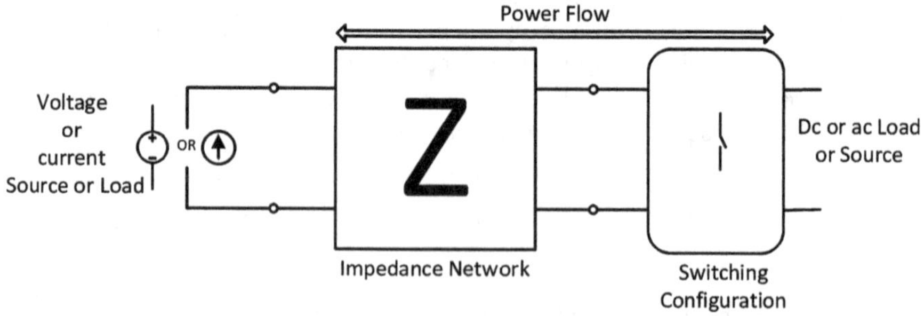

Figure 7.1 A general circuit configuration of an impedance-sourced network for power electronic conversion

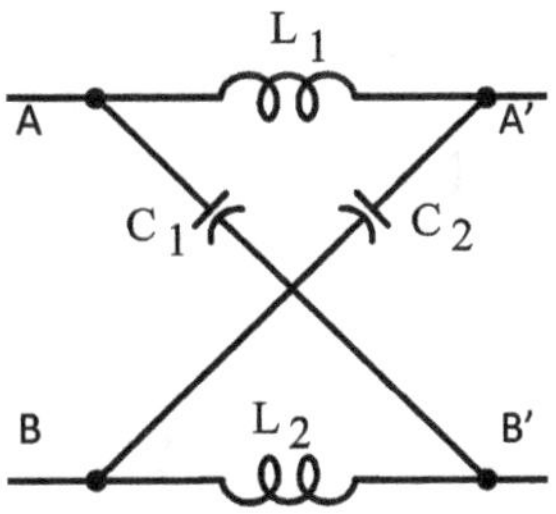

Figure 7.2 Basic Z-source impedance network with two inductors and capacitors

The impedance-sourced converter overcomes the above-mentioned conceptual and theoretical barriers and limitations of the classical voltage source inverter and current source inverter and it provides a novel power-conversion concept. The major advantage of this topology is that it can operate as a V-source or an I-source depending on the application and needs, and the output voltage can ideally be varied from $0 \sim \infty$. Since the publication of the first impedance source network, called a "*Z-source network*," in the year 2002 [1], many modified topologies with improved modulation and control strategies have been proposed and published to improve the performance in various applications [14]. Figure 7.2 shows the basic Z-source impedance network, which consists of the inductors L_1 and L_2 and capacitors C_1 and C_2 connected at both ends (Z-shape) and acts as a buffer between load and source (voltage source or current source).

According to the conversion functionality, impedance Z-source power converter can be divided into four main categories: dc–dc converters, dc–ac inverters, ac–ac converters, and ac–dc rectifiers. A further breakdown leads to two-level and multilevel, ac–ac and matrix converters, and non-isolated and isolated dc–dc converters. From the Z-source network topology standpoint, it can be voltage-fed or current-fed. Furthermore, impedance networks can be divided based on the magnetics used in the impedance source network, i.e., non-transformer-based [15–28] and transformer or coupled inductor-based [29–36]. Each topology has its own unique features and applications to which it is best suited. There is no one-size-fits-all solution. It is expected that new Z-source topologies will continuously be put forth to meet and improve converter performance in different applications.

7.1.2 Basic principles

The concept of the impedance-source network can be applied to any dc-to-dc, ac-to-ac, ac-to-dc, and dc-to-ac power conversion. The dc source and/or load can be a voltage or current source and/or a load. A Z-source impedance network is used as an example to briefly illustrate the operating principle and control of the impedance source network. Figure 7.3(a) shows the circuit diagram of the Z-source converter. Figure 7.3 (b) illustrates its equivalent circuit during the shoot-through state. During this state, the

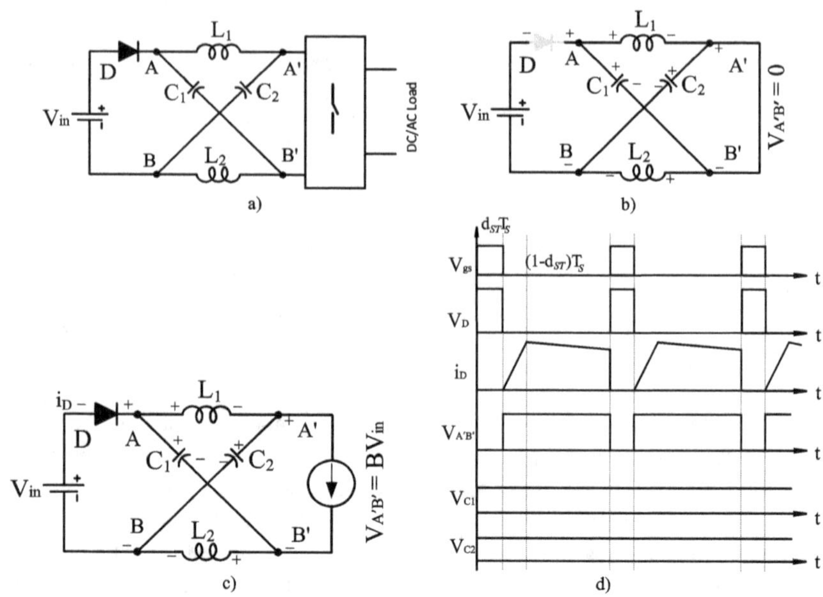

Figure 7.3 (a) Voltage-fed Z-source converter, (b) its equivalent circuit during the shoot-through state, (c) its equivalent circuit during the active state, and (d) corresponding current and voltage waveforms

output terminals of the impedance network A′ and B′ are short-circuited by a switch or a combination of switches which will, in turn, cause diode D in the network to reverse bias. The energy stored in the inductor and capacitor during this shoot-through state is transferred to the load during the next active state, in which the diode D is returned to conduction. The switching circuit viewed from the dc side during the active state is equivalent to a current source as shown in Figure 7.3(c). Averaging of these two switching states results in an expression to calculate the peak dc-link voltage $\widehat{v}_{A'B'}$ across terminals A′ and B′, in terms of its input voltage V_{in} given as

$$\widehat{v}_{A'B'} = \frac{1}{1-\beta d_{ST}} V_{in} \tag{7.1}$$

$$= BV_{in} \tag{7.2}$$

where d_{ST} is the fractional shoot-through time assumed in a switching period, and $\beta \geq 2$ is a factor determined by the impedance network chosen, e.g., for Z-source Inverter and quasi-Z-source Inverter (ZSI and qZSI), $\beta = 2$. Equating the denominator of the boost factor (B) to zero then results in the permissible range of d_{ST} as $0 \leq d_{ST} < \frac{1}{\beta}$, where the upper limit corresponds to an infinite gain. Waveforms of current and voltage considering a single switch at the dc-link (A′ and B′) are illustrated in Figure 7.3(d).

A three-phase voltage-fed Z-source inverter, as shown in Figure 7.4, is used as an example to briefly illustrate the operating principle as described in Figure 7.3

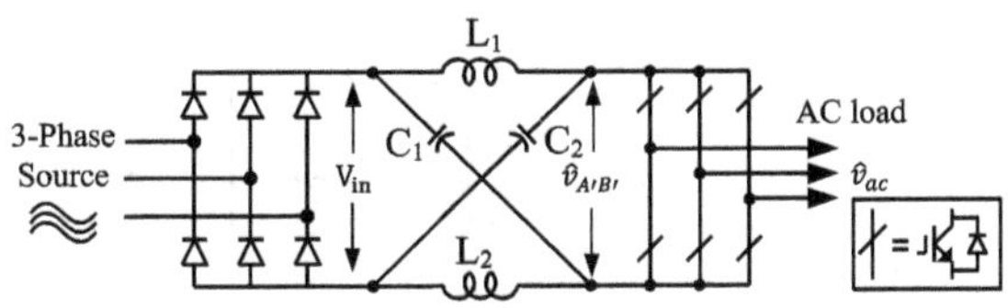

Figure 7.4 Voltage-fed Z-source 3-phase inverter—an example of a Z-source inverter

The three-phase Z-source inverter bridge has nine permissible switching states (six active states, two zero states, and one shoot-through state), unlike the traditional three-phase Voltage Source Inverter (VSI), which has eight (six active states, two zero states). During zero states, the upper three or lower three switches of the inverter bridge are turned on simultaneously, thus shorting the output terminals of the inverter and producing zero voltage across the load. During one of the six active states, the dc voltage is impressed across the load, positively or negatively. However, during the shoot-through states, the load terminals are shorted through both the upper and lower devices of any phase leg, any two-phase legs, and all three-phase legs [1] producing zero voltage across the load. This shoot-through state has the same effect, i.e., producing zero voltage across the load as the traditional zero states. However, these shoot-through states can boost the output voltage. The shoot-through state is forbidden in the traditional voltage-source inverter because it would cause a short circuit across the dc-link and damage the converter. The Z-source network and the shoot-through zero states provide a unique buck-boost capability for the inverter by varying the shoot-through time period and modulation index (M) of the inverter. Theoretically, the output voltage of the inverter can be written as

$$\widehat{v}_{ac} = \frac{MB}{2} \tag{7.3}$$

$$= M[1 - 2d_{ST}]^{-1} V_{in}/2) \tag{7.4}$$

The output voltage in (4) can be set to any value between 0 and ∞ by varying the value of M and d_{ST} However, some practical aspects and performance of the converter need to be considered for large voltage buck or boost operation in order to avoid exceeding the device limitations. Some of the waveforms during the boost operation are illustrated in Figure 7.5.

All traditional pulse width modulation (PWM) schemes can be used to control the impedance-source converter, and their theoretical input–output relationships still hold true. However, in addition to all states in the traditional modulation techniques, a new state called a "*shoot-through state*" is introduced and embedded in the modulation strategy for the impedance network-based power converter without violating the volt-sec balance in the operating principle. With the unique feature of these shoot-through states, several new PWM methods modified from the traditional techniques are prevalent in the literature. Figure 7.6 shows a broad

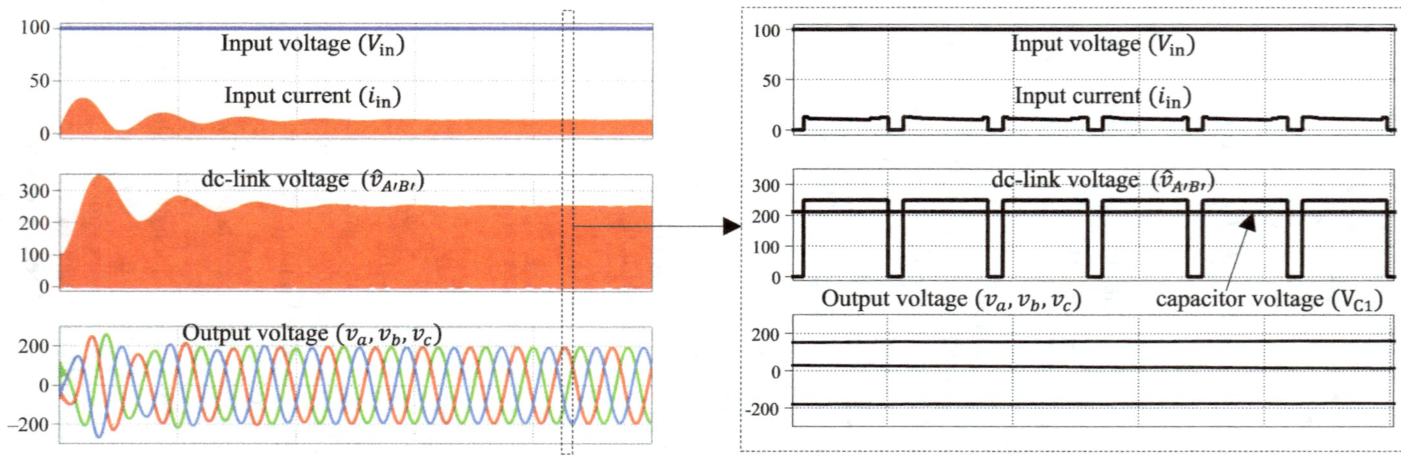

Figure 7.5 Illustration of some of the voltage and current waveforms in a voltage-fed Z-source inverter

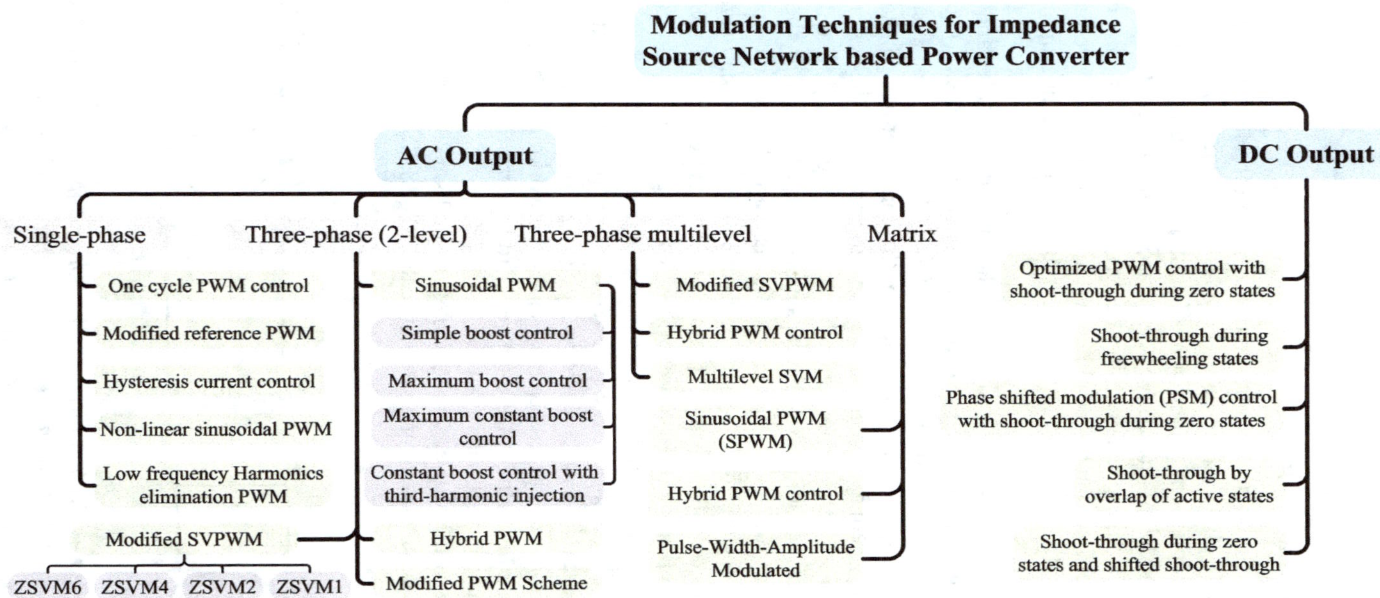

Figure 7.6 Categorization of modulation techniques for impedance-source network-based power converters—both AC and DC outputs

categorization of the impedance-network-based power converter using different switch configurations [37].

7.1.3 Modeling and control

A Z-source network shows non-minimum phase behavior due to the presence of zero in the right half-plane which could impose a limitation on the controller design. In order to implement a good control strategy, it is imperative to have a detailed dynamic model of the converter. Various small-signal analyses and mathematical models are presented in the literature to study the dynamic behavior of the system, which then can be implemented in different closed-loop control strategies with different complexities based on various applications [38–45].

To derive an accurate small-signal model, various state variables are selected, such as the input current $(i_{\rm in}(t))$, inductor currents $(i_{\rm L1}(t), i_{\rm L2}(t), \ldots)$, capacitor voltages $(v_{\rm C1}(t), v_{\rm C2}(t), \ldots)$, and load currents $\left(i_{\rm L}(t), i_{\rm d}(t), i_{\rm q}(t)\right)$. The small-signal model provides the required transfer function for the controller design and gives a detailed view of the system dynamics, which helps to understand the system limits, and provides guidelines for system controller design. In general, M and D_{st} are considered as control variables and the capacitor voltage $(v_{\rm C}(t))$ or the dc-link voltage $(v_{\rm PN}(t))$ and the load voltage $(v_{\rm x}(t))$ as variables to be controlled. Figure 7.7(a) shows a simplified Z-source converter model for small-signal analysis, where $v_{\rm C}(t)$ is controlled using D_{st} as a control variable (control switch $S_{\rm D}$). This is the most simplified model; however, it does not guarantee tight control of $v_{\rm x}(t)$, which requires an additional control variable M (control switch $S_{\rm M}$) as shown in Figure 7.7(b) [38]. In addition to the state variables, the parasitic resistance of the inductor (r) and the equivalent series resistance (ESR) of the capacitor (R) also influence the dynamics of the impedance-source networks and hence are also considered during the modeling of the converter to analyze the sensitivity of the circuit under parameter variations [39–41].

Based on some of the above state variables, several small-signal models have been proposed for symmetrical or asymmetrical ZSI [38,42,43] and qZSI [39].

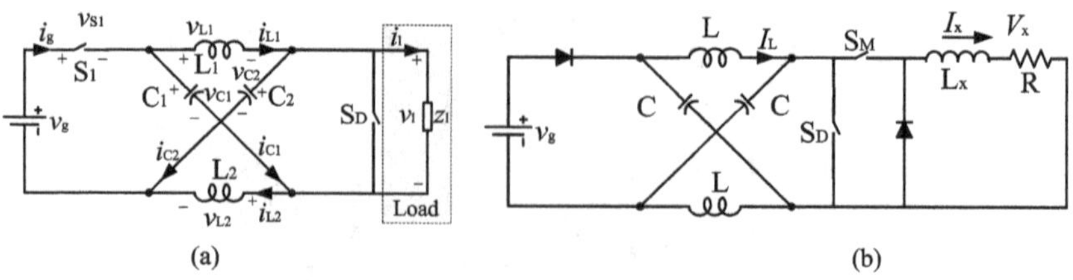

Figure 7.7 Simplified equivalent circuit of Z-source converter for small-signal modeling: (a) D_{st} (control S_1) as a control variable and (b) D_{st} (control S_D) and M (control S_D) as a control variable

Considering the symmetry of the network (using $v_{C1}(t) = v_{C2}(t) = v_C(t)$ and $i_{L1}(t) = i_{L2}(t) = i_L(t)$), a simplified small-signal model is presented in [40,42] for ZSI, where the load current is represented by a constant current source. However, such a model describes only the dynamics of the impedance network and fails to describe the dynamics of the ac load. To overcome this disadvantage, a third-order model is presented in [38] using $v_{C1}(t) = v_{C2}(t) = v_C(t)$, $i_{L1}(t) = i_{L2}(t) = i_L(t)$ and $i_l(t)$ as state variables. In this model, the ac side of the inverter is referred to as the dc side with an R_L load and taking its current as a third state variable. A similar third-order small-signal model is presented in [38], which considers the dynamics of the input-side current. In this, the current-fed qZSI is analyzed using $v_C(t)$, $i_L(t)$, and $i_{in}(t)$ as state variables to demonstrate the transient response of the inverter during the motoring and regeneration modes of operation for application in electric vehicles. Subsequent fourth- and higher-order small-signal models are also presented for inverters [44] and rectifiers [45] to better understand the dynamics of the input/output (load/source) and the impedance network. However, the complexity of formulating the small-signal model and the control-loop design increases with the increase in state variables. To simplify this, various assumptions (symmetry in impedance network, balanced load) and simplifications (representation of ac load/source by its equivalent dc load/source) are prevalent in the literature without loss of generality and changes to the dynamic performance.

The state-space-averaged small-signal modeling provides a derivation of various control-to-output $\left(G_{d_0}^{v_c}(s)\right)$ and disturbance-to-output $\left(G_{v_{in}}^{v_c}(s), G_{i_{load}}^{v_c}(s)\right)$ transfer functions, which helps to predict the system dynamics under the influence of various parameter changes. The root locus of the control-to-output transfer function in the s-domain gives a clear map of the converter dynamics. In addition, predicting a Right-Half-Plane (RHP) zero in the control-to-output transfer function is a major advantage of small-signal modeling. The presence of RHP zeros indicates that the non-minimum phase undershoots (the controlled capacitor voltage dips before it rises in response to a D_{st} increase), generally tend to destabilize the wide-band feedback loops, implying high gain instability and imposing control limitations. This means that the design of a feedback loop with an adequate phase margin becomes critical when RHP zeros appear in the transfer function. Various analyses of the pole-zero location and the impact of parameter variations on the converter dynamics are studied considering the wide operating ranges of different sources, e.g., fuel cells and photovoltaics. Figure 7.8 shows the locus of the poles and zeros with changes in various parameters such as L, C, D_{st}, R, and r. The impact of these parameter variations on the converter dynamics is summarized in Table 7.1.

The impact of parameter variations can provide direction to designers beforehand to choose component values whilst considering the design constraints, such as feedback control bandwidth, ripple content, size and cost of components, damping factor, resonant frequency, and overshoot/undershoot in the desired output.

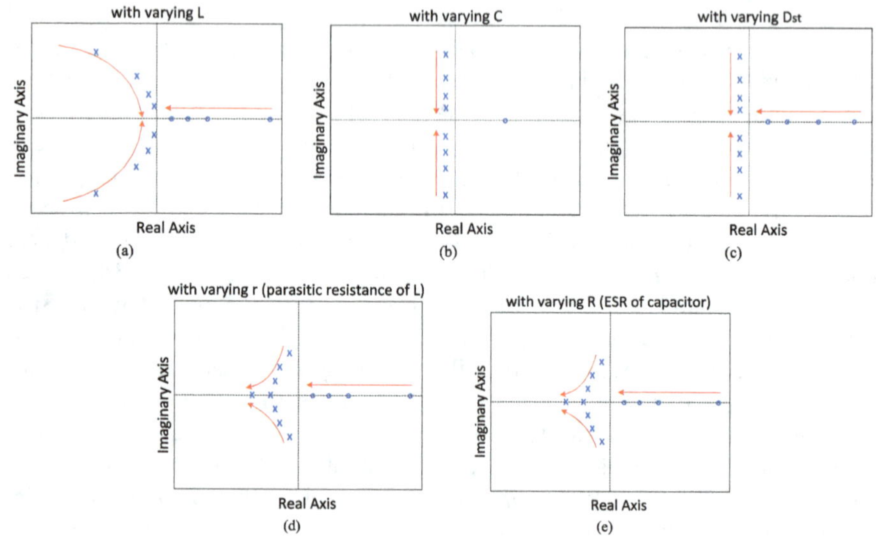

Figure 7.8 Locus of poles and zeros of control-to-output transfer function with varying (a) inductance (L), (b) capacitance (C), (c) shoot-through duty cycle (D_{st}), (d) parasitic resistance of inductor ®, and (e) equivalent series resistance (ESR) of the capacitor (R). The direction of the arrow indicates an increasing value of the parameter studied.

When considering the effect of parameter variations and the effect of poles and RHP zero, several closed-loop control methods are proposed in the literature to achieve the desired performance and to control the dc-link voltage and the ac output voltage of the impedance-source converter [41,46–49]. In all these control methods, there are two control degrees of freedom (D_{st} and M). The dc-link voltage is controlled by the shoot-through duty ratio (D_{st}) and the output voltage is controlled by the modulation index M.

The dc-link voltage across the inverter bridge can be controlled both directly and indirectly. In the direct dc-link voltage control method [46], the voltage across the inverter bridge is sensed directly by special sensing and scaling circuits as shown in Figure 7.9(b). This improves the transient response, enhances the disturbance rejection capability, and simplifies the controller design process. In the indirect method, the capacitor voltage at the impedance network is typically sensed and compared with the desired voltage as shown in Figure 7.9(a), (c), and (d). There are two control methods in this category: (1) measurement of capacitor voltage (V_C) in the impedance source network as shown in Figure 7.9(a) [44,47], and (2) measurement of capacitor voltage (V_C) and input voltage (V_{in}) to estimate the peak dc-link voltage as shown in Figure 7.9(c) and (d) [48].

Table 7.1 Summary of the impact of parameter variations on the Z-source converter dynamics (See Figure 7.8)

Parameter	**Change**	**Effect on the position of**		**Impact on system dynamics**
		Conjugate poles	**RHP zeros**	
Inductance (L)	Increasing	Move toward the imaginary axis	Move toward the imaginary axis	• Increase non-minimum phase undershoot • Increase settling time • Increase oscillatory response • Decrease damping ratio • Decrease natural frequency
Capacitance (C)	Increasing	Move toward the real axis	Constant	• Increase system damping • Increase rise time • Increase system settling time • Decrease natural frequency
Shoot-through duty cycle (d_{ST})	Increasing	Move toward the real axis	Move toward the imaginary axis	• Increase non-minimum phase undershoot • Increase system settling time • Decrease natural frequency
Equivalent series resistance (ESR) of capacitor (R)	Increasing	Move toward the real axis	Move toward the imaginary axis	• Increase system damping • Increase non-minimum phase undershoot • Increase current ripple through C
Parasitic resistance of inductor (r)	Increasing	Move toward the real axis	Move toward the imaginary axis	• Increase system damping • Increase non-minimum phase undershoot • Increase voltage ripple across L

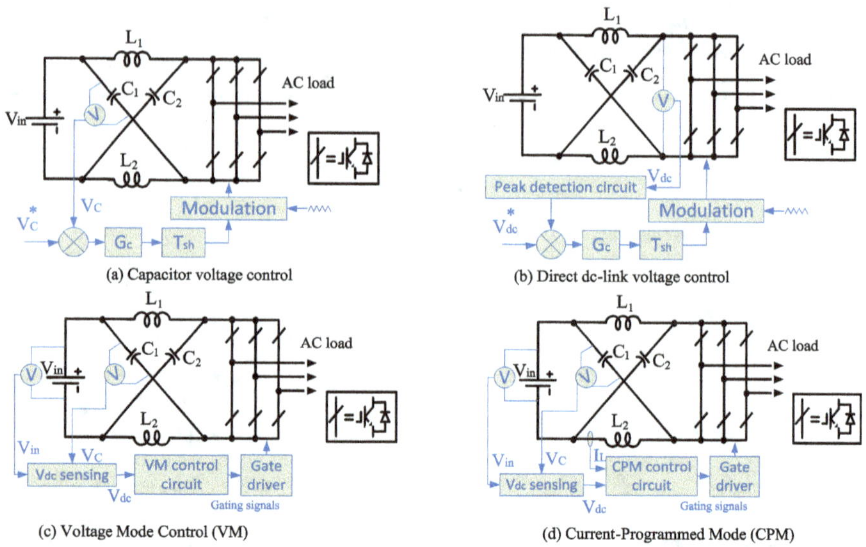

Figure 7.9 Control of dc-link voltage by indirect ((a), (c), and (d)) and direct control methods (b) for Z-source inverters

In the indirect control methods, the peak dc-link voltage becomes uncontrollable, while regulating the shoot-through duty cycle for fast changing the input voltages. This effect is unacceptable as it affects the output voltage, which forces a change in the modulation index. This may result in higher semiconductor stress and increase the harmonic distortion [47,48] in the output waveforms. The peak dc-link voltage is kept constant in the direct measurement technique; however, the control scheme becomes more complex, with additional circuitry. To resolve this limitation, the peak dc-link voltage is estimated by measuring the input voltage and capacitor voltage as shown in Figure 7.9(c) (voltage mode (VM) and Figure 7.9(d) (current programmed mode (CPM)). However, additional voltage/current sensors are required.

Figure 7.10 shows a block diagram of two popular closed-loop voltage control methods of ZSI, which consist of both dc-link control and ac output control [41,49]. T_{sh} appearing in Figure 7.10(a) is the shoot-through time; T_{sh} denotes the peak output voltage; and P_m is the modulation index. In both control methods, the capacitor voltage is controlled by regulating the shoot-through duty cycle (D_{st}) and the output voltage is controlled by regulating the modulation index (M), by using separate control loops with P and/or PI controllers. However, both control parameters are dependent on each other, as a change in one parameter imposes a limitation on the changeability of the other due to the insertion of a shoot-through time inside the null period. Putting a maximum limit on the control variable could mitigate this limitation.

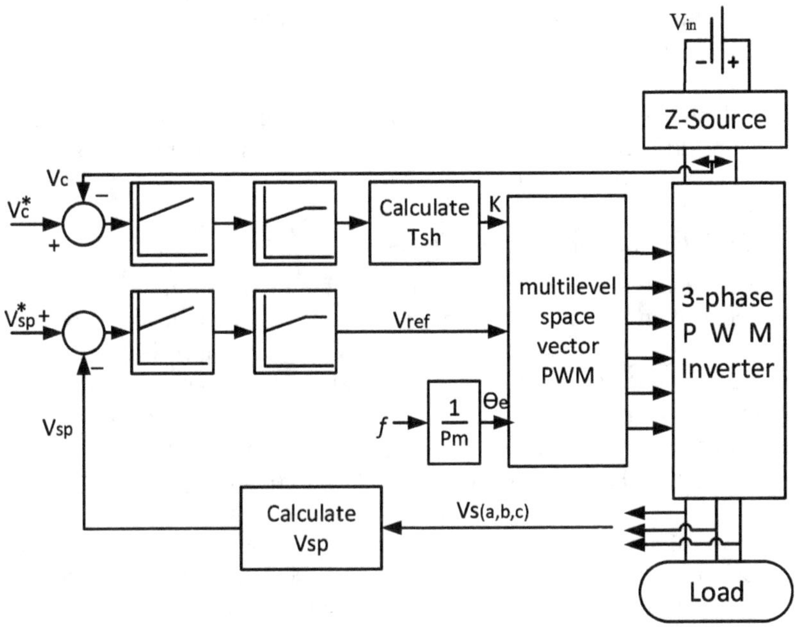

(a)

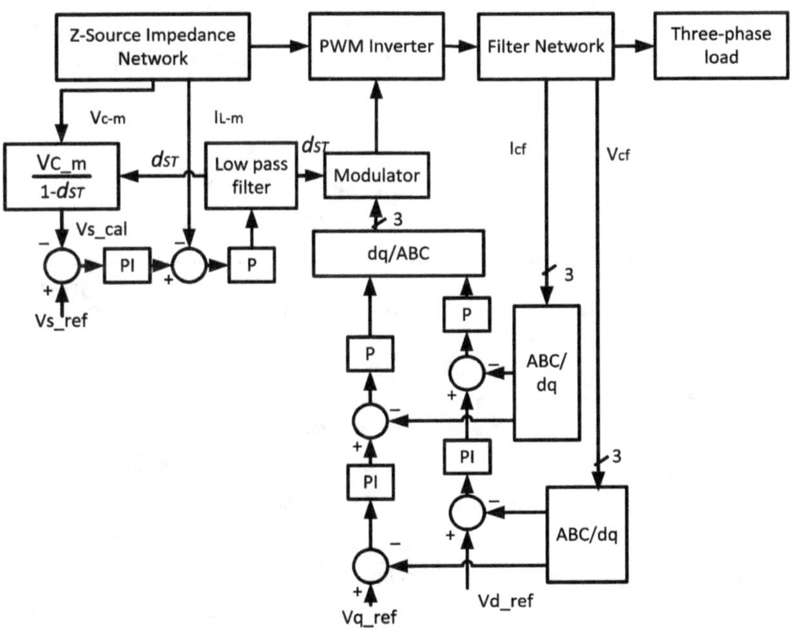

(b)

Figure 7.10 Indirect dc-link voltage controller modules with two degrees of freedom: (a) controller for both dc boost and ac output voltage of Z-source Inverter [49] and (b) multi-loop closed loop controller [41]

7.2 Categories of impedance source power converters based on conversion functionality

Figure 7.11 shows a broad categorization of the impedance source power converters using different switching configurations, which will be discussed in the following sub-sections in detail.

7.2.1 DC–DC converter topologies

Various isolated/non-isolated dc–dc converters are proposed in the literature with different control and modulation techniques. For example, a dc–dc converter realized using a quasi-Z-source Inverter (qZSI) with 2- or 3-leg H-bridge switching topology is proposed for distributed generation [50,51] as shown in Figure 7.12(a). A new quasi-Z-source push-pull converter topology with a reduced number of switching devices is also proposed in [52] as shown in Figure 7.12(b). It has the same gain as in [50,51], however, with reduced complexity in the gate circuit design.

A family of four-quadrant dc–dc converters using a Z-source or a quasi-Z-source network with a minimal number of switches and passive devices is presented in [53]. The converter provides four-quadrant operation using four-quadrant switches. Two basic converter topologies derived from ZSI and qZSI are shown in Figure 7.13. The converter has both buck/boost characteristics in the 0–1 range of the duty cycle. This feature along with changing the polarity of the load voltage by just controlling the duty cycle of the switch makes the converter very simple and more economical for many applications, e.g., dc-drives and other renewable energy systems.

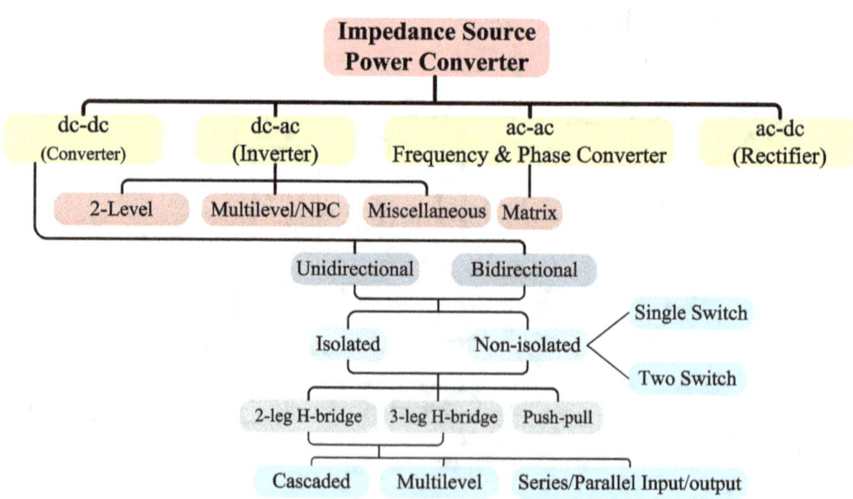

Figure 7.11 Categorization of impedance source network-based power converters

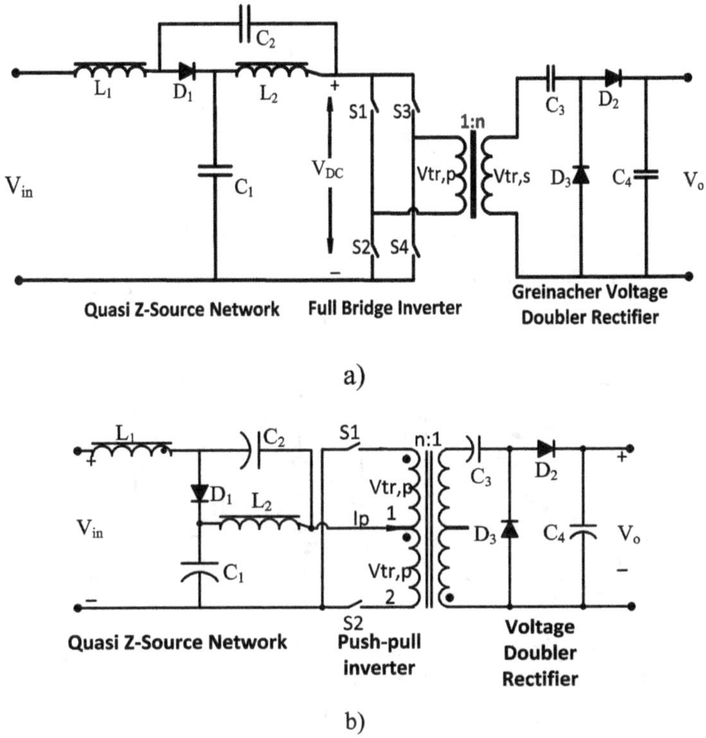

Figure 7.12 qZSI-based isolated dc–dc converter with (a) intermediate H-bridge switching topology [51] and (b) push-pull topology [52]

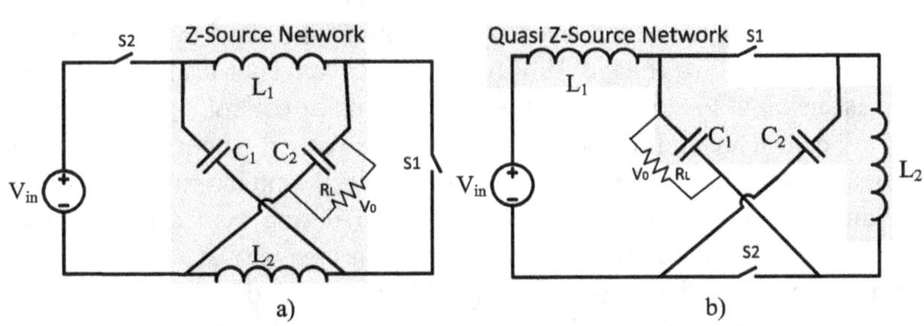

Figure 7.13 Family of four-quadrant dc–dc converters using (a) Z-source and (b) quasi-Z-source network [53].

A new boost dc–dc converter topology was proposed in [54] with three coupled inductors and called a Y-source converter as shown in Figure 7.14. It can achieve a very high boost at a lower shoot-through duty cycle of the switch and has one more degree of freedom to choose the voltage boost. The converter can be designed compactly with integrated magnetics, and fewer components are required to attain a high voltage boost. An isolated high-voltage boost converter using a

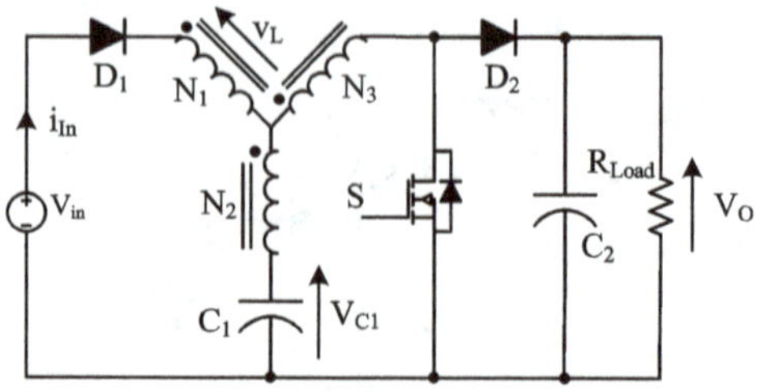

Figure 7.14 Y-source boost dc–dc converter [54]

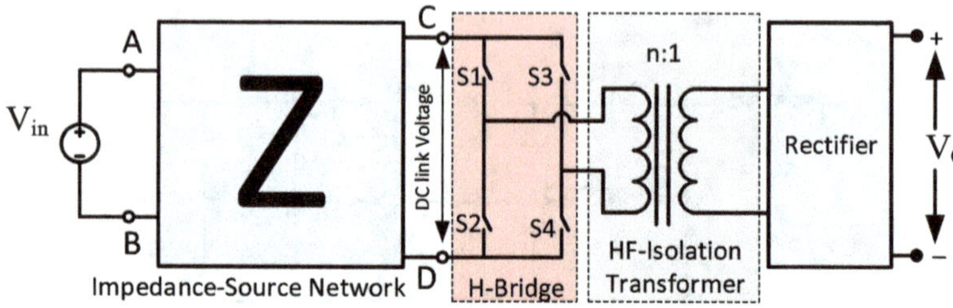

Figure 7.15 A general circuit configuration of an impedance-source network for dc–dc power conversion with intermediate H-bridge topology using various impedance-source networks [55]

Y-source impedance source network is also proposed in [55], which reduced the number of switching devices and the corresponding switching complexities.

Modulation of single-switch [54] and two-switch [55] dc–dc converter topologies is fairly simple and can be achieved by controlling the duty cycle of the switch depending on the dc-link voltage. However, a dc–dc converter using an intermediate H-bridge as shown in Figure 7.15 involves more complex control, as four switches are required to be switched optimally in order to get the desired output and performance.

Various modulation techniques are proposed in the literature, including shoot-through during freewheeling states, shoot-through during zero states, phase shift modulation (PSM) control with shoot-through during zero states [51,56] shoot-through by overlap of the active states [51,57] and shifted shoot-through [58], as shown in Figure 7.16. Due to the insertion of shoot-through in PWM, the switches in the H-bridge are compelled to commute at 2-3 times higher than the switching frequency. In shoot-through during freewheeling states, the top-side and bottom-side switches of the inverter bridge operate at three times the switching frequency ($f_{sw,top} = 3f_s$ and $f_{sw,bottom} = 3f_s$) as shown in Figure 7.16(a). This leads to high switching losses. Similarly, the number of commutations of the bottom-side switches remains the same while reducing the switching frequency of the top-side switch ($f_{sw,top} = f_s$ and $f_{sw,bottom} = 3f_s$) in the shoot-through during zero states modulation techniques (see Figure 7.16(b)). Phase Shift Modulation (PSM) control with shoot-through during zero states (see Figure 7.16(c)) equalizes the switching losses of the

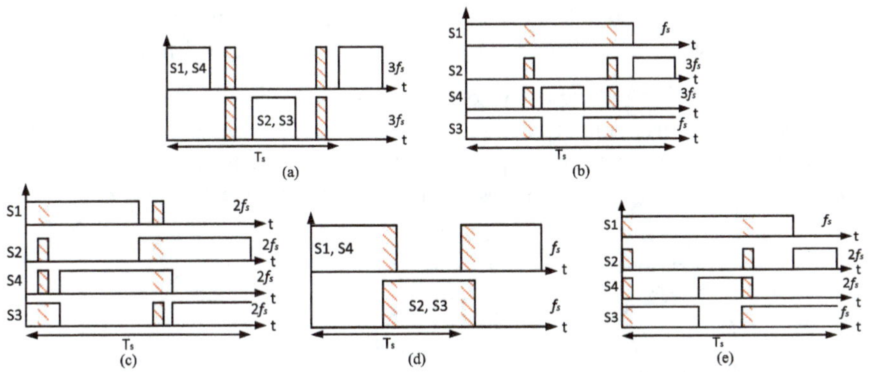

Figure 7.16 Some of the modulation techniques used for a dc–dc converter having an intermediate H-bridge are shown in Figure 7.15 (a) shoot-through during freewheeling states, (b) shoot-through during zero states, (c) phase-shift modulation (PSM) control with shoot-through during zero states, (d) shoot-through by the overlap of active states, and (e) shifted shoot-through (////, →Shoot-through)

top-side and bottom-side switches ($f_{sw,top} = 2f_s$ and $f_{sw,bottom} = 2f_s$), but this method is not effective in reducing the commutation time of the switches. The switching loss is minimized in the shoot-through by the overlap of active states ($f_{sw,top} = f_s$ and $f_{sw,bottom} = f_s$) as shown in Figure 7.16(d); however, the active-state and shoot-through-state duty cycles are not independently controllable. The inter-dependency of the active-state and shoot-through-state duty cycle could cause problems in the output-voltage compensation and also for systems that require independent control of the active and shoot-through state. A shifted shoot-through modulation technique (see Figure 7.16(e)) could reduce the switching frequency of the switches but it is complex and difficult to implement particularly in a microcontroller due to large number of comparator requirements. It also requires additional external components (logic gates, etc.) for implementation. The major disadvantage of this modulation technique is the loss of full soft-switching properties.

Every addition of the shoot-through state increases the commutation time of the semiconductor switches and so increases the switching loss in the system. Hence, minimization of the commutation time by the optimal placing of the shoot-through state in a switching time period is necessary in order to minimize the switching loss. In addition, the following points should be considered while placing the shoot-through state in the PWM control system for a qZSI-based dc–dc converter:

1. The maximum shoot-through duty cycle should never exceed 0.5; otherwise, the system may become unstable.
2. The minimum number of shoot-through states per switching period is two. One shoot-through state per period causes a discontinuous current and the converter will behave abnormally.

3. The shoot-through state should be in the zero state, i.e., the intact active state.
4. The active state and the shoot-through state should be independently controllable.

A new modulation technique called "*optimized PWM control with shoot-through during zero-states*" incorporating the shoot-through state to minimize the number of commutations in semiconductor switches for any impedance network-based dc–dc converter with intermediate H-bridge switching topology is presented in [50]. Figure 7.17

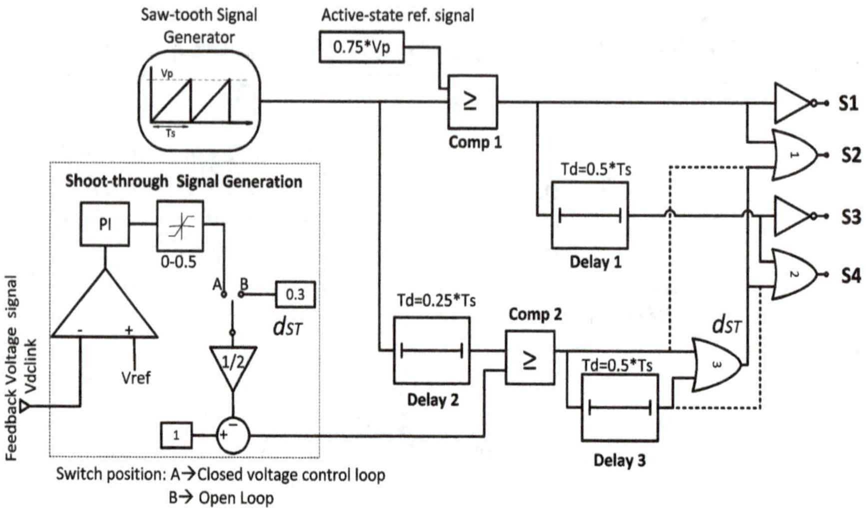

(a)

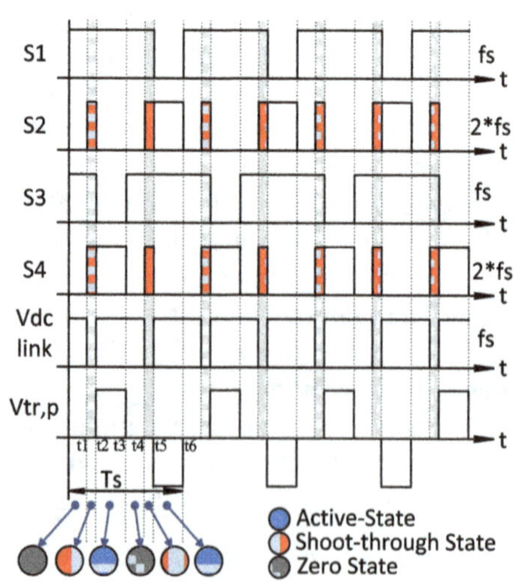

Figure 7.17 Optimized PWM control with shoot-through during zero states illustrating (a) a block diagram and (b) switching sequence for a dc–dc converter [50]

shows a block diagram of the optimized PWM control with shoot-through during zero-states and the corresponding switching sequence.

7.2.2 DC–AC inverter topologies

Different dc–ac topologies inspired from traditional converter topologies are implemented to utilize the basic properties of the impedance source network, e.g., 2-level H-bridge, multilevel/neutral point clamped, as well as the dual bridge. This will be discussed in detail in the following sub-sections.

7.2.2.1 2-Level H-bridge topologies

The impedance source network was originally implemented with a 3-phase H-bridge switching topology as a voltage source and a current source to demonstrate its superiority over the traditional VSI and CSI. Both single-phase as well as three-phase inverters can be implemented with an impedance source using the H-bridge circuit. Most of the topologies derived for 2-level voltage inversion have the general topology shown in Figure 7.18.

A standard carrier-based PWM as shown in Figure 7.19 is used for a single-phase H-bridge topology [59]. A shoot-through state is placed instead of a null state

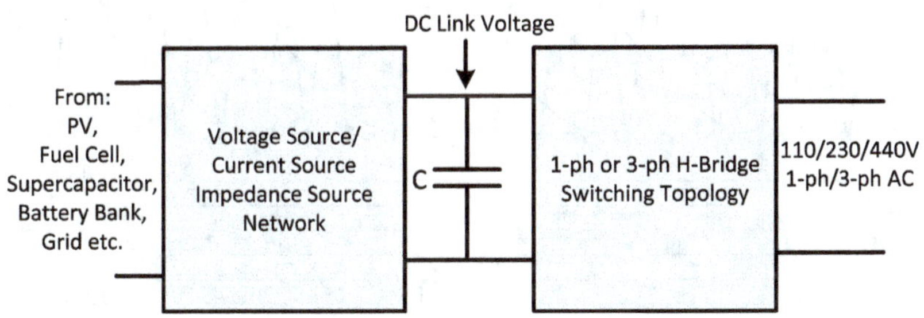

Figure 7.18 General impedance source inverter for 1-phase or 3-phase

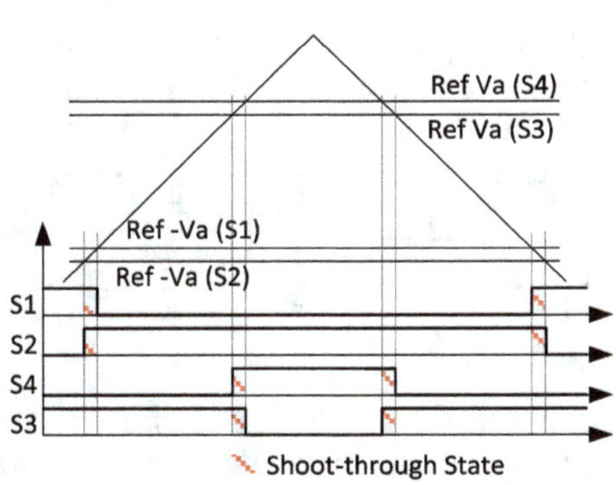

Figure 7.19 Generation of switching signal (shoot-through and PWM) for 1-ph Z-source inverter [59]

without altering the normalized volt-sec average voltage. The duration of each active state in a switching cycle is kept the same as in the traditional Sine PWM (SPWM). Therefore the output waveforms are still sinusoidal; however, they are boosted to the desired level by properly controlling the shoot-through time period. The paper extends the modulation concepts to the more complex three-phase H-bridge and four-phase H-bridge topologies for a voltage-fed Z-source inverter in both continuous and discontinuous modes.

For three-phase H-bridge topologies (2-level), SPWM and space-vector PWM (SVPWM) are used. SPWMs include simple boost control, maximum boost control, maximum constant boost control, and constant boost control with third-harmonic injection [60–62]. Simple boost control is the most basic and it is derived from the traditional sinusoidal PWM where a carrier triangular signal is compared to the three-phase reference signal for sinusoidal output voltage and two straight lines (V_p and V_n) to create shoot-through for voltage boost as shown in Figure 7.20(a). The disadvantage of this modulation technique is a decrease in the modulation index with an increase in the shoot-through range. The maximum shoot-through duty ratio of the

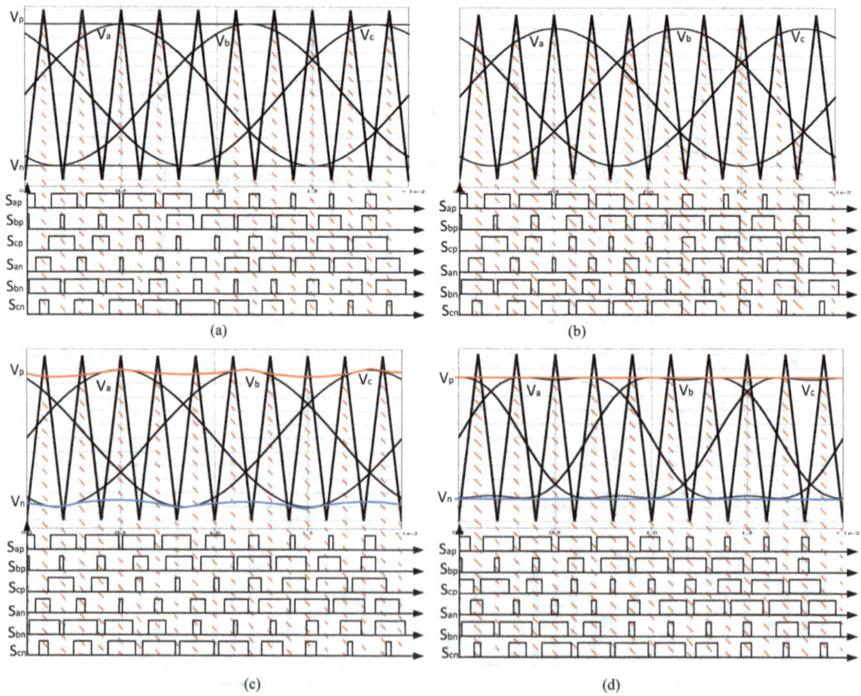

Figure 7.20 Sine-wave PWM. (a) Simple boost control. (b) Maximum-boost control. (c) Maximum-constant-boost control. (d) Constant-boost control with 1/6th of third harmonic injection (//// →Shoot-through period).

simple boost control is limited to $D_{st,\max} = (1 - M)$, which limits the boost factor to $B = [2M - 1]^{-1}$. As a result the device stress increases for the application, which requires a higher voltage boost. To address this issue a maximum-boost PWM control method is presented in [60]. This modulation technique maintains six active states unchanged from those of the traditional carrier-based PWM method; however, it utilizes all zero states to make shoot-through states as shown in Figure 7.20(b). These increase the range of the boost factor $B = \pi[3\sqrt{3}M - \pi]^{-1}$ compared to using a simple boost, which reduces the device stress.

However, due to the variable shoot-through time intervals, low-frequency ripple components are present in the capacitor voltage and inductor current, which increases the size and cost of the components in the impedance network. To achieve a constant shoot-through duty ratio and a maximum boost factor, a maximum constant-boost PWM control method is proposed in [61,62], which eliminates the low-frequency harmonic component in the impedance-source network. Figure 7.20(c) illustrates the switching waveforms of the maximum-constant-boost PWM control. The range of the modulation index is extended from 1 to $2/\sqrt{3}$ by injecting a third-harmonic component with 1/6 of the fundamental component magnitude to the three phase-voltage references. In this modulation technique, two straight lines V_p and V_n are required to generate a shoot-through time period as shown in Figure 7.20(d).

Besides SPWM, space vector pulse width modulation (SVPWM) has similarly been proven to be an effective modulation technique for traditional inverter topologies as it effectively reduces the commutation time of the switches, reduces the harmonic content in the output voltage/current and better utilizes the dc-link voltage, and consequently reduces the voltage stress and switching loss. This benefit encourages researchers and engineers to retrofit SVPWM for various impedance source inverters. However, proper insertion of the shoot-through state in the switching cycle without altering the volt-sec balance is crucial to reduce the additional commutation time of the switches and corresponding switching loss. Various SVPWM techniques are presented in the literature, e.g., ZSVM2 [63], ZSVM4 [64], and ZSVM6 [65]. Modified SVPWM techniques are also being used to reduce the common mode voltage and leakage currents for photovoltaic systems [66,67] and motor drives [68].

7.2.2.2 Multi-level/neutral point clamped (NPC)

The concept of a multilevel approach is adopted with an impedance source network to overcome some of the disadvantages of the classical multilevel topologies. Various impedance-sourced multilevel topologies with special modulation techniques are proposed to control the multilevel converter to get reduced harmonic distortion and device commutations. Figure 7.21(a) shows a three-level **neutral point clamped (NPC)** converter using two ZSIs [69–72]. The neutral point is connected to a common point of the two ZSIs and it is grounded. There are other topologies implemented with a reduced number of passive [73] and active devices [74] to reduce the size and cost of the system (see Figure 7.21(b)).

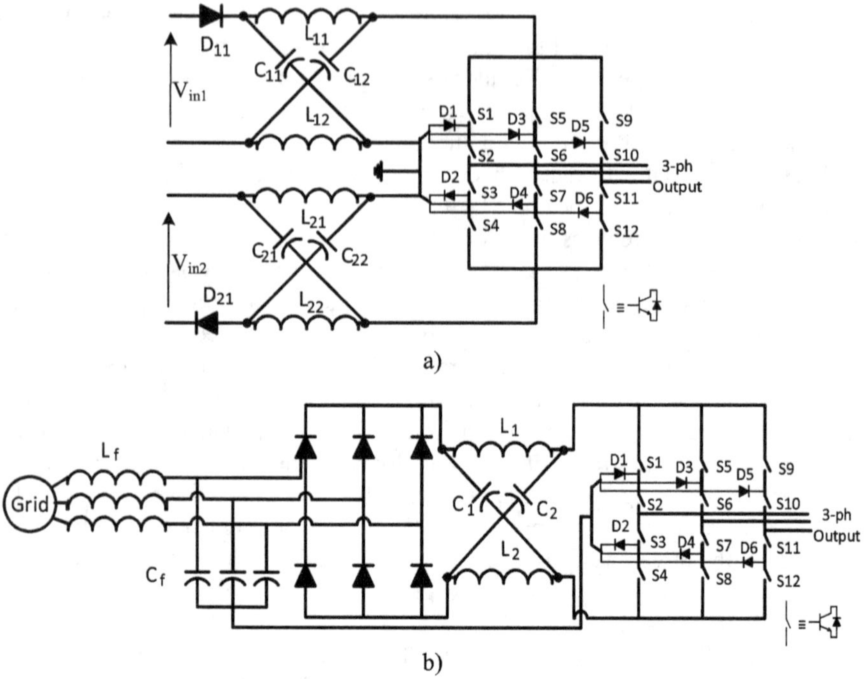

Figure 7.21 Various impedance network source multilevel converter topologies: (a) double-sourced NPC and (b) single-sourced NPC

The PWM schemes for Z-source NPC inverters are developed from classical three-level SVPWM modulation concepts using two-dimensional vectorial representations with "origin shifting." However, correct integration of the shoot-through state sequence with the classical PWM is essential for proper Z-source NPC operation, as some of the vectors can cause a short circuit across the full dc-link, which then results in zero voltage output. It is important to maintain the normalized volt-sec balance while sequencing the shoot-through states to accurately reproduce the desired three-phase sinusoidal voltages. In addition, careful integration of shoot-through with the conventional switching sequences is required to achieve maximal voltage-boost, minimal harmonic distortion, lower semiconductor stress, and a minimal number of device commutations per switching cycle. To achieve this, there are various continuous and discontinuous PWM schemes reported in [72] for controlling the Z-source NPC inverter with two impedance networks at its input side. The former modulation scheme is divided into two, i.e., continuous edge insertion (EI) PWM with symmetrical voltage boost and continuous modified reference (MR) PWM with minimal device commutation, and the latter modulation scheme is also divided into two, i.e., conventional 60°-discontinuous PWM and origin-shifted 60°-discontinuous PWM with reduced common mode switching. The device commutation count with EI PWM is the maximum of eight, which reduces to six with the continuously modified reference technique. The number further reduces to a minimum of four when the discontinuous PWM technique is opted for. However, with the same shoot-through duty ratio, the reduced number of shoot-throughs per switching cycle in the discontinuous scheme will produce lower common-mode voltage but higher inductor current ripples, which significantly increase the size of the passive components in the impedance network. A detailed comparison of continuous and discontinuous PWM schemes is provided in [72].

A nearest-three-vector (NTV) modulation principle and reduced common-mode switching (RCM) are proposed in [69] to minimize harmonic distortion, device commutation, and common mode voltage of the inverter. A hybrid PWM strategy similar to [75] for a two-level Z-source inverter is implemented for the Z-source NPC topology [76] to reduce its algorithm calculation by combining the theory of SVPWM and triangular-comparison PWM. A reduced component count Z-source NPC converter with a modified modulation technique is reported in [73] using a single Z-source network. This topology reduces the requirement of an additional impedance network to create a neutral point as explained in [69,72,76]. The modulation scheme is modified to create a full DC-Link (FDCL) and a partial DC-link (PDCL) shoot-through state to boost the output voltage without increasing the commutation time as in the conventional NPC modulation techniques [69,72]. An effective control method for a cascaded quasi-Z-source inverter using multilevel space vector modulation (MSVM) is presented in [77] for a single phase and in [78] for three phases to generate seven-level voltage. This control scheme achieves an independent control of the maximum power point technique (MPPT) for each photovoltaic panel and also balances the dc-link voltage across each H-bridge inverter to accomplish premium power quality for grid integration of photovoltaic panels for low switching frequency design.

7.2.3 AC–AC converter topologies

A matrix converter consists of nine bidirectional switches that allow any output phase to be connected to any input phase. It is a direct ac–ac converter with sinusoidal input/output waveforms and a controllable input power factor.

The maximum voltage gain of a traditional matrix converter does not exceed 0.866. In addition, the switches are more vulnerable to shoot-through on the same output phase leg. These limitations are overcome by the impedance source matrix converter [79] as shown in Figure 7.22. A traditional matrix converter with an impedance network can buck-boost the voltage as well as the frequency for the ac load requiring adjustable voltage and frequency, e.g., motor drives. Many modulation and control methods, e.g., simple maximum boost control, maximum boost control, maximum gain control, and hybrid minimum stress control have been proposed to improve the performance and reliability of the converter [80,81].

With some modifications made to the conventional modulation techniques (SVPWM, carrier-based PWM, PWAM, etc.) to incorporate the shoot-through

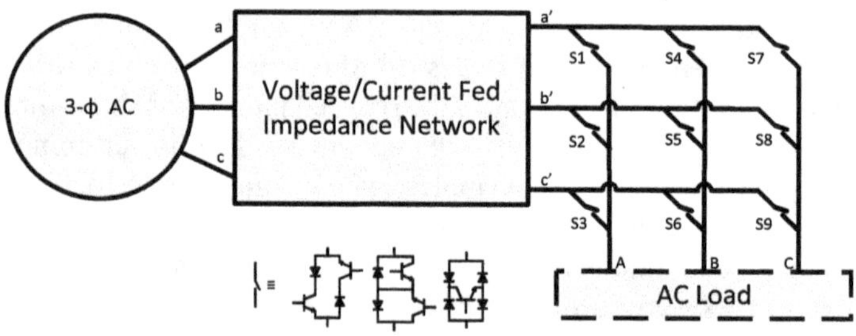

Figure 7.22 Impedance source network-based matrix converter [79]

state, various modified modulation techniques are implemented to control direct and indirect impedance-network-based matrix converters. The traditional carrier-based sinusoidal PWM (SPWM) [60–62] is applied to control and modulate various Z-source and quasi-Z-source direct matrix converters with a few modifications, e.g., four control strategies: simple maximum-boost control, maximum-boost control, maximum-gain control, and hybrid minimum-stress control are proposed in [79,81]. In the simple maximum-boost control, the modulation index is limited to $M = 0.5$ which means that the maximum voltage gain can only go up to 0.944. Maximum boost control utilizes all the zero states as shoot-through states. The range of the modulation index is extended to 0.866 by injecting 1/6 of the third harmonic signal in to the reference signal. The maximum-gain control method can obtain the maximum gain at the same modulation index among all four techniques, and the hybrid minimum-voltage-stress control can obtain the minimum voltage stress at the same voltage gain. In terms of the total harmonic distortion (THD) at the output, the maximum-constant-boost control method is effective in eliminating the low-order harmonics compared to the simple boost and maximum boost controls. A comparison of various control methods is also presented in [81].

Pulse-width-amplitude-modulation (PWAM) with a maximum-constant-boost shoot-through control strategy is also implemented in [82] to control a voltage-fed quasi-Z-source direct-matrix converter. This modulation technique reduces the switching frequency of the matrix converter by 1/3 compared to the SVPWM, which helps to reduce the switching losses by more than 50% compared to the SVPWM and 87% compared to the SPWM.

7.2.4 AC–DC converter topologies

Rectifiers based on impedance source networks have the capability to both buck and boost the input voltage in a single stage compared to a traditional rectifier. In addition, they can provide good input power factor, low line-current distortion, regeneration, and improved reliability. Figure 7.23 shows Z-source [83,84] and quasi Z-source [45] based rectifier topologies. A new concept of a bidirectional converter based on a matrix converter is also presented in [79]. The advantages of the ac–dc matrix converter are controllable input power factor, tight dc voltage

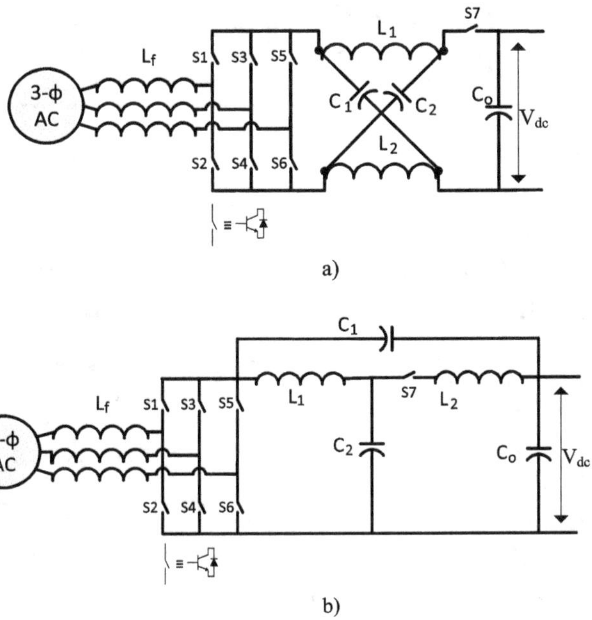

Figure 7.23 AC–DC rectifier topologies based on (a) Z-source [83,84] and (b) quasi Z-source network [45]

regulation, wide bandwidth with quick response to load variations, and single-stage buck voltage ac–dc power conversion.

7.3 Impedance-source network topologies

Impedance source networks are mainly derived from the Z-source network by modifying the original impedance network, or by rearranging the connections of inductors and capacitors [15–36]. New topologies are still being developed, mainly for four reasons: (1) reduction of the Z-source network component count and rating; (2) extension of voltage gain range; (3) achieving higher power density; and (4) application-oriented optimization and improvement.

The impedance source network is broadly classified into two categories based on magnetics: (a) non-transformer based and (b) coupled or transformer based line shown in Figure 7.24. Each topology has distinct features and advantages, which will be discussed briefly in the following sub-sections.

7.3.1 Non-transformer based

7.3.1.1 Z-source/quasi Z-source

Z-source converters are broadly classified into two types, voltage-fed and current-fed. However, unlike the traditional voltage-fed/current-fed inverter, the impedance

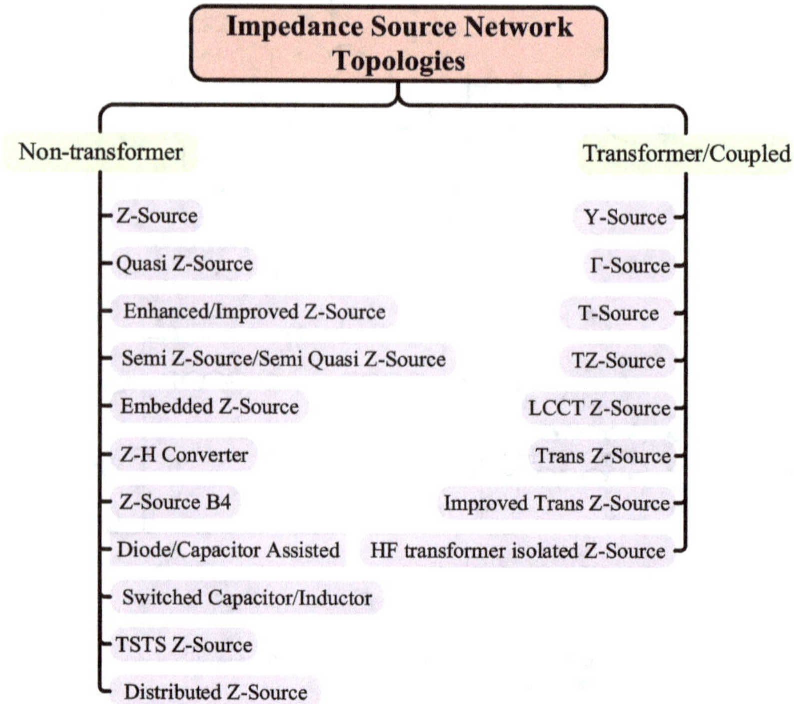

Figure 7.24 Impedance source network topologies [14]

source network provides a buffer between the source and the inverter bridge and facilitates a short- and an open-circuit at any time depending on the mode of operation. Traditional voltage-fed/current-fed Z-Source impedance networks suffer from problems like discontinuous input current in the boost mode for the voltage-fed ZSI (Figure 7.25(a)) and high current stress on the inductor in the current-fed ZSI (Figure 7.26(a)). Various voltage- and current-fed topologies derived from ZSI and qZSI (Figures 7.25(b) and (c) and 7.26(b) and (c)) with improved performance were proposed in [15] to solve the problem of ZSIs [1]. It should be noted that all three current-fed ZSIs, Figure 7.26(a)–(c), are capable of bidirectional power flow and buck-boost operation, although the switches have to be reverse-blocking devices. These new benefits extend this topology to applications of renewable energy generation and motor drives. The enhanced-boost Z-source inverter is also proposed in [16] with alternate-cascaded switched and tapped-inductor cells using some lower-rated components. Similarly, an improved Z-source [17] and an improved trans-Z-source [18] are proposed respectively to reduce the capacitor voltage stress. However, additional components are required to implement the circuit. This makes the system more costly and also decreases the power density of the converter.

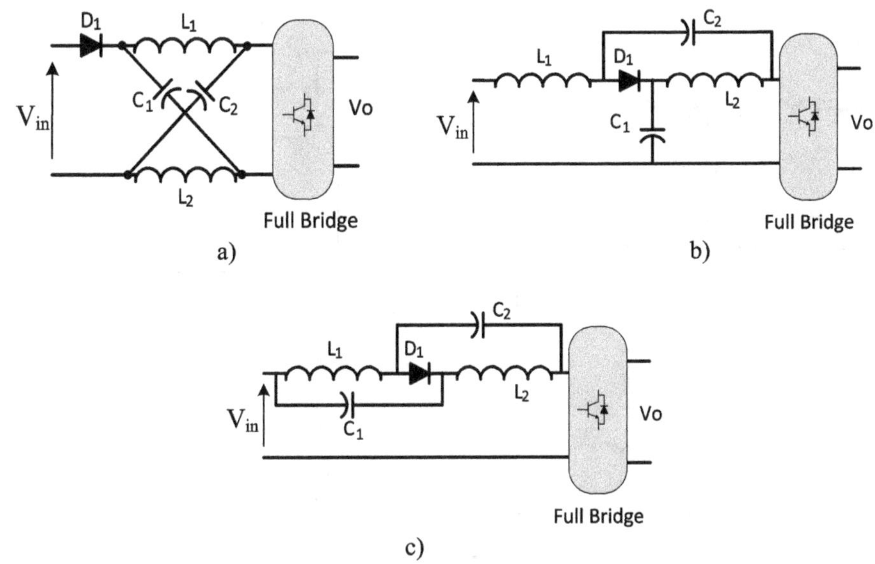

Figure 7.25 Various voltage-fed ZSI topologies. (a) ZSI with discontinuous input current, (b) qZSI with continuous input current, and (c) qZSI with discontinuous input current

7.3.1.2 Semi-Z-source/semi-quasi Z-source

Semi-Z-source inverters were proposed (see Figure 7.27) to achieve low cost and high efficiency in applications such as single-phase grid-tied PV power systems. A semi-Z-source inverter with only two active switches has a voltage boost function and a double-ground feature (both the PV panel and ac output can be grounded) that eliminates the need to float/isolate PV panels without leakage current and improves the safety [19]. Unlike the traditional ZSI/qZSI, a shoot-through state is not applicable to a semi-Z-source inverter. An improved non-linear sinusoidal PWM method is used to get the desired duty cycle to generate a sinusoidal output.

The advantage of the semi-Z-source and semi-quasi-Z-source inverters [20] is that they can be implemented using fewer switches compared to traditional ZSI and qZSI, but unfortunately, the voltage stress on the switching devices is high. This topology is suitable for a grid-connected micro-PV inverter using high-voltage SiC devices.

Two modulation techniques are prevalent in the literature to control and modulate the two switches of a single-phase Z-source/quasi-Z-source to get the desired output voltage, namely one-cycle control [21] and non-linear sinusoidal pulse-width modulation (SPWM) [19].

The voltage gain of the semi-Z-source/quasi-Z-source converter is not a straight line as with a full-bridge inverter. So, instead of a sinusoidal reference signal($v = V\sin\omega t$), a non-linear sinusoidal reference signal $v = [2 - M\sin\omega t]^{-1}$ is compared with the carrier signal to generate the gate drive signal for the two switches as shown in Figure 7.28(a).

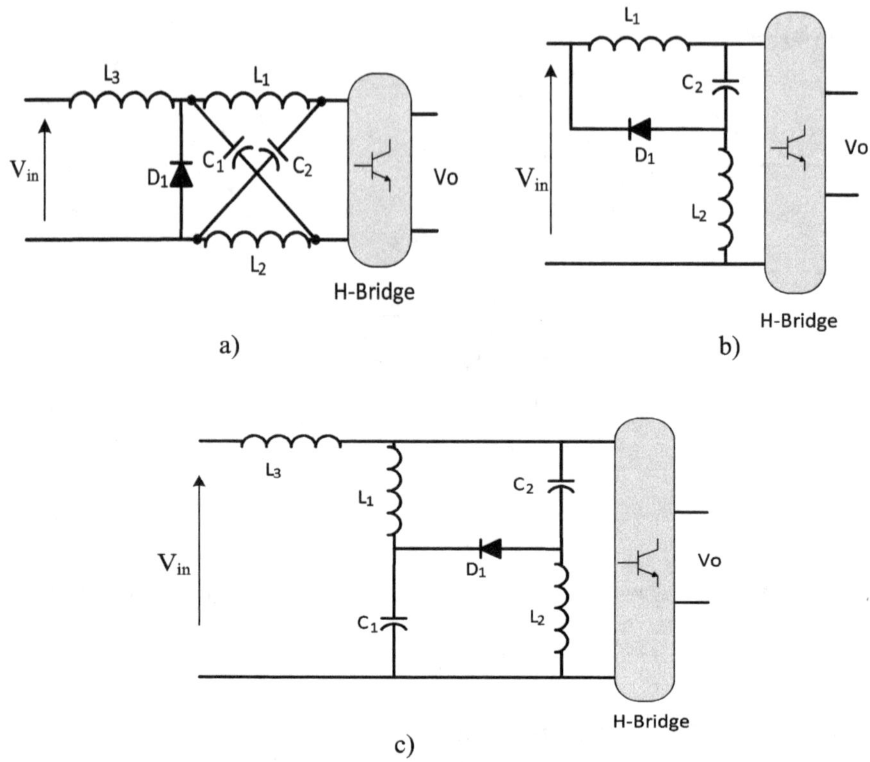

Figure 7.26 Various current-fed ZSI topologies (switches have to be reverse-blocking devices). (a) ZSI with continuous input current, (b) qZSI with discontinuous input current, and (c) qZSI with continuous input current.

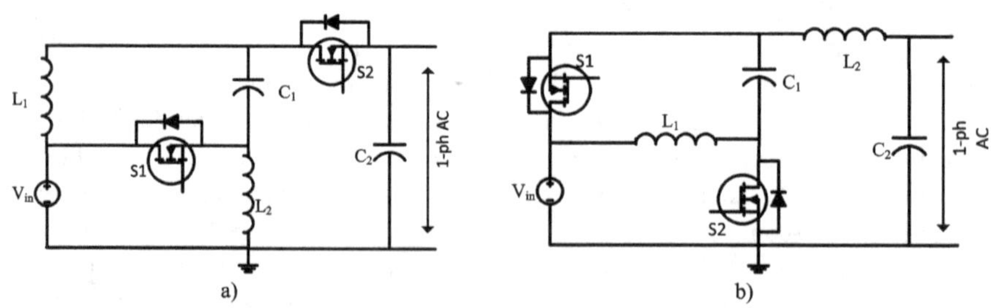

Figure 7.27 Semi-Z-source topologies. (a) Semi-Z-source inverter and (b) semi-quasi-Z-source inverter.

A one-cycle control method is adopted to control a single-phase semi-Z-source topology in [21]. In this control method, two switches work in a complementary fashion where the clock signal (CLK) is used to turn-on any one switch. The

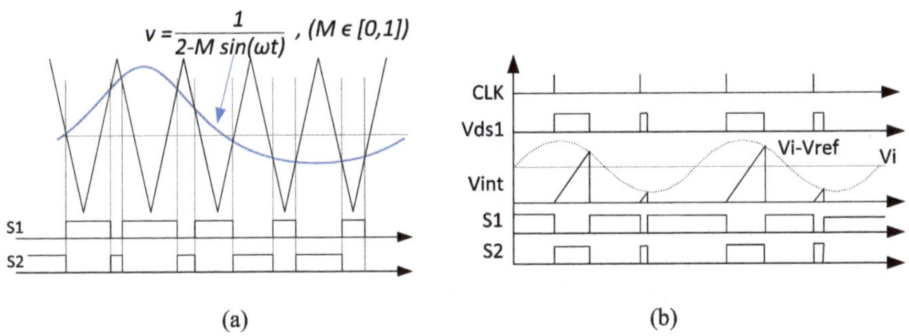

Figure 7.28 Generation of switching signal for 1-ph Z-source inverter using (a) non-linear sinusoidal pulse width modulation (SPWM) [19] and (b) one cycle control method [21]

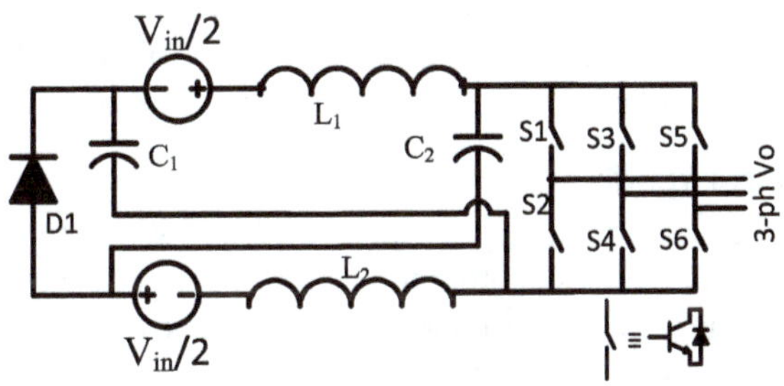

Figure 7.29 Two-level embedded Z-source inverter [22]

turn-on time of the switch is determined by the integrated voltage across the switch, and when it reaches the sinusoidal signal$(v_i - v_{ref})$, the integrator is reset and the switch turns off as shown in Figure 7.28(b). This control method has the ability to reject input perturbations and is robust to the system model, which provides a high-efficiency constant-frequency control.

7.3.1.3 Embedded Z-source

The embedded Z-source was proposed to achieve continuous input current and lower capacitor voltage rating, and its multi-source feature is especially suitable for PV power generation [22]. Figure 7.29 shows the circuit topology of a two-level embedded Z-source inverter. There are other similar embedded topologies with one or two DC sources suitable for battery storage systems.

7.3.1.4 Z-source B4 converter

Inspired by the traditional B4 VSI, a Z-source B4 topology is proposed in [23] with a reduced number of active components, having enhanced reliability, and lower cost. Figure 7.30 shows the Z-source B4 converter topology for a three-phase power conversion.

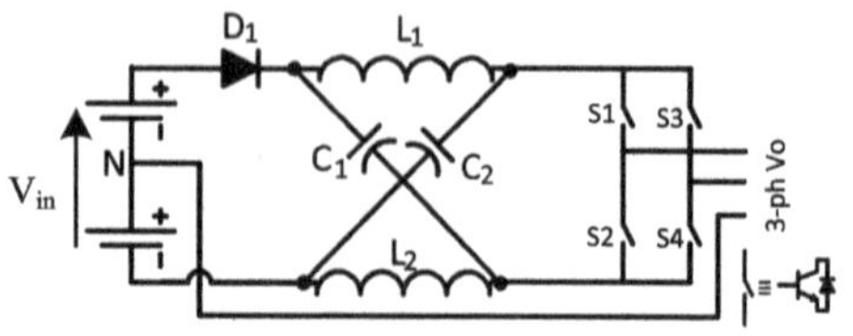

Figure 7.30 B4 Z-source inverter topology [23]

7.3.1.5 Switched inductor/capacitor

Extra inductors and capacitors have been added to the Z-source and quasi-Z-source impedance network, with the aim of improving the boost capability of the circuit. Many topologies are presented in the literature to reduce the stress on the passive components and also to eliminate the start-up inrush current. A switched inductor/capacitor ZSI/qZSI provides continuous input current and reduced voltage stress on the capacitor [24]. An embedded Z-source with a switched inductor combines the advantages of both topologies, e.g., high boost ratio, reduced capacitor voltage stress, and low input ripple current [25]. However, this switched inductor/capacitor topology needs a large number of passive devices, which increases the cost and volume of the converter.

7.3.1.6 Capacitor/diode assisted

The voltage-boost capability of the Z-source and quasi-Z-source is extended with the aid of capacitors and diodes in order to meet the needs of applications requiring very high voltage boost [26]. The impedance network can be extended by using diodes and/or capacitors as shown in Figure 7.31. The advantages of this topology are high voltage gain and reduced capacitor stress; however, this is at the cost of extra passive components.

7.3.1.7 TSTS Z-source

Three-switch three-state single-phase Z-source inverters (TSTS-ZSIs) were proposed a in [27], and they are classified into two groups, boost-TSTS-ZSI and buck–boost-TSTS-ZSI as shown in Figure 7.32.

This topology can be assembled using fewer switches compared to the traditional impedance source topologies, so higher power density can be achieved. In addition, it has a lower voltage stress and dual grounding, which makes it suitable for PV power generation.

7.3.1.8 Distributed Z-source

Distributed impedance networks such as transmission lines and hybrid inductor - capacitor (LC) components can be used for a Z-source network [28] as shown in Figure 7.33. These distributed Z-source networks are difficult to implement, but a distributed Z-source inverter does not need any extra diode or switch to achieve the voltage boost function, thus having the minimum component count. This topology could open a door for impedance source networks to radio-frequency (RF) power converter design by utilizing the distributed inductance and capacitance prominent at higher frequencies.

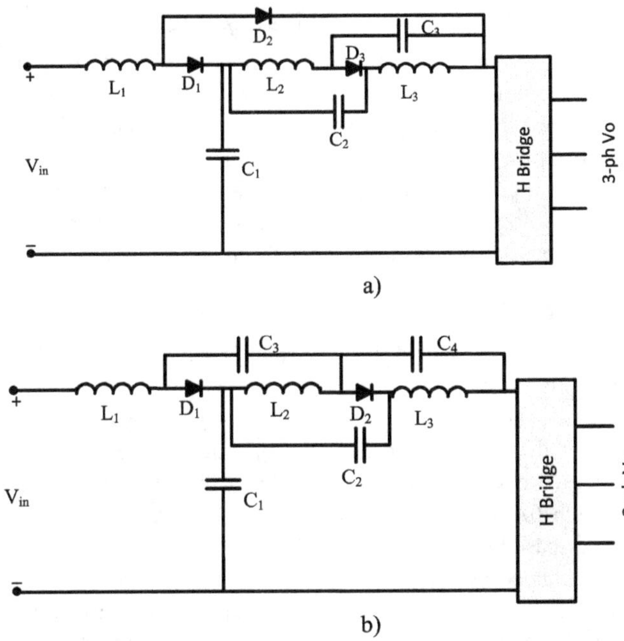

Figure 7.31 Assisted qZSI topologies. (a) Diode assisted and (b) capacitor assisted [26].

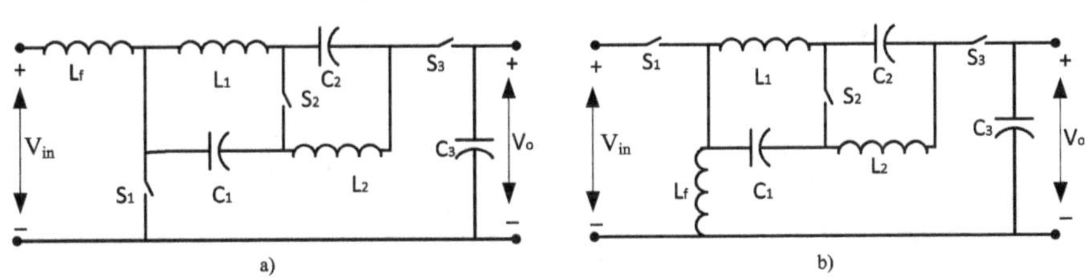

Figure 7.32 Three-switch three-state single-phase Z-source inverter (TSTS-ZSIs). (a) Boost topology and (b) Buck-boost topology [27].

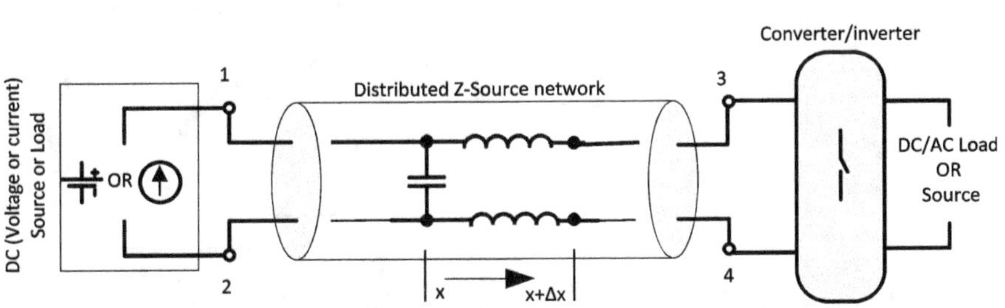

Figure 7.33 A general topology of distributed Z-source converter [28]

7.3.2 With transformer or magnetic coupling

Magnetically coupled inductors and transformers find a niche in impedance networks to improve the voltage boost capability as well as the modulation index. In addition, they reduce the number of passive components needed in the network, which improves the power density and reduces the cost of the system. A generic method is presented in [29] to illustrate the derivation of some of the magnetically coupled network topologies. The following sub-sections describe the impedance network topologies based on the transformer or coupled inductor.

7.3.2.1 Y-source

A unique impedance source network called the "*Y-source network*" topology was proposed in [53] using coupled inductors with three windings (N_1, N_2, and N_3) having versatile characteristics and features (see Figure 7.14). The gain of the converter is presently not matched with existing networks operated at the same duty ratio. The proposed converter has one more degree of freedom (three windings and shoot-through duty cycle of switch (d_{ST})) to choose the voltage boost, as compared to a classical impedance network-based boost converter [30]. Theoretically, any magnitude of voltage boost can be obtained by adjusting the turns ratio and shoot-through duty cycle of the switch. Additionally, the Y-source network can be used to derive other two-winding magnetically coupled impedance source (MCIS) networks by simply eliminating and/or rearranging the windings as shown in Figures 7.34–7.37 [31].

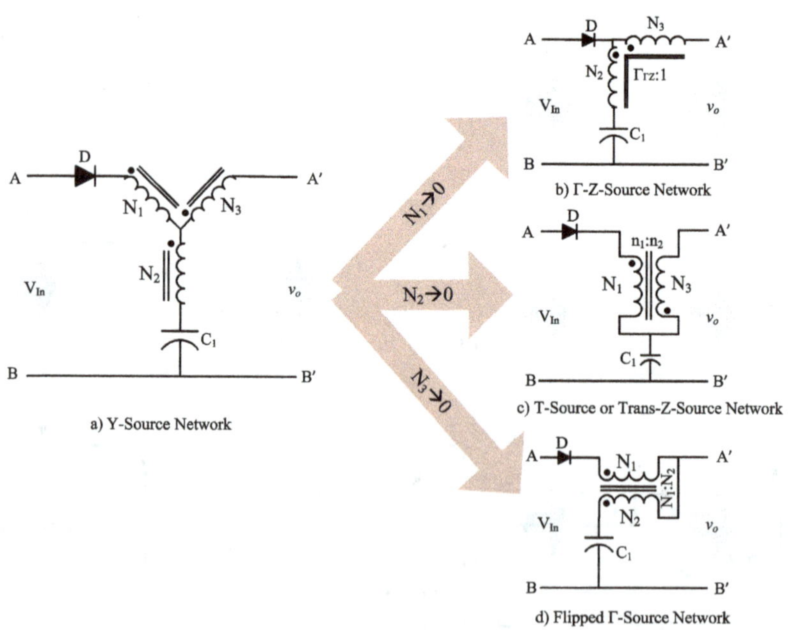

Figure 7.34 Derivation of conventional two-winding MCIS networks with discontinuous input currents: (b) **Γ***-Z-source network, (c) T-source or Trans-Z-source network, and (d) flipped* **Γ***-source network from (a) Y-source network [31]*

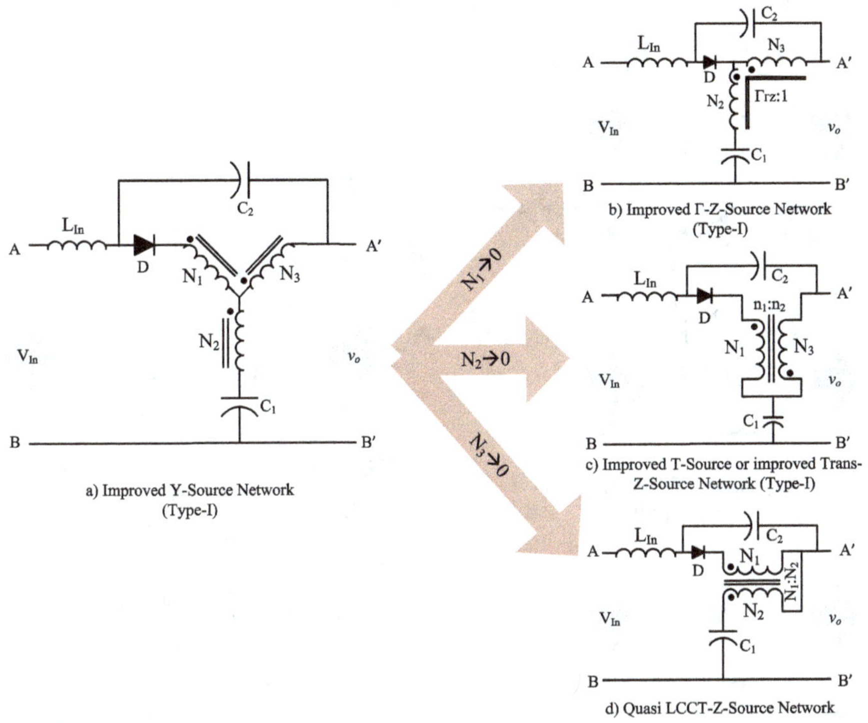

Figure 7.35 Derivation of conventional Type-I two-winding improved-MCIS networks with continuous input currents: (b) improved **Γ***-Z-source network, (c) improved T-source or Trans-Z-source network, and (d) quasi LCCT-Z-source network from (a) improved Y-source network [31]*

7.3.2.2 A-source

The A-source impedance network uses an autotransformer for realizing converters for any application that demands a very high dc voltage gain [32]. The network utilizes a minimal turns ratio compared to other magnetically coupled impedance source (MCIS) networks to attain a high voltage gain as shown in Figure 7.38. In addition, the converter draws a continuous current from the source, and hence it is suitable for many types of renewable energy sources.

7.3.2.3 Γ-Z-source

A unique Γ-shape two-winding coupled transformer is also implemented in an impedance network [33] to increase the gain and modulation ratio simultaneously while reducing the component count. Unlike other transformer-based impedance networks whose gains increase with an increased turns ratio, e.g., T-source, Trans-Z-source, TZ-source, or LCCT Z-source, the Γ-Z-source gain increases

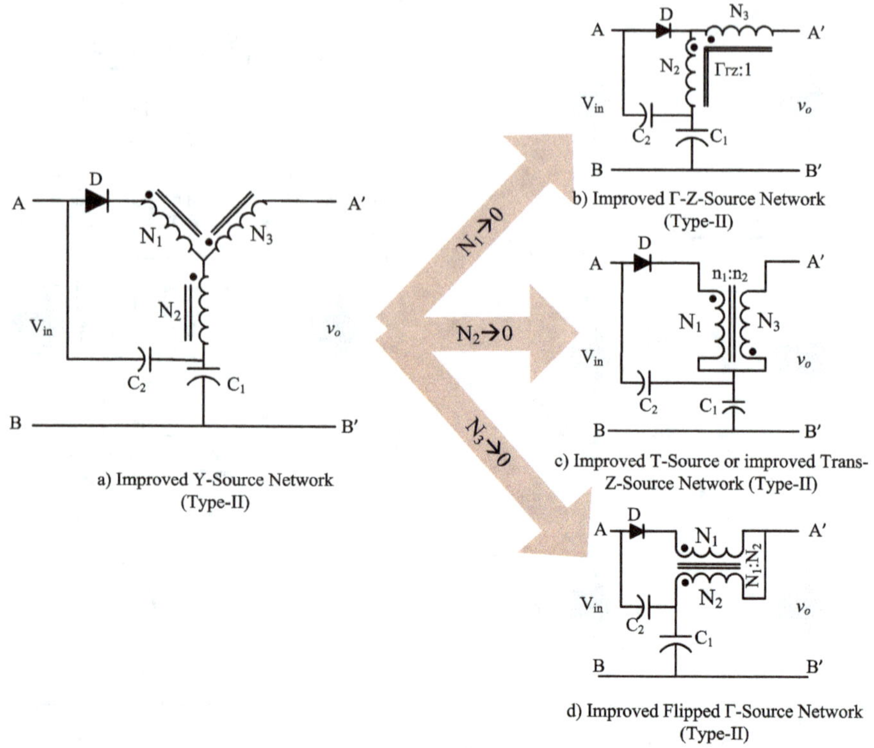

Figure 7.36 Derivation of conventional Type-II two-winding improved-MCIS networks with continuous input currents: (b) improved **Γ**-*Z-source network, (c) improved T-source or trans-Z-source network, and (d) quasi LCCT-Z-source network from (a) improved Y-source network [31]*

with a decrease in the turns ratio. Figure 7.39 shows the Γ-Z-source network topology for an inverter.

7.3.2.4 T-source

The T-source inverter also utilizes a two-winding coupled inductor and one capacitor [34] as shown in Figure 7.40. The gain of the converter can be set higher than the traditional ZSI and qZSI using a transformer turns ratio larger than 1. This topology is suitable for an NPC converter as it shares a common voltage source for both the passive arrangement and the converter circuit.

7.3.2.5 Trans Z-source

Theoretically, the original Z-source, quasi-Z-source, and embedded Z-source all have unlimited voltage gain. Practically, however, a high voltage gain (>2), will result in a high voltage stress imposed on the switches. Trans-Z-source (two

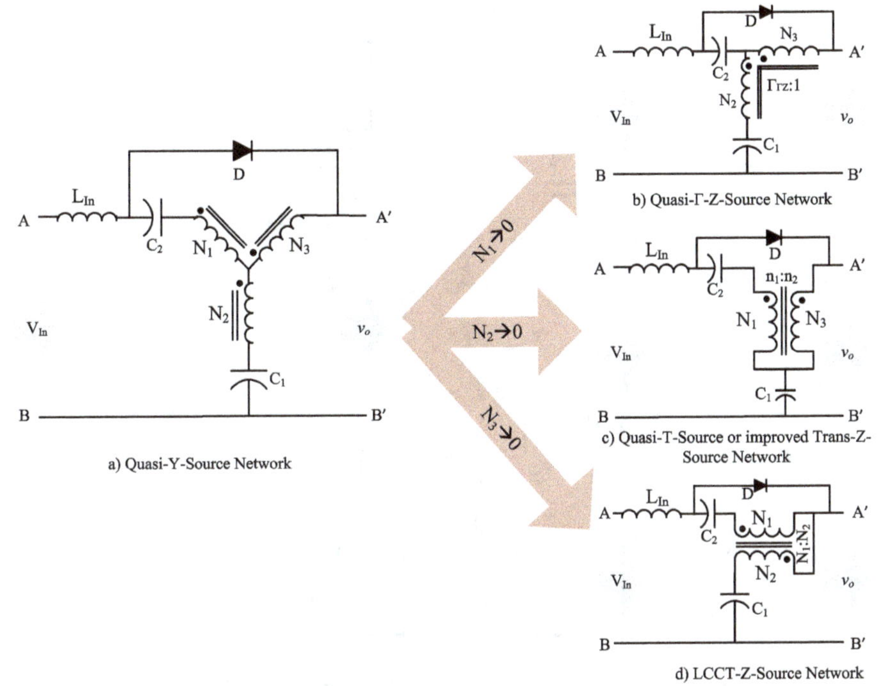

Figure 7.37 Derivation of two-winding quasi-MCIS networks with continuous input currents: (b) quasi-Γ-Z-source network, (c) quasi-T-source or quasi-Trans-Z-source network, and conventional (d) LCCT-Z-source network from (a) quasi-Y-source network [31]

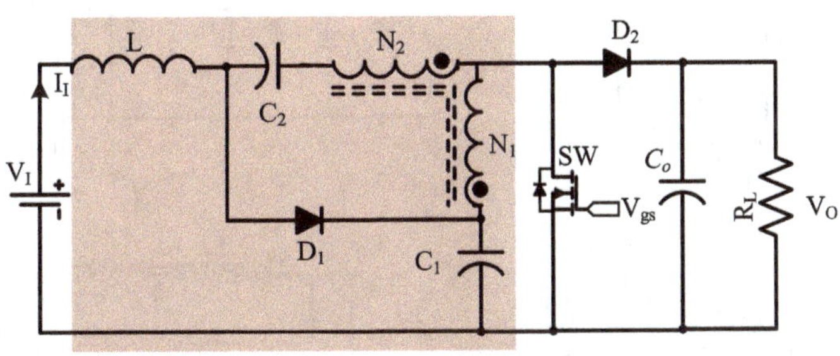

Figure 7.38 A-source network topology [32]

voltage-fed and two current-fed) inverters were proposed to have higher voltage gains while keeping voltage stress low and reducing the Z-source network to one transformer (or one coupled inductor) and one capacitor [35] as shown in Figure 7.41.

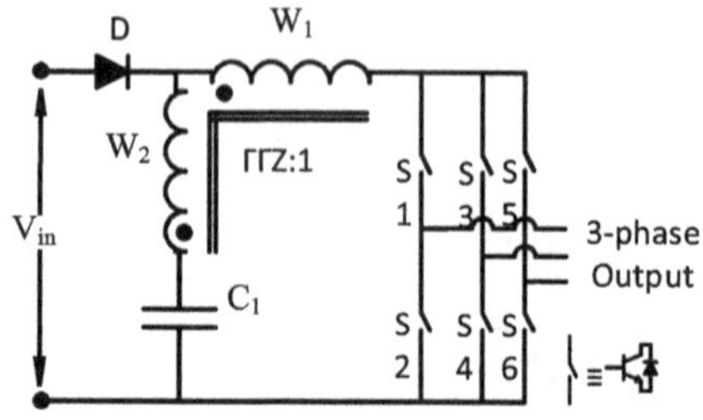

Figure 7.39 **Γ**-*Z-source network topology [33]*

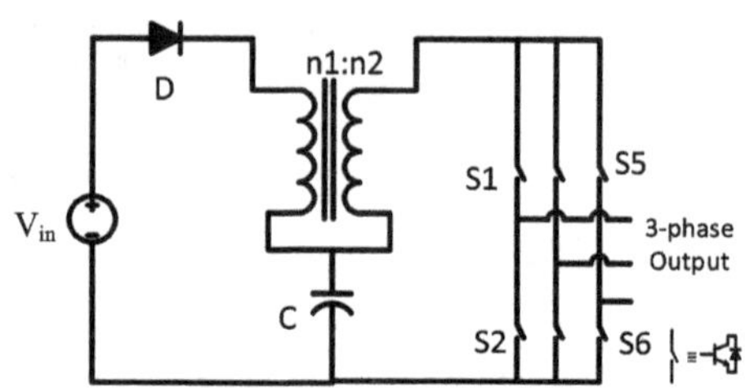

Figure 7.40 T-source network topology [34]

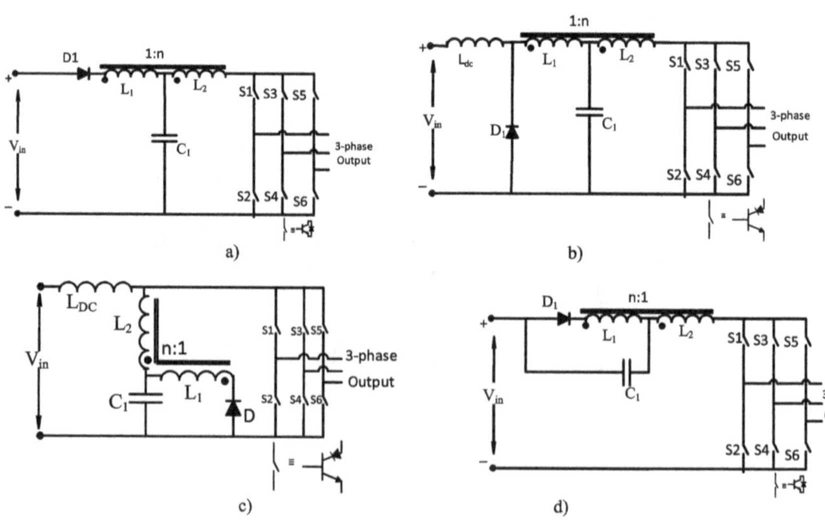

Figure 7.41 Trans-Z-source inverter. (a) Voltage-fed trans-Z-source; (b) current-fed trans-Z-source; (c) current-fed trans-quasi-Z-source, and (d) voltage-fed trans-quasi-Z-source [35]

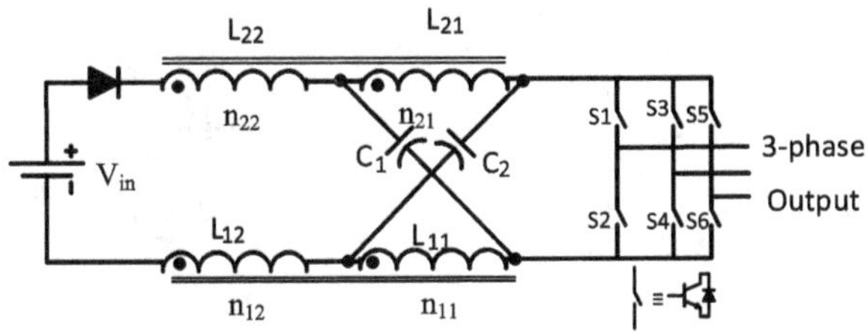

Figure 7.42 TZ-source network topology [36]

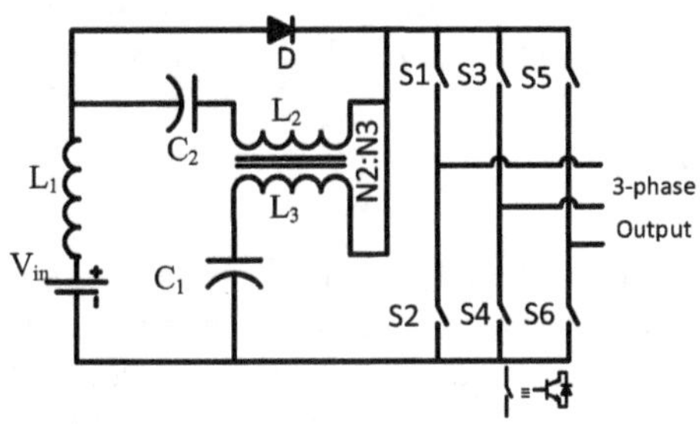

Figure 7.43 LCCT network topology [85]

7.3.2.6 TZ-source

The TZ-source also achieves high voltage gain by setting the turns ratio of the transformer to greater than 1 [36]. Comparatively, it requires a lower transformer turns ratio than the trans-ZSI; however, it requires four coupled inductors as well as the same number of other passive components as the traditional ZSIs. So this topology is not very effective in reducing the number of components and size. Figure 7.42 shows the circuit topology of the TZ-source network for a 3-phase inverter.

7.3.2.7 LCCT Z-source

With the inductor and a transformer integrated into a common core, the inductor-capacitor-capacitor-transformer Z-source inverters (LCCT-Z-source) as shown in Figure 7.43 achieve higher voltage gains and modulation index [85]. This topology maintains a continuous input current even at a light load and also filters out high-frequency ripples from the input current.

7.3.2.8 HF transformer isolated Z-source/quasi Z-source/ trans-Z-source

A family of impedance source networks with intermediate high frequency (HF) isolation is presented with different topologies for applications requiring isolation for safety reasons [86]. These topologies inherit all the benefits of the Z-source networks along with a higher boost ratio and lower device stress. However, this topology increases the number of active and passive components. In addition, the

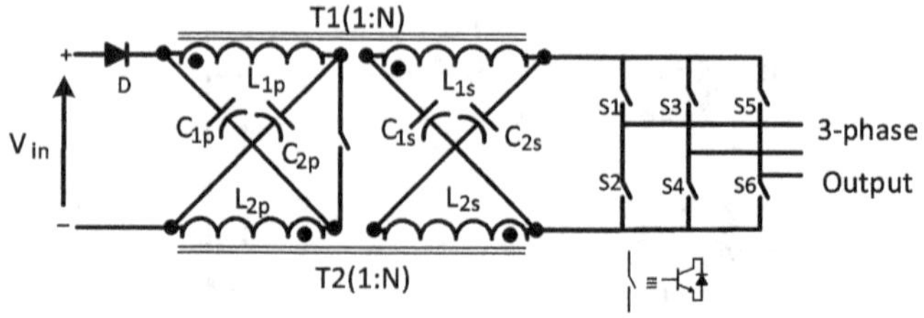

Figure 7.44 HF transformer isolated Z-source inverter [86]

coupled transformer must be designed properly to minimize leakage inductance. One example of a voltage-fed HF-isolated ZSI is shown in Figure 7.44.

7.4 Conclusions

Impedance source networks have added a new chapter in the field of power electronic topologies with their unique features and properties that overcome most of the problems faced by traditional converter topologies. Since the publication of the first Z-source network, there have been numerous contributions in the literature modifying the basic topology to suit the needs of many applications. In principle, it overcomes the conceptual and theoretical barriers and limitations of the traditional VSI and CSI and provides a novel power conversion concept. The superior performance of the impedance source network to design more robust and versatile converter topologies for various applications attracts researchers and designers from both academia and industry to explore it in more depth. Many topologies were developed to overcome the limitations and disadvantages of the traditional impedance source network. The number of combinations is large and the topologies might in some cases be confusing. In this chapter, an attempt is made to provide a global picture of the impedance source networks proposed in the literature to comprehend and identify their pros and cons. A broad classification of the topologies and modulation techniques with further sub-classification aims to provide an easy selection of control and modulation techniques for appropriate topologies for particular applications.

References

[1] F. Z. Peng, "Z-Source Inverter," *IEEE Trans. Ind. Applicat.*, vol. 39, no. 2, pp. 504–510, 2003.

[2] F. Z. Peng, X. Yuan, X. Fang and Z. Qian, "Z-Source Inverter for Adjustable Speed Drives," *IEEE Trans. Power Electron. (Letter)*, vol. 1, no. 2, pp. 33–35, 2003.

[3] Z. J. Zhou, X. Zhang, P. Xu and W. X. Shen, "Single-Phase Uninterruptible Power Supply Based on Z-Source Inverter," *IEEE Trans. Ind. Electron.*, vol. 55, no. 8, pp. 2997–3004, 2008.

[4] Y. Li, S. Jiang, J. G. Cintron-Rivera and F. Z. Peng, "Modeling and Control of Quasi-Z-Source Inverter for Distributed Generation Applications," *IEEE Trans. Ind. Electron.*, vol. 60, no. 4, pp. 1532–1541, 2013.

[5] B. Ge, H. Abu-Rub, F. Z. Peng, *et al.*, "An Energy Stored Quasi-Z-Source Inverter for Application to Photovoltaic Power System," *IEEE Trans. Ind. Electron.*, vol. 60, no. 10, pp. 4468–4481, 2013.

[6] L. Yushan, G. Baoming, F. Z. Peng, A. R. Haitham, A. T. de Almeida and F. J. T. E. Ferreira, "Quasi-Z-Source Inverter based PMSG Wind Power Generation Systems," in *Proc. ECCE 2011*, pp. 291–297, 17–22 Sept. 2011.

[7] C. J. Gajanayake, D. M. Vilathgamuwa and P. C. Loh, "Development of a Comprehensive Model and a Multiloop Controller for Z-Source Inverter DG Systems," *IEEE Trans. Power Electron.*, vol. 22, no. 4, pp. 1453–1463, 2007.

[8] J. Liu, S. Jiang, D. Cao and F. Z. Peng, "A Digital Current Control of Quasi-Z-Source Inverter with Battery," *IEEE Trans. Ind. Informatics*, vol. 9, no. 2, pp. 928–936, 2013.

[9] F. Z. Peng, M. Shen and K. Holland, "Application of Z-Source Inverter for Traction Drive of Fuel Cell-Battery Hybrid Electric Vehicles," *IEEE Trans. Power Electron.*, vol. 22, no. 3, pp. 1054–1061, 2007.

[10] A. S. Khlebnikov and S. A. Kharitonov, "Application of the Z-Source Converter for Aircraft Power Generation Systems," in *Proc. EDM 2008*, pp. 211–215, 2008.

[11] S. J. Amodeo, H. G. Chiacchiarini and A. R. Olivia, "High-Performance Control of a DC/DC Z-Source Converter Used for an Excitation Field Driver," *IEEE Trans. Power Electron.*, vol. 27, no. 6, pp. 2947–2957, 2012.

[12] J. C. R. Caro, F. Z. Peng, H. Cha and C. Rogers, "Z-Source-Converter-Based Energy-Recycling Zero-Voltage Electronic Loads," *IEEE Trans. Ind. Electron.*, vol. 56, no. 12, pp. 4894–4902, 2009.

[13] K. A. Corzine and R. W. Ashton, "A New Z-Source DC Circuit Breaker," *IEEE Trans. Power Electron.*, vol. 27, no. 6, pp. 2796–2804, 2012.

[14] Y. P. Siwakoti, F. Z. Peng, F. Blaabjerg, P. C. Loh and G. E. Town, "Impedance Source Network for Electric Power Conversion—Part I: A Topological Review" *IEEE Trans. Power Electron.*, vol. 30, no. 2, pp. 699–716, 2015.

[15] J. Anderson and F. Z. Peng, "Four Quasi-Z-Source Inverters," in *Proc. PESC 2008*, pp. 2743–2749, 2008.

[16] D. Li, P. C. Loh, M. Zhu, F. Gao and F. Blaabjerg, "Enhanced-Boost Z-Source Inverters with Alternate-Cascaded Switched-and Tapped-Inductor Cells," *IEEE Trans. Ind. Electron.*, vol. 60, no. 9, pp. 3567–3578, 2013.

[17] Y. Tang, S. Xie, C. Zhang and Z. Xu, "Improved Z-Source Inverter with Reduced Z-Source Capacitor Voltage Stress and Soft-Start Capability," *IEEE Trans. Power Electron.*, vol. 24, no. 2, pp. 409–415, 2009.

[18] M. K. Nguyen, Y. C. Lim and S. J. Park, "Improved Trans-Z-Source Inverter with Continuous Input Current and Boost Inversion Capability," *IEEE Trans. Power Electron.*, vol. 28, no. 10, pp. 4500–4510, 2013.

[19] D. Cao, S. Jiang, X. Yu and F. Z. Peng, "Low-Cost Semi-Z-Source Inverter for Single-Phase Photovoltaic Systems," *IEEE Trans. Power Electron.*, vol. 26, no. 12, pp. 3514–3523, 2011.

[20] H. Haimovich, R. H. Middleton and L. De Nicolo, "Large-Signal Stability Conditions for Semi-Quasi-Z-Source Inverters: Switched and Averaged Models," in *Proc. 52nd Conference on Decision and Control 2013*, Dec. 2013.

[21] Y. Tang, S. Xie and C. Zhang, "Single-Phase Z-Source Inverter," *IEEE Trans. Power Electron.*, vol. 26, no. 12, pp. 3869–3873, 2011.

[22] P. C. loh, F. Gao and F. Blaabjerg, "Embedded EZ-Source Inverter," *IEEE Trans. Ind. Appl.*, vol. 46, no. 1, pp. 256–267, 2010.

[23] P. C. Loh, N. Duan, C. Liang, F. Gao and F. Blaabjerg, "Z-Source B4 Inverter," in *Proc. PESC 2007*, pp. 1363–1369, 2007.

[24] D. Li, P. C. Loh, M. Zhu, F. Gao and F. Blaabjerg, "Generalised Multicell Switched-Inductor and Switched-Capacitor Z-Source Inverters," *IEEE Trans. Power Electron.*, vol. 28, no. 2, pp. 837–848, 2013.

[25] H. Itozakura and H. Koizumi, "Embedded Z-Source Inverter with Switched Inductor," in *Proc. IECON 2011*, pp. 1342–1347, 2011.

[26] C. J. Gajanayake, F. L. Luo, H. B. Gooi, P. L. So and L. K. Siow, "Extended-Boost Z-Source Inverters," *IEEE Trans. Power Electron.*, vol. 25, no. 10, pp. 2642–2652, 2010.

[27] L. Huang, M. Zhang, L. Hang, W. Yao and Z. Lu, "A Family of Three-Switch Three-State Single-Phase Z-Source Inverters," *IEEE Trans. Power Electron.*, vol. 28, no. 5, pp. 2317–2329, 2013.

[28] S. Jiang and F. Z. Peng, "Transmission-Line Theory Based Distributed Z-Source Networks for Power Conversion," in *Proc. APEC 2011*, pp. 1138–1145, 2011.

[29] P. C. Loh and F. Blaabjerg, "Magnetically Coupled Impedance-Source Inverters," *IEEE Trans. Ind. Applicat.*, vol. 49, no. 5, pp. 2177–2187, 2013.

[30] Y. P. Siwakoti, P. C. Loh, F. Blaabjerg, S. J. Andreasen and G. E. Town, "Y-Source Boost DC/DC Converter for Distributed Generation," *IEEE Trans. Ind. Electron.*, vol. 62, no. 2, pp. 1059–1069, 2015.

[31] Y. P. Siwakoti, F. Blaabjerg and P. C. Loh, "New Magnetically Coupled Impedance (Z-) Source Networks," *IEEE Trans. Power Electron.*, vol. 31, no. 11, pp. 7419–7435, 2016.

[32] Y. P. Siwakoti, F. Blaabjerg, V. P. Galigekere, A. Ayachit and M. K. Kazimierczuk, "A-Source Impedance Network," *IEEE Transactions Power Electron.*, vol. 31, no. 12, pp. 8081–8087, 2016, doi:10.1109/TPEL.2016.2579659.

[33] P. C. Loh, D. Li and F. Blaabjerg, "Γ-Z-Source Inverters", *IEEE Trans. Power Electron. (Lett.)*, vol. 28, no. 11, pp. 4880–4884, 2013.

[34] R. Strzelecki, M. Adamowicz, N. Strzelecka and W. Bury, "New Type T-Source Inverter," in *Proc. CPE 2009*, pp. 191–195, 2009.

[35] W. Qian, F. Z. Peng and H. Cha, "Trans-Z-Source Inverters," *IEEE Trans. Power Electron.*, vol. 26, no. 12, pp. 3453–3463, 2011.

[36] M. K. Nguyen, Y. C. Lim and Y. G. Kim, "TZ-Source Inverters," *IEEE Trans. Ind. Electron.*, vol. 60, no. 12, pp. 5686–5695, 2013.

[37] Y. P. Siwakoti, F. Z. Peng, F. Blaabjerg, P. C. Loh, G. E. Town and S. Yang, "Impedance Source Network for Electric Power Conversion—Part II: Review of Control and Modulation Techniques" *IEEE Trans. Power Electron.*, vol. 30, no. 4, pp. 1887–1905, 2015.

[38] M. Shen, Q. Tang and F. Z. Peng, "Modeling and Controller Design of the Z-Source Inverter with Inductive Load," in *Proc. IEEE PESC 2007*, pp. 1804–1809, 2007.

[39] Y. Li and F. Z. Peng, "AC Small Signal Modeling, Analysis and Control of Quasi-Z-Source Converter," in *Proc. IEEE ECCE 2012*, pp. 1848–1854, 2012.

[40] C. J. Gajanayake, D. M. Vilathgamuwa and P. C. Loh, "Small-Signal and Signal-Flow-Graph Modeling of Switched Z-Source Impedance Network," *IEEE Trans. Power Electron. (Lett.)*, vol. 3, no. 3, pp. 111–116, 2005.

[41] C. J. Gajanayake, D. M. Vilathgamuwa and P. C. Loh, "Modelling and Design of Multi-Loop Closed Loop Controller for Z-Source Inverter for Distributed Generation," in *Proc. IEEE PESC 2006*, pp. 1–7, 2006.

[42] P. C. Loh, D. M. Vilathgamuwa, C. J. Gajanayake, Y. R. Lim and C. W. Teo, "Transient Modeling and Analysis of Pulse-Width Modulated Z-Source Inverter," *IEEE Trans. Power Electron.*, vol. 22, no. 2, pp. 498–507, 2007.

[43] V. P. Galigekere and M. K. Kazimierczuk, "Small-Signal Modeling of PWM Z-Source Converter by Circuit-Averaging Technique," in *Proc. IEEE ISCAS 2011*, pp. 1600–1603, 2011.

[44] F. Guo, L. Fu, C. H. Lin, C. Li and J. Wang, "Small Signal Modeling and Controller Design of a Bidirectional Quasi-Z-Source Inverter for Electric Vehicle Applications," in *Proc. IEEE ECCE 2012*, pp. 2223–2228, 2012.

[45] Q. Lei, S. Yang, F. Z. Peng and R. Inoshita, "Steady State and Transient Analysis of a Three Phase Current-Fed Z-Source PWM Rectifier," in *Proc. VPPC 2009*, pp. 426–432, 2009.

[46] X. Ding, Z. Qian, S. Yang, B. Cui and F. Z. Peng, "A Direct Peak DC-link Boost Voltage Control Strategy in Z-Source Inverter," in *Proc. IEEE APEC 2007*, pp. 648–653, 2007.

[47] X. Ding, Z. Qian, S. Yang, B. Cui and F. Z. Peng, "A Direct DC-Link Boost Voltage PID-Like Fuzzy Control Strategy in Z-Source Inverter," in *Proc. IEEE PESC 2008*, pp. 405–411, 2008.

[48] G. Sen and M. E. Elbuluk, "Voltage and Current-Programmed Modes in Control of Z-Source Converter," *IEEE Trans. Ind. Applicat.*, vol. 46, no. 2, pp. 680–686, 2010.

[49] Q. V. Tran, T. W. Chun, J. R. Ahn and H. H. Lee, "Algorithms for Controlling Both the DC Boost and AC Output Voltage of Z-Source Inverter," *IEEE Trans. Ind. Electron.*, vol. 54, no. 5, pp. 2745–2750, 2007.

[50] Y. P. Siwakoti and G. Town, "Improved Modulation Technique for Voltage Fed Quasi-Z-Source DC/DC Converters," in *Proc. APEC 2014*, pp. 1973–1978, 2014.

[51] D. Vinnikov and I. Roasto, "Quasi-Z-Source-Based Isolated DC/DC Converters for Distributed Power Generation," *IEEE Trans. Ind. Electron.*, vol. 58, no. 1, pp. 192–201, 2011.

[52] Y. P. Siwakoti, F. Blaabjerg, P. C. Loh and G. E. Town, "High Voltage Boost Quasi-Z-Source Push-Pull Isolated DC/DC Converter," *IET Power Electron.*, vol. 7, no. 9, pp. 2387–2395, 2014.

[53] D. Cao and F. Z. Peng, "A Family of Z-Source and Quasi-Z-Source DC–DC Converters," in *Proc. APEC 2009*, pp. 1097–1101, 2009.

[54] Y. P. Siwakoti, P.C. Loh, F. Blaabjerg and G. Town, "Y-Source Impedance Network," *IEEE Trans. Power Electron. (Lett.)*, vol. 29, no. 7, pp. 3250–3254, 2014.

[55] Y. P. Siwakoti, P. C. Loh, F. Blaabjerg and G. E. Town, "Magnetically Coupled High Gain Y-Source Isolated DC/DC Converter" *IET Power Electron.*, vol. 7, vol. 11, pp. 2817–2824, 2014.

[56] D. Vinnikov, T. Jalakas, I. Roasto, H. Agabus and K. Tammet, "Method of Shoot-Through Generation for Modified Sine Wave Z-Source, Quasi-Z-Source and Trans-Z-Source Inverters," *Patent Owner: Tallinn University of Technology, World Intellectual Property Organization*, Publication number WO2012/116708 A2, Publication date 2012.

[57] H. Lee, H. G. Kim and H. Cha, "Parallel Operation of Trans-Z-Source Network Full Bridge DC/DC Converter for Wide Input Voltage Range," in *Proc. IEEE ECCE Asia 2012*, pp. 1707–1712, 2012.

[58] I. Roasto, D. Vinnikov, J. Zakis and O. Husev, "New Shoot-Through Control Methods for qZSI-Based DC/DC Converters," *IEEE Trans. Ind. Inform.*, vol. 9, no. 2, pp. 640–647, 2013.

[59] P. C. Loh, D. M. Vilathgamuwa, Y. S. Lai, G. T. Chua and Y. Li, "Pulse-Width Modulation of Z-Source Inverters," *IEEE Trans. Power Electron.*, vol. 20, no. 6, pp. 1346–1355, 2005.

[60] F. Z. Peng, M. Shen and Z. Qian, "Maximum Boost Control of the Z-Source Inverter", *IEEE Trans. Power Electron.*, vol. 20, no. 4, pp. 833–838, 2005.

[61] M. Shen, J. Wang, A. Joseph, F. Z. Peng, L. M. Tolbert and D. J Adams, "Constant Boost Control of the Z-Source Inverter to Minimize Current Ripple and Voltage Stress," *IEEE Trans. Ind. Applicat.*, vol. 42, no. 3, pp. 770–778, 2003.

[62] M. Shen, J. Wang, A. Joseph, F. Z. Peng, L. M. Tolbert and D. J Adams, "Maximum Constant Boost Control of the Z-Source Inverter," in *Proc. IEEE IAS Annual Meeting 2004*, 2004 pp. 142–147.

[63] U. S. Ali and V. Kamaraj, "A Novel Space Vector PWM for Z-Source Inverter," in *Proc.1st Int. Conf. on Electr. Energy Systems*, pp. 82–85, 2011.

[64] O. Ellabban, M. J. Van and P. Lataire, "Experimental Study of the Shoot-Through Boost Control Methods for the Z-Source Inverter," in *Proc. Eur. Power Electron. Drives Assoc.*, vol. 21, no. 2, pp. 18–29, 2011.

[65] Y. Liu, B. Ge, F. J. T. E. Ferreira, A. T. de Almeida and A. A. Rub, "Modeling and SVPWM Control of Quasi-Z-Source Inverter," in *Proc. IEEE EPQU 2011*, pp. 1–7.

[66] Y. Siwakoti and G. Town, "Three-Phase Transformerless Grid Connected Quasi Z-Source Inverter for Solar Photovoltaic Systems with Minimal Leakage Current," in *Proc. IEEE PEDG 2012*, pp. 368–373, 2012.

[67] F. Bradaschia, M. C. Cavalcanti, P. E. P. Ferraz, F. A. S. Neves, E. C. dos Santos, Jr. and J. H. G. M. da Silva, "Modulation for Three-Phase Transformerless Z-Source Inverter to Reduce Leakage Currents in Photovoltaic Systems," *IEEE Trans. Ind. Electron.*, vol. 58, no. 12, , pp. 5385–5395, 2011.

[68] Y. Siwakoti and G. Town, "Common-Mode Voltage Reduction Techniques of Three-Phase Quasi Z-Source Inverter for AC Drives," in *Proc. IEEE APEC 2013*, pp. 2247–2252, 2013.

[69] P. C. Loh, F. Gao, F. Blaabjerg, S. Y. C. Feng and N. J. Soon, "Pulsewidth-Modulated Z-Source Neutral-Point-Clamped Inverter," *IEEE Trans. Ind. Applicat.*, vol. 43, no. 5, pp. 1295–1308, 2007.

[70] P. C. Loh, F. Gao and F. Blaabjerg, "Topological and Modulation Design of Three-Level Z-Source Inverters," *IEEE Trans. Power Electron.*, vol. 23, no. 5, pp. 2268–2277, 2008.

[71] F. Gao, P. C. Loh, F. Blaabjerg and D. M. Vilathgamuwa, "Performance Evaluation of Three-Level Z-Source Inverters Under Semiconductor-Failure Conditions," *IEEE Trans. Ind. Applicat.*, vol. 45, no. 3, pp. 971–981, 2009.

[72] P. C. Loh, F. Blaabjerg and C. P. Wong, "Comparative Evaluation of Pulsewidth Modulation Strategies for Z-Source Neutral-Point-Clamped Inverter," *IEEE Trans. Power Electron.*, vol. 22, no. 3, pp. 1005–1013, 2007.

[73] P. C. Loh, F. Gao, P. C. Tan and F. Blaabjerg, "Three-Level AC–DC–AC Z-Source Converter Using Reduced Passive Component Count," *IEEE Trans. Power Electron.*, vol. 24, no. 7, pp. 1671–1681, 2009.

[74] F. B. Effah, P. Wheeler, J. Clare and A. Watson, "Space-Vector-Modulated Three-Level Inverters with a Single Z-Source Network," *IEEE Trans. Power Electron.*, vol. 28, no. 6, pp. 2806–2815, 2013.

[75] E. C. dos Santos Jr., E. P. X. P. Filho, A. C. Oliveria and E. R. C. da Silva, "Hybrid Pulse Width Modulation for Z-Source Inverters," in *Proc. IEEE ECCE 2010*, pp. 2888–2892, 2010.

[76] J. H. G. Muniz, E. R. C. da Silva and E. C. dos Santos Jr., "A Hybrid PWM Strategy for Z-Source Neutral-Point-Clamped Inverter," in *Proc. IEEE APEC 2011*, pp. 450–456, 2011.

[77] Y. Liu, B. Ge, H. A. Rub and F. Z. Peng, "An Effective Control Methods for Quasi-Z-Source Cascade Multilevel Inverter Based Grid-Tie Single Phase Photovoltaic Power System," *IEEE Trans. Ind. Inform.*, vol. 10, no. 1, pp. 399–40, 2014.

[78] Y. Liu, B. Ge, H. A. Rub and F. Z. Peng, "A Modular Multilevel Space Vector Modulation for Photovoltaic Quasi-Z-Source Cascaded Multilevel Inverter," in *Proc. IEEE APEC 2013*, pp. 714–718, 2013.

[79] B. Ge, Q. Li, W. Qian and F. Z. Peng, "A Family of Z-Source Matrix Converters," *IEEE Trans. Ind. Electron.*, vol. 59, no. 1, pp. 35–46, 2012.

[80] X. Liu, P. C. Loh, P. Wang and X. Han, "Improved Modulation Schemes for Indirect Z-Source Matrix Converter with Sinusoidal Input and Output

Waveforms," *IEEE Trans. Power Electron.*, vol. 27, no. 9, pp. 4039–4050, 2012.

[81] Q. Lei, B. Ge and F. Z. Peng, "Hybrid PWM Control for Z-Source Matrix Converter," in *Proc. ECCE 2011*, pp. 246–253, 2011.

[82] Q. Lei, F. Z. Peng and B. Ge, "Pulse-Width-Amplitude-Modulated Voltage-Fed Quasi-Z-Source Direct Matrix Converter with Maximum Constant Boost," in *Proc. APEC 2012*, pp. 641–646, 2012.

[83] X. Liu, P. C. Loh, F. Z. Peng and P. Wang, "Optimal Modulation of Indirect Z-Source Matrix Converter," in *Proc. IPEC 2010*, pp. 3049–3056, 2010.

[84] X. Ding, Z. Qian, Y. Xie and F. Z. Peng, "A Novel ZVS Z-Source Rectifier," in *Proc. APEC 2006*, pp. 951–955, 2006.

[85] M. Adamowicz, R. Strzelecki, F. Z. Peng, J. Guzinski and H.A. Rub, "New Type LCCT-Z-Source Inverters," in *Proc. EPE 2011*, pp. 1–10, 2011.

[86] S. Jiang, D. Cao and F. Z. Peng, "High Frequency Transformer Isolated Z-Source Inverters," in *Proc. APEC 2011*, pp. 442–449, 2011.

Chapter 8

DC/DC converters and building blocks for power electronics

Danillo Borges Rodrigues[1], Luiz Carlos Gomes Freitas[2] and Gustavo B Lima[2]

8.1 Introduction

In the long past, linear power regulators operated on the principle of voltage or current division; however, their inefficiency has relegated them to use in only basic circuits with low power requirements in the present day. The modern power electronic systems must be small, lightweight, reliable, and efficient, and, therefore, most computer-based automation, industrial, commercial, and residential equipment mostly adopts switching power supplies. Linear power supplies are limited to output voltages less than the input voltage, their power density is low because they require low-frequency (50 or 60 Hz) line transformers and filters. However, linear regulators may provide a very high-quality output voltage with little noise and no effects of electromagnetic interference (EMI) or EMC. Their main area of application is at low power levels as low dropout voltage regulators.

Semiconductor components in linear regulators operate in their active (linear) modes. Switching regulators use power electronic semiconductor switches to be controlled on and off, and the power loss, determined by multiplying the conduction voltage drop with the current passing through, is minimal. If the transition times between on/off and off/on are rapid, the switching losses can also be negligible, enabling switching regulators to attain high conversion efficiency. Modern power electronic switches can operate at high frequencies. The higher the operating frequency, the smaller and lighter the transformers, filter inductors, and capacitors. In addition, the dynamic characteristics of converters improve with increasing operating frequencies. The bandwidth of a control loop is usually determined by the corner frequency of the output filter. Therefore, high operating frequencies allow for achieving a fast dynamic response to rapid changes in the load current and/or the input voltage.

[1]Departamento de Engenharia Elétrica, Universidade Federal do Triângulo Mineiro, Brazil
[2]Faculdade de Engenharia Elétrica, Universidade Federal de Uberlândia, Brazil

Two main types of dc–dc converters can be distinguished: hard-switching pulse width modulated (PWM) converters and resonant and soft-switching converters. This chapter describes the PWM converters which have been very popular for the last few decades. They are widely used at all power levels. Topologies and properties of PWM converters are well understood and described in the literature [1–6]. The advantages of PWM converters include low component count, high efficiency, constant frequency operation, relatively simple control and commercial availability of integrated circuit controllers, and the ability to achieve high conversion ratios for both step-down and step-up applications. A disadvantage of PWM dc–dc converters is that PWM rectangular voltage and current waveforms cause turn-on and turn-off losses in semiconductor devices which limit practical operating frequencies to a megahertz range. Rectangular waveforms also inherently generate high levels of EMI. The functions of dc–dc converters are:

- to convert a dc input voltage V_{IN} into a dc output voltage V_{OUT};
- to regulate the dc output voltage against load and line variations;
- to reduce the ac voltage ripple on the dc output voltage below the required level;
- to provide isolation between the input source and the load (isolation is not always required);
- to protect the supplied system and the input source from electromagnetic interference (EMI);
- to satisfy various international and national safety standards.

PWM dc–dc converters may operate in three conduction modes: continuous conduction mode (CCM), critical conduction mode, and discontinuous conduction mode (DCM). In DCM, the inductor current drops to zero, remaining null for some time and starts increasing only at the beginning of the next switching cycle. The disadvantage of this mode is a high current stress on the converter's components, generating significant levels of EMI due to high current ripple at the input, and the necessity of employing additional input filters to ensure a high power factor. The advantages of DCM are simpler control, as the current in the inductor starts from zero in each cycle and a lower size of the converter's inductor. On the other hand, in CCM, the ripples in the input current are smaller, avoiding the need to use filters at the input to ensure a high power factor and reduce EMI.

That said, the analysis of the converters in this chapter is done under the assumption that the current in the inductor is never zero, which is the way to CCM operation. The critical conduction mode is a boundary between the other two and is obtained when the current in the inductor just drops instantaneously to zero and immediately starts rising, the key factor for CCM is that there is no period in the inductor current that it stays at zero because in that period both voltage and current would be zero, making the inductor to not have any differential equation at all to add to the circuit operation.

Typical inductor current waveforms for the three conduction modes are presented in Figure 8.1.

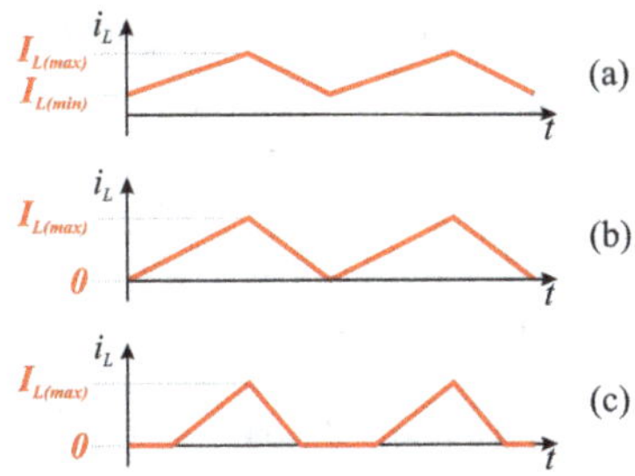

Figure 8.1 Power converter inductor current for various conduction modes: (a) continuous conduction mode, (b) critical conduction mode, and (c) discontinuous conduction mode

In this context, this chapter starts with a section on non-isolated dc–dc converters operating in CCM. Four basic dc–dc converter topologies are presented in Sections 8.2.1–8.2.5: buck, boost, buck–boost, integrated buck and boost, and three topologies of fourth-order dc–dc converters—SEPIC, Cuk, and Zeta. In Section 8.2.6, the concept of a quadric converter is presented. Non-idealities or parasitic of practical devices and components are briefly described in Section 8.3. Popular isolated converter topologies are discussed in Sections 8.4.1–8.4.4: flyback, forward, half-bridge, and full-bridge converters. In Section 8.5 ordinary rectifier types are presented and in Section 8.6 the concept of power factor corrector is presented. In Section 8.7 the concept of hybrid rectifiers is explained under ideal component and semiconductor device assumptions. In Section 8.8, examples of bidirectional power converter topologies are presented. Section 8.9 discusses the concept of interleaving multiple converters. Section 8.10 reviews the control principles of PWM dc–dc converters. Two main control schemes, voltage-mode control and current-mode control, are described, as well as the application of the state-space average modeling for dc–dc converters.

8.2 Non-isolated converter topologies

Converters discussed in this chapter will use pulse-width-modulation (PWM) to control the output, and a mathematical relationship that maps the duty cycle to the output voltage must be developed in order to make possible the design of a closed-loop control system. The PWM signal is obtained with an analog or a digital controller that commands the transistor, which is turned-on for time t_{ON} and then turned-off for time t_{OFF}. The process of turning on and off is repeated continuously with a period $T = t_{ON} + t_{OFF}$. Hence, the switch duty ratio D could be defined as

$$D = \frac{t_{on}}{t_{on} + t_{off}} \tag{8.1}$$

It can be seen from the above equation that the theoretical range of duty ratio D is from 0 to 1.

8.2.1 Buck converter

Figure 8.2 shows a step-down dc–dc converter, commonly known as a buck converter. The buck converter consists of dc input voltage source V_{IN}, controlled switch S, diode D, filter inductor L, filter capacitor C, and load resistance R_L. Typical waveforms in the converter are shown in Figure 8.3 under the assumption that the inductor current is always positive. It can be seen from the circuit that when switch S is driven to the on state, the diode D is reverse biased.

When the switch S is off, the diode conducts to support an uninterrupted current in the inductor. The relationship among the input voltage, output voltage, and the switch duty ratio D can be derived from, for instance, the waveform of the voltage across the inductor L, which is a difference between the output voltage V_{OUT} and the voltage across the diode D. The voltage across the inductor is equal to V_{IN} V_{OUT} when the switch S is on and equal to V_{OUT} when the switch is off.

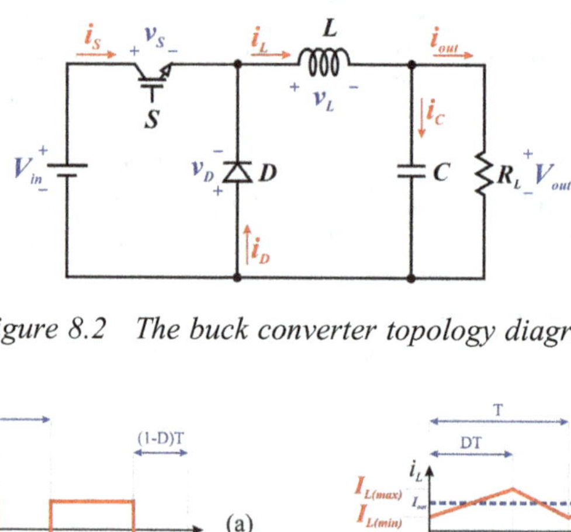

Figure 8.2 The buck converter topology diagram

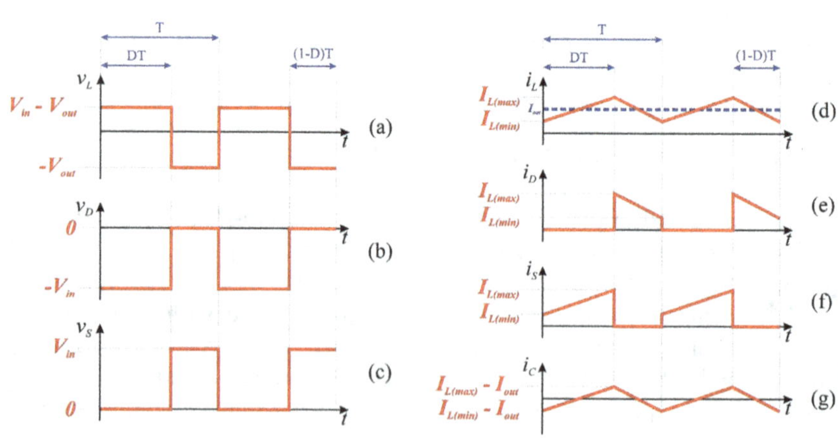

Figure 8.3 Typical voltage and current waveforms in the buck converter: (a) inductor voltage; (b) diode voltage; (c) switch voltage; (d) inductor current; (e) diode current; (f) switch current; (g) capacitor current

According to Faraday's law, the inductor volt–second product over a period of steady-state operation is zero. For the buck converter

$$(V_{in} - V_{out})DT = V_{out}(1 - D)T \tag{8.2}$$

Hence, the dc voltage transfer function, defined as the ratio of the output voltage to the input voltage, is

$$M_V = \frac{V_{out}}{V_{in}} = D \tag{8.3}$$

It can be concluded from the above equation that the output voltage of the buck converter is within the range from 0 to V_{IN}.

The filter inductor current i_L consists of a dc component I_O with a superimposed triangular ac component. This ac component forms the inductor current ripple. Its peak-to-peak value can be calculated as

$$I_{Lr} = \frac{V_{out}t_{off}}{L} = \frac{V_{out}(1 - D)T}{L} \tag{8.4}$$

The maximum and minimum values of the inductor current i_L are

$$I_{Lmax} = I_O + \frac{V_{out}(1 - D)T}{2L} \tag{8.5}$$

$$I_{Lmin} = I_O - \frac{V_{out}(1 - D)T}{2L} \tag{8.6}$$

In most design cases, the filtering inductance value L is selected to keep the current ripple below a certain limit I_{Lrmax}. Thus, the inductance value can be calculated as

$$L_{min} = \frac{V_{out}(1 - D)T}{I_{Lrmax}} \tag{8.7}$$

Almost the entire ac component of the inductor current flows through the filter capacitor as a current i_C. Current i_C causes a small voltage ripple across the dc output voltage V_{OUT}. To limit the peak-to-peak value of the ripple voltage below a certain value V_r, the filter capacitance C must be greater than

$$C_{min} = \frac{V_{out}(1 - D)}{8V_rLf^2} \tag{8.8}$$

Equations (8.7) and (8.8) are the key design equations for the buck converter. The input and output dc voltages (hence, the duty ratio D), and the range of load resistance R_L are usually determined by preliminary specifications. The designer needs to determine the values of passive components L and C and of the switching frequency f. The value of the filter inductor L is based on the desired level of inductor ripple current using (8.7). The value of the filter capacitor C is obtained from the voltage ripple condition (8.8). For the compactness and low conduction losses of a converter, it is desirable to use small passive components. Equations (8.7) and (8.8) show that it can be accomplished with a high switching frequency *f*. The switching frequency is limited, however, by the type of semiconductor switches used and by switching losses. It should be also noted that the values of *L* and *C* may be altered by the effects of parasitic components in the converter, especially by the equivalent series resistance (ESR) of the capacitor. The issue of parasitic components in dc–dc converters is discussed in Section 8.3.

8.2.2 Boost converter

Figure 8.4 presents a step-up or a PWM boost converter. It is composed of dc input voltage source V_{IN}, boost inductor L, controlled switch S, diode D, filter capacitor C, and load resistance R_L. Typical waveforms in the converter are shown in Figure 8.5 under the assumption that the inductor current is always positive. When the switch S is in the on state, the current in the boost inductor increases linearly.

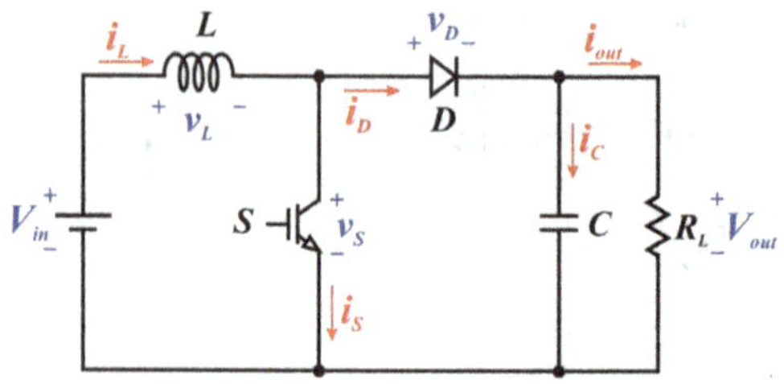

Figure 8.4 The boost converter topology diagram

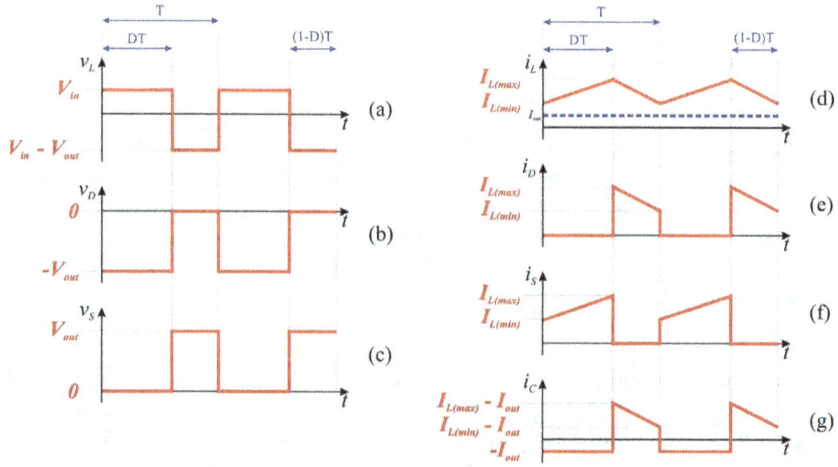

Figure 8.5 Typical voltage and current waveforms for the boost converter: (a) inductor voltage; (b) diode voltage; (c) switch voltage; (d) inductor current; (e) diode current; (f) switch current; (g) capacitor current

The diode D is off at that time. When the switch S is turned off, the energy stored in the inductor is released through the diode D to the output RC circuit. The voltage across the inductor is equal to V_{IN} when the switch S is on and equal to V_{IN} V_{OUT} when the switch is off. Since the average inductor voltage is zero for periodic operation

$$V_{in}DT = (V_{out} - V_{in})(1 - D)T \tag{8.9}$$

from which the dc voltage transfer function could be derived as

$$M_V = \frac{V_{out}}{V_{in}} = \frac{1}{1 - D} \tag{8.10}$$

It can be concluded from (8.10) that the output voltage of the boost converter is within the range from V_{IN} to infinity. Thus, a duty ratio limitation or over-voltage protection at the output is strongly recommended as the boost converter can generate very high voltage at the output.

The peak-to-peak value of the inductor current ripple can be calculated as

$$I_{Lr} = \frac{V_{in}t_{on}}{L} = \frac{V_{in}DT}{L} \tag{8.11}$$

The maximum and minimum values of the inductor current i_L are

$$I_{Lmax} = I_O + \frac{V_{in}DT}{2L} \tag{8.12}$$

$$I_{Lmin} = I_O - \frac{V_{in}DT}{2L} \tag{8.13}$$

In most cases, the filtering inductor value L is selected to keep the current ripple below a certain limit I_{Lrmax}. Thus, the inductor value can be calculated as

$$L_{min} = \frac{V_{in}DT}{I_{Lrmax}} \tag{8.14}$$

In the boost converter, the current is delivered to the output RC circuit only when diode D is conducting. Thus, this current is discontinuous which implies a larger filter capacitor requirement compared to that in the buck-derived converters to limit the output voltage ripple. The filter capacitor must provide the output dc current to the load when diode D is not conducting. The minimum value of the filter capacitance that results in the voltage ripple V_r is given by

$$C_{min} = \frac{V_{out}D}{V_r R_L f} \tag{8.15}$$

8.2.3 Buck–boost converter

Figure 8.6 depicts a buck–boost converter. It is comprised of dc input voltage source V_{IN}, inductor L, controlled switch S, diode D, filter capacitor C, and load

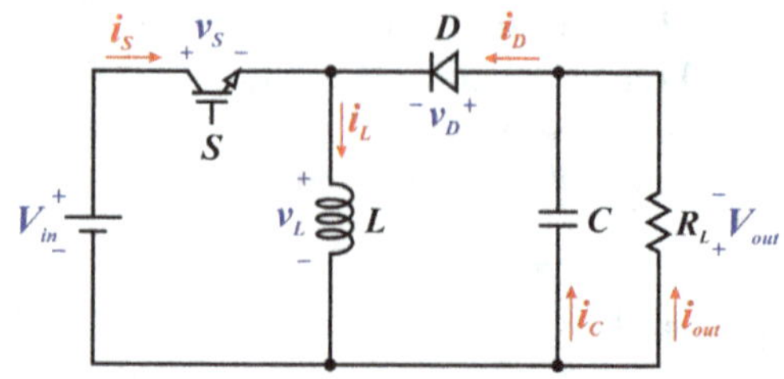

Figure 8.6 The buck–boost converter topology diagram

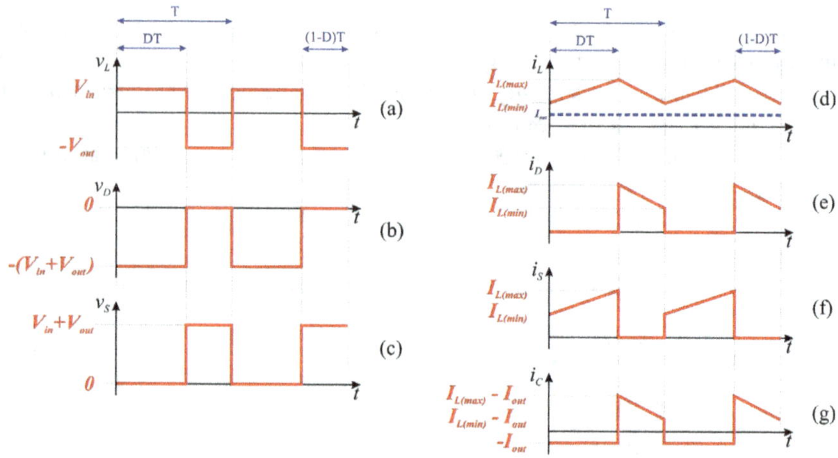

Figure 8.7 Typical voltage and current waveforms for the buck–boost converter: (a) inductor voltage; (b) diode voltage; (c) switch voltage; (d) inductor current; (e) diode current;(f) switch current; (g) capacitor current

resistance R_L. Typical waveforms in the converter are shown in Figure 8.7 under the assumption that the inductor current is always positive. When the switch S is in the on state, the current in the inductor increases linearly. The diode D is off at the time. When the switch S is turned off, the energy stored in the inductor is released through the diode D to the output RC circuit. The voltage across the inductor is equal to *V*IN when the switch S is on and equal to *V*OUT when the switch is off. Since the average inductor voltage is zero in steady-state operation

$$V_{in}DT = V_{out}(1 - D)T \tag{8.16}$$

from which the dc voltage transfer function turns out to be

$$M_V = \frac{V_{out}}{V_{in}} = \frac{D}{1 - D} \tag{8.17}$$

Equation (8.17) shows that the output voltage of the buck–boost converter is within the range from 0 to negative infinity. Hence, a duty ratio limitation or over-voltage protection at the output is strongly recommended since the buck–boost converter can generate a very high voltage at the output.

The output voltage V_{OUT} is negative with respect to the ground. Its magnitude can be either greater or smaller (equal at $D = 0.5$) than the input voltage as this converter name implies.

The structure of the output part of the converter is similar to that of the boost converter (reversed polarities being the only difference). Thus, the same equation for current ripples, inductor L value, and filtering capacitor C value can be applied.

8.2.4 Integrated buck and boost converter

Figure 8.8 depicts an integrated buck and boost converter. It is comprised of dc input voltage source V_{IN}, inductor L, two controlled switches S_1 and S_2, two diodes D_1 and D_2, filter capacitor C, and load resistance R_L. This converter has two operation modes: buck and boost. In the buck mode, the switch S_2 is off and the switch S_1 is being turned on and off with duty ratio D and frequency f. Assuming a positive current in the inductor L, diode D_2 always conducts in this operation mode. Thus, the converter works as a regular buck converter and all related equations are applicable.

In the second operation mode, the converter behaves as a boost converter. In this case, switch S_1 is always on and diode D_1 does not conduct. The switch S_2 is being turned on and off with duty ratio D and frequency f. Thus, the converter works as a regular boost converter and all related equations apply.

The integrated buck and boost converter is used where the output voltage needs to be regulated in a range that extends from below to above the input voltage. Such regulation is not achievable by the buck or boost converter alone. It can be achieved by a Single-ended Primary-inductor Converter (SEPIC) or Ćuk converter which are, however, higher order converters and their analysis and control are more complex. Thus, the integrated buck and boost converter is preferred by the industry in most such applications.

8.2.5 Fourth-order converters

Fourth-order non-isolated converters are known for having four energy storage elements, two inductors, and two capacitors. Thus, they present four state variables

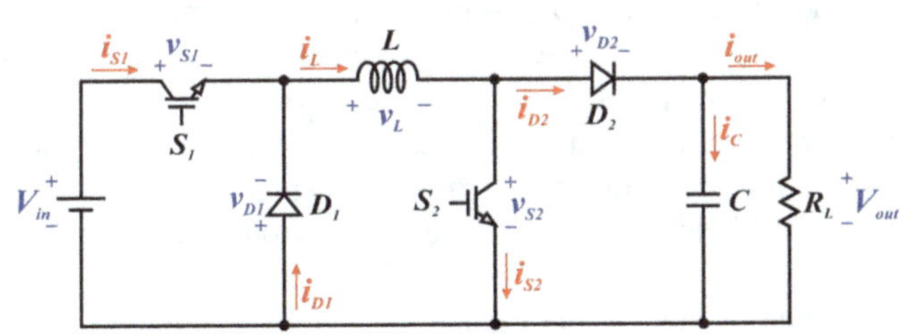

Figure 8.8 The integrated buck and boost converter topology diagram

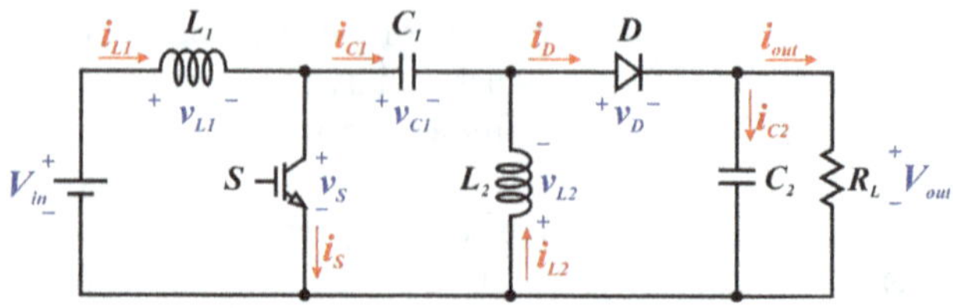

Figure 8.9 The SEPIC converter topology diagram

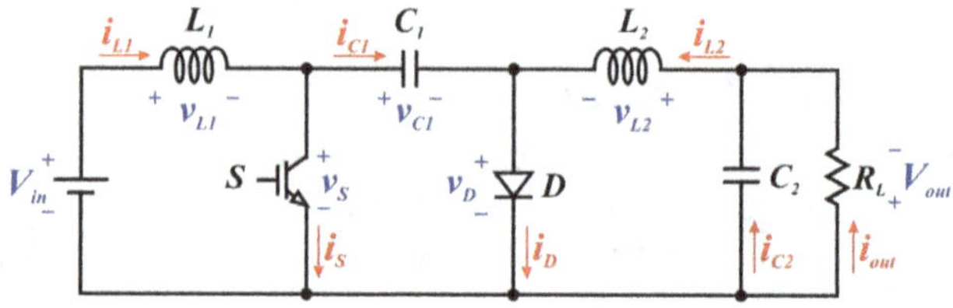

Figure 8.10 The Ćuk converter topology diagram

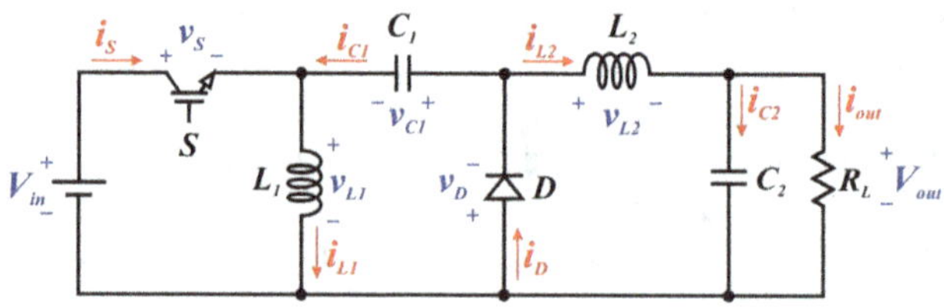

Figure 8.11 The Zeta converter topology diagram

and, consequently, fourth-order system matrices for each operating stage (closed or open switch). Figures 8.9, 8.10, and 8.11 illustrate the fourth-order SEPIC, Ćuk, and Zeta converters, respectively.

The SEPIC, Ćuk, and Zeta converters consist of a power supply V_{in}, one switch S, one diode D, two inductors L_1 and L_2, two capacitors C_1 and C_2, and load resistance R_L. During the stage in which the switch S is on, the diode D remains reverse biased, the power supply V_{in} supplies energy to the inductor L_1 and the capacitor C_1 (charged during the stage in which the switch S is open) supplies energy to inductor L_2. In the stage where the switch S is off, the inductor L_1 supplies energy to capacitor C_1, and the diode D is forward-biased to support an uninterrupted current in the inductors L_1 and L_2. Typical waveforms in these converters are shown in Figure 8.12, under the assumption that the inductor currents are always positive.

Using the Faraday's law for the inductors

$$V_{in}DT = V_{out}(1 - D)T \tag{8.18}$$

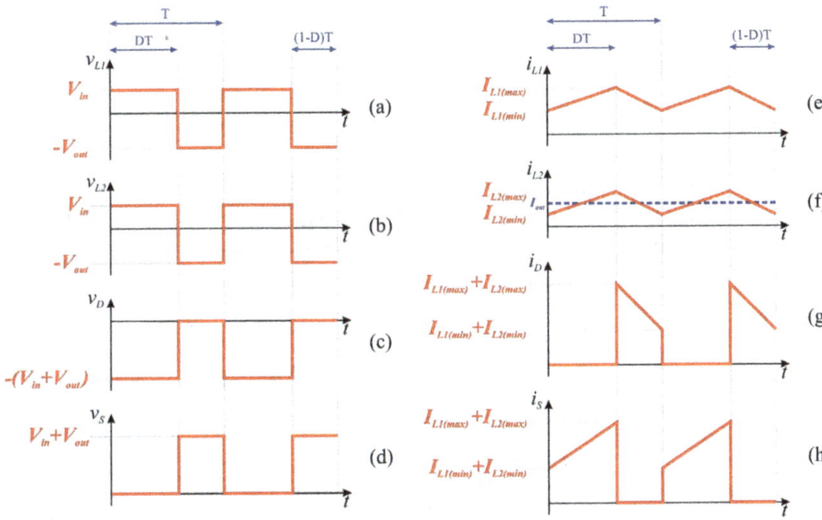

Figure 8.12 Typical voltage and current waveforms in SEPIC, Ćuk, and Zeta converters: (a) inductor L_1 voltage, (b) inductor L_2 voltage, (c) diode voltage, (d) switch voltage, (e) inductor L_1 current, (f) inductor L_2 current, (g) diode current, and (h) switch current

from which the dc voltage transfer function could be derived, which demonstrates that fourth-order non-isolated converters are step-down and step-up converters.

$$M_V = \frac{V_{out}}{V_{in}} = \frac{D}{1 - D} \tag{8.19}$$

For L_1, the average inductor current is given by (8.20), which is also the average source current. For L_2, since the average current in each capacitor is zero, the average current is given by (8.21).

$$I_{L1} = \frac{V_{out}D}{(1 - D)R_L} \tag{8.20}$$

$$I_{L2} = \frac{V_{out}}{R_L} \tag{8.21}$$

The filtering inductors L_1 and L_2 values are selected to keep the current ripple below a certain limit I_{Lrmax}. Thus, the inductor values can be calculated as

$$L_{1min} = \frac{V_{in}DT}{I_{L1rmax}} \tag{8.22}$$

$$L_{2min} = \frac{V_{out}(1 - D)T}{I_{L2rmax}} \tag{8.23}$$

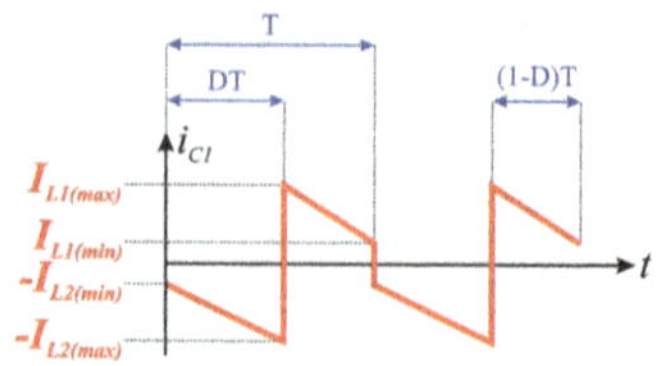

Figure 8.13 Waveforms of current for capacitor C_1 for SEPIC, Ćuk, and Zeta converters

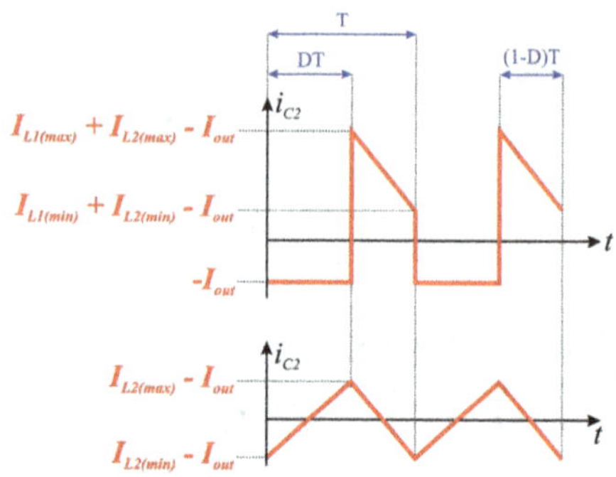

Figure 8.14 Waveforms of current for capacitor C2: (a) SEPIC converter (b) Ćuk and Zeta converters

Analyzing the equivalent circuits when the switch is on and when the switch is off, the current in the capacitor C_1 can be obtained as depicted in Figure 8.13. Therefore, the minimum value of capacitor C_1 to limit the peak-to-peak value of the ripple voltage below a certain value V_{C1rmax} can be given by (8.24).

$$C_{1min} = \frac{V_{out}DT}{R_L V_{C1rmax}} \tag{8.24}$$

Analyzing the equivalent circuits when the switch is on and when the switch is off, the current in the capacitor C_2 can be obtained as depicted in Figure 8.14. Therefore, the minimum value of capacitor C_2 to limit the peak-to-peak value of the ripple voltage below a certain value V_{C2rmax} can be given by (8.25), for SEPIC converter, and by (8.26) for Ćuk and Zeta converters.

$$C_{2min} = \frac{V_{out}DT}{R_L V_{C2rmax}} \tag{8.25}$$

$$C_{2min} = \frac{V_{out}(1-D)T^2}{8L_2 V_{C2rmax}} \tag{8.26}$$

8.2.6 Quadratic converters

Quadratic non-isolated converters are dc to dc converters that provide high conversion rates by adjusting only the operating cycle ratio, without the need to use high-frequency step-down or step-up transformers [7–11]. Among the topologies of traditional quadratic converters, the Quadratic Buck and Quadratic Boost converters stand out. These converters consist of V_{in}, one switch S, three diodes (D_1, D_2, and D_3), two inductors (L_1 and L_2), two capacitors (C_1 and C_2), and load resistance R_L.

8.2.6.1 Quadratic Buck converter

Figure 8.15 illustrates the Quadratic Buck converter topology. During the stage in which the switch S is on, diodes D_1 and D_3 remain reverse biased while diode D_2 is forward biased to form the current path of capacitor C_1. The power supply V_{in}, in turn, supplies power to inductor L_1 while capacitor C_1 (charged during the stage when switch S is off) supplies power to inductor L_2. In the stage when switch S is off, diode D_2 is reverse biased and diodes D_1 of D_3 are forward biased to provide the paths for the discharge currents of inductors L_1 and L_2, respectively. Typical voltages and current waveforms for the Quadratic Buck converter are depicted in Figure 8.16, under the assumption that the inductor currents are always positive.

Through the analysis of the circuit depicted in Figure 8.15, when the switch S is on, the voltage across the inductor L_1 is

$$v_{L1} = V_{in} - V_{C1} \tag{8.27}$$

When the switch S is off, the voltage across the inductor L_1 is

$$v_{L1} = -V_{C1} \tag{8.28}$$

Knowing that the average inductor voltages are zero and that the average capacitor currents are zero for steady-state operation, one can conclude that, for the Quadratic Buck converter, the average voltage across the capacitor C_1 is given by (8.30).

$$(V_{in} - V_{C1})DT = V_{C1}(1 - D)T \tag{8.29}$$

$$V_{C1} = DV_{in} \tag{8.30}$$

$$(DV_{in} - V_{out})DT = V_{out}(1 - D)T \tag{8.31}$$

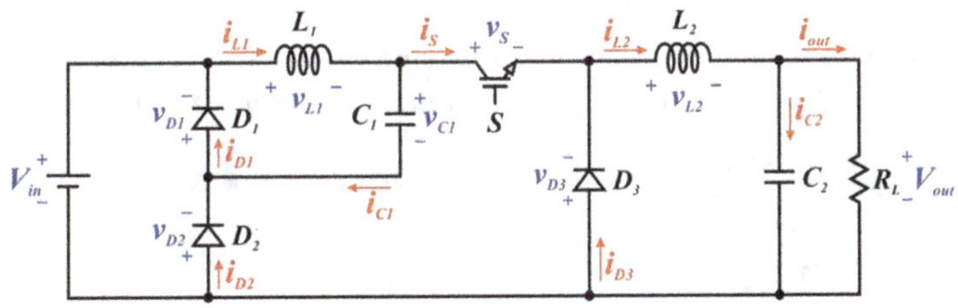

Figure 8.15 The Quadratic Buck converter topology diagram

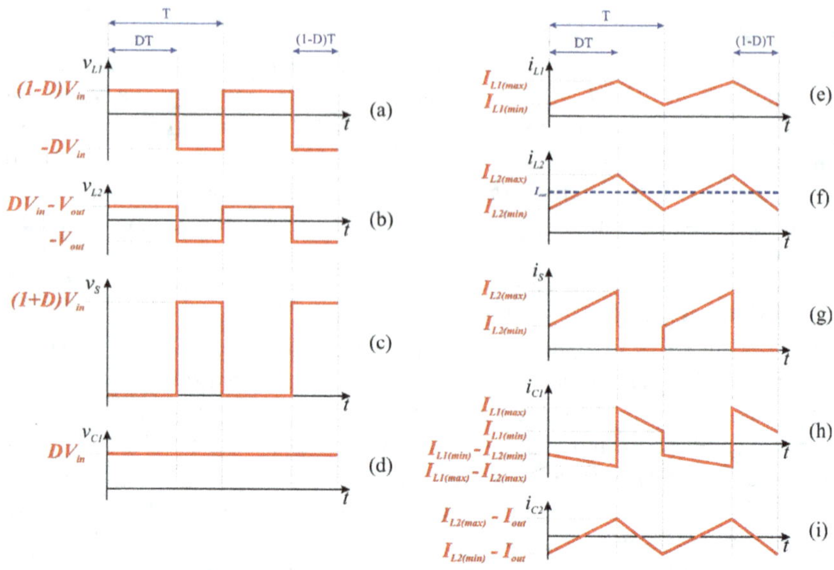

Figure 8.16 Typical voltage and current waveforms in Quadratic Buck converter: (a) inductor L_1 voltage, (b) inductor L_2 voltage, (c) switch S voltage, (d) capacitor C_1 voltage, (e) inductor L_1 current, (f) inductor L_2 current, (g) switch S current, (h) capacitor C_1 current, and (i) capacitor C_2 current

Hence, it can be concluded from the above equations that the dc voltage transfer function for the Quadratic Buck converter is

$$M_V = \frac{V_{out}}{V_{in}} = D^2 \tag{8.32}$$

The average inductor currents are

$$I_{L1} = \frac{V_{out}D}{R_L} \tag{8.33}$$

$$I_{L2} = \frac{V_{out}}{R_L} \tag{8.34}$$

In order to guarantee that the ripples of currents i_{L1} and i_{L2} do not exceed the limits defined by I_{L1max} and I_{L2max}, the minimum inductor sizes are given by

$$L_{1min} = \frac{V_{in}(1-D)DT}{I_{L1rmax}} \tag{8.35}$$

$$L_{2min} = \frac{V_{out}(1-D)T}{I_{L2rmax}} \tag{8.36}$$

The minimum values of capacitors C_1 and C_2 to ensure that voltage ripples v_{C1} and v_{C2} do not exceed the limits defined by V_{C1rmax} and V_{C2rmax} are given by (8.37) and (8.38), respectively.

$$C_{1min} = \frac{V_{out}(1-D)DT}{R_L V_{C1rmax}} \tag{8.37}$$

$$C_{2min} = \frac{V_{out}(1-D)T^2}{8L_2 V_{C2rmax}} \tag{8.38}$$

8.2.6.2 Quadric Boost converter

Figure 8.17 demonstrates the circuit of a Quadratic Boost converter. During the stage in which the switch S is on, diodes D_1 and D_3 remain reverse biased while diode D_2 conducts to form the load current path of inductor L_1. Power supply V_{in} supplies power to inductor L_1 while capacitor C_1 (charged during the stage when switch S is off) supplies power to inductor L_2. In the stage where the switch S is off, the diode D_2 is reverse biased and the diodes D_1 of D_3 conduct to provide the paths of circulation of the discharge currents of the inductors L_1 and L_2, respectively. Typical voltages and current waveforms for the Quadratic Boost converter are depicted in Figure 8.18, under the assumption that the inductor currents are always positive.

Through the analyses of the circuit depicted in Figure 8.17, when the switch S is on, the voltage across the inductor L_2 is

$$v_{L2} = V_{C1} \tag{8.39}$$

When the switch S is off, the voltage across the inductor L_2 is

$$v_{L2} = V_{C1} - V_{out} \tag{8.40}$$

Knowing that the average inductor voltages are zero and that the average capacitor currents are zero for steady-state operation, one can conclude that, for the

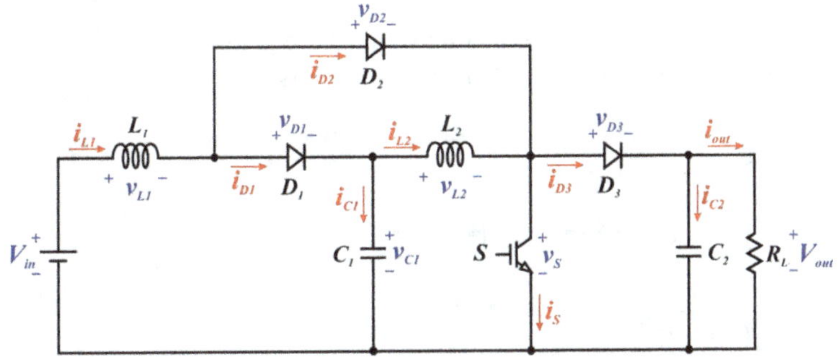

Figure 8.17 The Quadratic Boost converter topology diagram

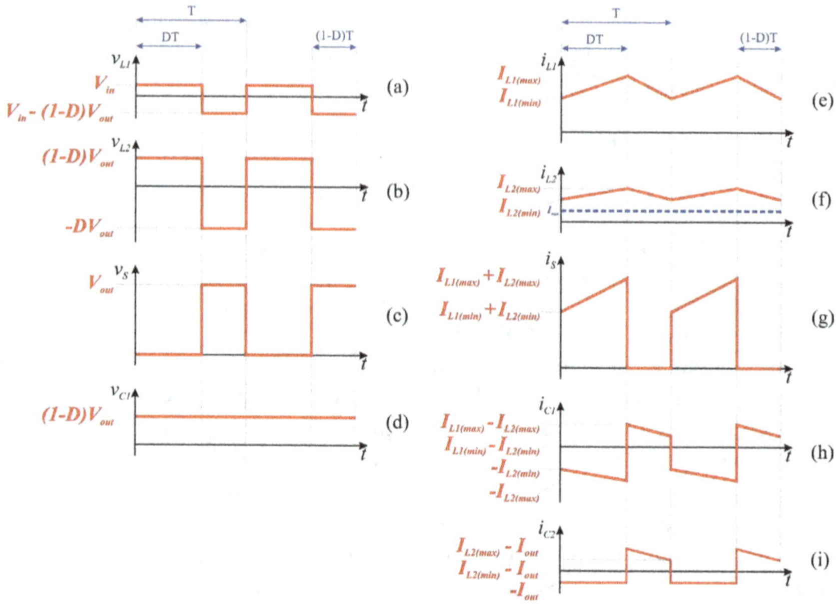

Figure 8.18 Typical voltage and current waveforms in Quadratic Boost converter: (a) inductor L_1 voltage, (b) inductor L_2 voltage, (c) switch S voltage, (d) capacitor C_1 voltage, (e) inductor L_1 current, (f) inductor L_2 current, (g) switch S current, (h) capacitor C_1 current, and (i) capacitor C_2 current

Quadratic Boost converter, the average voltage across the capacitor C_1 is given by (8.42).

$$V_{C1}DT = -(V_{C1} - V_{out})(1 - D)T \tag{8.41}$$

$$V_{C1} = V_{out}(1 - D) \tag{8.42}$$

$$V_{in}DT = -[V_{in} - V_{out}(1 - D)](1 - D)T \tag{8.43}$$

Hence, it can be concluded from the above equations that the dc voltage transfer function for the Quadratic Boost converter is

$$M_V = \frac{V_{out}}{V_{in}} = \frac{1}{(1 - D)^2} \tag{8.44}$$

The average inductor currents are

$$I_{L1} = \frac{V_{out}}{R_L(1 - D)^2} \tag{8.45}$$

$$I_{L2} = \frac{V_{out}}{R_L(1 - D)} \tag{8.46}$$

In order to guarantee that the ripples of currents i_{L1} and i_{L2} do not exceed the limits defined by I_{L1max} and I_{L2max}, the minimum inductor sizes are given by

$$L_{1min} = \frac{V_{in}DT}{I_{L1rmax}} \tag{8.47}$$

$$L_{2min} = \frac{V_{out}(1 - D)DT}{I_{L2rmax}} \tag{8.48}$$

The minimum values of capacitors C_1 and C_2 to ensure that voltage ripples v_{C1} and v_{C2} do not exceed the limits defined by V_{C1rmax} and V_{C2rmax} are given by (8.49) and (8.50), respectively.

$$C_{1min} = \frac{V_{out}DT}{(1 - D)R_L V_{C1rmax}} \tag{8.49}$$

$$C_{2min} = \frac{V_{out}DT}{R_L V_{C2rmax}} \tag{8.50}$$

8.3 Parasitic in dc–dc converters

The analysis of converters presented in Section 8.2 has been performed under the ideal switch, diode, and passive component assumptions. Non-idealities or parasitic of practical devices and components may, however, greatly affect some performance parameters of dc–dc converters. In this section, modeling and effects of parasitic on output voltage ripple, efficiency, and voltage transfer function of converters are discussed.

An ideal and a more realistic model of a capacitor are shown in Figure 8.19. The realistic model consists of a series connection of capacitance C and resistance r_C. The resistance r_C is called an ESR of the capacitor and is due to losses in the dielectric and physical resistance of leads and connections. Recall (8.8) which provided a value of the filter capacitance in a buck converter that limits the peak-to-peak output voltage ripple to V_r. The equation was derived under the assumption that the entire triangular ac component of the inductor current flows through a capacitance C. It is, however, closer to reality to maintain that this triangular component flows through a series connection of capacitance C and resistance r_C. Thus, the ripples across the resistance r_C add to the ripples across the capacitor.

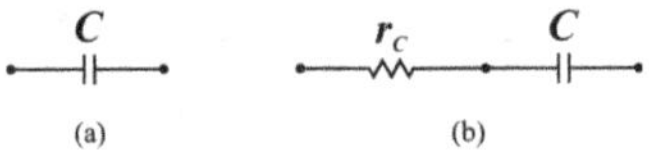

Figure 8.19 Capacitor model: (a) ideal one and (b) with series resistance

The peak-to-peak voltage ripple value across the resistance r_C can be calculated as

$$V_{rC} = r_C I_{Lr} = \frac{r_C V_{out}(1-D)T}{L} \tag{8.51}$$

where I_{Lr} is the peak-to-peak current ripple value in the buck converter inductor.

Depending on the used capacitor type and size, one of the ripples can be dominant. In most cases, especially when electrolytic capacitors are employed, the ripples over resistance r_C are significantly larger than ripples over a capacitance itself. Thus, many times for the ripple requirements, only the series resistance effect can be considered.

Ideal and realistic models of an inductor, a diode, and a transistor switch are presented in Figures 8.20–8.22. The efficiency of a dc–dc converter is defined as the ratio of the output power to input power and is commonly expressed in percent. All the parasitics shown in part (b) of the figures in this section decrease converter efficiencies. In well-designed practical converters, efficiencies range from high eighties to medium nineties percent. Parasitic components also lower the voltage transfer functions of practical converters resulting in maximum values of about ten instead of theoretical limits of infinity for boost or buck–boost converters.

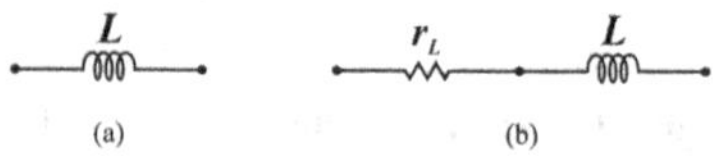

Figure 8.20 Inductor model: (a) ideal one and (b) with series resistance

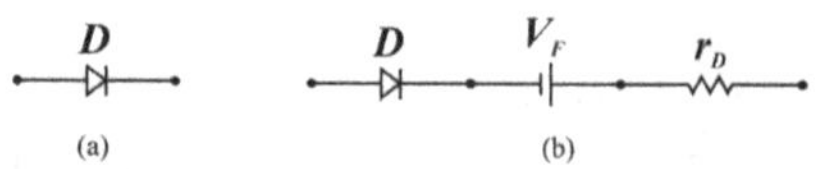

Figure 8.21 Diode model: (a) ideal one and (b) with series resistance and constant forward voltage

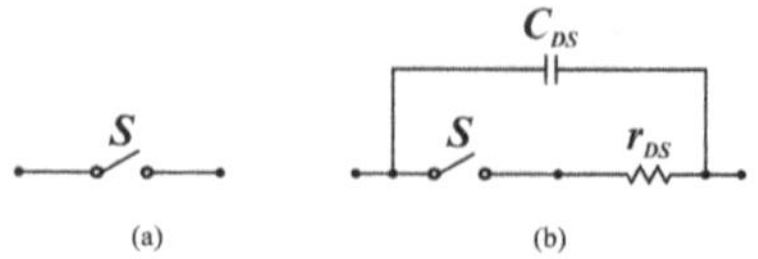

Figure 8.22 Switch model: (a) ideal one and (b) with series resistance and parallel capacitance

8.4 Isolated converters topologies

8.4.1 Flyback converter

A PWM flyback converter is a very practical isolated version of the buck–boost converter. The circuit of the flyback converter is presented in Figure 8.23. The inductor of the buck–boost converter has been replaced by a flyback transformer.

The input dc source V_{IN} and switch S are connected in series with the primary transformer. The diode D and the RC output circuit are connected in series with the secondary of the flyback transformer. Refer to Figure 8.23 for the converter operation. Typical voltages and current waveforms for the Flyback converter are depicted in Figure 8.24, under the assumption that the inductor currents are always

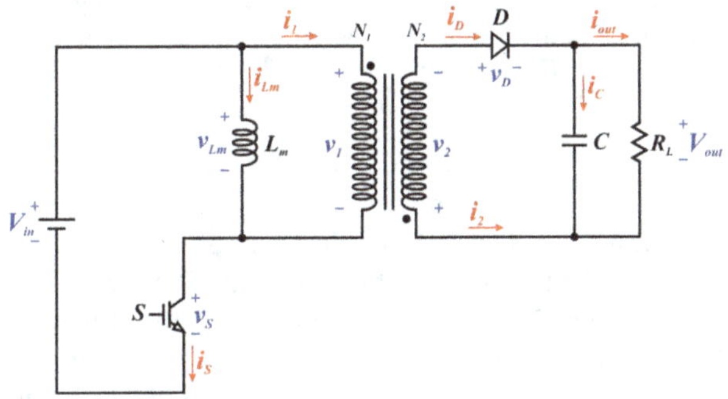

Figure 8.23 The flyback converter topology diagram

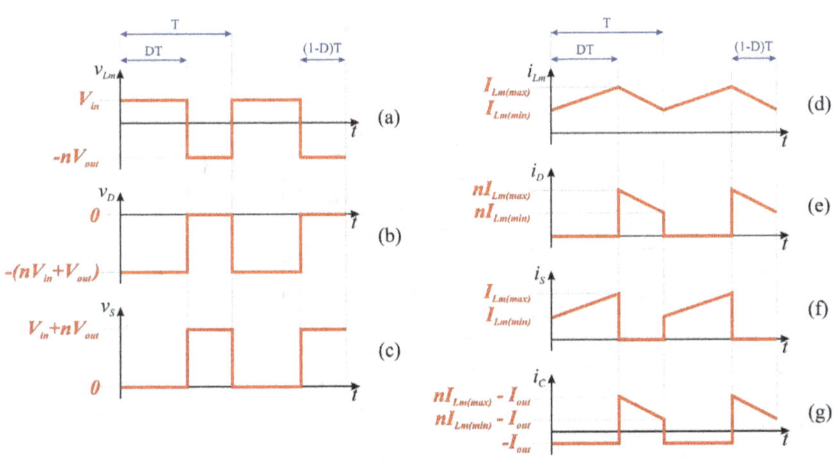

Figure 8.24 Typical voltage and current waveforms in Flyback: (a) magnetizing inductance L_m voltage, (b) diode D voltage, (c) switch S voltage, (d) magnetizing inductance L_m current, (e) diode D current, (f) switch S current, and (g) capacitor C current

positive. When the switch S is on, the current in the magnetizing inductance of the transformer increases linearly. The diode D is off and there is no current in the secondary side winding of the transformer. When the switch is turned off, the magnetizing inductance current is discharged through the diode D, and the transformed magnetizing inductance current is supplied to the RC load. The dc voltage transfer function of the flyback converter is

$$M_V = \frac{V_{out}}{V_{in}} = \frac{D}{n(1-D)} \tag{8.52}$$

It differs from the buck–boost converter voltage transfer function by the turns ratio factor $n = N_1/N_2$. A positive sign has been obtained by an appropriate coupling of the transformer windings.

Unlike in other isolated dc–dc converters, the magnetizing inductance L_m of the flyback transformer is an important design parameter, as it works also as the filtering inductor.

8.4.2 Forward converter

The circuit diagram of a forward converter is depicted in Figure 8.25. When the switch S is on, diode D_1 conducts and diode D_2 is off. The energy is transferred from the input, through the transformer, to the output filter. When the switch is off, the state of diodes D_1 and D_2 is reversed. Thus, the output filter operates exactly like in the buck converter. When the switch S is on, the voltage across the inductor is equal to V_{IN}/n V_{OUT}, and when the switch is off, the inductor voltage is equal to V_{OUT}. The transformer turns ratio $n = N_1/N_2$. Thus, the dc voltage transfer function

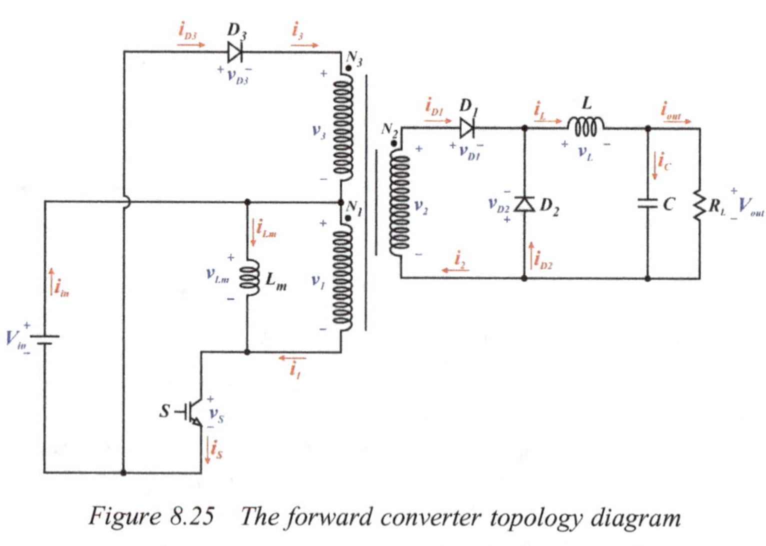

Figure 8.25 The forward converter topology diagram

of the forward converter is

$$M_V = \frac{V_{out}}{V_{in}} = \frac{D}{n} \tag{8.53}$$

In the forward converter, the energy-transfer current flows through the transformer in one direction. Hence, an additional winding with diode D_3 is needed to bring the magnetizing current of the transformer to zero. This prevents transformer saturation. The turns ratio N_1/N_3 should be selected in such a way that the magnetizing current decreases to zero during the time interval when the switch is off. In the most common configuration, the duty ratio D is limited to 0.5 and the mentioned turns ratio $N_1/N_3 = 1$.

A disadvantage of this topology is a high-voltage stress across the switch S. In the most common configuration, the voltage across the switch S when it is open is equal to $2V_{IN}$. This affects the efficiency of the converter and makes it unsuitable for high- or medium-power applications. As the structure of the converter is relatively simple and only one transistor is employed, the forward converter is commonly used for low-power applications.

The high-voltage stress problem of the basic forward converter topology could be solved by the two-switch topology variant, which is presented in Figure 8.26. The operation principle is very similar. Both switches S_H and S_L are turned on and off at the same time. When the switches are on, diodes D_H and D_L do not conduct. On the secondary side, diode D_2 is off and diode D_1 conducts delivering current to the output filter. When switches S_H and S_L are off, diodes D_H and D_L conduct allowing the transformer to demagnetize before the next cycle. Diode operation at the secondary is reversed, as in the basic forward converter described above. Thus, the output filter design is identical for both of them. In the two-switch forward converter, duty ratio D must be limited to 0.5 to allow demagnetization of the transformer.

Figures 8.27 and 8.28 illustrate the main differences in the waveform amplitudes of the forward and two-switch forward converters. The time interval ΔT_x for

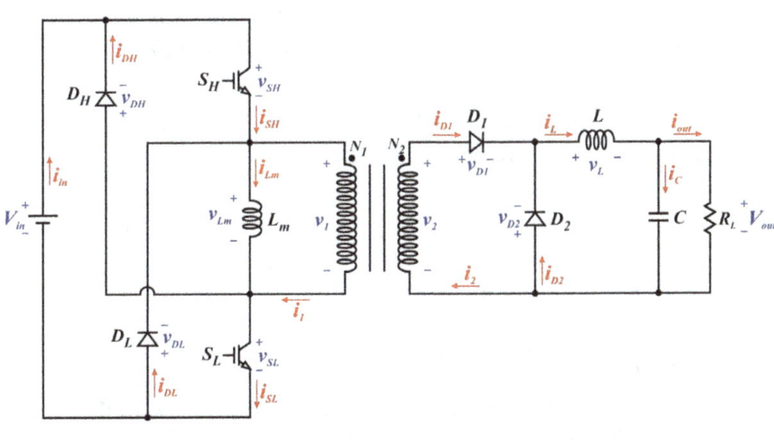

Figure 8.26 The two-switch forward converter topology diagram

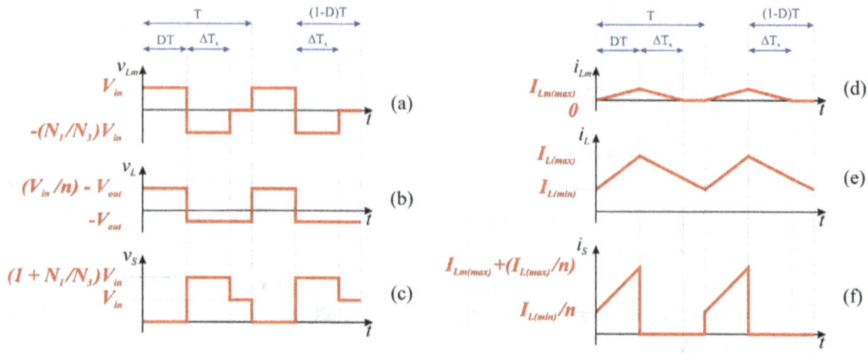

Figure 8.27 *Typical voltage and current waveforms in forward: (a) magnetizing inductance L_m voltage, (b) inductor L voltage, (c) switch S voltage, (d) magnetizing inductance L_m current, (e) inductor L current, and (f) switch S current*

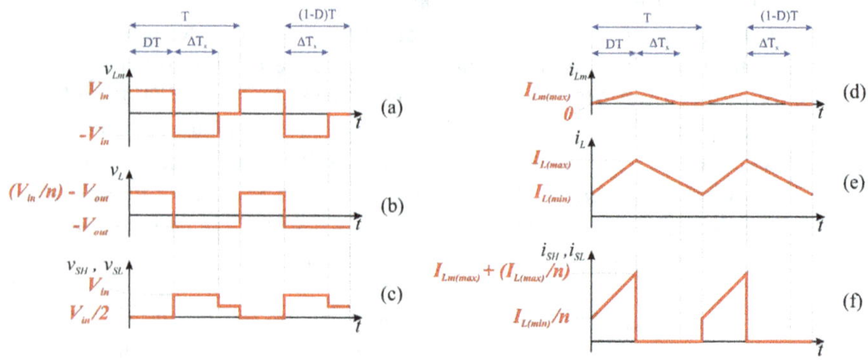

Figure 8.28 *Typical voltage and current waveforms in two-switch forward: (a) magnetizing inductance L_m voltage, (b) inductor L voltage, (c) switches SH and SL voltage, (d) magnetizing inductance L_m current, (e) inductor L current, and (f) switches SH and SL current*

the forward converter is determined by (8.54) and for the two-switch forward converter, it is determined by (8.55).

$$\Delta T_x = D \cdot T_{ch} \cdot \frac{N_T}{N_P} \tag{8.54}$$

$$\Delta T_x = D \cdot T_{ch} \tag{8.55}$$

8.4.3 Half-bridge converter

The PWM dc–dc half-bridge converter is shown in Figure 8.29. In the half-bridge converter, two switches generate a rectangular voltage waveform in a range from 0

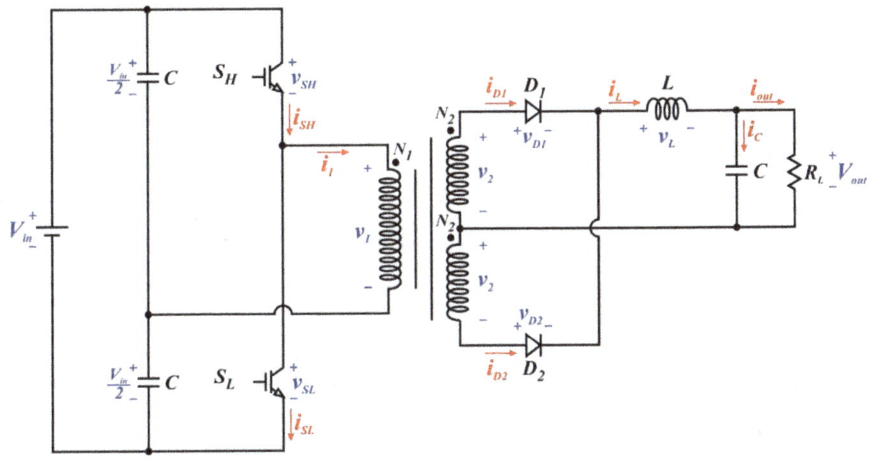

Figure 8.29 The half-bridge converter topology diagram

to V_{IN}. To avoid saturation of the transformer, a dc offset is cut off by the capacitor divider to which the transformer is connected. The switches S_H and S_L operate shifted in phase by $T/2$ with the same duty ratio D. The duty ratio must be smaller than 0.5. When switch S_H is on, diode D_1 conducts and diode D_2 is off. Diode states are reversed when switch S_L is on. When both controllable switches are off, the diodes are on and share the filter inductor current equally. Typical waveforms in the converter are shown in Figure 8.30 under the assumption that the inductor current is always positive. The dc voltage transfer function of the half-bridge converter is

$$M_V = \frac{V_{out}}{V_{in}} = \frac{D}{n} \tag{8.56}$$

where $D < 0.5$ and n is the turns ratio equal to N_1/N_2. The inductor current ripple is calculated in a similar manner as in the buck converter. Its peak-to-peak value is

$$I_{Lr} = \frac{V_{out}(1-2D)T}{L} \tag{8.57}$$

The maximum and minimum values of the inductor current i_L are

$$I_{Lmax} = I_O + \frac{V_{out}(1-2D)T}{2L} \tag{8.58}$$

$$I_{Lmin} = I_O - \frac{V_{out}(1-2D)T}{2L} \tag{8.59}$$

The filtering inductor value L is typically selected to keep the current ripple below a certain limit I_{Lrmax}. Thus, the inductor value can be calculated as

$$L_{min} = \frac{V_{out}(1-2D)T}{I_{Lrmax}} \tag{8.60}$$

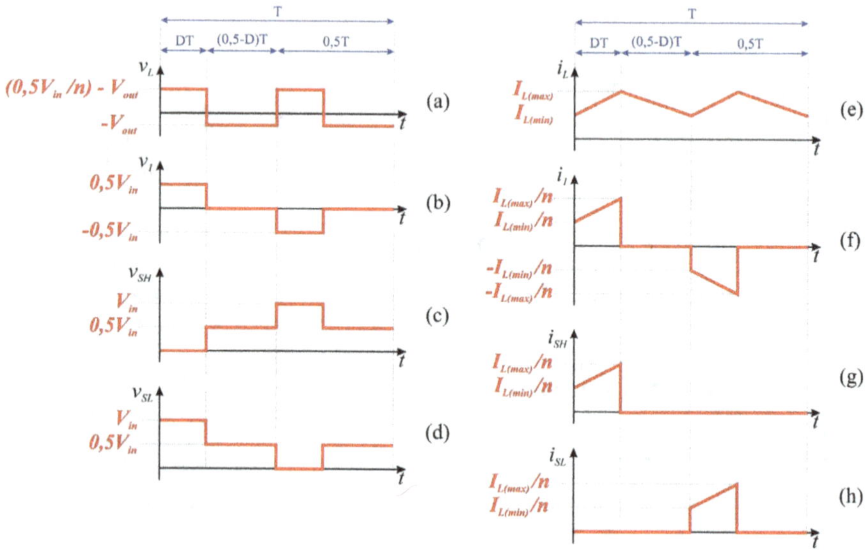

Figure 8.30 Typical voltage and current waveforms for the half-bridge converter: (a) inductor L voltage, (b) transformer primary voltage, (c) switch S_H voltage, (d) switch S_L voltage, (e) inductor L current, (f) transformer primary current, (g) switch S_H current, and (h) switch S_L current

Almost all of the inductor current ripple flows through the filter capacitor as current i_C. Current i_C causes a small voltage ripple at the dc output voltage V_{OUT}. To limit the peak-to-peak value of the ripple voltage below a certain value V_r, the filter capacitance C must be greater than

$$C_{min} = \frac{V_{out}(1-2D)}{32V_r L f^2} \tag{8.61}$$

8.4.4 Full-bridge converter

The PWM dc–dc full-bridge converter is shown in Figure 8.31. Its comparison to the half-bridge converter reveals that the capacitive voltage divider has been replaced by two controllable switches. The controllable switches are operated in pairs. When S_{H1} and S_{L2} are on, voltage V_{IN} is applied to the primary of the transformer and diode D_1 conducts. With S_{L1} and S_{H2} on, there is voltage V_{IN} across the primary transformer and diode D_2 is on. With all controllable switches off, both diodes conduct, similarly as in the half-bridge converter. The dc voltage transfer function of the full-bridge converter is

$$M_V = \frac{V_{out}}{V_{in}} = \frac{2D}{n} \tag{8.62}$$

Filtering components and current ripples are calculated in an identical manner as for the half-bridge converter.

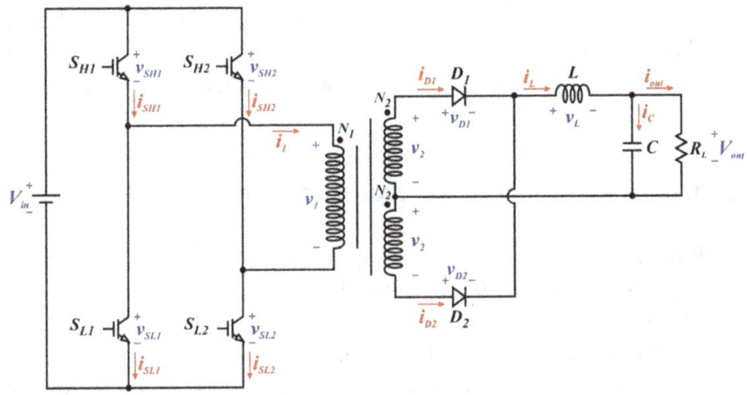

Figure 8.31 The full-bridge converter topology diagram

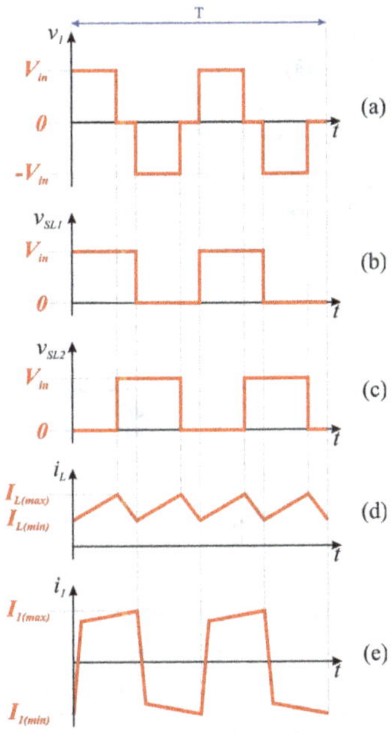

Figure 8.32 Typical voltage and current waveforms for the phase-controlled full-bridge converter: (a) transformer primary voltage, (b) switch S_{L1} voltage (c) switch S_{L2} voltage (d) inductor current, and (e) transformer primary current

The full-bridge topology allows also for another very popular control strategy, called phase-shift control. In this scenario, each half-bridge generates a square wave with the duty cycle equal to 0.5. The regulation is obtained by the phase shift of those two square waveforms as shown in Figure 8.32. The phase-shift control allows for obtaining zero voltage switching conditions for all switches. At the secondary, this control strategy provides similar waveforms as in a regular half- or full-bridge converter. The phase shift of 0 is equivalent to the duty cycle equal to zero, and the phase shift of 180 is equivalent to the duty cycle of 0.5.

8.5 Rectifiers in the switching power supply topologies

There are three basic rectifier topologies applicable for half- and full-bridge converter topology: full-bridge rectifier, center-tap rectifier, and current-doubler rectifier. The full-bridge rectifier is presented in Figure 8.33. It consists of four diodes D_1–D_4, inductor *L*, capacitor *C*, and load resistance R_L. The rectifier is supplied by the single winding of the transformer TR. The advantage of the full-bridge rectifier is that the voltage stress across diodes is equal to the peak input voltage to the rectifier. It also works with a simple two-winding type of transformer. A disadvantage of the full-bridge rectifier is that the inductor current always flows through two diodes which increases conduction losses. This makes this type of rectifier not suitable for low-voltage applications.

The center-tap rectifier is presented in Figure 8.34. It is composed of two diodes D_1 and D_2, inductor *L*, capacitor *C*, and load resistance R_L. The rectifier is supplied by the center-tapped secondary winding of the transformer TR. The advantage of this rectifier is that the inductor current flows through only one diode at a time, which makes it more suitable for low-voltage applications than a full-

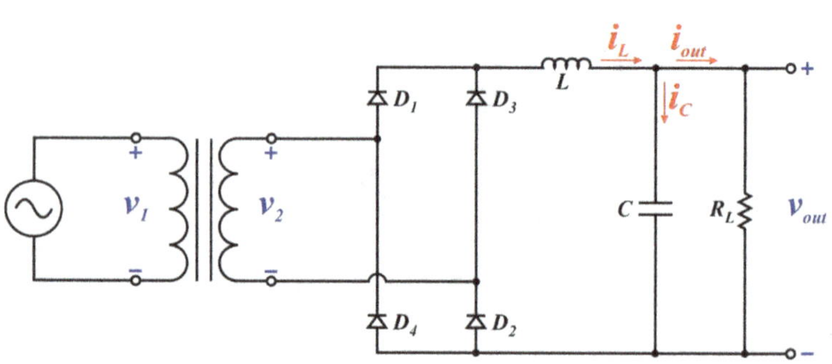

Figure 8.33 The full-bridge rectifier diagram

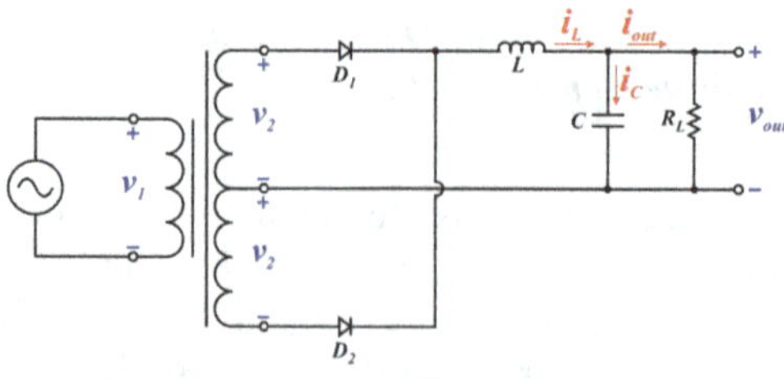

Figure 8.34 The center-tap rectifier diagram

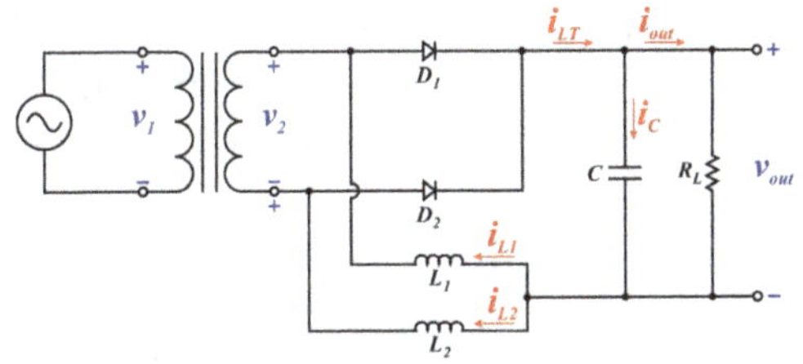

Figure 8.35 The current-doubler rectifier diagram

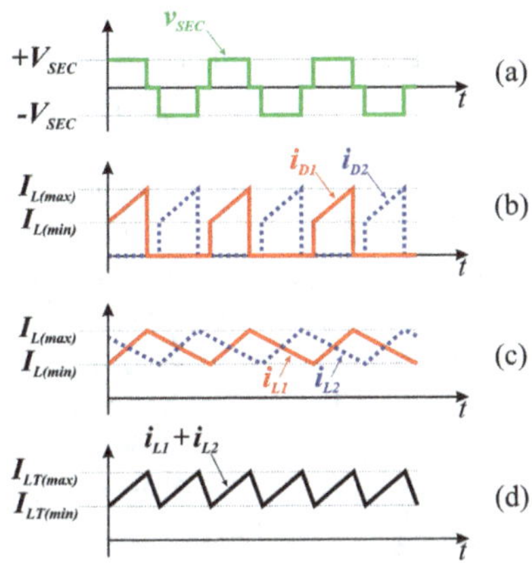

Figure 8.36 Typical voltage and current waveforms for the current-doubler rectifier: (a) transformer primary voltage, (b) diode currents, (c) inductor current, and (d) sum of inductor currents (not in scale)

bridge rectifier. The disadvantage of this rectifier is the requirement for more complicated transformer winding and high-voltage stress for the diodes. In this kind of rectifier, diodes experience two times higher voltage compared to an equivalent case with the full-bridge rectifier topology.

The current-doubler rectifier is presented in Figure 8.35. It comprises two diodes D_1 and D_2, two inductors L_1 and L_2, capacitor C, and load resistance R_L. The rectifier is supplied by the single winding of the transformer TR. The advantage of this rectifier is that the inductor current flows through only one diode at a time, which makes it more suitable for low-voltage applications than a full-bridge rectifier. The output current is shared by the two inductors which simplifies design for high-current applications. Also, the transformer winding in this case provides two times higher voltage, but there is two times lower current flowing through it. Thus, the transformer winding is greatly simplified, especially for high-current and low-voltage applications. The disadvantage of this rectifier is a high-voltage stress across the diodes. In the current-doubler rectifier, diodes are subject to two times higher voltage in comparison with an equivalent case of the full-bridge rectifier topology. Typical voltage and current waveforms for this type of rectifier are presented in Figure 8.36. In the current-doubler rectifier, both inductors operate at the same frequency as the transformer but the current in them is shifted in phase by 180. Thus, after the summation of those currents, there is a partial current ripple cancelation at the output.

8.6 Power factor correction

Typical power supplies take the ac voltage from the grid (50 or 60 Hz) as an input. Thus, the power supply usually consists of a low-frequency rectifier followed by a dc–dc converter as shown in Figure 8.37(a). In recent years, regulations and standards have put tight restrictions on the quality of input current of power supplies. In many cases, a sinusoidal input current with a low harmonic distortion level is required [12–14]. Thus, the power factor corrected pre-regulators are used instead of traditional rectifiers as shown in Figure 8.37(b).

The power factor corrector (PFC) consists usually of a rectifier (single-phase or three-phase) followed by the dc–dc boost converter. Other topologies, including bridgeless variants, are also used. The primary objective of the PFC is to provide a sinusoidal input current. Thus, by employing suitable control techniques, an input ac current can be imposed with a sinusoidal waveform following the input voltage as a reference, as portrayed in Fig. 8.38. Due to the switching at high frequency, the input ac current is composed of a fundamental component (50 Hz or 60 Hz) added to high-frequency harmonic components, which are multiple of the switching frequency. As the switching frequency is established in the range of tens of kHz, these

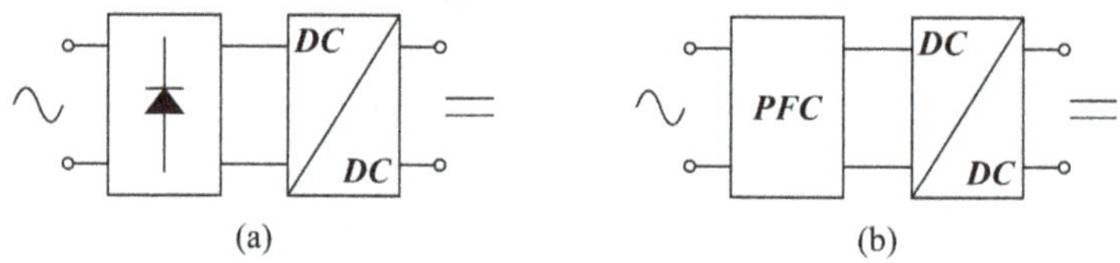

Figure 8.37 Block diagrams of ac–dc converters: (a) with a traditional rectifier and (b) with a power factor corrector

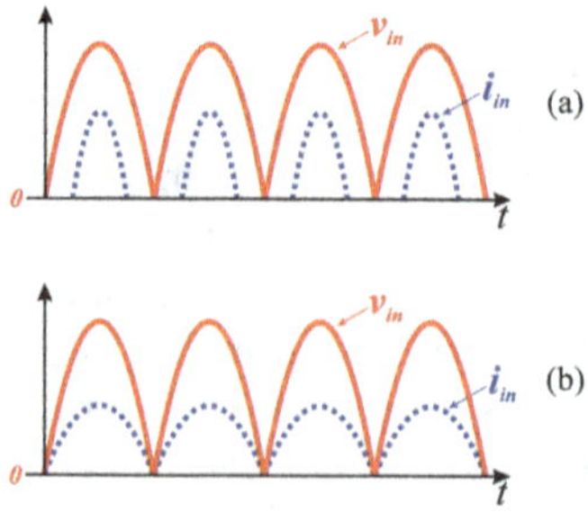

Figure 8.38 Input voltage and current of an ac–dc converter: (a) with a traditional rectifier and (b) with a power factor corrector

harmonic current components can be easily filtered through the use of small LC filters positioned before the input Boost inductors. The input current to the power supply with and without PFC is shown for comparison in Figure 8.38, neglecting distortions that typically happen at current values close to zero.

The power factor correction is classified according to the waveform of the Boost inductor current, which can be continuous or discontinuous. Related to the Boost pre-regulator used to obtain optimal operation of converters with high power factor, some of the main control strategies for CCM are briefly presented below.

8.6.1 Peak current modulation

Operating in continuous conduction mode, this converter allows the regulation of voltage on the dc-bus simultaneously with the power factor correction. This is possible through a control circuit with the feedback of the input voltage, the dc-bus voltages, and the input current. As the input current does not go to zero, in every switching period it is possible to make the input current follow the rectified input voltage by varying the duty ratio with an appropriate control strategy, as portrayed in Figure 8.39. This method is based on monitoring the peak current of a signal operating with a fixed frequency until a reference current is reached. The input current is sampled through a current sensor, and when this current reaches the reference current, the switch is turned off. The closing of the switch is determined by the switching frequency.

8.6.2 Hysteresis modulation

The principle of hysteresis modulation is based on monitoring the input current within a certain range of variation, called the range of hysteresis. This control creates two sinusoidal reference currents in phase with the input voltage, but with different amplitudes, thus generating the hysteresis band, as portrayed in Figure 8.40. When the input current tends to be greater than the maximum value of the hysteresis band, the switch is turned off and the current decreases until it reaches the lower value of the hysteresis band. At this moment, the switch is turned on, causing the input current to increase again. Hysteresis current control method has the disadvantage of operating with variable switching frequency, which can

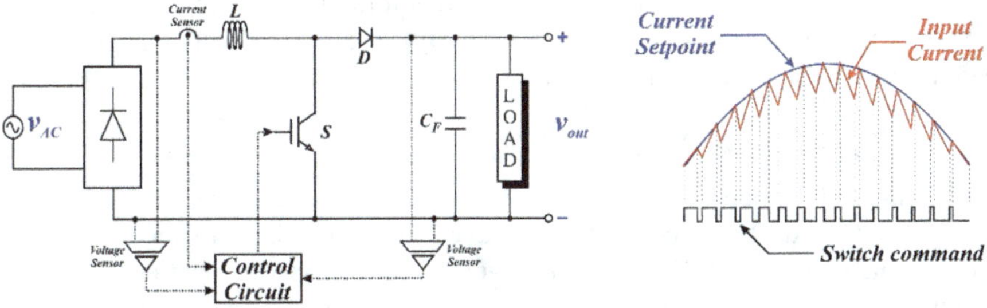

Figure 8.39 Single-phase Boost pre-regulator with power factor correction by peak current modulation

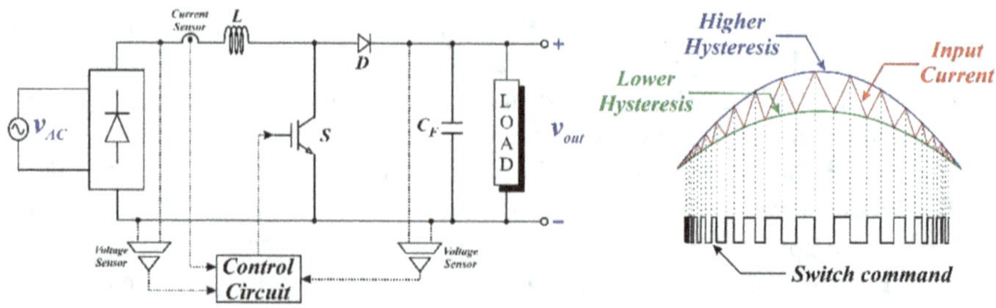

Figure 8.40 Single-phase Boost pre-regulator with power factor correction by hysteresis modulation

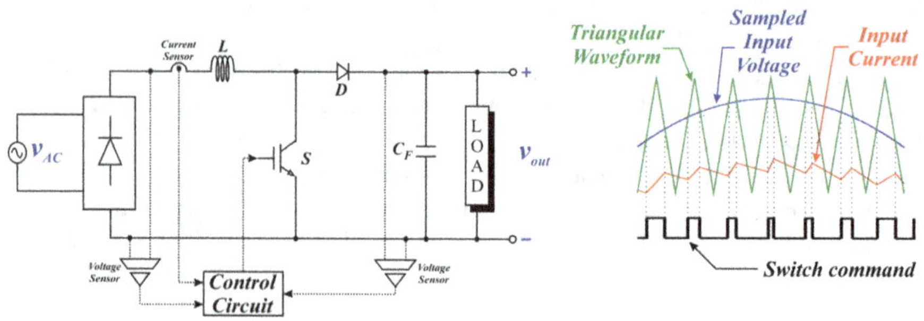

Figure 8.41 Single-phase Boost pre-regulator with power factor correction by sinusoidal PWM

lead to larger input filters. On the other hand, it provides a fast transient response for a wide load variation range.

8.6.3 Sinusoidal pulse width modulation

With this control technique, the gate-drive signal of the switch is obtained by comparing a triangular or sawtooth waveform with a fixed frequency, with a sinusoidal waveform with grid frequency, causing the duty ratio to vary sinusoidally and imposing a sinusoidal input current waveform, as portrayed in Figure 8.41. With this technique, the converter operates with a constant switching frequency leading to smaller input filters.

8.6.4 Three-phase PFC topologies

For three-phase systems, one can outline the Three-phase Diode-bridge rectifier followed by dc–dc boost converter portrayed in Figure 8.42. In medium power applications (up to 10 kW), even operating at continuous conduction mode, the structure's performance is not satisfactory since all the power is processed by a single dc–dc converter. Operating in discontinuous conduction mode, it presents as its main disadvantage the high-current stress on the converter's components, which

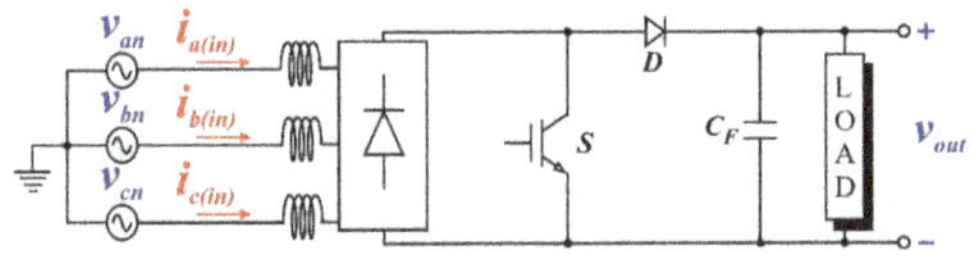

Figure 8.42 Three-phase Diode-bridge rectifier followed by dc–dc boost converter

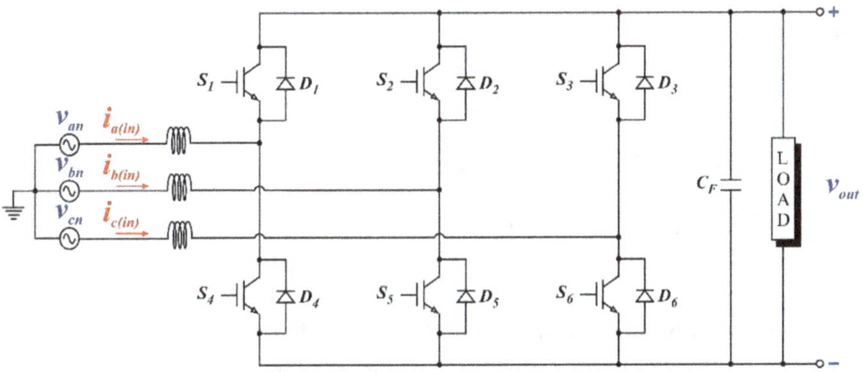

Figure 8.43 Classic three-phase PWM rectifier

limits application below a dozen kW. This is a common characteristic of the three-phase rectifiers associated with dc–dc converters.

That said, PWM techniques, applied to fully controlled three-phase rectifiers, have become unanimous for medium and high-power applications, ensuring high input power factor, and reduced harmonic distortion of current. Although they are more expensive when compared to ac–dc converters that employ conventional controls, the use of ac–dc converters controlled by PWM techniques has the following main advantages: operation with fixed switching frequency; precise design of high-frequency LC filters; reduced harmonic distortion rate of the ac input current (THD_I); high power factor; operation at high switching frequencies, increasing the converter's power density.

The most common topology is illustrated in Figure 8.43. This topology is one of the best known due to its use as a three-phase voltage source inverter (VSI). Nonetheless, bidirectional converters have their reliability compromised due to the risk of short circuits on the DC-bus, leading to greater complexity of the project.

In applications where bidirectional power flow is not necessary, topologies with unidirectional power flow (they employ fewer switches) become more attractive due to their robustness, reduced costs, and lower complexity of the project. In this context, several topological structures have been presented in the literature, for example, a classic topology of three-phase unidirectional PWM rectifier is portrayed in Figure 8.44.

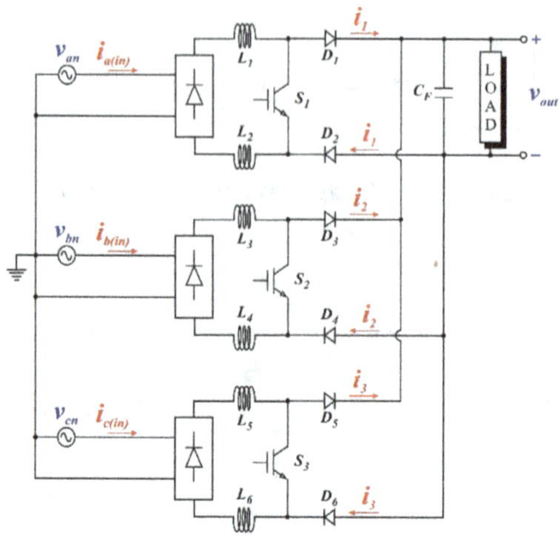

Figure 8.44 Classic three-phase PWM rectifier

It is noteworthy that all the power that is transferred to the load is actively processed, Therefore, making it unfeasible to use this structure in high-power applications (tens of kW) since the overall efficiency of the structure can be compromised due to the high-current stress to which semiconductor devices are subjected. Therefore, despite the inherent advantages of three-phase unidirectional PWM rectifiers, the use of these structures presents as its main disadvantage its limitation for applications in medium power (up to 10 kW). In conclusion, in applications where higher levels of dc power (or dc voltage) are required, the so-called Unidirectional Three-phase Rectifiers appear at three levels [15,16], and some Multipulse Rectifiers (MPR) [17,18] and Hybrid Rectifiers (HR) [19–21].

8.7 Hybrid rectifiers

The transition towards augmenting electrical energy systems requires the transformation of today's electric power system to a smarter future grid composed of different energy sources such as photovoltaic and wind. Thus, there is a trend to use direct current (dc) power distribution systems in order to reduce weight, size, and losses, while increasing the levels of transmitted power. In this context, Multipulse Rectifiers (MPR) have become more widely used. These are powered by special transformer units which can be divided into two groups, Transformer Rectifier Unity (TRUs) and Autotransformer Rectifier Unity (ATRUs). In order to guarantee the best performance of the MPR with transformers (MPTRU) and with autotransformers (MPATRU), Interphase Transformers (IPTs) and Interphase Reactors (IPRs) are used to absorb the instantaneous voltage differences of each rectifier group. The TRUs provide electrical isolation between the generator and the loads and process all required energy. Therefore, their weight and volume are higher when compared to ATRUs, which are not galvanically isolated. Both devices are robust and promote reduced voltage and current harmonic distortion while ensuring high efficiency and a high power factor. However, ATRUs stand out in terms of the reduced portion of power for which they are designed. Such an operational characteristic is known as kVA rating, which is a preponderant factor in the choice of MPR used in on-board systems such as More Electric Aircrafts (MEA).

As an alternative solution, one can outline the Three-phase Hybrid Rectifiers (TPHR) [19]. The main characteristic of TPHR is its association with switched converters connected in parallel to Diode-bridge Rectifier Units (DBRU), single-phase or three-phase, in such a way that the power processed by the switched converters is only a small fraction of the rated power. Thus, TPHR can prove to be an attractive solution for high-power high-density power electronics converters to present greater robustness.

Two TPHR topologies are presented in Figures 8.45 and 8.46, which are composed of a DBRU associated with single-phase switched converters connected in parallel. In Figure 8.45 the topology diagram of the TPHR with modified SEPIC converters is presented, and in Figure 8.46 the topology diagram of the TPHR with Boot converters supplied through isolated transformers is presented. These structures are both able to operate as an MPR without using TRUs or ATRUs. The magnitude and current waveform imposed upon the switched converter determine

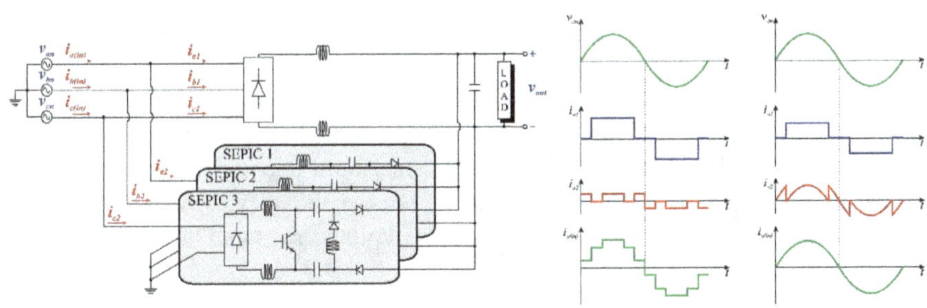

Figure 8.45 Three-phase hybrid rectifier using modified SEPIC converter

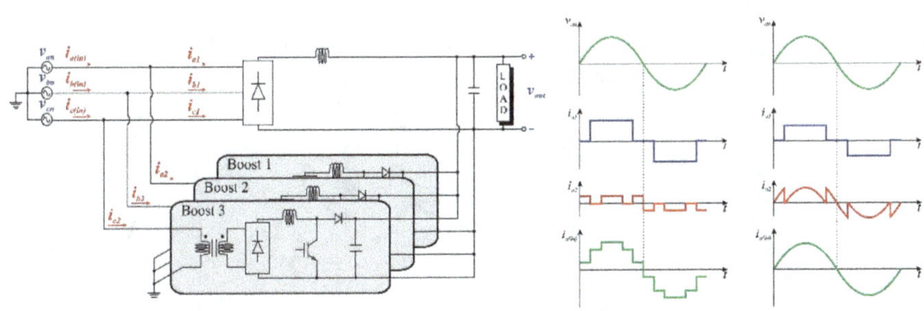

Figure 8.46 Three-phase hybrid rectifier using Boost converter

its power contribution and the final THD_I. To obtain a 12-pulse current waveform, the switched converters process less than 20% of the total active power delivered to the load. One therefore concludes that these structures associate the robustness, simplicity, and reliability of the DBRU with the high-frequency operation of switched converters, giving, therefore, a considerable reduction in volume, weight, and size of the RU. Therefore, a considerable reduction in costs can be achieved, as 80% of the power delivered to the load is processed by a simple DBRU. It is also highlighted here that retrofitting to existing installations is also feasible since the parallel path can be easily controlled by integration with the existing DC-link. To opt for Boost converters over SPEIC converters, transformers or autotransformers must be employed to ensure that the input voltage of Boost converters remains consistently lower than the DC-link voltage, which makes the use of SEPIC converters a more attractive proposition.

New experiments were performed aiming at imposing input sinusoidal currents (THD_I less than 5%) and reaching PF greater than 0.998. However, it becomes necessary to increase to 40% the power which is processed by the switched converters. In this manner, the control technique originally developed was improved, giving an appreciable reduction of the magnetic elements, maintaining therefore the division of power between the rectifier groups as desired, 40% to the switched converters and 60% to the DBRU. In addition, due to the repositioning of the filter inductors used, problems associated with the switch converters' current return path, were no longer observed, thus eliminating the need to use modified SEPIC converters, providing a reduction in passive elements and semiconductors, and leading to the structure illustrated in Figure 8.47.

Although guaranteeing low THD_I, high PF, and high-power density, it is not possible to regulate the DC-link voltage, with its value being defined by the value of the mains grid. This operational characteristic could become very attractive for applications where a tight DC-link voltage is not mandatory, however, it is susceptible to disturbances at the AC side, such as voltage sags and swells, that could

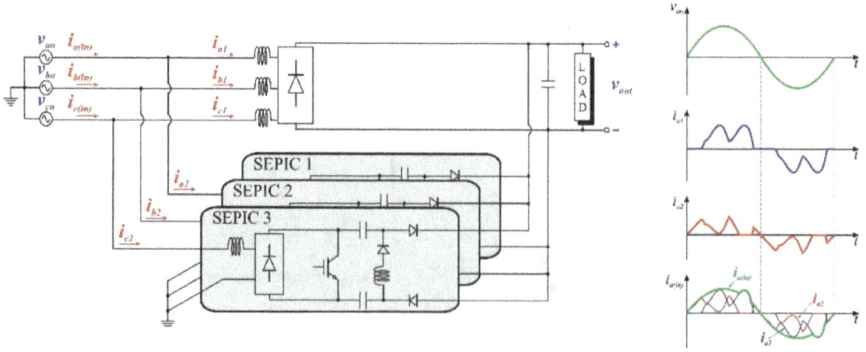

Figure 8.47 Three-phase hybrid rectifier using modified SEPIC converters with reduced volume and weight of magnetic elements

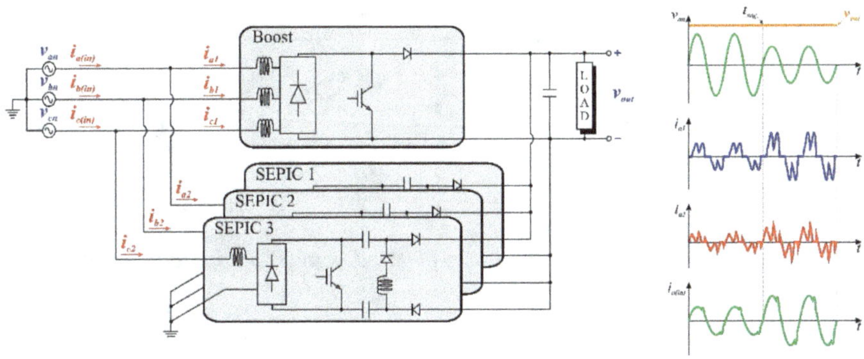

Figure 8.48 Three-phase hybrid rectifier using DBRU + cascaded Boost converter and modified SEPIC converters

lead to malfunctioning of power electronic converters connected to the DC side. With this in mind, a new structure TPHR was obtained through the addition of a switch and a diode between the DBRU and the DC-link capacitor (DBRU + Boost converter) making the structure immune to the disturbances at the ac side (Figure 8.48). Therefore, when voltage sag is detected on the mains grid, the developed control strategy goes into action with the aim of compensating the disturbance maintaining the voltage on the DC-link with the desired value. It is highlighted here that these new semiconductors are triggered when a fluctuation at the ac side occurs, as the SEPIC converters connected in parallel remain activated even during the sag occurrence. It should be emphasized here that the DBRU + Boost converter needs to be designed for a maximum 60% of the total output power. This operational characteristic avoids any oversizing of the SEPIC converters.

8.8 Bidirectional converters

Some applications, for instance, battery storage converters, require a bidirectional power flow capability of the converter [22–25]. In non-isolated converters, it can be achieved by applying synchronous rectification and proper control methods. Figure 8.49 shows an example of such a converter. It can be observed that after replacing a diode with a controlled switch, buck and boost converter topologies are identical, only the power flow is reversed. Thus, the converter can operate in one direction as a buck, in reverse as a boost. The integrated buck and boost converter topology gives much more flexibility in bidirectional operation in terms of the input and output voltage ratio.

For applications where galvanic isolation between two sides is required, the dual active bridge (DAB) converter can be used. Its block diagram is presented in Figure 8.50. It consists of two full bridges, a transformer and an inductor. As an inductor, the leakage inductance of the transformer could be used. Since this

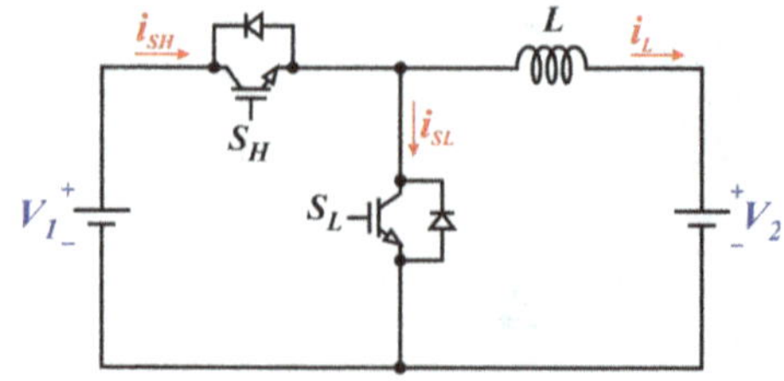

Figure 8.49 The bidirectional non-isolated converter topology diagram

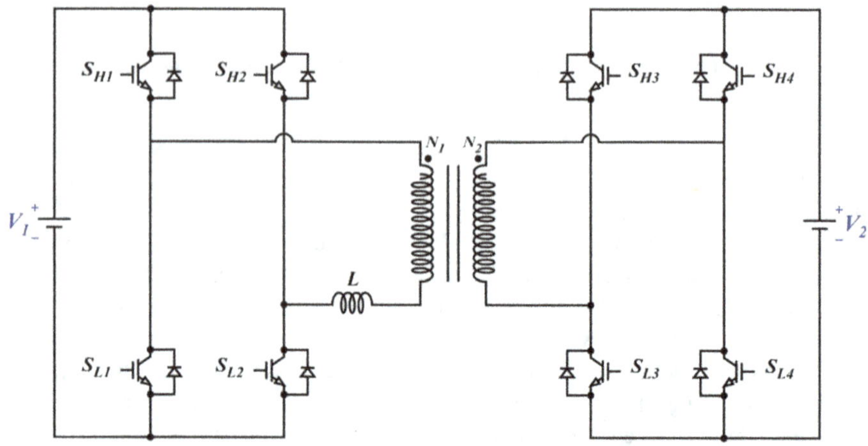

Figure 8.50 The dual active bridge converter topology diagram

converter has a symmetrical structure, it naturally allows for bidirectional operation. The DAB is usually controlled using the phase-shift control approach.

8.9 Interleaving

In recent years, the concept of interleaving has become very popular. It is related to the growing requirements on the power quality and significant amount of power (tens to hundreds of watts) which implies very high currents [26–29]. To meet those requirements, multiple converters are connected in parallel to split the current among them. In addition, a phase-shift between converters is introduced to allow for ripple cancelation. An example of a two-phase interleaved buck topology is presented in Figure 8.51. Typical voltage and current waveforms are shown in Figure 8.52. The two paralleled converters are controlled to share the current equally to minimize losses. The introduced phase-shift allows for partial ripple cancelation. Thus, interleaving facilitates an increase in the efficiency of converters, reducing the filtering components' size, and improving the output power quality.

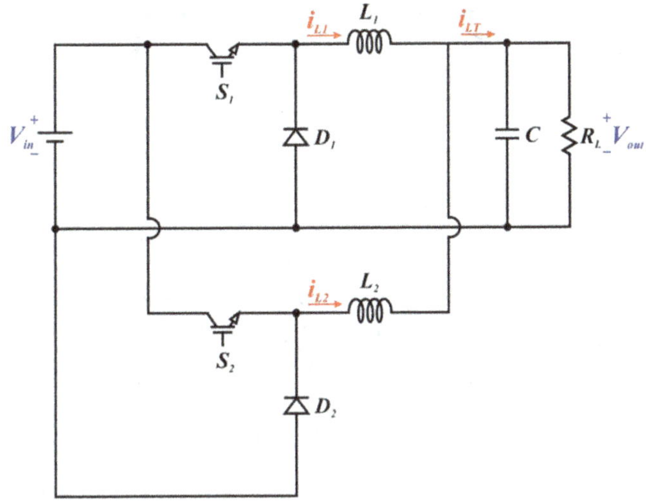

Figure 8.51 The interleaved buck converter topology diagram

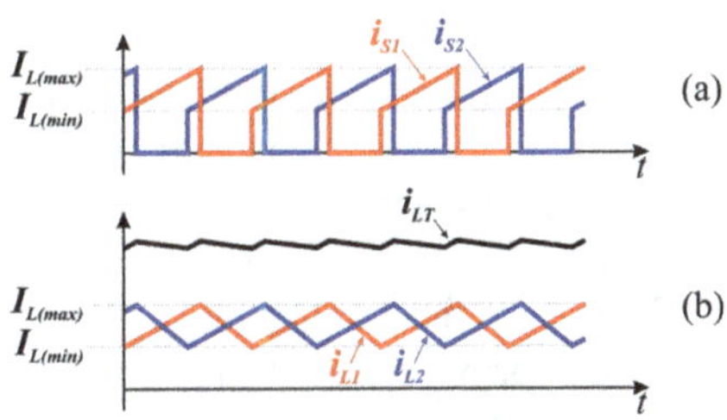

Figure 8.52 Typical voltage and current waveforms for the interleaved buck converter: (a) switch current and (b) inductor current

8.10 Control principles

A regulated dc output voltage must be provided by a dc–dc converter under varying load and input voltage conditions. The converter component values also change with time, temperature, pressure, etc. Hence, the control of the output voltage should be performed in a closed-loop manner using principles of negative feedback. The two most common closed-loop control methods for PWM dc–dc converters, namely, the voltage-mode control and the current-mode control, are presented schematically in Figure 8.53.

In the voltage-mode control scheme shown in Figure 8.53(a), the converter output voltage is sensed and subtracted from an external reference voltage in an error amplifier (EA). The EA produces a control voltage that is compared to a constant-amplitude sawtooth waveform. The comparator produces a PWM signal which is fed to drivers of controllable switches in the dc–dc converter. The duty

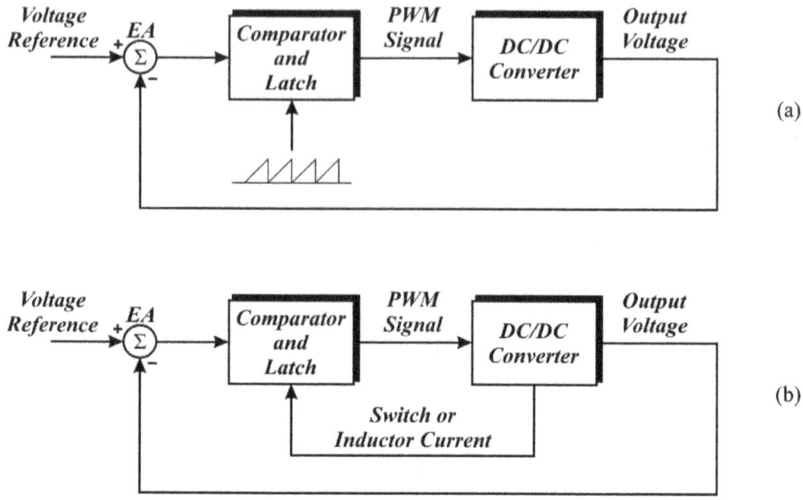

Figure 8.53 Main control schemes for dc–dc converters: (a) voltage-mode control and (b) current-mode control

ratio of the PWM signal depends on the value of the control voltage. The frequency of the PWM signal is the same as the frequency of the sawtooth waveform. An important advantage of the voltage-mode control is its simple hardware implementation and flexibility.

The EA in Figure 8.53(a) reacts fast to changes in the converter output voltage. Thus, the voltage-mode control provides good load regulation, that is, regulation against variations in the load. Line regulation (regulation against variations in the input voltage) is, however, delayed because changes in the input voltage must first manifest themselves in the converter output before they can be corrected. To alleviate this problem, the voltage-mode control scheme is sometimes augmented by a so-called voltage feedforward path. The feedforward path affects directly the PWM duty ratio according to variations in the input voltage. As will be explained below, the input voltage feedforward is an inherent feature of current-mode control schemes.

The current-mode control scheme is presented in Figure 8.53(b). An additional inner control loop feeds back an inductor current signal. This current signal, converted into its voltage analog, is compared to the control voltage. This modification of replacing the sawtooth waveform of the voltage-mode control scheme with a converter current signal significantly alters the dynamic behavior of the converter. The converter takes on some characteristics of a current source. The output current in PWM dc–dc converters is either equal to the average value of the output inductor current (buck-derived and Ćuk converters) or is a product of an average inductor current and a function of the duty ratio. In practical implementations of the current mode control, it is feasible to sense the peak inductor current instead of the average value. Since the peak inductor current is equal to the peak switch current, the latter can be used in the inner loop, which often simplifies the current sensor. Note that the peak inductor (switch) current is proportional to the input voltage. Hence, the inner loop of the current-mode control naturally accomplishes the input voltage feedforward technique. Among several current-mode control versions, the most popular is the constant-frequency one which requires a clock signal. Advantages of the current-mode control include input voltage feedforward, a limit on the peak switch current, equal current sharing in modular converters, and reduction in the converter dynamic order. The main disadvantage of the current-mode control is its complicated hardware, which includes a need to compensate the control voltage by ramp signals (to avoid converter instability).

Among other control methods of dc–dc converters, a hysteretic (or bang-bang) control is very simple for hardware implementation. The hysteretic control results, however, in variable frequency operation of semiconductor switches. Generally, a constant switching frequency is preferred in power electronic circuits for easier elimination of EMI and better utilization of magnetic components.

Application-specific integrated circuits (ASICs) are commercially available that contain main elements of voltage- or current-mode control schemes. On a single 14- or 16-pin chip, there is an EA, comparator, sawtooth generator or sensed current input, latch, and PWM drivers. The switching frequency is usually set by an external RC network and can be varied from tens to hundreds of kilohertz. The controller has an oscillator output for synchronization with other converters in modular power supply systems. A constant voltage reference is generated on the chip as well. Additionally, the ASIC controller may be equipped with various diagnostic and protection features: current limiting, overvoltage and undervoltage protection, soft start, dead time in case of multiple PWM outputs, and duty ratio limiting. In several dc–dc converter topologies, e.g., buck and buck–boost, neither control terminal of semiconductor switches is grounded (so-called high-side switches). ASIC controllers are usually designed for a particular topology and their PWM drivers may be able to drive high-side switches in low-voltage applications. In high-voltage applications, external PWM drivers must be used. External PWM drivers are also used for switches with high input capacitances. To take full advantage of the input–output isolation in transformer versions of dc–dc converters, such isolation must be also provided in the control loop. Signal transformers or optocouplers are used for isolating feedback signals.

Another approach to dc–dc converter control implementation is digital control. It is usually realized using microcontrollers (MCU), digital signal processors (DSPs), or programmable logic, for instance, field programmable gate arrays (FPGAs) [30]. In the digital controller, all analog signals, like voltages and currents, are converted to digital signals by analog-to-digital converters (ADCs). An ADC sampling frequency is usually synchronized with the converter switching frequency. The obtained digital signal corresponding to the output voltage is subtracted from the reference value to obtain an error signal. The error signal is processed by the digital implementation of the controller, for instance, a proportional-integral-derivative (PID) controller. The digital controller recalculates its output every time when new sample is obtained. In other words, it is synchronized with the ADC sampling frequency. The calculated controller output is then passed to a digital pulse width modulator (DPWM). The DPWM operates in a very similar way to a voltage-mode controller. The input signal is compared with a ramp (sawtooth) and proper pulse width is generated. In the digital domain, both signals are digital and have discrete values. The ramp is usually generated by the counting-up counter, which is reset periodically in order to obtain the desired switching frequency. Thus, the resolution of the DPWM is proportional to a ratio of the counterclock frequency to the switching frequency. There are techniques, commonly used in dedicated power converter DSPs, which allow to increase the resolution of the DPWM in some limited range. The other limitation of digital controllers is computation time. The dynamics of the digital controller are naturally limited by the sampling frequency. The sampling frequency can be easily increased, but it may lead to overloading an MCU or a DSP. It means that the update of the controller output will not be calculated before the next cycle is started. It can be overcome by using FPGA and paralleling calculations, but it is a much more complex and expensive solution. The biggest advantage of digital controllers is their flexibility. They allow modification of the parameters of the controller by reprogramming, which can be even done on the fly, while the converter is running. Moreover, it allows for the implementation of extensive diagnostic and protection algorithms that could be not available or hard to implement in analog controllers. Depending on the application, a low-cost MCU, powerful DSP, or FPGA can be used. All of them offer various performances and costs. In recent years, there have been also available dedicated ASICs with implemented digital controllers together with gate drivers and other necessary circuits for low- and medium-power applications.

8.10.1 State-space average modeling for dc–dc converters

Dynamic characteristics of closed-loop controlled dc–dc converters must fulfill certain requirements. To simplify analysis, these requirements are usually translated into desired properties of the open loop. The open loop should provide a sufficient (typically, at least 45) phase margin for stability, high bandwidth (about one-tenth of the switching frequency) for good transient response, and high gain (several tens of decibels) at low frequencies for small steady-state error.

The open-loop dynamic characteristics are shaped by compensating networks of passive components around the EA. Second- or third-order RC networks are commonly used. Since the converter itself is a part of the control loop, the design of compensating networks requires knowledge of the small-signal characteristics of the converter. There are several methods of small-signal characterization of PWM dc–dc converters. The most popular methods provide average models of converters under high switching frequency assumption. The averaged models are then linearized at an operating point to obtain small-signal transfer functions. Among analytical averaging methods, state-space averaging has been popular since the late 1970s. Circuit-based averaging is usually performed using a PWM switch or direct replacement of semiconductor switches by controlled current and voltage sources [30–32]. All these methods can take into account converter parasitic.

Mathematically, one can model and analyze any physical system with multiple inputs and outputs as an invariant linear system in time representing it in the space of states through the following matrix relationship:

$$\dot{x}(t) = A \cdot x(t) + B \cdot u(t) \tag{8.63}$$

$$y(t) = C \cdot x(t) + D \cdot u(t) \tag{8.64}$$

where $x(t)$ is the vector of states; $\dot{x}(t)$ is the derivative of the state vector relative to time; $y(t)$ is the output vector; $u(t)$ is the input vector; A is the system matrix; B is the input matrix; C is the output matrix; and D is the action matrix onwards.

Traditionally, the number of state variables sufficient for complete system representation coincides with the number of energy storage elements present in the circuit. In the case of dc–dc converters, the current in the inductor and the voltage in the capacitor are chosen as state variables. In the case of dc–dc converters operating in continuous conduction mode and containing only one switch, there are only two possible circuit configurations, one for the ON period of the switch and one for the OFF period of the switch. Therefore, it is necessary to determine the state matrix for both configurations and then calculate the average of those matrices. For this reason, the dc–dc converters are analyzed for an operating point, which typically corresponds to the average values of currents and voltages according to the design specifications.

Assuming that the dc–dc converter is operating in the continuous conduction mode, the system, input, output, and forward action matrix by average values, over one switching period, is given by:

$$A_{avg} = A_{on} \cdot D_{avg} + A_{off} \cdot \left(1 - D_{avg}\right) \tag{8.65}$$

$$B_{avg} = B_{on} \cdot D_{avg} + B_{off} \cdot \left(1 - D_{avg}\right) \tag{8.66}$$

where that A_{avg} is the system matrix by average values; B_{avg} is the input matrix by average values; A_{on} is the system matrix for the ON period of the switch; B_{on} is the input matrix for the ON period of the switch; A_{off} is the system matrix for the OFF period of the switch; B_{off} is the input matrix for the OFF period of the switch; and is D_{avg} the average duty cycle, determined by the static gain of the converter.

It is worth noting that the derivatives of the state variables in terms of average values are null. Thus, by replacing the terms of (8.63) with average values and making $\frac{dX_{avg}}{dt}$ equal to zero, it is possible to determine the average values of the state variables as

$$\frac{dX_{avg}}{dt} = A_{avg} \cdot X_{avg} + B_{avg} \cdot U_{avg} = 0 \tag{8.67}$$

$$X_{avg} = A_{avg}{}^{-1} \cdot B_{avg} \cdot U_{avg} \tag{8.68}$$

where X_{avg} is the vector of average values of the state variables and is U_{avg} the average value of the input voltage of the dc–dc converter.

The objective of determining the transfer function of a dc–dc converter is to produce a mathematical model capable of providing information about the behavior of the converter from small disturbances around the operating point. For this, the small-signal model is applied linearizing the system of state equations around this point.

It is considered that the state variables of the dc–dc converter are able to suffer a small variation given a disturbance in the operating duty cycle:

$$x(t) = X_{avg} + \hat{x}(t) \tag{8.69}$$

$$d(t) = D_{avg} + \hat{d}(t) \tag{8.70}$$

Using the matrix expression of the state equations taking into account the disturbances around the operating point, one has

$$\begin{aligned}\frac{d\left(X_{avg} + \hat{x}(t)\right)}{dt} &= \left\{A_{on} \cdot \left[D_{avg} + \hat{d}(t)\right] + A_{off} \cdot \left[1 - D_{avg} - \hat{d}(t)\right]\right\} \cdot \left[X_{avg} + \hat{x}(t)\right] \\ &+ \left\{B_{on} \cdot \left[D_{avg} + \hat{d}(t)\right] + B_{off} \cdot \left[1 - D_{avg} - \hat{d}(t)\right]\right\} \cdot U_{avg}\end{aligned} \tag{8.71}$$

Thus,

$$\begin{aligned}\frac{d\left(X_{avg} + \hat{x}(t)\right)}{dt} &= \left[A_{on} \cdot D_{avg} + A_{off} \cdot \left(1 - D_{avg}\right)\right] \cdot X_{avg} + \left[B_{on} \cdot D_{avg} + B_{off} \cdot \right. \\ &\left(1 - D_{avg})\right] \cdot U_{avg} + \left[A_{on} \cdot D_{avg} + A_{off} \cdot \left(1 - D_{avg}\right)\right] \cdot \hat{x}(t) \\ &+ \left[A_{on} \cdot \hat{d}(t) - A_{off} \cdot \hat{d}(t)\right] \cdot X_{avg} \\ &+ \left[A_{on} \cdot \hat{d}(t) - A_{off} \cdot \hat{d}(t)\right] \cdot \hat{x}(t)\end{aligned} \tag{8.72}$$

$$\begin{aligned}\frac{d\left(X_{avg} + \hat{x}(t)\right)}{dt} &= A_{avg} \cdot X_{avg} + B_{avg} \cdot U_{avg} + A_{avg} \cdot \hat{x}(t) \\ &+ \left[A_{on} \cdot \hat{d}(t) - A_{off} \cdot \hat{d}(t)\right] \cdot X_{avg} \\ &+ \left[A_{on} \cdot \hat{d}(t) - A_{off} \cdot \hat{d}(t)\right] \cdot \hat{x}(t) \\ &+ \left[B_{on} \cdot \hat{d}(t) - B_{off} \cdot \hat{d}(t)\right] \cdot U_{avg}\end{aligned} \tag{8.73}$$

Taking into account the property of the derivative of the sum, the equation (8.69) that informs that the derivative of the average matrix is null and that the multiplication of two perturbations, by generating a very small value, can be disregarded, it is thought that the following simplification

$$\frac{d\widehat{x}(t)}{dt} \approx A_{avg} \cdot \widehat{x}(t) + \left[\left(A_{on} - A_{off} \right) \cdot X_{avg} + \left(B_{on} - B_{off} \right) \cdot U_{avg} \right] \cdot \widehat{d}(t) \tag{8.74}$$

Making,

$$B_d = \left(A_{on} - A_{off} \right) \cdot X_{avg} + \left(B_{on} - B_{off} \right) \cdot U_{avg} \tag{8.75}$$

Thus,

$$\frac{d\widehat{x}(t)}{dt} = A_{avg} \cdot \widehat{x}(t) + B_d \cdot \widehat{d}(t) \tag{8.76}$$

The expression (8.76) represents a system of differential equations of 1st order, linear and invariant in time, and describes the behavior of the dc–dc converter for small disturbances around the operating point defined by X_{avg} and by D_{avg}.

By applying the Laplace Transform in the above equation one can get:

$$s \cdot I \cdot \widehat{X}(s) + \widehat{x}(0) = A_{avg} \cdot \widehat{X}(s) + B_d \cdot \widehat{D}(s) \tag{8.77}$$

where I is the identity matrix of order equal to the number of state variables and $\widehat{x}(0)$ is the vector of initial values of the disturbances of the state variables.

Considering zero initial conditions, it is possible to develop the equation (8.77) to $\widehat{X}(s)$:

$$\widehat{X}(s) = \left(s \cdot I - A_{avg} \right)^{-1} \cdot B_d \cdot \widehat{D}(s) \tag{8.78}$$

Finally, the matrix of transfer functions that relate the responses of the state variables to the duty cycle disturbance around the operating point is determined as

$$\widehat{G}_{xd}(s) = \frac{\widehat{X}(s)}{\widehat{D}(s)} = \left(s \cdot I - A_{avg} \right)^{-1} \cdot B_d \tag{8.79}$$

Therefore, in general terms, the modeling described above can be used to obtain the model in state variables of dc–dc converters that contain only one switch. Therefore, one must determine the state (A) and output (B) matrix in each step of the converter operation. With these matrices, it is possible to calculate, for a given operating point, the dynamic transfer function of the state variables of the converter as a function of the duty cycle (d). This function is traditionally used for designing controllers (PI, PID, P-Res, etc.) for voltage, current, and/or both loops for dc–dc converters that contain only two operating stages.

8.10.2 Transfer functions for dc–dc converters

This section will present an example of the application of the methodology presented in Section 8.10.1 in obtaining the matrix of transfer functions of the dc–dc Buck and Boost converters studied in Section 8.2.

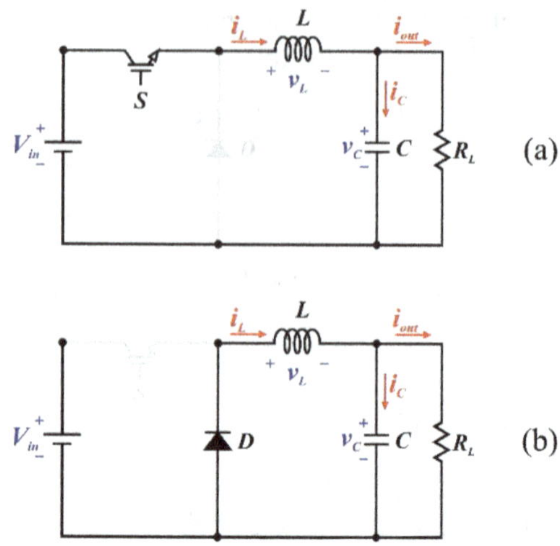

Figure 8.54 Buck converter equivalent circuit: (a) switch ON (b) switch OFF

Figure 8.54 shows the equivalent electrical circuits of the Buck converter for the conditions of the ON and OFF periods of the switch.

Thus, for the ON period of the switch (see Figure 8.54(a)) and knowing that the state variables of this system will be the current in the i_L inductor and the voltage in capacitor v_C, a simple analysis of the mesh containing the voltage source V_{in}, the switch S, the inductor L and the capacitor C, and an analysis of the capacitor connection node to the inductor and load, allows obtaining the representation of the Buck converter in the state space in the matrix form for the ON period of the switch, as given by (8.80).

$$\begin{bmatrix} \frac{di_L}{dt} \\ \frac{dv_C}{dt} \end{bmatrix} = \begin{bmatrix} 0 & -\frac{1}{L} \\ \frac{1}{C} & -\frac{1}{R_L \cdot C} \end{bmatrix} \cdot \begin{bmatrix} i_L \\ v_C \end{bmatrix} + \begin{bmatrix} \frac{1}{L} \\ 0 \end{bmatrix} \cdot v_{in} \tag{8.80}$$

For the OFF period of the switch (see Figure 8.54(b)), a mesh analysis containing diode D, inductor L, and capacitor C, and again an analysis of the capacitor connection node to the inductor and load, are able to provide the representation of the Buck converter in the state space in matrix form for the OFF period of the switch, as given by (8.81).

$$\begin{bmatrix} \frac{di_L}{dt} \\ \frac{dv_C}{dt} \end{bmatrix} = \begin{bmatrix} 0 & -\frac{1}{L} \\ \frac{1}{C} & -\frac{1}{R_L \cdot C} \end{bmatrix} \cdot \begin{bmatrix} i_L \\ v_C \end{bmatrix} + \begin{bmatrix} 0 \\ 0 \end{bmatrix} \cdot v_{in} \tag{8.81}$$

With mathematical models in the state space for the ON and OFF periods of the switch, it is possible to use equations (8.65), (8.66), (8.68), (8.75), and (8.76) to determine the matrix of transfer functions for the Buck converter following the average model in the state space as given by (8.82).

$$\widehat{G}_{xd}(s) = \begin{bmatrix} \dfrac{\widehat{I}_L(s)}{\widehat{D}(s)} \\ \dfrac{\widehat{V}_C(s)}{\widehat{D}(s)} \end{bmatrix} = \begin{bmatrix} \dfrac{V_{in}}{L} \cdot \dfrac{s + \dfrac{1}{C \cdot R_L}}{s^2 + \dfrac{1}{C \cdot R_L} \cdot s + \dfrac{1}{C \cdot L}} \\ V_{in} \cdot \dfrac{\dfrac{1}{C \cdot L}}{s^2 + \dfrac{1}{C \cdot R_L} \cdot s + \dfrac{1}{C \cdot L}} \end{bmatrix} \tag{8.82}$$

Figure 8.55, in turn, displays the equivalent electrical circuits of the Boost converter for the ON and OFF periods of the switch.

For the ON period of the switch (see Figure 8.55(a)), a mesh analysis containing *the voltage source* V_{in}, the L inductor and the switch *S* key, and an analysis of the capacitor connection node to diode *D* and load, allow to obtain the representation of the Boost converter in the state space in matrix form for ON period of the switch, as given by (8.83).

$$\begin{bmatrix} \dfrac{di_L}{dt} \\ \dfrac{dv_C}{dt} \end{bmatrix} = \begin{bmatrix} 0 & 0 \\ 0 & -\dfrac{1}{R_L \cdot C} \end{bmatrix} \cdot \begin{bmatrix} i_L \\ v_C \end{bmatrix} + \begin{bmatrix} \dfrac{1}{L} \\ 0 \end{bmatrix} \cdot v_{in} \tag{8.83}$$

For the OFF period of the switch (see Figure 8.55(b)), a mesh analysis containing *the voltage source* V_{in}, the L inductor, the D diode and capacitor *C*, and again an analysis of the capacitor connection node to the D diode and load, are able

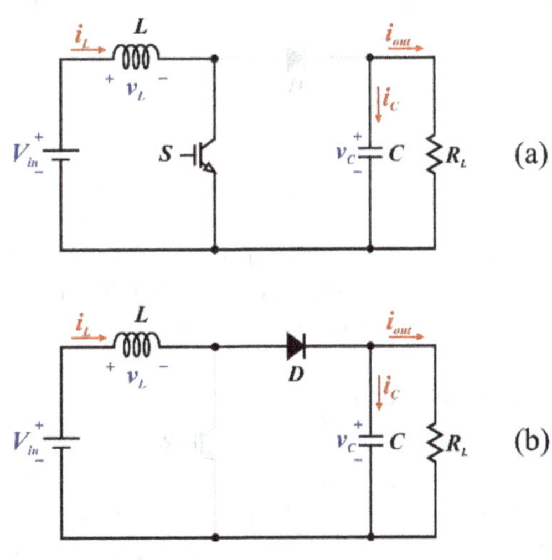

Figure 8.55 Equivalent circuit of the Boost converter: (a) switch ON (b) switch OFF

to provide the representation of the Boost converter in the state space in the matrix form for the OFF period of the switch, as given by (8.84).

$$\begin{bmatrix} \frac{di_L}{dt} \\ \frac{dv_C}{dt} \end{bmatrix} = \begin{bmatrix} 0 & -\frac{1}{L} \\ \frac{1}{C} & -\frac{1}{R_L \cdot C} \end{bmatrix} \cdot \begin{bmatrix} i_L \\ v_C \end{bmatrix} + \begin{bmatrix} \frac{1}{L} \\ 0 \end{bmatrix} \cdot v_{in} \tag{8.84}$$

In possession of mathematical models in the state space for the ON and OFF periods of the switch, one can use the modeling methodology presented in Section 8.10.1 to determine the matrix of transfer functions to the Boost converter following the small signal model using (8.87).

$$G_{xd}(s) = \begin{bmatrix} \frac{I_L(s)}{D(s)} \\ \frac{V_C(s)}{D(s)} \end{bmatrix}$$

$$= \begin{bmatrix} \frac{v_{in}}{L \cdot (1 - D_{avg})} \cdot \frac{s + \frac{2}{C \cdot R_L}}{s^2 + \frac{1}{C \cdot R_L} \cdot s + \frac{(1 - D_{avg})^2}{C \cdot L}} \\ -\frac{v_{in}}{C \cdot R_L \cdot (1 - D_{avg})^2} \cdot \frac{s - \frac{R_L \cdot (1 - D_{avg})^2}{L}}{s^2 + \frac{1}{C \cdot R_L} \cdot s + \frac{(1 - D_{avg})^2}{C \cdot L}} \end{bmatrix} \tag{8.85}$$

Since the current in the L inductor corresponds to the input current and voltage in capacitor C at the output of the Boost converter, these equations can be used for the design of the current and voltage controllers for the Boost converter applied to PFC (Power Factor Correction) systems.

It should be noted that, unlike the buck converter, the transfer functions of the Boost converter depend on the D_{avg}, which is the average duty cycle at the operating point. This means that these transfer functions represent medium models of small signals and are therefore valid only for small variations around the nominal operating point, i.e., for small steps around the average duty cycle. Therefore, these models are not valid for major disturbances in the duty cycle.

8.10.3 Considerations about the design of controllers for dc–dc converters

The most important small-signal characteristic is the control-to-output transfer function. Other converter characteristics that are investigated include the input-to-output (or line-to-output) voltage transfer function, also called the open-loop dynamic line regulation or the audio susceptibility, which describes the input–output disturbance transmission; the open-loop input impedance; and the open-loop dynamic load regulation. Buck-derived, boost, and buck–boost converters are second-order dynamic systems; Ćuk and SEPIC converters are fourth-order systems. The characteristics of buck and buck-derived converters are similar to each other. Another group of converters with similar small-signal characteristics is formed by boost, buck–boost, and flyback converters. Among parasitic components, the ESR of the filter capacitor r_C introduces additional dynamic terms into transfer functions. Other parasitic resistances usually modify slightly the effective value of the load resistance.

Buck-derived converters can be easily compensated for stability with second-order controllers. The boost converter (as well as buck–boost and flyback converters) is a non-minimum phase system. Non-minimum phase dc–dc converters are typically compensated with third-order controllers. Step-by-step procedures for the design of compensating networks are usually given by manufacturers of ASIC controllers in application notes.

Finally, the behavior of dc–dc converters in distributed power supply systems should be considered. An important feature of closed-loop regulated dc–dc converters is that they exhibit a negative input resistance. As the load voltage is kept constant by the controller, the output power changes with the load. With slow load changes, an increase (decrease) in the input voltage results in a decrease (increase) in the input power. This negative resistance property must be carefully examined during the distributed supply system design to avoid conditions that may lead to instabilities.

8.11 Conclusions

This chapter presented the analysis and design procedure of non-isolated dc–dc converters operating in CCM. Seven basic dc–dc converter topologies were analyzed: buck, boost, buck–boost, integrated buck and boost, SEPIC, Cuk, and Zeta. The basic concept of quadric converters was presented, and the Quadratic Buck and Quadratic Boost converters were analyzed. These converters can provide high conversion rates by adjusting only the operating cycle ratio, without the need to use high-frequency step-down or step-up transformers.

Non-idealities or parasitic of practical devices and components are presented since they can greatly affect some performance parameters of dc–dc converters. Modeling and effects of parasitic on output voltage ripple, efficiency, and voltage transfer function of converters were discussed. For applications where galvanic isolation is needed, isolated converters such as flyback, forward, half-bridge, and full-bridge were analyzed.

Ordinary rectifier types as well as power factor corrector converters for sinusoidal input current imposition were presented. The advantages and disadvantages of each topology are discussed, and the concept of hybrid rectifiers for high-power applications is explained. Special topologies such as bidirectional power converters and interleaved converters were also covered. Finally, for closed loop operation, basic control principles of PWM dc–dc converters were presented and two main control schemes, voltage-mode control and current-mode control, were described, including the application of the state-space average modeling technique.

References

[1] R. Erickson and D. Maksimovic, *Fundamentals of Power Electronics*, 2nd ed., Norwell, MA: Kluwer Academic, 2001.

[2] N. Mohan, T. M. Undeland, and W. P. Robbins, *Power Electronics: Converters, Applications and Design*, 3rd ed., New York, NY: John Wiley & Sons, 2002.

[3] M. K. Kazimierczuk, *Pulse-Width Modulated DC–DC Power Converters*, New York, NY: John Wiley & Sons, 2008.

[4] I. Pressman, K. Billings, and T. Morey, *Switching Power Supply Design*, 3rd ed., New York, NY: McGraw-Hill, 2009.

[5] D. W. Hart, *Power Electronics*, New York, NY: McGraw-Hill, 2010.

[6] M. H. Rashid, *Power Electronics: Circuits, Devices, and Applications*, 4th ed., Upper Saddle River, NJ: Pearson Prentice Hall, 2013.

[7] Forouzesh, M., Siwakoti, Y.P., Gorji, S.A., *et al.*, "Step-up DC–DC converters: a comprehensive review of voltage-boosting techniques, topologies, and applications," *IEEE Trans. Power Electron.*, 2017, 32, (12), pp. 9143–9178.

[8] Gorji, S.A., Mostaan, A., Tran My, H. and Ektesabi, M. (2019), Non-isolated buck–boost dc–dc converter with quadratic voltage gain ratio. *IET Power Electron.*, 12: 1425–1433. https://doi.org/10.1049/iet-pel.2018.5703.

[9] Hasanpour, S., Siwakoti, Y. and Blaabjerg, F. (2020), New single-switch quadratic boost DC/DC converter with Low voltage stress for renewable energy applications. *IET Power Electron.*, 13: 4592–4600. https://doi.org/10.1049/iet-pel.2020.0580

[10] Y. Wang, Y. Qiu, Q. Bian, Y. Guan and D. Xu, A single switch quadratic boost high step up DC–DC converter. *IEEE Trans. Ind. Electron.*, 6 (6): 4387–4397, 2019, doi:10.1109/TIE.2018.2860550.

[11] Chavoshipour Heris, P., Saadatizadeh, Z. and Rostami, N., (2019), Transformerless quadratic-based high step-down DC–DC converter with wide duty cycle range. *IET Power Electron.*, 12: 368–382. https://doi.org/10.1049/iet-pel.2018.5504.

[12] R. S. Maciel, L. C. de Freitas, E. A. A. Coelho, J. B. Vieira and L. C. G. de Freitas, "Front-end converter with integrated PFC and DC–DC functions for a fuel cell UPS with DSP-based control," in *IEEE Trans. Power Electron.*, vol. 30, no. 8, pp. 4175–4188, 2015, doi: 10.1109/TPEL.2014.2359891.

[13] Bodetto, M., Marcos-Pastor, A., El Aroudi, A., Cid-Pastor, A. and Vidal-Idiarte, E., "Modified Ćuk converter for high-performance power factor correction applications," in *IET Power Electron.*, vol. 8: pp. 2058–2064, 2015, https://doi.org/10.1049/iet-pel.2014.0967

[14] T. Meng, H. Ben and X. Wang, "A passive flyback auxiliary circuit with integrated transformer suitable for three-phase isolated full-bridge boost PFC converter," in *IEEE Trans. Power Electron.*, vol. 31, no. 7, pp. 4995–5003, 2016, doi:10.1109/TPEL.2015.2478896.

[15] Cittanti, D., Gregorio, M., Bossotto, E., Mandrile, F., and Bojoi, R. "Three-level unidirectional rectifiers under non-unity power factor operation and unbalanced split DC-link loading: analytical and experimental assessment," in *Energies* 2021, vol. 14, p. 5280. https://doi.org/10.3390/en14175280

[16] L. Dorn-Gomba, J. Ramoul, J. Reimers and A. Emadi, "Power electronic converters in electric aircraft: Current status challenges and emerging technologies," *IEEE Trans. Transport. Electrific.*, vol. 6, no. 4, pp. 1648–1664, 2020.

[17] S. Yang, J. Wang and W. Wang, "A novel 24-pulse diode rectifier with an auxiliary single-phase full-wave rectifier at DC side," *IEEE Trans. Power Electron.*, vol. 32, pp. 1885–1893, 2017, doi:10.1109/TPEL.2016.2560200.

[18] J. Chen, H. Bai, J. Chen and C. Gong, "A novel parallel configured 48-pulse autotransformer rectifier for aviation application," *IEEE Trans. Power Electron.*, vol. 37, pp. 2125–2138, 2022, doi:10.1109/TPEL.2021.3106322.

[19] Gonçalves, J.T., Valtchev, S., Melicio, R., Gonçalves, A., and Blaabjerg, F. Hybrid three-phase rectifiers with active power factor correction: A systematic review. *Electronics* 2021, 10, 1520. https://doi.org/10.3390/electronics10131520

[20] Freitas, L. C. G., Simões, M.G., Canesin, C.A., and de Freitas, L.C. Programmable PFC based hybrid multipulse power rectifier for ultra clean power application. *IEEE Trans. Power Electron*. 2006, 21, 959–966. Doi:10.1109/TPEL.2006.876859

[21] Freitas, L. C. G., Simões, M.G., Canesin, C.A., and de Freitas, L.C. Performance evaluation of a novel hybrid multipulse rectifier for utility interface of power electronic converters. *IEEE Trans. Ind. Electron.* 2007, 54, 3030–3041. Doi: 10.1109/TIE.2007.907004

[22] M. Safayatullah, M. T. Elrais, S. Ghosh, R. Rezaii, and I. Batarseh, "A comprehensive review of power converter topologies and control methods for electric vehicle fast charging applications," *IEEE Access*, 2022, vol. 10, pp. 40753–40793, doi:10.1109/ACCESS.2022.3166935.

[23] P. He and A. Khaligh, "Comprehensive analyses and comparison of 1 kW isolated DC–DC converters for bidirectional EV charging systems," *IEEE Trans. Transp. Electrific.*, vol. 3, no. 1, pp. 147–156, 2017, doi:10.1109/TTE.2016.2630927

[24] Muchina, E.G., Masike, L. and Njoroge Gitau, M. (2019), High boost-ratio bidirectional converter for interfacing low-voltage battery energy storage system to a DC bus. *IET Power Electron.*, 12: 2372–2381. https://doi.org/10.1049/iet-pel.2018.6083

[25] Singh, A.K. and Pathak, M.K. (2018), Single-phase bidirectional ac/dc converter for plug-in electric vehicles with reduced conduction losses. *IET Power Electron.*, 11: 140–148. https://doi.org/10.1049/iet-pel.2016.0899

[26] Hussein, B., Abdi, N. and Massoud, A., Development of a three-phase interleaved converter based on SEPIC DC–DC converter operating in discontinuous conduction mode for ultra-fast electric vehicle charging stations. *IET Power Electron.* 14, 1889–1903 (2021). https://doi.org/10.1049/pel2.12157

[27] Mousavinezhad Fardahar, S. and Sabahi, M. (2020), High step-down/high step-up interleaved bidirectional DC–DC converter with low voltage stress on switches. *IET Power Electron.*, 13: 104–115. https://doi.org/10.1049/iet-pel.2018.6164

[28] B. N. Alajmi, M. I. Marei, I. Abdelsalam and N. A. Ahmed, "Multiphase interleaved converter based on cascaded non-inverting buck-boost converter," in *IEEE Access*, 2022, vol. 10, pp. 42497–42506, doi:10.1109/ACCESS.2022.3168389.

[29] Choi, H., Jang, M., Ciobotaru, M. and Agelidis, V.G. (2016), Performance evaluation of interleaved high-gain converter configurations. *IET Power Electron.*, 9: 1852–1861. https://doi.org/10.1049/iet-pel.2015.0644

[30] D. B. Rodrigues, A. D. Costa, G. B. Lima, *et al.*, "DSP-based implementation of control strategy for sinusoidal input line current imposition for a hybrid three-phase rectifier," in *IEEE Trans. Ind. Inform.*, 2013, vol. 9, no. 4, pp. 1947–1963, doi: 10.1109/TII.2012.2227268.

[31] P. Azer and A. Emadi, "Generalized state space average model for multi-phase interleaved buck, boost and buck-boost DC–DC converters: Transient, steady-state and switching dynamics," in *IEEE Access*, 2020, vol. 8, pp. 77735–77745, doi:10.1109/ACCESS.2020.2987277.

[32] Ding, X., Wang, W., Zhou, M., *et al.*, Feedback control strategy for state-of-charge balancing and power sharing between distributed battery energy storage units in DC microgrid. *IET Power Electron.* (2023), 00, 1–14. https://doi.org/10.1049/pel2.12450

Chapter 9
Smart power electronic modules

Dorin O. Neacşu[1]

9.1 Introduction

With the simultaneous advent of power semiconductor devices and microcontroller platforms in the late 1980s and early 1990s, the circuit topologies and the principle of power electronics conversion have matured. Certain solutions have emerged as winners of worldwide acceptance. This allowed the transition into the next technological phase with a paradigm shift towards continuous improvement of the way power converters are actually built. New packaging methods emerged for the hardware and novel control architectures and platforms gained popularity.

Initiated in late 1994, the Power Electronics Building Blocks (PEBB) program, conceived in the Office of Naval Research (ONR, a top research sponsorship institution of the US Department of Defense), "Ship Hull, Mechanical and Electrical Systems Science and Technology Division", has sponsored and led a series electronics, materials, and manufacturing techniques. Another major program of ONR Division 334, the "Ship Hull, Mechanical, and Electrical Systems Science and Technology Division" was the Advanced Electrical Power Systems (AEPS) [1]. Both PEBB and AEPS programs included academia and industry, with notable contributions from CPES at Virginia Tech and SatCon Technology Corporation.

In addition to the actual R&D effort, these programs set the basis for standardization in power electronics converters of medium power (tens to hundreds kW). Similar programs were carried out within Oak Ridge National Laboratory (ORNL) [2,3] and major power converter manufacturers. Later on, ABB Corporation actually extended the PEBB program into the multi-megawatt range. Even more recently, the PEBB became a technology of choice for the development of future more-electric aircraft systems.

Springing off this R&D effort, a series of principles responded to the modularity demands of the production lines and after-market service expectations. The effort was continued by the industry in the mid-2000s with the implementation of hybrid integrated circuits, especially for the low kW range. While it is hard to draw

[1]Applied Electronics Department, Technical University of Iaşi, Romania

a line and identify the precise moment the first smart power module was created, it certainly came as a natural technological evolution and it has been adopted by power semiconductor manufacturers and system integrators alike. Figure 9.1 illustrates the technology evolution in low-voltage, medium-power range.

An explosive development in all semiconductor device technologies occurred after 1980. According to Moore's law, the number of transistors on the same integrated circuit chip has continuously doubled every 2 years. This increasing technological capacity combined with the advances in computer-aided design has allowed the creation of new very large-scale integrated circuits able to achieve impressive performance in signal processing and size reduction in electronic equipment. Integrate circuits spread into various fields of application including power electronics conversion.

While the main issue against smaller packaging of the power electronics equipment relies on the limited heat extraction capability, progress has been achieved in system-level packaging of the gate driver, sensing logic, and power semiconductor (Figure 9.2). A new set of devices has hence emerged and various manufacturers called them either "intelligent power modules", "smart power modules", or "advanced power converters". For the sake of uniformity, this chapter addresses these devices as Smart Power Modules or SPM [4–11].

"In 2020, motor drives represented the most significant power module market with a value of $1.6 billion," says Yole's Shalu Agarwal, "however, by 2026, EV/HEVs will become the most significant power module market, representing a market value of almost $3.6 billion. In addition, the power module packaging raw materials market is growing and will reach $3.5 billion by 2026" [12]. EV/HEV stands for Electric Vehicle and, respectively, Hybrid Electric Vehicle.

By other sources [13], the global intelligent power module market is anticipated to grow from USD 1.8 billion in 2022 to USD 3 billion by 2027, at a CAGR (Compound Annual Growth Rate) of 10.7%. Major drivers for this impressive growth come from the rising focus of Governments on generating power using renewable sources, the

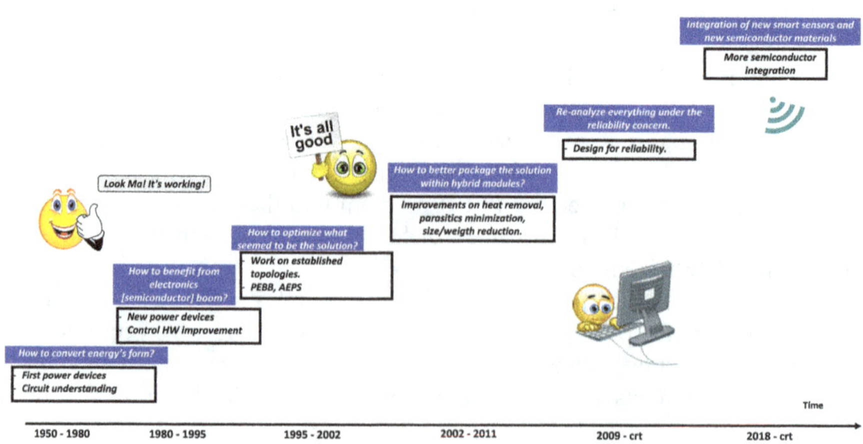

Figure 9.1 Time evolution of technology foci

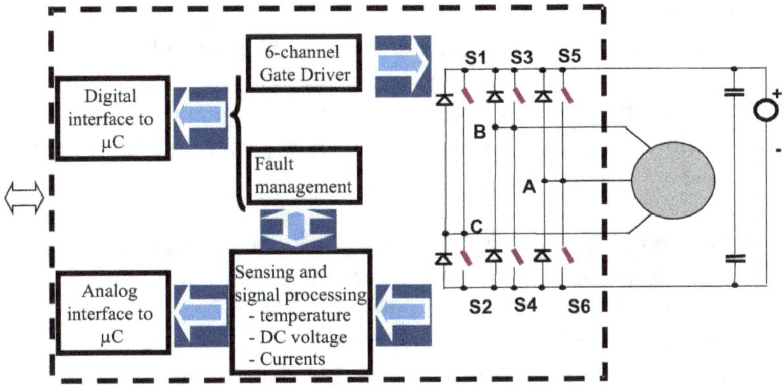

Figure 9.2 Typical architecture of a SPM

surging global electric vehicle demand, and the growing need for improving system reliability and performance. This market is currently dominated by 600 V devices and by silicon-based insulated gate bipolar transistor (IGBT) technology.

Advantages of the SPM-type devices consist of

- Reduction of system packaging space and weight.
    - Recent achievements in SiC technologies increased the power density for the complete three-phase inverter in the 10 kW range, up to 70 kW/l in volumetric, 50 kW/kg in gravimetric [14]. This can be compared to the US Department of Energy's minimal targets for 2020, of 14 kW/l or 14 kW/kg for the traction inverter in the 10 kW range. The milestones decrease with the installed power. However, the Silicon Carbide (SiC) or Gallium Nitride (GaN) inverters still do not meet the US government cost targets.
    - Easier debugging and field repair of the overall electronics. Less packaging speeds up the replacement of damaged components, and smaller package simplifies storage.
    - Simplified power connection (VDC+, VDC-, A, B, C) and controller connection through logic-level inputs.
    - A better thermal design and an optimized layout, both affect the system's reliability.
        - Using a power module supplied by the manufacturer rather than individual components yields hence recommended for the inverter application [15–18].
        - Improved reliability and lifetime [15–18]. This topic will be expanded later on, in this chapter.
        - Better power cycling capability than conventional devices [15–18];
    - Lower circuit inductance than discrete solutions.
        - Benefits in voltage spike reduction and operation at higher switching frequency with lower switch loss.
        - Improved reliability and lifetime [15–18].

- Expanding possibilities within control integrated circuits.
  - Propagation delays for all low-side and high-side IGBTs are matched within the same implementation platform.
  - Protection from over-current and over-temperature.
  - More sensing and measurement of temperature, currents, and voltages.

The initial limitations of using SPM devices are now mitigated. The first SPM devices were dedicated strictly to horsepower range applications due to the limited thermal capabilities of the existing packaging. After December 2011, SPM products were extended to currents like 300 A, covering basically the entire range of medium power applications. Another initial concern related to maintaining the switching frequency under 10 kHz due to the same limited thermal capabilities. Even if the switching frequencies have been increased over the years, the use of these SPM devices in motor drive applications justifies a limitation in switching frequency.

Despite all the merits of SPM technology, it does not eliminate the need for cooling and DC bus filtering. These two components are occupying the largest volume within the power conversion unit.

The success of SPM devices would have not been possible without the technological achievements related to IGBT device technology, packaging materials, and improved gate driver control. Their importance is outlined in the next section.

9.2 Technology background

While the goal of this chapter is not to provide a comprehensive review of power semiconductor technologies, the most important trends that encouraged the development of SPM are herein described. This is especially important since the richness of advanced technological equipment allowed a wide variety of power semiconductor devices with a clear trend for specialization with application. The IGBT used within SPM devices is highly specialized for medium-power motor drives.

The performance of modern power switches is quantified

- with reduction in power loss through lowest $V_{CE(ON)}$, tentatively with zero temperature coefficient,
- with controlled or softer transients able to reduce EMI [19–21].

9.2.1 IGBT device technologies and their performance

The SPM market is dominated in 2020 by over 95% of products in Silicon-based technologies. Even the positive prognosis from Yole Development's 2021 Annual Report expects SiC-based device revenue to reach 16% of the market while the GaN-based device revenue to reach 4.2% of the market by 2026 only. For such reasons, this chapter's main focus remains on Silicon-based IGBT.

The historical evolution of IGBT technologies after 1990, and their advantages to the creation of a smart or intelligent power module worth be briefly mentioned.

Several major device structures marked the evolution of IGBT devices: punch-through, non-punch-through, and more recently, field-stop. A possible classification can be further extended with the possible gate structures: planar or trench. Without entering technology details, Figure 9.3 illustrates the internal structures for each class of IGBT power devices.

Punch-through planar IGBT devices are mostly recommended for voltages in the range of 250 V–1,200 V and represent today a cost-effective technology, optimized for speed or short-circuit rating up to 10 μs. The non-punch-through planar IGBT devices are mostly recommended for higher voltages, like 600 V–1,200 V, for their ruggedness, with short-circuit tolerance of up to 10 μs, and optimized for high switching frequency. A more recent option, the field-stop trench IGBT is recommended for lower voltages, like 350 V–650 V, optimized for both conduction and switching performance, with a rated 5 μs short-circuit capability, and able to allow higher current rating in smaller packages. Alternatively, some rare punch-through trench IGBT devices may be used for 600 V, and they are a cost-effective technology at low switching frequencies, below 5 kHz, with a 3 μs short-circuit capability.

The current trend is the specialization of the IGBT chip depending on application and power level. This leads to products ranging from very low power levels, under 100 W, to tens of kW for a motor drive application. From an application point of view, most SPM products are created for the appliance market, where the application is a motor drive with a switching frequency below 5 kHz, with

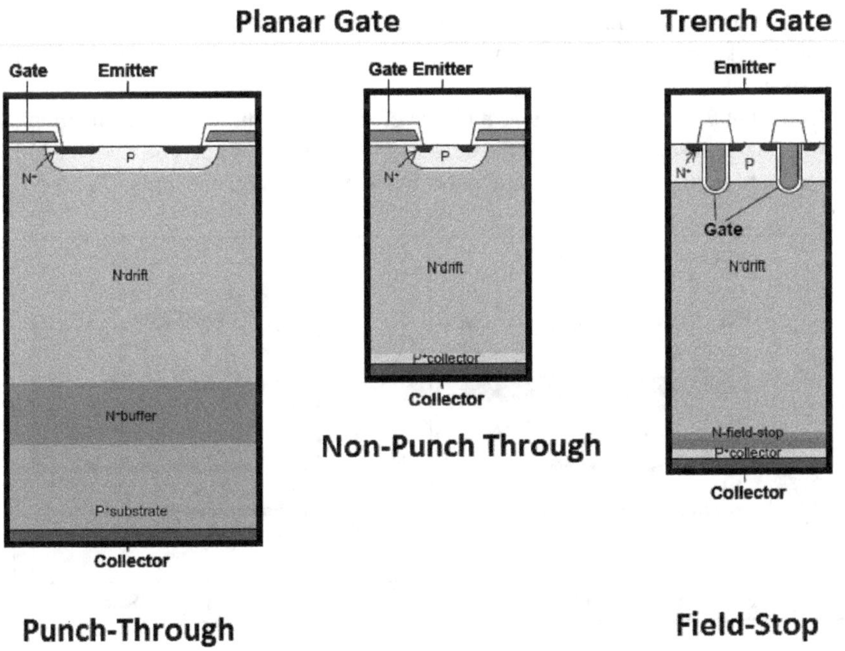

Figure 9.3 Typical IGBT device structures

transitions generally below 5 kV/ms (or 5 V/ns) to contain EMI emissions, and with a focus of IGBT design on reduction of the conduction loss. This is mostly the case of a Smart Power Module detailed herein.

Recent developments in power semiconductors for SPM modules worth including the RC-IGBT, a Reverse Conducting IGBT device, where the two chips for an IGBT and a reverse diode are included within a single chip. The two individual devices work alternately, which means that the single-chip RC-IGBT can better dissipate power while allowing a reduction in the number of chips inside the power module.

9.2.2 Super-Junction MOSFET

While conventional power MOSFET (D-MOS) devices allow a higher switching frequency than IGBTs, in the hundred-of-volt range, they also feature a larger voltage drop in conduction. The ON-state voltage drop depends on the drain to source resistance $R_{ds(on).}$ In order to reduce this resistance, new technologies have been introduced. Among them, the most successful relates to the Super-Junction MOSFET (SJ-MOS) device.

Figure 9.4 illustrates this technology. SJ-MOS has a pillar-shaped P layer deep into the N layer. P and N layers are also aligned alternately. The depletion layer spreads in the N-layer by applying V_{DS}, but the way it spreads in SJ-MOS is different from that in general D-MOS.

In the case of a conventional D-MOS device, the electric field intensity is the strongest at the P/N-layer interface and fades with distance. When the electric field intensity exceeds the limit of the silicon material, breakdown occurs, and this limits the voltage. To satisfy the voltage requirement, a compromise is thus made for the R_{ds} resistance.

Conversely, in the case of SJ-MOS, the electric field intensity is uniform in the N-layer. As a result, SJ-MOS can be designed with an N-layer that has lower resistance, realizing lower ON-state resistance devices. In other words, the SJ-MOS

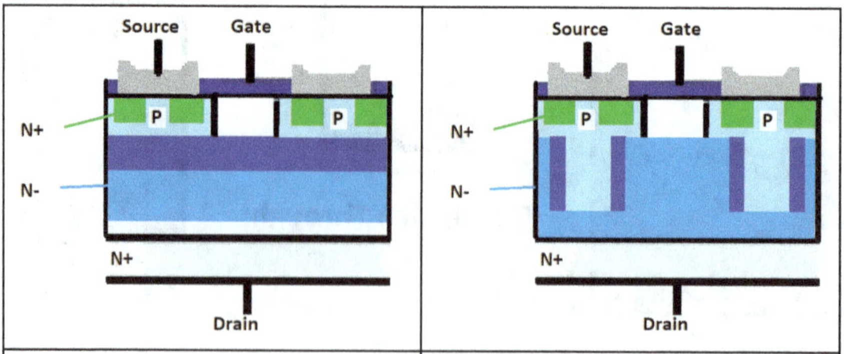

(a) Structure of a DMOS device | (b) Structure of an SJ-MOS device

Figure 9.4 Typical power MOSFET semiconductor structures: (a) Structure of a DMOS device; (b) Structure of a SJ-MOS device

can offer the same ON-state resistance as that of D-MOS but with a smaller chip size than D-MOS, or around 50% reduction in resistance for the same die size.

Another advantage of this technology refers to the C_{oss} capacitance of the MOSFET, which for the SJ-MOS device is nearly 40% lower, leading to reduced stored energy and faster switching, and consequently to lower loss.

One possible drawback, still under continuous improvement, shows that SJ-MOS has a larger p-n junction area than a planar MOSFET, leading to a faster reverse recovery time t_{rr}, but with a larger reverse recovery current i_{rr}.

9.2.3 Impact of wide-gap semiconductor material

The characteristics of Si-based power semiconductor devices are approaching the theoretical limits for the material. The SiC and GaN-based power semiconductor devices are attracting attention as next-generation materials. The most important advantages relate to higher breakdown voltage and higher thermal conductivity. A complete comparative analysis can be found in various chapters related to the properties of semiconductors. The application to the power semiconductor market can be systematized in Figure 9.5. Because of the peculiar properties of devices required for the SPM market, products are designed for the 600 V or 1,200 V ranges and rather follow MOSFET structures.

While discrete power switches or power modules are already in fabrication based on these new materials, the advances are limited for the SPM market. The SPM devices are intended for markets where the size and simplicity of the system are important. Counter-intuitively, the SPM based on SiC or GaN-based power semiconductor devices would actually require the highest level of system changes because the size optimization goes also inside and outside the power module.

For this reason, the current SPM products which include new wide-gap semiconductors are introduced in the very same packages as their IGBT counterparts,

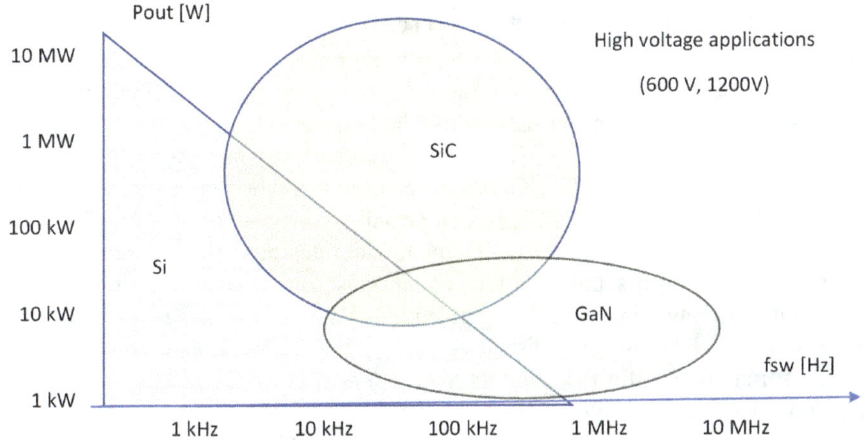

Figure 9.5 Useability of power semiconductor technologies

even if this is not a fully optimized solution. This approach allows the system integrator to use the size optimization already designed outside of the SPM.

Examples demonstrating this delay in the adoption of wide-gap semiconductors in SPM products are sparse. Mitsubishi Electric works currently (November 2022, [22]) on the PMF75CGA120 (1,200 V, 75 A), a SiC-based SPM comprising of their 2nd generation SiC MOSFET and SiC Schottky Barrier Diodes (SBD), with on-chip current and temperature sensing for highly accurate monitoring and fast protection.

In recent years (2019–2022), Infineon Corporation has launched a portfolio of highly integrated, compact power modules called CIPOS (Control Integrated Power Systems) designed to drive motors in applications ranging from home appliances, fans, and pumps to general-purpose drives. This platform started with Trenchstop™ IGBTs and includes today Cool-SiC™ MOSFET switches. A recent example from this family is the CIPOS™ Maxi 1,200 V, 20 A three-phase Intelligent Power Module [23].

For a complete image of the future of wide-gap-based SPMs, the GaN technology is seeing a possible employment with the joint effort between Sensitron Corporation and EPC Corporation to introduce a High-Power Density 350 V GaN SPM module that yields 60% smaller than comparable Si solution [24].

9.2.4 Gate driver technology

9.2.4.1 Variable gate resistance

The performance of the IGBT chip is further delineated with the optimization of the gate driver [25,26] through the inclusion of all its requirements within high-voltage integrated circuits. The effect of the two obvious extreme values for the gate resistance (small and large) on various system performance indices is shown in Figure 9.6, while Figure 9.7 illustrates the usage of different gate resistors on the ON and OFF transitions.

9.2.4.2 Mitigation of the Miller current

Other advanced gate drivers (like ST Microelectronics' TD351 [27]) mitigate the effect of the Miller current during the turn-off process. At the turn-off of any IGBT, the gate-collector capacitance is subject to a large dv/dt able to create a gate current often referred to as Miller current. This parasitic current may in certain circumstances trigger the gate and switch the IGBT into conduction. The conventional solution, mostly used by industry consists in using a negative voltage to bring the IGBT to the OFF state. This is not very easy to implement with integrated circuits since it would require a second power supply. Various compromises are herein possible with local power supplies converting the existing positive voltage into a smaller, local negative power supply, or the use of a discharge capacitor in between gate and emitter to provide a friendly path for the Miller current. Unfortunately, the use of an additional capacitor may interfere with the ability of the driver to control the gate.

Advanced gate drivers like TD351 [27] use an additional path for the Miller current during the OFF state of the IGBT as shown in Figure 9.8. A low impedance

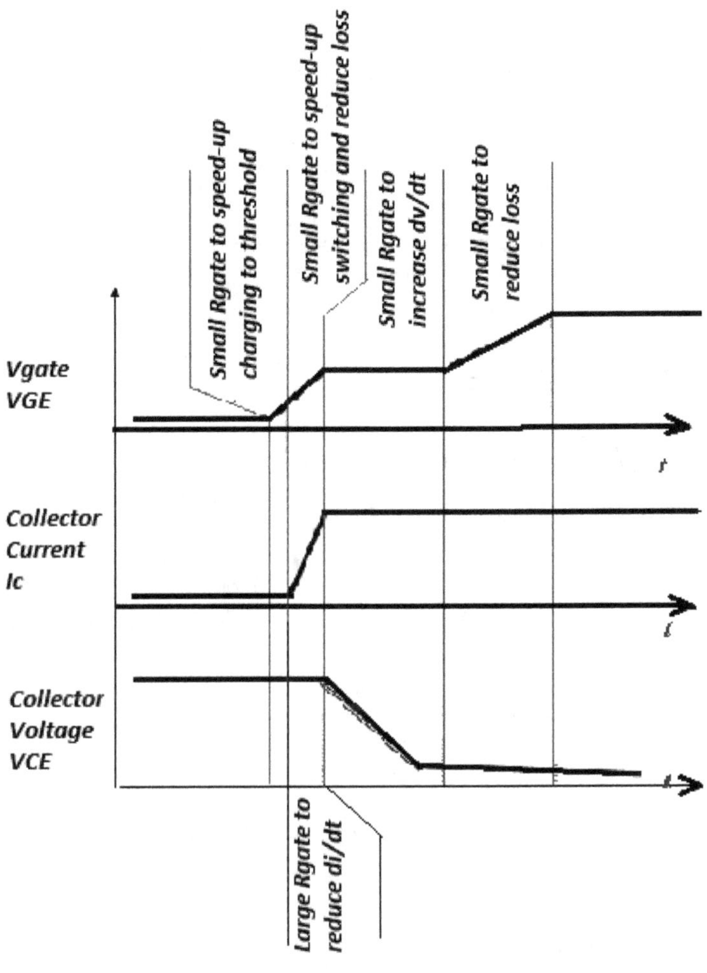

Figure 9.6 Effect of a small or large gate resistance on each transition interval

path is established between the IGBT's gate and emitter to carry the Miller current. It is important to mention that this solution does not affect the IGBT's turn-off characteristics while it rather keeps the gate at a low level during the entire off-time.

9.2.4.3 Reduction of the IGBT's tail current and minimum dead time

Another important improvement for the gate circuit requirements refers to the reduction of the IGBT's tail current through an optimized design of the power semiconductor. This has effects on the reduction of the required dead time from something around 2.4 μs to a value under 1 μs for kW range drives, where most SPMs are working.

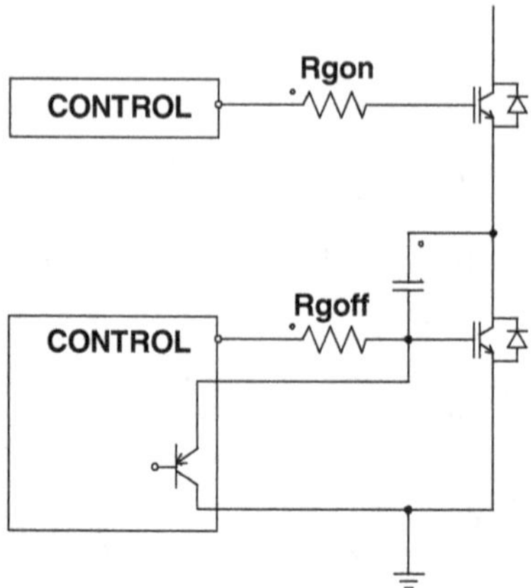

Figure 9.7 Implementation of the nonlinear gate resistance [18] shown with a PSIM drawing

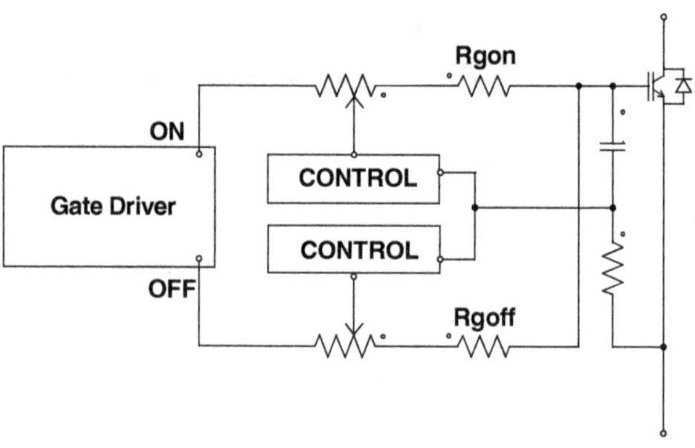

Figure 9.8 Mitigation of the Miller current

9.2.5 Sensor technologies

Smart Power Modules (SPM) incorporate power semiconductors and protection circuits. The most important sensor and protection circuit relates to temperature and overheating. Fuji Electric Corporation [14] was the first manufacturer to introduce chip overheat protection in SPMs. This feature continuously has been improved within the production of all manufacturers. The newest technology trends in SPM devices consist of the advancement of integrated sensor technologies.

Conventional temperature measurement is achieved with negative temperature coefficient (NTC) thermistors located near the power semiconductor chip. Most SPMs in the market integrate thermistor monitoring of temperature.

Given the proximity of the control IC and power semiconductor devices within the SPM, an affordable solution embraced by many manufacturers relates to temperature sensing inside the gate driver IC by using semiconductor sensing.

Both the NTC thermistor and the gate driver IC methods for temperature sensing introduce a small time lag in measurement and can be seen as a measurement of the case temperature rather than the junction temperature of the power semiconductors.

A novel integrated solution places a sensing diode (pn junction) on the same semiconductor die with the power devices. Monolithic integrated temperature sensing in bare-die IGBTs is accomplished by measuring the forward voltage drop (VF) of a string of polysilicon diodes, which are fabricated monolithically on the same die as the IGBT. The resulting voltage drop must be sensed and conditioned via an interface circuitry. Since the temperature diode is placed monolithically on silicon along with the power device, it is capacitive-coupled to the high-voltage switching node of the IGBT. Hence, the interface circuit must be designed to read the small temperature-dependent VF value, reject the switching voltages, and pass the signal across an isolation barrier.

For a sample achievement, this accurate temperature measurement allows Fuji's SPM devices to increase by 13% the allowable current with a reduction in safety margin [28]. When comparing for the same allowable current, the size of the device with a diode sensor yields considerably smaller. This advantage is very compelling for automotive applications.

Benefiting from accurate temperature sensors for junction temperature rather than case temperature, the risk for thermal destruction can be managed better and most products are now rated at 175°C instead of the former 150°C.

Other exotic methods for temperature sensing consider variations in semiconductor parameters with temperature. These include:

- Miller plateau detection;
- Switching rise or fall time intervals;
- Voltage drop measurement.

Despite the academic or laboratory attractivity, these methods are jeopardized by the strong dependency of these parameters on aging or production dispersion. Conversely, their application is rather for condition monitoring, as an early detector, or aging or damage. This is for the future because no current SPM includes such features.

Besides the temperature measurement, most SPMs offer current sensing for both control and protection. There are several main methods for current sensing and measurement:

- Integrated current shunt, with a calibrated series resistor and a voltage measurement of its voltage drop;
- Open-loop or closed-loop magnetic field sensing, also called Hall sensors require an additional integrated circuit which may not be very favorable within SPM modules;
- Rogowski sensor requires a magnetic setup around the SPM;
- Auxiliary IGBT current mirror sensing offers a pure semiconductor option (monolithic), which is very desirable within the SPM environment.

- Detection of IGBT desaturation can only be used for protection purposes, not for control.

Results from all these methods are currently influenced by temperature. The integrated current shunt and the auxiliary IGBT solutions are most used with SPM, with the resistive shunt being the most accurate and the auxiliary IGBT offering a rather coarse measurement. Among them, the newest one, the monolithic integrated current sensing is accomplished by measuring the current of a small IGBT in parallel with the main IGBT and then multiplying it by a known scaling factor. This method can offer unidirectional current measurement, so its use depends on the application. In most SPM implementations, IGBTs consist of thousands of cells in parallel via a metallization area on the top of the device. The small IGBT represents a portion of these cells that have been left disconnected from the rest of the cells and is made available as a current mirror. The emitter current for this small IGBT is further amplified and processed. The current sense ratio is temperature-dependent, but this can be corrected when combined with junction temperature sensing. Furthermore, the measurement can be compensated for the current dependency. This causes a limitation of the accuracy at a low current under 10% of the full current range.

9.2.6 Packaging technologies

Most importantly, the technology of package fabrication has seen several evolutionary steps. First, the transfer mold technology based on copper lead frames has been a good candidate for power semiconductor devices. For larger power levels, the requirements for heat extraction increase and the heat transfer through the copper leads is not enough. Hence, the integration of some cooling structures like mold resin and aluminum heat-sink provided a performance leap. Since the SPM devices expanded towards 300 A, even higher power levels require even more heat extraction which is now possible with integrated ceramic substrates. The role of the substrate in a power module is to provide the circuit connections as well as cooling. Different from low-power IC technology, the substrate of a power module must carry higher currents, provide larger voltage insulation, and deal with increased amounts of power loss and heat. It is common to face operations up to 150°C or 200°C. The most used substrate technologies are the Direct Bonded Copper (DBC) substrate, and the Insulated Metal Substrate (IMS).

The DBC substrates are composed of a ceramic tile (commonly alumina) and a sheet of copper bonded to one or both sides. The copper and ceramic tile are heated to a controlled temperature, in an atmosphere of nitrogen and oxygen, until a copper-oxygen eutectic forms which bonds successfully both to copper and the oxides used as substrates. The top copper layer can use printed circuit board technology to form an electrical circuit, while the bottom copper layer is usually kept plain for further attachment to a heat spreader.

The DBC substrates offer a low coefficient of thermal expansion, capable of good thermal cycling performances (up to 50,000 cycles). Other advantages include excellent electrical insulation and good spreading heat characteristics. For an SPM device, the power semiconductor can be packaged together with the gate driver integrated circuits (ICs) on the DBC substrate (Figure 9.9).

The ceramic material used within DBC substrates can be:

- alumina (Al_2O_3)
  - Widely used because of its low cost despite the limited thermal performance (24–28 W/mK) and despite being somewhat brittle.
  - The thickness of this substrate was reduced from 0.63 to 0.38 mm over the past years in order to reduce the thermal resistance (Rth) from the chip to the heat sink.
  - aluminum nitride (AlN)
    - more expensive, but with better thermal performance (>150 W/mK).
  - beryllium oxide (BeO)
    - good thermal performance, but somewhat toxic.

IMS represents a newer technological solution. It starts with a metal baseplate like aluminum, covered by a thin layer of dielectric (usually an epoxy-based layer), followed by a layer of copper (35 μm to more than 200 μm thick). Due to its structure, the IMS is a single-sided substrate. Thus, it can accommodate components on the copper side only.

The SPM can be designed with or without a baseplate. If the baseplate is missing, the DBC substrate is directly placed on the heat sink. Sometimes the copper baseplate is replaced with composite materials such as Al-SiC or Cu-Mo due to their low thermal conductivity.

An important aspect related to packaging concerns the terminal connection. The conventional solution of soldering joints has been proven critical to the device's lifetime. Modern solutions propose the replacement of solders with ultrasound welding [29].

Alternative to the bond wires used in conventional power modules to connect the top sides of the chips, Mitsubishi Electric's J-Series and J1-Series of power

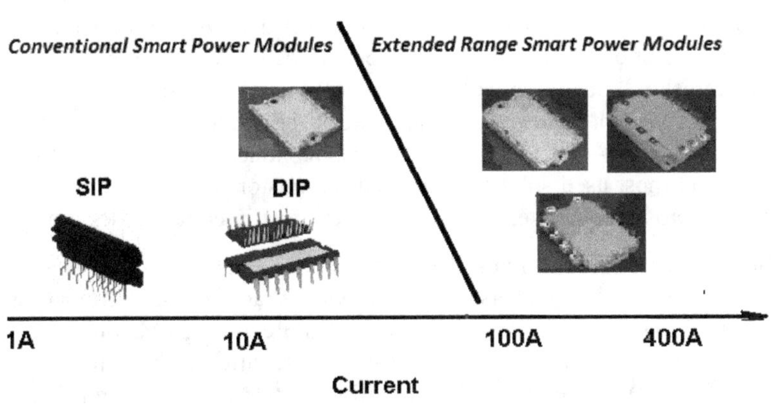

Figure 9.9 Packages for SPM devices

modules (2015) feature the Direct-Lead-Bonding (DLB) technology which replaces the bond wires by copper lead frames to conduct the high currents. While bond wires are prone to lift-off under power cycling, this failure mechanism cannot be seen in the DLB technology, and this leads to a significantly higher power cycling lifetime.

The pressure-based assembly is another direction in module packaging design. As mentioned, in a conventional power module system, the failure at the joints is a very common failure mode. In a press-pack power module, joints are formed by pressure-based bonding, which is simpler and more reliable compared to other connection methods, such as wire bonding, soldering, or sintering. On the downside, the electric and thermal conductivity of this type of joint is highly pressure-sensitive.

This brief review of technologies incorporated within a Smart Power Module allows the understanding of the efforts that made possible the integration of the entire power inverter within a single highly integrated module. These technologies fit into a complete three-phase inverter.

9.3 Basic usage

SPM has been introduced as a compact solution for large-volume production motor drives. Hence, a series of precautions have to be considered before usage.

9.3.1 Protection

All Smart Power Modules have internal protection of the IGBT against:

- under-voltage lockout—it triggers when the low voltage (~15 V) required to supply the gate drivers and miscellaneous electronics falls below a certain threshold when the gate driver IC ceases operation.
- over-current—it triggers when a DC or phase current within the power stage gets above a certain value. The simplest version is a simple desaturation circuit that monitors the IGBT's collector-emitter voltage and triggers if such voltage is above a threshold (usually 7 V) for a longer than usual time interval. Most complex solutions include a current sensing device and a comparison with a fixed threshold. Such a circuit does not fully sense the current and there may not be any current information available for the microcontroller.
- over-temperature—it triggers when the semiconductor die gets above a certain level. The most used solution implies a thermistor. However, monitoring the voltage across another semiconductor junction is in certain cases used.

The phase or DC bus currents, the DC bus voltage, and the semiconductor die's temperature are sometimes measured and reported back to the microcontroller as feedback information. This is not always a feature of a SPM, only the newer devices with advanced mixed-mode ICs have this feature implemented.

Just a few SPM devices allow an isolation barrier for either digital or analog signals sent to the microcontroller. Most SPM devices work directly on the grounding provided by the high-voltage DC bus. This, in turn, requires that the microcontroller is separated from any other circuit, or the external communication channels (RS232, RS485, SPI, IC2, CAN, and so on) are isolated. Commonly, the isolation requirements for grid-connected low-voltage drives are under 2,000 V [rms].

9.3.2 Bootstrap power supply

Most SPM are designed for low-kW power ranges, with phase currents less than 30 A. The power supply required for the control of each high-side IGBT travels together with the controlled device from the ground of the DC bus voltage to the positive DC bus voltage. This means hundreds of volts, and the local power supply needs to be isolated from the low-side electronics. The cheapest solution possible, at such power levels, consists of a bootstrap power supply as suggested in Figure 9.10.

Obviously, the newer SPM operated at currents towards 300 A requires proper power supplies for the high-side IGBTs. They are usually included within the hybrid integrated circuit.

A typical bootstrap circuit is shown in Figure 9.10. When the low-side IGBT turns-on, a current path is created from the power supply VDC (usually 15–18 VDC) through the $R_{bootstrap}$, $D_{bootstrap}$, $C_{bootstrap}$. The capacitor charges towards the VDC. When the low-side IGBT turns-off and the high-side IGBT turns-on, its gate circuitry is supplied from the capacitor $C_{bootstrap}$. The diode needs to be designed to carry the initial charging current and to withstand the entire bus voltage as reverse voltage.

The series resistor is used to limit the initial charging current, and it is missing in low-power applications. The capacitor needs to be selected to maintain the voltage decrease within an allowable band while the IGBT is kept ON. The initial charging of the capacitor is usually achieved with a special charging procedure where the microcontroller turns-on the low-side IGBT for a time long enough to

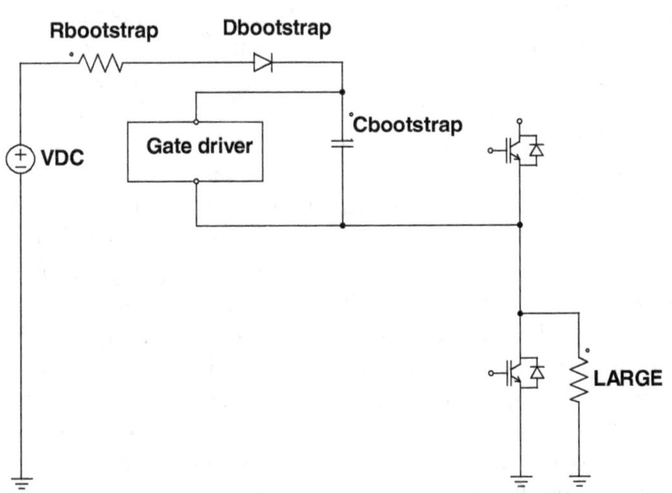

Figure 9.10 Bootstrap power supply

allow the voltage to rise from zero to VDC. Alternatively, the initial charging current is limited with a duty cycle control of the low-side IGBT.

In order to avoid the initial struggle with a proper microcontroller program, some low-power systems use a low-power resistor (like 100 kOhm, 1–3 W) in parallel with the low-side IGBT, and simply wait for several seconds before starting the actual pulse width modulation (PWM) generation. Unfortunately, it is all a qualitative process since it is virtually impossible to monitor the high-side gate circuit voltage. Ultimately, an Under Voltage Lock-Out protection prevents the IGBT against a very low high side gate power supply, usually around 8 V. Alternatively, the ST Microelectronics' products consider an active-low state for the low-side IGBTs and an active-high state for the high-side IGBTs. This allows a quick charging of the bootstrap power supply at initial power-up, when the low-side IGBTs turn-on and the high-side IGBT stay off.

The bootstrap capacitor is usually a ceramic Multilayer Ceramic Capacitor (MLCC) able to behave well in high frequency while storing some energy. Whilst traditionally a combination of ceramic and electrolytic capacitors was of choice, recently the expansion of ceramic capacitors into the large value domain would limit the bootstrap capacitor to a single ceramic capacitor of 1 to 47 μF, at 25 VDC.

The bootstrap diode needs to withstand 1.5 times the bus voltage, which is usually a requirement for a 1,000 VDC diode for a conventional 600 V IPM. Moreover, it has to be a fast recovery diode, 50 ns or better, at 1–2 A.

An alternative solution would obviously be the independent supply of each gate driver. This is done in industrial systems with special configurations of flyback power supplies, having a large number of secondary circuits and counting on asymmetrical loading of the gate driver circuits. Hence, the specific of such power supplies is that the installed power is considerably higher than the actual power, allowing thus the proper operation of the flyback converter. Interesting advances on the flyback control IC are also seen recently. They include advanced start-stop, provisions for quasi-resonant flyback, valley switching for low EMI, stand-by requirements, meeting the no-load power requirements (usually less than 300 mW), green mode status indicator, burst operation at low load, pulse skipping or pulse density modulation. So much on the same integrated circuit platform as the high-power gate drivers.

A more modern highly integrated alternative suggests directly using a switch-mode unregulated dc/dc converter, which is an ideal circuit substitute for the bootstrap diode. An example is shown in Figure 9.11. The RECOM™ dc/dc switch-mode converter comes in a small SIP4 package, works without any external components and any heat sink, and offers unregulated 15 V/15 V conversion, 1 kV isolation, 2 W output (133 mA) at an efficiency of 80%–85%. It can accommodate a capacitive load of up to 680 μF, more than enough for a large majority of gate driver applications. For instance, considering a maximum gate current request of 4.5 A for 500 ns, the minimum supply capacitor required to minimize the voltage drop at 1 V yields 2.25 μF.

A ceramic capacitor MLCC of 10 μF at 25 V, in a surface mount 1206 package can be adopted for any IRF IRAMS (another name for SPM). Alternatively, a 22 μF or 47 μF are also available in ceramic technology.

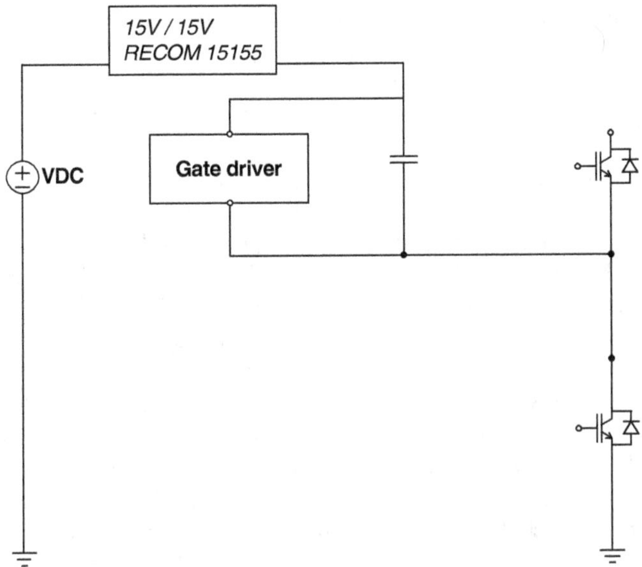

Figure 9.11 Using the unregulated 15/15 V power supply, with 1,000V isolation

Obviously, the use of a dc/dc hybrid converter is more expensive, but it solves a series of problems with bootstrap power supply and it can provide a great laboratory setup needed to develop new control algorithms, like a PWM with 60° no-switching intervals.

9.3.3 Digital interface

Ideally, the SPM device is directly connected to the microcontroller through proper selection of the logic family (like 3.3 V or 5 V). Even modern microcontrollers with a core operated at 1.2 V or 1.8 V must have a higher voltage interfacing at 3.3 V or 5 V. It is important to pay attention to the ON-state logic of these signals, since most European and Japanese companies use an active-low signal for turning ON the IGBTs, while certain American companies use an active-high signal for turning ON the IGBT. As already mentioned, ST Microelectronics has a special convention of using active-high signals for control of the high-side IGBT, and active-low signals for turning on the low-side IGBTs. Apparently, this helps with the start-up of the bootstrap power supply before the microcontroller starts the PWM. Additionally, these logic inputs may benefit from a Schmidt or hysteresis trigger.

9.4 Reliability

Since the power converter structure and the core technologies are well-known, contemporary efforts are more and more targeting lifetime and reliability improvement.

The advent of SPM subscribes to this desiderate and it can already be concluded that the implementation of a three-phase power inverter with an SPM, rather than discrete power devices, improves reliability. This is possible through a series of technological improvements, mostly leading to loss reduction and better thermal aspects.

Obviously, modern research approaches to reliability improvement include online junction temperature monitoring [30], or active gate drivers [25,26].

Even before applying such approaches, the power modules are inherently more reliable since they are built through the paralleling of power IGBT devices. It has been demonstrated in [31] that the power loss within a 1,200 A IGBT is higher than when using three parallel 400 A IGBT devices. This observation in addition to the inherent redundancy in operation comes to improve the reliability somewhat against the common wisdom that a circuit with more components is less reliable.

Hence, IGBTs are used with a parallel connection for the large-capacity inverters [29]. This translates into a set of requirements suitable for paralleling. The voltage drop during the conduction state of both the IGBT and anti-parallel diode must have a positive coefficient to avoid the concentration of current within any of the paralleled chips. Furthermore, it is necessary that the transition times are matched for the switching current balance. A typical figure of merit represents achieving a current balance as close as 2% of each other [29].

Reliability and lifetime of SPM are characteristics strongly depending on the thermal performance of the module. Hence, the rating of power converters in the low-kW range is mostly depending on thermal constraints. Given the complex structure of an SPM, the precise thermal modeling is difficult. It is therefore very common for the manufacturer to design and launch preliminary SPM devices ("engineering samples") without complete power or thermal rating. That is a blank space in the rating section, to be filled after more testing or customer assessment.

For most tested devices, empirical thermal models are made available to account for the internal losses without physical modeling. An empirical model for an International Rectifier's IRAMS SPM device [32] considers each switch individually and calculates the power loss with the following set of equations on dependence with the tested current.

$$E_{ON} = (h_1 + h_2 \cdot I^x) \cdot I^k = \left[(7.69e-4) + (2.99e-2) \cdot I^{-1.159}\right] \cdot I^2 \qquad (9.1)$$

$$E_{OFF} = (m_1 + m_2 \cdot I^y) \cdot I^n = \left[(1.76e-2) + (4.34e-2) \cdot I^{-0.492}\right] \cdot I^1 \qquad (9.2)$$

$$V_{CEON} = V_T + a \cdot I^b = 0.51 + 0.46 \cdot I^{0.649} \qquad (9.3)$$

The results of this method used for a conventional motor drive application built with a 20 A IRAMS IPM (another name for SPM) module show 2.3 W power loss per switch, and 14.1 W per entire package when operated in ambient temperature and trying to prevent a junction temperature close to 125°C.

Other manufacturers provide loss curves for their SPM.

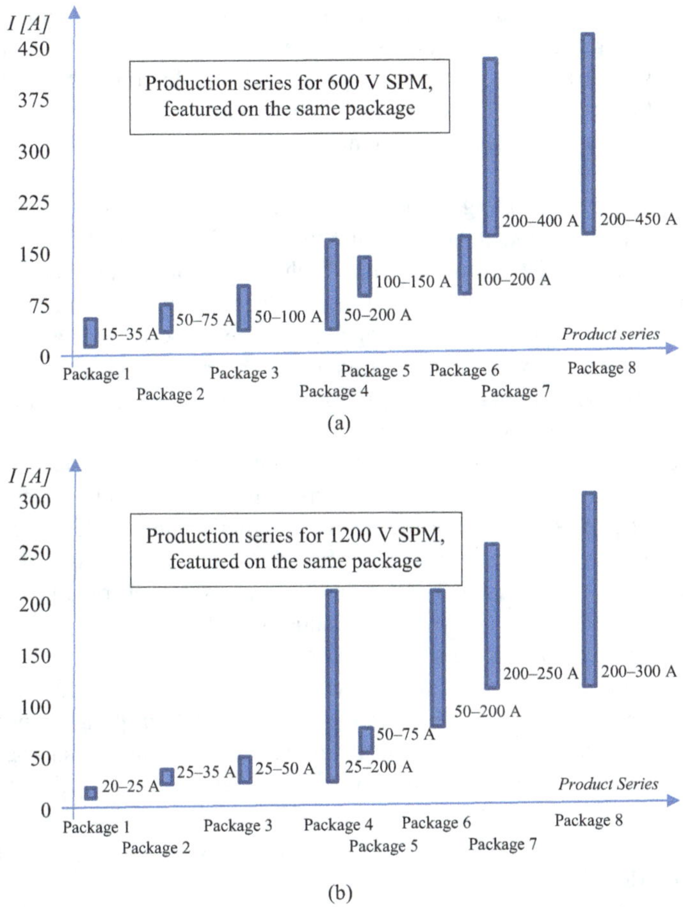

Figure 9.12 Examples of SPM devices by production series under the same package: (a) Production series for 600 V SPM, featured on the same package; (b) Production series for 1200 V SPM, featured on the same package

9.5 Variety of products

All semiconductor manufacturers [32–37] have lines of products dedicated to IPM or SPM. Instead of a comprehensive list of products, Figure 9.12 illustrates the variety of packages and products on the market. This ultimately allows us to outline the technology's performance saturation curve.

9.6 Emerging applications

In addition to the conventional usage of SPM, already discussed in Section 9.3, other applications of this class of devices are reviewed in this section. The inclusion of an entire inverter circuitry within a single hybrid module is similar to the moment of the transition from discrete devices to integrated circuits in analog electronics. An entire inverter becomes a single Bill-of-Materials' item, with a cost close to what a single IGBT device used to be. Hence complex structures can be built with multiple SPM devices, solving some of the limitations of the current implementation.

This concept of building a complex power structure with a multitude of smaller, more intelligent switches was first introduced to the academic community by Professor Ramu Krishnan of Virginia Tech (USA), at a round table about the future of power electronics, held in 1999 in San Jose, at IECON Conference. While at the time, the concept seemed way ahead of time, it became a favorite for the future usage of Smart Power Modules. It is one of the cases when power electronics meets high-tech, with high-scale integration of semiconductor-based functions to reduce the usage of passive components.

This idea proposes a paradigm shift from the conventional design sequence which starts with a topology selection, its simulation-based improvement, and the look for proper parts to build. For the SPM implementation, the design starts from the amazing features already existing on the SPM and—if they are not enough—multiple SPM devices are used. The system designer is thus absolved of a series of concerns that move into the power semiconductor courtyard.

The emergence of a new generation of converters offers the potential for either achieving system performance that meets the rigorous standards set by the European Union and other governments or delivering high-performance, highly reliable, and redundant custom converters tailored for specific special applications.

Independent efforts have been seen over the last years in studying multi-module converters [1], either as multi-level converter structures, or complex parallel and/or interleaved two-level inverters. A series of conferences' special sessions occur on the topic. More interestingly, new topologies are now reported when considering the entire six-switch inverter as a singular integrated circuit, without showing the drawing of the complete schematic. Some results are next briefly reviewed from both industry and academia to illustrate this topic of R&D.

9.6.1 Grid interface with multiple power modules

A large number of three-phase power converters are dedicated to setting up grid interfaces with reduced harmonics. This is mostly the result of the multitude of standards and regulation efforts undertaken by most countries in the world. This effort is even more important at higher power levels where the same percentage loss in harmonics represents more power loss in watts.

The three-phase AC/DC conversion employing IGBTs can be achieved with a full six-switch power converter, or with a six-switch power converter used as an active filter in addition to a higher power diode rectifier. Any of these two solutions can benefit from an SPM with a direct implementation of the power stage through an SPM.

The power levels for these solutions tend to be limited. Hence, the opportunity to propose a novel power conversion principle [32,38] is able to augment the power capability of an SPM device with a diode rectifier, or reverse the logic, to correct the input harmonics of a diode rectifier with an SPM device. The principle of this new conversion system can be understood from the current waveforms shown in Figure 9.13, with the converter circuit from Figure 9.14.

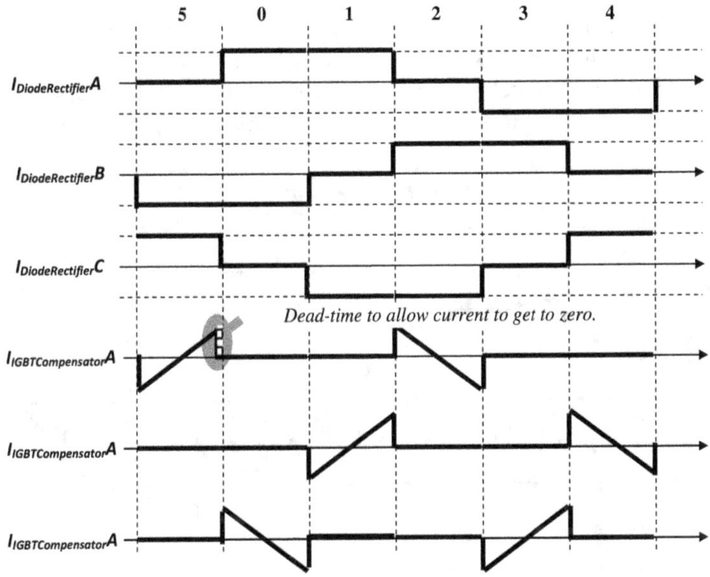

Figure 9.13 Waveforms defining the novel multi-module grid interface

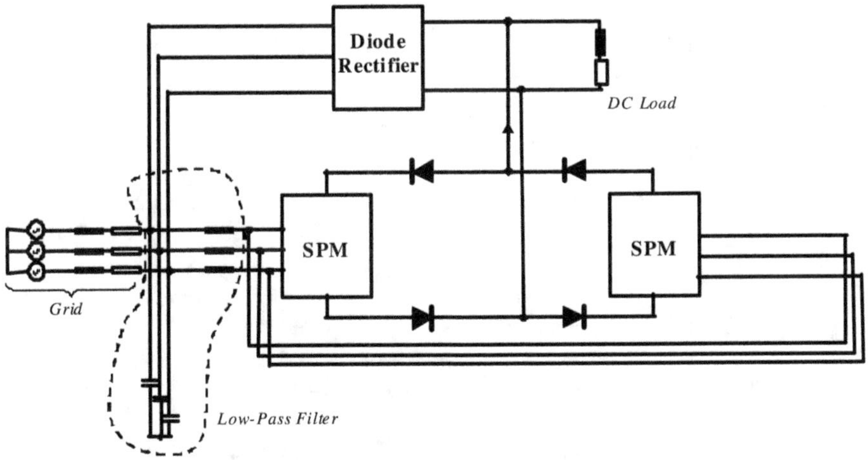

Figure 9.14 Novel multi-module grid interface targeting SPM implementation of the IGBT compensator

The conventional operation of the diode rectifier sets the proper current references for the SPM multi-converter module during intervals with no diode conduction. The current generated by the IGBT compensator converter closes through the two ON-state diodes and back to the power converter. This will overload a little the diode rectifier but this is really not important (cost-wise) in this application. The current circulation provides an improved grid current shown in Figure 9.15. In addition to the proper synchronization of the waveforms, a closed-loop current controller ensures the current waveforms.

The bidirectional switches are implemented with IGBT through two SPM modules as shown in Figure 9.16. The advantage of this setup is demonstrated in

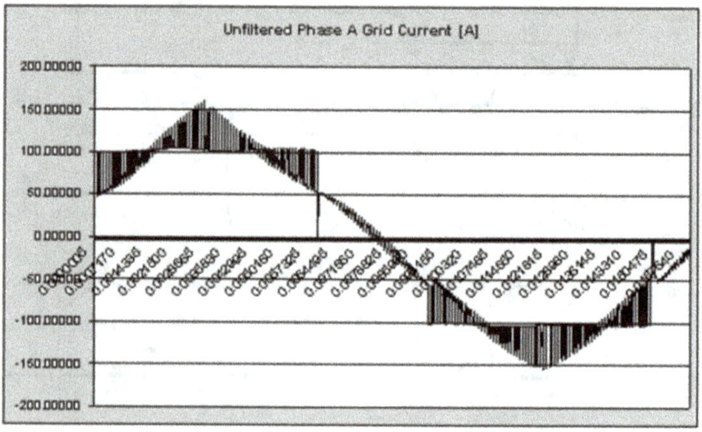

Figure 9.15 Grid phase current based on the converter from Figure 9.14 and current control

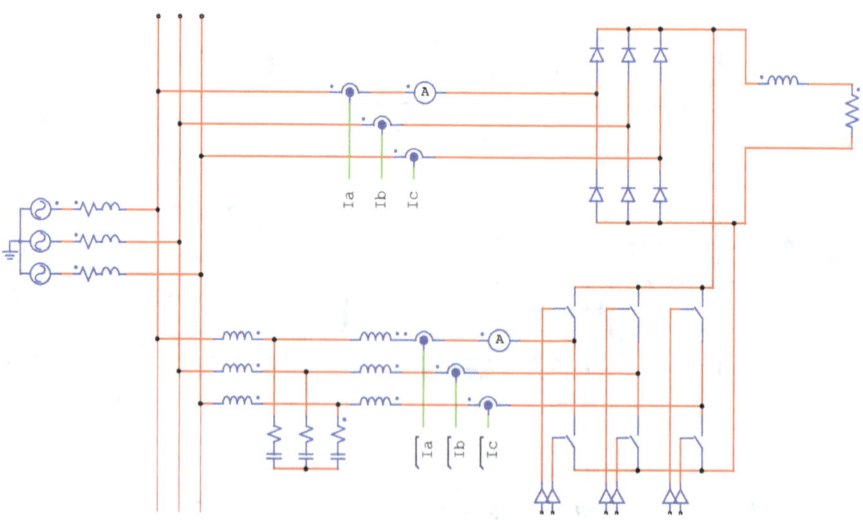

Figure 9.16 SPM implementation

Table 9.1 Comparative results for various grid interfaces

Topology	**Direct IGBT bridge**	**Active filter**	**New method**
Max. load current (DC)	Irectifier = 6.5 Adc	Irectifier = 20 Adc	**Irectifier = 50 Adc**
IGBT operation	Iph = 5.0 Arms, m = 0.8 Unity Power Factor	Iph = 22 Arms Waveform = harmonic difference	Iph = 20 Arms Pwr Cycle = 0.33
SPM's PWM Frequency	Fsw = 6.00 kHz	Fsw = 12.5 kHz	Fsw = 12.5 kHz
Pswitching (Switch + D)	0.87 W	1.68 W	2.42 W
Pcond, switch	3.42 W	2.99 W	1.46 W
Pcond, diode	0.80 W	0.35 W	1.06 W
Ploss, IPM module	**30.54 W**	**30.12 W**	**29.64 W**

Table 9.1 where a comparison is made while considering the same (numerically close enough) power loss. It is proven that more load current can be processed with the novel solution, which results in better utilization of both diodes and IGBTs.

Using highly integrated power modules provides herein an opportunity for reducing the count of passive components, with advantages in size, weight, efficiency, and reliability. In other words, we trade passive components for installed semiconductor power. Overall, a very low-cost solution provides a low component count (three active and three passive components apart from the microcontroller setup), without additional power supplies or DC link capacitors, with fully integrated, fully-protected electronics within the SPM.

9.6.2 Matrix converter with SPM devices

A three-phase AC–AC matrix converter [39,40] is made of nine bidirectional switches which allow any output terminal to be connected to any input terminal. Any symmetrical three-phase output voltages can be obtained from a set of input three-phase voltages by a suitable switching of the matrix. The PWM operation needs to prevent a short circuit between the input-side terminals, and each output must be connected so that provides a path for the inductive load current. These constraints require special precautions at switching off the power stage.

The major historical drawback with matrix converters lies in their rather difficult implementation. Various solutions for bidirectional power switches have been presented in the literature. Figure 9.17 provides an alternative based on the use of SPM devices. The solution assumes that the bidirectional switches can be spread onto two Current Source-type Converters. Each Current Source Converter is implemented with an SPM as shown in Figure 9.17(a). Repeating this for all three phases leads to the converter architecture from Figure 9.17, where the energy conversion is achieved with an all-semiconductor solution.

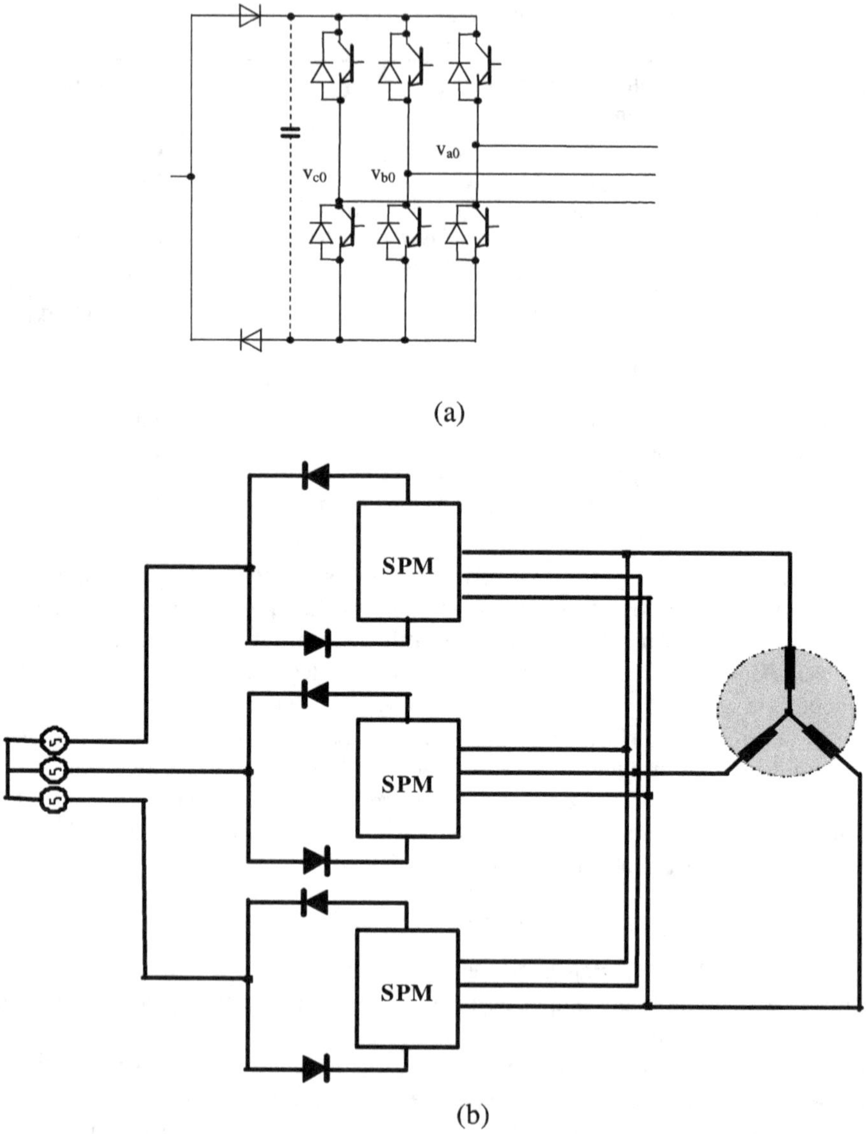

Figure 9.17 *(a) Realization of each Current Source Inverter with conventional three-phase power modules. (b) Implementation of a matrix converter with SPM devices (passive or EMI filters not shown)*

9.6.3 Multilevel converter with SPM devices

Multilevel converters are seeing increasing attention in the academic environment, mostly because of the challenges related to the definition of the PWM algorithms.

Thus, these can make up for an excellent student project. The original use of multilevel converters came from the need to provide electronic solutions to medium and high-voltage applications where power semiconductor devices were not available. Multilevel converters allow the use of devices rated at lower voltages within applications where the DC bus voltage is higher.

For instance, grid-level converters can be implemented with low-voltage integrated circuit technologies (under 40 V) [41]. SPM devices can be further used herein to build complex power converter structures able to reduce the count of passive components. The first solution is shown in Figure 9.18 [42].

A second approach for using SPM in multilevel converters has been the subject of a PhD thesis at the University of Bologna [43]. The base schemes are shown in Figure 9.19 for series or parallel connection of SPM.

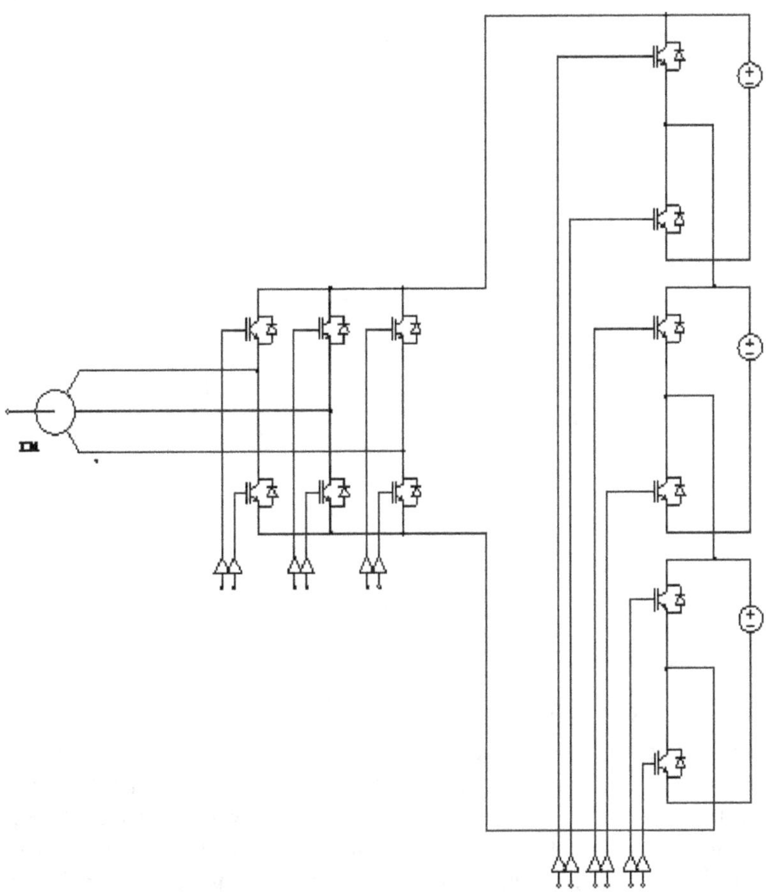

Figure 9.18 Novel multilevel topology built of Fuji IPM modules [34]

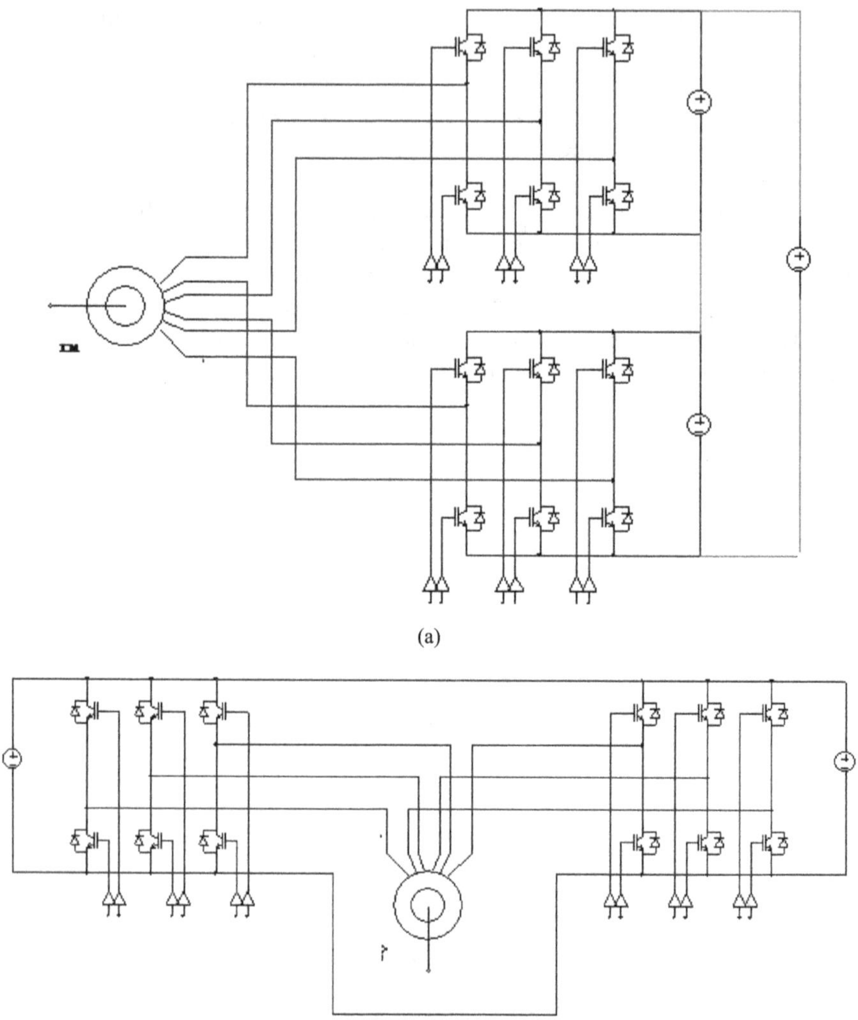

(a)

(b)

Figure 9.19 (a) Series and (b) parallel dual-SPM-based multilevel converters

9.6.4 A different direct converter with SPM devices

Modern reliability concerns have encouraged the study of new converter topologies. Among them, the direct conversion from an AC source to an AC load gains the most interest since it does not need the cumbersome DC capacitor bank.

Despite the important academic interest in the 9-switch AC/AC matrix converters, they are facing several major problems: the unconventional converter structure, the somewhat limited output voltage, and the need for additional external components for clamping the reactive energy.

An alternative to the conventional matrix converters was first proposed in 1976 [44] and it did not get too much attention for some difficulties of implementation at the time. However, the control challenges were solved many years later with the advent of microcontrollers [3,46–48], while the hardware implementation became attractive after the success of SPM [49].

Operation of this converter is derived from the backward operation of a Current Source Rectifier as shown in Figure 9.20. Conventionally, the operation of this power converter assumes a large inductance on the DC side. The control system is sensing the grid voltage through a low-power diode rectifier. The phase currents are also sensed for closed-loop control. The PWM controller follows the 120° program which is based on the Space Vector Modulation principle outlined within Figure 9.21.

Designing the PWM algorithm can be made using the current space vectors corresponding to the input phase currents or based on voltage vectors with constant magnitude equaling the DC side voltage. This allows us to convert the conventional current vector control onto a voltage vector control.

The operation of a Current Source Rectifier requires only one high-side switch and one low-side switch must be turned-on at a time. This yields nine possible combinations for the "ON"- switches. Each pair of two switches turned-on determines a specific state of the converter. A space vector can be associated with each such state to represent the system of line input currents in the complex plane. During such a conduction state, the output rectified voltage coincides with any of the line-to-line voltages or can be zero. Only seven distinct positions of the space vector can be obtained: I_1 - I_6 and zero I_0, and they are shown in Figure 9.21.

A desired position of the current space vector I is always placed between two neighboring vectors I_a and I_b, a,b = 1...6 which represent the two active states considered within the switching process. The sampling interval is completed with a zero state that can be obtained by turning-on the switches on the same leg so that

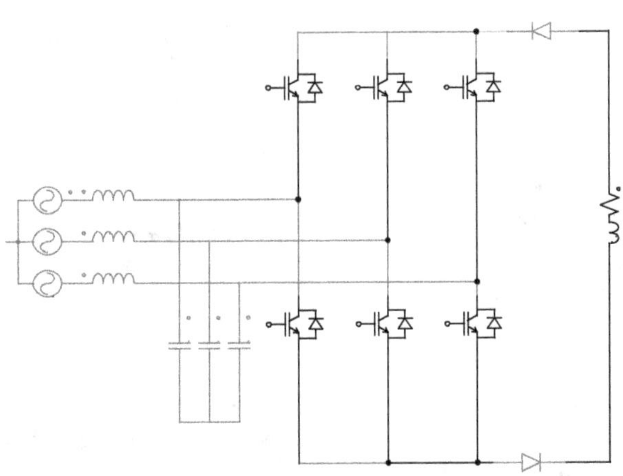

Figure 9.20 Implementation of a grid-side Current Source Converter with SPM

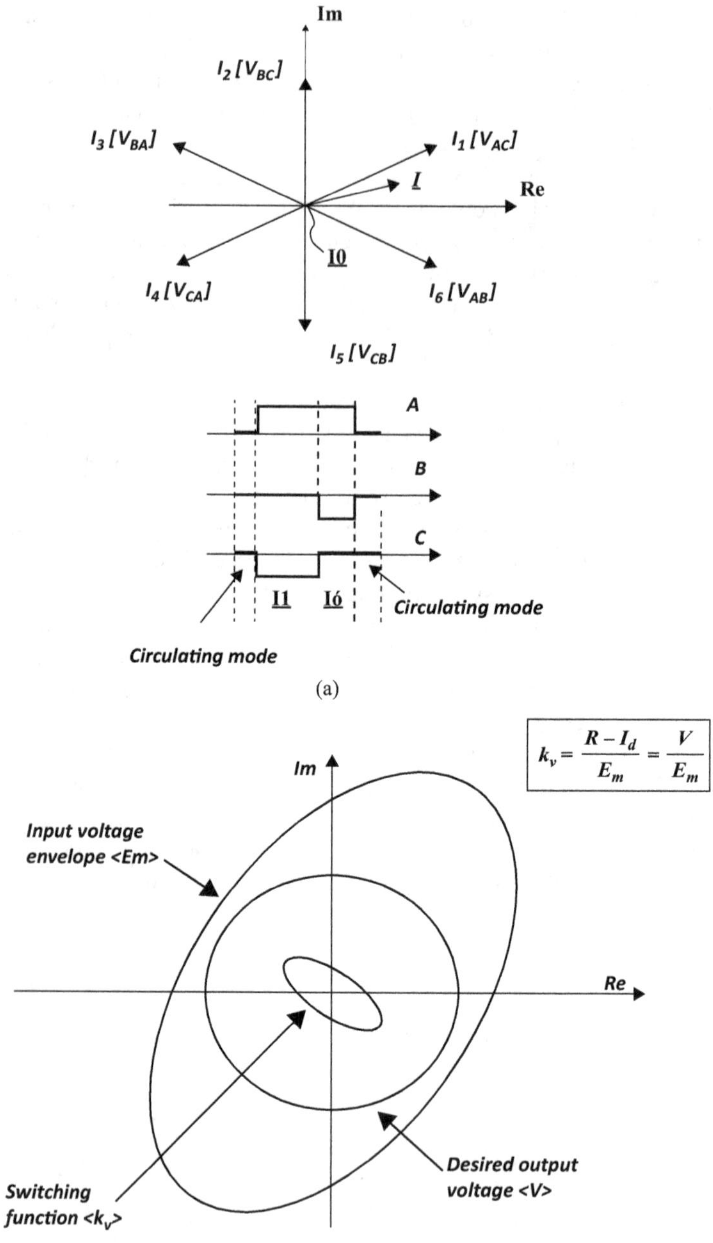

Figure 9.21 Space Vector Modulation for a Current Source Converter: (a) relationship between vectors and pulses, (b) trajectory in the complex plane

there is always a current path for the output inductive current. Writing the appropriate average relationship yields:

$$\bar{I}_a \cdot t_a + \bar{I}_b \cdot t_b + \bar{I}_0 \cdot t_0 = \bar{I} \cdot T \tag{9.4}$$

where T is the sampling interval; t_a is the time assigned for the state I_a; t_b is the time assigned for the state I_b; t_0 is the time assigned for the state I_0.

Similar reasoning is possible using voltage vectors from Figure 9.21, where each voltage vector is defined with a line-to-line combination from the input voltages. Designing the PWM circuitry based on voltage vectors allows for neglect of the load character and alleviates the need for a large inductance on the DC side. Moreover, there is no need for a closed-loop control of the output voltage. A PLL loop is used as a frequency multiplier locked to the grid frequency in order to produce the desired sampling frequency. The influence of the supply frequency variations can be thus reduced.

Observing the Space Vector Modulation algorithm allows an AC voltage reference for the voltage on the conventional "DC side". A direct AC/AC converter can thus be derived.

A second advantage derived from the application of the Space Vector Modulation algorithm consists of the opportunity to compensate for voltage harmonics or other grid-side distortion. Considering E_m the envelope of the rectified input voltage [50] and V the desired voltage vector at the output of the converter, the space vector theory can be re-written in voltage terms:

$$\begin{aligned} t_a &= T \cdot k_v \cdot \sin(60 - \alpha) \\ t_b &= T \cdot k_v \cdot \sin(\alpha) \\ t_0 &= T - t_a - t_b \\ k_v &= \frac{R \cdot I_d}{E_m} = \frac{V}{E_m} \end{aligned} \tag{9.5}$$

where V is one of the desired references for the three phases (v_{1d}, v_{2d}, v_{3d}).

Any unbalance of the grid voltage is reflected in the real-time envelope of the grid voltage E_m. The above equations allow for the proper adjustment of the pulse widths to take into account any distortion or unbalance within this envelope. This method has the drawback of limiting the maximum available voltage to the minimum value of E_m. Using the conventional Current Source Rectifier as a voltage AC/AC direct converter faces the same drawbacks as any Current Source Converter, that is not yet available as a power module product.

Figure 9.23 expands this concept to a three-phase system and illustrates the implementation with SPM. The PWM algorithms previously defined for Current Source Rectifiers cannot work for the hardware solution of Figure 9.23 since the previous solutions assume shorting the DC bus during the zero-states. Such operation is prevented by the internal operation of the SPM and a short dead time is generally introduced by such module to prevent shoot-through. Additional requirements for the bootstrap power supply should be met, that is frequent enough low-side switch-mode operation.

A new PWM algorithm is herein considered to use opposite active vectors during the zero states in order to avoid shoot-through and to comply with the usage of SPM modules. This is shown in Figure 9.22. Instead of producing a shoot-through during the zero states, the new algorithm uses two opposite vectors to compensate each other within the average vector equation used for Space Vector Modulation generation. The vector applied during the first zero state is selected to be the same as the closest active vector, and the vector applied during the second zero state is selected as the opposite of the vector applied during the first zero state.

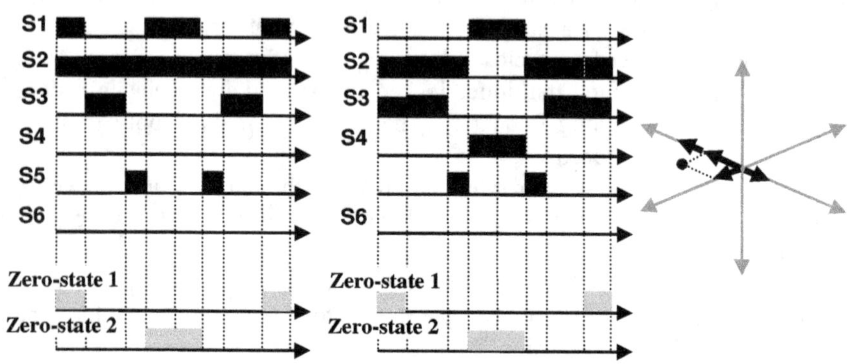

Figure 9.22 (a) Conventional CSI algorithm, (b) Adaptation for Smart Power Modules, (c) Vectorial representation for the new solution shown in (b)

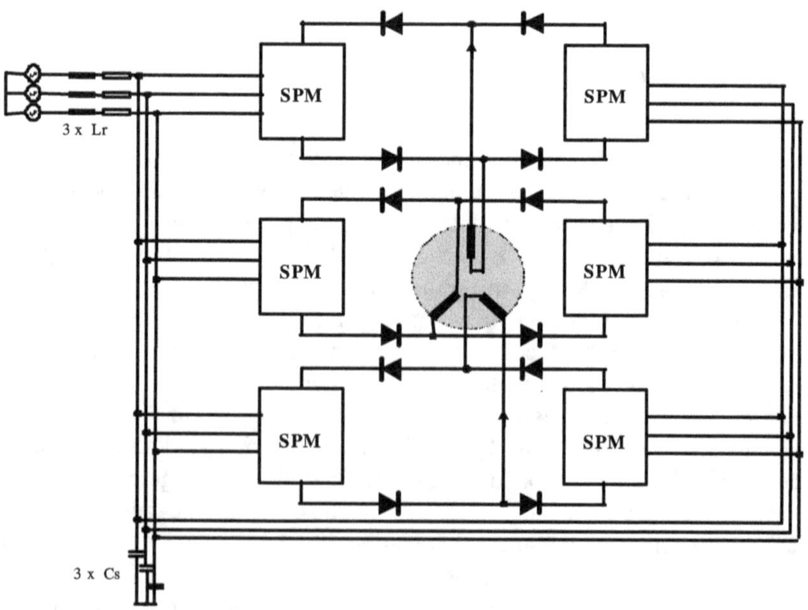

Figure 9.23 Implementation with SPM

It is important to understand that this converter behaves as a voltage source on the output side. Hence, the overall system represents a three-phase voltage source compatible with the conventional back-to-back converter. This novel converter is able to provide $\sqrt{3}$ times more load-side phase voltage than a conventional matrix converter without any special over-modulation algorithm, and with considerably fewer passive components. On the grid side (or input side), each phase current is a summation of six currents, yielding a high content in fundamentals. The grid-side currents can be improved with an interleaved carrier placement of the modules contributing to each phase current.

Harmonic results are shown in Figure 9.24 and they are superior to the conventional back-back converter when switched at the same frequency. Among the multiple coefficients available for harmonic characterization of power converters, we have chosen herein the Total Harmonic Distortion (THD) Coefficient:

$$\mathrm{THD}(\%) = \frac{100}{V_{(1)}}\sqrt{\sum\nolimits_{n=2}^{\infty}\left[V_{(n)}\right]^2} \tag{9.6}$$

Since the grid-side highly inductive impedance is basically a low pass filter (LPF) for currents, the higher-order current harmonics will be attenuated. The remaining spectrum of the current will be different from one PWM method to another, and from one switching frequency to another. A coefficient regarding current harmonics would better define the performance of a PWM method. Such a coefficient is called Harmonic Current Factor (HCF), and it can be expressed also based on the voltage harmonics at the converter output:

$$\begin{aligned}\mathrm{HCF}(\%) &= \frac{100}{I_{(1)}}\sqrt{\sum\nolimits_{n=5}^{\infty}\left[I_{(n)}\right]^2} = \frac{100}{\left(\frac{V_{(1)}}{\omega L}\right)}\sqrt{\sum\nolimits_{n=5}^{\infty}\left[\frac{V_{(n)}}{n\omega L}\right]^2}\\ &= \frac{100}{V_{(1)}}\sqrt{\sum\nolimits_{n=5}^{\infty}\left[\frac{V_{(n)}}{n}\right]^2}\end{aligned} \tag{9.7}$$

This analysis shows that a converter system comprised of six power modules IPM/SPM/IRAMS setup as shown in Figure 9.23 can drive a 2 kW motor, in similar operation and thermal conditions as a back-to-back dual SPM module would drive a 1 HP = 745 W motor.

The power density is also competitive as the entire power stage (with straight-pin mounting, six individual heat sinks, passive LC filtering, and power connectors) accounts for (2 in × 2 in × 7.5 in =) 0.49 l for 2 kW delivered power (that is 4.1 kW/l).

This should compare to contemporary industry goals of 4 kW/l, for this class of converters [3] for low kW-range, required to switch below 10 kHz due to motor requirements.

The advantages over the conventional back-back solution consist of:

- Longer lifetime, improved reliability;
- Considerable smaller package, improved power density;
- Fully standard compatible for both grid and EMI.

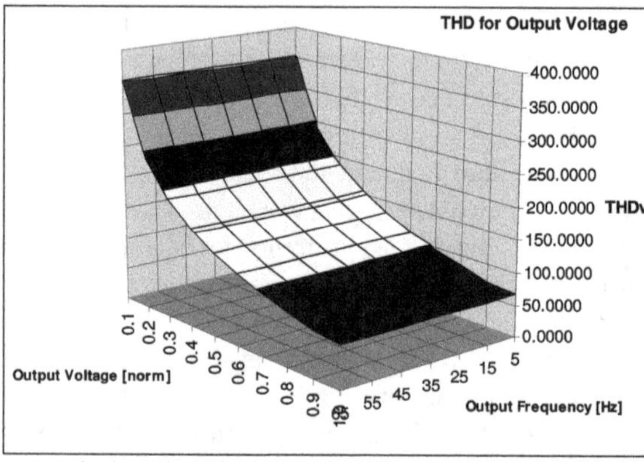

V[RMS]		17.9	35.9	54.0	72.2	90.4	108.6	126.2	145.4	162.2	171.6
m		0.1	0.2	0.3	0.4	0.5	0.6	0.7	0.8	0.9	1.0

(a)

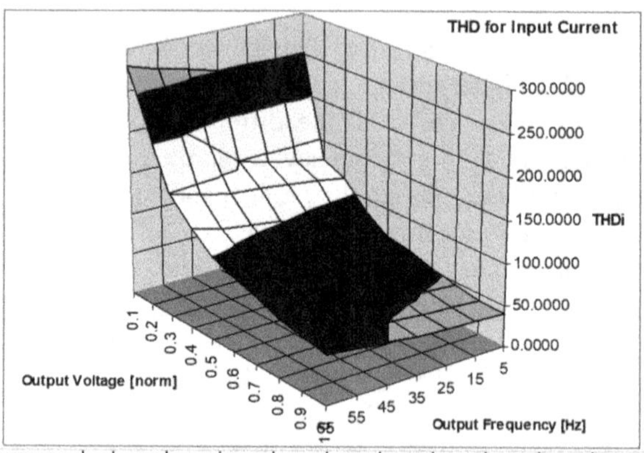

V[RMS]		17.9	35.9	54.0	72.2	90.4	108.6	126.2	145.4	162.2	171.6
m		0.1	0.2	0.3	0.4	0.5	0.6	0.7	0.8	0.9	1.0

(b)

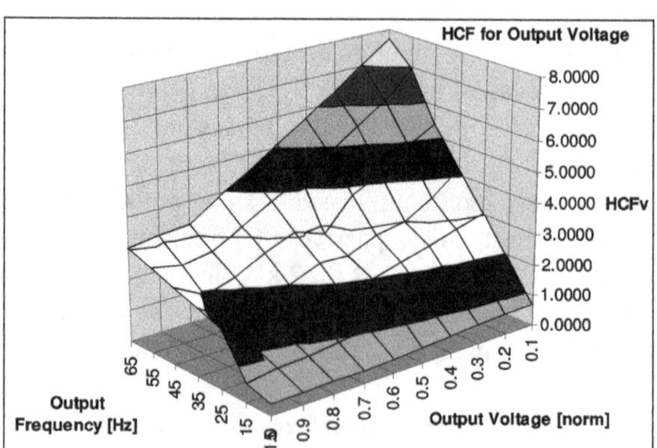

V[RMS]		17.9	35.9	54.0	72.2	90.4	108.6	126.2	145.4	162.2	171.6
m		0.1	0.2	0.3	0.4	0.5	0.6	0.7	0.8	0.9	1.0

(c)

Figure 9.24 Performance evaluated with harmonic results

The drawbacks are:

- Slightly higher cost of the system with current costs;
- Limited choice of power ratings available for power modules [15–18].

9.7 Conclusion

The advent of power semiconductor technology allowed the integration of semiconductor power devices and their control integrated circuits within the same hybrid device. This chapter has briefly reviewed the advantages of this integration, the technology options, and provided a historical perspective on the topic.

Finally, the conventional usage of SPM has been extended with a new class of multi-module power converters based on a network of switches implemented with multiple SPMs and ready to provide even better performance.

References

[1] Neacşu D.O. 'Current Advanced Power Conversion Concepts with Multiple Power Modules', Habilitation Thesis, Romanian Ministry of Education, May 2012.

[2] Tolbert L.M. *Smart Integrated Power Module*, 2012 U.S. DOE Hydrogen and Fuel Cells Program and Vehicle Technologies Program Annual Merit Review and Peer Evaluation Meeting, Project ID: APE046, May 15, 2012.

[3] Staunton RH, Ozpineci B, Theiss TJ, and Tolbert LM, 'Review of the State-of-the-Art in Power Electronics suitable for 10kW Military Power Systems', *ORNL/TM-2003/209 Annual Report*, 2003.

[4] Trzynadlowski A. *Introduction to Modern Power Electronics*. New York: Wiley, 2010.

[5] Wintrich, A., Nicolai, U., Tursky, W., and Reimann, T., 'Application Manual – Power Semiconductors', *SEMIKRON Internal Documentation*, 2011.

[6] Neacşu, D.O., 'Principle of a Novel Component Minimized Active Power Filter for High-Power Magnet Supplies', *IEEE Industrial Electronics Conference IECON*, Montreal, Canada, 2012.

[7] Fujihira T., Miyasaka T., and Ikawa O., 'Power Semiconductors: Current Status and Future Outlook', *Fuji Electric Review*, vol. 65, no. 4, 2019, pp. 202–209.

[8] Onishi Y., Miyasaka T., and Ikawa O., 'Power Semiconductors: Current Status and Future Outlook', *Fuji Electric Review*, vol. 66, no. 4, 2020, pp. 191–197.

[9] Onishi Y., Miyasaka T., and Ikawa O., 'Power Semiconductors Contributing to Vehicle Electrification and Energy Management', *Fuji Electric Review*, vol. 67, no. 4, 2021, pp. 224–231.

[10] Y. Yang, L. Dorn-Gomba, R. Rodriguez, C. Mak, and A. Emadi, 'Automotive Power Module Packaging: Current Status and Future Trends', *IEEE Access*, vol. 8, 2020, pp. 160126–160144.

[11] Cyril Buttay, 'Review of Intelligent Power Modules', *IWIPP 2021 Webinar*.

[12] David Manners, 'Power Module Packaging Market on a 12.5% CAGR 2020–6', *Electronics Weekly*, Nov. 2021, (https://www.electronicsweekly.com/news/business/power-module-packaging-market-12-5-cagr-2020-6-2021-11/)

[13] Anon., 'Global Intelligent Power Module Market Report 2022–2027: Increasing Demand for Intelligent Power Modules in Consumer Electronics and Industrial Verticals to Drive Growth', Research and Markets (https://www.globenewswire.com/en/news-release/2022/07/27/2486765/28124/en/Global-Intelligent-Power-Module-Market-Report-2022-2027-Increasing-Demand-for-Intelligent-Power-Modules-in-Consumer-Electronics-and-Industrial-Verticals-to-Drive-Growth.html), 2022.

[14] Yamaguchi K., "Design and evaluation of SiC-based High Power Density Inverter, 70 kW/liter, 50 kW/kg," *IEEE Applied Power Electronics Conference and Exposition (APEC)*, 2016, pp. 3075–3079.

[15] Motto E. *Application Specific Intelligent Power Modules – A Novel Approach to System Integration in Low Power Drives*, PCIM Conference, 1998.

[16] Bhalerao P., and Wiatr R. 'New Intelligent Power Module Series', *PCIM Europe Journal,* vol. 15, no. 2, 2008, pp. 21–22.

[17] Lee F.C., 'Evolution of Power Electronics Technologies – A CPES's Perspective'. Presented at *IEEE 21st International Symposium on Industrial Electronics*, Hangzhou, China, 2012.

[18] Neeb C., Boettcher L., Conrad M., and DeDoncker R.W., 'Innovative and Reliable Power Modules: A Future Trend and Evolution of Technologies'. *IEEE Industrial Electronics Magazine,* vol. 8, no. 3, 2014, pp. 6–16.

[19] Lutz J., Schlangenotto H., Scheuermann U., and DeDoncker R. *Semiconductor Power Devices*. Berlin: Springer; 2011.

[20] Hosen, T., and Yanagisawa, K., 'Fuji Electric's Semiconductors: Current Status and Future Outlook', *Fuji Electric Review*, vol. 57, no. 3, pp. 68–71, 2011.

[21] Gorgerino A, 'New-Generation IGBTs Increase Motor Drive Efficiency in Appliances', *ECN Magazine*, Jul. 2012.

[22] Anon., https://meus-semiconductors.com/products/power-devices/pmf75cga120 (accessed November 2022).

[23] Anon., https://www.infineon.com/cms/en/product/power/intelligent-power-modules-ipm/cipos-maxi/ (accessed November 2022).

[24] Anon., https://www.businesswire.com/news/home/20220517005327/en/Sensitron-and-EPC-Collaborate-to-Introduce-a-High-Power-Density-350-V-Gallium-Nitride-GaN-Half-Bridge-Intelligent-Power-Module-IPM-That-is-60-Smaller-Than-Comparable-Silicon-Solutions-and-Lower-Cost (accessed November 2022).

[25] Motto E.R., and Donlon J., 'Speed Shifting Gate Drive for Intelligent Power Modules'. *IEEE APEC Conference*, Dallas, TX, USA, 2006.

[26] Neacşu, D.O., and Nguyen H.H. 'Gate Drive Circuit with Feedback-Controlled Active Resistance', US Patent #6,459,324, 2002.

[27] Anon., AN2123 Application Note, 'TD351 Advanced IGBT Driver – Principles of Operation and Application', *ST Microelectronics Application Note*, 2005.

[28] Fujihira T., Miyasaka T., and Ikawa O., 'Power Semiconductors: Current Status and Future Outlook', *Fuji Electric Review*, vol. 64, no. 4, 2018, pp. 168–175.

[29] Igarashi, S., Kakiki, H., Nishimura, Y., and Goto, T., 'Design of High Reliability Packaging for Fuji High Power Module', *International Conference on Electrical Machines and Systems (ICEMS)*, 2011, pp. 1–6.

[30] Baker, N., Liserre, M., Dupont, L., and Avenas, Y., 'Improved Reliability of Power Modules: A Review of Online Junction Temperature Measurement Methods', *IEEE Industrial Electronics Magazine*, vol. 8, no. 3, 2014 , pp. 17–27.

[31] Neacşu, D.O., Wagner, E., and Borowy, B., 'A Simulation Benchmark for Selection of the PWM Algorithms for Three-Phase Interleaved Converters', *IEEE Transactions on Industrial Electronics*, 2008, vol. 55, pp. 1628–1636

[32] P. Wood, M. Battello, N. Keskar, and A. Guerra., IPM Application Overview – Integrated Power Module for Appliance Motor Drives. International Rectifier AN-1044; 2002.

[33] Fairchild Semiconductor. FSBS10CH60-*Smart Power Module.* Datasheet; 2005.

[34] International Rectifier. *IRAMS10UP60B-iMotion Series*, Integrated Power Hybrid IC for Appliance Motor Applications. Datasheet; 2008.

[35] ST Microelectronics. STGIPS10K60A – IGBT intelligent power module. Datasheet; 2010.

[36] Volke, A., and Hornkamp, M., IGBT Modules – Technologies, Driver, Applications, *Infineon Documentation*. Second edition. 2012.

[37] Shimizu N., Karasawa T., and Takagiwa K., 'New Lineup of Large-Capacity "V-Series" Intelligent Power Modules', *Fuji Electric Review,* vol. 58, no. 2, 2012, pp. 65–69.

[38] Neacşu D.O. *Switching Power Converters – Medium and High Power.* Boca Raton, FL: CRC Press, 2013.

[39] Friedli, T., and Kolar, J.W., 'Milestones in Matrix Converters', *IEEJ Journal of Industry Applications*, vol. 1, no. 1, 2012, pp. 12–14.

[40] Huber, L., Borojevic, D., and Burany, N., 'Analysis, Design and Implementation of the Space Vector Modulator for Forced-Commutated Cycloconverters', *IEE Proceedings Series B*, vol. 139, no. 2, 1992, pp. 103–113

[41] Neacşu, D.O., 'Fault-Tolerant Isolated Converter in Low-Voltage Technology for Automotive AC Auxiliary Power', *IEEE Industrial Electronics Conference IECON*, Vienna, Austria, Nov. 10–13, 2013, pp. 8184–8189.

[42] Su, G.J., and Adams, D., 'Multilevel DC Link Inverter for Brushless Permanent Magnet Motors with Very Low Inductance', *IEEE IAS Annual Meeting*, 2001.

[43] Lega, L., 'Multilevel Converters: Dual Two-Level Inverter Scheme', PhD Dissertation, University of Bologna, 2009.

[44] Jones, J., and Bose, B.K., 'A Frequency Step-Up Cycloconverter Using Power Transistors in Inverse Series Mode', *International Journal of Electronics*, vol. 41, no. 6, 1976, pp. 573–587.
[45] Neacşu, D.O., 'Effectiveness of AC/AC Current-Controlled Matrix Converter to Open-Winding Induction Machine Drive', *IEEE-CAS International Symposium SCS99*, Iasi, Romania, Jul. 7–9, 1999, pp. 379–382.
[46] Neacşu, D.O., 'Controlled AC/AC Voltage Source Matrix Converter for Open-Winding Induction Machine Drives', *IEEE Industrial Electronics Conf. IECON 1999*, San Jose, CA, USA, 1999.
[47] Neacşu, D.O., 'IGBT- Based "Cycloconverters" Built of Conventional Current Source Inverter Modules', *IEEE-CAS International Symposium SCS2003*, vol. 1, Iasi, Romania, 10–11, 2003, pp. 217–220
[48] Neacşu, D.O., 'Analysis and Design of IGBT-based AC/AC Direct Converters built of Conventional Current Source Inverter Module', *IEEE IAS 2004*, vol. 3, Seattle, WA, 2004, pp. 1824–1831.
[49] Neacşu, D.O., 'Towards an All-Semiconductor Power Converter Solution for the Appliance Market', *IEEE Industrial Electronics Conference IECON*, Glendale, AZ, USA, 2010.
[50] Enjeti, P., and Xie, B., 'A New Real Time Space Vector PWM Strategy for High Performance Converters', *IEEE/IAS Annual Meeting Conference Record 1992*, pp. 1018–1025.

Chapter 10
Converters in a power grid

Giri Venkataramanan[1], Fernando Mancilla-David[2] and Mahima Gupta[3]

10.1 Introduction

The electrical power system forms the essential backbone of the energy system that enables modern livelihood and supports a comfortable lifestyle for people who have access to it. It is commonly segmented into generation, transmission, distribution, and utilization systems. While electrical power converters have a role to play across all of these segments, their application in generation and utilization systems is rather specific in nature, as has been discussed in various other chapters in this book. The focus of discussion in this chapter is on their application in electrical transmission and distribution systems, which are often referred to together as the power grid. Since the power grid is largely operated as a three-phase ac system, the discussion in this chapter is mainly focused on three-phase power converters. An outline of different power converters that are used in the power grid classified into different categories is presented in Section 10.2. In the discussion, the focus is mainly placed on the circuit configuration of the constituent components, and not on their control, modulation, switching strategy, or design aspects. Such a discussion is beyond the scope of this volume and may be found in specialized texts, monographs, and references. Various specific applications of these converters in transmission and distribution systems are illustrated in Section 10.3. A brief description of the application is restricted to the functional details. Again, a detailed discussion of the operation and capabilities of the converters may be found in reference documents. The summary in the concluding section includes a brief discussion of the state of the technology and emerging trends, followed by a list of references.

[1]Department of Electrical and Computer Engineering, University of Wisconsin-Madison, USA
[2]Department of Electrical Engineering, University of Colorado Denver, USA
[3]Department of Electrical and Computer Engineering, Portland State University, USA

10.2 Power converter topologies

The feature of power converters used in the power grid that distinguishes them from industrial, residential, and commercial applications, is their power level, which may be reached up to thousands of megawatts. This large amount of power throughput places a large demand on the switching power semiconductor devices in terms of voltage and current ratings. This leads to certain topological considerations in the power converters that are often superseded at lower power applications. Various converters that fall under the common ac–dc and ac–ac categories are discussed briefly in this section, followed by a more detailed discussion of two new families of power converters that are emerging to find a unique place in power grid applications.

10.2.1 AC–DC converters

AC–DC converters often form a building block to realize an intermediate dc stage in several power grid applications. They are typically capable of bidirectional power flow and may form a current stiff (inductive) dc link or voltage stiff (capacitive) dc link, as described further in the following subsections.

10.2.1.1 Current source converters

Current source converter (CSC) technology used in power grids is relatively mature today [1]. The circuit consists of a three-phase bridge network with unidirectional current carrying and bidirectional voltage-blocking switching devices connected between each dc terminal and the ac terminal as illustrated in Figure 10.1. The switches used in the converter are silicon control rectifiers (SCRs or thyristors). Since SCRs do not have inherent turn-off capability, these converters employ the voltage reversals of the ac line to commute. Hence, they are termed line-commutated converters. The stiff current requirement on the dc side may be ensured using an explicit inductor or in some cases part of the dc side circuit

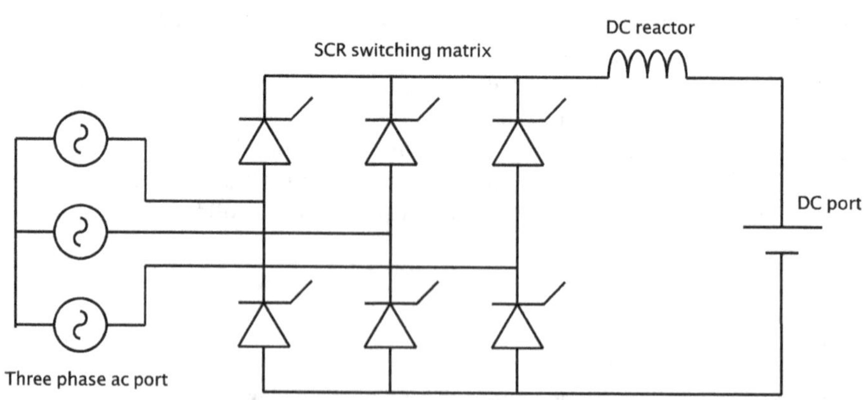

Figure 10.1 Schematic diagram of a current source converter with two dc levels

inductance. Although the figure indicates a single SCR switching device that interconnects each ac terminal with each dc terminal, a large number of series-connected devices are used to meet voltage-blocking requirements and also allow the provision of redundancy against device failures in the short circuit mode. Details of such interconnections and snubber circuits to ensure voltage distribution among series connected devices under the rubric of "valve design", the application of ac filter devices and dc filter devices are not shown in the figure. Such design details along with a detailed discussion of the operation and control of the converters may be found elsewhere. In most cases, it is common to use two power converters along with an appropriately designed wye-delta transformer to get phase multiplication on the ac side, and series connection on the dc side as shown in Figure 10.2.

While the technology of the line-commutated CSCs is mature, the lack of turnoff controllability of the conventional thyristor results in (i) poor power factor and considerable waveform distortion; (ii) need for a relatively stiff ac voltage supply; and (iii) inability to provide independent control of the active and reactive powers, and often superseded by converters that use gate-turn-off devices [2]. Although it is possible to realize CSCs with gate-turn-off devices, voltage source converters (VSCs) using gate-turn-off devices discussed in the following subsection have enjoyed a more dominant development due to trends in power semiconductor availability and various other application engineering issues [3].

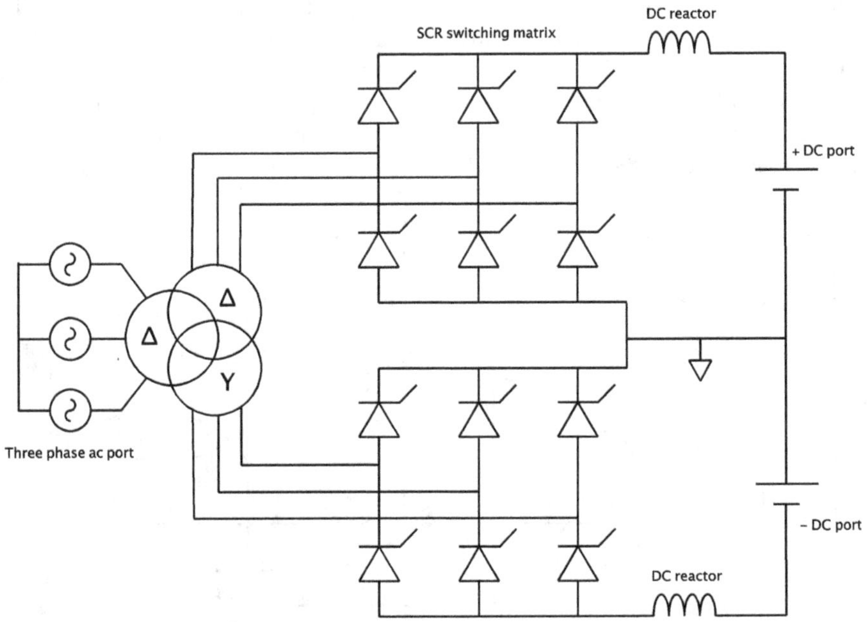

Figure 10.2 Schematic diagram of two current source converter series connected to form a bipolar or three-level dc system

10.2.1.2 Voltage source converters

The technology of VSC that employs a stiff voltage on the dc side realized using a capacitor bank is quite mature, and enjoys a broad and growing installation base for applications in motor drives, uninterruptable power systems, and wind and solar power systems. Figure 10.3 illustrates the power circuit of a two-level dc–three phase/ac VSC [4]. While the bridge circuit topology of the VSC is similar to that of the CSC, the switching elements in the VSC have unidirectional voltage blocking and bidirectional current carrying capability. With the advent of high-power/high-voltage switching devices such as the integrated gate-commutated thyristor (IGCT) and insulated-gate bipolar transistor (IGBT), the realization of VSC for application in the power grid has become practical.

Due to considerations of waveform quality, symmetry, and semiconductor switch utilization, often a (diode clamped) three-level converter topology (illustrated in Figure 10.4), which is also called a neutral-point clamped converter is used in power grid applications [5].

10.2.2 AC phase controllers

AC phase-controlled devices correspond to the first generation of solid-state power electronic devices built using SCRs [6,7]. While they are most commonly known for lamp dimming in domestic applications, in power grids, they are applied to realize an electronically regulated inductance, often called the thyristor-controlled reactor (TCR), whose power circuit is illustrated in Figure 10.5. The three-phase TCR consists of three reactors, each of them connected in series with two SCRs that are connected antiparallel, to allow current flow in both directions, as well as voltage standoff in both directions. The firing angle of each SCR, in relation to the zero crosses of the line voltage, determines the current flowing through the SCR and hence provides continuous control of fundamental reactance all the way up to infinity (when the SCRs are not conducting at all), down to the full value of the

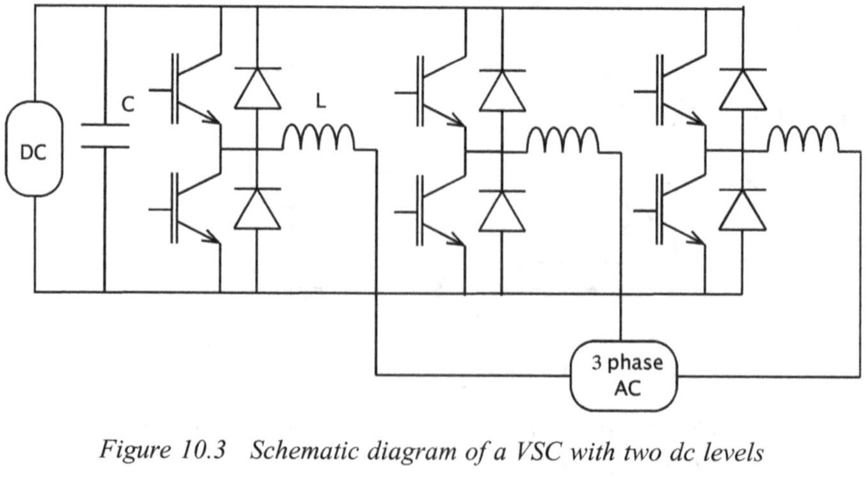

Figure 10.3 Schematic diagram of a VSC with two dc levels

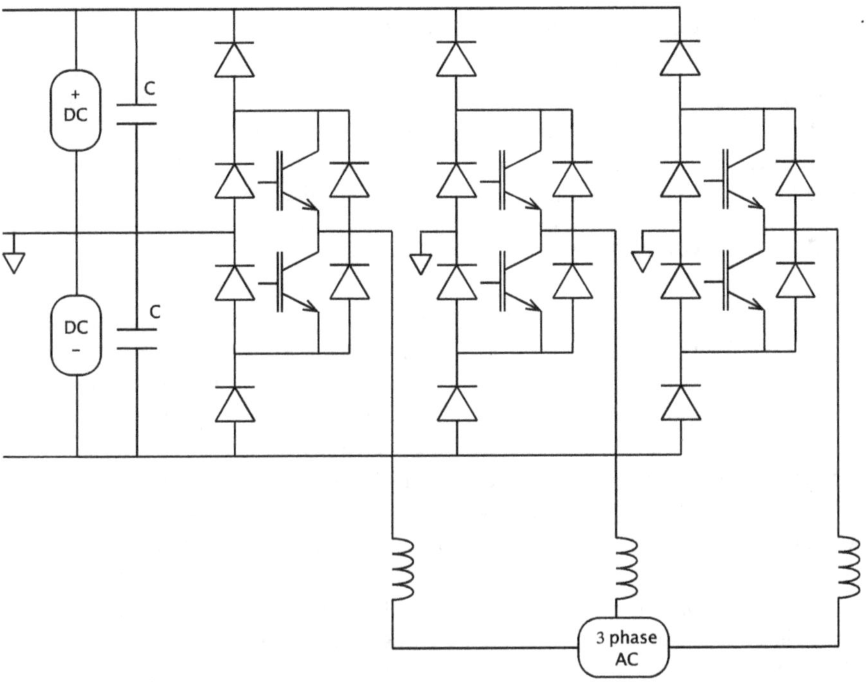

Figure 10.4 Schematic diagram of a VSC with three dc levels

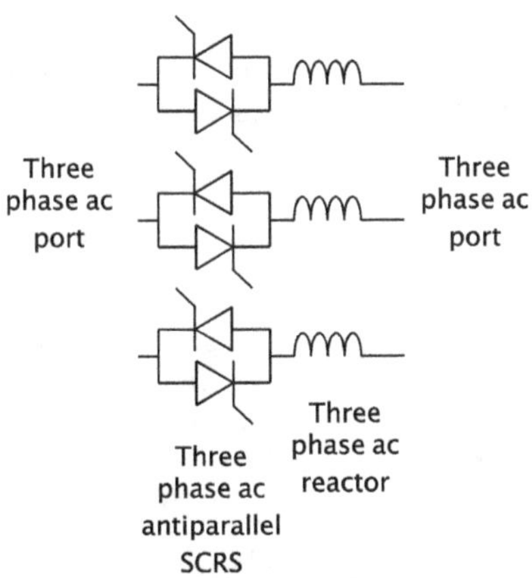

Figure 10.5 Schematic diagram of a thyristor-controlled reactor

reactance (when the SCRs are conducting completely). While a variable reactance by itself may not appear to be of such a high value in power grids, they are interconnected appropriately with capacitors to realize various applications, which will be described in Section 10.3.

10.2.3 AC pulse width modulated controllers

The principles of operation of three-phase ac pulse width modulated (PWM) converters have been described in detail [8,9]. Although their realization and applications for ac power flow control can take various forms, herein the focus is on the main features that elucidate their operating features. Since their realizations are not widely known, the operation is developed from primitive principles.

10.2.3.1 Converter circuit

A simplified schematic diagram of an N-throw single-pole three-phase ac PWM converter is illustrated in Figure 10.6. The system synthesizes an adjustable three-phase pole voltage ($V_{P(A-C)}$) by switching among N stiff three-phase voltage sources ($V_{T1(A-C)}, VT_{2(A-C)}, \ldots, V_{TN(A-C)}$). It is important to notice that in ac systems, the designation of the stiff voltage and stiff currents is arbitrary. Stiff voltage sources and stiff current sources can be transformed either by adding series inductors or shunt capacitors, respectively, to fit application considerations [10].

The throws of the switches as well as the sources are assumed ideal for the discussion herein, as is common in preliminary functional analysis of switching power converters. Furthermore, the nominal frequency of the three-phase voltages and currents (power frequency) are assumed identical, in the absence of which there could be no net steady-state power transfer among them.

As indicated in Figure 10.6, the three poles are ganged together, i.e., they switch concurrently between the throws to which they are connected. This can be mathematically expressed as $t_{iA} = t_{iB} = t_{iC} = t_i$, for $i = 1, 2, \ldots, N$.

10.2.3.2 Converter model

The transfer properties can be represented using the vector equations consisting of three components for each of the terminal quantities. In this case, the vectors representing the

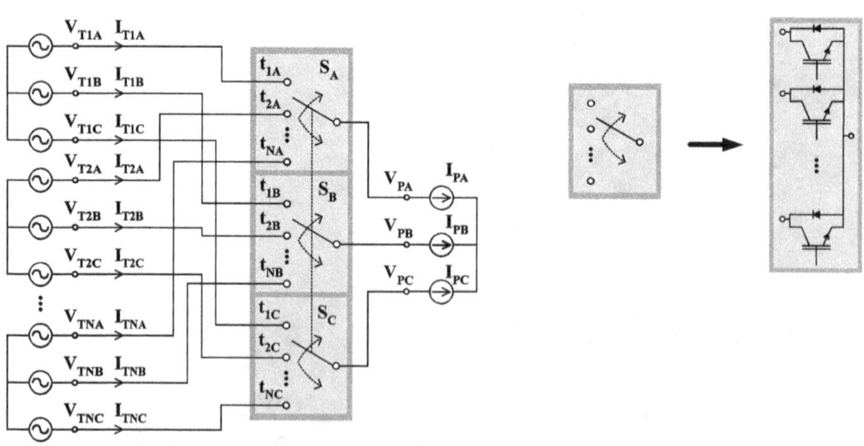

Figure 10.6 Schematic diagram of a three-phase PWM ac converter synthesizing an adjustable pole voltage obtained from N throws

throw voltages, throw currents, pole voltages, and pole currents become $V_{Ti} = \begin{bmatrix} V_{TiA} & V_{TiB} & V_{TiC} \end{bmatrix}^T$, $I_{Ti} = \begin{bmatrix} I_{TiA} & I_{TiB} & I_{TiC} \end{bmatrix}^T$, $V_P = \begin{bmatrix} V_{PA} & V_{PB} & V_{PC} \end{bmatrix}^T$, $I_P = \begin{bmatrix} I_{PA} & I_{PB} & I_{PC} \end{bmatrix}^T$, respectively, for $i = 1, 2, \ldots, N$. Thus, the pole voltage and throw currents can be expressed as

$$V_P(t) = \sum_{t=1}^{N} H_t . V_{St} \tag{10.1}$$

$$I_{Tt}(t) = H_t . I_P \tag{10.2}$$

where

$$H_t(t) = \begin{cases} 1 & \text{if } t_t \text{ is closed,} \\ 0 & \text{otherwise.} \end{cases}$$

is the switching function of a throw connecting voltage VTi to the current IP. When the repetition frequency of the switching function (or simply the switching frequency) is large with respect to the power frequency, net power transfer between the voltage ports and the current ports arises from the average value (dc component) of the switching functions. The dc component of the switching functions may be readily represented by the duty ratio of the particular throw. The transfer relationships (10.1) and (10.2) may be simplified as

$$V_P(t) = \sum_{t=1}^{N} d_t(t) . V_{Tt} \tag{10.3}$$

$$I_{Tt}(t) = d_t(t) . I_P \tag{10.4}$$

where the duty ratio of the ith throw is defined as

$$d_t(\tau) = \frac{1}{T} \int_{\tau}^{\tau+T} H_t(t) . \mathrm{d}t \tag{10.5}$$

and T is the switching period. As may be evident from (10.5),

$$\sum_{t=1}^{N} d_t(t) = 1 \tag{10.6}$$

Since the operating principle of these converters is based on controlling the connectivity between various three-phase ac voltage and/or current vectors by switching among their components concurrently, these converters have been termed vector switching converters (VeSCs). The realization of each of the multiple-throw single-pole switches using real semiconductors is shown in Figure 10.6. Each IGBT–diode pair may be replaced by an IGCT–diode or any other device with similar voltage blocking and current conducting capabilities with control in one direction [11].

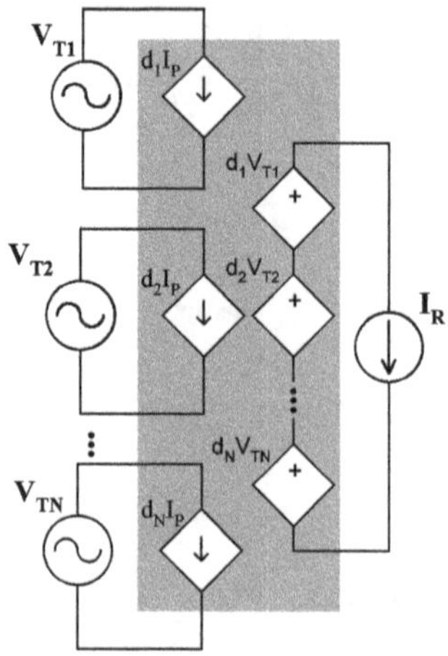

Figure 10.7 PWM ac VeSC equivalent circuit shown using duty ratio controlled coupled current and voltage sources

10.2.3.3 Equivalent circuit

Relationships (10.3) and (10.4) indicate a reciprocal input–output transfer property similar to that of a transformer. Therefore, the fundamental component averaged vector (or single phase) equivalent circuit of the converter system may be represented as shown in Figure 10.7, which describes the input–output reciprocal relationships explicitly as dependent sources. As is evident from Figure 10.7, the net pole voltage, V_R, depends on the value (magnitude and phase) of the throw voltages as well as the corresponding duty ratios. This feature becomes the defining element for the application of these converters in the power grid as will be described in a later section in this chapter.

10.2.4 Modular multilevel converter

The most recent addition to the family of power converters, which has today found a niche in power grid applications, is the modular multilevel converter (MMC) described in this subsection. Several topologies of power converters that use this approach have been presented and demonstrated in recent years. These include three-phase/ac–three-phase/ac matrix converters, single-phase/ac–three-phase/ac dual bridge converters, and dc-three-phase/ac inverters/rectifiers [12–16]. Due to the relative novelty of these converters, the topological structure is developed here from the first principles, starting from the description of a particular sub-module (SM) that constitutes the building block of this converter.

10.2.4.1 SM structure

A simplified schematic diagram of an SM in full-bridge configuration is shown in Figure 10.8(a). Bulk energy storage (CS) in the capacitor is typically in the form of dc, thereby providing an appropriate bias voltage or current for the switching

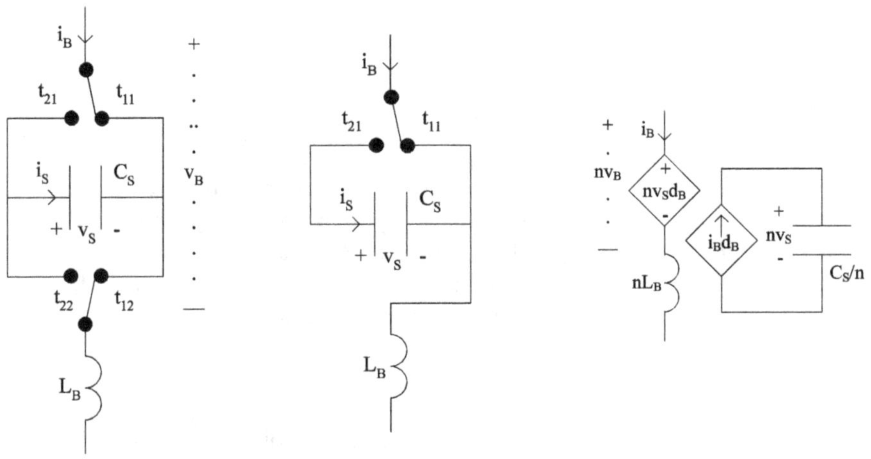

Figure 10.8 Schematic diagram of the introduction of sub-modules in MMC as: (a) functional diagram of symmetrical full-bridge configuration, (b) functional diagram of single-ended half-bridge configuration, and (c) averaged model of n series connected sub-modules

devices that constitute the bridge circuit. The presence of L_B indicates the stiffness of the bridge terminal current, which may be provided by an incidental amount of inductance in series with the bridge, with a minimal amount of energy storage in it. The voltage and current at the terminals and the dc capacitor along with the constituent switch throws of the SM may be defined as illustrated in Figure 10.8(a).

The realization of the throws of the SM depends on the polarity of terminal currents and the polarity of the bias voltage. The assumption of unipolar bias voltage v_S and bidirectional terminal current i_B leads to a classical H-bridge consisting of gate-turn-off semiconductors such as IGBTs with anti-parallel diodes. The full-bridge topology enables bipolar values for v_B and i_B through appropriate control of switch states. On the other hand, a reduced switch topological structure using a single-ended half-bridge may be realized as illustrated in Figure 10.8(b), allowing bidirectional values of i_B, and unidirectional values of v_B. The throws constituting the SM may be typically operated under repetitive switching in a duty ratio controlled mode, or generate an ensemble waveform to follow a command. The duty ratio of the throws t_{xy} is denoted as d_{xy}, where x and y represent the throw and pole designations of the switches in the bridge as illustrated. Such a primitive description of the operation of this topology enables the analysis of the power transfer mechanism and convenient application of modeling tools.

10.2.4.2 Averaged model

For sufficiently large switching frequencies and energy storage elements, the terminally stiff electrical variables in switching power converters may be described using their average values. The average transfer properties of the SM may be represented as

$$\begin{bmatrix} v_B \\ i_S \end{bmatrix} = \begin{bmatrix} d_B & 0 \\ 0 & d_B \end{bmatrix} \begin{bmatrix} v_S \\ i_B \end{bmatrix} \tag{10.7}$$

where $d_B = d_{21} - d_{22}$ for the full bridge and $d_B = d_{21}$ for the half bridge. As may be observed from (10.7), the energy storage capacitor current i_S and the SM terminal voltage v_B feature a reciprocity relationship with respect to the SM terminal current i_B and energy storage capacitor voltage v_S, determined by the "bridge duty ratio" d_B, where the bridge duty ratio d_B may take a value between -1 and $+1$ for the full bridge and a value between 0 and $+1$ for the single-ended half bridge.

10.2.4.3 String of SMs

An arbitrary number (n) of SMs either in the full-bridge or in the single-ended half-bridge configuration may be connected in a series string. Typically, these bridges would consist of identical modules for the sake of simplicity and convenience in packaging, design, control, and realization. Since they are connected in a series string, they carry the same current and under most circumstances, their dc bias voltage level will be nominally identical if their respective bridge duty ratios are also identical. Although the duty ratio among these modules may be identical, the switching signals may have phase shifts between them, thereby providing one of the benefits of multilevel power conversion. Furthermore, imperfections due to parasitic effects may be compensated through second-order perturbations in the control signals. In total, a series connected string of n SMs may be represented as a simplified equivalent circuit as illustrated in Figure 10.8(c).

10.2.4.4 Topology example

Two systems of power sources I and O (and/or loads, e.g., a dc voltage, three-phase ac system, etc.) may be said to be interconnected in a symmetrical bridge configuration if each terminal of System I is connected to each terminal of System O through an identical electrical branch. If each of the interconnection branches itself is a series connected string of SMs, an MMC or bridge-of-bridge (BoB) converter network is formed. BoB converter networks may be configured and operated in a manner to realize particular power conversion functions between different systems of power sources [12]. A dc to single-phase ac power converter using the BoB approach, along with an averaged equivalent circuit is illustrated in Figure 10.9 and Figure 10.10 respectively. The dc voltage and the ac voltage at the input and output terminals are defined with respect to their neutral point voltages in Figure 10.9, for the sake of notional convenience. The duty ratio of the different bridges is modulated to follow the relationship

$$D_{Bij} = \frac{V_{Ii} - V_{Oj}}{nV_S} \tag{10.8}$$

where the indices i and j correspond to the location of voltages and the bridge position. Through appropriate duty ratio control of current and voltage using a

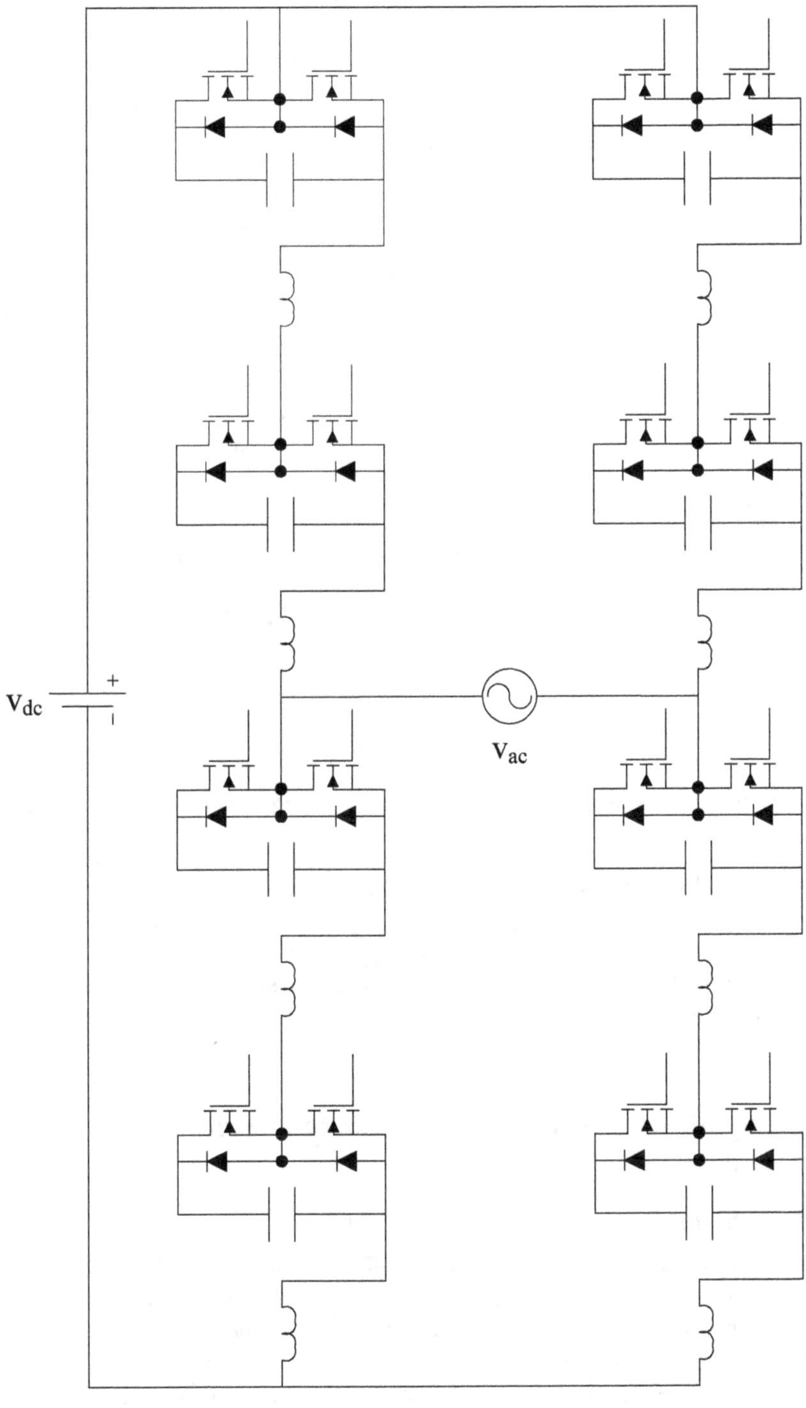

Figure 10.9 Power circuit of a single-phase-dc BoB converter

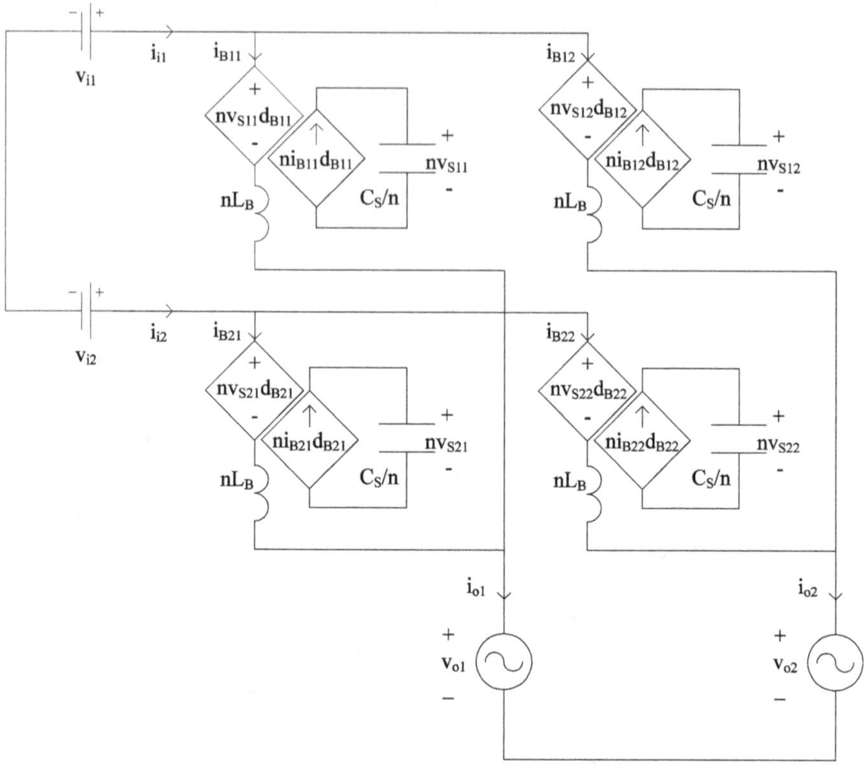

Figure 10.10 Averaged equivalent circuit of a single-phase-dc BoB converter

feedback mechanism, the BoB converter may be used to realize various power conversion functions in a versatile manner [15]. More interestingly, the approaches may also be extended to the operation and modulation of bridges in staircase or step waveform modulation as opposed to PWM operation, since the power levels in grid applications may preclude PWM operation with high switching frequency (Figure 10.10).

10.3 Application examples of power converters in a power grid

10.3.1 Shunt compensation

Shunt compensation is used in power systems to provide voltage support and overcome the impact of lagging power factor loads in transmission systems as well as in distribution systems. While fixed capacitors may be employed for the same purpose, they would not have the required dynamic capacity to react to varying loading conditions. Such applications have become more prevalent with the introduction of wind energy generation systems that have a high degree of variability.

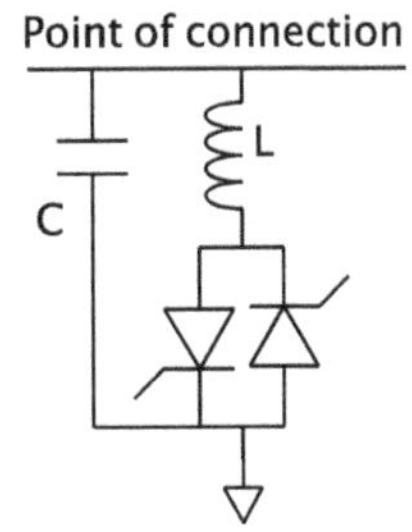

Figure 10.11 Single-phase schematic diagram of a static var compensator

Shunt compensation may use ac phase control, ac–ac power converters, or ac PWM control. Examples of each of these applications are illustrated in this section.

10.3.1.1 Static var compensator

A static var compensator (SVC) is realized using a TCR in parallel with a capacitor as illustrated in Figure 10.11. While the TCR is able to vary the reactance between a fixed value, up to infinity, the parallel connected capacitor provides a negative reactance bias, so that the net reactance of the SVC may be varied between a fixed negative value (X_C) when the SCRs are completely off, and a fixed positive value ($X_L - X_C$), corresponding to the condition when the SCRs are completely conducting. Thus through appropriate choice of C and L, the dynamic range of an SVC may be tailored to meet particular operating conditions. SVCs are utilized in transmission systems for voltage regulation. A detailed presentation of the operation of SVCs along with the TCRs, and their controls may be found in [17–20].

An Intellivar is topologically equivalent to the SVC, operated in distribution systems as opposed to the transmission system. The Intellivar has been implemented for flicker mitigation purposes with ratings up to 85 MVA. However, due to its low speed of response, reduction for frequencies above 5 Hz has not been very successful, and a substantial amount of harmonics is introduced in the system.

While the use of ac phase control utilizing SVC or Intellivar provides an economic way of realizing dynamic var compensation, it suffers from undesirable terminal current harmonics.

10.3.1.2 STATCOM

STATic COMpensator (STATCOM) is one of the first power electronic devices that use a VSC technology to be introduced under the rubric of flexible ac transmission system (FACTS) in the transmission system [17,21]. It consists of a three-phase VSC that is connected in shunt with the power system through a coupling transformer as shown in Figure 10.12. The transformer provides voltage matching, harmonic cancelation (see Figure 10.17), and a finite amount of inductance between the output of the STATCOM and the point of common coupling (PCC) connection. At steady-state operating conditions, the voltage output of the STATCOM is locked in phase with voltage at PCC. Under this condition, the

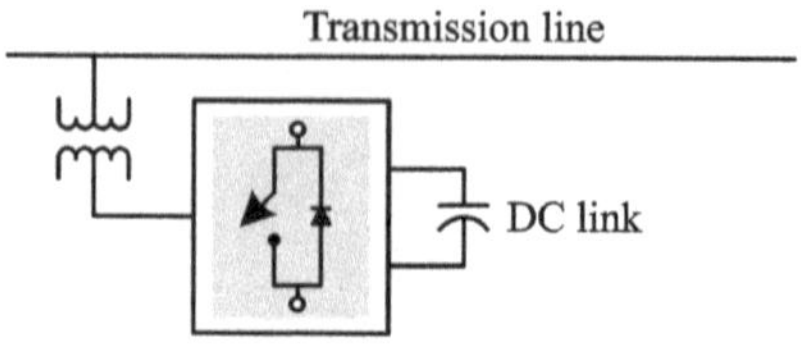

Figure 10.12 Block schematic diagram of a STATCOM

current injected into the line may be chosen to provide positive or negative vars depending on the magnitude of the STATCOM terminal voltage. When the STATCOM terminal voltage is higher than the PCC voltage, it supplies capacitive vars. When the STATCOM terminal voltage is lower than the PCC voltage, it draws inductive vars. Since the current capacity of the system is decided by the converter components and design, the amount of vars in the positive and negative direction is symmetrical. The VSC used for the system may be of two or three levels and may be modulated using PWM or multi-pulse in angle-controlled operation.

The Distribution STATic COMpensator (D-STATCOM) refers to the realization of STATCOM at distribution voltage levels, under various commercial trade names [22].

10.3.1.3 PWM ac VeSC shunt compensator

Voltage fluctuation at a load location becomes a severe power quality problem as it leads to light flicker, fluctuating torque in machinery loads, and the malfunction of protection devices and process control devices. This problem has its origins in loads such as arc furnaces and electric welders, and varying power sources such as certain engine-driven generators and wind turbines. The flicker source draws a current waveform that contains a 60 Hz fundamental component along with random sub-harmonic content in the range of a few Hz to approximately 30 Hz. The sub-harmonic currents propagate into the network and cause voltage fluctuation at the PCC at the corresponding frequencies [23].

As a result of the voltage fluctuations at PCC, other loads connected to the substation would experience the power quality problems described above. Among them, the light flicker is by far the most serious as it affects humans, producing eye irritations, headaches, and migraines. Ideally, it may be desired that the variations in the voltage at the PCC at a load location (DVPCC) be as close to zero as possible to eliminate flicker.

The classical approach to mitigate these voltage fluctuations has been to increase the size and number of generating units or to make the network more densely interconnected. However, this method has proven to be very expensive and insufficiently effective. More recently, techniques based on reactive power management have become more widespread. The concept of reactive power management is established by sizing the power system according to the maximum demand of real power and managing the reactive power by means of compensators and other equipment. This approach may often be more practical and economical.

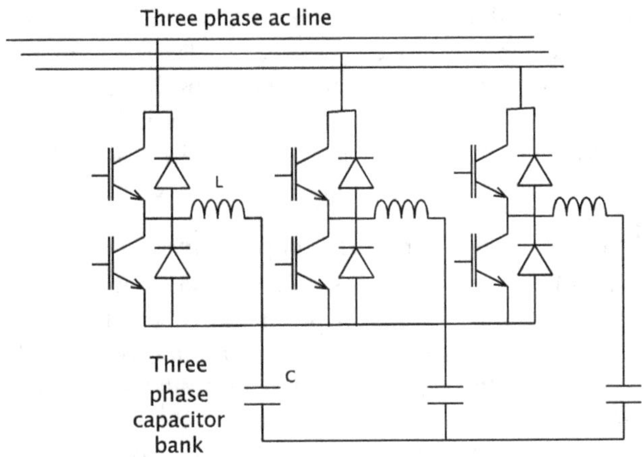

Figure 10.13 PWM ac VeSC shunt controller suitable for the realization of flicker control

Within this framework, shunt reactive power compensators can be used to counterbalance the effect of dynamic loads locally. By appropriately injecting a controlled amount of reactive current at the PCC, the effect of active and reactive components of the current in the line impedance can be controlled. In general, the compensation system's most important function is to maintain a substantially steady voltage profile therefore maintaining the relative voltage change (DVPCC) within a predetermined margin, over a specified bandwidth [24,25].

The approach of PWM ac VeSC-based controller may be used to inject a fast-acting variable amount of capacitance in parallel with the flicker source, as illustrated in Figure 10.13 [23]. The most critical feature of a reactive power compensator to be effective in flicker reduction is the speed response. Typically, if a compensator has a control time delay of 10 ms, no matter what the voltage rating it can reduce flicker, but not by a significant amount. On the other hand, if the delay is greater than 20 ms flicker can become accentuated for several frequencies. Therefore, the application of SVC for this approach is generally limited. On the other hand, the switching times of state-of-the-art semiconductors used in PWM applications have increased dramatically, therefore decreasing the response time and amount of harmonics. These advances in technology have broadened the possibilities of devices such as the PWM converters in applications such as flicker control, with bandwidths reaching several kilohertz, if necessary.

10.3.2 Series compensation

It has been long recognized that the most effective way to externally manipulate such a power flow in a transmission line is by injecting some form of additional series reactance with the line, generically known as series compensation. The first attempts at series compensation were carried out by inserting fixed capacitors in series with the line to decrease the effective impedance of the line and therefore increase the power flow through it, with the risk of certain undesirable dynamic phenomena such as sub-synchronous resonance. The development of semiconductors with turn-on capability has allowed the utilization of this approach to perform continuous control with faster dynamic properties to realize this without the concomitant dynamic problems. Various approaches for applying power converters to realize controllable series compensation are described in this subsection.

10.3.2.1 Thyristor-controlled series capacitor

The thyristor-controlled series capacitor (TCSC), sometimes referred to as the advanced series compensator (ASC) in the literature has been used to provide precisely controllable series compensation. Similar to an SVC, this device consists of a TCR in parallel with the capacitor to provide continuous control of the reactance. The entire device is connected in series with the transmission line as illustrated in Figure 10.14. While the principle of the device is rather simple, based on phase angle control of the reactance, the operational dynamic of the device may become rather complex. This is due to the possible resonances between the reactance of the line, the reactance of the TCR, and the reactance of the parallel capacitor. Since TCR is a variable reactance, as the control angle of the SCRs is varied across its dynamic range, the effective resonant modes of the system also shift. Since the operation of TCR also inherently produces several significant harmonics, the possibility of exciting such harmonics becomes real.

A detailed discussion of such issues including design considerations, dynamic operating range, control approach, etc. may be found in [26–28].

10.3.2.2 Static synchronous series compensator

Similar to the STATCOM, a VSC may be used to provide var compensation, in series with the line, and is generally termed a static synchronous series compensator (SSSC). Since transmission systems operate at voltages beyond the reach of

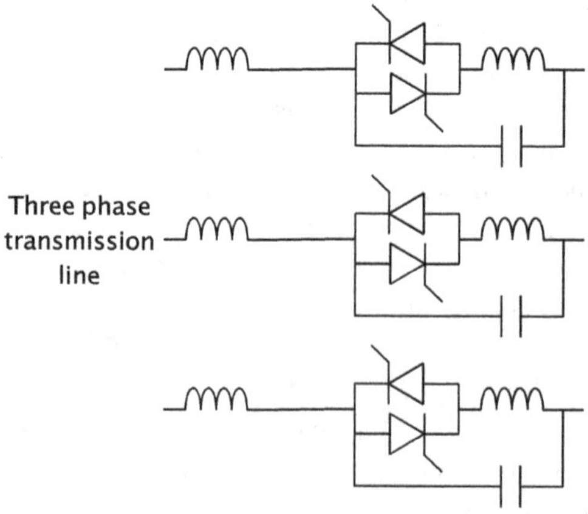

Figure 10.14 Schematic diagram of a three-phase thyristor-controlled series capacitor

any of the power semiconductor switching devices, coupling of the power converter has to be necessarily made through a series transformer as shown in Figure 10.15 [29,30].

While it may be noted that all three stages, the dc bus, the inverter, and the injection transformer, can be realized in several ways, a realization of a 48-pulse operation is described further. The realization of the VSC-based SSSC consists of three main stages, namely a dc bus, an inverter, and an injection transformer as illustrated in Figure 10.16. As can be seen, the dc bus is realized using two series connected dc capacitors with access to their mid-point N. This allows three voltage levels ($V_{C1}, V_{C2}, 0$). Four independent diode-clamped three-phase three-level converters are connected to the dc bus. The voltage synthesis for each of the phases has a conduction period 172.5° for each half cycle. Each three-level converter is then connected to a magnetic structure based on Y-zigzag and D-zigzag phase-shifting transformers (PSTs) that allow the creation of a 48-pulse waveform. Thus, the lowest characteristic harmonic numbers present in the output waveforms are 47 and 49.

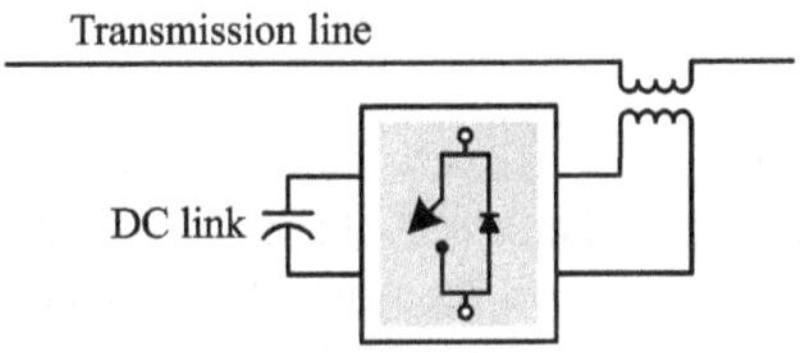

Figure 10.15 Block schematic diagram of a dc link solid-state series capacitor

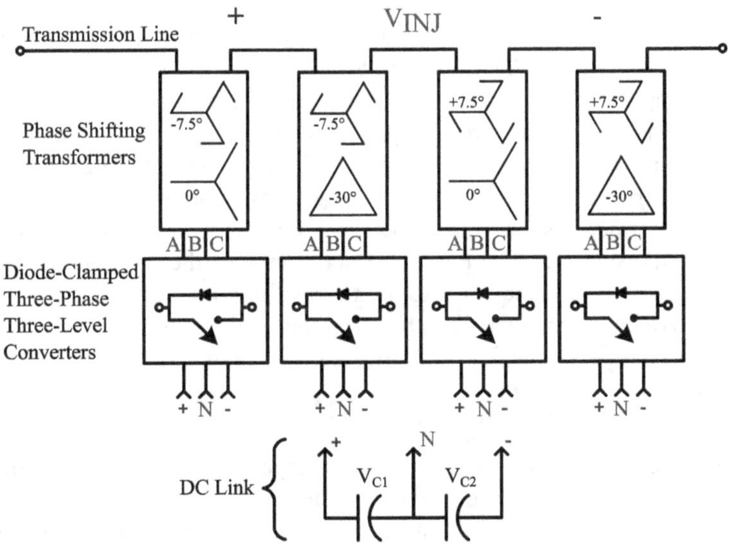

Figure 10.16 Transformer connection details of the 48-pulse SSSC

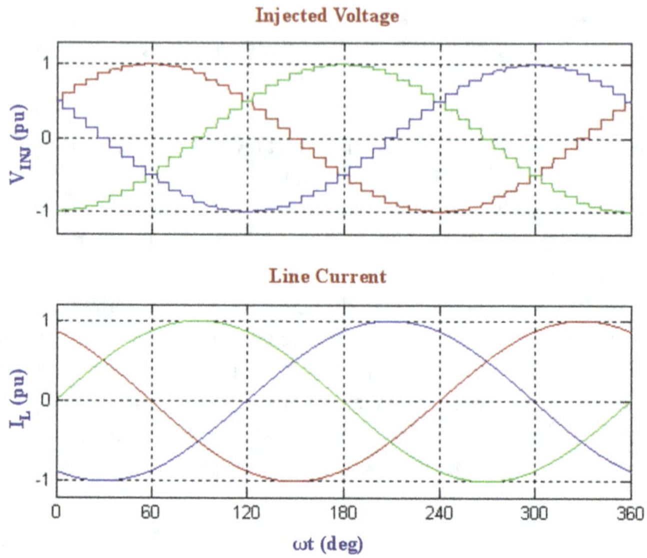

Figure 10.17 Three-phase output voltage waveforms from 48-pulse SSSC

Figure 10.17 shows typical switched three-phase voltage waveforms along with the corresponding line current. It can be seen that the injected voltage is of high harmonic quality and thus the SSSC realized in this manner can operate with no output filters.

10.3.2.3 PWM ac VeSC series controller

The approach of PWM ac VeSC-based power control may also be used for controlled series compensation [9,11]. The schematic diagram of a power circuit is illustrated in Figure 10.18. The switches SA, SB, and SC that are ganged together form a PWM ac VeSC converter, which operates in a duty ratio controlled mode. The output of the PWM converter is injected in series with the line through a series coupling transformer. When the switches are in the top position, the capacitors are introduced in series with the line. When the switches are in the bottom position, the coupling transformer is shorted, and the capacitors are isolated from the line. The duty ratio of the switches determines the percentage of the compensation capacitance that is introduced in series with the transmission line.

The equivalent impedance (X_{eq}) between the sending and receiving end of the transmission system may be derived using state-space averaging techniques as

$$X_{eq} = X_L - n^2(1 - D)^2 X_C \tag{10.9}$$

where X_C is the reactance of the compensation capacitors, n is the turns ratio of the transformer, X_L is the reactance of the inductors line (which may include the leakage reactance of the coupling transformers), and D is the duty ratio of the top throws of the switches. The performance of the controller from simulations during a typical case is illustrated in Figure 10.19.

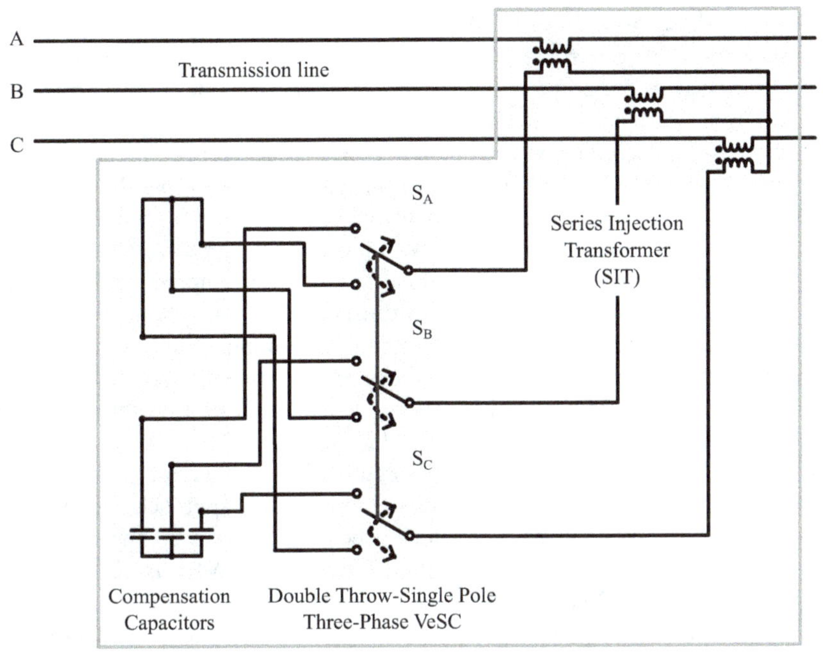

Figure 10.18 Schematic diagram of a three-phase PWM-controlled series capacitor

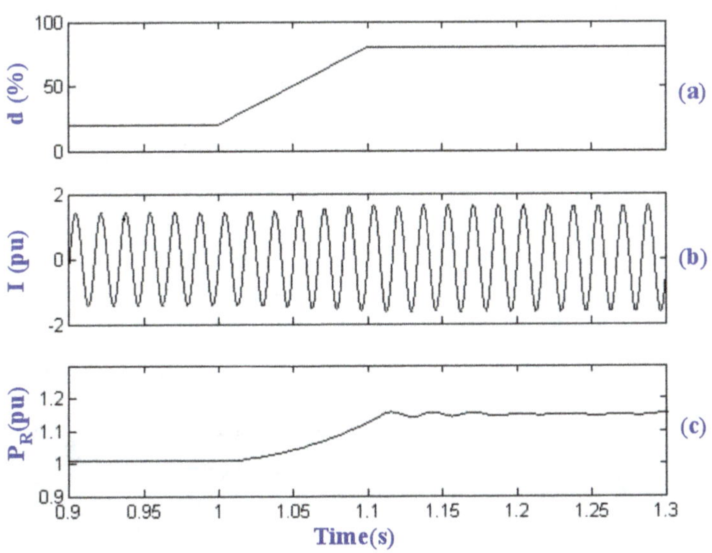

Figure 10.19 A plot of (a) duty ratio, (b) current, and (c) power throughput using a PWM-controlled series compensator obtained from simulation of a transmission system case

10.3.2.4 Voltage sag correction

The proliferation of electronic equipment results in increasing sensitivity of industry processes to the power quality. On the other hand, the more competitive market environment leads to reduced profit margins, which is the primary reason why the process shutdown caused by the power quality problem is intolerable. The power quality problem of most concern is the voltage quality problem, such as voltage sags. Voltage sags are very commonly caused by power system faults, most of which are single line-to-ground short circuit faults. The short circuit faults can be caused by lighting strokes, storms, animals, and other unpredictable factors. Although the voltage quality can be improved at the power system level, rather often the solution is too costly to implement and it is impossible to eliminate the voltage problem. Alternatively, utilities and customers turn to the practical solution of feeding premium power to sensitive loads, which is realized by the mitigation equipment at the distribution feeder close to the customer [31].

In such cases, if the system is strong (measured by short circuit capacity being much larger than unity) at PCC, the series compensation approach requires less rating size for the power converter. A block schematic diagram of such a system, commonly referred to as a dynamic voltage restorer (DVR) is illustrated in Figure 10.20. While VSC-based DVRs can be realized using two-level converters, which are well suited for 480 V systems, for systems operating at distribution power levels a multilevel converter is a more attractive solution. Moreover, since voltage sags occurring in such systems are due to single line-to-ground faults, it is necessary for a DVR to be able to handle imbalances in sags, as well as zero sequence type of voltages. Therefore, a power converter topology with independent control of each phase as illustrated in Figure 10.21 is suitable for such applications.

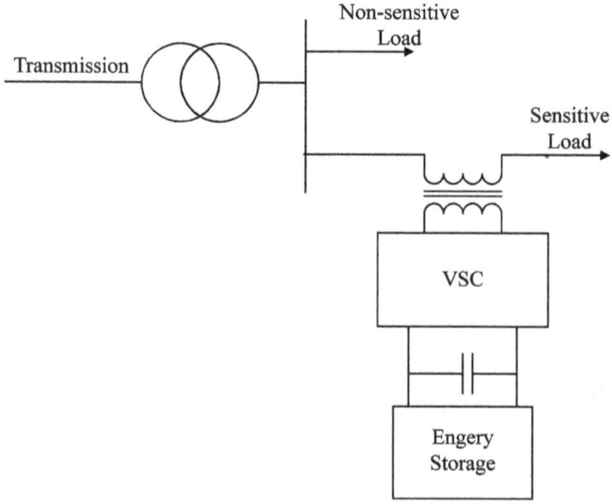

Figure 10.20 Single-line schematic diagram of a dynamic voltage restorer illustrating a series connection

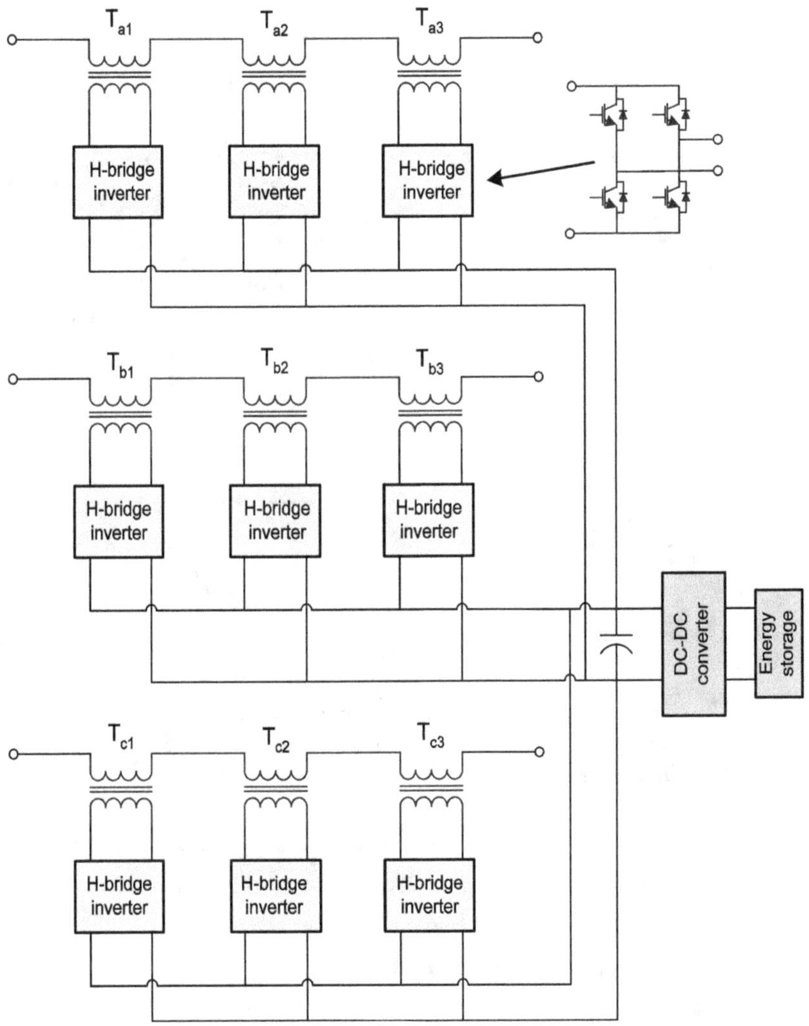

Figure 10.21 Realization of the three-phase dynamic voltage restorer using multilevel converters operating as independent three-phase converters

Moreover, due to the use of transformers with voltage-adding capability, the output waveform may be of a multilevel nature. In this case, the converters may employ a phase-shifted PWM strategy among them to provide excellent quality output waveforms at the output (Figure 10.22), thereby requiring minimal filtering components.

10.3.3 Shunt-series compensation

The shunt-series power flow controller is a device introduced under the rubric of FACTS, that combines the features of shunt compensation such as the STATCOM,

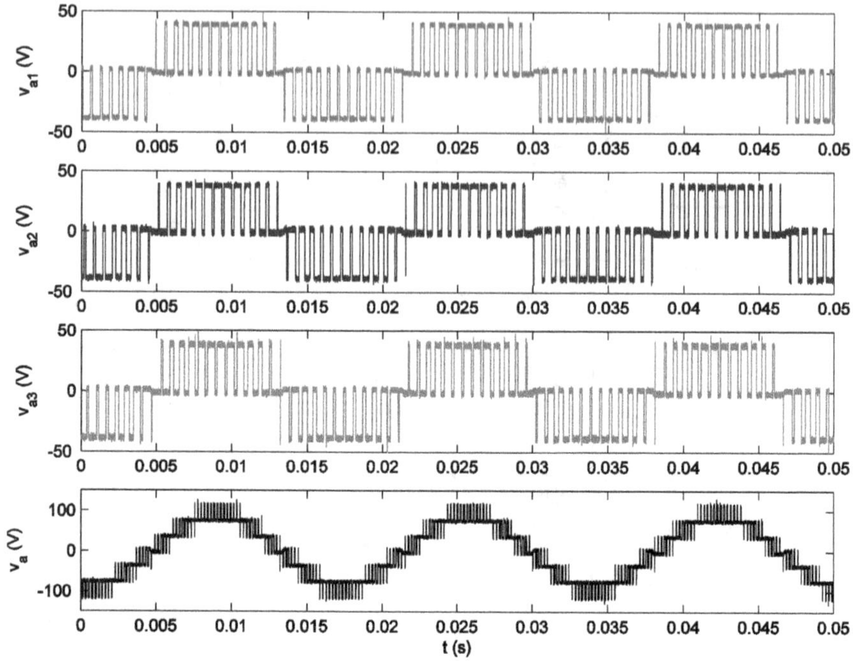

Figure 10.22 Waveforms from three series connected H-bridge single-phase inverters with phase-shifted PWM generating a seven-level output voltage in a DVR

and series compensation such as the SSSC together by interconnecting the two converters. VSC-based converters and PWM ac controller-based approaches are discussed in this section.

10.3.3.1 Unified power flow controller

An arrangement of the shunt series power flow controller under the title of unified power flow controller (UPFC) provides a means for controlling the active and reactive power flows in a power system through the synthesis of a fully adjustable series voltage [32]. This topology, based on a dc link between a shunt VSC (a STATCOM) and a series VSC (a SSSC) connected in a back-to-back configuration, has certain inherent degrees of freedom determining its control capabilities (Figure 10.23). The STATCOM and the SSSC can develop arbitrary terminal voltages and phase angles (four degrees of freedom) within the bounds dictated by their shared dc link voltage. In the case of applying a PWM control scheme to these VSCs, the dc link voltage has to be maintained constant. Therefore, the sum of the active power injections of the STATCOM and SSSC must be zero. Taking this into account, the initial four degrees of freedom of the UPFC are reduced to three, making it possible to regulate the active and reactive power flow between the interconnected power systems and the voltage at the PCC of the

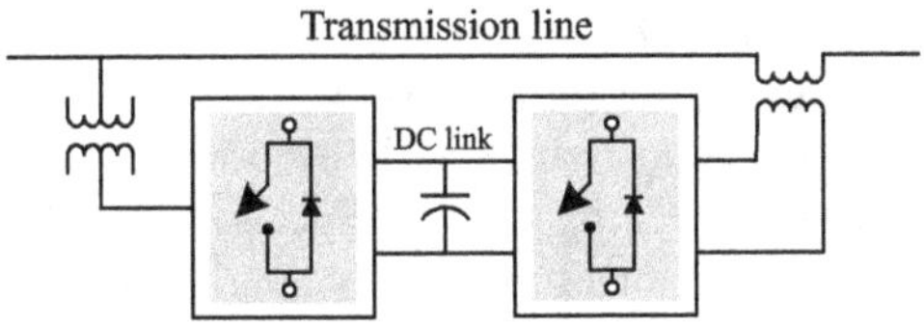

Figure 10.23 Block diagram of a UPFC using shunt and series connected VSCs with a common dc link

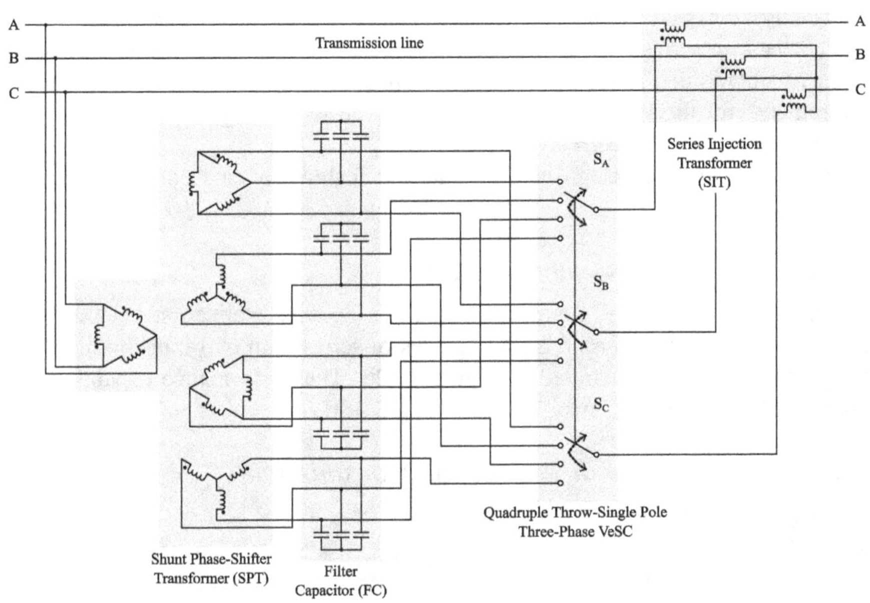

Figure 10.24 Realization of a shunt-series power flow controller using a PWM ac converter

STATCOM. Each of the STATCOM and SSSC may be realized using multi-pulse converters described in the previous sections.

10.3.3.2 PWM ac VeSC shunt/series power flow controller

The schematic diagram of the PWM ac power converter applied as a shunt/series controller is illustrated in Figure 10.24 [33,34]. The central feature of the approach is to control the power flow along the transmission line by injecting an ac voltage with controllable magnitude and phase angle in series with the line. The system is configured by including a shunt phase-shifter transformer (SPT), filter capacitors (FCs), a quadruple-throw single-pole three-phase PWM VeSC, and a series injection transformer (SIT). This system may be located at any point throughout the line.

The SPT (with four secondary winding sets) has the double function of lowering the impressed voltage to match the capability of the semiconductor switches and simultaneously phase shift the impressed voltage by 0°, 90°, 180°, and 270°, respectively, to obtain four sets of three-phase voltage vectors lined up along four cardinal directions. The capacitor bank FC is used to absorb the high-frequency currents introduced due to switching from permeating into the system across the transformer. The VeSC uses a quadruple-throw single-pole three-phase switch to synthesize a controllable voltage at the pole terminals by adjusting the duty ratios d_1, d_2, d_3, and d_4. Given the throw voltages generated by the SPT, the range of values of the pole voltage can be regulated in magnitude and phase angle, both of which can be independently controlled through duty ratios d_1, d_2, d_3, and d_4.

Finally, the SIT accommodates the voltage to an appropriate level to be injected back in series with the transmission line and also provides the necessary electrical isolation. It is important to notice that the stiff voltage and stiff current requirements for the PWM ac converter are given by the capacitor bank and SIT leakage inductance, respectively.

The performance, design, and operational features of this converter may be found in [33,34].

10.3.4 Series–series compensation

The interline power flow controller (IPFC) is another realization of a FACTS device, where two VSC-based SSSC devices in series with two transmission lines share their dc link as illustrated in Figure 10.25. The performance capability and design details of the IPFC may be found in [35,36].

10.3.5 High-voltage direct current transmission

10.3.5.1 Conventional CSC

High-voltage direct current (HVDC) based on CSCs has been in use for 50 years, starting evolving from systems using mercury arc valves to modern systems using light-fired thyristors. These systems are generally used for bulk power transmission over long distances using overhead lines (500 km or longer) or undersea (50 km or longer in many cases) or to interconnect systems operating out of synchronism or at different frequencies. These systems operate with voltages up to 600 kV to ground and carry currents up to 3,000 A (Figure 10.26).

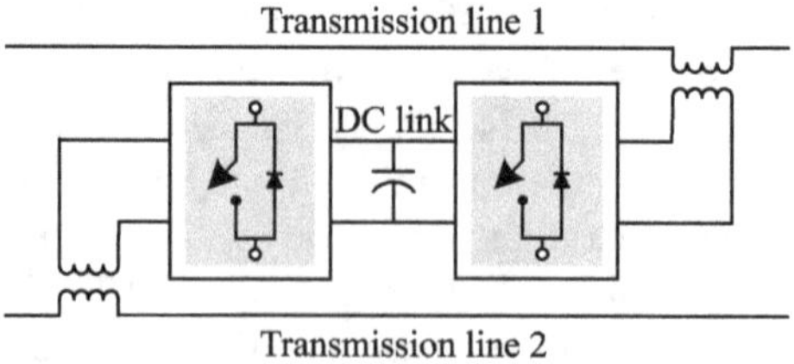

Figure 10.25 Block diagram of an interline power flow controller using interconnected VSCs

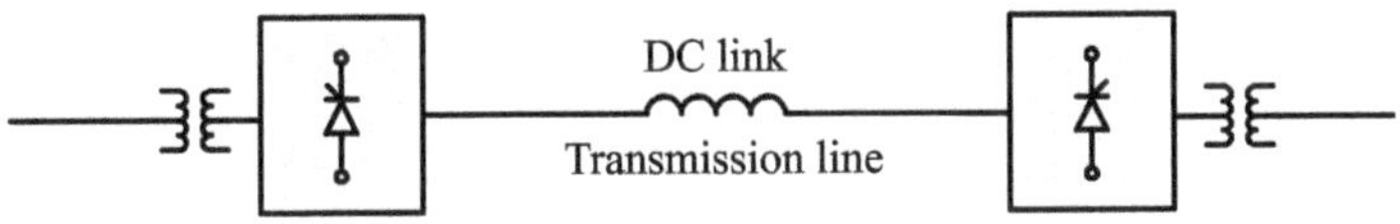

Figure 10.26 Block diagram of an HVDC transmission system

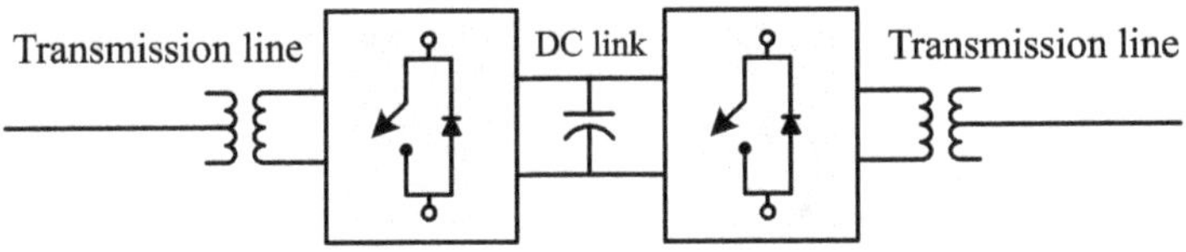

Figure 10.27 Block diagram of VSC-based HVDC system

The lack of turn-off controllability of the thyristors used in the line-commutated converter results in: poor power factors and considerable waveform distortion, the need for a relatively stiff ac voltage supply, and the inability to provide independent control of the active and reactive powers. Moreover, it requires fast communication channels between the two stations to ensure power balance and appropriate controls [1].

10.3.5.2 VSC-HVDC

Over the last decade, VSC-based HVDC transmission has started to come into use (Figure 10.27) [37]. These systems, referred to as HVDC lights by vendors, are based on a modular design, reducing the installation time. VSC transmission is often used in cases where an ac system is not practical, generally due to long underground or undersea cables, and excessive losses. One common application is supplying power to isolated island systems without adequate local generation to provide reactive support for conventional HVDC converters. The converters in VSC transmission are able to supply power to weak or passive ac load systems. Another application where VSC-HVDC transmission is often used is the transmission of power from off-shore wind farms to the mainland. The converters used in this system may be two-level or three-level topologies.

In contrast to CSC systems, the newer VSC-based technology (i) does not require any communications links between the different converter stations; (ii) does not need to rely on the ac network's ability to keep the voltage and frequency stable; and (iii) offers independent control of active and reactive power [16].

10.3.5.3 MMC-HVDC

The most recent development in the field of HVDC systems is the application of the MMC or BoB converter to realize HVDC systems [38–40]. A schematic diagram of the system is illustrated in Figure 10.28. The first field application of the BoB approach is the Trans Bay Cable Project which was commissioned in 2010 under the product name Siemens HVDC Plus. The SMs are made up of IGBT switches with anti-parallel diodes.

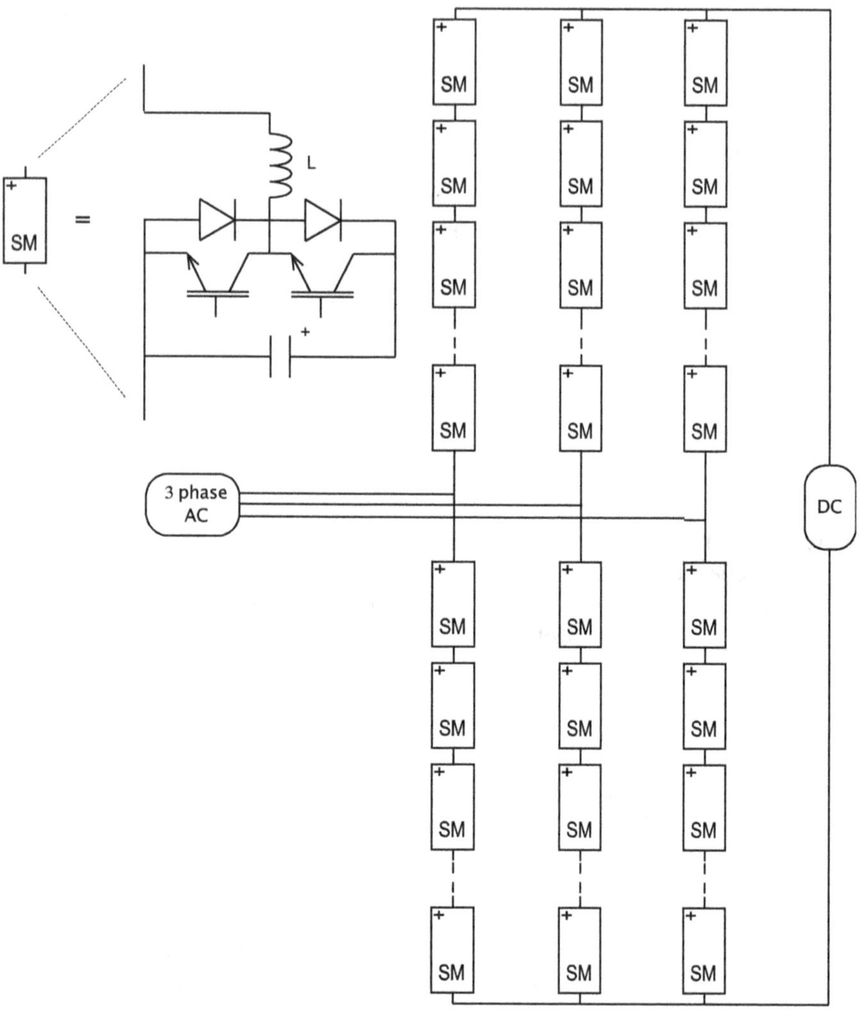

Figure 10.28 Schematic diagram of a three-phase ac-to-dc MMC or BoB converter for HVDC converter stations

10.3.6 Low-frequency high-voltage ac transmission

The low-frequency high-voltage ac (LF-HVAC) transmission approach is a promising alternative to increase power transfer capability and improve voltage profile through the transmission line [41]. Due to the reduction in frequency for transmission, the line reactance is also decreased leading to an improved voltage profile along the line and a larger transient stability limit. Past propositions to introduce this technology have been mainly based on cycloconverters which lead to high levels of voltage and current harmonic distortion levels. LF-HVAC systems have been utilized so far for (i) the grid connection of large wind farms at a fractional

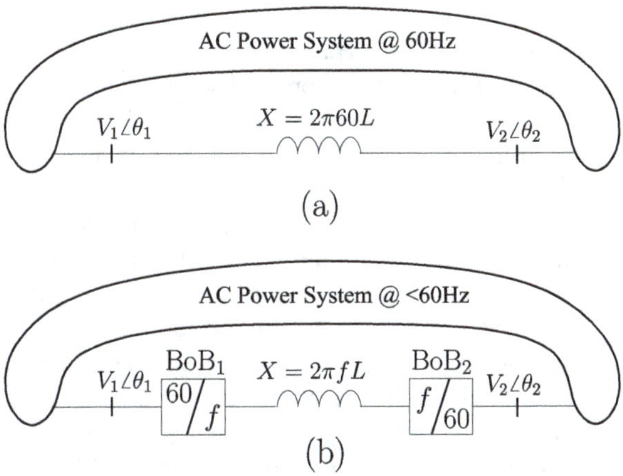

Figure 10.29 Simplified schematic diagram of an ac system illustrating (a) a transmission line and (b) the transmission operating at a lower frequency

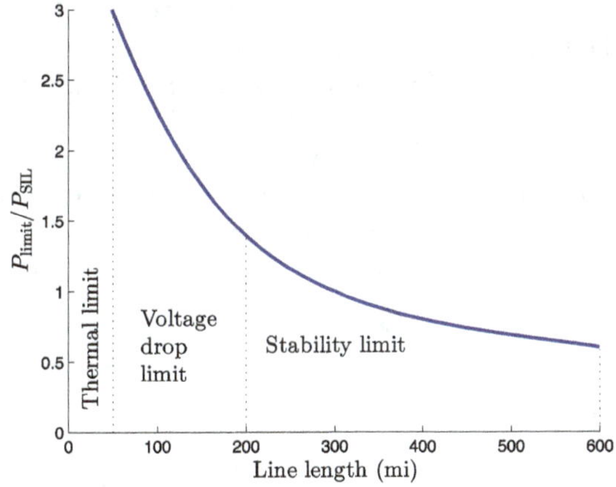

Figure 10.30 Loadability curve of a transmission line as a function of the line length

frequency of 50/3 Hz, and have been termed fractional frequency transmission systems [42], and for (ii) cable transmission systems which are limited in transmission length when operated at conventional frequencies [43].

Referring to Figure 10.29(a) the power throughput through a transmission line is roughly proportional to the angle between the voltage at the two ends of the line, and inversely proportional to the impedance of the line. It is well understood in the power industry that viable approaches to control the line throughput power include adjusting the transmission angle and the line impedance. On conventional transmission lines, the line impedance is fixed at $X = 2\pi 60L$ and the transmission angle is always kept below about 40° due to transient stability considerations. The latter is especially critical for long transmission lines, as transient stability considerations limit the throughput power loading. Figure 10.30 illustrates a typical loadability curve for a transmission line, where the limit is expressed as a fraction of the surge impedance loading (SIL). As may be seen, transient stability becomes the predominant limitation for lengths longer than 200 miles (mi). This leads to very inefficient utilization of the conductors—as suggested in the figure a line of 600 mi can be loaded to less than 30% of its thermal limit. Since the transient stability limit of the line is fixed by the transmission angle, an alternative and convenient means to increase the transient stability limit is to reduce the line's reactance.

Figure 10.29(b) shows the implementation of LF-HVAC transmission lines through embedding frequency converters at each end of a number of lines. This would allow the transmission line to be operated at a frequency $f < 60$ Hz. Operational benefits of LF-HVAC transmission lines clearly include the capability for asynchronous operation with full control of the throughput active power and reactive power at each end of the transmission line. Furthermore, since the line at both ends is terminated by power converters interfacing to the rotating machine and load centers, the classical transient stability considerations leading to a nominal limit of 40° are no longer valid. Moreover, power converters have some built-in storage capacity and active control over the power throughput across the line under transient conditions. This may permit the transmission angle to be increased well beyond the nominal limit of 40°, even up to 90°. Thus, by properly selecting the operating frequency, the line reactance can be decreased so that the line capacity can be utilized up to its thermal limit.

Although the previous discussion has focused on overhead transmission lines, similar arguments may be brought forth regarding underground cables. Furthermore, the same ideas may be applied to converting an entire area of a transmission system into LF-HVAC.

Figure 10.31 illustrates a 60 Hz to LF-HVAC converter realized using the BoB converter approach. In this case, all the bridge SMs will be realized using the full bridge topology.

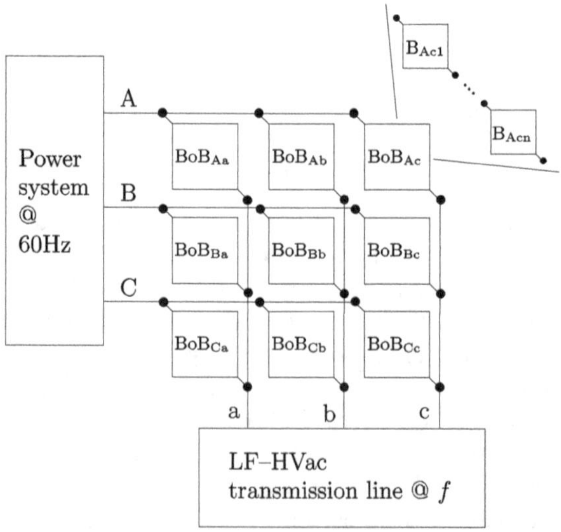

Figure 10.31 Illustration of a BoB converter-based frequency conversion for realizing LF-HVAC systems

10.3.7 Solid-state power transformer

The concept of a solid-state power transformer (SSPT) is not recent, but with the advancement of high-power semiconductors, the SSPT may soon become a practical reality [44]. Figure 10.32 shows a possible realization of an SSPT. As the figure suggests, an SSPT is made up of the cascaded connection of a VSC acting as a rectifier; a dc link; a VSC acting as an inverter; a high-frequency ac transformer; a VSC acting as a rectifier; a dc link; and a VSC acting as an inverter. Variations of this topology have been presented in the literature, but all of them follow more or less the same approach. Differences arise from the realization of the various VSC stages.

The main advantage of such a configuration is that it provides the benefits of an ordinary transformer—namely, electrical isolation and a potentially large transformation ratio, but, utilizing a reduced amount of magnetic material. The high-frequency transformer processes ac power at a frequency of a few tens of kilohertz, and therefore can be realized utilizing less magnetic material, making it lighter, cheaper, and easier to transport when compared to a standard 50 or 60 Hz standard transformer rated at the same throughput power. Furthermore, due to the presence of VSCs at the input and output, the SSPT can interconnect two ac systems operating at arbitrary frequencies (e.g., 50 and 60 Hz), decoupling the dynamics and providing independent power factor compensation for both systems. The VSCs can also provide active filtering functionality, and the intermediate dc link can serve as a point of interconnection for dc power sources (e.g., photovoltaic arrays). Because of all of these features, SSPTs are considered by some authors as an important building block for the smart grid of the future. Applications of SSPTs have been mostly reported for medium- to low-power systems in the context of ac and dc microgrid networks; for variable-speed ac drives; and for the realization of power transformers in distribution systems.

The main drawback of SSPTs stems from the usage of a large number of power semiconductors, leading to a decrease in efficiency, control complexity, and potential failures associated with the semiconductors. Research to improve efficiency has been reported in terms of replacing standard silicon-based switches with silicon carbide and other wide band gap semiconductors.

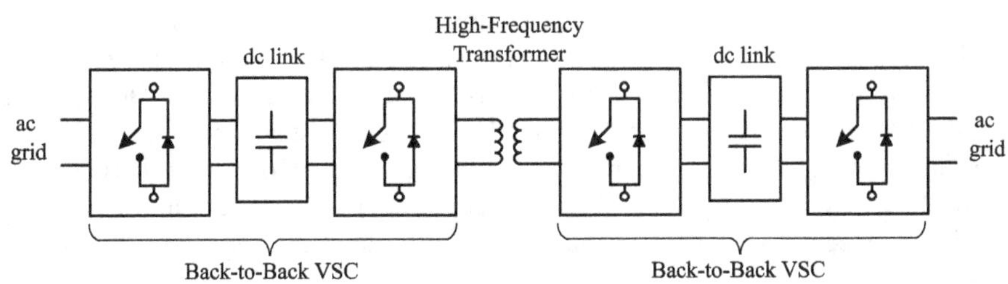

Figure 10.32 Illustration of a solid-state transformer using two dc links and a high-frequency transformer

10.3.8 Multi-port hybrid converters for renewable energy integration

The increasing share of intermittent sources of energy such as solar and wind intensifies the need for energy storage technologies such as batteries and pumped-storage hydropower. Battery energy storage systems are growing fast due to their high energy densities and smaller size and weight, which leads to added flexibility and ease of installation [45,46]. More recently, the falling costs of photovoltaic (PV) and battery energy storage (BES) technologies have raised interest in the creation of hybrid *PV+BES* power plants. These technologies are enabled by the development of advanced power electronics with unprecedented functionality, efficiency, reliability, and power density critical for an electrified world economy [47].

Currently, hybrid PV+BES power plants can be broadly classified as either AC grid-coupled or converter-coupled systems based on their point of common coupling. AC grid-coupled systems do not share any components and the storage systems can act independently of the PV system. From the perspective of power converter architectures, AC grid-coupled systems are expected to have similar characteristics to independently operated PV plants and BES plants. On the other hand, in converter-coupled systems, the PV and BES share the grid-connected DC/AC or the AC/AC converter. This section focuses on converter-coupled configurations for multi-port hybrid PV+BES power plants because of the distinct characteristics and the expected advantages.

While the growth of hybrid plants is driven by significant declines in project costs due to cost savings by sharing equipment, cutting interconnection and permitting costs, and leveraging federal tax credits, there are several other technical advantages as well. First, converter-coupled systems permit the energy storage of otherwise clipped energy that occurs when the inverter loading ratio (ILR) exceeds 1. Typically, ILR varies between 1 and 2. Note that an ILR of 1.3 is equal to the average value for utility-scale PV systems in the US [48]. Further, since the batteries are directly charged from the PV system, converter-coupled systems have higher charging efficiency due to the lower number of power electronic stages. Lastly, independent of the point of coupling, hybridization of the photovoltaic power plants with battery systems broadens the plant's grid services including fast dispatch flexibility, frequency regulation, and voltage-ampere reactive support.

Figures 10.33 and10.34 illustrate the simplified block diagrams of converter-coupled hybrid PV+BES system architectures. The PV and the BES power plants share a common AC grid-connected converter that connects the two DC systems (i.e., the PV and the BES) to the AC grid. The two DC systems interface with each other either at the DC node or at the AC node as illustrated in Figure 10.33 and Figure 10.34, respectively. Based on the coupling node configuration, galvanic isolation between the DC sources and the AC grid may be at the AC-grid or at the converter level.

Referring to the notations in the figures, the hybrid power transfer capability of the PV+BES plant to the AC grid is P_{inv}. In other words, the shared DC/AC or the

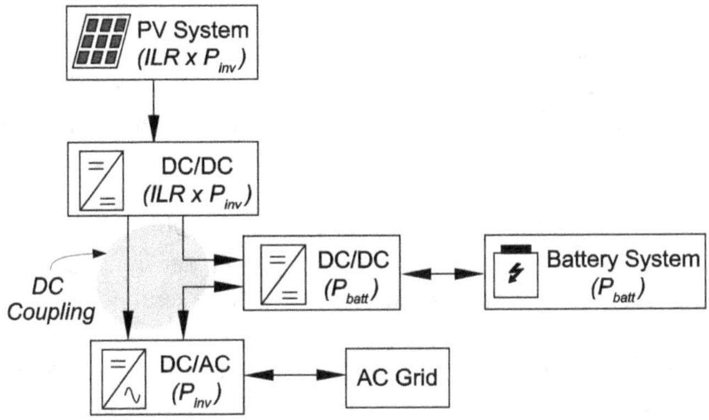

Figure 10.33 Block diagram of converter-coupled photovoltaic and battery energy storage power plants with DC-coupled configuration

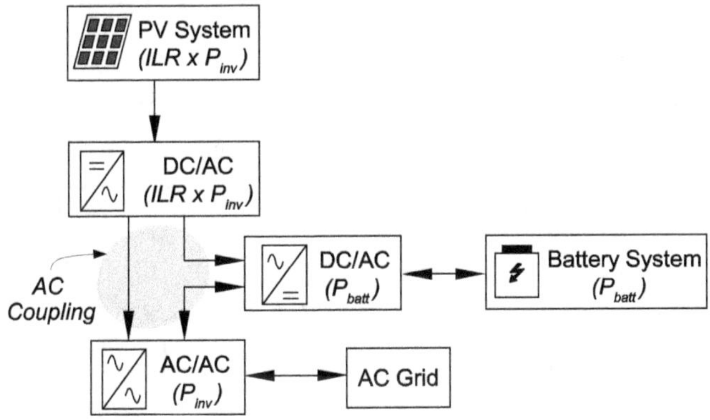

Figure 10.34 Block diagram of converter-coupled photovoltaic and battery energy storage power plants with AC-coupled configuration

AC/AC conversion system is rated for P_{inv}. Hence, the PV power plant's rated power is $P_{pv} = ILR \times P_{inv}$. The BES-rated power is assumed to be P_{batt}. Typically, $P_{batt} \approx 0.5 \times P_{inv}$ in today's hybrid power plants [48]. In summary, the photovoltaic and the battery converter are rated for P_{pv} (=$ILR \times P_{inv}$) and P_{batt}, respectively. As hybrid PV+BES power plants become common, the average inverter loading ratio is expected to increase since more clipped energy can be stored in the BES [48]. The remaining section presents some example topologies of converter-coupled multi-port hybrid converters. Note that the list is not exhaustive and many other configurations are feasible [47].

10.3.8.1 DC stacked converters

Three-phase two-level VSCs have traditionally been ubiquitous and are the workhorse topology in many applications such as traction inverters, industrial motor drive systems, uninterruptible power supplies, etc. [49]. In fact, two-level VSCs have dominated industrial-scale grid-connected energy storage projects until the present times [50]. Further, VSCs offer an architectural solution for *hybrid* utility-scale power plants. The connection of the low-voltage battery and photovoltaic systems to the medium voltage utility grid is enabled by (a) series connection of several low-voltage systems, and (b) utilization of a low-frequency step-up transformer, illustrated with turns ratio of "*t*" in the figure, which may be a solid-state based transformer for increased efficiency and power density. Figure 10.35 illustrates the circuit schematic for the solution along with several possible DC coupling configurations.

The DC coupling configurations can be determined by the voltage class of the photovoltaic panels and batteries. These DC energy sources are all series and/or parallel connections of a basic *cell*. These cells operate at a low voltage, ranging from less than 1 V to 4 V [51,52]. These low-voltage systems do not interface well with higher-voltage systems and hence several PV or battery cells are connected in series to form a PV module or a battery system. Considering the PV module

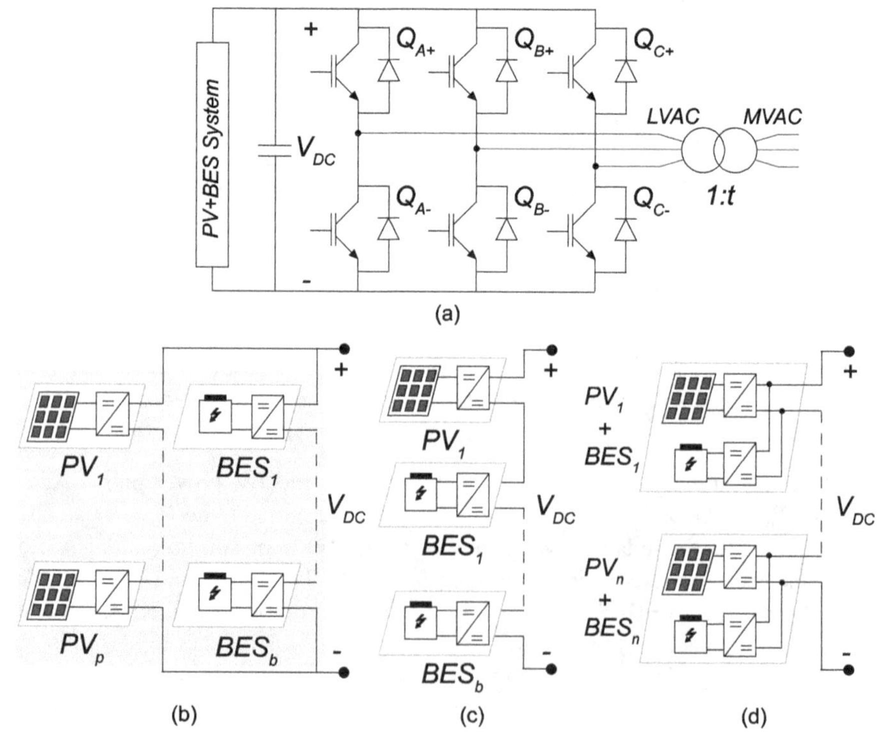

Figure 10.35 (a) Converter topology of a three-phase two-level voltage source converter with PV and BES systems cascaded to the DC link. Figures (b)–(d) illustrate several possible DC coupling configurations for the PV+BES systems.

technologies, PV modules can be categorized into three categories: 1) high voltage ($\approx$ 240 V) amorphous silicon modules, 2) medium voltage ($\approx$ 60 V) amorphous silicon modules, and 3) low voltage multi-crystalline silicon modules ($\approx$ 30 V) [53]. A utility-scale PV power plant can comprise several PV modules that connect in-series and/or in-parallel to form PV strings or PV arrays. Traditional photovoltaic power plants feature a central DC/DC converter that cascades to a grid-connected DC/AC inverter for power transport. The DC/DC converter implements the maximum power-point tracking algorithm (MPPT) to maximize the energy yield. In contrast to central DC/DC converters, distributed DC/DC converters are growing fast due to several advantages including energy yield improvements during shading, module mismatch, elimination of single-point failure, and added monitoring and diagnostics. Similarly, based on the voltage class of the battery energy storage, their DC/DC converters can feature either boost or buck-boost configuration. Many multi-port PV+BES hybrid configurations are possible. Figure 10.35(b)–(d) presents three scenarios.

Figure 10.35(b) illustrates the case where several PV arrays connect to the DC port via several DC/DC converters. In parallel, several BES systems connect to the shared DC port via their own DC/DC converters. In Figure 10.35(c) configuration, several PV arrays connect in series with several BES systems to the DC port via DC/DC converters. In contrast to these configurations, each PV system may be connected to a smaller BES system at the DC port, also illustrated in Figure 10.35(d). In simpler configurations, DC batteries may directly connect to the DC link without a DC/DC converter. In the case of hybrid PV+BES systems, DC/DC converters interface each of the DC systems with each other and the AC grid plays a distinct role in the control of power flow. Since the common point of coupling between the PV, BES, and AC grid converters is DC, this topology is a DC-coupled configuration of Figure 10.33.

Considering the DC/DC converter design, in theory, any unidirectional and bidirectional DC/DC converter can be used for the integration of the photovoltaic systems and the battery systems respectively. Boost or buck-boost configurations are most compatible with hybrid PV+BES systems due to their voltage class in comparison to the AC grid. Availability of boost mode of operation helps reduce the number of series connected PV modules or battery systems, thus potentially extending the life expectancy and mismatch issues [54]. Several power conversion topologies are proposed in the literature [47]. Figure 10.36 illustrates the circuit schematics of the boost and buck-boost capable single-ended primary-inductor converter (SEPIC) topologies with PV and BES systems connected in Figure 10.35 (d) configuration.

While a simple boost converter features the lowest voltage (V) X current (A) rating for the power semiconductor devices, it will not offer the buck mode of operation. In contrast, buck-boost converters enable buck mode of operation with device VA rating as a trade-off. Hence, unless required by the PV, battery, or DC link specifications, boost configuration may be an ideal choice since buck-boost configurations suffer from additional device stresses [55,47].

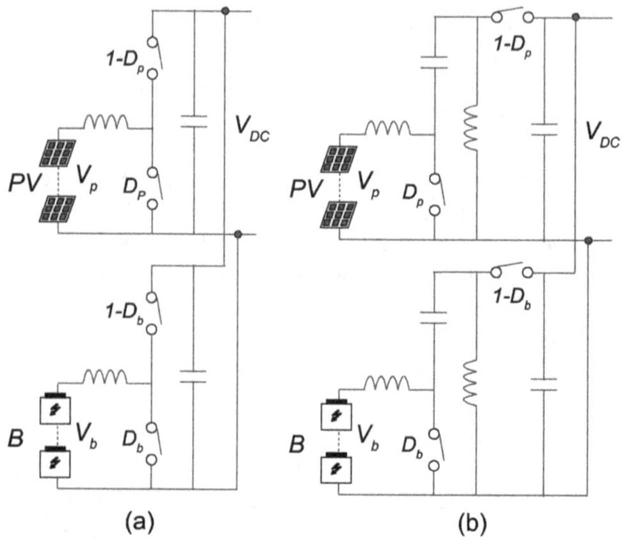

Figure 10.36 Common DC/DC converter topologies to control the power flow between the PV modules, batteries, and the DC/AC grid-tied inverter. (a) Boost configuration, and (b) Buck-Boost configuration featuring single-ended primary-inductor converters

10.3.8.2 Cascaded bridge converters

In contrast to DC stacked topologies wherein the DC sources/loads are connected in series and interfaced to the AC grid using a single inverter, modular topologies feature smaller inverters that interface the several distributed DC sources/loads to the AC grid. To provide galvanic isolation, the cascaded bridge architecture can feature a low-frequency transformer at the AC grid or the module level. This section presents two such approaches: one DC-coupled configuration and one AC-coupled configuration.

Figure 10.37 illustrates the circuit topology featuring *n* PV systems and battery systems. The several DC systems connect to the AC grid via several non-isolated DC/DC and DC/AC converters and enable voltage compatibility via a series connection of smaller converters. While an AC grid transformer at the grid level may provide a step-up function (not shown in the figure), the primary purpose is to provide galvanic isolation between the DC sources and the AC grid. Since the common point of coupling between the PV, BES, and AC grid converters is DC, this topology is a DC-coupled configuration of Figure 10.33. Note that while the DC-coupling configuration of Figure 10.35(d) is presented in Figure 10.37, other configurations including those presented in Figure 10.35 are also feasible.

In contrast to the above topologies, Figure 10.38 features a high-frequency transformer as a part of the design architecture which provides galvanic isolation between the PV systems, battery systems, and the AC grid at the module level.

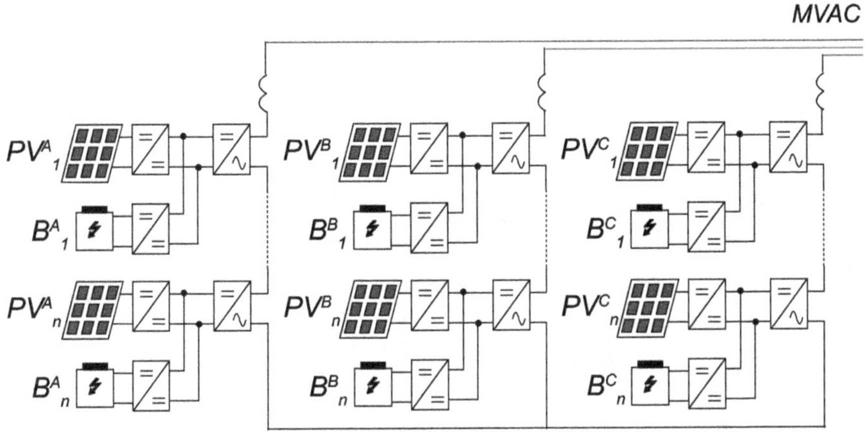

Figure 10.37 Converter topology of a three-phase modular converter with both battery systems and PV systems connected to modular DC/AC converters via their DC/DC converters

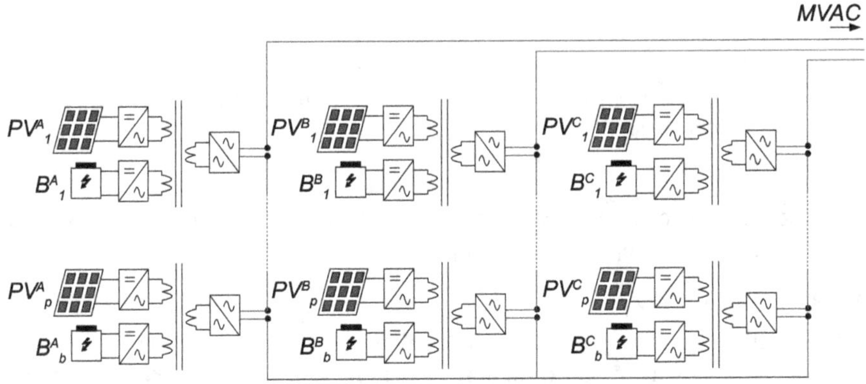

Figure 10.38 Converter topology of a three-phase modular converter with both battery systems and PV systems connected to modular DC/AC/AC converters via a three-winding transformer for galvanic isolation

Since the common point of coupling between the PV, BES, and the AC grid converters is AC, this topology is an AC-coupled configuration of Figure 10.34.

Figure 10.39 presents a converter architecture to realize the multi-port DC/AC/AC converters to connect the PV+BES systems to the AC ports. The schematic features a triple-active bridge converter where Bridge 1 converts DC waveforms to high-frequency AC waveforms. Consequently, bridges 2 and 3 are utilized to convert high-frequency AC to grid-frequency AC waveforms. Triple active bridge converters are expected to operate at high efficiencies due to the zero-voltage

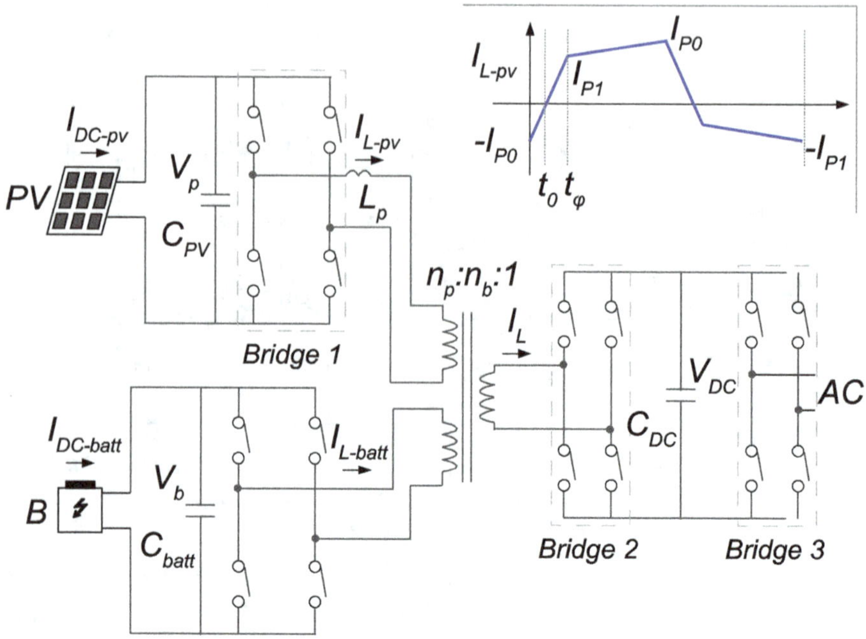

Figure 10.39 Converter topology of a triple active bridge converter connecting the PV and BES systems to the AC output ports to build a modular converter

switching mode of operation. More details on the operation and modulation strategies of the converter can be found in [56–58].

10.3.8.3 Summary

More recently, the falling costs of photovoltaic (PV) and battery energy storage (BES) technologies have raised interest in the creation of hybrid *PV+BES* power plants. Together with increasing energy storage capacity by storing clipped energy, hybrid plants broaden a power plant's grid services by featuring fast dispatch flexibility and voltage-ampere reactive support. While many proposed photovoltaic plants are being developed with co-located batteries, the dominant architecture of the utility-scale PV+BES power plants is uncertain. This section presents various suitable candidates for hybrid PV+BES power plants. Note that the power architectures are expected to evolve over time based on future trends to facilitate the necessary grid services.

10.4 Conclusions

While power converters applied in the power grid are varied, a majority of them can be classified into a few categories that are illustrated in Section 10.2. Among these, the SCR-based technologies have in general been the most mature owing to their early date of introduction, and also the availability of power semiconductor technology that reaches the power levels warranted by the power grid. While continuing advances in dc link VSC-based technologies are becoming viable in the power grid, many of the devices introduced under the rubric of FACTS remain dormant due to considerations of economic scale. On the other hand, the recent introduction of MMCs for realizing high-power converters using gate-turn-off switch technologies appears to be a game changer in displacing SCR-based converter technologies. However, the technology of PWM ac power flow control using VeSCs and solid-state transformers continues to stay within the realm of academic research, in the absence of commercial interests required to promote them. Emerging advances in power semiconductors such as silicon carbide and gallium nitride materials may tip the economies of scale to make such newer technologies more attractive in the future. This chapter has provided a rather concise outline of different power converters used in the power grid, along with selected application examples, chosen to illustrate the diversity of approaches in use. The discussion is necessarily brief due to the nature of this volume and the readers may refer to the primary literature for additional details.

References

[1] Hingorani NG. Power Electronics in Electric Utilities: Role of Power Electronics in Future Power Systems. *Proceedings of the IEEE*. 1988;76(4):481–482.

[2] Cuzner RM, and Venkataramanan G. Current Source Rectifiers in Discontinuous Conduction Modes of Operation. *IEEE Transactions on Industry Applications*. 2015;51(1):470–478.

[3] Cuzner R, Drews D, and Venkataramanan G. Power Density and Efficiency Comparisons of System-Compatible Drive Topologies. *IEEE Transactions on Industry Applications*. 2015;51(1):459–469.

[4] Yazdani A, and Iravani R. *Voltage-Sourced Converters in Power Systems: Modeling, Control, and Applications*. Hoboken, NJ: Wiley-IEEE Press; 2010.

[5] Parkhideh B, and Bhattacharya S. Vector-Controlled Voltage-Source-Converter-Based Transmission Under Grid Disturbances. *IEEE Transactions on Power Electronics*. 2013;28(2):661–672.

[6] Hingorani N, and Gyugyi L. *Understanding FACTS: Concepts and Technology of Flexible AC Transmission Systems*. Piscataway, NJ: IEEE Press; 2000.

[7] Hingorani NG. Flexible AC Transmission. *IEEE Spectrum*. 1993;30(4): 40–45.

[8] Venkataramanan G. Three-Phase Vector Switching Converters for Power Flow Control; 2004.

[9] Venkataramanan G, and Johnson BK. Pulse Width Modulated Series Compensator; 2002.

[10] Garcia-Vite PM, Mancilla-David F, and Ramirez JM. Per-Sequence Vector-Switching Matrix Converter Modules for Voltage Regulation. *IEEE Transactions on Industrial Electronics*. 2013;60(12):5411–5421.

[11] Mancilla-David F, Bhattacharya S, and Venkataramanan G. A Comparative Evaluation of Series Power-Flow Controllers Using DC- and AC-Link Converters. *IEEE Transactions on Power Delivery*. 2008;23(2):985–996.

[12] Ludois DC, Reed JK, and Venkataramanan G. Hierarchical Control of Bridge-of-Bridge Multilevel Power Converters. *IEEE Transactions on Industrial Electronics*. 2010;57(8):2679–2690.

[13] Reed J, Venkataramanan G, and Martínez F. Complex Phasor Modeling and Control of Modular Multilevel Inverters. In: 2011 *IEEE Energy Conversion Congress and Exposition*; 2011. p. 4013–4020.

[14] Ludois DC, and Venkataramanan G. Simplified Terminal Behavioral Model for a Modular Multilevel Converter. *IEEE Transactions on Power Electronics*. 2014;29(4):1622–1631.

[15] Akagi H. Classification, Terminology, and Application of the Modular Multilevel Cascade Converter (MMCC). *IEEE Transactions on Power Electronics*. 2011;26(11):3119–130.

[16] Kawamura W, Hagiwara M, and Akagi H. Control and Experiment of a Modular Multilevel Cascade Converter Based on Triple-Star Bridge Cells. *IEEE Transactions on Industry Applications*. 2014;50(5):3536–3548.

[17] Gyugyi L. Reactive Power Generation and Control by Thyristor Circuits. In: 1976 *IEEE Power Electronics Specialists Conference*; 1976. p. 174–184.

[18] Mendoza-Araya P, Munoz Castro J, Cotos Nolasco J, *et al.* Lab-Scale TCR-Based SVC System for Educational and DG Applications. *IEEE Transactions on Power Systems*. 2011;26(1):3–11.

[19] Manjrekar MD, and Venkataramanan G. Control Strategies for a Hybrid Static Reactive Compensator. *Proceedings of 1996 Canadian Conference on Electrical and Computer Engineering*. 1996;2:834–837 vol.2.

[20] Sapkota B, and Vittal V. Dynamic VAr Planning in a Large Power System Using Trajectory Sensitivities. *IEEE Transactions on Power Systems*. 2010; 25(1):461–469.

[21] Gali OMA, and Mancilla-David F. Realization of an AC-Link Pulse Width Modulated Shunt Converter for STATCOM Applications. *2012 North American Power Symposium (NAPS)*. 2012;p. 1–6.

[22] Lee TL, Hu SH, and Chan YH. D-STATCOM with Positive-Sequence Admittance and Negative-Sequence Conductance to Mitigate Voltage Fluctuations in High-Level Penetration of Distributed-Generation Systems. *IEEE Transactions on Industrial Electronics*. 2013;60(4):1417–1428.

[23] Gutierrez M, Venkataramanan G, and Sundaram A. Solid State Flicker Controller Using a Pulse Width Modulated AC–AC Converter. In: *Conference Record of the 2000 IEEE Industry Applications Conference. Thirty-Fifth IAS Annual Meeting and World Conference on Industrial Applications of Electrical Energy (Cat. No. 00CH37129)*. vol. 5; 2000. p. 3158–3165, vol. 5.

[24] Rahmani S, Hamadi A, Al-Haddad K, *et al.* A Combination of Shunt Hybrid Power Filter and Thyristor-Controlled Reactor for Power Quality. *IEEE Transactions on Industrial Electronics*. 2014;61(5):2152–2164.

[25] Bilgin HF, and Ermis M. Design and Implementation of a Current-Source Converter for Use in Industry Applications of D-STATCOM. *IEEE Transactions on Power Electronics*. 2010;25(8):1943–1957.

[26] Johansson N, Angquist L, and Nee HP. An Adaptive Controller for Power System Stability Improvement and Power Flow Control by Means of a Thyristor Switched Series Capacitor (TSSC). *IEEE Transactions on Power Systems*. 2010;25(1):381–391.

[27] Li K, Zhao J, Zhang C, *et al.* Dynamic Simulator for Thyristor-Controlled Series Capacitor. *IEEE Transactions on Industry Applications*. 2010;46 (3):1096–1102.

[28] Daneshpooy A, and Gole AM. Frequency Response of the Thyristor Controlled Series Capacitor. *IEEE Transactions on Power Delivery*. 2001;16 (1):53–58.

[29] Alvarez HJ, Ramirez J, Zuniga P, *et al.* Static Synchronous Series Compensator for active power flow control during unconventional operating conditions. In: 2014 *IEEE International Autumn Meeting on Power, Electronics and Computing (ROPEC)*; 2014. p. 1–6.

[30] Mancilla-David F, and Venkataramanan G. Modeling and Control of the Static Synchronous Series Compensator under Different Operating Modes. In: 2007 *IEEE Power Electronics Specialists Conference*; 2007. p. 2443–2449.

[31] Milanovic JV, and Zhang Y. Modeling of FACTS Devices for Voltage Sag Mitigation Studies in Large Power Systems. *IEEE Transactions on Power Delivery*. 2010;25(4):3044–3052.

[32] Gyugyi L, Schauder CD, Williams SL, *et al.* The Unified Power Flow Controller: A New Approach to Power Transmission Control. *IEEE Transactions on Power Delivery*. 1995;10(2):1085–1097.

[33] Mancilla-David F, and Venkataramanan G. Realisation of an AC Link Unified Power Flow Controller. *IET Generation Transmission & Distribution*. 2012;6:294–302.

[34] Barragán-Villarejo M, Venkataramanan G, Mancilla-David F, *et al.* Dynamic Modelling and Control of a Shunt-Series Power Flow Controller Based on AC-Link. *IET Generation Transmission & Distribution*. 2012; 6:792–802.

[35] Gyugyi L, Sen KK, and Schauder CD. The Interline Power Flow Controller Concept: A New Approach to Power Flow Management in Transmission Systems. *IEEE Transactions on Power Delivery*. 1999;14(3):1115–1123.

[36] Jiang S, Gole AM, Annakkage UD, *et al.* Damping Performance Analysis of IPFC and UPFC Controllers Using Validated Small-Signal Models. *IEEE Transactions on Power Delivery*. 2011;26(1):446–454.

[37] Venkataramanan G, and Johnson BK. A Superconducting DC Transmission System Based on VSC Transmission Technologies. *IEEE Transactions on Applied Superconductivity*. 2003;13(2):1922–1925.

[38] Saeedifard M, and Iravani R. Dynamic Performance of a Modular Multilevel Back-to-Back HVDC System. In: 2011 *IEEE Power and Energy Society General Meeting*; 2011. p. 1-1.

[39] Qin J, and Saeedifard M. Predictive Control of a Modular Multilevel Converter for a Back-to-Back HVDC System. *IEEE Transactions on Power Delivery*. 2012;27(3):1538–1547.

[40] Sekiguchi K, Khamphakdi P, Hagiwara M, *et al.* A Grid-Level High-Power BTB (Back-to-Back) System Using Modular Multilevel Cascade Converters Without Common DC-Link Capacitor. *IEEE Transactions on Industry Applications*. 2014;50:2648–2659.

[41] Carrasco M, Mancilla-David F, Venkataramanan G, *et al.* Low Frequency HVac Transmission to Increase Power Transfer Capacity. In: *2014 IEEE PES T&D Conference and Exposition*; 2014. p. 1–5.

[42] Wang X, Wei X, and Meng Y. Experiment on Grid-Connection Process of Wind Turbines in Fractional Frequency Wind Power System. *IEEE Transactions on Energy Conversion*. 2015;30(1):22–31.

[43] Miura Y, Mizutani T, Ito M, *et al.* Modular Multilevel Matrix Converter for Low Frequency AC Transmission. In: 2013 *IEEE 10th International Conference on Power Electronics and Drive Systems (PEDS)*; 2013. p. 1079–1084.

[44] She X, Huang AQ, and Burgos R. Review of Solid-State Transformer Technologies and Their Application in Power Distribution Systems. *IEEE Journal of Emerging and Selected Topics in Power Electronics*. 2013; 1(3):186–198.

[45] Horiba T. Lithium-Ion Battery Systems. *Proceedings of the IEEE*. 2014;102 (6):939–950.

[46] Lawder MT, Suthar B, Northrop PWC, *et al.* Battery Energy Storage System (BESS) and Battery Management System (BMS) for Grid-Scale Applications. *Proceedings of the IEEE*. 2014;102(6):1014–1030.

[47] Gupta M. An Examination of Power Converter Architectures for Utility-Scale Hybrid Solar Photovoltaic and Battery Energy Storage Systems: The Features of Several Power Conversion Architectures. *IEEE Industry Applications Magazine*. 2023;29(2):12–31.

[48] Denholm P, Eichman J, and Margolis R. Evaluating the Technical and Economic Performance of PV Plus Storage Power Plants. National Renewable Energy Laboratory; 2017. NREL/TP-6A20-68737. Contract No. DE-AC36-08GO28308.

[49] Gupta M, and Venkataramanan G. DC Bus Regulation in Cascaded Three Phase AC Power Converters with only Decoupling Capacitors. In: 2018 *IEEE Energy Conversion Congress and Exposition (ECCE)*; 2018. p. 4948–4955.

[50] Wang G, Konstantinou G, Townsend CD, *et al.* A Review of Power Electronics for Grid Connection of Utility-Scale Battery Energy Storage Systems. *IEEE Transactions on Sustainable Energy*. 2016;7(4):1778–1790.

[51] Bauer A, Hanisch J, and Ahlswede E. An Effective Single Solar Cell Equivalent Circuit Model for Two or More Solar Cells Connected in Series. *IEEE Journal of Photovoltaics*. 2014;4(1):340–347.

[52] Liu X, Ai W, Marlow MN, *et al.* The Effect of Cell-to-Cell Variations and Thermal Gradients on the Performance and Degradation of Lithium-Ion Battery Packs. *Applied Energy*. 2019;248:489–499.
[53] Elasser A, Agamy M, Sabate J, *et al.* A Comparative Study of Central and Distributed MPPT Architectures for Megawatt Utility and Large Scale Commercial Photovoltaic Plants. *In: IECON 2010 – 36th Annual Conference on IEEE Industrial Electronics Society*; 2010. p. 2753–2758.
[54] Fantham TL, and Gladwin DT. Impact of Cell Balance on Grid Scale Battery Energy Storage Systems. Energy Reports. 2020;6:209–216. *4th Annual CDT Conference in Energy Storage & Its Applications*.
[55] Mohan N. *Power Electronics: A First Course*. 1st ed. John Wiley & Sons; 2011.
[56] Channegowda P and Venkataramanan G. Comparative Evaluation of Capacitor-Coupled and Transformer-Coupled Dual Active Bridge Converters. In: 2016 *IEEE Energy Conversion Congress and Exposition (ECCE)*; 2016. p. 1–8.
[57] Xu W, Guo Z, Vetrivelan A, *et al.* Hardware Design of a 13.8-kV/3-MVA PV Plus Storage Solid-State Transformer (PVS-SST). *IEEE Journal of Emerging and Selected Topics in Power Electronics*. 2022;10(4):3571–3586.
[58] Majmunović B, Mukherjee S, Martin T, *et al.* 1 kV, 10-kW SiC-Based Quadruple Active Bridge DCX Stage in a DC to Three-Phase AC Module for Medium-Voltage Grid Integration. *IEEE Transactions on Power Electronics*. 2022;37(12):14631–14646.

Chapter 11

Permanent magnet synchronous motor drives

Luciano Coutinho Gomes[1] *and Luiz Carlos Gomes Freitas*[1]

11.1 Introduction

The control of electrical drives is a paradigm where a system must have dynamic stability under rather a variability of parameters, there are also innovations pushing the boundaries, with some technological constraints, and market requirements for cost and demand. Permanent magnet synchronous motor (PMSM) drives are well-suited for the evolving demands of modern industry, where energy efficiency, reliability, high performance, affordability, and connectivity are essential. Their adoption is expected to continue growing as industries seek ways to optimize their processes and reduce their environmental impact. PMSMs are used in electric vehicles, due to the required peak power rating, which ranges from less than 10 kW in mild hybrid configurations to 175 kW and above in high-performance strong hybrid and battery-electric drive trains. Quite surprisingly, permanent magnet (PM) synchronous motors show a continuous expansion also in harsh environments with important cost limitations such as in-home appliances [1]. The range of application of PM motors is also enhanced by the recent introduction of new constructive topologies, which are essentially the migration, at different degrees, of isotropic PM motors to pure synchronous reluctance (SynR) motors. The addition of a correct number of PMs to the synchronous reluctance rotor yields a higher power factor and improved efficiency, with limited back electromotive force (EMF) even in the flux-weakening speed range and without a necessary change in the stator design [1]. Figure 11.1 shows a sketch of two structures, among the several possible ones, and the relative axes definition. PMSMs have been widely used in the industry due to the advantages of high power density, high efficiency, and a rapid control response [2–8]. The purpose of this chapter is to provide an overview of PMSM drives and to highlight their importance in the power electronics community.

Section 11.2 presents some research in sensorless control for PMSM, with the special target of highlighting the delicate connections between any theoretical algorithm and its implementation in the power electronic converter. With the same critical sensibility, Section 11.3 illustrates the direct torque control (DTC) and the

[1]Faculdade de Engenharia Elétrica, Universidade Federal de Uberlândia, Brazil

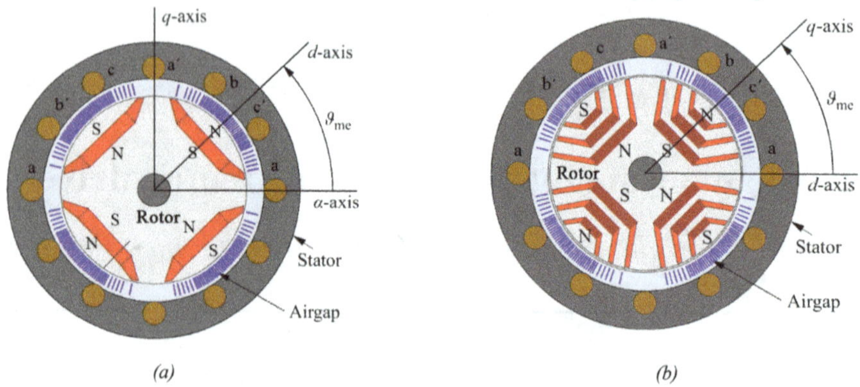

Figure 11.1 Cross-sectional views of four-pole PM motors. (a) Internal PM (IPM) motor. (b) PM-assisted SynR (PMASR) motor.

model predictive control (MPC), reporting a selected example of multi-object, energy-efficient control techniques application for PMSM drives.

11.2 PMSM sensorless control techniques

The delicate position sensors, inherent in every PMSM control, increase the cost of the drive while reducing its reliability. A consolidated trend is the reconstruction of the position through the measurement of electrical quantities only. Although sensorless control has been a major research focus in the last three decades, it still represents a challenge for scientists and currently forms a major part of the literature on PMSM machine control. Even when the position sensor cannot be completely removed, e.g., for safety reasons, the sensorless control is used to provide a backup for the rotor position information in case of a sensor failure. Most commercial motor drives now have a sensorless option embedded in the control system [9].

The extensive study over the years has almost completely defined the most interesting control techniques. The challenge now is to meet demanding criteria such as accuracy, robustness, swiftness, and capability of working over the entire range of motor operations. Among the few possible hardware improvements, there are:

- The direct measurement of the motor voltages. It will profitably substitute the use of the reference voltages and the associated nonideality compensation techniques [10], increasing the robustness and accuracy of practically all the sensorless techniques.
- The development of special PM motors, for example, PM-assisted synchronous reluctance (PMASR) motors, designed to account for certain sensorless machine characteristics [1]. The saliency ratio, the level of cross-saturation, and the eventual presence of secondary saliencies, over a broad load range, are all design aspects that may heavily affect the performance of the sensorless

drive, and they still deserve the attention of the researchers. Particular care will be taken to avoid the saturation of L_q under load, preventing a zero-saliency condition. To this aim, an interesting solution is proposed in [11], where a V-shaped PM array and an eccentric rotor are manufactured on purpose, by deriving a formula for the design of the air gap shape, based on the magnetic circuit model of the interior permanent magnet (IPM) motor. That solution is proven to minimize the effect of loading over the position estimation accuracy. The control technique includes a fast initial position estimation using the magnetic hysteresis phenomenon. Equally, it is expected that special winding configurations with inherent self-sensing features will be further investigated [12].

As regards software solutions, two distinct techniques are established, depending on the dominant speed range of the application: (i) high-frequency signal (HF) injection methods for operations at a standstill and low speed (normally below 20%–30% of the nominal speed), and (ii) back-EMF or flux linkage estimation for middle and high speed. The former technique is based on tracking the position of asymmetries through the injection of high-frequency (HF) voltage signals, and by measuring and postprocessing the related HF currents [13–17]. The latter technique is suited when the speed is high enough to have a favorable signal-to-noise ratio. The techniques used for estimating the back electromotive force and flux linkage can be divided into open-loop methods and closed-loop methods. Some topologies of model-based sensorless control methods with observers can be found in [11,18–25]. For applications where low-speed operation is only transient, an e-EMF-based sensorless controller coupled with a simple start-up procedure offers an attractive low-cost alternative [26].

11.2.1 Key factors for sensorless controls evaluation

In the following, a brief description is given of the different groups in which the schemes can be categorized. Then, the key factors of each sensorless solution are presented. Although these factors may not cover all aspects, they are indeed important considerations to guide the choice of the most appropriate methodology for each application:

- physical principle and dependence on the special design of the motor;
- additional hardware and computational requirements;
- dependence on the parameters and sensitivity to their variations;
- complexity of tuning and dependence on pre-measured values;
- completeness of theoretical analysis, e.g., convergence domain and dynamics;
- inclusion of nonlinearity, as additional saliency and saturation.

In the higher power range (medium-voltage drives), additional peculiarities must be taken into account when proposing a sensorless technique in place of a standard motor drive [19]:

- difficult measurement and conditioning of the DC bus voltage;
- necessarily low switching frequency and consequent high current ripple;
- large temperature range.

A thorough and complete technology status review of sensorless control solutions for PMSM is provided in the excellent papers [19,27,28].

11.2.2 A glance at HF injection methods

Figure 11.2 shows a complete general block schematic diagram of the inner current loop of an HF injection-based sensorless drive and how it interacts with the power inverter. The output of the current regulators, issued by two proportional-integral (PI) blocks in the synchronous (d, q) reference frame, is summed up to HF voltage signals $u_{\alpha}^{\mathrm{h}}, u_{\beta}^{\mathrm{h}}$. The resulting stator voltage references u_{α}^{*} and u_{β}^{*} constitute the input of a space vector modulation (SVM) inverter that computes the switching pattern for the six switches $\mathrm{S}_1-\mathrm{S}_6$ of the voltage inverter. The phase currents (normally two out of three) are measured and post-processed to get either the fundamental frequency estimation (FFE block) or the high frequency estimation (HFE block). The latter is exploited to estimate the rotor electro-mechanical position θ_{me}. When a rotating carrier-signal voltage at HF $\omega_h = \mathrm{d}\theta_{\mathrm{h}}/\mathrm{d}t$ as

$$\mathbf{u}_{\alpha\beta}^{\mathrm{h}} = U_{\mathrm{h}} \begin{bmatrix} \cos(\omega_{\mathrm{h}} t) \\ \sin(\omega_{\mathrm{h}} t) \end{bmatrix} \tag{11.1}$$

is applied to a PMSM, it produces a current vector whose negative sequence component is influenced by the saliency-dependent HF inductances:

$$\mathbf{i}_{\mathrm{n},\alpha\beta}^{\mathrm{h}} = \frac{U_h}{\omega_h} \frac{\left(L_d^h - L_q^h\right)}{2 L_d^h L_q^h} \begin{bmatrix} \sin(\theta_{\mathrm{h}} + 2\theta_{\mathrm{me}}) \\ -\cos(\theta_{\mathrm{h}} + 2\theta_{\mathrm{me}}) \end{bmatrix} \tag{11.2}$$

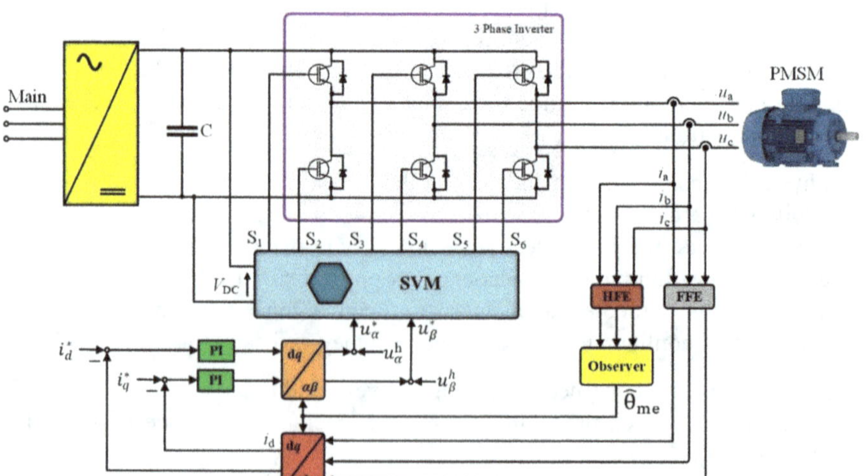

Figure 11.2 General block schematic diagram of an HF injection-based PMSM drive

where L_d^{h}, L_q^{h} are in general the HF differential inductances of the direct and quadrature axis, evaluated by linearization around the working point, according to the small signal theory. Further insights can be found in [22,24,29–31].

The negative sequence current vector (11.2) can be profitably used for rotor position detection [28]. As an alternative, a pulsating carrier voltage can be injected into the g_d- or g_q-axis of a tentative synchronous frame (g_d, g_q):

$$\mathbf{u}_{\alpha\beta}^{\mathrm{h}} = U_{\mathrm{h}}\cos\left(\omega_{\mathrm{h}}t\right)\begin{bmatrix}\cos\left(\widehat{\theta}_{\mathrm{me}}\right)\\ \sin\left(\widehat{\theta}_{\mathrm{me}}\right)\end{bmatrix} \tag{11.3}$$

where $\widehat{\theta}_{\mathrm{me}}$ is the estimated position of (g_d, g_q), with respect to the stationary α-axis. It is possible to demonstrate that, like (11.2), the resulting carrier current space vector is composed of two components, relative to positive and negative sequence currents, both containing the position information [24].

The high-frequency signal injection method is divided into rotation signal injection and pulse vibration signal injection. In general, pulsating carrier observers accomplish an accurate position estimate, but they are less robust and have worse transient response than those based on rotating carriers. Furthermore, the injection of an oscillating signal on the (estimated) flux axis minimizes torque ripple, vibration, and noise [28]. The rotating HF injection is suitable in applications at zero speed under load conditions [32]. The measured signals provided by signal injection in the synchronous reference frame are modulated by the carrier frequency, while the stationary injection methods are already demodulated around the DC level. A detailed comparison of these two approaches has been presented in [29], which analyses the effects of multiple saliencies, cross-saturation, resistive effects, and inverter-induced voltage distortions. It is worth pointing out that HF injection-based sensorless techniques have reached scientific maturity, and now the focus is on the best implementation practice.

11.2.3 An HF sensorless technique for IPM and PMASR motors

Besides the injection of a pulsating carrier in the stationary reference frame and of a rotating carrier in the synchronous reference frame, the third way for sensorless HF control is the injection of a pulsating HF carrier in the stationary reference frame, as proposed by Liu and Zhu [20]. The technique is accurate as a rotating carrier method, but the position is retrieved from the amplitude-modulated carrier current response, which is as simple as the pulsating carrier method. The technique is based on an HF magnetic model of the motor that includes the cross-coupling between the orthogonal axes in the synchronous reference frame, represented by a cross-coupling differential inductance L_{dq}^{h} in the HF inductance matrix $\mathbf{L}_{dq}^{\mathrm{h}}$ [29]. Furthermore, the HF model neglects the speed-related terms, provided that the frequency of the injected voltage is sufficiently higher than the fundamental excitation. Under those hypotheses, an α − axis pulsating voltage excitation

$$\mathbf{u}^{\mathrm{h}}_{\alpha\beta} = U_{\mathrm{h}} \begin{bmatrix} \cos(\omega_{\mathrm{h}} t) \\ 0 \end{bmatrix} \tag{11.4}$$

produces the HF carrier's current response

$$\mathbf{i}^{\mathrm{h}}_{\alpha\beta} = \begin{bmatrix} I_{\mathrm{p}} + I_{\mathrm{n}}\cos(2\theta_{\mathrm{me}} + \theta_{\mathrm{m}}) \\ I_{\mathrm{n}}\sin(2\theta_{\mathrm{me}} + \theta_{\mathrm{m}}) \end{bmatrix} \sin(\omega_{\mathrm{h}} \mathrm{t}) \tag{11.5}$$

where $I_{\mathrm{p}} = U_{\mathrm{h}}/(\omega_{\mathrm{h}} L_{\mathrm{p}})$, $I_{\mathrm{n}} = U_{\mathrm{h}}/(\omega_{\mathrm{h}} L_{\mathrm{n}})$ and θ_{m}, L_{p}, L_{n} are scalar functions of the elements of the HF inductance matrix $\mathbf{L}^{\mathrm{h}}_{dq}$.The HF currents are demodulated by a multiplication by $2\sin(\omega_{\mathrm{h}}\mathrm{t})$ followed by a low-pass filter (LPF), so that

$$\mathbf{i}^{\mathrm{hLPF}}_{\alpha\beta} = \begin{bmatrix} I_{\mathrm{p}} + I_{\mathrm{n}}\cos(2\theta_{\mathrm{me}} + \theta_{\mathrm{m}}) \\ I_{\mathrm{n}}\sin(2\theta_{\mathrm{me}} + \theta_{\mathrm{m}}) \end{bmatrix} \tag{11.6}$$

The retrieval of the position information is relatively simple, once I_{p} and θ_{m} are pre-detected and eliminated from (11.6). The mechanism of rotor position detection deserves careful attention. The conventional inverse-tangent function is substituted by a two-phase-type phase-locked loop, which is effective at low speed, but that entails the risk of divergence after abrupt speed changes, that must be avoided.

The experimental results reported in [20] show excellent precision and good dynamic performance under different load conditions. When evaluated by the assessment grid outlined in Section 11.2.1, it is easy to see that the method does not require additional hardware and that the computational requirement is not excessive. Both I_{p} and ϑ_{m} depend on motor parameters, but they can be derived from some preliminary measurements, that could be even substituted by a proper finite element analysis (FEA) model of the motor. Also, a complete theoretical analysis is somewhat missing, while the magnetic polarity detection, a classic ambiguity in every reluctance-based technique, is solved by a smart start-up procedure. In synthesis, the technique well represents a good starting point for the next development in the specific sector of HF injection methods. Some issues are still open to improvements, but the more and more common availability of FEA models and online parameter estimation techniques [33] should rapidly complete the task. For example, Reference 22 presents a variant of the method, based on the injection of a square wave pulsating signal. The bandwidth of the position estimation is significantly improved, and the influence of the rotor resistance, which can cause problems, as shown by Ghazi Moghadam and Tahami [23], is further reduced with respect to the case of sinusoidal HF injection.

11.2.4 An HF injection method for surface-mounted permanent magnet motors

Isotropic PM motors still represent the largest share of industrial applications. Therefore, even if the trend is toward new PM motor topologies that embed sensorless-oriented features, the study of techniques that extend their validity to surface-mounted permanent magnet (SPM) motors remains a meaningful task. As regards the most suitable method, the HF injection using a pulsating voltage vector seems to yield better results for SPM motors, where the machine saliency is given by the magnetic saturation and not by the rotor anisotropy [1].

An interesting example is given by Zaim *et al.* [21]. The estimation of the rotor position is obtained by injecting a voltage pulse u_{gd} along the direct axis of a *tentative* synchronous rotating reference frame (g_d, g_q). It is supposed that the g_d-axis is aligned with the real d-axis at the start-up. The effect of the current adds up to that of the PM, saturating the stator iron and creating a virtual anisotropy. In the presence of a mismatch φ between the real and the tentative reference frames, the anisotropy creates a cross-coupling between the tentative axes, so that the injected voltage pulse (along g_d) generates a current i_{gq} along the axis g_q. A simple PI controller is then used to eliminate the mismatch. The domain of convergence and the choice of the PI constants are well discussed in [21]. The study also determines the minimum virtual saliency $(L_q - L_d)/L_q$ required to satisfy the observability conditions on the rotor position. The availability of an FEA model of the motor facilitated the computation of practical values for the amplitude u_{gd} and the related application time T_{gd}. This computation is essential to generate the required virtual saliency and enable position estimation.

According to the checkpoints listed in Section 11.2.1, the technique has light hardware and computational requirements, and little tuning complexity, while it reveals a quite heavy dependence on the motor structure, due to the need to know the behavior of the virtual anisotropy as a function of the injected currents. Concerning the completeness of information, further analysis of the dynamics and its connection to the frequency of applying voltage pulses along the g_d-axis would benefit from additional exploration.

Yang and Lorenz [34] show an example of a different innovative point of view. It presents a synchronous-frame HF model of the SPM motor in which the eddy currents induced in both the stator and the rotor are represented through a couple of additional stator resistances along the (d, q) axes. The idea is that injection of a pulsating HF signal as (11.3) can enhance the intrinsic asymmetry of the two resistances, enabling the identification of the rotor position. An in-depth comparison of injection methods, highlighting the pros and cons of both, is carried out in [35].

11.2.5 A glance at extended EMF-based methods

The sensorless methods based on spatial inductance variations are suitable at low speed and standstill, while at high speed they generally suffer the effects of backEMF harmonics and voltage limitation. They also introduce detrimental effects such as additional noise and losses. When required by the application, the full operating speed range is obtained by considering the addition of extended EMF-based techniques, which are complementary as regards speed range [36].

In recent times, in correspondence to an increasing interest in alternative PM rotor structures, the concept of back-EMF has been extended to include any possible anisotropy effect. Because the resulting e-EMF model is of general validity for all synchronous motors including SPM, IPM, and PMASR types, the study of unified sensorless control techniques valid for all the synchronous motors at high speeds becomes possible [15,16,33].

Adopting the space vector notation, the voltage balance equation can be written highlighting the symmetrical voltage components, and both the terms due to the PM and the anisotropy:

$$\mathbf{u}_{dq} = R\mathbf{i}_{dq} + L_d \frac{d\mathbf{i}_{dq}}{dt} + \omega_{\text{me}} L_q \begin{bmatrix} 0 & 1 \\ -1 & 0 \end{bmatrix} \mathbf{i}_{dq} + \mathbf{e}_{dq}^{\text{ext}} \tag{11.7}$$

where the e-EMF vector $\mathbf{e}_{dq}^{\text{ext}}$ is given by:

$$\mathbf{e}_{dq}^{\text{ext}} = \mathbf{e}_{dq} + \begin{bmatrix} 0 \\ (L_d - L_q)\left(\omega_{\text{me}} i_d - \frac{di_q}{dt}\right) \end{bmatrix} \tag{11.8}$$

The same model can be easily back-transformed into the stationary reference frame while maintaining the synchronous inductances among the parameters. By applying the inverse Park transformation to (11.7) the voltage balance becomes:

$$\mathbf{u}_{\alpha\beta} = R\mathbf{i}_{\alpha\beta} + L_d \frac{d\mathbf{i}_{\alpha\beta}}{dt} + \omega_{\text{me}} (L_d - L_q) \begin{bmatrix} 0 & 1 \\ -1 & 0 \end{bmatrix} \mathbf{i}_{\alpha\beta} + \mathbf{e}_{\alpha\beta}^{\text{ext}} \tag{11.9}$$

$$\mathbf{e}_{\alpha\beta}^{\text{ext}} = \left((L_d - L_q)\left(\omega_{\text{me}} i_d - \frac{di_q}{dt}\right) + \Lambda_m \omega_{\text{me}} \right) \begin{bmatrix} -\sin(\theta_{\text{me}}) \\ \cos(\theta_{\text{me}}) \end{bmatrix} \tag{11.10}$$

Similarly, to the HF injection methods, the e-EMF-based techniques are well established. Effective implementations of the most conventional schemes are extensively reported in the literature [37], and now the research is focusing on the relevant aspects that still afflict the performances. Notably, some open issues are (a) the presence of LPFs in the estimation chain; (b) the extraction of the position and the speed from the e-EMF quantity; and (c) the dependence on the motor parameters.

The e-EMF vector is traditionally obtained by either a full or a reduced-order observer. The observer may include the inverse magnetic model, so that the real PM motor acts as a reference model, while the adaptive model consists of the voltage model (that computes the flux linkages) and the inverse magnetic model, which returns the stator currents. The error between the estimated currents and the measured ones is returned to the adaptive model to correct the estimated flux components. In light of the evaluation scheme of Section 11.2.1, the reduced-order observers are simpler and easier to tune than the full-order observers. On the other hand, some studies indicate that the latter can be made less sensitive to model parameter uncertainties and noise, with good dynamic performance. A comprehensive analysis and comparison were carried out by Hinkkanen *et al.* [13,38].

11.2.6 Sliding mode observers for the extended EMF

A modem trend is the substitution of the conventional observer with a sliding mode observer (SMO) to get the back-EMF signals. This is testified by many recent publications [15,39–41] that highlight the advantages in terms of reduced order, simple implementation, and inherent robustness against parameter variations. In a conventional application to SPM motors, the sliding surface is selected as

$$\sigma = \hat{\mathbf{i}}_{\alpha\beta} - \mathbf{i}_{\alpha\beta} = 0 \tag{11.11}$$

Where $\mathbf{i}_{\alpha\beta}$ is the stator current vector in the stationary reference frame and $\hat{\imath}_{\alpha\beta}$ is its estimated value. The SMO vector equation is

$$L_s \frac{\mathbf{di}_{\alpha\beta}}{dt} = \mathbf{u}_{\alpha\beta} - R_{s\hat{\imath}_{\alpha\beta}} - k\Gamma\left(\hat{\imath}_{\alpha\beta} - \mathbf{i}_{\alpha\beta}\right) \tag{11.12}$$

where k is the observer gain. In principle, r(x) should be the sign function, but the trend is to use a sigmoid function to reduce the chattering phenomenon:

$$\Gamma(\mathbf{x}) = \frac{2}{1+e^{-a\mathbf{x}}} - 1 \tag{11.13}$$

and a is a variable parameter that is used to change the sigmoid profile, for example, according to speed [40,41]. Based on the *equivalent control* method, when the system on the sliding surface (11.11) comes true, and an estimation of the e-EMF is obtained as

$$\hat{\mathbf{e}}_{\alpha\beta}^{ext} = k\Gamma\left(\hat{\imath}_{\alpha\beta} - \mathbf{i}_{\alpha\beta}\right) \tag{11.14}$$

that contains the information about the rotor position θ_{me}, as shown by (10.10). The continuous sigmoid function (11.13) has a more favorable harmonic content with respect to the sharp sign function, and this eases the extraction of the low frequency EMF signal from (10.14). As an example of contribution to this topic, Zhao *et al.* [15] proposed an adaptive line enhancer to filter the estimated e-EMF without introducing any phase delay between the original and filtered e-EMF components.

In the case of IPM or PMASR motors, the expression of the e-EMF is much more complex with respect to an SPM motor, since the magnitude of the e-EMF is a function of rotor speed, stator current, and derivative of stator current, which means that the vector is load dependent, as shown by (11.10). High-speed and heavy loads applications can exhibit a relevant distortion of the e-EMF. On the other hand, the use of IPM and PMASR motors is destined to grow, and very likely, this will encourage the study of suitable EMF observer structures and lag-less speed and position estimators. A first good example is given by Zhao *et al.* [42] who developed an e-EMF-based quasi-SMO in the discrete-time domain, proposing an online parameters adaptation mechanism. The result is a rotor position observer that is highly robust to load variations. The paper accomplishes most of the requirements listed in Section 11.2.1 as regards completeness of analysis, innovation, and inclusion of several technical aspects. The computational requirement is not a hindrance, especially in view of the likely availability of low-cost field programmable gate arrays (FPGA) and floating-point microcontrollers.

11.3 MPC in PMSM

A PMSM drive system is a typical nonlinear control system. To have high-performance motor drives with ever faster responses, more centralized control structures, with fewer nested control loops should be adopted. For example, DTC schemes, or simpler feedback loops on centralized multivariable controllers. MPC approach has proven to be the most promising candidate for current and future developments [43–53].

11.3.1 Key factors for MPC

In MPC, a mathematical model of the controlled system is used to predict its future evolution over a receding horizon of finite length. The control sequence to apply at any instant is determined according to optimal criteria, formulated as the maximization or minimization of a certain cost function that depends on the system evolution over the prediction horizon. Constraints on the state and input variables can be included in the optimization problem so that the feasibility of the control action is automatically guaranteed by design. This is one of the salient features that differentiate MPC from other control techniques, where constraint satisfaction is typically managed after the completion of the control design. Moreover, MPC is inherently oriented to the design of multi-variable control laws with a centralized architecture. Unfortunately, these advantages come at the price of an increased computational effort [54,55], since in principle the controller output must be obtained by solving online, at every sampling instant, the constrained optimization problem underlying the MPC formulation. This fact imposes some limitations on both the model complexity and the prediction horizon length that can be effectively considered in a practical MPC implementation. Anyway, the advent of faster microcontrollers and FPGA [56] will undoubtedly improve the situation. The research in MPC is very active, and new fundamental results are expected from the advancement of both the theoretical foundations, and the methods used for its design, analysis, and implementation. The potentialities of a new contribution should be evaluated according to the following aspects, in addition to those listed in Section 11.2.1:

easy formulation and management of soft and hard constraints;
possibility of inclusion of limits and constraints in the design phase;
limited use of approximated linear models;
direct control of the inverter switching states, bypassing the modulator.

11.3.2 Direct torque and flux control

Although there is still no common agreement in the terminology, the DTC conventionally identifies any control scheme where the torque and flux are controlled in a closed loop, by acting directly on the inverter switching commands, and without resorting to any intermediate current control loop to speed up the torque response [57].

In *nonlinear* DTC schemes, the first to be introduced in the mid-1980s, bangbang controllers based on hysteresis comparators are used to keep the torque and flux regulation errors within specified error bands. Several efforts have been made over time to improve the overall performances of the basic scheme, trying to overcome its intrinsic issues, such as the increased torque and current ripples, the variable frequency of the switching commands, and the difficulty of operating at low-speed regimes. Research is still active nowadays, especially in the context of brushless DC motors, where the major concern remains the reduction of the torque ripple [58,59].

In *linear* DTC schemes, the torque and possibly also the stator flux are controlled by conventional PI controllers that calculate, at any instant, the required average voltage vector to be applied over a fixed sampling period. The voltage vector is then synthesized by a pulse-width modulator (PWM), which in most cases exploits the SVM technique. A recent contribution to the design of linear DTCs for IPM motors is reported in [60]. The solution comprises both torque and flux limiters so that the operation in the flux-weakening region can be easily established, as demonstrated in the complementary paper [61].

11.3.3 MPC method for IPM motors

Future advancements in the field of DTC are expected by the gradual replacement of conventional linear and nonlinear schemes with more advanced control techniques, among which MPC represents the most promising candidate. A recent contribution in this sense is given by Preindl and Bolognani [62,63], and sketched in Figure 11.3.

It consists of a DTC for IPM motors based on a finite control set MPC, which is designed for optimal torque generation within the current and voltage limitations imposed by the drive. The control input set is constituted by all the admissible switching states of a two-level, three-phase voltage source inverter (operating in three-phase conduction mode). Accordingly, the cost function of the MPC problem

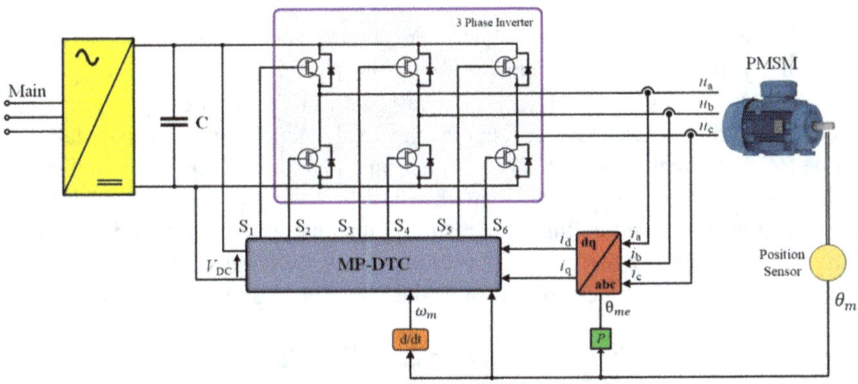

Figure 11.3 General block schematic diagram of an MP-DTC for a PMSM drive

is formulated as the sum of a term $c_T(i_d, i_q)$ that penalizes any deviation from the specified torque reference τ_{m}^*, i.e.,

$$c_T(i_d, i_q) = \left[\tau_{\text{m}}(i_d, i_q) - \tau_{\text{m}}^*\right]^2 \tag{11.15}$$

and a term $c_A(i_d, i_q)$ that penalizes any solution violating the optimal operating condition for the given motor speed and torque request. Below the motor base speed, when operations are only limited by the drive-rated current I_N, i.e.,

$$g_i(i_d, i_q) = \sqrt{{i_d}^2 + {i_q}^2} - I_N \leq 0 \tag{11.16}$$

a maximum-torque-per-ampere (MTPA) criterion is adopted to get minimum copper losses. Accordingly, the cost term $c_{A,1}$ is chosen equal to

$$c_{A,1}(i_d, i_q) = \left(i_d + \frac{L_d - L_q}{\Lambda_{\text{mg}}} + \left({i_d}^2 - {i_q}^2\right)\right)^2 \tag{11.17}$$

At higher speeds, the MTPA criteria can be satisfied up to the torque request that violates the constraint imposed by the drive-rated voltage U_N, i.e.,

$$g_v(i_d, i_q) = \sqrt{\left(\frac{L_q}{L_d} i_q\right)^2 + \left(i_d + \frac{\Lambda_{\text{mg}}}{L_d}\right)^2} - \frac{U_N}{L_d \omega_{\text{me}}} \leq 0 \tag{11.18}$$

$$c_{A,2}(i_d, i_q) = g_v^2(i_d, i_q) \tag{11.19}$$

To allow operation in such a flux-weakening regime, the extra cost is introduced to penalize the solutions off the boundary of (11.18). For sufficiently large motor speeds, the MTPA operation becomes incompatible with the voltage limitation, and therefore the optimal operating points, where a certain torque request is satisfied within the voltage constraints with the minimum amount of current, are always selected on the maximum voltage contour line, up to the point where the maximum torque is delivered, on the maximum-torque-per-voltage (MTPV) characteristic.

The operational constraints (11.16) and (11.18) are included in the MPC formulation by introducing two barrier functions (different from the simple *cost* functions) $c_{L,1}(i_d, i_q)$ and $c_{L,2}(i_d, i_q)$. An extra barrier function $c_{L,3}(i_d, i_q)$ is introduced to avoid solutions located beyond the MTPV curve in the flux-weakening regime [62,63]. Finally, an overall cost function c is defined as the weighted sum of all the cost and barrier functions cited above.

Since the control set is finite, the resulting unconstrained optimization problem is solved by evaluating the cost c for all the admissible control sequences over the prediction horizon, and then by selecting the one yielding the minimum cost.

11.3.4 A perspective on centralized MPC-based structures

The on-board computational power in most of the current electric drives is still insufficient for running a full-order MPC at the sampling rates (order of tens of kilohertz) usually required to get satisfactory control performances. The trend that can be noticed in recent contributions is either to find a closed-form solution to the MPC problem [64,65] or to reduce the number of admissible control values involved in the online optimization process [66,67].

In [64], it is proposed a continuous control set nonlinear model predictive control scheme for the simultaneous speed and *d*-current regulation in PMSMs. The design is based on a novel predictive output tracking control technique for nonlinear systems developed by the authors. The technique consists of approximating the evolution of the system output over the prediction horizon with Taylor's expansion to the time variable. The approximation is used to reformulate the optimal output tracking problem, specified over the prediction horizon, as a parametric optimization problem for the control input and its derivatives. Since the solution of the problem can be specified in closed form, it follows that no online optimization is indeed required, and an explicit analytical expression (depending on the system state and input disturbance) of the optimal nonlinear predictive control law can be provided. To the general evaluation criteria outlined in Section 11.3.1, a point that surely needs attention is the inclusion of the constraints, which at present is missing.

A promising approach seems to be that proposed in [66,67], which is based on an MPC finite control set similar to that adopted in [62,63]. The two contributions differ mainly on how the constraints are managed: while in [66] the constraints are converted into penalty terms to be added to the cost function, in [67] the constraints are used to select beforehand the feasible control sequences that will be taken into account by the optimization process. Moreover, the cost function in [67] has a hierarchical structure, which allows to set priorities in the control goals. Some efforts are still required to solve the two main problems that undermine the applicability of the MPC, namely

- the increased torque ripple compared to the use of a voltage modulator,
- the limited prediction horizon achievable in practice with current technology.

The first issue is fixed in [66], by adding an extra term to the cost function, that weights a high-pass filtered version of the torque command, while in [67] it is suggested to increase the number of voltage vectors in the control set. The second problem, instead, still requires further investigation, even though it is believed that it will be easily overcome with the advent of more powerful computational units.

11.4 Energy efficiency in PMSM drives

Optimizing power conversion efficiency, especially in variable speed applications utilizing technologies such as PMSMs, is critical to reducing energy costs, mitigating environmental impact, promoting a green economy, and complying with regulations. As the demand for energy-efficient solutions continues to grow, it drives innovation and increases our ability to address global energy and environmental challenges. Especially for small and medium power variable speed applications, PMSMs represent the ideal choice due to their inherent high efficiency across all speed and load torque ranges. Predicted values of efficiency are about 95%–97% for a 10–100 kW PMSM. Over the past years, thanks to the ever-decreasing cost of variable frequency AC power supplies enabled by the remarkable advances in power switching devices and digital control units, this improvement has been mostly achieved by the introduction of adjustable speed drives in applications where constant speed drives were generally used, e.g., pumps, fans, air conditioners, heat pumps, etc.

The major losses in PMSMs occur on the stator, either as joule losses on the stator windings or as core losses in the stator teeth and yoke. Apart from machine design considerations, the stator losses, being controllable, can be reduced by a proper loss minimization control strategy. This is especially true for IPM and PMASR motors that present a degree of freedom in the composition of the electromagnetic torque. These motors are becoming increasingly popular in traction drives for electric and hybrid vehicles [68], owing especially to their capability of operating over a wide speed range, with a constant power speed range of up to five times the base speed, and under intermittent overloading conditions, with up to almost twice the rated torque, at least for short durations [69]. Below the base speed, copper losses can be effectively reduced by an MTPA control strategy, which is a common trend. Available methods are either model-based, comprising parameter adaptation for enhanced robustness and capability to cope with magnetic saturation effects [70], or model-free, mostly perturbation-and-observe methods, based on the extremum seeking control working principle [71].

From a close examination of the solutions proposed so far in the literature concerning the efficient control of PMSMs, it can be noticed that most of them are still focused exclusively on optimizing the operating condition of the machine, without concern about the additional switching and harmonic losses produced by the almost ever-present PWM voltage source inverter. This habit is likely to change since it may be a source of misunderstanding.

It is worth noting that, in general, the optimal operating conditions of the motor and the inverter do not coincide, and a trade-off must be found. In this sense, better results are expected if the efficiency optimization is performed at a "system level", in which the tight interaction between the motor and inverter is explicitly taken into account. By proceeding in this way, it becomes also easier to consider additional optimization objectives (multi-objective optimization), e.g., the minimization of the number of inverter switching [72], and operational constraints, e.g., the thermal limits of the power components (power switches and freewheeling diodes) in the inverter [73]. It is believed that even from the energy efficiency perspective, the advent of smarter drives, with more computational power on board, will help to popularize an integrated optimization approach, especially when embedded within a more general centralized MPC-based architecture.

11.5 Conclusions

The use of PMSMs is widespread, and the associated power electronic control has reached a high level of maturity. New hybrid (reluctance and PM) motor topologies are studied to provide motors with improved efficiency and self-sensing features. The use of these motors complicates the control algorithms and makes the availability of a complete and reliable magnetic model unavoidable. Therefore, the modeling of complex systems will probably take a large part of research activity in the field. Further technological steps will be the use of more powerful computing devices, as FPGAs, this, in turn, will facilitate both direct voltage measurement and the implementation of more advanced signal filtering techniques. Finally, the emerging energy awareness is going to draw new paradigms, based on multivariable control, for the contemporary optimization of different power electronics and drive aspects. In perspective, the best candidate is the MPC, as soon as its online implementation becomes technically feasible.

Acknowledgments

The author gratefully acknowledges the contributions of Mauro Zigliotto from the University of Padova, who started this work.

References

[1] K. Rajashekara, "Present status and future trends in electric vehicle propulsion technologies," *Emerging and Selected Topics in Power Electronics, IEEE Journal of*, vol. 1, pp. 3–10, 2013.

[2] X. Sun, F. Cai, Z. Yang, and X. Tian, "Finite position control of interior permanent magnet synchronous motors at low speed," *IEEE Transactions on Power Electronics*, vol. 37, pp. 7729–7738, 2022. https://doi.org/10.1109/TPEL.2022.3146841

[3] Z. Novak, and M. Novak, "Adaptive PLL-based sensorless control for improved dynamics of high-speed PMSM," *IEEE Transactions on Power Electronics*, vol. 37, pp. 10154–10165, 2022. https://doi.org/10.1109/TPEL.2022.3169708

[4] X. Sun, C. Hu, G. Lei, Y. Guo, and J. Zhu, "State feedback control for a PM hub motor based on gray wolf optimization algorithm," *IEEE Transactions on Power Electronics*, vol. 35, pp. 1136–1146, 2020. https://doi.org/10.1109/TPEL.2019.2923726

[5] T.-I. Yeam and D.-C. Lee, "Design of sliding-mode speed controller with active damping control for single-inverter dual-PMSM drive systems," *IEEE Transactions on Power Electronics*, vol. 36, pp. 5794–5801, 2021. https://doi.org/10.1109/TPEL.2020.3028601

[6] T. Li, X. Sun, M. Yao, D. Guo, and Y. Sun, "Improved finite control set model predictive current control for permanent magnet synchronous motor with sliding mode observer," *IEEE Transactions on Transportation Electrification*. 2023. https://doi.org/10.1109/TTE.2023.3293510

[7] A.T. Woldegiorgis, X. Ge, H. Wang, and Y. Zuo, "An active flux estimation in the estimated reference frame for sensorless control of IPMSM," *IEEE*

Transactions on Power Electronics, vol. 37, pp. 9047–9060, 2022. https://doi.org/10.1109/TPEL.2022.3156726

[8] L. Chen, H. Xu, and X. Sun, "A novel strategy of control performance improvement for six-phase permanent magnet synchronous hub motor drives of EVs under new European driving cycle," *IEEE Transactions on Vehicular Technology*, vol. 70, pp. 5628–5637, 2021. https://doi.org/10.1109/TVT.2021.3079576

[9] S. Vaez-Zadeh, *Control of Permanent Magnet Synchronous Motors*, Oxford University Press, Oxford, 2018. DOI 10.1093/oso/9780198742968.001.0001

[10] L. Peretti and M. Zigliotto, "FPGA-based voltage measurements in ac drives," Electrical Machines (ICEM)*, 2010 XIX International Conference on*, September 2010, pp. 1–6.

[11] S. Murakami, T. Shiota, M. Ohto, K. Ide, and M. Hisatsune, "Encoderless servo drive with adequately designed IPMSM for pulse-voltage-injection-based position detection," *Industry Applications, IEEE Transactions on*, vol. 48, no. 6, pp. 1922–1930, 2012.

[12] A. Faggion, E. Fomasiero, N. Bianchi, and S. Bolognani, "Sensorless capability of fractional-slot surface-mounted pm motors," *Industry Applications, IEEE Transactions on*, vol. 49, no. 3, pp. 1325–1332, 2013.

[13] T. Tuovinen, M. Hinkkanen, L. Hamefors, and J. Luomi, "Comparison of a reduced-order observer and a full-order observer for sensorless synchronous motor drives," *Industry Applications, IEEE Transactions on*, vol. 48, no. 6, pp. 1959–1967, 2012.

[14] R. Leidhold, "Position sensorless control of PM synchronous motors based on zero-sequence carrier injection," *Industrial Electronics, IEEE Transactions on*, vol. 58, no. 12, pp. 5371–5379, 2011.

[15] Y. Zhao, W. Qiao, and L. Wu, "Improved rotor position and speed estimators for sensorless control of interior permanent-magnet synchronous machines," *Emerging and Selected Topics in Power Electronics, IEEE Journal of*, vol. 2, no. 3, pp. 627–639, 2014.

[16] K. Lu, X. Lei, and F. Blaabjerg, "Artificial inductance concept to compensate nonlinear inductance effects in the back EMF-based sensorless control method for PMSM," *Energy Conversion, IEEE Transactions on*, vol. 28, no. 3, pp. 593–600, 2013.

[17] F. Genduso, R. Miceli, C. Rando, and G. Galluzzo, "Back EMF sensorless-control algorithm for high-dynamic performance PMSM," *Industrial Electronics, IEEE Transactions on*, vol. 57, no. 6, pp. 2092–2100, 2010.

[18] S. Morimoto, K. Kawamoto, M. Sanada, and Y. Takeda, "Sensorless control strategy for salient-pole PMSM based on extended EMF in rotating reference frame," in *Industry Applications Conference, 2001. Thirty-Sixth IAS Annual Meeting. Conference Record of the 2001 IEEE*, vol. 4, September 2001, pp. 2637–2644.

[19] M. Pacas, "Sensorless drives in industrial applications," *Industrial Electronics Magazine, IEEE*, vol. 5, no. 2, pp. 16–23, 2011.

[20] J. Liu and Z. Zhu, "Novel sensorless control strategy with injection of high-frequency pulsating carrier signal into stationary reference frame," *Industry Applications, JEEE Transactions on*, vol. 50, no. 4, pp. 2574–2583, 2014.

[21] S. Zaim, B. Nahid-Mobarakeh, and F. Meibody-Tabar, "Robust position sensorless control of nonsalient PMSM at standstill and low speeds," *Emerging and Selected Topics in Power Electronics, JEEE Journal of*, vol. 2, no. 3, pp. 640–650, 2014.

[22] J. Liu and Z. Zhu, "Sensorless control strategy by square-waveform high-frequency pulsating signal injection into stationary reference frame," *Emerging and Selected Topics in Power Electronics, IEEE Journal of*, vol. 2, no. 2, pp. 171–180, 2014.

[23] M. Ghazi Moghadam and F. Tahami, "Sensorless control of PMSMs with tolerance for delays and stator resistance uncertainties," *Power Electronics, IEEE Transactions on*, vol. 28, no. 3, pp. 1391–1399, 2013.

[24] A. Accetta, M. Cirrincione, M. Pucci, and G. Vitale, "Sensorless control of PMSM fractional horsepower drives by signal injection and neural adaptiveband filtering," *Industrial Electronics, IEEE Transactions on*, vol. 59, no. 3,pp. 1355–1366, 2012.

[25] Y. Xu, M. Yao, and X. Sun. "Overview of position-sensorless technology for permanent magnet synchronous motor systems," *World Electric Vehicle Journal*, vol. 14, no. 8: p. 212, 2023. https://doi.org/10.3390/wevj14080212

[26] Z. Wang, K. Lu, and F. Blaabjerg, "A simple startup strategy based on current regulation for back-EMF-based sensorless control of PMSM," *Power Electronics, IEEE Transactions on*, vol. 27, no. 8, pp. 3817–3825, 2012.

[27] I. Boldea, "Control issues in adjustable speed drives," *Industrial Electronics Magazine, IEEE*, vol. 2, no. 3, pp. 32–50, 2008.

[28] F. Briz and M. Degner, "Rotor position estimation," *Industrial Electronics Magazine, IEEE*, vol. 5, no. 2, pp. 24–36, 2011.

[29] D. Raca, P. Garcia, D. Reigosa, F. Briz, and R. Lorenz, "Carrier-signal selection for sensorless control of PM synchronous machines at zero and very low speeds," *Industry Applications, IEEE Transactions on*, vol. 46, no. 1, pp. 167–178, 2010.

[30] P. Guglielmi, M. Pastorelli, and A. Vagati, "Cross-saturation effects in IPM motors and related impact on sensorless control," *Industry Applications, IEEE Transactions on*, vol. 42, no. 6, pp. 1516–1522, 2006.

[31] N. Bianchi and S. Bolognani, "Influence of rotor geometry of an 1PM motor on sensorless control feasibility," *Industry Applications, IEEE Transactions on*, vol. 43, no. 1, pp. 87–96, 2007.

[32] R. Bojoi, M. Pastorelli, J. Bottomley, P. Giangrande, and C. Gerada, "Sensorless control of PM motor drives: A technology status review," in *Electrical Machines Design Control and Diagnosis (WEMDCD)*, 2013 *IEEE Workshop on,* March 2013, pp. 168–182.

[33] Y. Inoue, Y. Kawaguchi, S. Morimoto, and M. Sanada, "Performance improvement of sensorless IPMSM drives in a low-speed region using online

parameter identification," *Industry Applications, IEEE Transactions on*, vol. 47, no. 2, pp. 798–804, 2011.

[34] S.-C. Yang and R. Lorenz, "Surface permanent magnet synchronous machine position estimation at low speed using eddy-current-reflected asymmetric resistance," *Power Electronics, IEEE Transactions on*, vol. 27, no. 5, pp. 2595–2604, 2012.

[35] S.-C. Yang and R. Lorenz, "Comparison of resistance-based and inductance-based self-sensing controls for surface permanent-magnet machines using high-frequency signal injection," *Industry Applications, IEEE Transactions on*, vol. 48, no. 3, pp. 977–986, 2012.

[36] M. Corley and R. Lorenz, "Rotor position and velocity estimation for a salient-pole permanent magnet synchronous machine at standstill and high speeds," *Industry Applications, IEEE Transactions on*, vol. 34, no. 4, pp. 784–789, 1998.

[37] S. Bolognani, S. Calligaro, and R. Petrella, "Design issues and estimation errors analysis of back-EMF-based position and speed observer for SPM synchronous motors," *Emerging and Selected Topics in Power Electronics, IEEE Journal of*, vol. 2, no. 2, pp. 159–170, 2014.

[38] M. Hinkkanen, T. Tuovinen, L. Hamefors, and J. Luomi, "A combined position and stator-resistance observer for salient PMSM drives: Design and stability analysis," *Power Electronics, IEEE Transactions on*, vol. 27, no. 2, pp. 601–609, 2012.

[39] T. Bemardes, V. Foletto Montagner, H. Grundling, and H. Pinheiro, "Discrete-time sliding mode observer for sensorless vector control of permanent magnet synchronous machine," *Industrial Electronics, IEEE Transactions on*, vol. 61, no. 4, pp. 1679–1691, 2014.

[40] Z. Qiao, T. Shi, Y. Wang, Y. Yan, C. Xia, and X. He, "New sliding-mode observer for position sensorless control of permanent-magnet synchronous motor," *Industrial Electronics, IEEE Transactions on*, vol. 60, no. 2, pp. 710–719, 2013.

[41] H. Kim, J. Son, and J. Lee, "A high-speed sliding-mode observer for the sensorless speed control of a PMSM," *Industrial Electronics, IEEE Trans actions on*, vol. 58, no. 9, pp. 4069–4077, 2011.

[42] Y. Zhao, W. Qiao, and L. Wu, "An adaptive quasi-sliding-mode rotor position observer-based sensorless control for interior permanent magnet synchronous machines," *Power Electronics, IEEE Transactions on*, vol. 28, no. 12, pp. 5618–5629, 2013.

[43] M. Abu-Ali, F. Berkel, M. Manderla, S. Reimann, R. Kennel, and M. Abdelrahem, "Deep learning-based long-horizon MPC: Robust, high performing, and computationally efficient control for PMSM drives," *IEEE Transactions on Power Electronics*, vol. 37, pp. 12486–12501, 2022. https://doi.org/10.1109/TPEL.2022.3172681

[44] X. Sun , T. Li, Z. Zhu, G. Lei, Y. Guo, and J. Zhu, "Speed sensorless model predictive current control based on finite position set for PMSHM drives,"

IEEE Transactions on Transportation Electrification, vol. 7, pp. 2743–2752, 2021. https://doi.org/10.1109/TTE.2021.3081436

[45] M. Wu, X. Sun, J. Zhu, G. Lei, and Y. Guo, "Improved model predictive torque control for PMSM drives based on duty cycle optimization," *IEEE Transactions on Magnetics*, vol. 57, p. 8200505, 2021. https://doi.org/10.1109/TMAG.2020.3008495

[46] S. Xu, Z. Sun, C. Yao, H. Zhang, W. Hua, and G. Ma, "Model predictive control with constant switching frequency for three-level T-type inverter-fed PMSM drives," *IEEE Transactions on Industrial Electronics*, vol. 69, pp. 8839–8850, 2022. https://doi.org/10.1109/TIE.2021.3114716

[47] C. Xue, D. Zhou, and Y. Li, "Finite-control-set model predictive control for three-level NPC inverter-fed PMSM drives with LC filter," *IEEE Transactions on Industrial Electronics*, vol. 68, pp. 11980–11991, 2021. https://doi.org/10.1109/TIE.2020.3042156

[48] X. Tian, Y. Cai, X. Sun, Z. Zhu, and Y. Xu, "A novel energy management strategy for plug-in hybrid electric buses based on model predictive control and estimation of distribution algorithm," *IEEE/ASME Transactions on Mechatronics*, vol. 27, pp. 4350–4361 2022. https://doi.org/10.1109/TMECH.2022.3156150

[49] X. Zhang, H. Bai, and M. Cheng, "Improved model predictive current control with series structure for PMSM drives," *IEEE Transactions on Industrial Electronics*, vol. 69, pp. 12437–12446, 2022. https://doi.org/10.1109/TIE.2021.3134088

[50] T. Xu, X. Wang, D. Xiao, X. Meng, Y. Mao, and Z. Wang, "A novel two-mode inverter-based open-winding PMSM drive and its modulation strategies," *IEEE Transactions on Power Electronics*, vol. 38, pp. 8762–8774, 2023. https://doi.org/10.1109/TPEL.2023.3267064

[51] L. Chen, H. Xu, X. Sun, and Y. Cai, "Three-vector-based model predictive torque control for a permanent magnet synchronous motor of EVs," *IEEE Transactions on Transportation Electrification*, vol. 7, pp. 1454–1465, 2021. https://doi.org/10.1109/TTE.2021.3053256

[52] Y. Luo, K. Yang, and Y. Zheng, "Luenberger observer-based model predictive control for six-phase PMSM motor with localization error compensation," *IEEE Transactions on Industrial Electronics*, vol. 70, pp. 10800–10810, 2023. https://doi.org/10.1109/TIE.2022.3229340

[53] X. Sun, T. Li, X. Tian, and J.G. Zhu, "Fault-tolerant operation of a six-phase permanent magnet synchronous hub motor based on model predictive current control with virtual voltage vectors," *IEEE Transactions on Energy Conversation*, vol. 37, pp. 337–346, 2022. https://doi.org/10.1109/TEC.2021.3109869

[54] Z. Zhu, Y. Li, D. Howe, C. Bingham, and D. Stone, "Influence of machine topology and cross-coupling magnetic saturation on rotor position estimation accuracy in extended back-EMF-based sensorless PM brushless ac drives," in *Industry Applications Conference, 2007. 42nd IAS Annual Meeting. Conference Record of the 2007 IEEE*, 2007, pp. 2378–2385.

[55] P. Cortes, M. Kazmierkowski, R. Kennel, D. Quevedo, and J. Rodriguez, "Predictive control in power electronics and drives," *Industrial Electronics, IEEE Transactions on*, vol. 55, no. 12, pp. 4312–4324, 2008.

[56] M. W. Naouar, E. Monmasson, A. Naassani, I. Slama-Belkhodja, and N. Patin, "FPGA-based current controllers for ac machine drives: A review," *Industrial Electronics, IEEE Transactions on*, vol. 54, no. 4, pp. 1907–1925, 2007.

[57] F. Betin, G.-A. Capolino, D. Casadei, *et al.*, "Trends in electrical machines control: Samples for classical, sensorless, and fault-tolerant techniques," *Industrial Electronics Magazine, IEEE*, vol. 8, no. 2, pp. 43–55, 2014.

[58] M. Masmoudi, B. El Badsi, and A. Masmoudi, "Direct torque control of brushless DC motor drives with improved reliability," *Industry Applications, IEEE Transactions on*, vol. 50, no. 6, pp. 3744–3753, 2014.

[59] M. Masmoudi, B. El Badsi, and A. Masmoudi, "DTC of B4-inverter-fed BLDC motor drives with reduced torque ripple during sector-to-sector commutations," *Power Electronics, IEEE Transactions on*, vol. 29, no. 9, pp. 4855–4865, 2014.

[60] Y. Inoue, S. Morimoto, and M. Sanada, "Control method suitable for direct torque-control-based motor drive system satisfying voltage and current limitations," *Industry Applications, IEEE Transactions on*, vol. 48, no. 3, pp. 970–976, 2012.

[61] Y. Inoue, S. Morimoto, and M. Sanada, "Comparative study of PMSM drive systems based on current control and direct torque control in flux-weakening control region," *Industry Applications, IEEE Transactions on*, vol. 48, no. 6, pp. 2382–2389, 2012.

[62] M. Preindl and S. Bolognani, "Model predictive direct torque control with finite control set for PMSM drive systems. Part 1: Maximum torque per ampere operation," *Industrial Informatics, IEEE Transactions on*, vol. 9, no. 4, pp. 1912–1921, 2013.

[63] M. Preindl and S. Bolognani, "Model predictive direct torque control with finite control set for PMSM drive systems. Part 2: Field weakening operation," *Industrial Informatics, IEEE Transactions on*, vol. 9, no. 2, pp. 648–657, 2013.

[64] R. Errouissi, M. Ouhrouche, W.-H. Chen, and A. Trzynadlowski, "Robust nonlinear predictive controller for permanent-magnet synchronous motors with an optimized cost function," *Industrial Electronics, IEEE Transactions on*, vol. 59, no. 7, pp. 2849–2858, 2012.

[65] R. Errouissi, M. Ouhrouche, W.-H. Chen, and A. Trzynadlowski, "Robust cascaded nonlinear predictive control of a permanent magnet synchronous motor with anti-windup compensator," *Industrial Electronics, IEEE Trans actions on*, vol. 59, no. 8, pp. 3078–3088, 2012.

[66] E. Fuentes, C. Silva, and J. Yuz, "Predictive speed control of a two-mass system driven by a permanent magnet synchronous motor," *Industrial Electronics, IEEE Transactions on*, vol. 59, no. 7, pp. 2840–2848, 2012.

[67] M. Carraro, M. Zigliotto, and L. Peretti, "FPGA-based hierarchical finite states predictive control for PMSM drives," *in Power Electronics, Machines and Drives (PEMD 2014), Seventh JET International Conference on,* April 2014, pp. 1–6.

[68] G. Pellegrino, A. Yagati, P. Guglielmi, and B. Boazzo, "Performance comparison between surface-mounted and interior PM motor drives for electric vehicle application," *Industrial Electronics, IEEE Transactions on*, vol. 59, no. 2, pp. 803–811, 2012.

[69] G. Pellegrino, A. Vagati, and P. Guglielmi, "Design tradeoffs between constant power speed range, uncontrolled generator operation, and rated current of IPM motor drives," *Industry Applications, IEEE Transactions on*, vol. 47, no. 5, pp. 1995–2003, 2011.

[70] Q. K. Nguyen, M. Petrich, and J. Roth-Stielow, "Implementation of the MTPA and MTPV control with on-line parameter identification for a high speed IPMSM used as traction drive," in *Power Electronics Conference (IPEC-Hiroshima 2014 – ECCE-ASIA), 2014 International*, 2014, pp. 318–323.

[71] R. Antonello, M. Carraro, and M. Zigliotto, "Maximum-torque-per-ampere operation of anisotropic synchronous permanent-magnet motors based on extremum seeking control," *Industrial Electronics, IEEE Transactions on*, vol. 61, no. 9, pp. 5086–5093, 2014.

[72] G. Prior and M. Krstic, "Quantized-input control Lyapunov approach for permanent magnet synchronous motor drives," *Control Systems Technology, IEEE Transactions on*, vol. 21, no. 5, pp. 1784–1794, 2013.

[73] J. Lemmens, P. Yanassche, and J. Driesen, "Optimal control of traction motor drives under electrothermal constraints," *Emerging and Selected Topics in Power Electronics, IEEE Journal of*, vol. 2, no. 2, pp. 249–263, 2014.

Chapter 12
Induction motor drives

Marcelo Godoy Simões[1,2]

12.1 Introduction

This chapter covers a brief introduction to induction machines, with their features and importance, and why such induction machines are so important in the modern age. Then some early developments are explained, after the birth of power electronics with the advent of thyristors, initially induction machines had programmable voltage starting and crude control with silicon-controlled rectifiers on natural commutation. The chapter discusses how variable speed induction motors are important for energy-saving purposes because most industrial applications would have centrifugal load torque applications, where the change is proportional to the square of speed, and consequently power changes in proportion with the cube of speed with energy varying at different speeds. Variable frequency drives manage energy consumption by changing the speed, and continuous speed control is desirable for a variety of applications such as moving air, fluids, heating and ventilation, lifting water, moving belts, and loads in container systems, providing energy efficiency, reliable dynamics, and cost efficiency.

Variable frequency drives can have speed feedback, speed control is important on a shaft where on the other side of the shaft there is some mover that impresses constant torque. On the other hand, if the device on the shaft imposes constant speed, the motor drive will mostly have to impress a torque control, and induction motor drives can be programmed for either torque or speed closed-loop control.

When McMurray invented the first forced-commutation based on thyristors and proposed for the first time a variable frequency operation with the proportionality of impressed voltage with frequency, i.e., a Volts/Hertz operation, the induction motor drive was essentially born, such operation was based on voltage-fed square-wave 6-step inverter, without modulation. During the 1960s and 70s, several theoretical and practical achievements furthered the motor drives. Induction machine control was enhanced when in 1969 Hasse introduced direct vector control (DVC), there was a decoupled system achieved with flux measurements in the

[1]Professor in Flexible and Smart Power Systems, Department of Electrical Engineering, University of Vaasa, Finland
[2]Professor Emeritus, Colorado School of Mines, USA

air-gap of the induction machine. Indirect vector control (IVC) was introduced by Blaschke in 1972. The analysis of the rotor d-q currents in the rotating reference frame made possible the vector control with analysis of the synchronous reference frame and stationary reference frame dynamic models of the machine, nowadays such IVC conditions are defined as "Blaschke equations". Those techniques are called field-oriented control (FOC) and are discussed in detail in this chapter.

During the 1980s, there were many seminal works. By that time, sinusoidal pulse width modulation had not yet been established with the at-the-time technology of bipolar transistors. Therefore, the current control of machines was dominated by the variable frequency operation of bang-bang controllers. In 1980, B.K. Bose made a significant contribution by constraining the variable frequency of those controllers in his paper published in the *IEEE Transactions on Industrial Electronics* entitled "An adaptive hysteresis-band current control technique of a voltage-fed PWM inverter for machine drive system", which was a breakthrough allowing possible better implementations of field-oriented control. Induction machines were just acceptable for fixed frequency operation, or as induction generators, and for the first time in 1983, J.M. Elder published an understanding on his paper entitled "The process of self excitation in induction generators", while the industrial automation field was for the first time understanding that asynchronous electrical machines, could be considered for "AC servo system for position control" as published by A. Vagati in 1984.

The years 1985 and 1986 were important for the publication of important books that became adopted, studied, and educated a generation of engineers, including the author of this chapter. In 1985, Werner Leonhard published by Springer-Verlag his book "Control of Electrical Drives", still some high power applications were developed at that time, and the book by G.K. Dubey was very important "Thyristorised Power Controllers", also in the same year D.W. Novotny started the idea of controlling voltage-source-converters in current mode control, giving the foundation for further modulation implementations with his paper in *IEEE Transactions on Industry Applications* "Current control of VSI-PWM inverters". In 1986 B.K. Bose published the first edition of his worldwide famous outstanding book, published by Prentice-Hall "Power Electronics and AC Drives". A very important book with abc, d-q, and symmetrical analysis book for computer-based implementation was published by P.C. Krause with McGraw-Hill Book "Analysis of Electric Machines".

T.M. Rowan published in the *IEEE Transactions on Industry Applications* "A new synchronous current regulator and an analysis of a current-regulated PWM inverter", furthering the initial work by D.W. Novotny into possible d-q quadrature reference frame-based controllers. In 1988 for the first time, there were implementations of pulse width modulation based on voltage space vectors. In this same year, two important new conceptions were made, the birth of stator flux orientation with the paper by Xu, DeDoncker, and Novotny in the IEEE IAS Annual Meeting of 1988, and control systems started to be touched by this effervescent area of induction motor drives when Verguese proposed for the first time approaches of observers for flux estimation in induction machines in the *IEEE Transactions on*

Industrial Electronics followed by C. Schauder with his splendid breakthrough paper on "Adaptive speed identification for vector control of induction motors without rotational transducers" presented at the IEEE IAS Annual Meeting in 1989. The 1990s witnessed the advancement of adaptive control and model reference control for induction machines. During this period, there was a focus on studying the behavior of the system when subjected to high-frequency signal injection. Induction machines, composed of copper, iron, and magnetic fluxes circulating in imperfect and potentially anisotropic materials, posed challenges. To address these challenges, researchers explored the use of a carrier signal superimposed on PWM waveforms. This approach aimed to improve frequency tracking, signal-to-noise ratio detection, and spectral classification, necessitating the development of advanced signal processing algorithms. In 1994 M.G. Simões proposed for the first time a neural network-based control for vector control signal estimation. After the 2000s there were still some developments of motor drives, parameter variation enhancements, self-commissioning, sensorless-based control, and auto-tuning of controllers in real-time, but eventually so much was done that the research area started to decrease novel implementations. With the advent of artificial intelligence and strong computer processing for real-time, and allowing the integration of motor drives and generator converters in smart-grids with multifunctionalities in several upper layers of energy management and forecasting with digital twin implementation, the area is again evolving in a fast pace.

This chapter shows the theoretical foundations of abc three-phase and d-q stationary and reference frame modeling and analysis to support the implementation of scalar control, indirect vector control based on rotor flux alignment, and direct vector control based on stator flux alignment. Scalar control is a V/f approach that may have slip frequency feedforward torque control, but they have somewhat inferior performance when compared to vector control, they are easy to implement and are very popular for pumping and industrial applications. The main constraint to using a scalar control method for induction motors/generators is related to the transient response. If shaft torque and speed are bandwidth limited and torque varies slowly to track required speed variations (within hundreds of milliseconds up to the order of almost a second), scalar control may work appropriately. Hydropower and wind power applications have slower mechanical dynamics than such a timing constraint. Therefore, scalar control may be a reasonable approach for renewable energy-type applications. Field-oriented control uses a dynamic model of the IM where the voltages, currents, and fluxes are expressed in space vector forms. Field-oriented control can achieve excellent performance in transient and steady-state conditions; two types of field-oriented control are defined based on the position of the rotor flux: indirect and direct. By adding the slip position to the measured rotor position, the flux position in indirect field-oriented control is obtained. The flux position in direct field-oriented control is calculated based on the terminal variables and rotor speed. Since field-oriented control stems from a frame transformation that requires rotor speed, the knowledge of rotor position needs to be acquired accurately to perform such transformation. Vector control relies on accurate machine parameter characterization, as well as robust sensor monitoring and feedback of essential machine variables. It requires a powerful microcontroller or a processor based on digital signal processors (DSP) for rapid real-time computation using efficient PWM methodology. Additionally, it necessitates reliable protection against over-currents or over-voltages. This setup enables an inverter to exert current-control capabilities within a closed-loop speed control or a closed-loop torque control for the induction machine.

12.2 Introduction and early developments of induction machine drives

Induction machines operating either as motors or generators, are very important for our modern society. Industry, rural, and household applications use induction motors for 85% or more applications in rotating systems. They dominate the market of electrical motors and consume more than 60% of total industrial electricity in industrialized countries. The main reasons are a very strong low-cost design with reliability and inexpensive maintenance. There are no permanent magnets, consequently, the manufacturability is only constrained by the availability of iron, copper, machinery for punching the magnetic steel sheets, and electromechanical people capable of understanding how to do winding and assembling. During the 21st century, there will be further concerns about energy savings, sustainability, and global economics. Therefore, the use of induction machines as either motors or generators becomes of primordial importance, regarding sustainability and widespread manufacturability. Until the 1980s there were concerns about starting up induction motors with direct connection to the utility grid, and also constraints on how much reactive power would be needed to maintain the magnetization of those machines. Those machines would operate at constant speed plus/minus the slip variation around the synchronous speed, so they were initially considered constant nominal speed operation. There were some developments in the 70s and 80s regarding soft-start control, controlling torque by the rotor winding (for slip-rings machines with rotor connection), with some promises with thyristor-based forced-commutation McMurray topologies. In the 1960 there was the flourishing beginning of power electronics. Professor B.K. Bose published a paper "Electronic Speed Control of Motors" *Journal of the Institution of Engineers* (India), pp. 172–182, Sept. 1962, which was essentially a dc-drive using thyratrons. General Electric had a lot of resources and devoted electrical engineers to work with the newly invented device silicon-controller rectifier (SCR) also called thyristor. In the early 1960s, induction motors could be controlled with thyristors (SCRs) by managing the impressed machine voltage through natural commutation. However, this approach resulted in numerous low-frequency harmonics that caused shaft torque pulsations. Furthermore, by solely controlling the voltage at a constant fundamental frequency, the machine flux was not sustained, risking a potential loss of electromagnetic stability. In 1961, William McMurray (who is considered to be the founding father and guru of power electronics) invented forced commutation techniques: the McMurray and the McMurray–Bedford inverter, later the ac-switched commutation in 1980. It is a fact that there has been no analytical paper nor any known report on how Volts/Hertz was initially proposed in the 1960s, so for all practical purposes, the V/f control of voltage-fed IM drive was invented by McMurray, although it is not mentioned clearly in literature. There were studies and implementations of cycloconverters and ac voltage control by thyristors in the 1960s literature.

The sinusoidal PWM (SPWM) technique was invented by Schonung and Stemmler in 1964, but around that time only thyristors were devices for high voltage, bipolar junction transistors were available only for radio applications, very small power rating, very low frequency of operation, so still took 20 more years until SPWM became applicable in transistors-based inverters. In 1965 McMurray deployed thyristors to essentially start the revolution toward variable-frequency motor drives, and the open-loop volts/hertz control technique became very popular; by keeping V/Hz constant, with some machine performance measurement to fine-tune a look-up table it could be possible to maintain the flux nearly constant for variable speed drive operation, greatly enhancing the bandwidth of the control dynamics, as well as keeping a good torque per ampere performance. It became possible to extend the speed range beyond the synchronous frequency by implementing flux weakening to increase speed. This was achieved by constraining the voltage at the machine terminals, typically through 120-degree phase-shifted square-wave operation on induction machines with a floating-neutral configuration. This approach aimed to prevent the circulation of 3rd harmonics and maintain some control over harmonics and subharmonics.

During the 70s closed-loop speed control with slip and flux regulation was implemented mostly on analog control circuits. William McMurray working for General Electric developed in the 60s the forced commutation of thyristors in voltage-fed inverters, and assuming square-wave 6-step waveforms for machines with floating neutral he developed for the first time induction motor drives in volts/Hz with speed control. During the 1960s and 1970s, the early stage of induction motor drives was dominated by voltage-fed square-wave 6-step inverter where forced commutation of thyristors was used and the frequency control was simple to implement in voltage-fed inverter. Before PWM was implemented in inverters, the dc-link voltage could be controlled by a front-end thyristor bridge rectifier.

Toward the end of the 80s and throughout the 1990s there were tremendous developments of power electronics-based control of induction motors, with further and better scalar and then vector control schemes, development of direct torque schemes, continuous applications of DSPs, ASICS, then FPGAs. There was a lot of R&D in advanced control techniques in adaptive control, state observers, flux observers, on-line estimation or parameters, sensorless control, neural network-based control, and high-frequency signal injection for parameter and robustness improvements. This chapter lays out some foundations, based on d-q instantaneous theory transient modeling of induction machines with their fundamental controls. The principles are explained to allow the implementation of scalar and vector control approaches for induction machines operating as motor drives.

The idea of Volt/Hz control is that flux is proportional to the integral of voltage minus leakages, for example, a stator-flux-oriented vector control of an induction motor drive will have the estimation of stator flux by the time integration of the terminal voltage subtracted of the stator resistance. In the magnetic modeling of an induction machine, the air-gap flux will be the stator flux subtracted by the stator leakage (across the magnetizing inductance), and the rotor flux will be further in the machine by subtracting the rotor leakage. Therefore, a simple dimension analysis gives a unit of [volts*seconds] for flux. It is simple to assume that Volts/Hertz means a constant volts*seconds, i.e., a constant flux. McMurray proposed a forced-commutation SCR-based circuit where the voltage would be programmed proportionally to the frequency variation (V/f operation) and with careful slowing down or ramping up frequency it could make possible the first variable frequency drives.

Park (in 1929) and later Kron and Stanley (1938) made pioneering contributions to the quadrature two-fluxes reaction theory of electrical machinery. However, significant advancements in the control of induction machines occurred in 1969 when Hasse introduced direct vector control (DVC). In his thesis, Hasse achieved a decoupled system through flux measurements in the air gap of the induction machine. The indirect vector-control (IVC) was introduced by Blaschke in 1972. The analysis of the rotor d-q currents in the rotating reference frame made possible the vector control with analysis of the synchronous reference frame and stationary reference frame dynamic models of the machine, nowadays such IVC conditions are defined as "Blaschke equations". Throughout the 70s only analog circuits with operational amplifiers were implemented for solving equations capable of controlling induction machines, and typically the power electronics modulation was based on hysteresis band controllers, which were easily implemented with Schmitt-trigger analog comparators. The first time the author of this chapter observed the implementation of an analog-based (with op-amps) vector control implementation for an induction machine was in 1988. In the USA there was a solid introduction of the Texas Instruments digital signal processors (DSPs), allowing fast calculations in real-time, the families C20 and C25 were integers and C30 was a floating-point DSP processor.

12.3 Variable speed drives features and enhancements

Variable-speed induction motors are important for energy-saving purposes. Most industrial applications will have centrifugal load torque applications, the change is proportional to the square of the speed, and consequently power changes in proportion to the cube of speed with energy varying at different speeds. Variable frequency drives effectively control energy consumption by adjusting speed, and maintaining continuous speed control proves advantageous in numerous applications such as air and fluid movement, heating and ventilation, water lifting, belt operation, and handling loads in container systems. This ensures energy efficiency, reliable dynamics, and cost-effectiveness in a diverse range of shaft-rotating applications.

Variable frequency drives can have speed feedback when the precision of speed measurement is fundamental, compounded machines for construction, transportation, and electric vehicles. Speed control is important on a shaft where on the other side of the shaft there is some mover that impresses constant torque. On the other hand, if the device on the shaft imposes constant speed, the motor drive will mostly have to impress a torque control, and induction motor drives can be programmed for either torque or speed closed-loop control.

The induction machine may also have estimation methodologies, for example, to avoid speed sensors, or some other variables may be measured to estimate other ones. This class of motor drives is normally defined as sensorless motor drive control. To have sensorless motor drives, the estimation techniques will depend on a mathematical model. The model will need most certainly voltage and current feedback, or only current signals, to estimate the speed and/or the torque. The model will influence the accuracy of the machine parameters, and then, measurement of motor parameters will be important, or some class of parameter identification will have to be implemented. The chain of the system is: a motor drive may depend on speed or torque measurement, which can be estimated with a model that depends on measuring voltage and or current variables, which will require parametric identification of the machine in advance, or in real-time. The overall stability of the induction motor drive scheme will be affected by how this control chain is implemented, and then advanced control (analog and/or digital) will be required for mathematical implementation in real-time.

In the 1990s, following the development of adaptive control and model reference control for induction machines, a period of study emerged to observe the system's behavior with high-frequency signal injection. The induction machine, constructed with copper, iron, and magnetic fluxes circulating in a real material that is not perfect and likely anisotropic, prompted the exploration of using a carrier signal superimposed on PWM waveforms. This approach could assist in frequency tracking, signal-to-noise ratio detection, and spectral classification, necessitating the implementation of advanced signal processing algorithms. At the time of the first decade of the 21st century, there was a lot of implementation of advanced motor drive control with higher layers of system management. For instance, an induction machine can function as a wind turbine or a hydropower turbine, converting energy from a prime mover fueled by various energy resources. Achieving power balance control from the mechanical side toward the drive, and potentially extending further to another inverter connected to a grid, would demand additional layers of control and signal processing compared to a straightforward motor drive. In recent years there have been "green energy policies" in several countries for the installation of small- or medium-power Distributed Energy Resources (DER), including renewable or alternative energy sources and storage elements. Those units will have power electronics interfaces and maybe some induction motors or induction generators connected to "intelligent power sources". Those intelligent power sources cooperate to meet the energy demand by exploiting renewable energy at the maximum extent, making possible smart grids where there is enhanced information technology throughout the whole system with capabilities, with impact on the environment, science and technology, economics, and lifestyle. The use and operation of induction machines as motors or generators have been very popular for renewable energy sources. In the past, there were concerns about reactive power consumption and poor voltage regulation under varying speeds for directly connected induction machines, but the advances and developments in power electronics, real-time control, and application of modeling and signal processing have been facilitating contemporary use and retrofit of modern adjustable speed and torque control of induction machines for smart-grid applications.

We are in the age of energy transformation, from fossil fuel to renewable energy, and the sustainability and circular economy of the whole society must be considered. Therefore, in such a perspective it is important to understand how the manufacturability of induction machines is accessible to nearly all, anywhere: they have simple and robust construction, the materials utilized are mostly iron and copper, aluminum, resins, and any industrial base of tools and shop supplies are mostly sufficient as raw resources. In induction machine applications, maintaining a strong foundation in electrical engineering is essential. Students equipped with an understanding of electromagnetism, energy conversion, and electrical circuits, along with hands-on skills, can learn everything necessary to design an induction machine from scratch. They can order magnetic steel lamination and, with the assistance of mechanical technicians and individuals trained in wiring transformers and electrical motors, build an induction machine using grassroots resources. This not only facilitates the creation of a machine but also contributes to local economic development, potentially generating new opportunities for labor and crafts. Therefore, induction machines are certainly a present and a future solution that should be a humanity treasure kept in our electrical engineering schools and communities as a sustainability cultural resource.

12.4 Electromechanical considerations

Induction machines have robust construction and relatively low manufacturing costs. Induction machines are more economical when compared to synchronous machines. For high-power applications, such a difference is less perceptive because high-power machines are typically custom-made. But for medium and small sizes the price difference is dramatic, reaching 80% of the difference, i.e., for most residential, rural, and some small-commercial sizing, induction machines are less expensive than any other machine. For the same kVA rating, induction machines are somewhat larger than synchronous machines, because the magnetizing current circulates through the stator, so a part of the magnetic iron is for excitation of the machine (while permanent magnet machines are of course more compact). In applications such as induction generators, when they are connected to the grid they have fewer constraints on the turbine speed control, and are naturally stable, with very low maintenance requirements. There are three constraints for grid-connected induction generators: (1) starting-up, (2) reactive power requirements to be provided by the grid connection, or from an inverter-based control, and (3) directly connected induction generators will not have a variable shaft speed to optimize any turbine mechanical performance (unless they are doubly-fed with converters on the rotor side). A long time ago the US National Electrical Manufacturers Association (NEMA) standardized the variations in torque-speed characteristics and frame sizes assuring physical interchangeability between motors of competing manufacturers, thereby making them a commercial success and available for integral horsepower ratings with typical voltages ranging from 110 V to 4,160 V. The only perceived weakness of induction machines would be lower efficiency (since the rotor dissipates power) and the need for reactive power in the stator. The induction machine is made up of two major components: (i) the stator consisting of steel laminations mounted on a frame in such a manner that slots are formed on the inside diameter of the assembly very similar to a synchronous machine assembly, and (ii) the rotor consisting of a structure of steel laminations mounted on a shaft with two possible configurations classified as wound rotor or cage rotor. For construction details, it is advisable to consult the book authored by the writer of this chapter, titled "Modeling and Analysis with Induction Generators – 3rd Edition", published by Taylor and Francis/CRC Press. The induction machine includes an external casing that forms the magnetic path for the three-phase stator circuits. Bearings offer mechanical support, maintaining shaft clearance (air gap) between the rotor and stator cores. In the case of a wound rotor, there are brush holders and carbon brushes arranged to connect with the rotor windings. The winding of a wound rotor could be a three-phase type with the same number of poles as the stator, generally connected in Y with three terminal leads connected to the slip rings through carbon brushes. Wound rotors are usually made for very large power machines, where power electronic circuits will control the rotor winding currents. For industrial and commercial applications typically squirrel cage rotors are used. Squirrel cage rotor windings consist of solid bars of conducting material embedded in the rotor slots and shorted at the two ends by conducting rings. In large machines the rotor bars are made of copper alloy brazed to the end rings, when rotors are sized up to about 20 inches in diameter they are usually stacked in a mold made of aluminum casting, enabling a structure combining the rotor bars, end rings, and cooling fan. Induction machines are very simple, rugged, and very reasonable in manufacturing costs. Some variations on the rotor design are used to alter the torque speed features, and when machines are used for induction generators they are optimized for better efficiency, less magnetizing circulating current, higher-leakage inductance (to boost their output voltage and filter out harmonics), and with cooling capable to keep low overall temperature of operation even for very low shaft speeds and internal and enclosed environments.

12.5 Three-phase to two-phase and reference-frame transformations

To start modeling an induction motor, some assumptions are necessary on the physical variables and mathematical transformations. Various options exist for defining the three-phase a-b-c circuits and the quadrature d-q circuits, involving the calculation of variable projections and mapping between them. Additionally, different power systems may adopt one notation, while alternative notations may be assumed in the realm of power electronics. Moreover, it is crucial to specify whether the modeler intends to maintain electrical power constant across transformations or, alternatively, keep the current invariant. This decision is contingent upon whether the closed-loop control system operates based on power or current. Power systems engineers generally favor power-invariant transformations, while power electronics engineers tend to prefer current-invariant transformations. In cases where the system is presumed to be balanced and symmetrical, specifying a normal component to the plane defined by d-q is unnecessary. However, for asymmetrical and imbalanced conditions, a zero-sequence component may manifest on a z-axis from d-q, making the notation of a *dq0* transform applicable. Figure 12.1(a) shows a possible configuration typically used in power systems applications where phase a is at 0°, phase b at $-120°$, and phase c at $+120°$. They are stationary and connected for example on a cross-sectional plane of a machine, where the machine shaft would be normal to such plane (crossing the origin of the axis), and another stationary quadrature frame in this figure is defined as alpha (⟨) placed at 0°, plus beta (®) at +90°. It is possible to have a geometrical and algebraic conversion, similar to a reflection and projection of vectors, from a-b-c to ⟨-®, such mapping could go both ways, having a transformation from three-phase to two-phase, or a transformation from two-phase to three-phase equivalent variables and equations. Figure 12.1(b) shows the diagram for three-phase a-b-c- to stationary $\alpha\beta 0$ reference frame and to rotating $dq0$ reference frame where the phase a and the q-axis are initially aligned, while Figure 12.1(c) is for phase a and the d-axis are initially aligned.

When a designer uses a simulation software with those transformations embedded in blocks or formulas, it is very important to first understand if the notation and terminology used in those block diagrams follow this convention or another one, because it would be possible that, for example, the three-phase system would be defined instead as phase a at 0°, phase c at $-120°$ and phase b at $+120°$, with several other possibilities for the ⟨-® axis, maybe alpha (®) placed at 0°, beta (⟨) at +90° or maybe (⟨) placed at 0°, and beta (®) at $-90°$ or any other configuration with possibly not even aligning one of those with phase a. For each configuration, the projection and reflection of variables will give a slightly different mapping and matrix calculations. When the reader studies this subject and starts to figure out the algebraic transformations, it is important to start from scratch and compare notes, because some textbooks will use different notations and formatting, some schools of thought in Europe will be different from others in North America, some tradition of power systems modeling versus power electronics modeling will differ as well. If the reader does all their own coding and designing from their own understanding of the differential equations through the mappings of three-phase to two-phase, and vice versa, writing their own equations, it will then work well. But it would be important to write comments or establish clearly such simple notation and definitions, particularly when that computer-based simulation is shared, distributed, and used by other people. With the previous step of transforming a-b-c

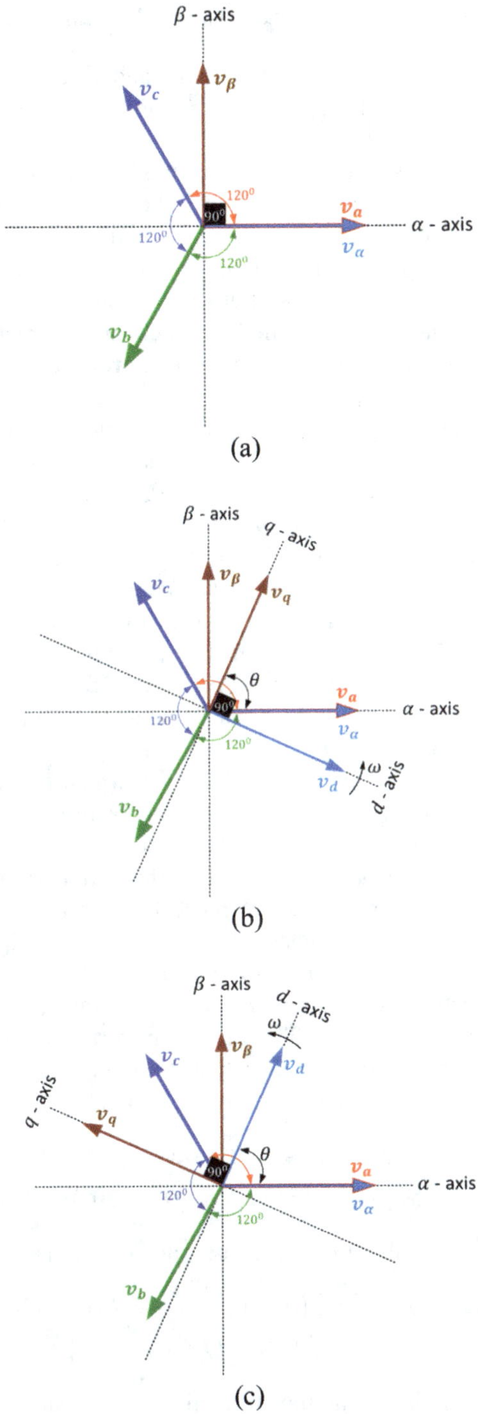

Figure 12.1 Diagram for three-phase ABC to stationary αβ0 reference frame and rotating dq0 reference frame. (a) ABC three-phase. (b) Phase a and the q-axis are initially aligned. (c) Phase a and the d-axis are initially aligned

three-phase models into ⟨-® two-phase systems (and vice versa), we have instantaneous sinusoidal equations, time-shifted by $\frac{2\pi}{3}$ rad (120°) into a sine and a cosine waveform, but all of those equations (for voltage and current, or maybe flux) will be time-varying. It is possible to have one more transformation, going to an equivalent dc (direct-current) circuit, or model, therefore, transients in the original three-phase will be exponential responses in the equivalent DC, and steady-state three-phase in the original system will be steady-state constant values on the equivalent DC circuit. By doing these transformations in real-time we would be transforming a three-phase, highly complex system of differential equations, into an equivalent virtual dc machine, maybe by just controlling such a virtual machine with a command for the torque, and another one for the flux (such as in any dc-motor), it would be possible to map back those commands to impress the right currents (or voltages) in the original three-phase induction machine (through a real-time calculation with fast microcontrollers or DSP, or ASIC hardware connected to a three-phase inverter). This is what is called vector control, or field-oriented control of induction motor (or generator) drives.

To achieve an equivalent DC virtual machine, a new rotating reference frame should be assumed to be spinning at the instantaneous electrical angular frequency of the variables ω (rad/s). This angular frequency could be synthesized for grid-connected inverters or power systems applications, for example, using a PLL (phase-lock loop) calculation. However, for induction machines (and also for permanent magnet synchronous machines), it is taken the approach of calculating the instantaneous position of the machine flux vector, which rotates at the electrical angular frequency ω_e rad/s with their instantaneous position θ_e in rad. The subscript e indicates the electrical framework when the variable has no subscript that should be either related to the mechanical framework or the stationary frames. As depicted already, Figure 12.1(b) has a rotating reference frame for phase a and the q-axis aligned at $t = 0$ and the case where phase a and the d-axis are aligned when $t = 0$, so proper calculation from the sequence a-b-c with the alphabet of Figure 12.1(a), should then be projected on such rotating reference frame, assuming an angular speed ω running counter-clockwise. The whole transformation from three-phase, to two-phase, to rotating reference frame, such as depicted in Figure 12.2, when there is phase a to q-axis alignment like in Figure 12.1(b) is:

$$T_{dq0} = k_1 \left[\sin\sin(\theta) \; \sin\sin\left(\theta - 120^0\right) \; \sin\sin\left(\theta + 120^0\right) \right. \\ \left. \cos\cos(\theta) \; \cos\cos\left(\theta - 120^0\right) \; \cos\cos\left(\theta + 120^0\right) \; k_0 \; k_0 \; k_0 \right] \tag{12.1}$$

and, for phase a to d-axis alignment like in Figure 12.1(c) is:

$$T_{dq0} = k_1 \left[\cos\cos(\theta) \; \cos\cos\left(\theta - 120^0\right) \; \cos\cos\left(\theta + 120^0\right) \right. \\ \left. -\sin\sin(\theta) \; \left(\theta - 120^0\right) \; \left(\theta + 120^0\right) \; k_0 \; k_0 \; k_0 \right] \tag{12.2}$$

When it is current invariant transformation $k_1 = \frac{2}{3}$ and $k_0 = \frac{1}{2}$ whereas for power invariant transformation $k_1 = \sqrt{\frac{2}{3}}$ and $k_0 = \sqrt{\frac{1}{2}}$. Those transformations can

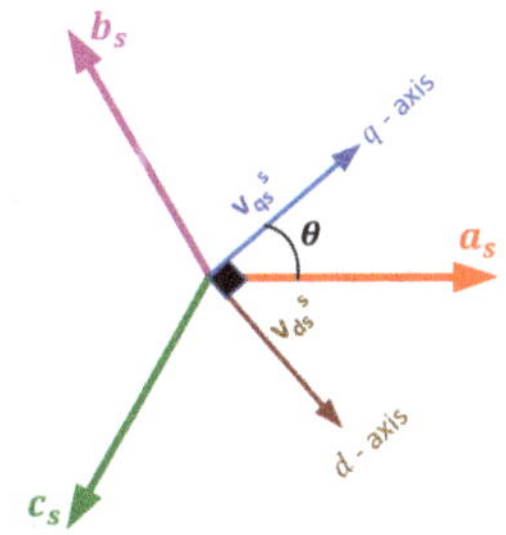

Figure 12.2 Vector representation for position of q-d stationary to ABC frame for vector control implementation

be valuable not only for power-systems power flow management, systems protection, and steady-state evaluation of synchronous generators but also in instantaneous power theory. Typically, induction motors have a few different notations, in this chapter it is followed as, i.e., Prof. B.K. Bose, and other US researchers and investigators adopted, assuming the three-phase axis a-b-c and the two-phase frame qs-ds (instead of alpha-beta), both stationary, as well as the rotating reference-frame (qe-de, but most classic publications and books dropped the e, keeping only q-d), superscript s means that the frame is stationary, it is made $\theta = 0$ to make the calculations easier. For an induction machine, it is required to impress currents and voltages on the machine in real time, with an inverter, so it is assumed current invariant transformation, but a scaling factor is applied to machine power to correct it to the right scale.

By expanding the matrix multiplication and working on the algebraic steps, two equations are developed to convert any three-phase system into two-phase equivalent:

$$\begin{bmatrix} vqs^S \\ vds^S \\ vo^S \end{bmatrix} = \begin{bmatrix} \cos(\theta) & \cos(\theta - 120^\circ) & \cos(\theta + 120^\circ) \\ \sin(\theta) & \sin(\theta - 120^\circ) & \sin(\theta + 120^\circ) \\ 1/2 & 1/2 & 1/2 \end{bmatrix} \begin{bmatrix} vas \\ vbs \\ vcs \end{bmatrix} \tag{12.3}$$

further manipulation and isolating the three-phase variables on the left-hand side, assuming the angle from axis a to axis q^s to be zero, then a reverse transformation that takes a two-phase stationary system into a three-phase stationary one is obtained:

$$\begin{aligned} vqs^S &= vas \\ vds^S &= \left(\frac{1}{\sqrt{3}}\right)(vcs - vbs) \end{aligned} \tag{12.4}$$

Those equations show voltages, in fact, any three-phase variables (voltages, currents, fluxes) could be translated from three-phases to two-phases, or vice versa, using Equations (12.4) and (12.5).

$$
\begin{aligned}
\text{vas} &= \text{vqs}^{\text{S}} \\
\text{vbs} &= \left(\frac{-1}{2}\right)\left(\text{vqs}^{\text{S}} + \sqrt{3}\left(\text{vds}^{\text{S}}\right)\right) \\
\text{vbs} &= \left(\frac{-1}{2}\right)\left(\text{vqs}^{\text{S}} - \sqrt{3}\left(\text{vds}^{\text{S}}\right)\right)
\end{aligned}
\tag{12.5}
$$

There is another transformation used for modeling the induction machine: mapping the stationary frame (d^s-q^s) into a rotating or reference frame (d-q), which is rotating at the electrical angular frequency ω_e. Figure 12.2 shows the relationship of those cartesian frames, the process of going from the stationary frame to the rotating reference frame is called Vector Rotation (VR) while going back is called Inverse Vector rotation (IVR). Looking at variables (voltages, currents, fluxes) on the stationary frame (d^s-q^s) shows sine waves apart by 90° i.e., typically defined by equations with cos and sin functions, when they are projected on a rotating reference frame (d-q), they will be DC variables; however, it would be important to define the two important features: (i) the initial conditions for θ_e, and this is decided on an induction machine in defining which magnetic flux (stator, air-gap, or rotor flux) would be aligned, for example with the axis d (it could be q, but in vector control and field-oriented control it is mostly assumed the d-axis to align with the flux, and by electromagnetics analysis, the torque variable would be aligned with the q-axis), as well as (ii) how the angle θ_e is generated, the terminology is, if the angle θ_e is generated directly from either a flux-measurement (with sensors on the machine), or from a flux-calculation that receives voltage and current, this is called Direct Vector Control (DVC), while if the angle θ_e is generated indirectly, with probably a feedforward calculation, or mapping variables and parameters to calculate the slip frequency $\omega_r = (\omega_e - \omega_{sl})$, i.e., with feedforward calculation of ω_{sl} and a speed sensor measurement of ω_r, both can be added and integrated on time to generate θ_e which can be used in the vector-rotation and inverse-vector-rotation, such strategy is called Indirect Vector Control, and it is typically used in high-performance induction motor drives. When the d-q strategy is used to control a grid-connected inverter, there is no magnetic flux, but the angle θ_e can be generated with a PLL connected to the point of common coupling of the inverter to the utility grid. Figure 12.3 shows the frames

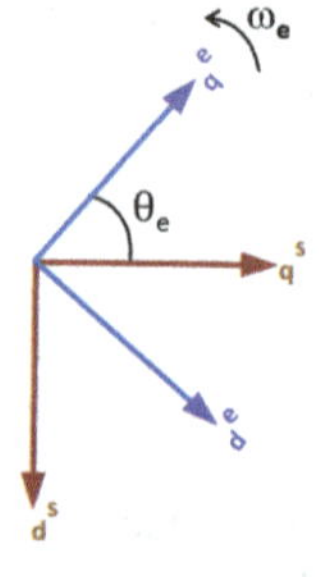

Figure 12.3 Vector representation for vector rotation

relationship adopted in the modeling of the induction machine for vector control in this chapter.

Using the cartesian frames as indicated in Figure 12.3, the process of Vector Rotation (VR) is implemented with the equation below, transforming the 90° degrees phase apart of variables in qs-ds into the rotating reference frame q-d (rotating at the electrical angular speed $\omega_e = \frac{d\theta_e}{dt}$ where of course $\theta_e = \int_0^t \omega_e dt + \theta_e(att = 0)$, since such integration may run for a long time, it is better to implement in the simulation a self-reset, i.e., every $\theta_e = 2$ the integration re-starts from zero, this could avoid numerical overflow in case of a long-run unbounded integration.

$$\begin{aligned} \mathrm{vqs} &= \mathrm{vqs}^{\mathrm{S}} \cos(\theta\mathrm{e}) - \mathrm{vds}^{\mathrm{S}} \sin(\theta\mathrm{e}) \\ \mathrm{vds} &= \mathrm{vqs}^{\mathrm{S}} \sin(\theta\mathrm{e}) + \mathrm{vds}^{\mathrm{S}} \cos(\theta\mathrm{e}) \end{aligned} \tag{12.6}$$

Although the rotating reference frame has a superscript e, to indicate the rotation at the electrical angular frequency of the machine flux, ame, the superscript is omitted for convenience, and particularly for writing a clear code, this transformation works for any variables such as voltages, currents, and fluxes. The equations needed for Inverse Vector Rotation (IVR) i.e., mapping from the reference frame to the stationary qs-ds frame are:

$$\begin{aligned} \mathrm{vqs}^{\mathrm{S}} &= \mathrm{vqs}\ \cos(\theta\mathrm{e}) + \mathrm{vds}\ \sin(\theta\mathrm{e}) \\ \mathrm{vds}^{\mathrm{S}} &= -\mathrm{vqs}\ \sin(\theta\mathrm{e}) + \mathrm{vds}\ \cos(\theta\mathrm{e}) \end{aligned} \tag{12.7}$$

Those equations can be applied to any input variable, it would be interesting to have functions or block diagram calculations to convert any input from three-phase and from stationary to rotating and back again, this is easily made in any contemporary block diagram-based system simulator, or with built-in functions. If the user decides on adopting ready-made blocks or functions embedded in the simulator, again it is very important to clearly check how those computations are made based on the adopted terminology and standard notation of the abc three-phases with the quadrature stationary and rotating reference frames, as well as making sure if the transformation is power-invariant or current-invariant, i.e., the assumptions made for constructing the available blocks and functions in the simulator system.

12.6 Induction motor model

It is expected that someone reading this chapter should have some understanding of Physics and Electrical Engineering foundations regarding electrical motors, mostly how transformers work, and how induction machines operate in steady-state using phasor analysis. However, to model an induction machine with ordinary differential equations (ODE) and in block diagrams, we can initially adopt the understanding of a per-phase equivalent circuit as indicated in Figure 12.4. The parameters in this per-phase model have components with an s-subscript for the stator, r relates to the rotor components, S (capital letter) is the slip (difference from electrical angular speed and the rotor speed in rad/s), so the rotor side will have zero current when there is no torque and a composite parallel connection with the magnetizing branch

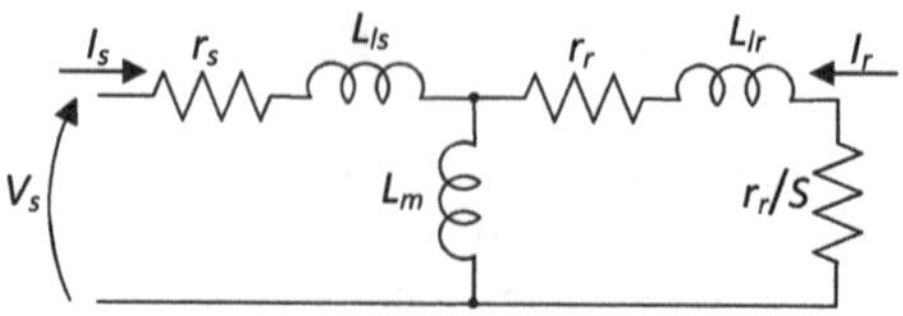

Figure 12.4 Per phase equivalent circuit for an induction machine

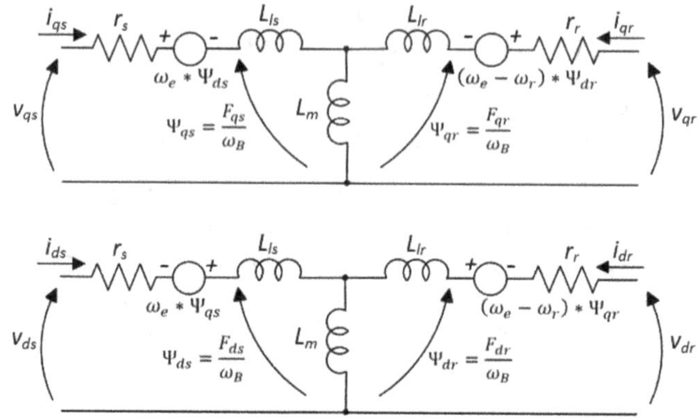

Figure 12.5 d-q rotating reference frame equivalent circuit of induction machine

for any other slip value. Such a simple circuit is very often used for sizing, dimensioning, and evaluating the steady-state response of induction machines, assuming perfectly balanced and symmetrical operation in all the three-phases, therefore, just one equivalent phase is used in the analysis.

For transient modeling, an equivalent qs-ds model can be constructed, but that is still a sinusoidal-quadrature operation, and then projecting such into the rotating reference frame q-d (rotating at the electrical angular speed ω_e), Figure 12.5 shows the quadrature equivalent q-d frame circuit for an induction machine. Transient analysis can be performed in this equivalent circuit and then transformed back to qs-ds and converted into abc framework. However, from a controller's perspective, we can operate fully in the q-d rotating equivalent frame, and by correctly imposing set-points of operation (in real-time), only the set-points commands will be converted back to the power electronics interface that will drive the induction machine.

Looking from the left side of the circuit inwards with a KVL approach and assuming that the voltage across the point from the leakage inductance (L_{ls}) to ground is represented by the derivative of the magnetic flux there, the fluxes variables are taken into consideration, i.e., the linked flux across the magnetizing inductance (L_m) and the stator leakage (L_{ls}), in the machine that is called Stator Flux, then the equations for both quadrature cartesian stationary reference frame (q^s-d^s) are given by Equations (12.8).

$$\begin{aligned} vqs^S &= Rs * iqs^S + \frac{d}{dt}\Psi qs^S \\ vds^S &= Rs * ids^S + \frac{d}{dt}\Psi ds^S \end{aligned} \tag{12.8}$$

and when using the rotating reference frame (q^e-d^e), dropping the superscript *e* and defining only as (q-d), there is rotational electromotive force (EMF) represented by the multiplication of electrical angular frequency ω_e with flux from the other axis, i.e., $\omega_e * \Psi_{ds}$ for the q-equivalent circuit and $\omega_e * \Psi_{qs}$ for the d-equivalent circuit, as indicated in (12.9), of course, in the (d-q) quadrature equivalent circuits the variables are considered to be DC with exponential decay transient responses, Equation (12.9) also shows the developed equations for the rotor circuit of the induction machine, i.e., by evaluating KVL from the right side of the equivalent circuit inwards. There the rotational electromotive force (EMF) is represented by the multiplication of slip angular frequency ω_{SL} with flux from the other axis, i.e., $\omega_{SL} * \Psi_{dr}$ for the q-equivalent circuit and $\omega_{SL} * \Psi_{qr}$ for the d-equivalent circuit.

$$\left.\begin{aligned} \mathrm{vqs} &= \mathrm{Rs} * \mathrm{iqs} + \frac{\mathrm{d}}{\mathrm{dt}}\Psi\mathrm{qs} + \mathrm{we} * \Psi\mathrm{ds} \\ \mathrm{vds} &= \mathrm{Rs} * \mathrm{ids} + \frac{\mathrm{d}}{\mathrm{dt}}\Psi\mathrm{ds} - \mathrm{we} * \Psi\mathrm{qs} \\ \mathrm{vqr} &= \mathrm{Rr} * \mathrm{iqr} + \frac{\mathrm{d}}{\mathrm{dt}}\Psi\mathrm{qr} + \omega_{SL} * \Psi\mathrm{dr} \\ \mathrm{vdr} &= \mathrm{Rr} * \mathrm{idr} + \frac{\mathrm{d}}{\mathrm{dt}}\Psi\mathrm{dr} - \omega_{SL} * \Psi\mathrm{qr} \end{aligned}\right\} \tag{12.9}$$

The next step would be to isolate the differential equations on the left end side, achieving a state-space formulation on a set of first-order differential equations, of course, several algebraic manipulations are necessary, the next few steps try to summarize the procedures, initially normalizing impedances by choosing a base electrical angular frequency, and by replacing the Greek symbol for flux with the letter F, to make equations easier to write, and allowing coding and functions written for simulation purposes; the normalized linkage fluxes would be redefined as algebraic formulations indicated in (12.10), where wb = 260 = 377 rad/s for a 60 Hz machine, or similar angular frequency normalization for 50 Hz.

$$\left.\begin{aligned} Fqs &= wb * \Psi_{qs} \\ Fds &= wb * \Psi_{ds} \\ Fqr &= wb * \Psi_{qr} \\ Fdr &= wb * \Psi_{dr} \\ Fqm &= wb * \Psi_{qm} \\ Fdm &= wb * \Psi_{dm} \\ Xls &= wb * L_{ls} \\ Xlr &= wb * L_{lr} \\ Xm &= wb * L_m \end{aligned}\right\} \tag{12.10}$$

Observing the d-q quadrature equivalent circuits with the normalized variables, it is possible to compute currents with the following calculations with (12.11); if a block diagram model is implemented, those current calculations will be necessary for machine torque instantaneous computation—the block would receive the linkages, dividing by the respective impedance—generating output machine currents in the q-d frame (iqs, ids, iqr, idq).

$$\begin{aligned}
\mathrm{iqs} &= \frac{\mathrm{Fqs} - \mathrm{Fqm}}{\mathrm{Xls}} \\
\mathrm{ids} &= \frac{\mathrm{Fds} - \mathrm{Fdm}}{\mathrm{Xls}} \\
\mathrm{iqr} &= \frac{\mathrm{Fqr} - \mathrm{Fqm}}{\mathrm{Xlr}} \\
\mathrm{idr} &= \frac{\mathrm{Fdr} - \mathrm{Fdm}}{\mathrm{Xlr}}
\end{aligned} \tag{12.11}$$

All the equations for currents in (12.11) can be substituted for their values in (12.8) and (12.9) and after some rearranging the following equations are developed for the stator:

$$\begin{aligned}
\frac{\mathrm{d}}{\mathrm{dt}}\mathrm{Fqs} &= \mathrm{wb}\left(\mathrm{Vqs} - \frac{\mathrm{we}}{\mathrm{wb}}\mathrm{Fds} - \frac{\mathrm{Rs}}{\mathrm{Xls}}(\mathrm{Fqs} - \mathrm{Fqm})\right) \\
\frac{\mathrm{d}}{\mathrm{dt}}\mathrm{Fds} &= \mathrm{wb}\left(\mathrm{Vds} + \frac{\mathrm{we}}{\mathrm{wb}}\mathrm{Fqs} - \frac{\mathrm{Rs}}{\mathrm{Xls}}(\mathrm{Fds} - \mathrm{Fdm})\right)
\end{aligned} \tag{12.12}$$

And for the rotor side, if the assumption is a squirrel cage machine, the rotor is shorted and both vqr and vdr are zero, so the differential equations are:

$$\begin{aligned}
\frac{\mathrm{d}}{\mathrm{dt}}\mathrm{Fqr} &= -\mathrm{wb}\left(\frac{\mathrm{w_{SL}}}{\mathrm{wb}}\mathrm{Fdr} - \frac{\mathrm{Rr}}{\mathrm{Xlr}}(\mathrm{Fqr} - \mathrm{Fqm})\right) \\
\frac{\mathrm{d}}{\mathrm{dt}}\mathrm{Fdr} &= -\mathrm{wb}\left(-\frac{\mathrm{w_{SL}}}{\mathrm{wb}}\mathrm{Fqr} + \frac{\mathrm{Rr}}{\mathrm{Xlr}}(\mathrm{Fdr} - \mathrm{Fdm})\right)
\end{aligned} \tag{12.13}$$

The unknown variables in those equations (Fqm and Fdm) are calculated as (12.14) and after rearranging one can get (12.15) and (12.16); a missing variable is the machine torque computation and a mechanical torque shaft computation will be required to calculate the shaft angular speed to give back to the model.

$$\mathrm{Fqm} = \mathrm{Xm}\left(\frac{(\mathrm{Fqs} - \mathrm{Fqm})}{\mathrm{Xls}} + \frac{(\mathrm{Fqr} - \mathrm{Fqm})}{\mathrm{Xlr}}\right) \tag{12.14}$$

$$\begin{aligned}
\mathrm{Fqm} &= \frac{\mathrm{Xm1}}{\mathrm{Xls}}\mathrm{Fqs} + \frac{\mathrm{Xm1}}{\mathrm{Xlr}}\mathrm{Fqr} \\
\mathrm{Fdm} &= \frac{\mathrm{Xm1}}{\mathrm{Xls}}\mathrm{Fds} + \frac{\mathrm{Xm1}}{\mathrm{Xlr}}\mathrm{Fdr}
\end{aligned} \tag{12.15}$$

$$\mathrm{Xm1} = \left(\frac{1}{\frac{1}{\mathrm{Xm}} + \frac{1}{\mathrm{Xls}} + \frac{1}{\mathrm{Xlr}}}\right) \tag{12.16}$$

The general equation for torque in an induction machine is the vector multiplication of the magnetizing flux with the space vector of the rotor current, as in (12.17), where P is the number of poles in the machine, and (3/2) is the correction factor required for a current-invariant transformation (power must be corrected). In

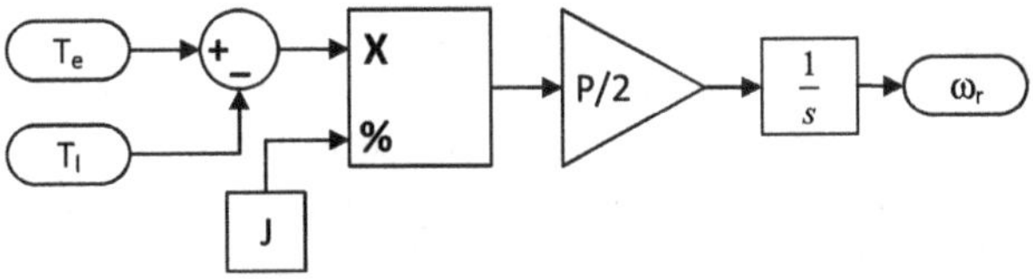

Figure 12.6 Block diagram for calculating the electrical angular speed for the machine model

the rotating reference frame, the torque equation is resolved as (12.18), but typically (12.19) is the most typical torque equation calculated with a q-d model, which can be easily incorporated in Figure 12.6 as input machine-developed electromagnetic torque. Using the normalized flux linkages the developed machine torque can be calculated by (12.20) and with a few extra blocks for multiplication, subtraction, and scaling factor. Typically, computer simulations employ a step size that may result in noisy signals when multiplying certain variables such as currents and fluxes. This noise may not pose significant issues with high mechanical inertia. However, introducing a low-pass filter in conjunction with the torque signal can effectively refine the waveform, ensuring a clear and accurate representation in an output graph or in the view of tools like MATLAB/Simulink oscilloscope.

$$\mathrm{Te} = \left(\frac{3}{2}\right)\left(\frac{\mathrm{P}}{2}\right)(\Psi\mathrm{m} \times \mathrm{Ir}) \tag{12.17}$$

$$\mathrm{Te} = \left(\frac{3}{2}\right)\left(\frac{\mathrm{P}}{2}\right)(\Psi\mathrm{dm} * \mathrm{iqr} - \Psi\mathrm{qm} * \mathrm{idr}) \tag{12.18}$$

$$\mathrm{Te} = \left(\frac{3}{2}\right)\left(\frac{\mathrm{P}}{2}\right)(\Psi\mathrm{ds} * \mathrm{iqs} - \Psi\mathrm{qs} * \mathrm{ids}) \tag{12.19}$$

$$\mathrm{Te} = \left(\frac{3}{2}\right)\left(\frac{\mathrm{P}}{2}\right)\left(\frac{1}{\omega_b}\right)(\mathrm{Fds} * \mathrm{iqs} - \mathrm{Fqs} * \mathrm{ids}) \tag{12.20}$$

The angular shaft speed of the machine will be dependent on the rotating shaft inertia J, impressed machine torque (T_e), and developed load (T_L) will have acceleration following Newton's law. The first-order differential equation as in Equation (12.21) notes that ω_m is the mechanical angular speed on the machine shaft, and that is related to the required electrical rotor angular speed ω_r by the multiplication factor of the number of poles of the machine (P) divided by 2, as indicated in (12.22); T_e is the torque applied by the motor, T_L is the load torque depending on the motion, J is the moment of inertia of the combined spinning shaft plus attached load, and ω_m is the shaft angular speed in rad/s, if the model requires RPM, another scaling factor must be imposed.

$$\mathrm{T_e} - \mathrm{T_L} = \mathrm{J}\frac{\mathrm{d}}{\mathrm{dt}}(\omega_m) \tag{12.21}$$

$$\omega_r = \left(\frac{\mathrm{P}}{2}\right) * \omega_m \tag{12.22}$$

The load torque depends on the application, it could be linear, constant, with a back-slash, but for most of the industrial applications, the assumption is a "fan load", i.e., the impressed load torque will be proportional to the square of the mechanical rotation with a factor B, as indicated in (12.23). If for any reason the machine will be controlled in position, then a mechanical angle θ_m will have to be computed by the time integration of ω_m, this may happen for robotic manipulators and conveyor belt positioning systems. In fact, most of the time induction motor drives are used in speed or torque control mode only. The mechanical equations above can be combined and a 5th ordinary differential equation can be added to the system of the induction machine, i.e. the four differential equations in the quadrature d-q frames depicted by (12.12) and (12.13), such a mechanical equation comes from the shaft dynamics in equation (12.24), where the ODE in (12.25) is for the electrical rotor angular speed.

$$\mathrm{T_L} = \mathrm{B} * (\omega_m)^2 \tag{12.23}$$

$$\mathrm{T_e} - \mathrm{B} * (\omega_m)^2 = \mathrm{J}\frac{\mathrm{d}}{\mathrm{dt}}(\omega_m) \tag{12.24}$$

$$\frac{\mathrm{d}\omega_r}{\mathrm{dt}} = \left(\frac{1}{\mathrm{J}}\right) * \left(\frac{\mathrm{P}}{2}\right) * (\mathrm{T_e} - \mathrm{T_L}) \tag{12.25}$$

Figure (12.6) shows a simple block diagram simulation implementation, whereas the complete block diagram to simulate an induction machine is portrayed in Figure (12.7). It is important to notice that this model still requires as input the electrical angular speed impressed by the input voltages, i.e., this is not a stand-alone and independent model, since the designer must connect with the previous voltage source, or the previous PWM inverter, where the instantaneous value ω_e will be defined. Typically this has no effect in most of the motor drives control simulations, but for example, using the model as an induction generator, the instantaneous value ω_e should be internally generated. The author of this chapter investigated several possibilities, and the easiest one is to convert this model that runs on a rotating reference frame into another one that rotates on the stationary frame, i.e., the variables would be sinusoidal, it is left as an exercise to the reader to simply make $\omega_e = 0$ in this current model, and the whole differential equations will be in the stationary frame (with the electromotive force calculations $\omega_e * \Psi_{ds}$ and $\omega_e * \Psi_{qs}$ naturally going to zero, and the other ones on the rotor circuit, namely $\omega_{SL} * \Psi_{dr}$ and $\omega_{SL} * \Psi_{qr}$ will change to their correct stationary reference frame ω_r multiplied by the corresponding d or q flux, since naturally the $\omega_e = 0$ makes the calculation of slip frequency to be the right ones).

For induction generators, it is recommended that the reader study the book by M.G. Simoes. The electrical model must have a starting-up, as in the real world there is always a remnant flux in the core for electric machines to start-up from blackout conditions. It is often better to simulate in an a-b-c framework instead of a d-q one. A final comment is that angular speed terminology uses the Greek letter but in writing code and computer language the letter w is used, which shows in tags and labels in Matlab/Simulink implementation. Figure 12.7 shows a block diagram for the computer-based simulation model of an induction machine, where the differential equations are based on the flux linkages. This model must receive the

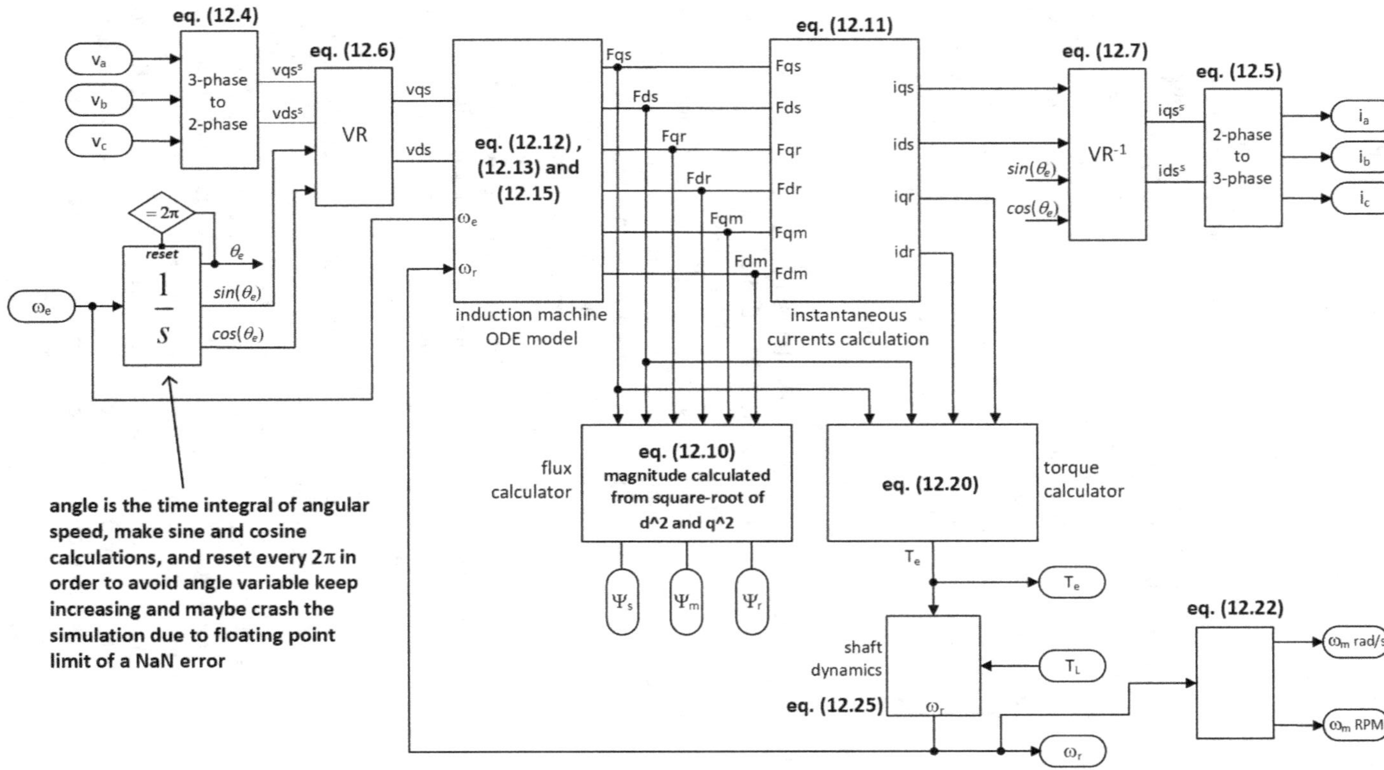

Figure 12.7 Diagram for induction machine with flux linkages based state-space simulation

impressed machine voltages (va, vb, vc) from either sinusoidal voltage sources or from a PWM converter, since the machine model is on the d-q rotating reference frame the angular electrical speed ω_e is also an input provided by the voltage generation of va, vb, and vc. Such input of electrical angular speed can be a drawback if the machine must self-start as a generator, and then the solution would be either making it speed $\omega_e = 0$ and converting all internal variables in the same model to be on the d-q stationary frame, or by entirely modifying the approach for the induction machine model to be on an abc three-phase frame. For most of the motor drive schemes, particularly for indirect vector control (IVC), the mode in Figure 12.7 is sufficiently good and well established and supported by electrical engineering mathematical analysis. This model calculates the d-q currents and transforms back the dynamical evolution of three-phase currents, so the outputs for such a model are three-phase machine currents (ia, ib, and ic), plus the machine torque (te), and mechanical angular speed ω_e in rad/s or maybe in RPM as well. In addition, the model can provide internal flux calculations (rotor, air-gap, and stator) as well as power losses, developed power, and mechanical load changes and variations.

12.7 Scalar control

Before the 1970s, variable frequency drives (VFD) as considered a modern concept, and by that time only Volts/Hertz was the adopted methodology. It was known that the speed of an induction motor could be managed by varying the frequency of a controlled three-phase supply at the same time maintaining a constant (rated) flux density. The McMurray forced commutation inverter was a contemporary approach. However, in reality, thyristors are not good at forced commutation, they are a better fit with natural commutation. Transistors-based switching must be used, and still toward the first few years of the 1980s, most of the control schemes for any motor drive would be analog, with operational amplifiers, maybe with some hard-wired digital logic gates. For such VFDs, above rated speed, the applied voltage is kept constant at the rated value, such an operation is referred to as constant power mode, while at lower frequencies (lower shaft RPM), the voltage would then be proportional. In the case of a wide-range VFD, during the initial start-up at very low frequencies, the system encounters challenges due to insufficient flux in the machine. To address this, compensation is necessary, as the stator resistance exhibits a significant voltage drop at high power. Non-linear boosting is implemented to overcome these issues. After a few Hertz of operation, the VFD transitions to Volts/Hertz mode, requiring variable voltage power electronics to ensure a proportionate voltage is applied to the induction machine terminals. At maximum speed, the switching power electronics device ceases to vary the voltage, maintaining it at the maximum level while decreasing the flux to sustain a constant power level in the machine. Until the late 1970s, only machines with open rotor windings allowed for flux weakening, as vector control remained a theoretical concept. It was not until the advent of microelectronics, with microprocessors and the latest digital signal processors, that field orientation of induction machines became feasible. Initially confined to research labs, this technology later revolutionized the entire motor-drive manufacturing industry around the 1990s, ushering in a wave of contemporary vector-controlled speed and torque motor drives for industrial applications. During that timeframe, IGBTs and Power-MOSFETs increasingly emerged as viable transistor options. The market for motor drives, ranging from a few hundred watts to several kilowatts, for medium voltage applications experienced exponential growth and widespread adoption.

The simplified per-phase equivalent model of Figure 12.4 shows how a steady-state formulation can be done by neglecting the resistive losses in the rotor $3 * R_r^{'} * |I_r|^2$. The electrical power transferred from the mechanical shaft to the rotor circuit is approximated by (12.26); the generator torque shaft can be calculated as the ratio of P_{mech} to the mechanical angular speed. For scalar voltage control, the angular shaft speed is leading the stator angular speed by $\omega_m = \frac{2}{P}\omega_e(1+s)$ and the developed electrical torque on the shaft is given by (12.27).

$$P_{elec} = 3 * R_r^{'} * \frac{\omega}{\omega_r}|I_r|^2 \approx P_{mech} \tag{12.26}$$

$$T_e = 3\frac{P}{2} * \frac{\Psi_m{}^2}{R_r^{'}} * \frac{\omega_r}{(\tau_r * \omega_r)^2 + 1} \tag{12.27}$$

For $\omega_r = \frac{1}{\tau_r}$ the critical slip is $S_{critical} = \frac{1}{\tau_r * \omega_e}$ and the maximum torque developed (peak) is $T_{m,\max} = 3\frac{P}{2} * \frac{\Psi_m^2}{L_{ls}}$; the machine will operate above (generator) and below (motor) the synchronous frequency ω_s with its operating slip well below the critical slip. Therefore, the denominator of (12.27) is close to unity and the torque is virtually proportional to the rotor angular frequency for constant flux, following that (12.28) has constant torque which can be impressed by flux $\Psi_m \approx \frac{V_{stator}}{\omega_e}$, Then, a family of curves for constant flux can be derived, and plotted as depicted in Figure 12.8, showing steady-state curves for operational conditions of variable frequency drives,

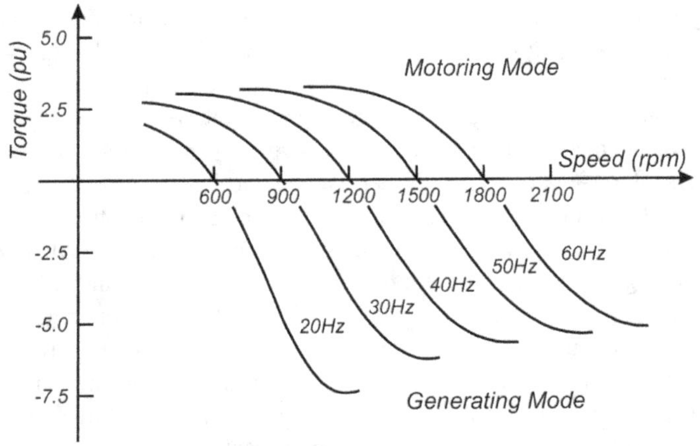

Figure 12.8 Variable frequency, quasi-steady state control with torque/speed performance for constant Volts/Hz

the motoring mode (motor drives) is on the positive torque half-plane, while for induction generator applications in wind and hydro turbines for example, the operating angular frequency must be above the synchronous speed, such a negative slip will give developed shaft torques as indicated in the negative half-plane.

$$T_e = 3\frac{P}{2 * r_r} * \Psi_m^2 * \omega_{sl} \tag{12.28}$$

Figure 12.8 indicates that for a fixed Volts/Hertz ratio the characteristic curve would be locked and the torque cross-over operating point changes in accordance with the impressed stator frequency as shown in Figure 12.8 and Equation (12.28). Depending on the shaft speed one point along the torque speed curve will be developed. For negative slip, the machine will develop negative torque and the operation will be a generator one. If variable speed is required, the V/Hz ratio must be maintained to keep the rated airgap flux. The firmness of airgap flux results in good utilization of the motor iron, a high torque per stator ampére and rated torque is possible for a wide speed range. It is important to note that neglecting the stator resistance and the saturation effect, both playing a very important role in a realistic quantitative evaluation, Equation (12.28) is not precise for real-life operation. For motoring mode, the stator resistance drop is opposed to the impressed voltage, and a boost scheme is required for low-frequency operation. For generating mode, the stator voltage is reversed resulting in an increased back-emf with increased air-gap flux, that is why the negative torque is higher than the positive torque in the family of curves. The torque initially rises as the speed decreases, but this increase is counteracted by the saturation of the magnetizing inductance.

Until the middle of the 1970s when the primitive VFD based on thyristors forced commutation was implemented, the induction motors would make a lot of audible noise, particularly because the switching was in low audible frequency, even bipolar transistors were very slow, and any low-frequency open-loop modulation would make torque pulsations on the shaft, and some harmonics and sub-harmonics may even create dangerous mechanical vibrations that could break the shaft. The machines at that time had very quick aging due to capacitive coupling of the variable high-frequency voltages and currents coupling of windings and the external iron case and base carcass, with the overall impact caused by electromagnetic interference. Such crude VFD schemes were not amenable for torque control—typically those drives were not used as electrical generators—but of course further mathematical analysis, and parametric case studies, allowed slip control, which is a quasi-stationary approximation of magnetizing flux, so eventually only a lower than average steady-state operation would be achieved, eventually having a very bad transient response. For many years and still today some industry people, plus power systems personnel, curse the bad performance of those machines based on such past experience. However, those are mostly uninformed people who have not been aligned with the latest technological implementations of advanced field orientation, signal processing, adaptive control, and modern techniques that can be applied and implemented in induction machines.

There was a technological revolution in the 1980s with new microprocessors that were developed for personal computers with associated microcontrollers that were launched yearly with better and faster performance than previous ones. Texas Instruments, Motorola, and Analog Devices started to manufacture and sell DSPs, which were better than microprocessors because they allowed the multiplication of variables in one computer cycle, and each new edition of a DSP always incorporated more features, generation of PWM, easier I/O, easier serial communication with other devices, better system compilers, typically using C language, and quickly there were new applications of real-time control of induction motors. This chapter discusses the next two baseline underpinnings of indirect vector control and direct vector control, but much more should be studied by the motivated reader, student, or engineer who wants to become an expert, it is recommended to follow all references indicated in the bibliography sections at the end of this chapter. The next sections will define the analysis and mathematics required to understand the electrical engineering technology for IVC and DVC.

12.8 Indirect vector control speed motor drive

During the 1980s the ideas and methods for induction machines modeling and control were very new to most people. Even the author of this chapter, after being introduced to the foundations of power electronics and some control methodologies learned the difficulties in implementing a variable frequency drive for three-phase induction motor control and decided to enroll in a graduate course during his Master degree period about "The General Theory of Alternating Current Machines" by B. Adkins and R.G. Harley. The book was initially published in 1957, it was a graduate course during the 1980s considered to be advanced. Computer-based simulations at that time were still on mainframes or on compiled languages such as Pascal and C for DOS personal computers, students and researchers would have to implement all the mathematical routines and recipes. The general theory of electrical machines was more conceptual and focused at that time on the approach of having a "Primitive Machine" i.e., a canonical model that could be handled to represent any type of machine, DC motors, asynchronous motors, synchronous motors, sometimes including damper windings. It was a general introduction to the two-axis theory, and most typically the computer models implemented were used to observe how start-up transients would develop in the electromechanical system, harmonics, oscillations, short-circuit conditions, and in fact, the on-going findings of vector control were not yet associated with these kinds of graduate studies. The use of the word "general" in the title of the book and associated courses was criticized, because the concepts were not comprehensive in the sense that every possible type of machine and problem could be dealt with, typically permanent magnet motors were not included, variable reluctance types, stepper-motors, and no imbalance of operation was included, and particularly no control approach was even employed. Vector control was coming out of theoretical foundations, in a few reports. Books mostly adopt space vector approaches which are very mathematical, fancy, and compact. Nevertheless, formulations of electrical machines using polar complex numbers and space vector algebra proved to be more obstructive than inspiring.

Vector-controlled induction motors can be implemented based on the machine stator flux, air-gap flux, or rotor flux. The d-q transformation is made such as to have an alignment of the rotating reference frame with the chose flux. For instance, ensuring that the d-axis of the rotating reference frame possesses the entire flux magnitude implies that any q-axis component would govern the torque. This orthogonality can be elucidated through the electromagnetism of machines, employing the general theory of electrical machines. Given that a q-axis component results in a flux vector projection of zero, any alteration in the q-axis solely impacts the machine's active power, consequently affecting the machine's torque in a constant speed operation. This supports the notion that a closed-loop speed control, incorporating a proportional-integral (PI) controller, will generate a reference signal usable for commanding the torque and, consequently, the q-axis current component in the machine. A hypothetical virtual d-q machine could be controlled on the rotating reference frame by guaranteeing (i) that the flux is aligned with one axis, (ii) the other axis will control torque, (iii) providing any mathematical feedforward or decoupling between the axis, and (iv) the induction machine would have a current-controller, specifically a power electronics inverter where the operation is to impress current, most of the time that is a voltage-source three-phase PWM system, with inner current feedback control. To impress currents on the machine terminals, as dictated by the hypothetical rotating reference model, would make such a three-phase induction motor or generator to follow the commands given on the hypothetical virtual d-q machine.

There are several possible compositions of reference frames in the literature, but in this chapter, we adhere to the vector rotation representation (Figure 12.3), with the d-q rotating reference frame for an induction machine (Figure 12.5). In the quadrature equivalent q-d frame circuit of an induction machine, the stator flux results from integrating the voltage across Lm and Lls. The air-gap flux arises from integrating the voltage across Lm alone, while the rotor flux originating from the stator side is generated by integrating the voltage across Lm and Llr. It is conceivable that in a steady state, these voltages lag from one to the next as we progress from the external terminals toward the inner equivalent circuit. Consequently, the fluxes would also experience a lag. Historically, direct-vector control was pioneered by Hasse, employing flux pick-up coil sensors on the air-gap. Naturally, specialized machines were constructed for this purpose. Subsequently, indirect vector control was introduced by Blaschke, featuring analytical calculations and feedforward control based on the rotor flux. Please refer to Figure 12.5, where the machine flux is directed toward the right side of the equivalent circuit (typically, the rotor is short-circuited in the case of considering squirrel-cage machines). In this case, the system is parameter sensitive on Rr (rotor resistance).

Then in 1988 Xu, DeDoncker, and Novotny published a paper where they implemented direct vector control based on stator flux. The identical voltage integration across Lm and Lls can be calculated by measuring the voltage terminals and currents of the machine, subtracting the voltage drop across the stator resistance, and integrating over time. In this approach, the performance parameter is sensitive to Rs. However, measuring Rs is easier than Rr, and thermal correction methods have been proposed in. The only practical concern is the necessity of a minimum shaft speed and frequency for the integration to yield satisfactory performance. Nevertheless, sensorless-based control systems typically have restrictions on minimum machine speed, susceptibility to temperature effects, or sensitivity to other parameters. The author of this chapter draws upon personal experience in this regard, having implemented various strategies. It is advisable to consider either (i) indirect vector control, which is based on rotor flux orientation, or (ii) direct vector control, which relies on stator flux orientation. There are some discussions on Universal Field Orientation schemes also proposed by DeDoncher and Novotny in the same year of 1988 where a general controller could be aligned to any flux, but typically such global-based field orientation is not adopted in real life. Vector control or field-oriented control became the de-facto industrial implementation and typically all manufacturers have modern drives based on these principles. Especially in the 21st century, there is a wealth of robust technology, including powerful microprocessors, microcontrollers, DSPs, and FPGAs, readily accessible for industrial, commercial, and even hobbyist applications. IGBT transistors now enable applications at scales of several kilovolts and kiloamps, with switching frequencies exceeding 25 kHz. These transistors come equipped with integrated gate-drives and power devices, ensuring highly compact implementations with top-notch EMI and EMC safety standards. Both IVC and DVC necessitate mathematical transformations, which will entail:

- Having either a d-q or a polar space vector formulation of an induction machine or maybe a three-phase (abc) sinusoidal approach.
- Establish a three-to-two-phase (abc) to (d^s-q^s) projection using Clarke transformation. Vector control implementations usually assume an ungrounded motor with balanced three-phase currents such that only two motor current phases need to be sensed. It is necessary to develop the backward two-to-three phase (d^s-q^s) to (abc) projection which will then assume a Space Vector PWM modulator, or an inverse Clarke transformation plus one among several possible sinusoidal PWM modulation schemes.
- Forward and backward two-to-two phase, (d-q) to (d^s-q^s) and (d^s-q^s) to (d-q) projections using the Park and inverse Park transformations, respectively.
- Depending on the chosen flux orientation (stator, air-gap, or rotor) the mathematical decoupling torque and field signals to be used in the control strategy plus any necessary sensor measurements for algorithm development.

The indirect vector control (IVC) is commonly used in industrial applications where a closed-loop speed control will easily operate the drive systems throughout a range from zero speed to maximum speed (with variable torque) then beyond where high-speed can be achieved at constant power and field-weakening mode. For such IVC the flux angle position (ω_e) is feedforwarded with internal calculations, the measurement of stator currents (which are also in an inner current-mode control loop for the inverter) with measurement of the rotor speed allows the instantaneous calculation of ω_e by summing the measured rotor angle corresponding to the rotor speed and the calculated reference value of the slip angle corresponding to the slip frequency ω_{SL}. The idea behind IVC is to use the location of the rotor, such angular speed will help to calculate the angle θ_r, then with proper calculation of the slip frequency a feedforward signal would generate $_{SL}$, both can be integrated as indicated by (12.29) and added to generate angle θ_e, which can then be used for vector-rotation (VR) and inverse-vector-rotation (IVR). The system can easily transition toward being torque control or speed control, with management of the proportional+integral (PI) initializations on the torque loop or speed loop control. An equation for calculating $_{SL}$ can be derived from the q-d rotating reference circuits in Figure 12.5 by examining the circuit from rotor side port. The resulting Equation (12.30) can then be formulated.

$$\begin{aligned} \theta_r &= \int \omega_r dt \\ \theta_{SL} &= \int \omega_{SL} dt \\ \theta_e &= \theta_r + \theta_{SL} = \int (\omega_r + \omega_{SL}) dt \end{aligned} \tag{12.29}$$

$$\begin{aligned} &\frac{\mathrm{d}}{\mathrm{dt}}\Psi\mathrm{qr} + \mathrm{Rr} * \mathrm{iqr} + \omega_{SL} * \Psi\mathrm{dr} = 0 \\ &\frac{\mathrm{d}}{\mathrm{dt}}\Psi\mathrm{dr} + \mathrm{Rr} * \mathrm{idr} + \omega_{SL} * \Psi\mathrm{qr} = 0 \end{aligned} \tag{12.30}$$

It is not practical to measure rotor currents, particularly when driving squirrel-cage induction machines (for doubly-fed IM this is very simple and possible). It is necessary to eliminate both iqr and idr from Equation (12.3). Again observing the rotating reference d-q circuits in Figure 12.5, the following relationships in (12.31) can be derived. Note that inductances in those circuits are self-inductance, there is no mutual effect from one to another. Isolating on the left-end side the Equation in (12.31) for iqr and idr and replacing them in (12.30) we obtain the differential Equation (12.32).

$$\begin{aligned} \Psi\mathrm{qr} &= \mathrm{Lm} * \mathrm{iqs} + \mathrm{L_R} * \mathrm{iqr} \\ \Psi\mathrm{dr} &= \mathrm{Lm} * \mathrm{ids} + \mathrm{L_R} * \mathrm{idr} \end{aligned} \tag{12.31}$$

$$\begin{aligned} &\frac{\mathrm{d}}{\mathrm{dt}}\Psi\mathrm{qr} + \frac{\mathrm{Rr}}{\mathrm{L_R}}\Psi\mathrm{qr} - \frac{\mathrm{Lm}}{\mathrm{L_R}}\mathrm{Rr} * \mathrm{iqs} + \mathrm{w_{SL}}\ \ \Psi\mathrm{dr} = 0 \\ &\frac{\mathrm{d}}{\mathrm{dt}}\Psi\mathrm{dr} + \frac{\mathrm{Rr}}{\mathrm{L_R}}\Psi\mathrm{dr} - \frac{\mathrm{Lm}}{\mathrm{L_R}}\mathrm{Rr}\ \ \mathrm{ids} + \mathrm{w_{SL}}\ \ \Psi\mathrm{qr} = 0 \end{aligned} \tag{12.32}$$

The control objective is to impose orthogonality i.e., the d-axis flux component must be zero on the q-axis, while the q-axis flux component must be zero on the d-axis. Therefore, the whole flux vector must be aligned with the d-axis. The vector controller must then impose alignment of θ_e with the d-axis in these circumstances the component $\Psi_{qr} = 0$ and of course, their derivative in time is also $\frac{d\Psi_{qr}}{dt} = 0$. Then the first equation in (12.32) can be solved for the slip frequency ω_{SL} giving the feedforward control condition in (12.33). The variables multiplying the torque current iqs are typically considered constant, particularly for the whole torque control mode range where the operational speed is less than the nominal speed rating. This factor that multiplies iqs is called *Slip Gain*. It is important to consider that rotor resistance Rr varies with temperature plus rotor skin effect as well as Lr may have temperature and saturation effects, Lm has saturation and core loss effects, and the d-component of flux can be assumed to be the whole flux magnitude but it is constant only for normal operation and not in constant power field weakening high-speed mode.

$$\omega_{SL} = \left(\frac{\mathrm{Lm} * \mathrm{Rr}}{\Psi\mathrm{dr} * \mathrm{L_R}}\right) * \mathrm{iqs} \tag{12.33}$$

Solving the second equation in (12.32) for flux, and assuming it to be constant, the relationship in (12.35) holds, again reminding us that there is no q-component of the rotor flux, so the whole magnitude of rotor flux is $\Psi_{dr} = \Psi_{rotor}$ which rotates at ω_e and theoretically fully aligned at θ_e. If the machine will go into field weakening mode, Equation (12.34) must be solved in real-time (it is a differential equation), or a look-up table can be used, since the rotor time constant for flux variation will be slow, and a quasi-steady-steady assumption can be made to avoid the numerical solution of (12.34) in real-time. Most of the time a simple look-up table with a low-pass filter in series will emulate the behavior of (12.34).

$$\Psi\mathrm{dr} = \mathrm{Lm} * \mathrm{ids} - \left(\frac{\mathrm{L_R}}{\mathrm{Rr}}\right) * \frac{d}{dt}\Psi\mathrm{dr} \tag{12.34}$$

$$\Psi\mathrm{dr} = \mathrm{Lm} * \mathrm{ids} \tag{12.35}$$

If the stator component ids is held constant, the flux will be constant. Then, for any torque current command, the slip frequency is easily calculated (*Slip-Gain*), it is an internal calculation in the control so there is no noise corruption of external variables. The slip speed is directly proportional to iqs in induction machines, the torque is proportional to the slip, and the torque is then proportional to iqs. Therefore, for constant flux speed control will operate to maximum rated speed, at maximum power rating. The whole motor drive can be fully in torque control mode. If induction generators are controlled, the torque command at the iqs closed-loop control can be easily changed from positive to negative, or vice versa, if the shaft is on a prime mover at a constant speed, it is very simple to make the system to be bidirectional in the power flow, so the power at the mechanical shaft will go backward in the drive system by just flipping the iqs set-point. Of course, the drive system must have a way to make it go to another converter at the dc-link. In industrial drives, a torque command typically incorporates an outer loop to regulate speed, as most industrial automation relies on speed control. However, if there is a need for position control, such as in robotic manipulators, pick-and-place actuators, or shaft positioning, an outer loop can be added to the speed control, which, in turn, is layered on top of the torque control. Nevertheless, the implementation of such layered controllers, progressing from inner to outer loops, requires careful simulation and study to ensure that all feedback controllers operate effectively within their linear range, avoiding saturation issues. If the speed control drive must go into field weakening, the whole differential Equation (12.34) must be implemented, or a look-up table with a low-pass filter may program the ids set-point to make the machine go into field weakening as the speed increases.

Current control plays an important role in power electronic circuits. Particularly current-regulated PWM inverters are widely applied for ac motor drives and continuous ac power supplies where the objective is to produce a sinusoidal ac output. In the past those inverters were hysteresis-band (bang-bang) controlled, but with fast digital control it became possible to have d-q PI control of current set-points compared to the actual measured currents, such controllers would then generate the set-point d-q voltages to be then either converted to abc and control the whole inverter in abc, or the PWM would receive such d-q commands for modulation of the inverter output voltages. The main task of the control system in current-regulated inverters is to force the current vector in the three-phase load according to a reference trajectory. To operate the three-phase induction motor in a vector-controlled mode, sensed three-phase stator currents are to be controlled by the three-phase reference currents. Such a control technique is called a current controller. However, independent three-phase sinusoidal current controllers are difficult to implement, because PI controllers do not work well with variable steady-state conditions. The PI+ resonant controllers can be implemented directly on the abc frame and do not require any transformation, but they work well for constant angular speed, so they are mostly recommended for grid-connected inverters.

In the indirect vector control technique the rotor flux vector position is computed from the speed feedback signal, IVC eliminates the need to use flux sensors or a flux model. It is necessary to have an accurate measurement of the shaft position to determine the precise position of the rotor flux vector. The block diagram for the IVC controller is shown in Figures 12.9 and 12.10; there are two main control loops, one for flux and another one for speed. For a motor drive working from zero speed to maximum rated speed, the rotor flux is kept constant, so ids

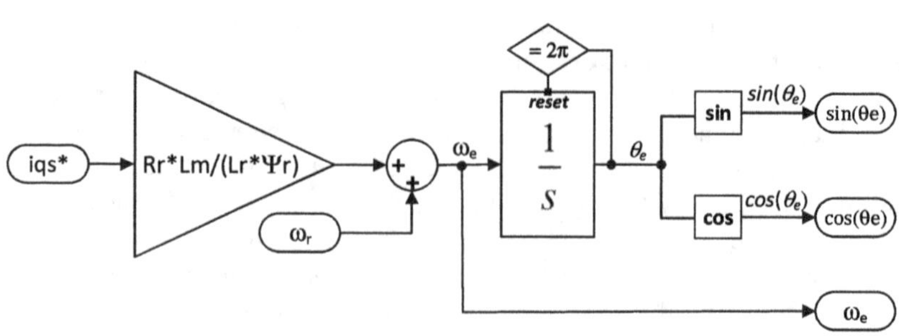

Figure 12.9 Computation of sine and cosine signals based on slip-gain command (ω_{SL}) added to the instantaneous shaft speed (ω_r) for having (ω_e) which after integration gives the flux angular position θ_e

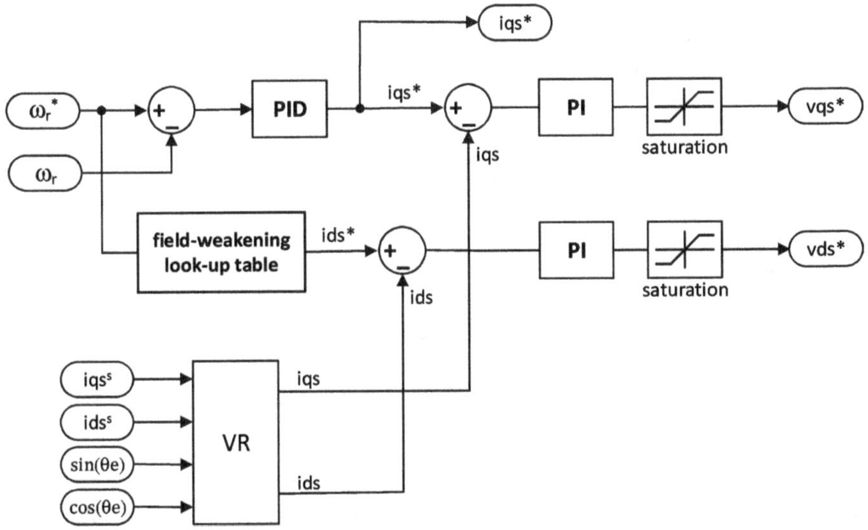

Figure 12.10 Closed-loop control for speed and inner controllers for quadrature current (iqs and ids loops) for quadrature voltage outputs (vqs and vds) to be sent for the transistor inverter

set-point does not change until the base speed. Typically for medium power induction machines the magnitude of the rotor flux is 0.8 Wb (or a little less) while the machine is in the constant flux and variable torque region. The controller computes the slip frequency that will be added to the shaft speed provided, which gives the angular speed of the rotor flux vector from which the flux angle can be computed. This angular speed is integrated to generate θ_e and reset every full rotation (2rads). Such angular position of the rotor flux has instantaneous calculations of sine and cosine for the vector-rotation (VR) and inverse-vector-rotation (VR^{-1}).

When the machine runs above or below the set point for the speed reference, the PI controller will generate the proper iqs* command, which in turn will be compared to the current measured and vector-rotated (iqs) of the motor. There are two inner PI controls for the q-axis and the d-axis, a proper voltage set-point vqs* will be generated with a signal provided by iqs*, and the d-axis will control ids at the machine, typically kept constant for nominal speed range operation. Those PI regulators have a saturation in the diagram, to typically normalize their signals to $[-1, +1]$ range, then they will feed the sinusoidal pulse-width modulator to generate the PWM patterns for the transistor circuit. Such a $[-1, +1]$ saturation serves as a limitation in case of transient response of the PI that may have some settling time during sharp transitions of the main speed control loop. Extensive and exhaustive simulation case studies must be conducted with careful observation if those PI controllers respond well, if there is not much saturation, and of course, if the steady-state error of the loops drives the error toward zero.

The d-axis quadrature control will assume that for the normal speed range (from zero to base speed) the flux is kept constant by maintaining ids* constant.

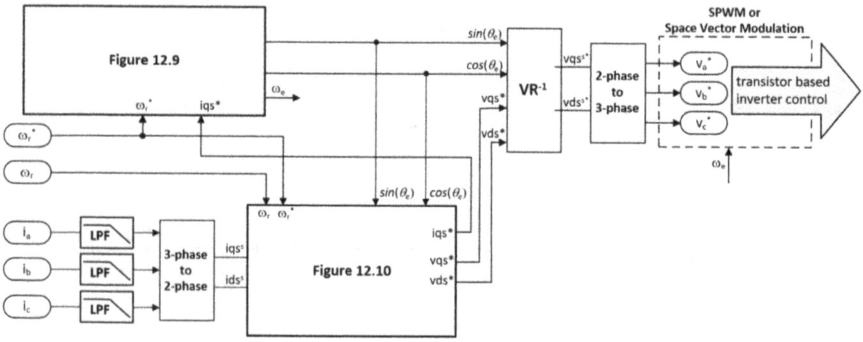

Figure 12.11 Complete Indirect Vector Control (IVC) induction motor drive

However, above the base speed, the flux field is weakened to keep the power constant, with a programmable look-up table as indicated in the block inserted from speed (ω_r^*) to ids*, sometimes for small machines it is important to have a low-pass filter after this look-up table to slow down the change, for larger machines typically the inertia will take care of flux change. If the angular speed set-point is set for a higher speed than the base, the overall behavior is to keep the machine in constant power variable (decreasing) flux, but there is a minimum flux to be maintained otherwise the whole functionality of the drive collapses, because a machine without flux has no field for any energy conversion, so the optimization of real-life induction motor drives in field weakening mode is very ad-hoc and dependent on experimental validation. The loop is closed with flux and sent through a PI to develop the vds*. After the signals vds* and vqs* are generated, they are sent through IVR and a two-phase to three-phase block to synthesize the output patterns for Va*, Vb*, and Vc*. These signals are sent through a PWM and inverter to create the voltages on the terminals of the machine. Combining all these blocks for the IVC speed controller there is a full-fledged speed control loop depicted in Figure 12.11. The controller inputs the currents and speed and outputs the reference voltages to control the machine to the specified speed. The input currents are filtered using a low pass filter because typically in real-life implementation there is high-frequency switching and noise on current measurements.

12.9 Direct vector control with stator flux orientation

A possible methodology for implementing direct vector control (DVC) is by aligning the stator flux along the d-axis as shown in Figure 12.12. By observing Figure 12.5 with the d-q rotating reference frame, d-q equivalent circuit of an induction machine, the stator flux is the one from the left port of the equivalent circuit after the stator resistance; therefore, it is possible to acquire the instantaneous voltages and currents, subtracting the voltage drop on the resistance, and through a time integration the stator flux can be synthesized from such measurements, this will be discussed later in

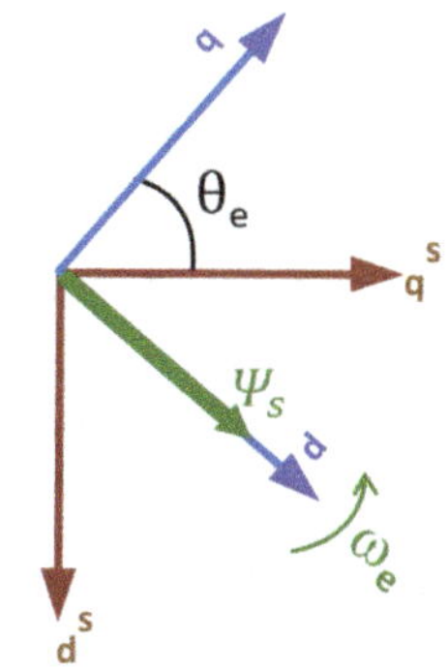

Figure 12.12 Vector representation for DVC stator flux

this section. At this time it is important to look to the right port of the d-q equivalent circuits in Figure 12.12 and try to reach the stator flux by making some algebraic derivation based on the Blaschke equations. Although they were made initially for IVC, through proper substitutions it is possible to arrive at the stator flux from the rotor flux, as indicated by Equation (12.36).

$$\begin{aligned}(1+\mathrm{sT_R})*\Psi_{\mathrm{dr}} &= \mathrm{L_M}*\mathrm{i_{ds}}+\omega_{SL}*\mathrm{T_R}*\Psi_{\mathrm{qr}}\\ (1+\mathrm{sT_R})*\Psi_{\mathrm{qr}} &= \mathrm{L_M}*\mathrm{i_{qs}}-\omega_{SL}*\mathrm{T_R}*\Psi_{\mathrm{dr}}\end{aligned} \tag{12.36}$$

Examining Figure 12.5, the d-q rotating reference frame equivalent circuit of an induction machine reveals that the q-axis stator flux is expressed by (12.37), and iqr is defined by (12.38). By combining these expressions and isolating iqr on the left-hand side, the result is Equation (12.39). Continuing the algebraic manipulations, we calculate both rotor flux components Ψ_{qr} and Ψ_{dr} using Equations (12.40) and (12.41), respectively, applying the Blaschke equations from formulation (12.42) with parameters specified in Equation (12.43).

$$\Psi_{qs} = \Psi_{qr} - L_{LR}*i_{qr} + L_{LS}*i_{qs} \tag{12.37}$$

$$i_{qr} = \frac{1}{L_M}*\Psi_{qs} - i_{qs}*\left(\frac{L_{LS}}{L_M}+1\right) \tag{12.38}$$

$$\Psi_{qs} = \frac{L_M}{L_R}\Psi_{qr} + i_{qs}*\left(\frac{L_{LR}*L_S+L_{LS}*L_M}{L_R}\right) \tag{12.39}$$

$$\Psi_{qr} = \frac{L_R}{L_M}*\Psi_{qs} - i_{qs}*\left(\frac{L_{LR}*L_S+L_{LS}*L_M}{L_M}\right) \tag{12.40}$$

$$\Psi_{dr} = \frac{L_R}{L_M}*\Psi_{ds} - i_{ds}*\left(\frac{L_{LR}*L_S+L_{LS}*L_M}{L_M}\right) \tag{12.41}$$

$$
\begin{aligned}
(1+\mathrm{sT_R}) * \Psi_{\mathrm{ds}} &= \mathrm{i_{ds}} * \beta + \omega_{SL} * \mathrm{T_R} * \left(\Psi_{\mathrm{qs}} - \frac{\alpha}{L_R} * i_{qs}\right) \\
(1+\mathrm{sT_R}) * \Psi_{\mathrm{qs}} &= \mathrm{i_{qs}} * \beta - \omega_{SL} * \mathrm{T_R} * \left(\Psi_{\mathrm{ds}} - \frac{\alpha}{L_R} * i_{ds}\right)
\end{aligned} \tag{12.42}
$$

$$
\begin{aligned}
\beta &= \left(L_S + \frac{s * T_R * \alpha}{L_R}\right) \\
\alpha &= L_{LR} * L_S + L_{LS} * L_M
\end{aligned} \tag{12.43}
$$

The following relationships for the stator flux can be defined for alignment of the magnitude of the flux along the d-axis: $\Psi_{qs} = 0$ and $\Psi_{ds} = \Psi_s$. Replacing the d-axis alignment definitions in Equation (12.42) the following dynamical relationships for the stator flux can be defined in (12.44).

$$
\begin{aligned}
(1+\mathrm{sT_R}) * \Psi_{\mathrm{s}} &= \mathrm{i_{ds}} * \beta - \omega_{SL} * \mathrm{T_R} * \frac{\alpha}{L_R} * i_{qs} \\
0 &= \mathrm{i_{qs}} * \beta - \omega_{SL} * \mathrm{T_R} * \left(\Psi_{\mathrm{s}} - \frac{\alpha}{L_R} * i_{ds}\right)
\end{aligned} \tag{12.44}
$$

There is a coupling between the stator flux and the torque component $i_{qs.}$ To decouple these signals, a stator flux feedback control can generate the set-point for ids with a PI controller plus a decoupler as indicated in Figure 12.13 and the corresponding Equation (12.45). Replacing Ψ_s and simplifying, Equation (12.46) is obtained.

$$i_{ds} = PI(s) * \left(\Psi_s^* - \Psi_s\right) + i_{dq} \tag{12.45}$$

$$\Psi_s * (1 + sT_R) = \beta * \left(PI(s) * \left(\Psi_s^* - \Psi_s\right)\right) + \beta * i_{dq} - \omega_{SL} * \frac{T_R}{L_R} * \alpha * i_{qs} \tag{12.46}$$

Therefore, to decouple the stator flux from the torque component i_{qs}, the last two terms must be added to zero, and solving for the decoupler current results in Equation (12.47) with the block diagram to calculate it in Figure 12.14.

$$i_{dq} = \left(\frac{T_R * \alpha}{L_R * \beta}\right) * \omega_{SL} * i_{qs} \tag{12.47}$$

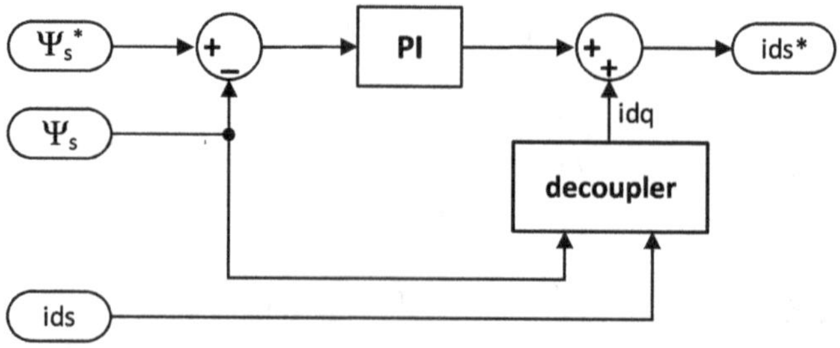

Figure 12.13 Decoupler circuit

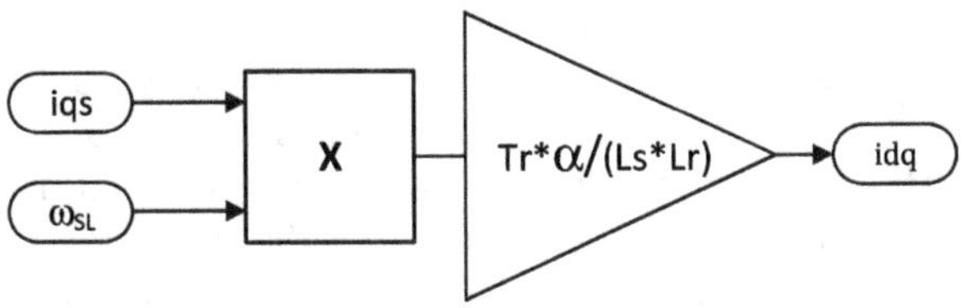

Figure 12.14 Block diagram for calculating decoupling current

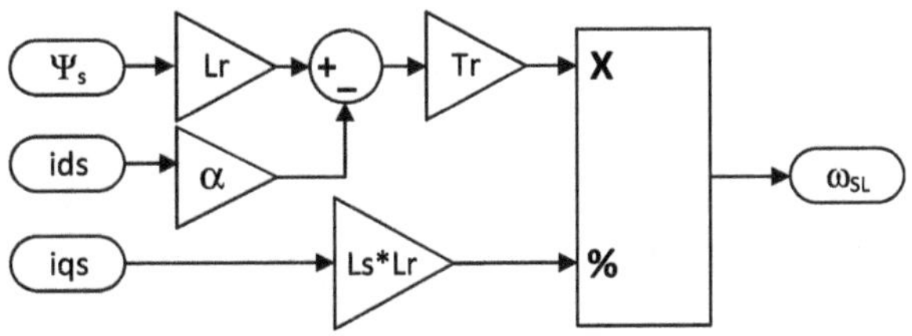

Figure 12.15 Real-time estimation for slip frequency ω_{SL}

Of course, the only unknown in (12.47) is the slip frequency, which can be calculated from (12.44), resulting in (12.48). However, this dynamic system becomes unwieldy for real-time applications due to the decoupler being a feed-forward addition to the flux loop control. The PI controller can compensate for the dynamic response caused by iqs, and only the steady-state solution for the decoupling factor is considered by imposing s = 0. The final equation for calculating the slip frequency is provided in (12.49), illustrated by a block diagram in Figure 12.15.

$$\omega_{SL} = \frac{i_{qs} * (L_S * L_R + sT_R\alpha)}{T_R * (\Psi_S * L_R - \alpha * \ i_{ds})} \tag{12.48}$$

$$\omega_{SL} = \frac{i_{qs} * L_S * L_R}{T_R * (\Psi_S * L_R - \alpha * \ i_{ds})} \tag{12.49}$$

The motor drive system must have as inputs a reference torque and a reference flux as set-points. To determine the value of the reference current I_{qs}, it is necessary to decouple torque and flux with some feedforward mechanism. The following relationship (12.50) is used, and assuming the stator flux is aligned with the d-axis, consequently $\Psi_{qs} = 0$ in (12.50), accordingly then the iqs* set-point for a given torque is calculated by (12.51).

$$\mathrm{T_e} = \frac{3\mathrm{P}}{4}\left(\psi_{\mathrm{ds}}\mathrm{i_{qs}} - \psi_{\mathrm{qs}}\mathrm{i_{ds}}\right) \tag{12.50}$$

$$\mathrm{i_{qs}}^* = \left(\frac{4}{3\mathrm{P}\Psi_{\mathrm{s}}^*}\right) * \mathrm{T_e^*} \tag{12.51}$$

The DVC system also needs vector rotation (VR) and inverse-vector-rotation (VR-1) where the angle θ_e must be calculated from the stator flux. The estimation of stator flux might be implemented by measuring the voltages and currents at the machine terminals and knowledge of the stator resistance. Through the integration on the quadrature-axis, as indicated by (12.52) the sine and cosine waves can be synthesized directly (instead of angle θ_e), by performing a normalization, i.e., a division of the instantaneous d and q fluxes by the magnitude of the flux, as depicted by (12.53). The magnitude of the flux can be also computed by the instantaneous evaluation of the square root of the addition of each d and q flux component squared as indicated by (12.54).

$$\begin{aligned} \Psi_{qs}^{s} &= \int \left(v_{qs}^{s} - R_S * i_{qs}^{s} \right) dt \\ \Psi_{ds}^{s} &= \int \left(v_{ds}^{s} - R_S * i_{ds}^{s} \right) dt \end{aligned} \tag{12.52}$$

$$\begin{aligned} \sin(\theta_e) &= \frac{\Psi_{qs}^{s}}{\Psi_s} \\ \cos(\theta_e) &= \frac{\Psi_{ds}^{s}}{\Psi_s} \end{aligned} \tag{12.53}$$

$$\Psi_s = \sqrt{\left(\Psi_{qs}^{s}\right)^2 + \left(\Psi_{ds}^{s}\right)^2} \tag{12.54}$$

Integrals are problematic for very low-frequency operation because if the voltage and current sensors have any small dc-offset, the integration will eventually saturate the output of analog operational amplifiers, and even when calculated internally in microcontrollers, the quadrature flux will build up an increasing variable that may hit some processing overflow, and during transient responses can lead to unstable operation. A low-pass filter with a very low corner frequency can be used to implement the integration in analog-based measurements. Such a low-pass filter might be a very good approximation for integration as long as the signal frequency to be integrated is higher than 10x the corner frequency of the low-pass filter. To understand this approximation it is suggested that the reader compare the Bode Diagram (magnitude and frequency response) for a pure integral $\frac{1}{s}$ with a low-pass filter $\frac{K}{s+\omega_c}$, and it is simple to observe that for high-frequency the low-pass filter will have a -90° phase-shift. It is possible to implement a gain correction provided by K and ω_c with a programmable analog amp-op-based low-pass filter or by digital filtering. The time-domain evaluation is shown in (12.55), rewriting the quadrature flux equations with a low-pass filter gives Equation (12.56). To define the corner frequency ω_c, the following relationship is used $\omega_c = \frac{1}{2\pi f_c}$ where f_c is the corner frequency of the low-pass filter, with simulation studies it can be chosen to be 0.0001 Hz, where the motor drive will work if the frequency of the machine is above 0.001 Hz, which is typically more than enough good approximation for any 60 Hz or 50 Hz induction machine. The block for calculating the stator flux with the low-pass filter as an integrator is shown in Figure 12.16 and the controller for torque control is depicted in Figure 12.17. The block diagram shows a coefficient K1 which must be considered if a low-pass filter is used instead of a

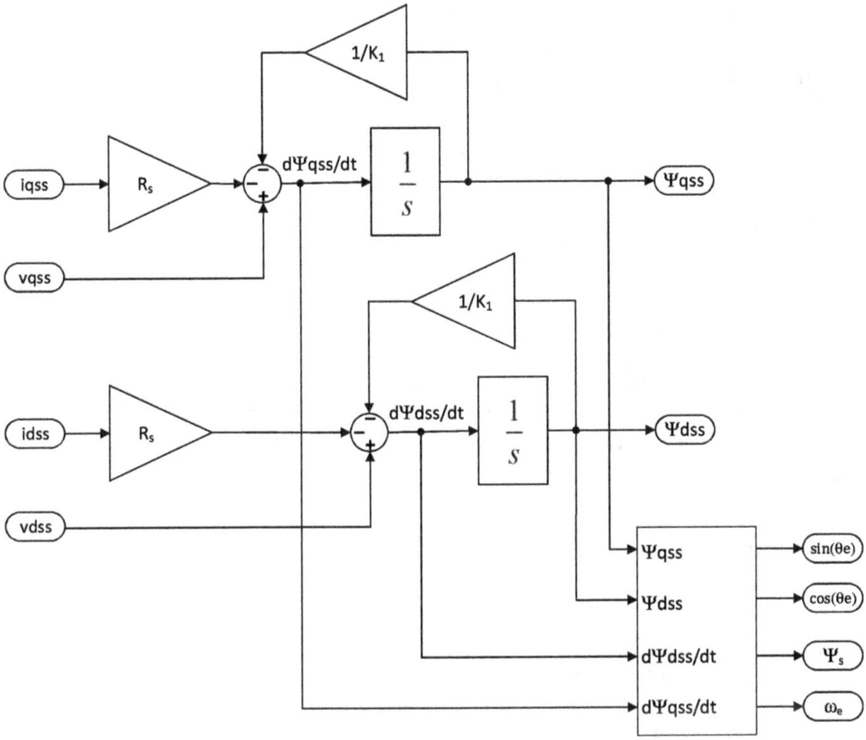

Figure 12.16 Block diagram for calculating stator flux

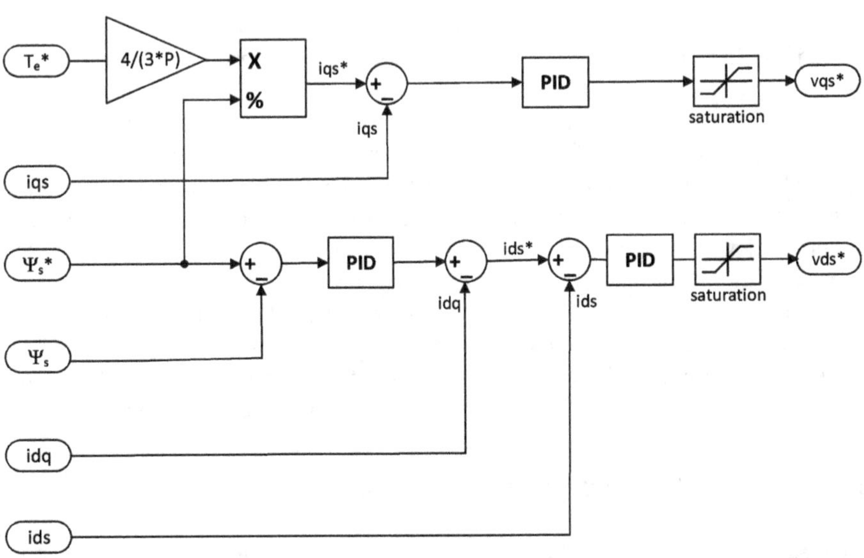

Figure 12.17 DVC stator flux torque controller

straightforward direct integration.

$$
\begin{aligned}
\Psi^s_{qs} &= \int \left(v^s_{qs} - r_S * i^s_{qs} - \omega_c * \Psi^s_{qs}\right) dt \\
\Psi^s_{ds} &= \int \left(v^s_{ds} - r_S i^s_{ds} - \omega_c * \Psi^s_{ds}\right) dt
\end{aligned}
\tag{12.55}
$$

$$
\begin{aligned}
\Psi^s_{qs}(s) &= \left(\frac{1}{s+\omega_c}\right) * \left(v^s_{qs}(s) - r_S * i^s_{qs}(s)\right) \\
\Psi^s_{ds}(s) &= \left(\frac{1}{s+\omega_c}\right) * \left(v^s_{ds}(s) - r_S * i^s_{ds}(s)\right)
\end{aligned}
\tag{12.56}
$$

The electrical frequency can be calculated, first starting by the definition of the flux angle position in (12.57), taking the derivative of the angle to get speed, as in (12.58), after some algebraic simplification, Equation (12.59) can be used to estimate the angular frequency ω_e in rad/s.

$$
\theta_e = \tan^{-1}\left(\frac{\Psi^s_{qs}}{\Psi^s_{ds}}\right)
\tag{12.57}
$$

$$
\frac{d\theta_e}{dt} = \omega_e = \frac{1}{\left(\frac{\Psi^s_{qs}}{\Psi^s_{ds}}\right)^2 + 1}\left(\frac{d}{dt}\left(\frac{\Psi^s_{qs}}{\Psi^s_{ds}}\right)\right)
\tag{12.58}
$$

$$
\omega_e = \frac{\Psi^s_{ds}\,\frac{d}{dt}\left(\Psi^s_{qs}\right) - \Psi^s_{qs}\,\frac{d}{dt}\left(\Psi^s_{ds}\right)}{\left(\Psi^s_{qs}\right)^2 + \left(\Psi^s_{ds}\right)^2}
\tag{12.59}
$$

An overall complete controller is shown in Figure 12.18 where the input voltages and currents from the machine will be fed to the stator estimation equations, and the reference torque with corresponding current torque will go to the vector controller. The input currents and voltages have been filtered in an experimental setup using an 8th-order low-pass Butterworth filter to obtain cleaner signals. The stator resistance for improved performance must be compensated for temperature variation, one possibility is with proper installation of thermistors on the machine, and some calibration. It is possible to develop a thermal model for the stator resistance variation with the outer temperature of the machine's exterior surface. The choice depends on the excellence of very high-performance motor drives that may go to extremely low-frequency operation. Typically it has been found if the motor drive operated above 5% of base speed and above, a DVC system would work with good to excellent performance, in addition, such a DVC approach is a very well fit for induction generator power electronics interfacing.

12.10 Universal field-oriented control

Universal Field Oriented (UFO) control was proposed for the first time by De Doncker and D.W. Novotny in 1988. The idea is to hybridize the operation of a

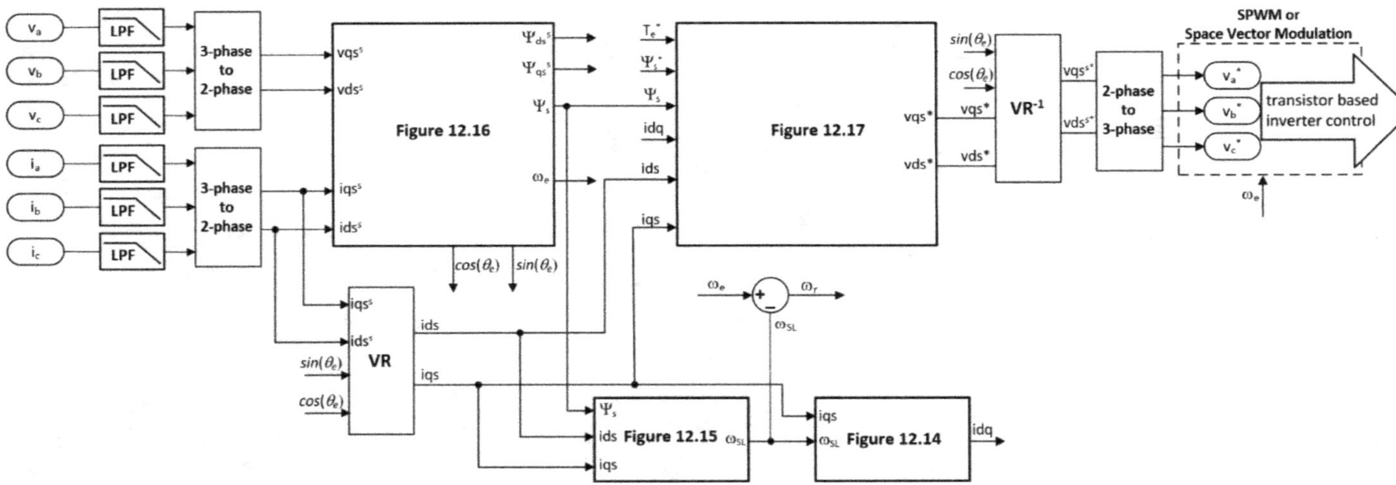

Figure 12.18 Stator flux oriented direct vector control (DVC) of induction motor drive

motor drive, for example with a stator flux orientation based on motor terminal quantities and the stator resistance combined with an indirect vector control, where only current and speed are measured with very high noise immunity (because the slip gain calculation is internal in the controller, with discussions of air-gap flux orientation for a particular torque/amp enhancement in those conditions). Probably the main concept was to have a wide-range induction motor drive, based on the fact that in the end, all those calculations are internal to any digital controller, and it could seamlessly transition from any alignment to any other one. Of course, there were additional calculations depending on the machine parameters for the decoupling equations of the controller based on the stator flux, and the air-gap flux. A current regulator was implemented with the transition from the linear PWM regulation to a square wave operation going to higher speeds with flux weakening region.

For the UFO approach, there is an indirect-vector control start-up mode, then around 5 Hz (when the fundamental component of the stator voltage becomes good noise immunity for sensors and data acquisition) the drive can transition to stator flux orientation, and the inner PI controllers must have bumpless implementation for switching back and forth. At base speed and beyond, the system can go into a 6-step square-wave mode operation. Similar principles have been used for other research and publications, and there was a research conducted by the author of this chapter with researchers at the University of Tennessee and General Motors that resulted in an issued patent for hybrid IVC/DVC start-up of motors for electric vehicles, where the integration of air-gap voltage for stator flux orientation was improved with programmable cascaded digital low-pass filters, with further improvements with neural networks based signal processing. It is possible to have a somewhat complex structure that allows all possible reference frames such as indicated in Figure 12.19. Such a structure has proper controller initialization and reset transition techniques to prevent large transients during the switching of the

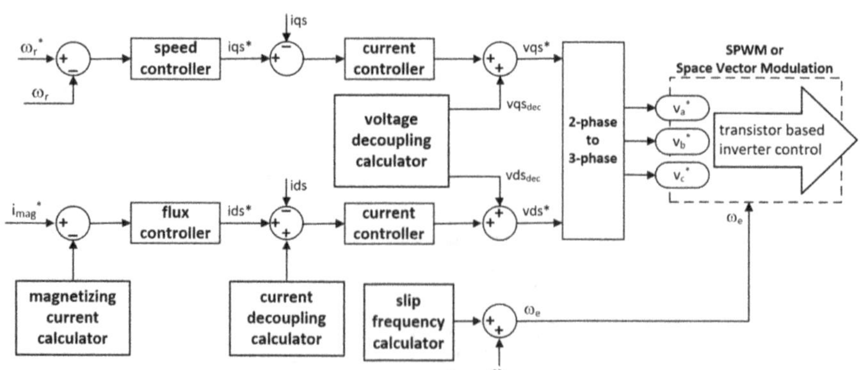

Figure 12.19 Universal Field Oriented Control, for induction machines, with current and voltage decoupling on any possible flux reference frame

reference frames, and it allows a mixed hybridization of controllers with an average weighting, also possible to implement as a fuzzy logic-based control management.

The choice of reference frame in a UFO has an important impact on the performance of the drive in practical realization. The classical IVC rotor flux as a reference frame has the parameters in the feedback path. Whereas the air-gap flux controller and the stator flux controller can be practically realized without any error in the flux position measurement, to avoid flux sensors, the stator flux position is calculated from the stator voltages and currents, and a DVC is only parameter dependent of the stator resistance. By changing dynamically the reference frame of the UFO controller without any change in the controller hardware enables the drive to operate over a very wide speed range. At low speeds, the UFO controller can be set to IVC with rotor or air-gap flux control, and at higher speeds the DVC with a stator flux control is preferable. Therefore, the best controller is used without any additional hardware and software. Compared to an IVC drive with an internal current regulator, it is required to add the stator voltage sensors for the stator flux vector in the DVC mode of operation. Adaptation schemes to compensate for temperature drift between the controller parameters and the machine parameters are well known when only stator resistance has to be adapted. The steady-state analysis for a DVC stator flux system always retains maximum torque capability no matter how large the error in the parameters is, but the DVC is somewhat sensitive to the variations in leakage and magnetizing inductances. An IVC with rotor flux alignment may have reduced torque capability and poorer torque/ampere ratio, or static instability (pull-out) for large rotor time constant detuning. In case of field weakening, as the machine is operated near its maximum capacity and at a reduced flux level, parameter sensitivity becomes more significant. In such a case stator field-oriented control shows a better performance than the rotor field-oriented control for field weakening. The problem associated with the stator field orientation is to take care of the feedforwarding of signals for proper decoupling. One major problem in implementing field-oriented control for induction motor drives is to develop the current regulator.

In most drives, the current control is realized by a high-gain current controller. But as the machine approaches the flux weakening region, the voltage margin goes to zero and it becomes difficult to maintain excellent transient response. The scheme described in the papers required a decoupling of the torque-producing component of the current and the stator flux. The PI regulators in both the torque and flux loops limited the transient response from this method, Profumo showed that by introducing direct self-control the problem would help to improve the regulation. Such direct self-control is based on the instantaneous error of each torque and flux control loop. This method was proved well suited to high-power drives with a low switching frequency that was implemented by Texas Instruments TI DSP C25 or C30. Currently, in the 21st century, it is certainly even easier to replicate the experimental work. It is evident that the UFO controller is applicable to all existing field-oriented controller schemes due to the generality of its reference frame. A single digital implementation of the UFO controller allows the designer to have the availability of six possible field orientation schemes with one single piece of hardware and software. The stator flux controller leads to a robust speed range drive that does not depend on the machine parameters during the steady-state operation and to obtain full dc bus utilization an improved current and voltage control is necessary for field weakening.

12.11 Conclusion

This chapter showed foundations of electrical engineering, physics, electromagnetism, and some basic control theory, for the analysis and modeling of induction machines, how their differential equations can be used for equivalent circuits in three-phase abc, as well as stationary d-q, and rotating d-q reference frames. The rotating reference frame allows a virtual dc-machine-like operation by imposing alignment of a machine flux (rotor, air-gap, or stator) with the d-axis, while the q-axis makes the torque control, and also closed-loop control possible. For parameter estimation, model-reference-control, and induction generator applications, typically the rotating reference frame does not capture the parametric requirements, and for those applications, it is better to use either stationary or abc frames. In terms of feedback signals, input currents and voltages are usually measured for the control loop. Speed is a measurement variable, but it is sometimes undesirable because speed sensors are fragile and may fail, so sensorless induction machine drives can be adopted. In the very long historical past, the first thyristor-based induction motor drive was typically implemented with analog op-amp-based control, where the variable frequency could be implemented with a voltage-oscillator control for frequency variation, and a proportional open-loop configuration would control the voltage to maintain a pretty much constant ratio of impressed voltage by frequency (Volts/Hz). When such SCR-based circuits were implemented in the 1960s, it was the birth of motor drive defined as Scalar Control. With the advent of transistor-based inverters with bipolar-junction-transistors and SPWM, it was then possible to have the implementation of scalar control of induction motors to achieve a suitable torque speed with impressed slip, since the slip frequency (angular speed of the machine flux subtracted by the equivalent shaft speed in the machine flux rotating reference frame) is proportional to torque, those were the slip controllers, still Scalar Control, but with a better and enhanced torque calculation. By then, the dynamics and bandwidth were very limited and the transient response of induction motor drives was not considered as fast as it would be required for industrial automation and even in electric transportation, so toward the 1970s and 1980s the armature control of dc-motor drives with rheostats were still common in electric buses and some trains. SCRs became enhanced with the invention of the GTOs but they were still used for very low-frequency operation.

Scalar control can be easily understood based on the fundamental principles of induction machine steady-state modeling. For generators where there is a power electronic interface that could be either in series connection with the machine for front-end control or in parallel for allowing a path reactive power for isolated operation, the induction generator could then be employed in wind turbines, or hydropower with scalar control by disregarding the coupling effect on the generator. However, for a full-fledged operation in a grid, it was not possible by that time to maintain the requirements of droop control. Flux and torque are functions of frequency and voltage, respectively. Hence, with the introduction, acceptance, and integration of microprocessors, DSPs, and microcontrollers in vector control, the realization of complete industrial automation became achievable. This extended to applications such as robotic manipulators, electric vehicles, transportation, and advanced motor drives utilizing induction machines. Additionally, these technologies found applications in the distributed generation of renewable energy sources connected to a microgrid. Field-oriented control has mathematical transformations from the machine abc three-phase model toward a rotating-reference-frame virtual DC model, then proper alignment of a chosen field orientation is made either on d or on the q axis. This chapter showed all the formulations based on the abc and quadrature notations. The indirect vector control with rotor flux, as well as the direct vector control with stator flux, were discussed, evaluated, and all the equations required for modeling, analysis, and control were described in this chapter.

The main constraint to using a scalar control method for induction motors/generators is related to the transient response—if shaft torque and speed are bandwidth limited and torque varies slowly to track required speed variations (within hundreds of milliseconds up to the order of almost a second), scalar control may work appropriately. Hydropower and wind power applications have slower mechanical dynamics than such a timing constraint. Therefore, it seems that scalar control may still be a good approach for renewable energy type applications, mostly for do-it-yourself and home-made solutions. Field-oriented control uses a dynamic model of the IM where the voltages, currents, and fluxes are expressed in space vector forms. Since the IM is described by differential equations, the model accounts for both steady state and transient dynamics of the IM. Therefore, field-oriented control can achieve excellent performance in transient and steady-state conditions. In the rotating rotor flux frame, quantities rotating at synchronous speed appear as DC quantities. With the flux aligned to the d axis of the reference frame, the d component of the stator current represents the flux and the q component of the stator current represents the torque. Therefore, the control of IM is reduced to a simple control scheme. The simplicity has resulted from the fact that torque and flux components are decoupled in the adopted reference frame. Field-oriented control is dependent on having good machine parameter characterization, in addition to excellent sensor monitoring and feedback of the required machine variables, plus a very capable microcontroller, or DSP-based processor, for very fast real-time computation with PWM methodology, good protection against over-currents or over-voltages, making possible that an inverter controlling the induction machine impress current-control capabilities in a closed-loop speed control, or in a closed-loop torque control. It is possible also to have universal field orientation, where any flux can be aligned to any rotating reference frame axis, but typically this subject is more academic than practical. Nowadays it is mostly common to have high-performance inverters because costs make it affordable for those complex controllers to be implemented, and typically outdated volts/Hertz inverters are still on the market because they became inexpensive, widely available, maybe obsolete as probably the major manufacturers will concentrate their production for mass market wholesale of advanced motor drives, which today became viable, thanks to so many creative researchers, academics, and R&D conducted by the motor drives industries.

Bibliography

B. K. Bose, "An adaptive hysterese-band current control technique of a voltage-fed pwm inverter for machine drive system," *IEEE Trans. Ind. Electron.*, vol. 37, pp. 402–408, 1980.

J. M. Elder, J. T. Boys, and J. L. Woodward, "The process of self excitation in induction generators," *Proc. Inst. Elect. Eng. B*, vol. 130, no. 2, pp. 103–108, 1983.

A. Vagati and F. Villata, "AC servo system for position control," *Proceedings of ICEM*, pp. 871–874, 1984.

M. Godoy Simões, and F. A. Farret, *Modeling and Analysis with Induction Generators* 3rd Edition, Taylor and Francis / CRC Press, 2014. ISBN 978-114-8224467-0.

B. K. Bose, *Power Electronics and AC Drives*, Prentice-Hall, Englewood Cliffs, New Jersey, 1986, 1st Edition.

D. W. Novotny, and T. A. Lipo, "Vector Control and Dynamics of AC Drives," Clarendon Press, 1st Edition, 1996.

P.C. Krause, *Analysis of Electric Machine*, McGraw-Hill Book Company, 1986, 1st Edition.

K. Hasse, "Zur Dynamik drehzahlgeregelter Antriebe mit Stromrichtergespeisten Asynchron Kurnchlu laufennaschinen" *Dannstadt, Techn. Hochsch., Dissertation*, 1969.

B. K. Bose, M. Godoy Simões, D. R. Crecelius, K. Rajashekara, and R. Martin, "Speed sensorless hybrid vector controlled induction motor drive" IEEE/IAS Annu. Meeting Conf. Rec., pp. 137–143, 1995.

D. R. Crecelius, "Implementation of a speed sensorless vector controlled induction motor drive with zero speed start-up," *M.S. Thesis*, University of Tennessee, Knoxville, 1994.

Patent disclosure letter, "Speed sensorless vector-controlled induction motor drive with zero speed operation," Nov/13/1993, GM H-181981, 1993.

R. W. De Doncker, and D. W. Novotny, "The universal field oriented controller," *IEEE Industrial Application Society Annual meeting*, pp. 450–456, 1988.

F. Profumo, A. Tenconi and R. W. De Doncker, "The universal field oriented (UFO) controller applied to wide speed range induction motor drives," PESC '91 Record 22nd Annual IEEE Power Electronics Specialists Conference, Cambridge, MA, USA, pp. 681–686, doi:10.1109/PESC.1991.162749, 1991.

R. W. de Doncker and F. Profumo, "The universal field oriented controller applied to tapped stator windings induction motors," *20th Annual IEEE Power*

Electronics Specialists Conference, Milwaukee, WI, USA, pp. 1031–1036 vol. 2, doi:10.1109/PESC.1989.48592, 1989.

A. Smith, S. Gadoue, M. Armstrong , *et al.* “Improved method for the scalar control of induction motor drives.” *IET Electric Power Applications* 7.6, 487–498, 2013.

T. A. Lipo. *Introduction to AC Machine Design*, Wisconsin Power Electronics Research Center, University of Wisconsin, 2004.

R. C. Bansal, T. S. Bhatti, and D. P. Kothari, “A bibliographical survey on induction generators for application of nonconventional energy systems,” *IEEE Trans. Energy Convers.*, vol. 18, no. 3, pp. 433–439, 2003.

M. Godoy Simões, B. K. Bose, and Ron Spiegel, “Fuzzy logic based intelligent control of a variable speed cage machine wind generation system,” IEEE Power Electronics Specialists Conference, pp. 389–395, 1995.

N. R. Zargari and G. Joos, “Performance investigation of a current-controlled voltage-regulated PWM rectifier in rotating and stationary frames,” *IEEE Trans. on Ind. Electr.*, vol. 42, no. 4, pp. 396–401, 1995.

M. G. Simões and B. K. Bose, “Neural network based estimation of feedback signals for a vector controlled induction motor drive,” IEEE/IAS Annu. Meeting Conf. Rec., pp. 471–479, 1994.

R. W. De Doncker and D. W. Novotny “The universal field oriented controller” *IEEE Trans. Ind. Appl.*, vol. 30, pp. 92–100, 1994.

M. Marchesoni, P. Segarich and E. Soressi, “A simple approach to flux and speed observation in induction motor drives,” IEEE/IECON Conf. Rec., pp. 305–310, 1994.

N. J. Kim and D. S. Hyun “Very low speed control of induction machine by instantaneous speed and inertia estimation,” IEEE/IECON Conf. Rec., pp. 605–610, 1994.

Patent disclosure letter, “Efficiency optimization and performance enhancement control of wind turbine generators,” Dec/21/1994, EPA 539-93, 1994.

SPWM integrated circuit Hanning Elektro-Werke GmbH, PWM 1/87 data sheet, 1993.

T. Ohtani, N. Takada, and K. Tanaka, “Vector control of induction motor without shaft encoder,” *IEEE Trans. on Ind. Appl.*, vol. 28, pp. 157–164, 1992.

T. G. Habetler, F. Profumo, M. Pastorelli, and L. M. Tolbert, “Direct torque control of induction machines using space vector modulation,” *IEEE Trans. On Industry Applications*, vol. 28, no. 5, pp. 1045–1053, 1992.

P. L. Jansen and R. D. Lorenz, “A physically insightful approach to the design and accuracy assessment of flux observers for field oriented induction machine drives,” IEEE/IAS Ann. Meet. Conf. Rec., pp. 570–577, 1992.

J. C. R. de Oliveira, B. de Fornel and J. C. Hapiot, “Microprocessor implementation of an adaptive estimator for induction motors,” IEEE/IECON Conf. Rec., pp. 127–132, 1992.

C. B. Jacobina, E. B. Souza Fl., A. M. N. Lima, and J. D. P. Rolim, “Current control for induction motor drives based on input-output dynamic discrete-model,” in Conf. Rec. IECON’92, 1992, pp. 33–137, 1992.

Werner Leonhard, "30 Years space vectors, 20 years field orientation, 10 years digital signal processing with controlled AC-drives, a review (part 1)" *EPE Journal*, 1:1, 13–19, DOI:10.1080/09398368.1991.11463257, 1991.

Werner Leonhard, "30 Years space vectors, 20 years field orientation, 10 year digital signal processing with controlled AC-drives, a review (part 2)" *EPE Journal*, 1:2, 89–101, DOI:10.1080/09398368.1991.11463267, 1991.

H. Kubota, K. Matsuse, and T. Nakano, "DSP-based speed adaptive flux observer of induction motor," IEEE/IAS Ann. Meet. Conf. Rec., pp. 380–384, 1991.

Y. Hori and T. Umeno, "Flux observer based field orientation controller for high performance torque control," Conf. Rec. of Intl'l Pow. Elec. Conf., Tokyo, pp. 1219–1226, 1990.

G. R. Siemon, "Modeling of induction machines for electric drives," *IEEE Trans. on Ind. App*, Vol. 2:5, No. 6, pp. 1126–1131, 1989.

C. Schauder, "Adaptive speed identification for vector control of induction motors without rotational transducers," IEEE/IAS Ann. Meet. Conf. Rec., pp. 493–499, 1989.

H. L. Nakra and B. Dubé, "Slip power recovery induction generators for large vertical axis wind turbines," *IEEE Trans. Energy Conversion*, vol. EC-3, pp. 733–737, 1988.

X. Xu, R. D. Doncker, and D. W. Novotny, "Stator flux orientation control of induction machines in the field weakening region," Conf. Proc. IEEE-IAS Annual Meeting, Pittsburgh, 1988, pp. 437–443.

X. Xu, R.W. De Doncker, and D. W. Novotny, "A stator flux oriented induction machine drive," IEEE/PESC Conf. Rec., pp. 870–876, 1988.

H. W. V. Broeck, H.-C. Skudelny, and G. V. Stanke, "Analysis and realization of a pulsewidth modulator based on voltage space vectors," *IEEE Trans. On Industry Applications*, vol. 24, no. 1, 1988, pp. 142–150.

X. Xu, R.W. De Doncker, and D. W. Novotny, "Stator flux orientation control of induction machines in the field weakening region," IEEE/IAS Ann. Meet. Conf. Rec., pp. 437–443, 1988.

G. C. Verghese, "Observers for flux estimation in induction machines," *IEEE Trans. on Industrial Electronics*, vol. 35, pp. 85–94, 1988.

T. M. Rowan and R. J. Kerkman, "A new synchronous current regulator and an analysis of a current-regulated pwm inverter," *IEEE Trans. Ind. Applicat.*, vol. 22, pp. 678–690, 1986.

G. K. Dubey, S. R. Doradia, A. Joshi, and R. M. K. Sinha, *Thyristorised Power Controllers*, John-Wiley, 1986.

Werner Leonhard, *Control of Electrical Drives*, Springer-Verlag, Berlin, 1985, 1st Edition.

D. M. Brod and D. W. Novotny, "Current control of VSI-PWM inverters," *IEEE Trans. Ind. Applicat.*, vol. 21, pp. 526–570, 1985.

B. K. Bose, "A microprocessor-based control system for a near term electric vehicle," *IEEE Trans. Ind. Appl.*, vol. 17, pp. 626–631, 1981 (complete EV control system).

B. K. Bose and H. Sutherland, "A microprocessor-based real-time feedback controller for an electric vehicle drive," IEEE IAS Conf. Rec. pp. 743–748, 1979 (power system control).

R. Gabriel and W. Leonhard, "Field oriented control of a standard AC motor using microprocessors," Conf. Rec. IEEFJIAS Annual Meeting, pp. 910–916, 1979.

F. Blaschke, "The principle of field orientation as applied to the new TRANSVECTOR closed loop control system for rotating field machines," *Siemens Review*, vol. 34, pp. 217–220, 1972.

P. C. Krause and C. H. Thomas, "Simulation of symmetrical induction machinery," *IEEE Trans. Power App. Syst.*, vol. PAS-84, no. 11, pp. 1038–1053, 1965.

D. C. White and H. H. Woodson, *Electromechanical Energy Conversion*, New York: Wiley, 1959.

A. Schonung and H. Stemmler, "Static frequency changers with 'subharmonic' control in conjunction with reversible variable speed a.c. drives. *Brown Boveri Rev.*, 555–577, 1964.

W. A. Lewis, "A basic analysis of synchronous machine – part I," *AIEE Transaction*, vol. 77, pp. 436–256, 1958.

E. Clarke, *Circuit Analysis of AC Power Systems – Symmetrical and Related Components*. John Wiley & Sons, 1943, vol. I.

R. H. Park, "Two-reaction theory of synchronous machines generalized method of analysis-part I," *AIEE Transactions of the*, vol. 48, no. 3, pp. 716–727, 1929.

R. E. Doherty and C. A. Nickle, "Synchronous machine I and II," *AIEE Transactions*, vol. 45, pp. 912–942, 1926.

A. Blondel. "The two-reaction method for study of oscillatory phenomena in coupled alternators," *Revue Générale de L'Electricité*, vol. 13, pp. 235–251, 515–531, 1923.

C. L. Fortescue, "Method of symmetrical co-ordinates applied to the solution of polyphase networks," *American Institute of Electrical Engineers, Transactions of the*, vol. XXXVII, no. 2, pp. 1027–1140, 1918.

Chapter 13

The four-quadrant operation of the switched reluctance machines

Paulo Robson Melo Costa[1], Marcelo Vinícius de Paula[2] and Tárcio André dos Santos Barros[1]

13.1 Introduction

This chapter is dedicated to the study of switched reluctance machines, one of the most important devices in electrical and mechanical engineering. The purpose of this chapter is to provide a comprehensive overview of the four-quadrant operation of switched reluctance machines, including its operating principles, performance characteristics, control techniques, and practical applications.

The switched reluctance machine is an electric machine that stands out for its simplicity, robustness, and energy efficiency. It is capable of operating in motoring, generating, or braking modes. Moreover, it is widely applied in the industry, including automotive, aerospace, marine, wind, and more.

This work is intended for engineers, researchers, and students who are interested in the field of electrical and mechanical engineering. We hope that it will serve as a valuable resource for anyone seeking to understand the switched reluctance machine and its potential applications. We will present simulation results to demonstrate the performance and efficiency of the machine in four-quadrant operation.

The Switched Reluctance Machine (SRM) is an electric machine that operates based on the magnetic reluctance principle. Unlike traditional electric machines, such as induction and synchronous machines, the SRM does not have a magnetic field in the rotor. Instead, an electric current passes through the stator windings, creating a magnetic field that interacts with the rotor material, inducing torque.

Switched Reluctance Motors have several advantages that make them suitable for various applications, such as: 1. Simplicity: simple construction with fewer parts than other electric machines, making them less complex and more reliable; 2. Cost-effectiveness: lower cost than other types of electric motors, due to their simple construction and lack of permanent magnets; 3. Robustness: high

[1]School of Electrical and Computer Engineering, University of Campinas, Brazil
[2]School of Mechanical Engineering, University of Campinas, Brazil

tolerance to mechanical and thermal stress, making them suitable for harsh environments and high-temperature applications; 4. High torque density: produce high torque in a small package, making them suitable for high-torque applications, such as electric vehicles and industrial drives; 5. No rare earth materials: do not require rare earth materials, making them a more environmentally friendly, sustainable option, and less dependent on rare earth price fluctuation; 6. Ease of maintenance: a simple design that allows easy maintenance and repair. Conversely, some drawbacks such as the nonlinear behavior, high electromagnetic torque ripple level, and elevated acoustic noise restrain the large-scale usage of these machines. These drawbacks and challenges may be overcome through adequate control strategies or machine design.

Some of the typical applications of SRM include: 1. Industrial drives: such as pumps, fans, conveyors, and compressors, due to their high torque density and ruggedness; 2. Electric vehicles: due to their high torque and efficiency, making it suitable for electric and hybrid vehicles; 3. Renewable energy: such as wind turbines and hydroelectric generators, due to their high efficiency and low maintenance requirements; 4. Household appliances: such as washing machines, refrigerators, and air conditioners, due to their high efficiency and low cost; 5. Aerospace and defense: such as unmanned aerial vehicles, due to their reliability, high torque density, and ruggedness; 6. Robotics: due to its high torque and reliability, making it suitable for tasks that require high torque and precise control.

The chapter is divided into nine sections. The first section provides an introduction to the SRM and the purpose of the chapter. The second section covers the SRM configuration and the operating principles. The four-quadrant model and ideal driving angles for SRM drive are presented in the third and fourth sections. The fifth and sixth section focuses on speed control and current control. The seventh section covers modern torque control methods used to operate the SRM efficiently, including Average Torque Control, Current Profiling through Torque Sharing Functions, and Direct Instantaneous Torque Control methods. The eighth section presents the simulation results of the four-quadrant operation. The final section concludes the chapter with a summary of key findings, future research directions, and recommendations for practical applications.

13.2 Switched reluctance machines operation principles

The SRM is composed of a Stator and a Rotor. Both structures are made of laminated ferromagnetic material, present salient poles, and are separated by a small air gap. A typical SRM is identified by the number of poles in the stator and the rotor. Thus, an SRM with four stator poles and two rotor poles is referred to as 4/2 SRM. A cross-section view of a two-phase 4/2 SRM is presented in Figure 13.1(b), while a three-phase 12/8 SRM and its parts are indicated in Figure 13.1(a).

Differently from other electric machines, there are not any windings or permanent magnets in the rotor. Hence, a Magnetomotive Force (MMF) is produced by the stator windings, which are allocated around the stator poles and divided into

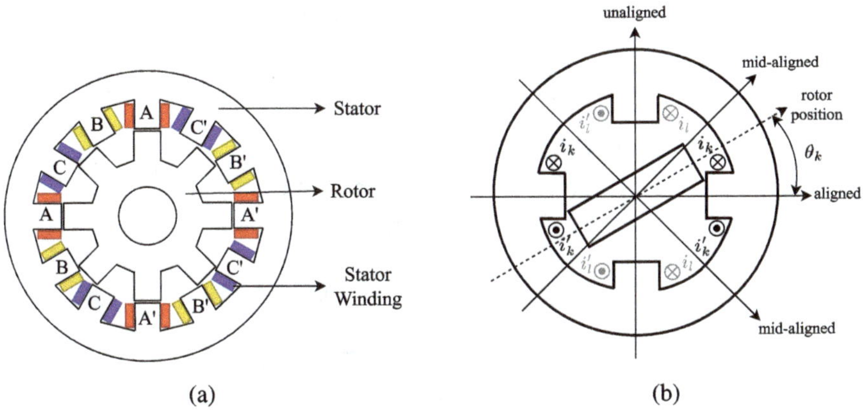

Figure 13.1 Cross-section view of two different SRM topologies: (a) three-phase 12/8 SRM and (b) two-phase 4/2 SRM

phases. Associating the stator and rotor cores, a magnetic path is established, which is variable according to the rotor position, θ. The minimum length of the magnetic path is achieved when the poles of the stator and rotor are fully aligned.

Once an electric current is applied to the stator windings, the Flux linkage (λ) is created and flows through the magnetic path with reluctance determined as:

$$\Re(\theta, i) = \frac{l(\theta)}{\mu(\theta, i) \cdot A(\theta)} \tag{13.1}$$

where l and A are the length and the cross-section area of the magnetic path, and μ is the magnetic permeability of the material. The geometric parameters of the reluctance, l and A, and the magnetization term, represented by μ, are variable according to the rotor position. Moreover, the magnetic permeability is reduced as the magnetic saturation occurs, i.e., it is dependent on the excitation current (i). As stated in (13.1), the reluctance is a function of rotor position and excitation current.

The inductance (L) is related to the capability to store energy. The greater the inductance, the greater the back-electromotive (EMF) force and the greater opposition to current changes. The inductance is calculated as the ratio of the produced flux linkage by the excitation current, see (13.2). Letting N be the number of turns in a winding, the inductance can be written as in (13.3). An idealized profile of the inductance and the reluctance for the two-phase 4/2 SRM relying on the rotor position is shown in Figure 13.2

$$\lambda(\theta, i) = L(\theta, i) \cdot i \tag{13.2}$$

$$L(\theta, i) = \frac{N^2}{\Re(\theta, i)} \tag{13.3}$$

Considering the k-phase of the SRM, the excitation voltage (v_k) is equal to the voltage drop at the phase resistance (R_k) added to the total induced back-EMF as

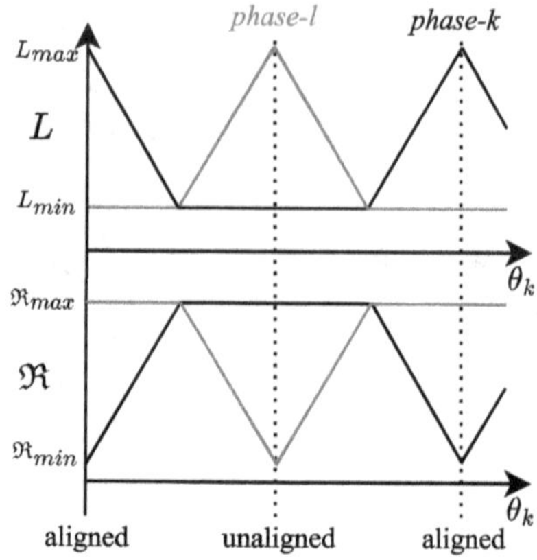

Figure 13.2 Idealized inductance and reluctance profile of two-phase 4/2 SRM

written in (13.4).

$$v_k = i_k R_k + \frac{\mathrm{d}\lambda(\theta_k, i_k)}{\mathrm{d}t} \tag{13.4}$$

Inserting (13.2) into (13.4), and using the product and chain rules for differentiation, we get:

$$v_k = i_k R_k + L(\theta_k, i_k)\frac{\mathrm{d}i_k}{\mathrm{d}t} + i_k\left(\frac{\mathrm{d}L(\theta_k, i_k)}{\mathrm{d}i}\frac{\mathrm{d}i}{\mathrm{d}t} + \frac{\mathrm{d}L(\theta_k, i_k)}{\mathrm{d}\theta}\frac{\mathrm{d}\theta}{\mathrm{d}t}\right) \tag{13.5}$$

To simplify, the magnetic saturation is neglected for now. That is, the inductance is no longer dependent on the phase current. Defining the rotational speed as $\omega = \mathrm{d}\theta/\mathrm{d}t$, we can write:

$$v_k = i_k R_k + L(\theta_k)\frac{\mathrm{d}i_k}{\mathrm{d}t} + i_k\omega\frac{\mathrm{d}L(\theta_k)}{\mathrm{d}\theta} \tag{13.6}$$

It can be noted that the rate of change of the flux linkage is separated into two components. The first term represents the inductive voltage drop. The second term is the speed-dependent EMF (e_k) produced due to the variation of the inductance over the position, enabled by the salient pole structure. This can be expressed as:

$$e_k(\theta_k) = \frac{\mathrm{d}L(\theta_k)}{\mathrm{d}\theta} i_k\omega \tag{13.7}$$

From (13.6), the equivalent model per phase of the SRM can be derived as shown in Figure 13.3. It is used to predict the performance of the motor, analyze its behavior under different operating conditions, and design the machine.

The per-phase equivalent circuit of an SRM consists of: 1. Voltage source: representing the voltage applied to the motor, this component models the voltage across the stator windings; 2. Resistance: representing the losses in the motor, this

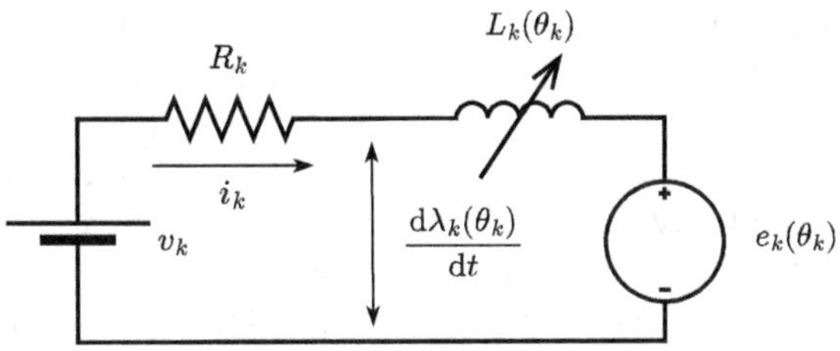

Figure 13.3 Equivalent circuit of one phase of SRM

component models the resistance of the stator windings and can include the iron losses in the magnetic material; 3. Inductance: representing the magnetic properties of the motor, this component models the magnetic energy stored in the magnetic field; 4. Back-electromotive force (EMF): this component models the voltage generated across the stator windings due to the movement of the rotor.

13.2.1 Power in SRM

The power in the k-phase of the SRM can be obtained by multiplying (13.6) by the phase current:

$$p_k = i_k v_k = i_k^2 R_k + i_k L(\theta_k) \frac{di_k}{dt} + i_k^2 \omega \frac{dL(\theta_k)}{d\theta} \tag{13.8}$$

To facilitate the physical interpretation of the second term on the right side of (13.8), some manipulation is necessary. Reusing the product rule for derivative, one can write:

$$\frac{d}{dt}\left(\frac{1}{2} L(\theta_k) i_k^2\right) = i_k L(\theta_k) \frac{di_k}{dt} + \frac{1}{2} i_k^2 \frac{dL(\theta_k)}{d\theta} \frac{d\theta}{dt} \tag{13.9}$$

Substituting (13.9) into (13.8), a meaningful expression is obtained:

$$p_k = i_k^2 R_k + \frac{d}{dt}\left(\frac{1}{2} L(\theta_k) i_k^2\right) + \frac{1}{2} i_k^2 \omega \frac{dL(\theta_k)}{d\theta} \tag{13.10}$$

This equation implies that the input power equals the sum of: the winding resistive losses, $i_k^2 R_k$; the rate of change of the magnetic stored energy, $\frac{d}{dt}\left(\frac{1}{2} L(\theta_k) i_k^2\right)$; and the air gap power, $\frac{1}{2} i_k^2 \omega \frac{dL(\theta_k)}{d\theta}$.

Once the air gap power is the product of electromagnetic torque (Te) and rotational speed (ω), the phase electromagnetic torque expression is given:

$$Te_k = \frac{1}{2} i_k^2 \frac{dL(\theta_k)}{d\theta} \tag{13.11}$$

It can be noted that the torque has a quadratic relation to the phase current and, hence, is independent of the current direction. The inductance change with the rotor position, caused by the salient geometry, is the torque production mechanism in an

SRM. The generated torque always acts to minimize the reluctance and maximize the inductance. In other words, if the phase is excited while the rotor goes from an unaligned position to an aligned position, a motoring torque is produced and contributes to the rotational movement. Conversely, a regenerative braking torque, that opposes the rotational movement, is delivered if the phase is excited while the rotor approximates the unaligned position. The rotating direction of an SRM can be changed by altering the phase sequence of the stator current. The sequence and timing of excitation phase currents are the key fundamentals for the four-quadrant operation of the SRM.

The torque expression from (13.11) is valid when the machine works in the linear magnetic region. To obtain the torque behavior for the entire magnetic range, it is necessary to study the magnetic characteristics of the machine and the energies involved in its driving as will be investigated hereafter.

13.2.2 Magnetic characteristics

The flux-current curve of a switched reluctance machine (SRM), shown in Figure 13.4(a) for different rotor positions, is a graphical representation of the relationship between the flux linkage in the machine and the exciting current. It is an essential tool for understanding the behavior of the SRM, and it is used to design the machine and to predict its performance. This curve is typically nonlinear and is divided into two regions: the linear region, where the magnetic flux increases linearly with the current, and the saturation region, where the magnetic flux reaches a maximum value and does not increase further, even if the current continues to increase. This means that the machine's torque capability increases in the saturation

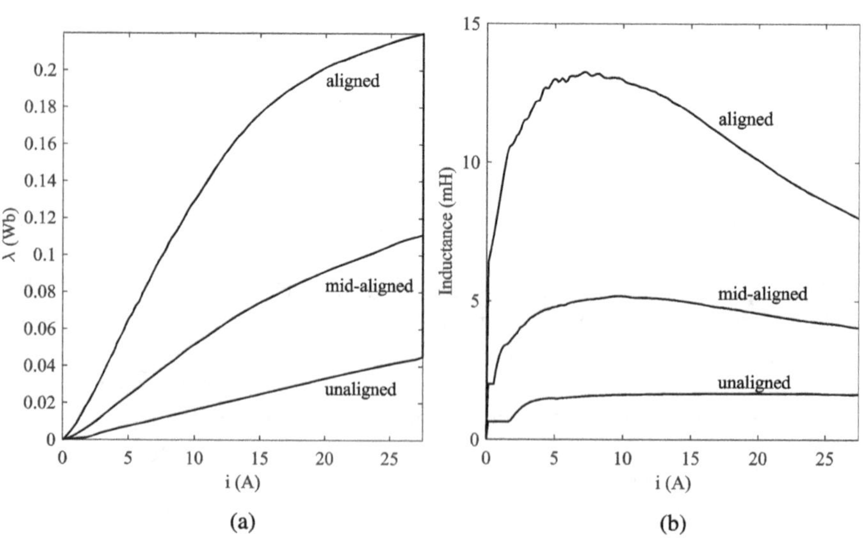

Figure 13.4 Phase flux linkage and inductance at different rotor positions and various current values: (a) flux linkage and (b) phase inductance

region. Notably, the air gap increases with the rotor from the aligned to the unaligned position, reducing the magnetic saturation effect. The maximum magnetic flux is a key parameter for determining the machine's torque capability and power density. As a consequence of the flux-current curve, the inductance-current curve at different rotor positions is obtained and shown in Figure 13.4(b).

Another perspective of the SRM characteristic can be delivered by the flux-position curve, given in Figure 13.5(a) for different excitation currents. As observed, the greater the current, the less sensitive the flux is to a change in current. The respective inductance-position profile is presented in Figure 13.5(b). This inductance profile obtained experimentally from a 12/8 SRM can be compared to the idealized profile from Figure 13.2. Three main regions can be highlighted: the negative slope region, proper for generating mode; the constant region with low inductance, suitable for fast magnetization or demagnetization; and the positive slope region, valid for motoring mode. Notice that the positive or negative slope region is dependent on the speed direction, namely, if the rotor position is increasing or decreasing.

13.2.3 Co-energy and torque production

In a rotating machine, the mechanical energy in terms of the electromagnetic torque and change in rotor position is defined by:

$$W_m = \int_{\theta_1}^{\theta_2} T_e \mathrm{d}\theta \tag{13.12}$$

Despising the power losses, the input electrical energy (W_e) is equal to the sum of the magnetic stored energy (W_f) and the energy converted into mechanical

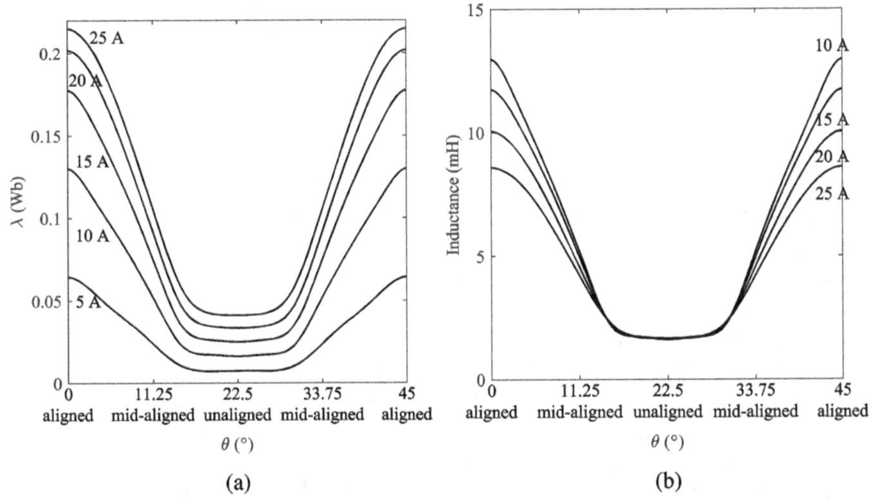

Figure 13.5 Phase flux linkage and inductance for one electrical cycle at different excitation current values: (a) flux linkage and (b) phase inductance

energy (W_m), as stated in (13.13). The incremental portions of these energies are written in (13.14).

$$W_e = W_f + W_m \tag{13.13}$$

$$\mathrm{d}W_e = \mathrm{d}W_f + \mathrm{d}W_m \tag{13.14}$$

Taking (13.4), despising the resistance and multiplying by the current, the change in the electrical energy can be stated by

$$\mathrm{d}W_e = i\mathrm{d}\lambda \tag{13.15}$$

Substituting (13.15) and using (13.12) into (13.14), the change in the magnetic stored energy is given by

$$\mathrm{d}W_f = i\mathrm{d}\lambda - T_e\mathrm{d}\theta \tag{13.16}$$

The graphical analysis of the energies involved in the electromechanical energy conversion is instrumental and allows the calculation of the torque produced for a saturated magnetic material. From a point $P1(I_1, \lambda_1)$ at the flux-current curve for a given position, see Figure 13.6(a), the magnetic energy stored in the magnetic field (W_f) is equal to the area above the curve and is limited due to the magnetic material saturation. It can be calculated as

$$W_f(\theta, i) = \int i\mathrm{d}\lambda \tag{13.17}$$

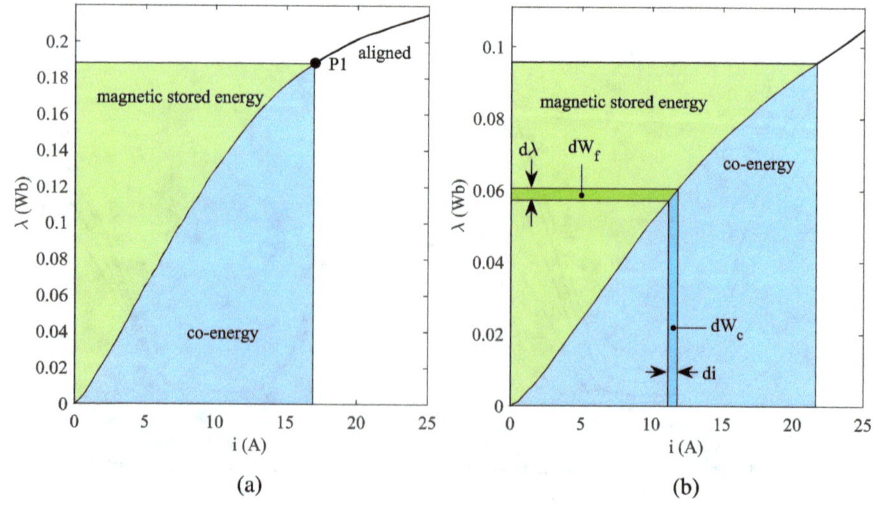

Figure 13.6 Flux-current curve and energies stored at a certain rotor position: (a) magnetic energy and co-energy stored and (b) stored energy representation

The area below the flux-current is defined as co-energy (W_c). It has no physical meaning, but a change in the co-energy due to the rotor position transition with constant current is equal to the mechanical work at the period. The co-energy is calculated by (13.18).

$$W_c(\theta, i) = \int \lambda \mathrm{d}i = i\lambda - W_f(\theta, i) \tag{13.18}$$

As shown in Figure 13.6(b), for a given rotor position and from (13.17) and (13.18), an infinitesimal change of magnetic stored energy ($\mathrm{d}W_f$) can be calculated by considering constant current for an infinitesimal change in flux linkage ($\mathrm{d}\lambda$). Likewise, a small change of the co-energy ($\mathrm{d}W_f$) can be obtained by an infinitesimal change in current ($\mathrm{d}i$) when the flux linkage is maintained constant.

By differentiating (13.18) and substituting the expression for incremental magnetic energy from (13.16), we can compute the incremental co-energy:

$$\mathrm{d}W_c(\theta, i) = \lambda \mathrm{d}i + T_e \mathrm{d}\theta \tag{13.19}$$

Since the co-energy is a function of flux-current characteristic and the rotor position, its incremental portion can be calculated as a function of partial derivatives of i and θ:

$$\mathrm{d}W_c(\theta, i) = \frac{\mathrm{d}W_c}{\mathrm{d}i}\mathrm{d}i\bigg|_{\theta=\mathrm{const}} + \frac{\mathrm{d}W_c}{\mathrm{d}\theta}\mathrm{d}\theta\bigg|_{i=\mathrm{const}} \tag{13.20}$$

Combining (13.19) and (13.20), results in the torque expression:

$$T_e = \frac{\mathrm{d}W_c}{\mathrm{d}\theta}\bigg|_{i=\mathrm{const}} \tag{13.21}$$

Figure 13.7 presents the amount of change in the co-energy (ΔW_c) when the rotor moves from a mid-aligned to an aligned position when a constant current is maintained. The area is highlighted and is equal to the mechanical energy developed in the period. For a generic SRM with N_ϕ phases, the total electromagnetic torque is

$$T_e = \sum_{k=1}^{N_\phi} \frac{\mathrm{d}W_{c,k}(\theta_k, i_k)}{\mathrm{d}\theta} \tag{13.22}$$

The supply source provides both magnetic energy and co-energy. In the linear magnetic region, these two components are the same. Consequently, double the mechanical work done must be transferred from the source.

The results of the integral of (13.17) and (13.18) at various rotor positions are presented in Figure 13.8. It becomes clear that the co-energy in the SRM increases with the saturation level, which leads to better torque output capability without the need for higher voltage. This results in better utilization of the power semiconductor devices in the drive converter and reduces the volt-ampere requirements.

Using (13.21) along with the flux-current curves at one electrical cycle shown in Figure 13.5(a) it is possible to obtain the electromagnetic torque profile depicted

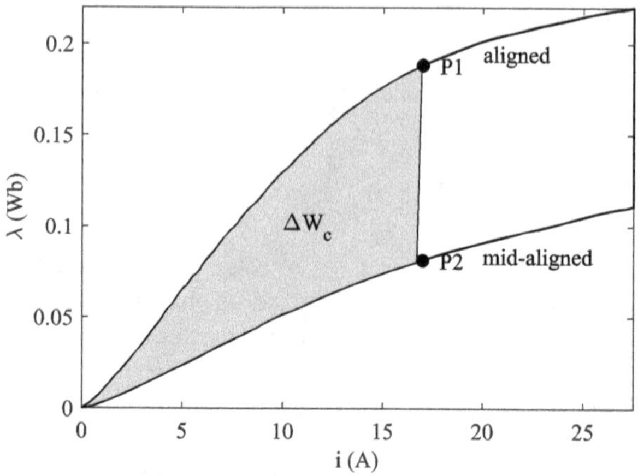

Figure 13.7   Co-energy change from mid-aligned to aligned position with constant current

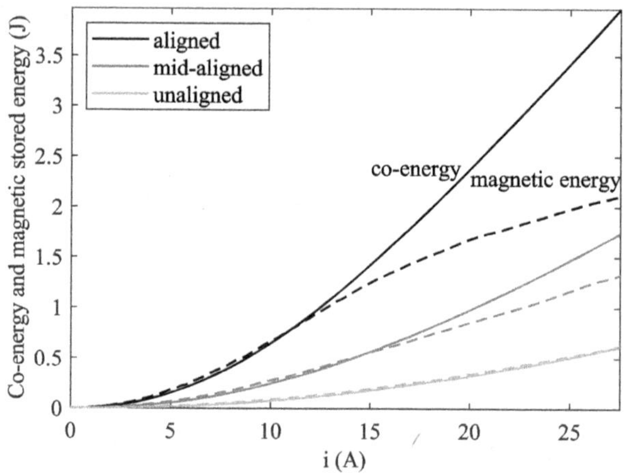

Figure 13.8   Co-energy and magnetic stored energy for various rotor positions and at different excitation current values

in Figure 13.9. It can be noted that, according to (13.11), positive or negative torque can be produced through the same current. Therefore, motoring or generating torque can be achieved by phase exciting during pole alignment or misalignment, respectively.

Figure 13.10(a) shows the flux linkage profile of the aforementioned 12/8 SRM obtained experimentally according to [1]. The profiles are stored as look-up tables, which are used for modeling and simulating the SRM in MATLAB/Simulink environment. Furthermore, the electromagnetic torque profile may also be obtained, as presented in Figure 13.10(b). The torque information is often embedded in the control strategy for online torque estimation or current profiling.

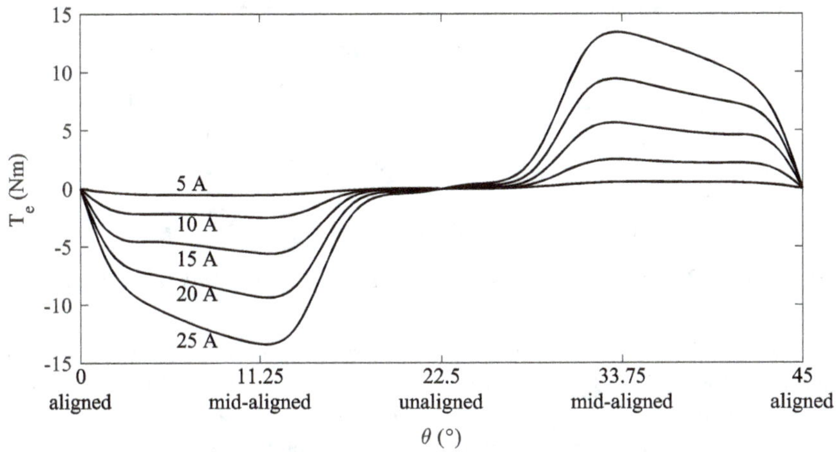

Figure 13.9 Phase electromagnetic torque for one electrical cycle at different excitation current values

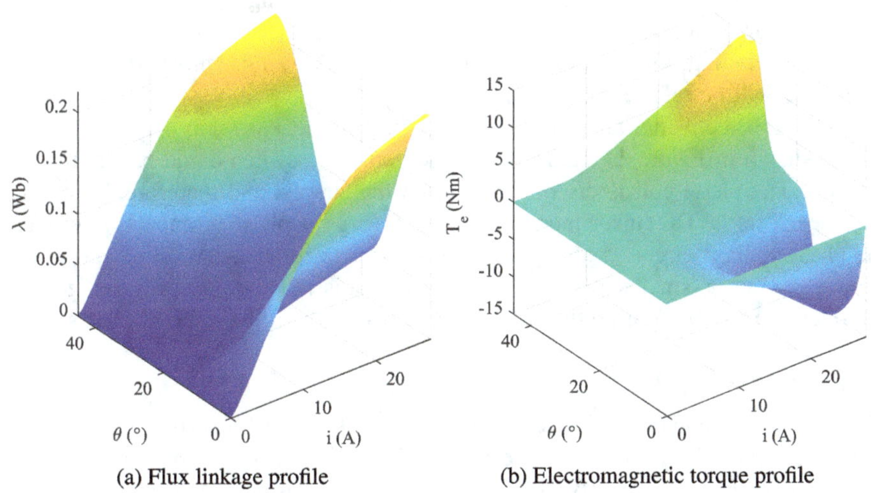

Figure 13.10 Profiles of the studied 12/8 SRM: (a) flux linkage profile and (b) electromagnetic torque profile

13.2.4 Asymmetrical half-bridge converter

To apply the phase excitation signals in SRMs, a power converter is required. One commonly used type of converter is the Asymmetric Half-Bridge Converter (AHB). As shown in Figure 13.11, this converter comprises two active switches and two diodes placed diagonally, giving it its "asymmetric" designation. In an SRM, torque is correlated to the square of the current, and therefore, independent of the phase current's direction. This property allows the asymmetric bridge converter

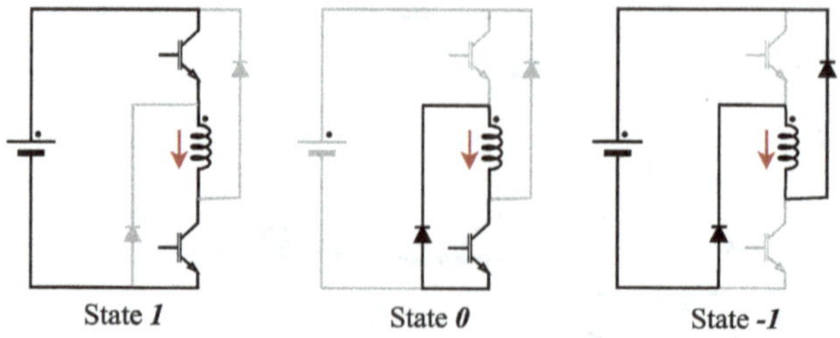

Figure 13.11 Possible states of asymmetrical half-bridge converter for one phase

to generate unidirectional current flow with a reduced number of active switches compared to the Full Bridge Converter. Since each phase of an SRM is electrically isolated from the others, an AHB leg is needed for each phase.

During phase excitation, the converter turns on the two switches as depicted in Figure 13.11, thereby applying the DC link voltage (V_{DC}) to the coils. This state will be referred to as state 1. When the phase excitation ends, the active switches are turned off, as seen in Figure 13.11. At this point, the stored energy in the phase causes the two diodes to become forward-biased, resulting in a negative DC link voltage being applied to the phase. The current then may continue to flow through the phase winding, via the diodes, and back to the DC link until it reaches zero. This state will be referred to as state -1.

The AHB converter also has a third operating mode, known as the freewheeling mode, shown in Figure 13.11. In this mode, only one switch-diode pair is in conduction. The phase coil is disconnected from the supply, and therefore, no energy flows back to the DC link. Since the voltage across the phase terminals is zero and the phase is short-circuited, the current decays due to the phase winding resistance. The decay rate is smaller than in the -1 state. This mode is typically utilized for phase current control during the conduction period and will be referred to as state 0.

13.2.5 Ideal operation

To achieve continuous rotation, each phase of an SRM must be driven sequentially, resulting in pulsed current and torque profiles. The relationship between the electrical angular frequency of the drive (ω_e) and the mechanical angular frequency (ω) is given by (13.23), where N_S and N_R are the number of poles on the stator and the rotor, respectively, and N_Φ is the number of machine phases. For example, in a three-phase 12/8 SRM, the electrical frequency is eight times the mechanical frequency.

$$\omega_e = \frac{N_S}{2}\frac{N_R}{2}\frac{1}{N_\Phi}\omega \tag{13.23}$$

The phase electrical angle (θ) can be related to the mechanical angle (θ_m) through the number of rotor poles (N_R):

$$\theta = \frac{1}{N_R}\theta_m \tag{13.24}$$

Therefore, in a three-phase 12/8 SRM, the phase electrical angle varies from 0 to 45 mechanical degrees.

Figure 13.12 shows waveforms for an idealized three-phase 12/8 SRM operating as a motor. The inductance profile and driving angles (θ_{on} and θ_{off}) are depicted. The conduction period is defined by the driving angles, at that period the enable signal for phase excitation receives a value 1. In motoring operation, the driving angles are chosen so that the phases are excited during the positive slope of the inductance. During this period, the current (i) is regulated at a reference value (i^*). The states 1 and 0 are used to regulate the current during the enable period, whereas the -1 state is used for phase demagnetization. The electromagnetic torque generated by each phase adds up to produce continuous rotation. However, two types of torque ripple are present. The first, and most significant one, occurs during phase commutation at a frequency of 24 times each rotation of the rotor for the three-phase 12/8 SRM. The second type, high-frequency torque ripple, arises due to switching from current regulation and has a frequency equal to the switching frequency and amplitude dependent on the current control method used.

The choice of driving angles can lead to negative torque production in the outgoing phase because of the current tail entering the negative inductance slope region, as presented in Figure 13.12. The negative torque production may diminish the efficiency of the SRM. On the other hand, in real machines, in which the

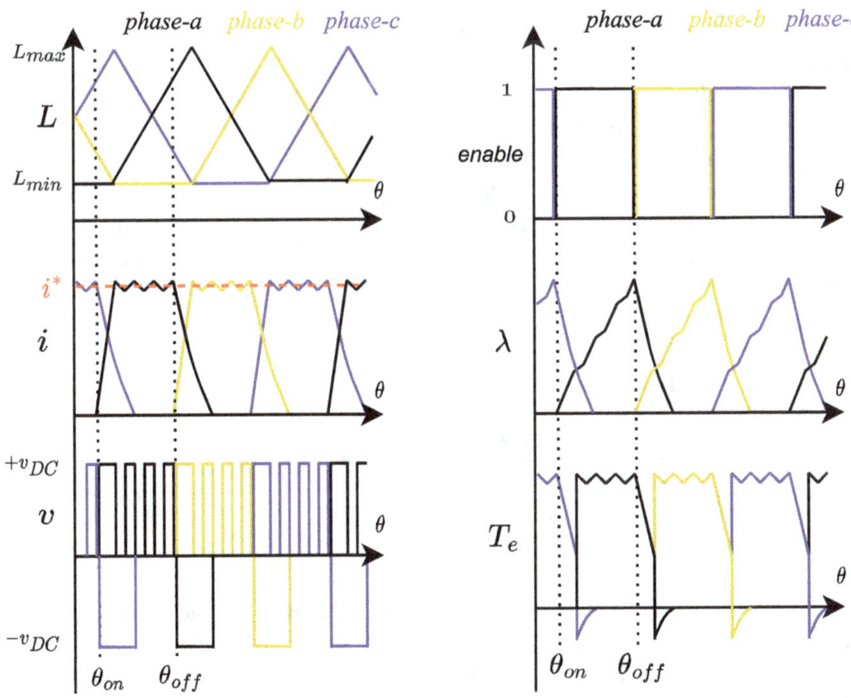

Figure 13.12 Idealized curves of three-phase 12/8 SRM during motor operation

inductance profile is not ideal, the torque production of the incoming phase may rise to values above the mean torque, causing the torque ripple to increase. In such cases, the negative torque produced during the outgoing phase can be utilized to offset any excess positive torque generated in the incoming phase, but this may come at the expense of efficiency.

During operation, two components of force act on the rotor: the radial force, which deforms the stator core and produces acoustic noise, and the tangential force, which delivers torque and some noise by lateral deformation of the stator pole [2]. The tangential force is an order of magnitude smaller than the radial force.

13.3 Four quadrant model

The SRM plant can be described by its two state variables, rotor angular speed (ω) and phase current (i), and two inputs, phase voltage (v) and load torque (T_l). To simplify the model, the inductance $L(\theta, i)$ and its derivative $dL(\theta, i)/d\theta$ are assumed to be constant, and calculated as the average value during the conduction period at rated current [3]. Applying the electromagnetic torque given by (13.11), the mechanical equation can be written as:

$$\frac{1}{2}i^2\frac{\mathrm{d}L(\theta)}{\mathrm{d}\theta} - T_l = B\omega + J\frac{\mathrm{d}\omega}{\mathrm{d}t}, \tag{13.25}$$

where T_l is the load torque, B is the viscosity friction coefficient, and J is the inertia moment.

However, due to the nonlinear product of state variables in the voltage equation (13.6), it is necessary to obtain a linearized plant to employ linear control techniques. Perturbing the system around a steady-state operation point with small signals, the system states and inputs become $i = \bar{i} + \tilde{i}$, $\omega = \bar{\omega} + \tilde{\omega}$, $v = \bar{v} + \tilde{v}$, and $T_l = \bar{T}_l + \tilde{T}_l$, respectively. The steady-state and small signal values are represented by $^{-}$ and $\sim$ indexes, respectively. Substituting the perturbed variables in (13.6) and (13.25), we can eliminate the steady-state terms and neglect the multiplication of small signal terms. The resulting linearized model is described by two equations, one for the phase current (13.26) and another for the rotor angular speed (13.27).

$$\frac{d\tilde{i}}{dt} = \left(-\frac{R}{L} - \frac{1}{L}\frac{dL}{d\theta}\bar{\omega}\right)\tilde{i} - \frac{1}{L}\frac{dL}{d\theta}\bar{i}\tilde{\omega} + \frac{1}{L}\tilde{v} \tag{13.26}$$

$$\frac{d\tilde{\omega}}{dt} = \left(\frac{1}{J}\frac{dL}{d\theta}\bar{i}\right)\tilde{i} - \frac{B}{J}\tilde{\omega} - \frac{1}{J}\tilde{T}_l \tag{13.27}$$

For later substitution, we define the variables $R_{eq} = R + \frac{dL}{d\theta}\bar{\omega}$ as the equivalent resistance, $K_b = \frac{dL}{d\theta}\bar{i}$ as the back EMF constant, and $\tilde{e}_b = K_b\tilde{\omega}$ as the induced EMF. Additionally, we can include the load torque in the model as incremental friction instead of an input, where $B_t = B + B_l$, and B_t, B, and B_l represent the total, machine, and load viscosity friction coefficients, respectively.

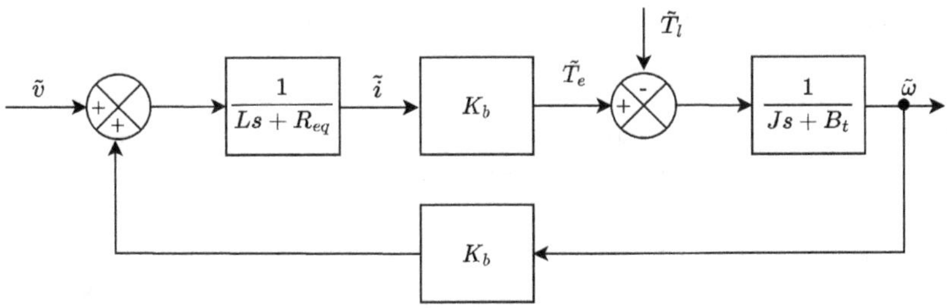

Figure 13.13 Block diagram of the linearized SRM model

It is important to note that when operating in motoring mode, $dL/d\theta > 0$, while for generating mode, $dL/d\theta < 0$. This indicates that the values of the equivalent resistance and back EMF constant will be different for motoring and generating operations. To differentiate between these modes, we will use the superscripts "m" and "g". The state space representation is shown below, where the superscript "x" can take on the values of either "m" or "g"

$$\dot{x} = \mathbb{A}x + \mathbb{B}uy = \mathbb{C}x + \mathbb{D}u \tag{13.28}$$

where

$$\mathbb{A} = \begin{bmatrix} -R_{eq}^x/L & -K_b^x/L \\ K_b^x/J & -B_t/J \end{bmatrix} \mathbb{B} = \begin{bmatrix} 1/L \\ 0 \end{bmatrix} \tag{13.29}$$

$$\mathbb{C} = [0 \quad 1] \mathbb{D} = [0] \tag{13.30}$$

$$x = \begin{bmatrix} \tilde{i} \\ \tilde{\omega} \end{bmatrix} u = [\tilde{v}]. \tag{13.31}$$

The system states are represented by the vector x and the inputs by u. The matrices $\mathbb{A}$, $\mathbb{B}$, $\mathbb{C}$, and $\mathbb{D}$ represent the system coefficients and are defined based on the linearized equations for the phase current and rotor angular speed. Finally, the system output y is given by the product of the system state and output matrices. The block diagram of the linearized SRM is depicted in Figure 13.13.

13.4 Optimization of the driving angles

During the design process of an SRM controller, it is essential to carefully select the driving angles, θ_{on} and θ_{off}, to maximize the machine's performance by achieving high efficiency and minimizing torque ripple. However, optimizing for these two objectives simultaneously is often not feasible, and it is necessary to select the optimal driving angles that prioritize one objective over the other or strike a balance between them. The specific application of the SRM will dictate which objective is more important at any given moment. Additionally, since the SRM's characteristics vary with speed and current, the optimal driving angles are also variable and must be determined for each operating point.

Selecting an overly advanced turn-on angle will cause the conduction period to extend into a region where the inductance profile is nearly constant, leading to an increase in copper losses without contributing to torque production. On the other hand, if the turn-on angle is too delayed, the phase current will not have enough time to reach its peak value, resulting in decreased torque production. Similarly, the optimal turn-off angle is chosen to occur before the inductance slope becomes negative, as this avoids the production of negative torque, which could increase torque ripple.

The adjustment of the turn-on and turn-off angles has a significant impact on the output torque, motor efficiency, and torque ripple. However, manual tuning of these angles for each operating point can be a time-consuming process. To expedite the tuning process, optimization algorithms, such as Genetic Algorithm (GA) [4], Particle Swarm Optimization (PSO) [5], Latin Hypercube Sampling (LHS) space-filling [6], etc., can be employed to determine the optimal driving angles quickly and efficiently.

In the context of four-quadrant operation, the selection of driving angles is crucial for achieving the desired torque output, either motoring (positive torque) or generating (negative torque), and the required speed direction, which can be counter-clockwise (positive speed) or clockwise (negative speed). Figure 13.14 illustrates how the driving angles are allocated in the inductance profile for each quadrant. Note that in negative speed quadrants (*Q*2 and *Q*3), the position θ is decreasing. The speed direction is determined by the order in which the angles are applied, while the torque direction is determined by the slope of the inductance curve for the position variation. Proper selection of the driving angles is essential to achieve optimal performance and efficiency of the SRM in all four quadrants of operation.

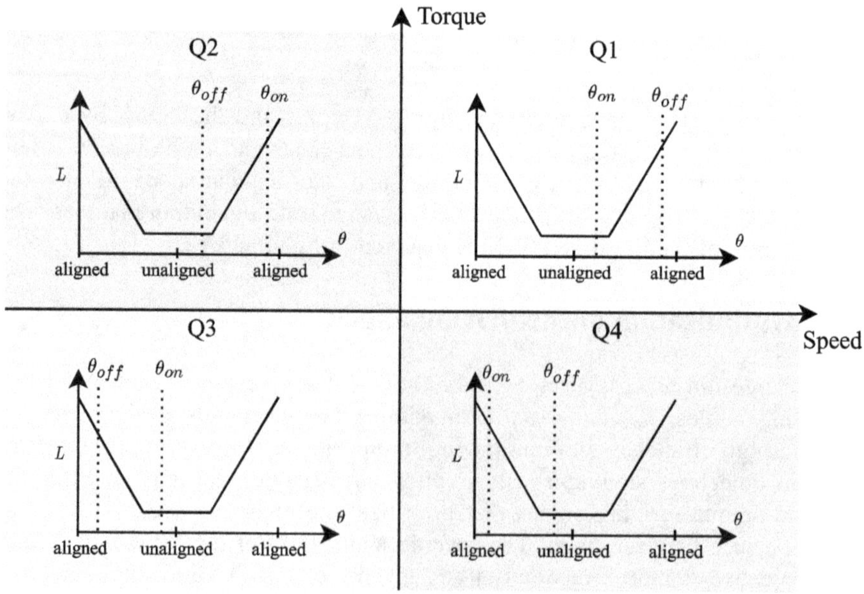

Figure 13.14 Driving angles in four-quadrant operation

13.5 Speed control

The speed control process of an SRM starts with the speed reference (ω^*) provided by the application system. The measured speed is compared to the speed reference, resulting in the speed error. Based on that information, the speed controller delivers the necessary current or torque to accelerate or decelerate the machine and drive the mechanical load attached to the shaft. It is important to emphasize that the current or torque command must be correctly comprised and agree with the compatible conduction period at the inductance profile to obtain the desired effect of accelerating or decelerating the machine.

A wide variety of controllers has been used for speed control, such as the Proportional–Integral Controller (PI) [7], Proportional–Integral–Derivative Controller (PID) [8], Fuzzy Controller [9], Dahlin Controller [10], and Sliding Mode Controller [11,12]. Some works have proposed speed-sensorless control, including the Generalized Proportional–Integral (GPI) Observers [13].

Figure 13.15 presents a PI speed control scheme for SRM. A speed-to-current controller ($PI_{\omega,I}$) or speed-to-torque controller ($PI_{\omega,T}$) can be employed. Its output is interpreted by the current or the torque controller. Generally, the current control inner loop is employed when the application requires low-cost or simple implementation. Conversely, a torque control inner loop is applied for high performance or precision.

Based on the linearized model from (13.27), a speed-current transfer function is obtained, which is useful for the controller design.

$$\frac{\tilde{\omega}(s)}{\tilde{i}}(s) = \frac{K_b}{B_t + sJ} \tag{13.32}$$

If an inner torque controller is employed, a speed-torque transfer function can be obtained through the mechanical equation (13.25) and is written as:

$$\frac{\omega(s)}{T_e(s)} = \frac{1}{B_t + sJ} \tag{13.33}$$

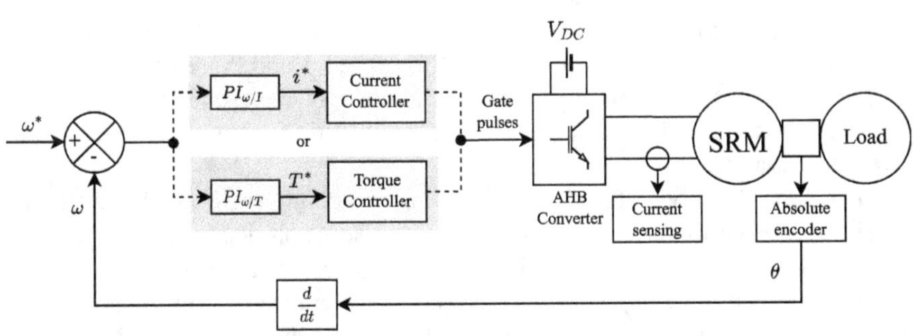

Figure 13.15 Speed control of SRM

The following sections present the main current and torque control strategies for SRM drive.

13.6 Current control

Current control is essential for SRM to achieve the desired torque and speed performance. The current control strategy determines the switching instants of the converter based on the reference current waveform and the feedback current measurement. The current control can be implemented in different ways, such as Hysteresis Control (HC), Pulse Width Modulation (PWM) control, or Model Predictive Control (MPC).

Hysteresis control is a simple and robust method that commutes the converter whenever the current error exceeds a predefined hysteresis band [14]. This method ensures fast current tracking, but it has a variable switching frequency and demands a high sampling frequency.

Alternatively, PWM control is a method that ensures fixed frequency switching with a variable duty cycle to regulate the current amplitude [15]. This method reduces switching losses and EMI, but it has a slower current response than hysteresis control.

Model predictive control is a method that predicts the future behavior of the SRM based on a mathematical model and selects the optimal switching state for each sampling period [16]. This method can achieve high performance in terms of current tracking, torque ripple reduction, efficiency improvement, and EMI suppression. However, it requires high computational power and accurate model parameters.

In motoring mode, there are mainly two switching logics that can be used for current control. In the *hard chopping* logic, the converter switches commute simultaneously, so the states 1 and -1 are used and the voltage applied to the phase is either $+v_{DC}$ or $-v_{DC}$. This strategy produces more current ripple and, consequently, more torque ripple. Another logic is the *soft chopping*, in which one converter switch is kept closed while the other is commuted. In this logic, the switching losses and the current ripple are reduced due to the use of the freewheeling state. The state -1 is used only for fast demagnetization after the conduction period. In Figure 13.12, a soft chopping logic was used for current control during motor operation. It can be seen that the states 1 and 0 are used for current regulation within the driving angles. After the turn-off angle, the state -1 is used and $-v_{DC}$ applied to the phase until the current extinguishes and the converter diodes stop conducting.

In generating mode, the *hard chopping* switching logic is similar to the motoring mode. However, *soft chopping* logic has some particularities, its idealized operation is illustrated in Figure 13.16. First, it is necessary to magnetize the phase, thus, the state 1 must be applied after θ_{on} until the reference current (i^*) is achieved for the first time. The magnetization interval is named *dwell* period and ends at θ_e position. After that, the current can be regulated by the converter states 0 and -1.

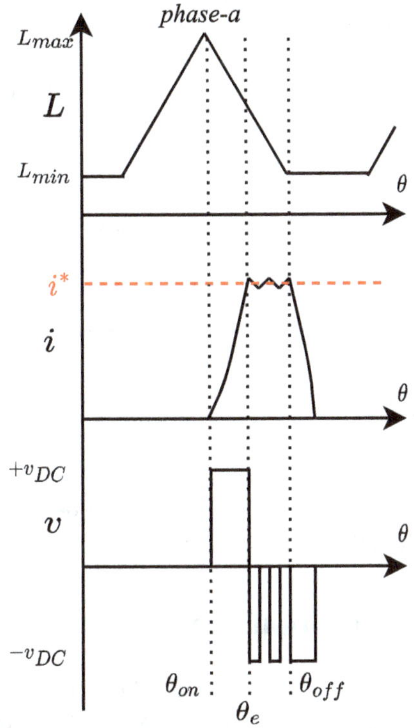

Figure 13.16 Idealized waveforms of a soft chopping current controller SRM for generator operation

A Hysteresis Current Controller with soft chopping logic for the motor and generator is summarized in Table 13.1. The logic is valid for increasing θ (positive speed, quadrants 1, and 4), but it can be easily deduced for decreasing θ (negative speed, quadrants 2, and 3).

In PWM control, a current controller provides the voltage necessary for current tracking. This voltage is then compared to the PWM carrier signal and the switching gates signals are given. Based on the linearized model from (13.26), a transfer function of (13.34) is obtained and is useful for the current controller design.

$$\frac{\tilde{i}(s)}{\tilde{v}}(s) = \frac{1}{R_{eq} + sL} \tag{13.34}$$

The choice of the current control method depends on various factors such as SRM characteristics, converter topology, application requirements, cost constraints, and others. In general, hysteresis control is suitable for low-power applications where fast response and low torque ripple are important; PWM control is suitable for medium-power applications where fixed switching frequency and low

Table 13.1 Switching logic by the soft chopping

Position	Current error	Applied state Motor	Generator
$\theta_{on} \leq \theta < \theta_e$[1]	–	n/a	1
$*\theta_e \leq \theta < \theta_{off}$	$e_I > HB$[2,3]	n/a	0
	$e_I < -HB$	n/a	-1
$*\theta_{on} \leq \theta < \theta_{off}$	$e_I > HB$	1	n/a
	$e_I < -HB$	0	n/a
$\theta_{off} \leq \theta$	–	-1	-1

[1] θ_e is the position for the first occurrence of $e_I < -HB$
[2] $e_I = i - i^*$
[3] HB is the hysteresis band for HC

EMI are important; predictive control is suitable for high-power applications where optimal performance and efficiency are important.

13.7 Electromagnetic torque control methods

This section will focus on the various methods of electromagnetic torque control in switched reluctance machines. It is important to state that, for all methods cited, a reference or a command torque is previously provided (usually by a speed controller). One of the most widely used techniques is Direct Average Torque Control (DATC). This technique operates by regulating the average torque over a specified period of time, rather than regulating the instantaneous torque. This method has the advantage of simplicity and achieves high power and high-speed operating points, but it requires a longer settling time resulting in slower torque response and can produce more torque ripple.

Another approach is Current Profiling through Torque Sharing Functions (TSF). The TSF splits the torque production among phases such that the total electromagnetic torque tracks a given reference. The torque reference is converted into a current reference based on a precise model of the SRM. Then, a feedback loop is used to adjust the current profile of each phase of the machine. This method allows for improved torque accuracy and reduced torque ripple, while also offering increased efficiency.

Finally, Direct Instantaneous Torque Control (DITC) is a method that can be used to regulate electromagnetic torque. This technique operates by directly measuring or estimating the instantaneous torque and adjusting the converter states to maintain the torque within a limited band. This method has the advantage of providing fast torque response and low torque ripple, but it requires a high sampling frequency.

Each of these methods has its own advantages and disadvantages, and the choice of control method will depend on the specific application requirements.

They can also be combined to operate in different operating points. Each one of these methods will be detailed below.

13.7.1 Direct average torque control

The DATC technique involves estimating the average torque produced by the machine and adjusting the phase currents to maintain the desired torque. This is achieved by monitoring the rotor position and the phase currents and adjusting the current to keep the average torque within a specified value. The main advantage of this technique is that it is relatively simple to implement and does not require complex sensing or control algorithms. However, it may not provide precise control of the torque and may result in elevated torque ripple due to the nonlinear behavior of the machine.

Measuring the average torque through torque transducers can be a significant financial investment, which is not always feasible. Therefore, it is important to estimate the average torque produced by the machine from the terminal electrical quantities.

To determine the average torque produced by the SRM, it is necessary to know the flux linkage. From (13.4), the time derivative of the flux linkage can be obtained by the electrical quantities as:

$$\frac{\mathrm{d}\lambda_k}{\mathrm{d}t} = v_k - i_k R_k \tag{13.35}$$

As the mechanical energy change is equal to the co-energy, or the difference between the initial and final input energy during one electrical cycle, we can write:

$$W_m = W_e(t = t_{ext}) = \int_{t_{on}}^{t_{ext}} \left(i_k \frac{\mathrm{d}\lambda_k}{\mathrm{d}t} \right) \mathrm{d}t \tag{13.36}$$

where t_{on} is the time at the phase magnetization start (occurs when $\theta = \theta_{on}$) and t_{ext} is the instant when the phase is fully demagnetized ($i = 0$ and $\lambda = 0$). The input energy is reset after demagnetization on each electrical cycle, i.e., $W_e(t = t_{on}) = 0$. For this calculation, the iron losses were despised.

The symmetry of the phases allows for using a single estimator for all phases of the machine, which reduces the computational cost of the controller. So, the average torque ($\bar{T}_e$) can be calculated as:

$$\bar{T}_e = \frac{N_\phi N_R}{2\pi} W_m \tag{13.37}$$

Figure 13.17 shows the scheme used to estimate the average torque in one of the SRM phases using only the voltage and current values in the phase. In a practical implementation of the average torque estimator, the phase resistance (R_k) is previously measured or online estimated, whereas the phase current (i_k) is easily sensed. The phase voltage (v_k) can be directly sensed or estimated based on the known DC link voltage and the converter switches states. The output of some PWM-based current controllers (with i_k/v_k transfer function) can also provide v_k.

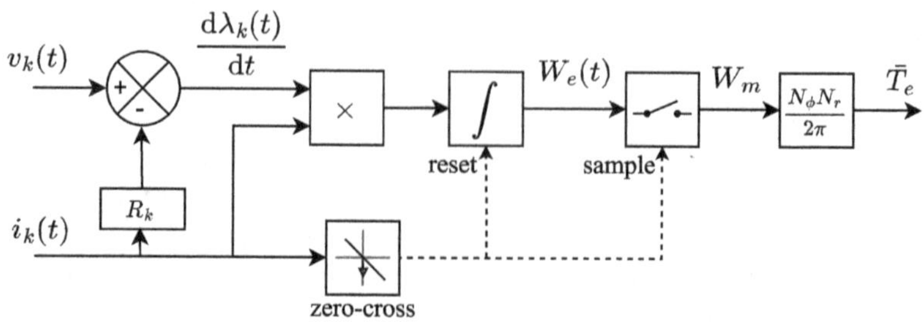

Figure 13.17 Scheme of the average torque estimator

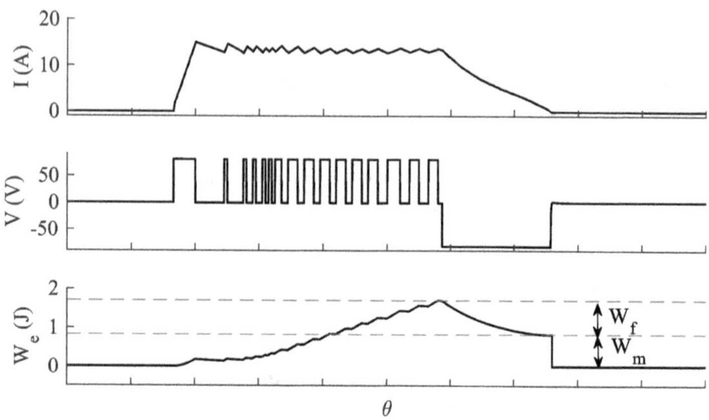

Figure 13.18 Main waveforms for the average torque estimator

In all cases, it is necessary to ensure that the obtained v_k is the mean value during the sampled time.

Figure 13.18 shows the phase current and voltage waveforms for one electrical cycle. During the enable time, the input energy (W_e) steadily increases until it reaches its peak at the point of demagnetization start. After this point, the magnetic energy stored in the machine is transferred back to the source and the input energy starts to decrease. The difference between the input energy and the energy returned to the source is equal to the mechanical energy output of the machine, which is used to evaluate the average torque.

The DATC method employs look-up tables that contain information about the optimal control parameters for each operating point. Each operating point is defined by a torque value and an instantaneous speed value. The control structure of the DATC method is depicted in Figure 13.19.

A closed-loop torque control system with feedforward is implemented in the control structure. An external speed loop generates the torque reference through the PI_ω controller. A PI torque controller (PI_T) processes the error between the

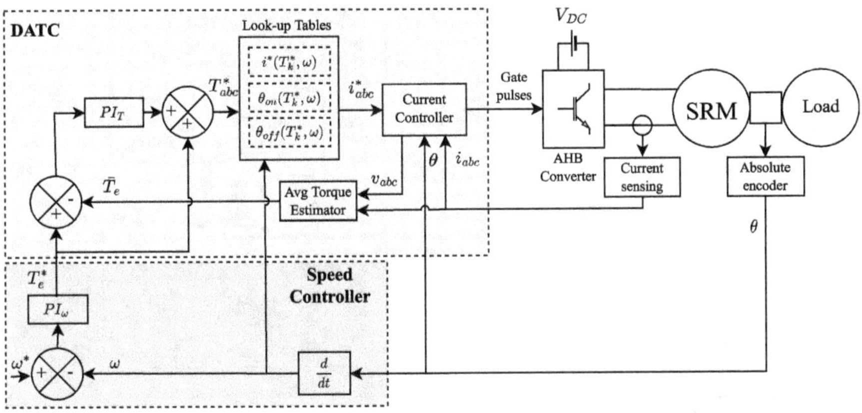

Figure 13.19 Scheme of the Direct Average Torque Control (DATC)

reference and the estimated average torque to generate the final torque reference. The look-up tables take this torque reference and the instantaneous speed as inputs and provide the optimal current reference, turn-on, and turn-off angles for each phase.

The current regulation is achieved by a hysteresis current controller, which utilizes the information from the lookup tables, as well as the rotor position and phase currents to synthesize the gate signals for the asymmetric half-bridge converter.

One notable advantage of the Direct Average Torque Control method is its low sampling rate requirement, which allows for a simpler and more cost-effective control system. This feature makes the DATC method particularly suitable for high-speed applications. However, due to the non-linearities of the SRM, DATC might produce a high torque ripple.

13.7.2 Current profiling through torque sharing functions

Differently from DATC, which uses a constant reference current during the electrical cycle, the Current Profiling through Torque Sharing Functions (CP-TSF), shown in Figure 13.20, shares the torque between phases and uses a current profile that compensates for the machine's non-linearities to achieve constant torque production.

After the torque reference (T_e^*) is obtained through a speed controller, it can be divided among the phases through the TSF, resulting in the instantaneous reference torque for each phase ($T_{e,k}^*(\theta)$) as a function of the rotor position (θ).

The TSF for the k-th phase is denoted by $f_k(\theta)$, where $0 \leq f_k(\theta) \leq 1$. We can write that:

$$T_{e,k}^*(\theta) = f_k(\theta) \cdot T_e^* \tag{13.38}$$

The sum of all TSFs is always equal to 1 to maintain the total instantaneous torque at a constant value, i.e.,

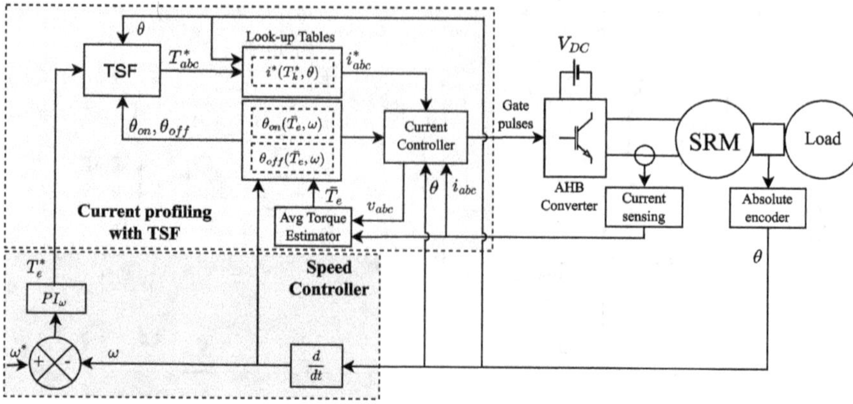

Figure 13.20 Scheme of the Current Profiling through Torque Sharing Function (CP-TSF) method

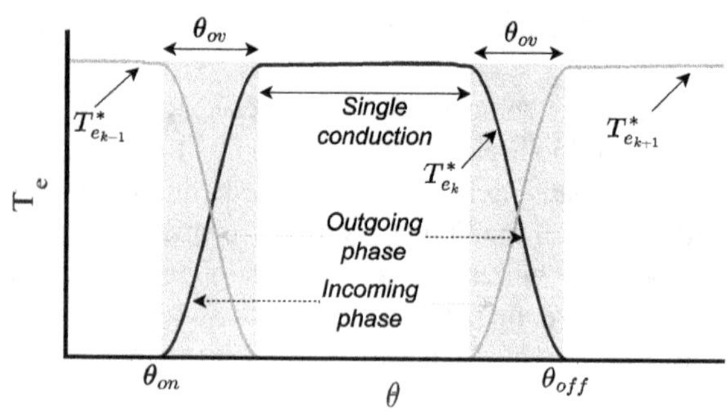

Figure 13.21 Cosine TSF

$$\sum_{k=1}^{N_\phi} f_k(\theta) = 1 \tag{13.39}$$

where N_ϕ is the number of SRM phases.

Various mathematical functions can be used as TSFs, such as linear, exponential, cubic, and cosine. The cosine TSF is illustrated in Figure 13.21, where θ_{on} represents the turn-on angle, θ_{off} denotes the turn-off angle, and θ_{ov} indicates the overlapping angle.

TSF is usually divided into three stages: incoming phase, single conduction, and outgoing phase. During phase commutation, there will be simultaneously an incoming phase and an outgoing phase. The overlapping angle determines the commutation period. For now, the cosine function will be considered for TSF and can be written as follows:

$$f_k(\theta) = \begin{cases} \dfrac{1}{2} - \dfrac{1}{2}\cos\dfrac{\pi}{\theta_{ov}}(\theta - \theta_{ov}), & \theta_{on} \le \theta < \theta + \theta_{ov} \\ 1, & \theta + \theta_{ov} \le \theta < \theta_{off} - \theta_{ov} \\ \dfrac{1}{2} + \dfrac{1}{2}\cos\dfrac{\pi}{\theta_{ov}}(\theta - \theta_{ov}), & \theta_{off} - \theta_{ov} \le \theta < \theta_{off} \\ 0, & others \end{cases} \tag{13.40}$$

where the first, second, and third expressions are valid for the incoming, single conduction, and outgoing stages, respectively.

The phase reference current (i_k^*) is crucial for regulating the torque of a motor. A popular approach for obtaining this information is the Current Profiling (CP) technique. In CP, the static torque profile from Figure 13.9 is inverted to obtain the current required for a particular rotor position and torque, i.e., $i(T_e, \theta)$. The resulting current profile can be observed in Figure 13.22. The reference current acquired is then compared to the phase current feedback signal, and any difference is regulated by a current controller. In practice, achieving certain torque values near the unaligned and aligned position may not be possible without requiring an excessively high current. To avoid such conditions, the current is saturated at the maximum allowable value, which is 27.5 A in this particular case.

The phase reference current obtained by inverting the phase reference torque using cosine TSF is presented in Figure 13.23 for different turn-on and turn-off angles. In this case, a total reference torque of 5 Nm was set, and the overlapping angle was 1°. It is noteworthy that proper driving angle selection can result in a minimum rms current. Additionally, the angles must be chosen in a way that avoids reference current saturation, which can cause the desired torque to be unavailable, resulting in torque sags, and thereby increasing torque ripple.

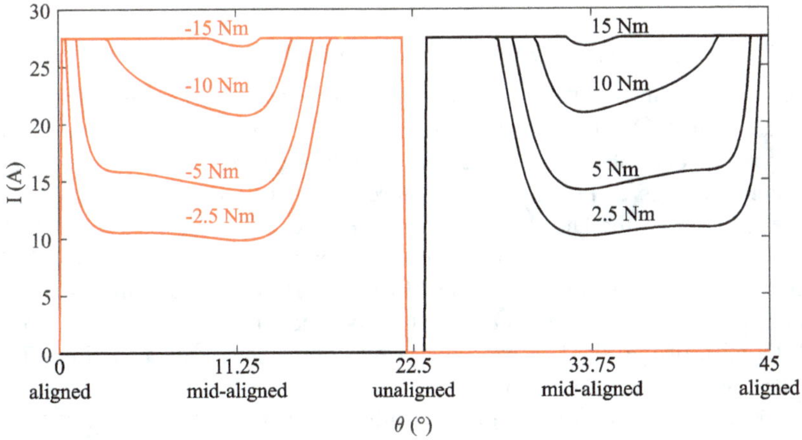

Figure 13.22 Required current profile for various torque values

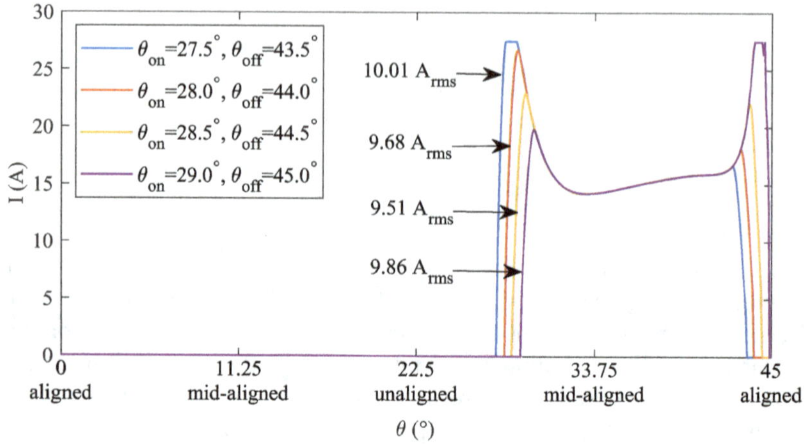

Figure 13.23 Required current profile produced by cosine TSF for 5 *Nm torque*

Accurately tracking current references is critical for minimizing torque ripple when using the CP-TSF technique, which offers low ripple and excellent dynamic response. However, achieving this requires a proper selection of driving angles and fast current regulation. The challenge lies in tracking the reference current at the start of the incoming stage when the rotor position is close to the misaligned position and the inductance derivative is at its minimum. At this point, producing sufficient torque can be difficult, and using a current profiling function may result in a steep current reference slope that the controller hardly regulates. Likewise, at the end of the outgoing stage, as the rotor approaches the aligned position, the inductance increases, and the current and torque can only decrease at a smaller rate than the reference value. In practice, the current might be extinguished after the turn-off angle.

13.7.3 Direct instantaneous torque control

The basic idea behind DITC is to estimate the instantaneous electromagnetic torque produced by the SRM and compare it with the desired torque. Based on the existing difference, the controller adjusts the converter states, hence the voltage applied to the stator windings to bring the actual torque closer to the desired torque. The scheme structure of the DITC can be seen in Figure 13.24.

To estimate the instantaneous electromagnetic torque, DITC uses a $T_e(i,\theta)$ function that relates the torque produced by the SRM to the stator current and rotor position, as depicted in Figure 13.10(b). This function is obtained based on the characteristics of the SRM, it can be a polynomial function or a Look-up Table. The phase torques add up to deliver the total torque. Once the instantaneous electromagnetic torque is estimated, the controller compares it with the desired torque to determine the torque error. Based on the torque error, a double-band hysteresis logic determines the drive signals applied to the driving converter.

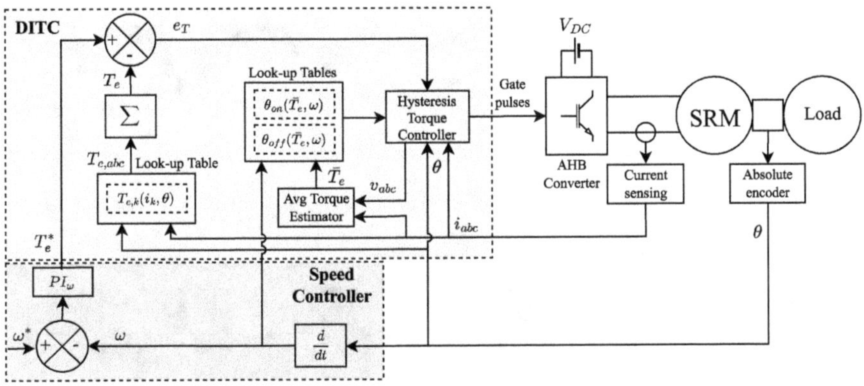

Figure 13.24 Scheme of the Direct Instantaneous Torque Control (DITC)

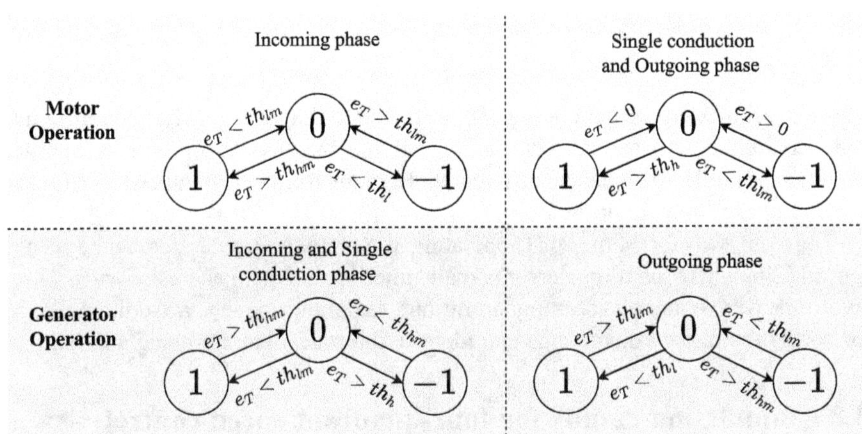

Figure 13.25 Hysteresis torque controller logic for motor and generator operation

The hysteresis logic used in DITC consists of four thresholds, th_{lm}, th_{hm}, th_h, and th_l, which represent the medium-low, medium-high, high, and low thresholds, respectively. The thresholds are used to determine the state of the SRM, which can be magnetization (1), demagnetization (−1), or free-wheeling (0). Each phase receives a signal that allows it to conduct, and depending on the torque error and phase condition, a different set of hysteresis is applied.

Figure 13.25 shows a state machine for DITC in motoring and generating modes, designed to minimize torque ripple. However, the logic can be modified to maximize efficiency. The driving angles are essential for determining phase stages and duration. In motor operation, states 1 and 0 are preferred in the incoming phase, while the hysteresis band formed by th_l and th_{lm} and state (−1) are rarely used. In the outgoing and single conduction phases, the free-wheeling and demagnetization

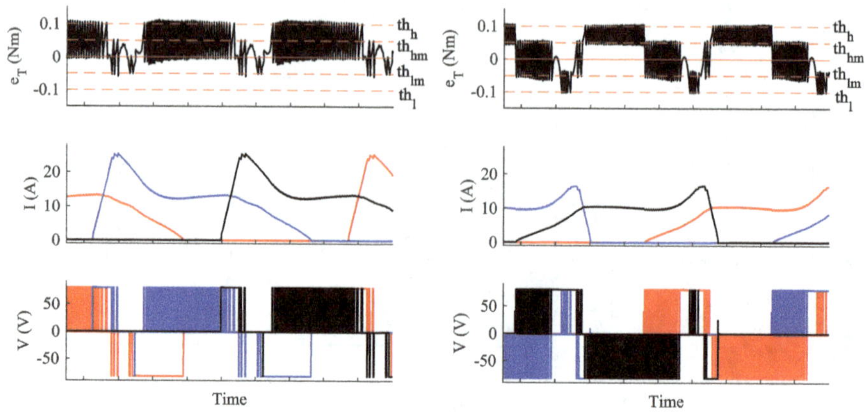

Figure 13.26 DITC waveform during operation: (a) motor operation and (b) generator operation

states are prioritized. In generating mode, the state diagram for the incoming and single conduction phases are the same, and the magnetization state is avoided whenever possible to enhance efficiency. If more torque is required during the outgoing phase, state 1 can be used.

The ideal waveforms for DITC operating as a motor and generator can be seen in Figure 13.26, where the torque error is maintained within the hysteresis bands. These waveforms were obtained assuming an infinite sampling frequency. Additionally, by monitoring the phase voltage, one can identify the states that are being applied.

13.8 Simulation results for four-quadrant speed control

The four-quadrant operation involves controlling the speed and torque in both the motoring and generating modes of operation and both directions. The SRM can operate as a motor in quadrants 1 and 3 and as a generator in quadrants 2 and 4. The four-quadrant speed control involves regulating the speed and torque in all four quadrants.

To verify the four-quadrant operation of the SRM with the aforementioned DITC method, a computational model can be developed using MATLAB/Simulink. The dynamic mathematical model of the SRM is based on the experimentally acquired magnetization curves, as proposed in [1], and the resulting Look-up Tables presented in Figures 13.10(a) and 13.10(b). In this chapter, the machine is a three-phase 12/8 SRM, with a nominal power of 1.5 kW, a nominal speed of 157 rad/s (1,500 rpm), and a base speed of approximately 96 rad/s (954.93 rpm). The inductance varies from 1.675 mH to 13.88 mH. The measured phase resistance is 0.3 Ω. During the simulations, the supply voltage is kept constant at 80 V for the sake of simplicity.

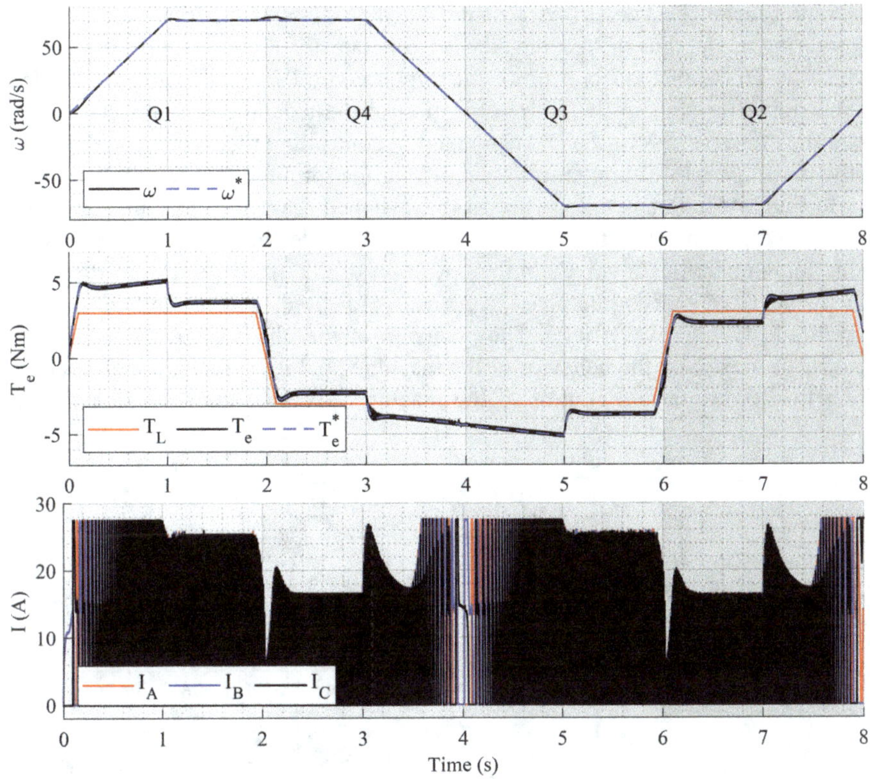

Figure 13.27 Four-quadrant speed control of SRM using DITC

The reference speed initially ramps up to 70 rad/s over a period of 1 s. It is then held constant before being decreased in a ramp to −70 rad/s. At the same time, a positive load torque of 3 Nm is gradually applied. After 2 s, the load torque is then inverted to −3 Nm. Finally, at 6 s, the load torque is inverted again. With these two profiles, the machine operates in all four quadrants.

In Figure 13.27, we can observe the reference speed (ω^*) and the actual speed (ω) of the switched reluctance machine. The speed is regulated by a Proportional-Integral (PI) controller that generates the reference torque (T_e^*). It is worth noting that the developed torque is higher during acceleration ramps, which is expected due to the need to overcome the inertia of the rotor. In motoring mode (quadrants $Q1$ and $Q2$), the electromagnetic torque produced by the machine is higher than the load torque due to the torque loss caused by the viscous friction (B). On the other hand, in generating mode (quadrants $Q2$ and $Q4$), the electromagnetic torque produced by the machine is lower than the load torque also due to the friction losses. In addition, the current waveform is presented. The results show that considering the same load torque (absolute value), the electric current reaches its maximum established value with a lower load torque in motoring mode compared to generating mode. This occurs because the machine supplies the friction torque only in motoring mode, thus the machine tends to operate at a higher magnetic saturation level. The DITC method is a powerful torque control method that provides low torque ripple throughout the entire operation of the machine. The results demonstrate the effectiveness of the DITC method in controlling the torque and the speed of the machine, while simultaneously reducing the torque and the current ripples.

In Figure 13.28, a detailed view of the switched reluctance machine operation is presented for one electrical cycle. The figure provides insights into the behavior of the main quantities involved in the operation of the machine. It is worth noting that for negative speeds, the phase sequence is inverted, which is reflected in the current and voltage waveforms. The inversion of the phase sequence can have a significant impact on the behavior of the machine, and it is crucial to consider this effect in the design and control of the machine. By analyzing the flux linkage

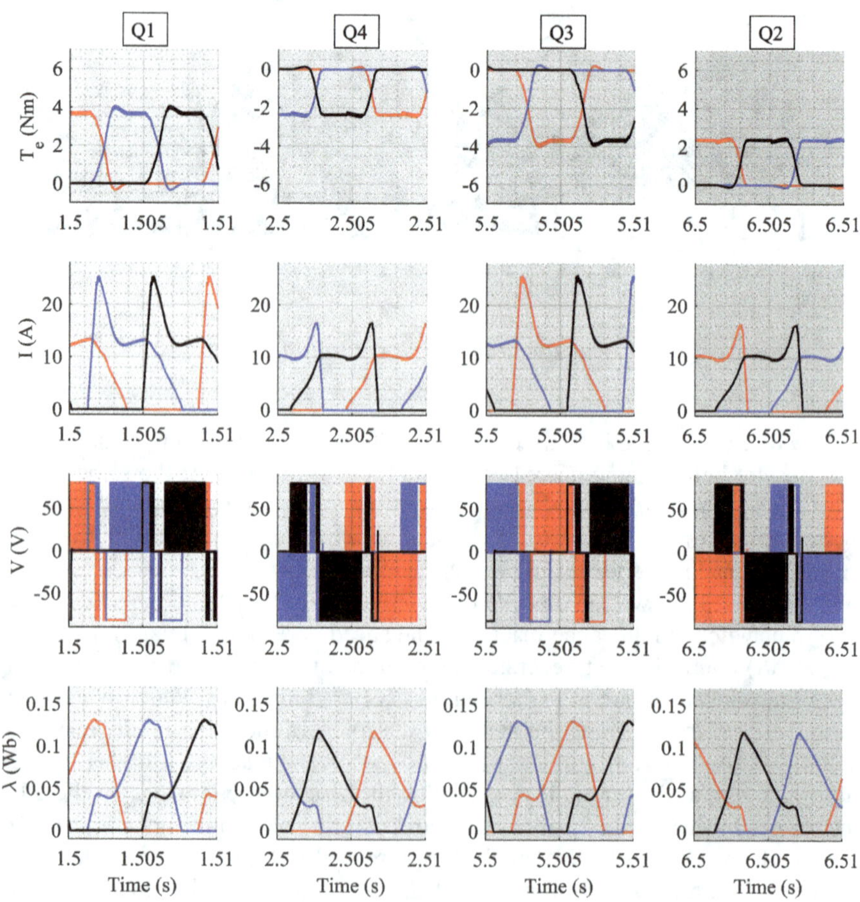

Figure 13.28 Details during the electrical cycle for each quadrant

waveforms, we can observe that the motoring torque is produced during the magnetization period, while the generating torque is produced during the demagnetization stage. This behavior is a consequence of the magnetic properties of the machine, which allows for the production of torque through the variation of the magnetic reluctance. Overall, Figure 13.28 provides valuable information about the operation of the SRM. These insights can be used to improve the design and control of the machine, leading to better performance and efficiency.

The simulation results show that the SRM can operate with low torque ripple in all four quadrants and that the presented DITC control scheme is able to achieve accurate speed and torque regulation.

Besides, the simulation results also demonstrate the effectiveness of the proposed four-quadrant speed control scheme for SRM and highlight the potential of these machines for a wide range of applications.

13.9 Conclusions

The four-quadrant operation of switched reluctance machines is a complex and challenging topic, but it can be attained with appropriate control. In this chapter, the operation principles of SRMs were presented, and the idealized and practical models for SRM operation were described. Different methods for current and torque control were discussed, such as Direct Average Torque Control, Current Profiling through Torque Sharing Functions, and Direct Instantaneous Torque Control.

Moreover, the optimization of driving angles and the speed control for the four-quadrant operation were also presented. The simulation results showed that the proposed methods can provide accurate and efficient control of SRMs. Nonetheless, it is worth noting that these results were obtained under ideal conditions, and practical implementations may require further adjustments and considerations.

In conclusion, the four-quadrant operation of SRMs presents many opportunities and challenges for research and development, while still being indicated for some industrial applications. Future work should focus on improving the accuracy and efficiency of control methods, as well as developing new techniques to overcome the limitations of SRMs, such as torque ripple and acoustic noise. Topics such as fault detection and operation and sensorless operation are also trending.

Acknowledgments

This book chapter was supported by the National Council for Scientific and Technological Development, CNPq under grants 402983/2021-1 and 308589/2022-0, by the Brazilian government in partnership with FUNDEP through the ROTA2030 program under grants 27192*38 and 27192*77.

References

[1] Barros TAS, Santos PJ, Paula MV, *et al.* Automatic Characterization System of Switched Reluctance Machines and Nonlinear Modeling by Interpolation Using Smoothing Splines. *IEEE Access*. 2018;6:26011–26021.

[2] Fang G, Scalcon FP, Xiao D, *et al.* Advanced Control of Switched Reluctance Motors (SRMs): A Review on Current Regulation, Torque Control and Vibration Suppression. *IEEE Open Journal of the Industrial Electronics Society*. 2021;2:280–301.

[3] Hannoun H, Hilairet M, and Marchand C. High performance current control of a switched reluctance machine based on a gain-scheduling PI controller. *Control Engineering Practice*. 2011;19(11):1377–1386.

[4] Jiang JW, Bilgin B, Howey B, *et al.* Design optimization of switched reluctance machine using genetic algorithm. In: 2015 *IEEE International Electric Machines & Drives Conference (IEMDC)*; 2015. p. 1671–1677.

[5] Ma C, and Qu L. Multiobjective Optimization of Switched Reluctance Motors Based on Design of Experiments and Particle Swarm Optimization. *IEEE Transactions on Energy Conversion*. 2015;30(3):1144–1153.

[6] Neto PJdS, Barros TAdS, de Paula MV, *et al.* Design of Computational Experiment for Performance Optimization of a Switched Reluctance Generator in Wind Systems. *IEEE Transactions on Energy Conversion*. 2018;33 (1):406–419.

[7] Song S, Hei R, Ma R, *et al.* Model Predictive Control of Switched Reluctance Starter/Generator with Torque Sharing and Compensation. *IEEE Transactions on Transportation Electrification*. 2020;6(4):1519–1527.

[8] da Cunha Reis MR, de Araujo WRH, Gomes VM, *et al.* Optimized techniques for driving and control of the switched reluctance motor to improve efficiency. *Control Engineering Practice*. 2019;90:1–18.

[9] Wang Y, and Deng Z. An Integration Algorithm for Stator Flux Estimation of a Direct-Torque-Controlled Electrical Excitation Flux-Switching Generator. *IEEE Transactions on Energy Conversion*. 2012 June;27(2):411–420.

[10] de Paula MV, dos S Barros TA, Moreira HS, *et al.* A Dahlin Cruise Control Design Method for Switched Reluctance Motors with Minimum Torque Ripple Point Tracking Applied in Electric Vehicles. *IEEE Transactions on Transportation Electrification*. 2020;7(2):730–740.

[11] de Paula MV, Dos Santos, and Barros TA. A Sliding Mode DITC Cruise Control for SRM with Steepest Descent Minimum Torque Ripple Point Tracking. *IEEE Transactions on Industrial Electronics*. 2021;69(1):151–159.

[12] De Paula MV, Williamson SS, and Barros TAS. Four-Quadrant Model Following Sliding Mode Cruise Control for SRM with DITC applied to Transportation Electrification. *IEEE Transactions on Transportation Electrification*. 2022;8(3):3090–3099.

[13] De La Guerra A, Arteaga-Pérez MA, Gutiérrez-Giles A, *et al.* Speed-sensorless control of SR motors based on GPI observers. *Control*

Engineering Practice. 2016;46:115–128. Available from: https://www.sciencedirect.com/science/article/pii/S096706611530037X.

[14] Walz-Lange A, and Schullerus G. Sensorless Control of a Switched Reluctance Machine Based on Switching Frequency Evaluation. *IEEE Transactions on Industry Applications*. 2022;58(4):4768–4777.

[15] Dhale SB, Mobarakeh BN, Nalakath S, *et al.* Digital Sliding Mode Based Model-Free PWM Current Control of Switched Reluctance Machines. *IEEE Transactions on Industrial Electronics*. 2022;69(9):8760–8769.

[16] Li X, and Shamsi P. Model Predictive Current Control of Switched Reluctance Motors With Inductance Auto-Calibration. *IEEE Transactions on Industrial Electronics*. 2016;63(6):3934–3941.

Chapter 14
Reliability in power electronics

Erick Matheus da Silveira Brito[1]*, João Marcus Soares Callegari*[2]*, Allan Fagner Cupertino*[3]*, Heverton Augusto Pereira*[4] *and Ariya Sangwongwanich*[5]

14.1 Introduction

Nowadays, the search for more efficient and reliable devices is present in different contexts of our lives. With technological advances, the average consumer wants to buy new electronic devices, such as smartphones, TVs, and other products with higher performance, warranties, and lower costs. Normally, only two of these three features can be achieved simultaneously. In terms of engineering is not different, although efficiency and product volume/weight are also a concern.

During the manufacturing process, there are different product requirements that must be prioritized. For example, in domestic applications, customers usually prefer low-cost and high-performance devices, since they expect to replace them with a newer version before the item's end-of-life. On the other hand, aerospace applications usually neglect the costs and prioritize high-efficient and high-reliable devices. We can say that industry applications are middle-term, where all requirements should be balanced in the design and operation of a certain product or service.

In the past decades, the industry has shown interest in designing electronic systems with longer lifetimes and more accurate and reliable components. Usually, when a product is launched in the market, the customer buys it accepting that it will fail at some time and when it occurs, they have to fix it (if possible) or replace it. As illustrated in Figure 14.1, if the failure occurs before the warranty period, the manufacturer is submitted to additional costs and the client remains for a certain time without its product until it is fixed/replaced. After the warranty period, the client bears the repair/replacement costs, but the manufacturer may lose credibility

[1]Institute of Exact and Technological Sciences, Universidade Federal de Viçosa, Brazil
[2]Graduate Program in Electrical Engineering, Universidade Federal de Minas Gerais, Brazil
[3]Department of Electrical Energy, Universidade Federal de Juiz de Fora, Brazil
[4]Department of Electrical Engineering, Universidade Federal de Viçosa, Brazil
[5]Department of Energy Technology, Aalborg University, Denmark

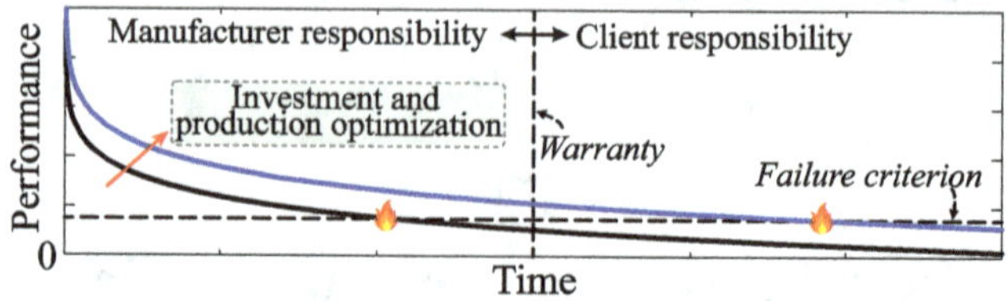

Figure 14.1 Theoretical product performance curve over time. The failure occurs when its criterion is achieved.

or reputation. As a consequence, early failures negatively impacted the industry, even occurring before or after the warranty period. From that point, new reliability-based approaches have emerged recently, aiming to find an optimal product design, taking into account the development improvements and warranty services costs.

14.1.1 Why study reliability in power electronics?

The answer is very simple, to reduce costs. In an industry-based context, every time a component or system fails, it should be fixed or replaced. In both cases, there are costs associated with maintenance, system downtimes, and possible emergencies caused by the failure. In this context, the prediction, as far as possible, of the time a certain item would fail can avoid or minimize these costs. There also could be secondary interests in increasing reliability, such as improving the product performance or prolonging its operation time before failure. However, making it work while reducing costs can be very challenging as well.

In practice, it is almost impossible to anticipate all the possible causes that lead a component or system to fail. There are several uncontrolled and unpredictable variables, characterized as catastrophic events, such as overcurrent or overvoltage, that would cause component failure. Therefore, it is common to address reliability as the probability of a component failing under a determined time, given its mission profile operation.

The IEEE Reliability Society defines reliability as "*the ability of an item to perform a required function under stated conditions for a stated period of time*". The expression lifetime or end-of-life refers to the time when a specific object (mechanism, component, converter, subsystem, or system) reaches its failure criteria. The increase in leakage current, the maximum variation of thermal resistance, or the increase in the voltage drop are examples of failure criteria [1]. The object does not necessarily explode, get destroyed, or instantly stop working when the failure criteria are achieved. It just means that it becomes inappropriate to continue operating under such conditions.

Reliability engineering in power electronics focuses on the study of the failure root cause, the failure mechanism behavior under a certain type of stress, and product design aspects that affect the component's lifetime. The comprehension of reliability metrics helps the engineers to consider the majority of the parameters and events that would wear-out the component over time.

14.1.2 Fundamentals of reliability engineering

Before talking about metrics of reliability, it is important to correctly denote the specific terms applied to this field. The basic concepts are described below:

- **Reliability**: as defined before, it describes the probability that an item will perform a required function without failure under stated conditions for a stated period of time.
- **Unreliability**: as the name suggests, it is the complement of reliability. In other words, unreliability describes the probability that an item reaches a failure under stated conditions for a stated period of time.
- **Robustness**: is the ability to withstand or overcome an adverse condition and keep working despite faults in its subsystems or parts.
- **Resilience**: is the ability to withstand or to recover quickly from an adverse condition within acceptable degradation parameters and a satisfactory timeframe.
- **Redundancy**: is the intentional duplication of critical components that are not strictly necessary, to improve the system's reliability.

Reliability (or unreliability) and robustness are the most common concepts applied to industrial power electronic systems. Their major role is to improve the system's performance. Recently, resilience is increasingly taken into account, which gives rise to the concept of R^3: Reliability, Robustness, and Resilience [2]. Finally, redundancy is applied to systems where reliability is crucial and/or components repair/replacement is very costly.

The reliability of a certain component is a function of time, denoted by $R(t)$. If t_f refers to the specific time when the failure occurs for a given time interval of $[0,T]$, then:

$$R(t) = P\{t_f > T\} \tag{14.1}$$

In other words, $R(t)$ is the probability of the failure occurring outside of the considered time interval, i.e. $t_f > T$. Following the same logic, the unreliability of a certain component is the opposite of the reliability, denoted by $U(t)$. Therefore:

$$U(t) = 1 - R(t) = P\{t_f \leq T\} \tag{14.2}$$

If, instead of having the specific failure time t_f, we have a failure probability function over time, $f(t)$, then:

$$U(t) = \int_0^T f(t)\mathrm{d}t \tag{14.3}$$

and

$$R(t) = 1 - \int_0^T f(t)\mathrm{d}t \tag{14.4}$$

Figure 14.2 demonstrates the differences between $f(t)$ (probability density function, or PDF) and $R(t)$ and $U(t)$ (cumulative density functions, or CDF). Note

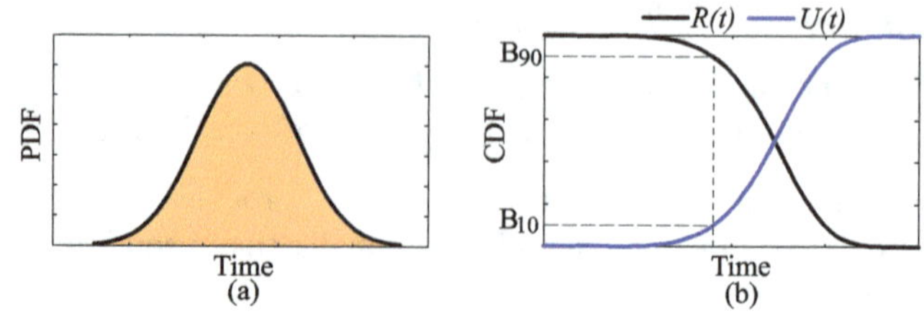

Figure 14.2 Probability functions: (a) is a PDF, or probability density function and (b) are both CDF, cumulative density functions.

that, since the component has only two states over time (before or after failure), at any point, $R(t) + U(t) = 1$, i.e., they are complementary.

Finally, the failure rate function $\lambda(t)$ can be described as:

$$\lambda(t) = \frac{f(t)}{R(t)} = \frac{f(t)}{1 - U(t)} \tag{14.5}$$

If a constant failure rate is considered, then we can calculate the Mean Time To Failure, or MTTF, given by:

$$MTTF = \frac{1}{\lambda} \tag{14.6}$$

The MTTF is usually provided in hours of utilization. Since this can be a huge number (ranging between 10^7 and 10^{10}) for electronic components in general, a unit called FIT (Failure in Time) was created. 1 FIT is equal to one failure per 10^9 hours. If the component under failure is repairable, other concepts can also be applied, such as Mean Time Between Failure (MTBF) and Mean Time to Repair (MTTR) [3].

From the early 1960s to recent decades, several handbooks with constant failure rates based on empirical and field data were released by the military and electronic industry. The most known was the Military Handbook, or just MIL-HDBK, released by the American Department of Defense in 1965 [4]. Despite receiving several updates to include variabilities in the constant failure rates, such as utilization, application, or environmental factors, this approach became inaccurate and inadequate, being canceled in 1995. In its place, modern reliability approaches focused on the root cause of the failure mechanisms had emerged and took place in academia and industry.

14.1.3 Failures in power electronics

A failure is defined as any kind of component degradation or parameter variation that causes an inappropriate operation of the system. The time when the first failure occurs can be defined as the system's lifetime (without considering repair). The objective evidence of a failure is called failure mode (such as open-circuit or short-circuit) and failure mechanism refers to the type of process that caused the failure (chemical, mechanical, electrical, thermal, etc.). According to [5], the failures can be classified according to:

1. The cause of occurrence:
   - **Misuse failures**: due to the application of stresses during use which exceeds the stated capabilities of the item;
   - **Primary failures**: where the direct or indirect cause of the failure is not due to the failure of another device;
   - **Secondary failures**: also known as induced failures, generated by the failure of another device.
2. The consequences as a result of the failure:
   - **Critical failures**: those can cause, with a high probability, damage to persons or material unacceptable to other parts of the system;
   - **Failures of primary importance**: those can reduce the functionality of a system, although different than those previously mentioned;
   - **Failures of secondary importance**: are those that do not reduce the functionality of a system.
3. The nature of the failure:
   - **Total failures**: when variations in the characteristics of the element are such to completely compromise its function;
   - **Partial failures**: when the variations of one or more characteristics of the element do not stop its complete functioning;
   - **Intermittent failures**: characterized by a succession, generally casual, of periods of normal operations with periods of failure breakdowns, without any maintenance operations, carried out on the device.
4. The time when the failure occurs:
   - **Early failures**: attributable to intrinsic weaknesses in the construction of the element and whose causes are normally identifiable during the manufacturing process or manifested during the initial use of the component;
   - **Random failures**: due to uncontrollable and unpredictable factors that occur during the "useful life" of the component and with a probability of occurrence independent of time;
   - **Wear-out failures**: generated by aging phenomena and with an increasing probability of occurring with the passage of time.

The last failure classification (i.e., over time of occurrence) can be arranged in a typical curve known as the "bathtub curve", due to its format with steep sides and a flat bottom, as shown in Figure 14.3. Since early failures, also known as burn-in of infant mortality, are identified during the manufacturing process or the first hours of operation, their failure rates sharply drop over time. On the other hand, random failures, caused by catastrophic events, are unpredictable and their failure rate is constant. Finally, the wear-out failures, caused by the components aging and degradation processes, become more relevant at the end of the system lifespan. The total failure rate over the operating time is the sum of three regions.

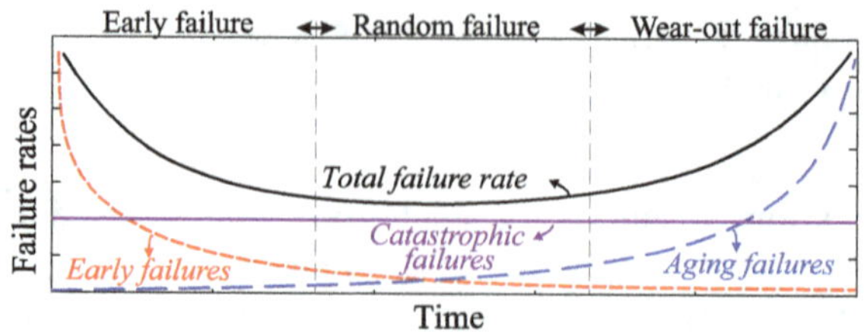

Figure 14.3 Hazard function of the conventional bathtub curve

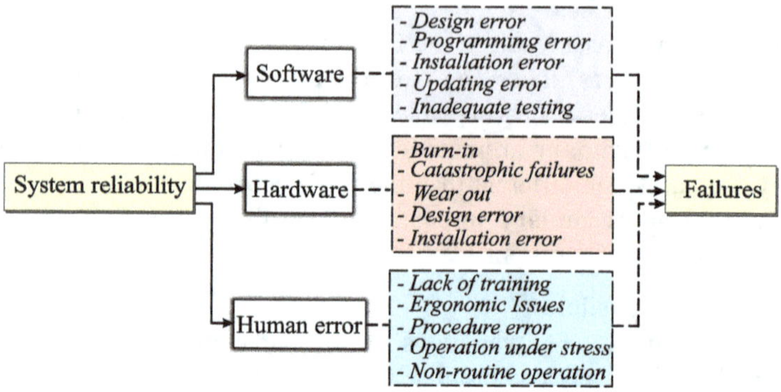

Figure 14.4 The classification of failure types in the power electronic system reliability

The most common failures that occur in power electronic systems are divided into three categories: software, hardware, and human error. More details about the causes of the failures are listed in Figure 14.4 [6]. Nevertheless, this chapter will focus on hardware failures, due to wear-out processes. An industry-based survey from 2011 showed that the most fragile components in a power electronic converter are the power devices (with 31% of the answers) and capacitors (around 17%) [7]. Other important components to be considered in a more complete reliability analysis are gate drivers, connectors, printed circuit boards (PCB), sensors, digital signal processors (DSP), etc. On the other hand, failures on resistors, inductors/magnetic components, and electromechanical components are quite rare and negligible in reliability analysis. More recently, a novel survey realized in 2018 showed that power semiconductors and capacitors are still the most critical components from industry experts' perspectives [8]. Therefore, this chapter focuses on the reliability evaluation of these two components.

14.1.4 Reliability-oriented design

During the early 1990s, military handbooks and methodologies based on constant failure rates began to decline due to their inaccuracy when applied to modern power electronics systems. By that time, approaches such as Physics-of-Failure (PoF) started to gain prominence in the reliability field. The PoF approach aims to investigate the failure root cause by analyzing the component strengths (SG) and the

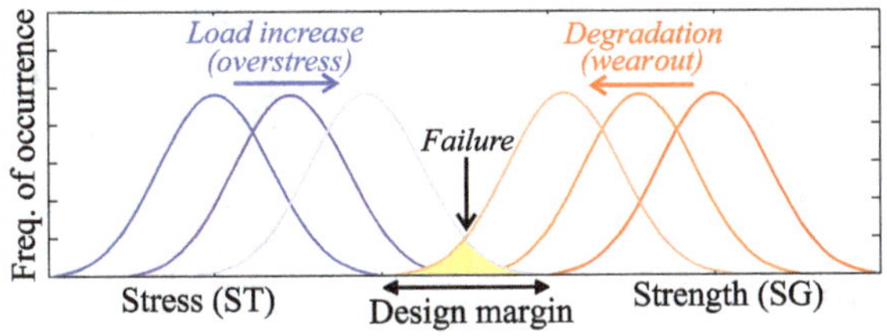

Figure 14.5 Stress-strength frequency of occurrence with failure over time due to overstress and wear-out. Available in [10].

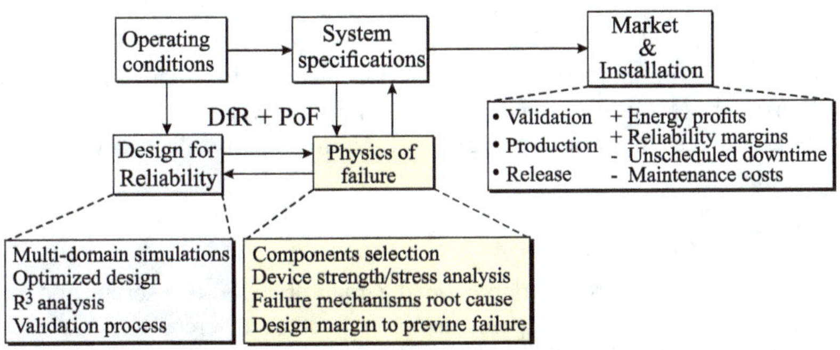

Figure 14.6 Design for Reliability flowchart for power electronic systems

stresses (ST) [9]. For example, given the PDF curves illustrated in Figure 14.5, the strength can be the capacitor's maximum voltage rating and the stress is the capacitor operation voltage. Under a population of various components, there might be variations in the component specification due to uncertainties during the manufacturing process. Therefore, instead of having only one value of the capacitor's maximum voltage rating, we have a probability distribution centered on the average value with a given standard deviation. The same concept can be applied to the capacitor operating voltage, caused by instabilities and operational variances. Nevertheless, the overstress on the operation increases the component stresses and the wear-out degradation reduces the component strength. At a certain point, defined by the failure criteria, both PDF curves will overlap, leading the weaker components to fail.

The previous analysis suggests that the failure can be avoided (or delayed) by either increasing the component strength or reducing its stresses. Therefore, there is a design margin where reliability engineering can act to reduce component failure and increase its lifetime. This is another modern concept, called Design for Reliability (DfR) [11]. By applying the DfR strategy, the system reliability is taken into account at the design stage, reducing the high costs of preventive and emergency maintenance. In addition, the DfR can address the most reliable components for a specific power electronic system under a specific operation, enhancing energy conversion efficiency and reducing costs with unscheduled maintenance and system downtime, as presented in Figure 14.6. The mission profile represents a set of conditions that define the system operation and the load stresses. For example, in the case of photovoltaic (PV) energy systems, the mission profiles are solar irradiance and the ambient temperature, while wind speed can also be considered for wind energy systems.

14.2 Reliability of power semiconductor devices

Semiconductor devices are fundamental in power electronics since they are the basic power conversion unit. Since power electronics became a part of our power system, many concerns regarding reliability have arisen. Due to the several types and structures of semiconductor devices available on the market, this is a broad topic to be approached in a single book chapter. For this reason, this section focuses on insulated-gate bipolar transistors (IGBT), metal-oxide-semiconductor field-effect transistors (MOSFET), and diodes, which are very common in variable-speed motor drives, renewable energy conversion systems, electric vehicles, and power supplies.

14.2.1 Technology overview

Semiconductor devices are available on the market in both discrete and power module packages. Figure 14.7 presents the aspect of discrete IGBT with free-wheeling diode and TO 247 package. Other discrete packages use a similar approach. Since vertical semiconductor devices are common in power electronics*, the connections are made at the top and at the bottom of the chip through bond wires and chip solder, respectively. After manufacturing the device, the epoxy package is built using a mold. This is especially important to protect the device against moisture.

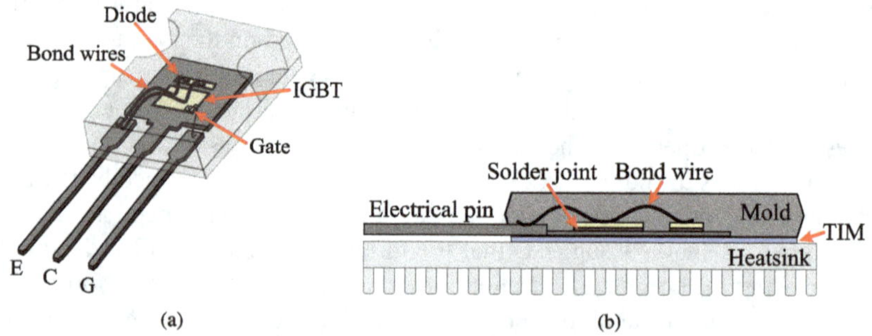

Figure 14.7 Structure of a discrete TO247 IGBT with freewheeling diode. (a) Internal bond wire connections; (b) cross-sectional view of the device.

*Vertical semiconductor devices are preferred in power electronics since this approach improves the heat transfer and current rating in comparison with lateral semiconductor devices.

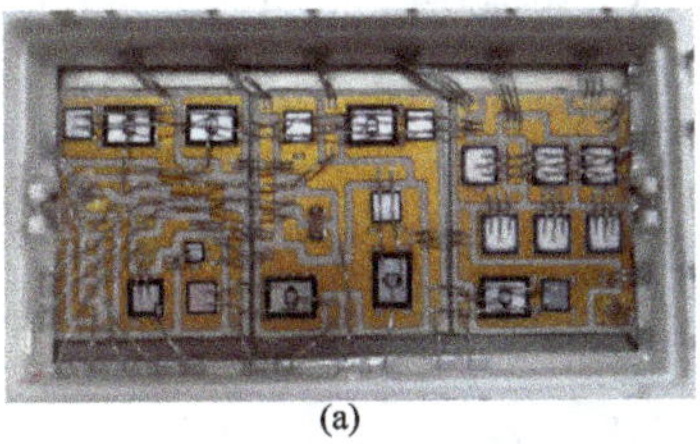
(a)

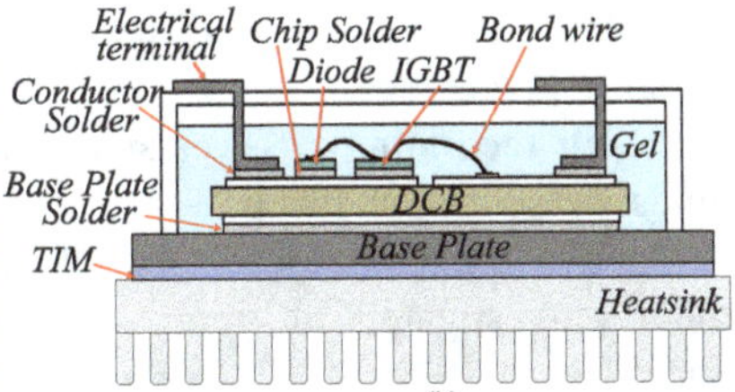

(b)

Figure 14.8 Structure of an IGBT power module. (a) Internal bond wire connections; (b) cross-sectional view of the device.

The aspect of an IGBT module is illustrated in Figure 14.8. Again, bond wires and solder joints are used to provide proper electrical connection. The novelty is the direct-bonded copper (DBC) ceramic substrate, where the devices are soldered. The DBC provides internal conduction paths in the power module (as in printed-circuited boards), and electrical insulation between the die and the heatsink. Then, DBC is soldered to the base plate, which is the surface in contact with the heatsink. The silica gel used in power modules increases the blocking voltage capability (which is significantly reduced if the gel is removed) and protects the devices against moisture.

Regarding the employed materials, aluminum bond wires are common in the industry. The chip solder is based on soft solders, such as tin-silver, indium, or tin-lead. Generally, a thermal interface material (TIM) is employed to fill the gap between the device and the heatsink, improving the heat transfer between two surfaces. That is the principal role of thermal grease. In some discrete power devices, the TIM should be also used to provide proper electrical insulation, since one side of the package is also an electrical connection (usually the collector for IGBTs). For this reason, thermal pads are used. It is also worth mentioning that proper mounting torque is required for efficient heat transfer.

As observed, semiconductor devices are made of several materials with different coefficients of thermal expansion (CTE). Therefore, the device joints (bond wires, chip solder, and base plate solder) are subjected to important thermo-mechanical stresses which can evolve into several failure mechanisms. For this reason, advanced packaging technologies have been developed to improve the power module performance in the following aspects:

- **Stray inductance**: controlling the stray inductance to low values (few nH) is very important since it affects the device voltage stress and contributes to the switching losses. Low-inductance connections play an important role in the performance of wide band-gap semiconductor devices.
- **Improved bond wires**: Aluminum ribbon wires can improve the conduction (electrical and thermal) and improve the surge capability. Another alternative is to employ copper bond wires. This approach can improve the conduction (electrical and thermal), surge capability, and power cycling capability of the

device. However, copper bond wires increase the cost and present a relatively high stray inductance.

- **Die-attach technology**: Silver sintering is an alternative to soft solder compounds, presenting better power cycling performance. Indeed, silver sintering and copper bond wires are interconnection technologies that have been strongly investigated to improve the reliability of silicon carbide power devices.
- **Double-side cooling**: In this case, the interconnection among the dies does not employ bond wires. Instead, two DCBs are employed and added to a transfer-molded package. This approach improves the power density at the system level through a more efficient heat transfer mechanism.
- **Press-pack modules**: This technology also does not employ bond wires. Instead, pressure contacts are adopted. In contrast with traditional power module technology, the press-pack packaging concept has the advantage of failing in short circuits and providing double-side cooling. The short circuit failure mode is useful in high-voltage applications, where several semiconductors are connected in series. In this case, the failure of a single device does not result in the loss of the stack, resulting in a fault-tolerant scheme. The cost and assembly complexity are the main drawbacks of this approach.

Manufacturing cost is the main challenge for the wide adoption of advanced packaging technologies. High-performance markets, such as electric mobility and avionics, have looked at these approaches due to their inherent requirements of high efficiency and power density.

14.2.2 Degradation and typical failure mechanisms

The failures in semiconductor devices can be related to several stressors. Obviously, more complex devices present a higher variety of failure mechanisms. Table 14.1 classifies and summarizes the main failure modes observed in traditional IGBT modules. Some of these failure mechanisms are also observed in MOSFETs and diodes.

Table 14.1 Classification of IGBT failure modes. Adapted from [12].

Classification	**Cause**	**Failure**
Chip-related failure modes	Overstress	Cosmic rays burnout gate overvoltage (voltage breakdown) overcurrent overvoltage (voltage breakdown)
	Wear-out	Electro-chemical migration hot-carrier injection time-dependent dielectric breakdown
Package-related failure modes	Overstress	Mechanical shock thermal runaway and flashovers thermal shock
	Wear-out	Electro and electro-chemical migration insulation degradation thermo-mechanical creep thermo-mechanical fatigue

Overstress failures are related to catastrophic events that lead to operating conditions beyond the device's safe operating area. On the other hand, wear-out failures are related to repetitive stresses that accumulate throughout the device's useful life. Different stressors can be identified. For example, the electro-migration has an important correlation with the current which flows through the device. On the other hand, the time-dependent dielectric breakdown is accelerated by the electric field in the device. In recent years, both academia and industry have conducted research on power semiconductor failure mechanisms. It was found that most of failures observed in semiconductor devices are related to thermo-mechanical stress, which depends on the device temperature profile.

The two most dominant wear-out failure mechanisms are bond wire and solder joint fatigue. Bond wire fatigue is caused by the CTE mismatch, where the thermo-mechanical stress results in plastic deformation during power cycling. This process results in lift-off and heel cracking. Figure 14.9 presents a scanning acoustic microscope (SAM) photo of bond wire indicating the lift-off and heel crack. It is important to note that the bond wire fatigue leads to an increase in the on-state chip voltage $v_{ce,on}$.

The solder joint fatigue is also caused by thermo-mechanical stresses. In this case, fractures, delamination, and voids are developed in the solder alloy. These micro defects can be observed between the die and DBC or DBC and baseplate. Figure 14.10 presents a SAM photo of IGBT chip solder edge delamination and cracks. It is worth mentioning that the solder joint fatigue mainly increases the device's thermal resistance.

It is important to mention that in some applications (e.g., photovoltaic inverters and wind turbines), the device junction temperature cycles present relatively low amplitude ($\Delta T_j < 40°C$). Under this scenario, bond wire fatigue is the dominant failure mechanism [16]. On the other hand, in applications such as avionics, electric vehicles, and locomotives, the power semiconductor device is submitted to high

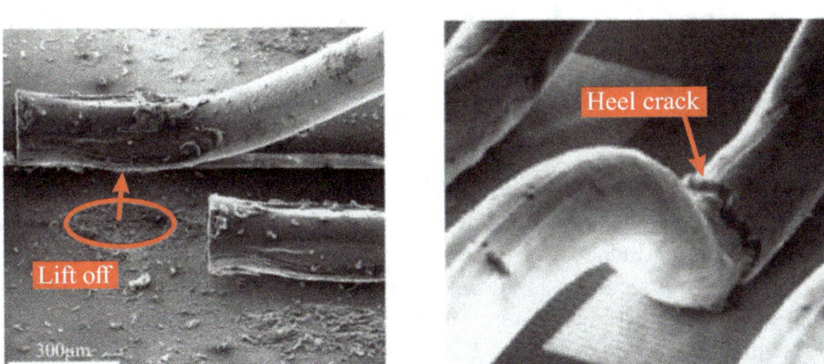

(a) (b)

Figure 14.9 SAM images of an IGBT bond wire: (a) lift-off and (b) heel crack, both available in [13].

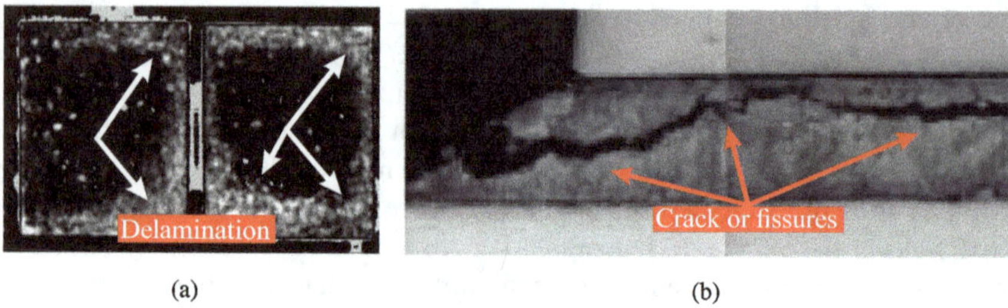

Figure 14.10 SAM images of an IGBT chip solder: (a) upper view of chip solder edge delamination, available in [14] and (b) cross-section of the solder crack propagation, available in [15].

junction temperature in tens of milliseconds. Under such conditions, the analysis of solder joint fatigue is mandatory in reliability analysis [17].

According to a recent industry survey, manufacturers expect power semiconductor devices to operate from 5 to 30 years [8]. In addition, manufacturers must provide warranties for newly launched products. Therefore, accelerated power cycling tests are a common tool to predict the device's lifetime. Basically, the tests accelerate the wear-out and identify the most common failures and their stressors. In addition, data from accelerated tests can be used to identify wear-out mechanisms and develop lifetime models. The setups for semiconductor endurance and reliability testing are defined by IEC 60747-9 and IEC 60747-34 standards.

Methods for monitoring and evaluating bond wire and solder joint fatigues in power semiconductor devices during power cycling tests are described in [18]. During the tests, failure criteria are defined to decouple different failure modes and generate distinct data for lifetime models. It is very important to obtain separate lifetime models for each failure mechanism in such a manner the critical and weak points of a component can be identified. Typically, the failure criteria applied to bond wire fatigues is the increase of 5%–20% in the on-state voltage drop $v_{ce,on}$. On the other hand, an increase in 20%–50% of junction-to-case thermal resistance is the assumed failure criterion for solder joints [1].

14.2.3 Electro-thermal modeling

As discussed above, the main failure mechanism of a power semiconductor device is related to fatigue induced by thermo-mechanical stress, where the temperature cycling during the operation is the main stress factor. Therefore, power loss and thermal modeling are crucial steps in the reliability and lifetime estimation of power semiconductor devices, e.g., IGBT.

Generally, power losses in the IGBT $P_{loss,IGBT}$ consists of conduction loss $P_{c,IGBT}$ and switching loss $P_{sw,IGBT}$ following:

$$P_{loss,IGBT} = P_{c,IGBT} + P_{sw,IGBT}, \tag{14.7}$$

$$P_{c,IGBT} = \frac{1}{T}\int_{t_0}^{t_0+T} v_{ce} i_c(t)\mathrm{d}t, \tag{14.8}$$

$$P_{sw,IGBT} = f_{sw}(E_{on} + E_{off}), \tag{14.9}$$

where t_0 is a given start time and T is the period of the signal, v_{ce} and i_c are the on-state collector-emitter voltage and collector current of the power device, respectively, f_{sw} is the switching frequency, E_{on} and E_{off} are the turn-on and turn-off energies, respectively.

The power loss of the diode can also be determined in a similar way as follows:

$$P_{loss,diode} = P_{c,diode} + P_{sw,diode}, \tag{14.10}$$

$$P_{c,diode} = \frac{1}{T}\int_{t_0}^{t_0+T} v_f i_f(t)dt, \tag{14.11}$$

$$P_{sw,diode} = f_{sw} E_{rec}, \tag{14.12}$$

where v_f and i_f are the forward voltage and current, respectively and E_{rec} is the reverse recovery energy. The turn-on energy can usually be neglected for the diode.

Notably, most of the key parameters in the power loss calculation are temperature-dependent. For instance, the on-state collector-emitter voltage v_{ce}, the collector current i_c, and the gate-emitter voltage v_{ge} of the IGBT are strongly dependent on the chip junction temperature $T_{j,IGBT}$. The same consideration also applies to the diode, where the forward voltage v_f is a function of the diode forward current i_f and chip junction temperature $T_{j,diode}$. The switching energies of the IGBTs E_{on} and E_{off} are also the function of chip junction temperature $T_{j,IGBT}$ in addition to the IGBT collector current i_c, gate resistance r_g, and collector-emitter blocking voltage v_{ce}. Finally, the diode reverse recovery energy E_{rec} is a function of the diode forward current i_f and chip junction temperature $T_{j,diode}$.

The thermal model can estimate the junction temperature of the power devices (e.g., IGBTs and diodes) from the calculated power loss. A typical thermal model is shown in Figure 14.11, which is based on an electro-thermal circuit proposed by [19]. In general, the thermal model is usually based on either the multi-layer Foster

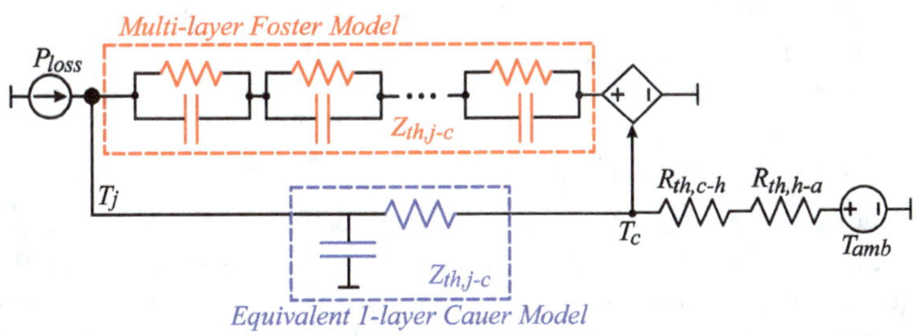

Figure 14.11 Electro-thermal circuit to estimate junction temperature in function of Foster and Cauer networks. Available in [19].

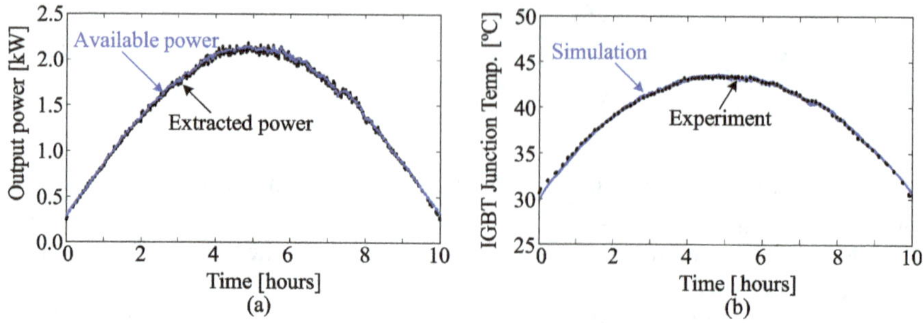

Figure 14.12 (a) One-day power profile applied to the PV inverter. (b) Junction temperature measured experimentally and junction temperature estimated via thermal model.

or Cauer thermal networks, which are employed to determine the transient response of the temperature in the power device for a given power loss condition. The main difference between the two models is that the Cauer model represents a physical structure inside the power device, but it requires the knowledge of material property and geometry for parameterization. On the other hand, the Foster model does not represent the physical structure of the component but its parameters can be obtained from the experimental test. Therefore, it is more common in the datasheet to provide the thermal model parameters based on the Foster model. The thermal impedance based on the Foster model Z_{th} can be analytically expressed as:

$$Z_{th}(t) = \sum_i R_{th,i}\left[1 - \exp\left(\frac{-t}{\tau_{th,i}}\right)\right], \tag{14.13}$$

where $\tau_{th} = R_{th}C_{th}$ and i is the number of nodes in the Foster network.

An example of a thermal stress profile of the IGBT used in a PV inverter is shown in Figure 14.12. In this case, a one-day output power mission profile shown in Figure 14.12(a) is applied to the PV inverter, and the estimated junction temperature is compared with the experimental measurement. From Figure 14.12(b), it can be noticed that the dynamic of the mission profile is clearly reflected in the dynamic of the thermal stress during the operation. Several thermal cycling can be clearly observed. Furthermore, the estimation of the junction temperature of the IGBT matches the experimental result well during the operation, which validates the accuracy of the electro-thermal model.

14.2.4 Examples of lifetime models

In the literature, two types of lifetime models can be found: empirical and physics-based lifetime models. Empirical models are based on a statistical analysis of accelerated power cycling tests. The drawbacks of empirical models include the need for separated models for each failure mechanism and extrapolation that is necessary when the operational conditions are beyond the tested conditions, leading to inaccuracies. On the other hand, physics-based lifetime models compute multi-domain failure mechanisms and provide a more realistic lifetime prediction. However, the complex implementation of these methods makes the empirical models the most popular.

The empirical models are expressed in terms of the number of cycles to failure N_f. Accelerated power cycling tests allow us to determine the dependence of N_f with the cycle parameters, such as junction temperature, heating periods, and chip technology. The most used empirical models for bond wire lift-off lifetime analysis are stated below:

- *Coffin–Mason model*: The simplest lifetime model is based on the Coffin–Mason strain law under the plastic regime. In this model, only the junction temperature amplitude ΔT_j is taken into account:

$$N_f = A(\Delta T_j)^{\beta}, \tag{14.14}$$

  where A and β are curving fitting parameters. The Coffin–Mason lifetime model is the basis of almost every empirical lifetime model available in the literature.
- *Arrhenius model*: This model includes the influence of the mean junction temperature T_{jm}. Accordingly:

$$N_f = A(\Delta T_j)^{\beta} \exp\left(\frac{E_a}{k_b T_{jm}}\right), \tag{14.15}$$

  where E_a is the activation energy and k_b is the Boltzmann constant.
- *Norris–Landzberg model*: This model extends the Coffin–Mason model, introducing the dependence of N_f on the thermal cycling frequency f_c:

$$N_f = A(\Delta T_j)^{\beta_1} f_c^{\beta_2} \exp\left(\frac{E_a}{k_b T_{jm}}\right). \tag{14.16}$$

  where β_1 and β_2 are curve fitting parameters.
- *Bayerer model*: Reference [20] developed a purely statistical lifetime model based on more than 150 accelerated power cycling tests, covering several Infineon package designs. Bayerer model added chip technology information to previous lifetime models, such as current per bond foot I_b, blocking voltage class V, and bond wire diameter D:

$$N_f = A(\Delta T_j)^{\beta_1} \exp\left(\frac{\beta_2}{T_{jm} + 273}\right) (t_{on})^{\beta_3} I_b^{\beta_4} V^{\beta_5} D^{\beta_6}, \tag{14.17}$$

  where β_i, $i = 1$ to 6, are curve fitting parameters. Also, t_{on} is the load pulse duration. The Bayerer model is widespread in the literature, also known as CIPS08, referring to the Conference where it was published.
- *Scheuermann model*: Reference [21] proposed a lifetime model based on power cycling tests performed on the Semikron SKiM63 power module family. The SKiM63 module is based on 1.2 kV solder-free chips, where silver

sintering technology is applied. Moreover, the bond wires are made of aluminum (Al) with 300 μm diameter:

$$N_f = A(\Delta T_j)^{\alpha} \exp\left(\frac{E_a}{k_b T_{jm}}\right)(ar)^{\beta_0+\beta_1 \Delta T_j}\left(\frac{C_d + (t_{on})^{\gamma}}{C_d + 1}\right) f_d, \tag{14.18}$$

where A, α, β_0, β_1, γ and C_d are curve fitting parameters. This model includes the influence of bond wire aspect ratio *ar* in the lifetime estimation. In contrast with other lifetime models, Scheuermann model also allows estimating the diode lifetime using the same equation through the factor f_d (1 for IGBT and 0.6204 for diode).

Regarding solder joint fatigue, the Coffin–Mason power law is often employed. This model is given by:

$$N_f = \frac{1}{2}\left(\frac{\Delta\alpha_s \Delta T_j L_s}{\gamma_s x_s}\right)^{c_s^{-1}}, \tag{14.19}$$

where $\Delta\alpha_s$ is the mismatched CTE between the upper and lower layer, ΔT_j is the temperature cycle amplitude, L_s is the lateral size of the solder joint, c_s is the fatigue exponent, x_s and γ_s are the thickness and the ductility factor of the solder, respectively [13]. Equation (14.19) shows that the fatigue on the solder joint can be minimized by applying materials with similar CTE or reducing the size of the solder joint.

As observed, Bayerer and Scheuermann are the most advanced lifetime models for bond wire lifetime prediction. However, the use of these models is not straightforward, device constructive characteristics are required. For example, the Scheuermann model requires a bond wire aspect ratio (see Figure 14.13), while the Bayerer model requires the number of bond wires and bond feet, information that is confidential and restricted. Therefore, it is important to ensure these parameters do not deviate from the range where the accelerated power cycling tests were

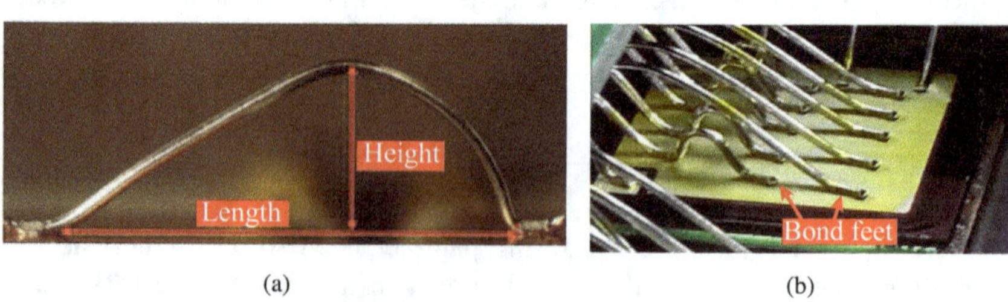

Figure 14.13 Device constructive parameters required by advanced empirical lifetime models: (a) bond wire aspect ratio ($ar = H_{ar}/L_{ar}$); (b) number of bond wires, number of bond feet, and bond wire diameter. Images available in [22] and [23].

Table 14.2 Parameters of Scheuermann lifetime model. Available in [21].

Parameter	Value	Experimental condition
A	2.4368×10^{14}	–
α	-4.923	$64\text{ K} \leq \Delta T_j \leq 113\text{ K}$
β_1	-9.012×10^{-3}	–
β_0	1.942	$0.19 \leq ar \leq 0.42$
C_d	1.434	–
γ	-1.208	$0.07\text{ s} \leq t_{on} \leq 63\text{ s}$
f_d	0.6204	–

Table 14.3 Parameters of Bayerer lifetime model. Available in [20].

Parameter	Value	Experimental condition
A	9.34×10^{14}	–
β_1	-4.416	$45\text{ K} \leq \Delta T_j \leq 150\text{ K}$
β_2	1285	$20°\text{C} \leq T_{jm} \leq 120°\text{C}$
β_3	-0.463	$1\text{ s} \leq t_{on} \leq 15\text{ s}$
β_4	-0.716	$3\text{ A} \leq I_b \leq 23\text{ A}$
β_5	-0.761	$6\text{ V} \leq V/100 \leq 33\text{ V}$
β_6	-0.5	$75\ \mu\text{m} \leq D \leq 500\ \mu\text{m}$

performed. Table 14.2 and 14.3 summarize the Scheuermann and Bayerer lifetime model parameters and their range of values, respectively.

Finally, it is worth mentioning that the presented lifetime models focus on thermo-mechanical stresses, i.e., other effects such as humidity or vibration are not taken into account.

14.3 Reliability of capacitors

Capacitors are one of the most used passive components in power electronic systems. The range of applications is very large, varying from pF (ceramic capacitors) up to kF for double-layer technology and supercapacitors. The operation voltage range is also very large, reaching thousands of volts in certain situations. The main role of the capacitors applied to power electronic systems is energy storage, input/output power balance, filtering, snubbers, etc. However, these components also have a downside, impacting significantly the converter volume and cost. In addition, capacitors are known to be one of the most failure-prone components in the system, which makes their reliability a very important topic to be considered in the Design of Reliability [7,8]. Therefore, this section will discuss in more detail the electro-thermal performance of the most used technologies of capacitors applied to power electronics and their lifetime models.

14.3.1 Technology overview

Among all capacitor applications in a power electronic system, the DC-link may be the one that imposes the highest stresses on these components. This is caused by the energy storage requirements, high current ripples from the switching power semiconductors nature, and elevated operating voltage. The most used technology of capacitors for DC-link applications are: aluminum electrolytic capacitors (Al-Caps), metalized polypropylene film capacitors (MKP-Caps), and multi-layer ceramic capacitors (MLC-Caps). An example of these three capacitor packaging technologies is shown in Figure 14.14.

Table 14.4, adapted from [24], shows a performance comparison of the three types of capacitors in different aspects. Clearly, each technology will have its own advantages and disadvantages among the others. Al-caps usually have good capacitance and energy density while benefiting from the lowest cost per joule. As a drawback, they have reliability issues related to high values of equivalent-series

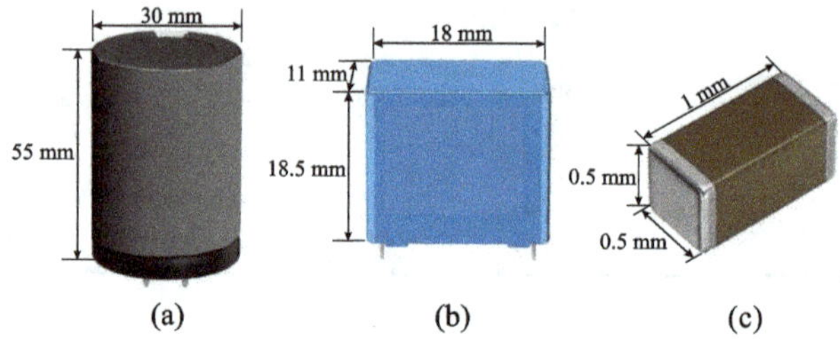

Figure 14.14 Example of three capacitor packaging technologies: (a) electrolytic aluminum snap-in capacitor, (b) metalized polypropylene film capacitors, and (c) multilayer ceramic chip capacitors.

Table 14.4 Performance comparison of three types of capacitors for DC-link applications. Adapted from [24].

Parameter	Al-Caps	MKP-Caps	MLC-Caps
Capacitance	+++	++	+
Voltage	++	+++	+
Volume	++	+	+++
Ripple current	+	+++	+++
ESR	+	+++	+++
Frequency range	+	++	+++
Capacitance stability	++	+++	+
Overvoltage capability	++	+++	+
Temperature range	++	+	+++
Energy density	+++	+	++
Useful life	+	+++	+++
Cost per joule	+++	++	+

resistance (ESR). On the other hand, one can say that MKP-Caps is the opposite, with good reliability but low energy density resulting in large and costly capacitor banks. The MLC-Caps are the smallest among the three technologies and can be used in a wider range of frequency and voltage. However, they suffer from temperature and mechanical stresses, in addition to a poor capacitance range and energy density.

14.3.2 Degradation and typical failure mechanisms

Despite the differences, all three technologies of capacitors are mainly stressed by the same parameters: operating voltage, temperature, and current ripples. The wear-out process occurs due to the power losses that are dissipated in the ESR. This creates internal hot-spot temperatures which can lead the components to failure over time. More details about the electro-thermal modeling of the capacitors will be addressed in the next subsection. However, there are some particularities between the technologies, for example, related to the dominant failure modes and mechanisms.

Aluminum electrolytic capacitors normally fail in open circuit mode, due to electrolytic loss or poor connection on the terminals. On the other hand, a short-circuit failure mode can also occur due to the dielectric breakdown of the oxide layer [25]. There is another common failure mode, known as an open vent, caused by the increase in internal pressure. Since Al-Caps suffer from high ESR values, the power losses dissipated in the component cause an increase in the internal temperature, leading to the apparition of hot spots that accelerate the degradation process of the electrolyte [26]. This electrochemical reaction intensifies the electrolyte vaporization, leading to an increase in the ESR and a reduction in the equivalent capacitance, which also accelerates the degradation process. According to the manufacturer, the failure is achieved when the capacitor reaches a reduction of 20% in the equivalent capacitance or an increase of two times the initial ESR.

The film capacitors typically fail in open circuits as well, due to self-healing dielectric breakdown, connection instability, and reduction in electrode area caused by oxidation (moisture) [27]. The absorption of humidity by the film can also cause a short-circuit failure [28]. It is important to note that MKP-Caps are stressed by humidity, so these kinds of components cannot be installed in very harsh environments without taking carefully this factor into account in the Design for Reliability. The failure criteria are defined by the manufacturers as a 5% deviation on the initial equivalent capacitance or an increase of three times the ESR.

Different from the other capacitors, MLC-Caps normally fail in short-circuit mode, caused by dielectric breakdown or damage on the component body due to mechanical vibration [29]. Therefore, DC-link applications using these capacitors must carefully investigate the bank's reliability to avoid dangerous operation due to its failure mode. Despite that, the dielectric of ceramic capacitors is expected to last for much more time than Al-Caps or MKP-Caps, thus its wear-out process is not much of an issue. Finally, the failure criteria defined by the manufacturers is a drift of 10% on the equivalent capacitance, an increase of two times the initial ESR, or a reduction on the equivalent parallel resistance R_p (related to leakage current).

Table 14.5 Comparison of failure types and critical stressors of three types of capacitors. Adapted from [24].

Capacitor type	Dominant failure mechanisms	Most critical stressors	Failure criteria
Al-Caps	Electrolyte vaporization, electrochemical reaction	T_{amb}, V_c, I_c	C: 20% reduction ESR: 2 times
MKP-Caps	Moisture corrosion, dielectric loss	T_{amb}, V_c, humidity	C: 5% reduction ESR: 3 times
MLC-Caps	Insulation degradation, flex cracking	T_{amb}, V_c, vibration	C: 10% reduction ESR: 2 times $R_P < 10^7 \Omega$

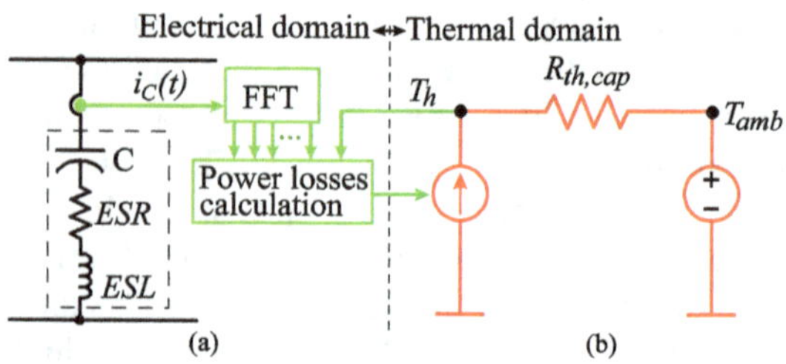

Figure 14.15 (a) Capacitor equivalent electrical circuit model. (b) Capacitor thermal model, fed by power losses computation using Fast Fourier Transform (FFT).

Table 14.5 provides a summarized table with the dominant failure mechanisms and most critical stressors, adapted from [24].

14.3.3 Electro-thermal modeling

Similar to the power semiconductor device, electro-thermal modeling is also essential for power loss and temperature estimation of capacitors. In the case of capacitors, the power loss is mainly generated from the ripple current. An equivalent electrical circuit model of the capacitor is shown in Figure 14.15(a), which consists of the primary capacitance *C*, the equivalent series resistance *ESR*, and the equivalent series inductance *ESL*. The power loss in the capacitor is mainly due to the equivalent series resistance *ESR* (resistance of the dielectric material, connection pins, and lead terminals). On the other hand, the equivalent series inductance *ESL* represents the capacitor inductance of the winding layers and connected power terminals [30]. The combination of these elements results in the frequency-dependent impedance characteristic of the capacitor, which can be determined as:

$$Z_{cap}(f) = ESR(f) + j\left(2\pi f ESL - \frac{1}{2\pi f C}\right) \tag{14.20}$$

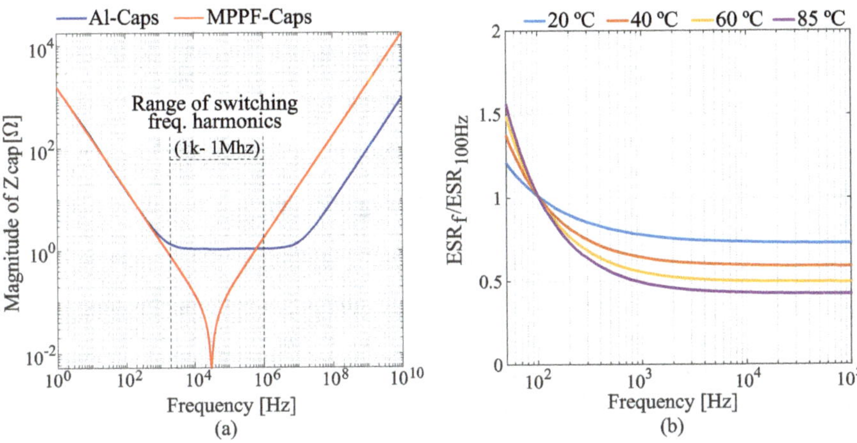

Figure 14.16 (a) Equivalent magnitude impedance comparison of an Al-Cap and MKP-Cap as a function of frequency. (b) Frequency and temperature characteristics of ESR from aluminum electrolytic capacitor part number TDK B43630.

It is worth mentioning that different capacitor types may have different impedance characteristics, as shown in Figure 14.16(a). In these curves, the Al-Cap is 100 μF/450 V (part number TDK B4360) and the MKP-Cap is 100 μF/450 V (part number TDK B32674). In the case of film capacitors, the equivalent series resistance *ESR* at the low-frequency region is usually higher than that of the electrolytic capacitors. Nevertheless, their influence on the total impedance is relatively small compared to the capacitive reactance, which makes Z_{cap} almost the same for both curves in the low-frequency region. On the other hand, film capacitors have lower Z_{cap} when compared to electrolytic capacitors in the medium-frequency region (in the range of switching frequency harmonics). Therefore, a hybrid combination of film and electrolytic capacitors is often used for the DC-link application to minimize power loss and enhance reliability.

Ripple current in the capacitor contributes to the power loss due to the equivalent series resistance *ESR*. The characteristic of the capacitor equivalent series resistance *ESR* is strongly affected by both frequency and temperature conditions. Therefore, the total power loss in the capacitor is the sum of the power loss generated by the ripple current at each frequency f (under a certain operating temperature *Th*) following [31]:

$$P_{loss,cap} = \sum_{i}^{h} ESR(T_h, hf) I_{cap,rms}(hf)^2. \tag{14.21}$$

The power loss model is then coupled with the thermal model as illustrated in Figure 14.15(b). Here, it can be seen that the hotspot temperature of the capacitor (e.g., the core temperature) is influenced by both the power loss in the capacitor,

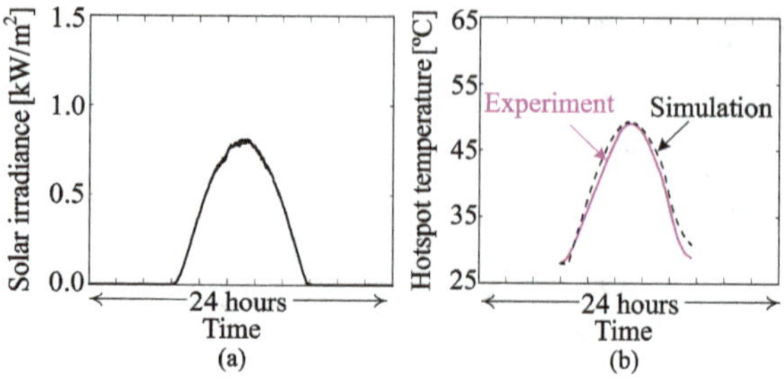

Figure 14.17 (a) One-day solar irradiance profile applied to the PV inverter. (b) Hot-spot temperature measured experimentally and hot-spot temperature estimated via a thermal mode.

the thermal resistance $R_{th,cap}$, and the ambient temperature T_{amb}

$$T_h = P_{loss,cap}R_{th,cap} + T_{amb}, \tag{14.22}$$

where h is the harmonic order of the fundamental frequency f_0 and $R_{th,cap}$ is the capacitor thermal resistance, defined by the function of the capacitor dimensions [30]. Usually, the feedback loop between the power loss and the hot-spot temperature calculation is required, to include the temperature-dependent *ESR* in the power loss calculation, as it is shown in Figure 14.16(b) for an aluminum electrolytic capacitor manufactured by TDK (part number B43630).

An example of the hotspot temperature estimation of an electrolytic capacitor used in DC-link for PV inverter is shown in Figure 14.17. In this case, a one-day mission profile is considered, as shown in Figure 14.17(a). It can be noticed in Figure 14.17(b) that the hotspot temperature of the capacitor has a much slower dynamic than the mission profile (e.g., power loading) of the component. This is mainly due to the relatively high thermal inertia of the capacitor packaging, where the time constant is in the range of hours. Nevertheless, the estimated hotspot temperature is well aligned with the experimental measurement, which verifies the accuracy of the model.

It should be pointed out that the electro-thermal circuit in Figure 14.15(b) can be applied for the power losses on electrolytic and film capacitors. However, for certain ceramic capacitors, i.e., class II ceramic capacitors, this model is inadequate due to their ultra-high dielectric constants at the cost of non-linear voltage-dependent capacitance value. Therefore, low-frequency large-signal excitation can cause substantial device heating, which is not represented in the circuit in Figure 14.15(b). For this type of capacitor, Menzi *et al.* [32] propose a new normalized Steinmetz model to predict more accurately the power losses on ceramic capacitors.

14.3.4 Examples of lifetime models

The most generic and accepted lifetime model applied to all three capacitor technologies discussed in this chapter is presented below, in the function of the temperature dependence and voltage stress:

$$L = L_0 \left(\frac{V_c}{V_0} \right)^{-n} \exp \left[\left(\frac{E_a}{k_b} \right) \left(\frac{1}{T_h} - \frac{1}{T_0} \right) \right], \tag{14.23}$$

where L is the capacitor lifetime, T_h is the hot-spot temperature, and V_c is the operating voltage. E_a is the activation energy, n is the voltage stress exponent, and k_b is the Boltzmann's constant (8.62×10^{-5}). The subscript index 0 in these variables denotes the rated values provided by manufacturers. For the sake of comparison, the rated useful life L_0 for TDK aluminum electrolytic capacitor (part number B43630) under nominal voltage V_0 and T_0 equal to 85°C is 2,000 h while TDK metalized polypropylene film capacitor (part number B32674) is 200,000 h. This shows how MKP-Caps have higher reliability when compared to Al-Caps, as previously mentioned. The coefficients E_a and n chosen are different for each capacitor technology, depending on the dielectric material and dimensions. A more simplified lifetime model can be described as follows, where $E_a = 0.94$ eV and n ranges from 7 to 9.4 for MKP-Caps and 3 to 5 for Al-Caps [33]:

$$L = L_0 \left(\frac{V_c}{V_0} \right)^{-n} 2^{\left(\frac{T_0 - T_h}{10} \right)} \tag{14.24}$$

The simplified model indicates that, for the same operating voltage, the capacitor lifetime is reduced by half for each temperature increase of 10°C. Additionally, this model considers only low-voltage stress [34]. It is important to note that the presented lifetime models take into account only the electro-thermal stresses. This means that other parameters such as mechanical vibration and humidity are not considered. For this purpose, a more complex multi-physics model is required to enhance reliability and accuracy.

14.4 Power converter reliability estimation

14.4.1 Damage accumulation models

The lifetime models presented in the previous sections compute the expected number of cycles (for semiconductors) or operating time (for capacitors). Supposing a mission profile with variable operating conditions, how should we proceed to compute the equivalent number of cycles/operating time? This question is answered through damage accumulation models.

In the case of semiconductors, the damage (or life consumption, LC) caused in the device for a known operating condition i is given by:

$$LC_s = \frac{n_i}{N_{f,i}} \tag{14.25}$$

where LC_s is the semiconductor life consumption, n_i is the number of thermal cycles at the condition i and $N_{f,i}$ is the number of cycles to failure computed through the lifetime model.

For capacitors, the damage caused by a known operating condition j is given by:

$$LC_c = \frac{t_j}{L_j} \tag{14.26}$$

where t_j is the capacitor operating time at the condition j and L_j is the time to failure computed through the lifetime model.

The damage definitions aforementioned allow us to conclude that when $LC_i = 1$, the device reaches the end of life. The challenge is how to combine the damages under different conditions. In the fatigue analysis literature, several damage accumulation models were developed [35]. The most simple approach is the *linear damage accumulation*, also known as the Palmgren–Miner rule. In this case, the total damage caused by a mission profile with sample time t_{mp} is computed as follows:

$$LC_s = \sum_i \frac{n_i}{N_{f,i}}, \tag{14.27}$$

$$LC_c = \sum_j \frac{t_{mp}}{L_i}. \tag{14.28}$$

However, one can ask about the validity of linear damage accumulation for semiconductors and capacitors. Recently, some studies have discussed this point. Reference [36] presents detailed experimental analyses for semiconductors while reference [37] presents a discussion on capacitors. Both references indicate that the Palmgren–Miner rule leads to conservative results, i.e., the expected damage in practice is lower than the estimation given by Equations (14.27) and (14.28).

At this point, the following question arises: the lifetime models for semiconductors require specific power cycle parameters such as average temperature T_j, cycle amplitude ΔT_j, and heating time t_{on}. In the general case, how can we obtain these parameters from a given mission profile? Indeed, due to the irregular thermal cycles presented in the junction temperature profile, a cycle counting algorithm must be used. The goal is to convert it into a set of cycles with constant parameters (average temperature, cycle amplitude, and heating time). Among the possible solutions, the Rainflow algorithm is commonly employed in fatigue analysis [38].

The Rainflow algorithm was first created for fatigue stress/strain computation in mechanical applications. In these applications, the resulting fatigue is unaffected by the period of time that a stress/strain force is applied. However, the heating time affects fatigue in power devices. Therefore, the traditional Rainflow algorithm cannot be used to characterize the thermal cycles, unless modifications are implemented. Reference [39] describes how to implement a modified Rainflow function to compute the effective heating time from the original temperature profile data.

14.4.2 Uncertainties and Monte Carlo simulation

The Palmgren–Miner rule estimates the accumulated damage for a given mission profile. Let us suppose a 1-year mission profile. Since the damage accumulation is linear, the inverse of the life consumption (LC) gives an estimation of a lifetime. For example, if the damage is 0.1, the estimated lifetime is 10 years. However, in practice, this reasoning leads to significant errors, because the stress in the devices is not identical due to the manufacturing process and stress variation. Therefore, one capacitor can present higher damage than another due to the tolerances. Thus, lifetime is usually expressed in terms of statistical values (probabilities) rather than a constant value.

Therefore, a statistical analysis is required in the lifetime evaluation procedure. For this task, three steps are necessary:

1. Computation of static lifetime parameters.
2. Monte Carlo Simulation.
3. Unreliability estimation.

For semiconductors, the stochastic parameters $T_{j,m}$, ΔT_j, and t_{on} are converted into equivalent deterministic static values, denoted by $T'_{j,m}$, $\Delta T'_j$, and t'_{on}. For capacitors, the stochastic parameters T_h, and V_c are translated to the static values T'_h, and V'_c. The translation procedure is not unique and we do recommend the methodology proposed by [40]. The physical interpretation of the static values for semiconductor devices is presented in Figure 14.18. The static value is a power cycling with constant parameters (amplitude, average value, and heating time) which provides the same LC computed by the Palmgren–Miner rule.

Once the equivalent static values have been obtained, the Monte Carlo simulation can be performed with a population of k samples. The Monte Carlo simulation computes the number of cycles to failure considering a normal distribution for the values of the lifetime model parameters and thermal loading parameters. For the latter, the mean value is its own static value. The standard deviation can be obtained from sensitivity analyses, as described in [40].

In the final step, the lifetime distribution obtained from the Monte Carlo simulation is fitted with the Weibull PDF $f(t)$, given by:

$$f(t) = \frac{\beta}{\eta^{\beta}} t^{(\beta-1)} \exp\left[-\left(\frac{t}{\eta}\right)^{\beta}\right] \tag{14.29}$$

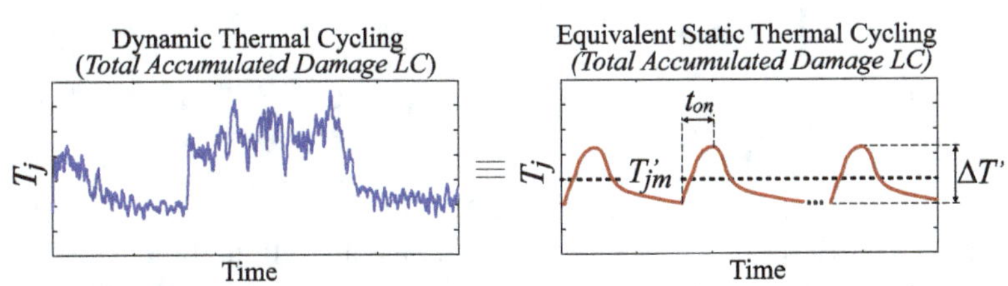

Figure 14.18 Translation from dynamic mission profile to static values for semiconductor devices

where β is the shape parameter, η is the scale parameter, and t is the operation time. Then, the cumulative density function (CDF) is computed by:

$$F(t) = \int_0^t f(t)\mathrm{d}t \tag{14.30}$$

In the context analyzed, the function $F(t)$ is also known as the unreliability function, i.e., $F(t) = U(t)$, and represents the proportion of population failure for a given time.

14.4.3 From component to the system

After computing all component individual reliability (or unreliability) functions, one can ask the obvious question. How to combine them to obtain system-level reliability? The answer, however, is not that simple. The system reliability function will depend on how the components are associated in a probability block diagram and also if there is any dependence or correlation between them. For example, after the failure of a certain component, the reliability of the others will remain unaffected or will they be more stressed? The more dependence on the component reliability, the more complex the system reliability is. From now on, we will assume that all component reliability functions are independent, for the sake of simplicity. In addition, similar components will have the same reliability function.

Typically, the components or subsystems that make part of a bigger system can be arranged in a probability diagram block in the form of: series associations, parallel associations, combined associations, or complex systems.

- **Series association**: is the simplest way to combine the components or subsystems to form a system. In this type of combination, the failure of one component, regardless of which one, is sufficient to cause the failure of the entire system. Figure 14.19(a) shows the referred block diagram. There is no redundancy and the order of component failure makes no difference. In addition, the more the number of series-associated components, the lower the system reliability (or the higher the system unreliability), as shown in Figure 14.20(a). In conclusion, the system reliability will always be lower than the reliability of the weaker component. The reliability R(t) and unreliability U (t) functions can be determined as follows:

$$R_{sys}(t) = \prod_{i=1}^{n} R_{comp,i}(t) \tag{14.31}$$

$$U_{sys}(t)= 1 - \prod_{i=1}^{n} \left(1 - U_{comp,i}(t)\right) \tag{14.32}$$

- **Parallel association**: is also known as redundant configuration, and it is used to increase the system reliability by duplicating the components that perform the same activity or function. Therefore, the components' individual reliability functions are associated in parallel in the probability block, as shown in

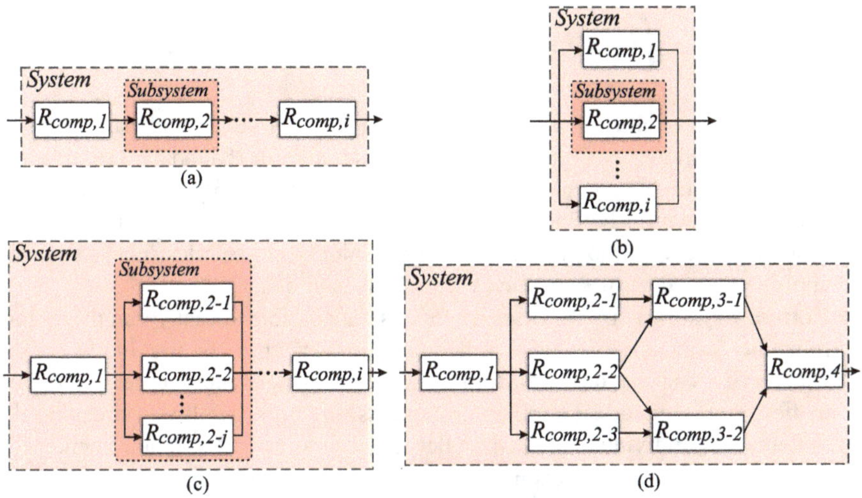

Figure 14.19 Probability block diagrams: (a) series association, (b) parallel association, (c) combined association, and (d) complex systems.

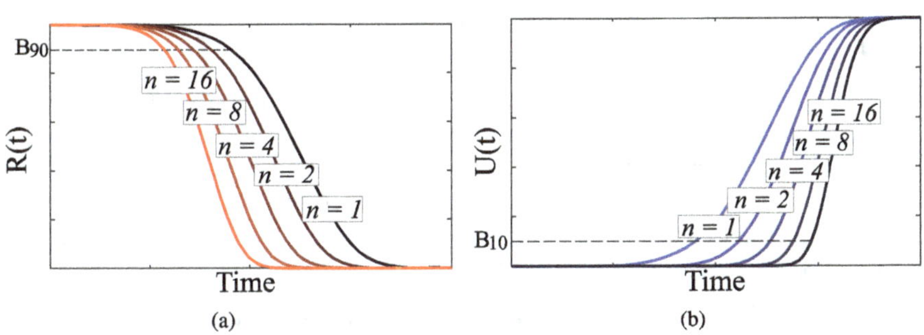

Figure 14.20 Reliability CDF with n components in the system: (a) series association and (b) parallel association.

Figure 14.19(b). Contrary to the series association, in this case, the failure of any duplicated component does not lead to the system failure, due to redundancy. The system will remain healthy until at least one component is still operative. Therefore, the reliability of the system increases with the number of parallel-associated components, as shown in Figure 14.20(b). Because of that, the system reliability will always be higher than the reliability of the stronger component. The reliability R(t) and unreliability U(t) functions can be determined as follows:

$$R_{sys}(t) = 1 - \prod_{i=1}^{n} \left(1 - R_{comp,i}(t)\right) \tag{14.33}$$

$$U_{sys}(t) = \prod_{i=1}^{n} U_{comp,i}(t) \tag{14.34}$$

- **Combined association**: are a combination of both series and parallel associations, as shown in Figure 14.19(c). In fact, adding redundancy to all components might be complex and expensive. Instead, only the most critical components would be doubled. The easiest way to find the reliability of the system is to divide the probability diagram block into smaller subsystems and apply the concepts of series and parallel associations.
- **Complex systems**: are the cases where it is not possible to easily find the system reliability using the previous methodology, as shown in Figure 14.19(d). The complexity can be derived from component dependency, order of failure mattering, non-identical similar components, the impossibility of breaking the system into individual series-parallel configurations, the requirement of a specific number of components succeeding to the system success, etc. In these cases, there are other ways to solve the problem, such as k-out-of-n parallel configurations, consecutive k-out-of-n redundancy, reliability of nonidentical components, decomposition method, event space method, path-tracing method, and so on.

14.4.4 Step-by-step procedure

This subsection will summarize all the steps to determine the system reliability of a generic power electronic converter, considering semiconductors and capacitors as the system components. If other components must be added to the reliability computation, the procedure is the same, except for the lifetime models that were not provided by this chapter. The step-by-step procedure is the following, according to Figure 14.21:

1. **Electrical simulation:** It is important to have an electrical simulation that represents the real system operation with good accuracy. Several engineering software can be used. The mission profiles are the input of the electrical simulation, and the power electronic converter electric variables (current, voltage, power losses, etc.) are the outputs.
2. **Power losses calculation:** As discussed in this chapter, power semiconductors (IGBT, MOSFET, or diodes) and capacitors (electrolytic or film) are the critical components in reliability analysis. In the case of power semiconductors, it is necessary to compute the conducting and switching losses. On the other hand, the power losses on the capacitors are dissipated on the equivalent series resistance as a function of the current and its harmonics. Power loss lookup tables can be an intelligent solution to interpolate different operating points. The dimensions of the lookups will depend on the number of input variables.
3. **Electro-thermal analysis:** This is an interface between the electrical and the thermal domain. The goal is to convert the electric power losses into the component thermal loading, which will be the input of the lifetime models in the next step. It is worth mentioning that the steady-state point of the electrical

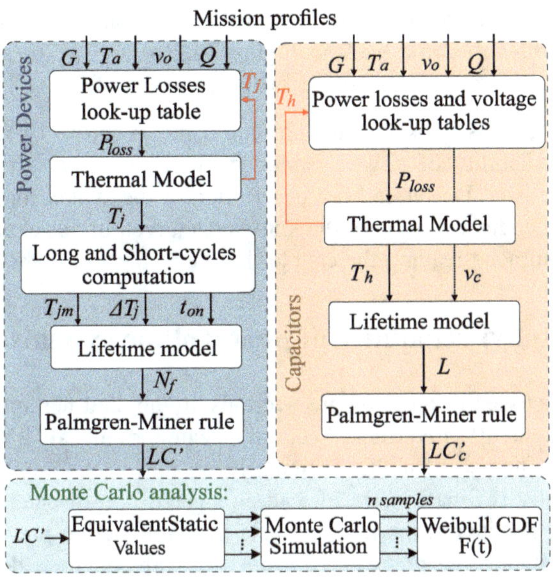

Figure 14.21 Flowchart for the reliability evaluation of power devices and capacitors

simulation is normally acceptable so the power losses would be stabilized. This can help reduce the electrical simulation complexity and time consumption.

4. **Static lifetime computation:** The thermal loading is translated to lifetime consumption by terminating the number of cycles to failure in the case of semiconductors and the number of operating hours in the case of capacitors. As a reminder, the semiconductors will experience short-cycle and long-cycle thermal loading, which means that their lifetime will be the sum of the damage, considering linear accumulation. We suggest using the Rainflow algorithm for long-cycle counting.
5. **Monte Carlo simulations:** The static lifetime computation considers that ideally, all similar components would fail at the same time. In practice, this is not true due to uncertainties and parameter variations during product fabrication. Therefore, parameter variations are inserted in the Monte Carlo simulations to obtain a failure probability density function and then integrate it into the component reliability (or unreliability) cumulative density function. Since we can have a large population sample, the Monte Carlo simulation must be performed with the equivalent deterministic static parameters.
6. **Component and system reliability:** Finally, the final step is to associate all components in a system-level probability diagram block (in series, parallel, or combined associations). After the system reliability function is obtained, a very widespread statistic metric can be applied: the Bx%. This means that the system lifetime can be determined by the time when x% of the samples have failed (in the case of unreliability function) or are still in operation (in the case of

reliability function). Normally B1, B5, or B10 are usually applied. The authors invite the reader to come back to Figure 14.2(b) and try to use the recently acquired knowledge to interpret better the concepts of Bx%.

The authors of this book chapter provide some generic MATLAB codes for calculating the reliability of power semiconductors and capacitors. They can be downloaded from [41]. Despite a brief description of the syntax for using the functions, using these codes will require the knowledge given in this chapter. Therefore, we strongly recommend reading the chapter before trying to use the codes.

14.5 Case study: reliability of photovoltaic systems

The concepts discussed in the previous sections are applied herein in a case study with multifunctional PV inverters. Accurately categorizing all PV inverter technologies is exceedingly difficult, particularly as new technologies may evolve. There are basically two categories of grid-connected non-dispatchable PV inverters: (*i*) standard PV inverter, which extracts available active power from the PV array and feeds it into the grid. The current firmware of these devices does not allow inverters to perform ancillary services; and (*ii*) multifunctional PV inverter, which extracts available active power from the PV array, and the idle power capability is used to perform ancillary services. Despite the non-dispatchability of active power, they can be equipped with communication infrastructure and can contribute to coordinated control strategies.

This section will outline the major effects of performing ancillary services by multifunctional PV inverters and then discuss the control-based strategies for PV inverter reliability improvement.

14.5.1 Effects of the ancillary services

The multifunctional PV inverters can provide multi-task ancillary services to the grid, such as reactive power [42] or power factor (PF) compensation [43], harmonic current compensation (HCC) [44], volt-var and volt-watt capabilities [45], and others. Previous studies already showed that the multifunctional operation shortens the PV inverter lifetime, due to increased electro-thermo-mechanical stress when providing ancillary services.

For illustrative purposes, Figure 14.22 shows the thermal loading of PV inverter components when performing reactive power compensation. Figure 14.22 (a) shows the top view of commercial PV inverter hardware. Initially, the PV inverter is processing 500 W. Afterwards, the inverter compensates 500 var, additionally to the previous 500 W. Figures 14.22(b) and (c) show infrared images under 500 W and 500+500 var loading conditions, respectively. As noted, the reactive power injection increases the average inverter's internal temperature. With this undesirable effect in mind, a case study applying the previous reliability assessment procedures will be conducted for the multifunctional two-stage PV system shown in Figure 14.23(a) performing reactive power compensation. The DC

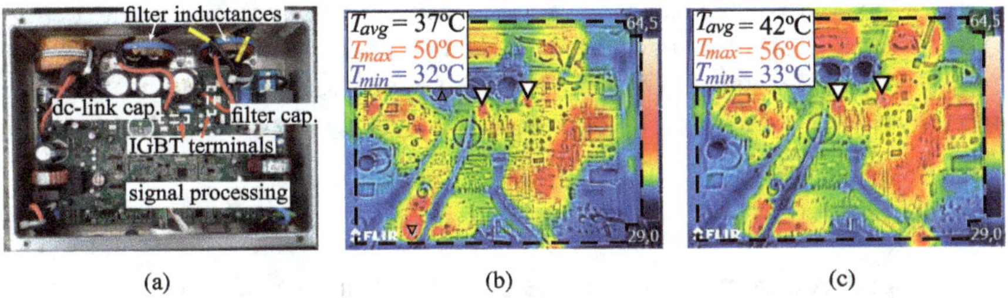

Figure 14.22 (a) Commercial PV inverter hardware. (b) Thermal loading when the inverter is processing 500 W, and (c) thermal loading when the inverter is processing 500 W + 500 var.

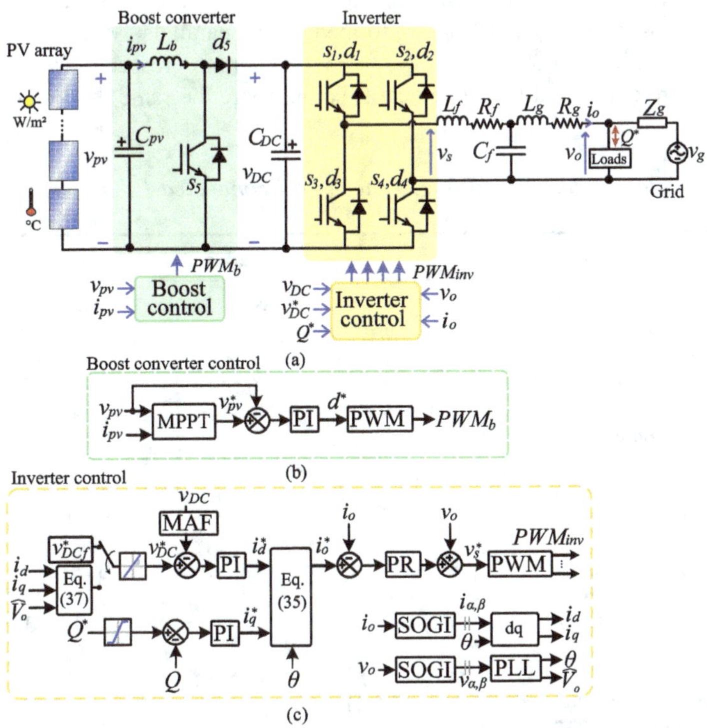

Figure 14.23 (a) Two-stage single-phase grid-connected PV system. (b) Control diagram of the boost converter. (c) Control diagram of the inverter.

stage is composed of a PV array and boost converter, where the maximum power point tracking (MPPT) perturb and observe (P&O) algorithm is responsible for maximizing energy extraction from the PV modules. Figure 14.23(b) shows the boost converter control diagram, with a voltage loop based on a proportional-integral (PI) controller. The second stage comprises a single-phase H-bridge inverter to suppress the harmonic components produced by the switching. Figure 14.23(c) shows the inverter control diagram, composed of cascaded loops: slow outer reactive power (Q) and DC-link voltage loops (v_{DC}); and fast inner current loop. The PI-based outer loops compute i_d^* and i_q^* current references applied to the inner loop. The inner loop current reference i_o^* is calculated by:

$$i_o^* = i_d^* \cdot \cos(\theta) - i_q^* \cdot \sin(\theta), \tag{14.35}$$

where the grid voltage phase angle θ is extracted using a SOGI-PLL structure. A proportional-resonant (PR) controller is used to track i_o^*. The parameters of the PV system employed in this and in Section 14.5.2 are shown in Table 14.6. The full-bridge inverter and boost converter in this section employ power devices rated at 20 A/600 V, manufactured by Infineon, part number IKW20N60T. Capacitor banks are composed of snap-in aluminum electrolytic capacitors manufactured by TDK, part number B43522. C_{pv} and C_{DC} are formed of two parallels 390 μF/420 V (1×2) and three parallels 1,000 μF/450 V (1×3) capacitors, respectively. All semi-conductors are coupled at the same heat sink and share the same thermal network.

Table 14.6 Parameters of the single-phase grid-connected PV system

Parameter	Label	Value
Converter rated power (Section 14.5.1, Section 14.5.2)	S_n	5 kVA, 3 kVA
Boost converter inductance	L_b	980 μH
Boost converter internal resistance	R_b	38.65 mΩ
Boost converter/inverter switching frequency	f_{sw}	20 kHz
Sampling frequency	f_s	40 kHz
LCL filter inductance	L_f, L_g	2 mH
LCL filter internal resistance	R_f, R_g	300 mΩ
LCL filter capacitance	C_f	0.1 μF
Grid voltage (RMS)	V_o	220 V
Grid fundamental frequency	f_n	60 Hz
Grid angular frequency	ω_n	377 rad/s
Fixed DC-link voltage (Section 14.5.1, Section 14.5.2)	v_{DCf}^*	400 V, 385 V
Power devices (IKW20N60T, Section 14.5.1)	s_{1-5}, d_{1-5}	20 A/ 600 V
Power devices (IKW40N60H3, Section 14.5.2)	s_{1-5}, d_{1-5}	40 A/ 600 V
DC-link capacitor (B43522, 1,000 μF/450V, Section 14.5.1)	C_{DC}	3.0 mF (1×3)
DC-link capacitor (B43630, 390 μF/450V, Section 14.5.2)	C_{DC}	1.17 mF (1×3)
Boost converter capacitor (B43522, 390 μF/ 420V, Section 14.5.1)	C_{pv}	780 μF (1×2)
Boost converter capacitor (B43630, 180 μF/ 450V, Section 14.5.2)	C_{pv}	180 μF (1×1)
Heatsink thermal resistance (Section 14.5.1, Section 14.5.2)	R_{ha}	0.3 K/W, 0.5 K/W

Safe operation is guaranteed through dynamic power saturation to avoid exceeding the inverter-rated power S_n. Also, the reactive power reference is limited according to:

$$Q^* \leq \sqrt{S_n^2 - P_{pv}^2}, \tag{14.36}$$

where P_{pv} is the intermittent active power available in the PV array. If $P_{pv} < S_n$, the PV inverter has the idle capability to compensate reactive power. On the other hand, $P_{pv} = S_n$ indicates that the multifunctional PV inverter operates as a standard device.

The increase in real-field experience and technological development of real-time monitoring systems allows the availability of mission profile data from electronic power systems. The use of real-field data allows for predicting the inverter lifetime more precisely. Figures 14.24 (a)–(c) show the adopted 1-year mission profiles of solar irradiance and ambient temperature sampled at 1 min from Aalborg–Denmark (AAL), Goiânia–Brazil (GOI), and Izaña–Spain (IZA), respectively. These three locations have different meteorological characteristics, with low temperature and solar irradiance profiles in AAL, high temperature and solar irradiance profiles in GOI, and low temperature and high solar irradiance profiles in IZA. The latter mission profile is considered ideal for PV system energy generation. Figures 14.24 (d) and (e) show the grid voltage and load-demanded reactive power mission profiles, employed in this and the next section.

Figure 14.21 shows the generic procedure adopted to translate the 1-year mission profile of ambient temperature (T_a), solar irradiance (G), and reactive power (Q) in device thermal loading, and then into system lifetime. The device

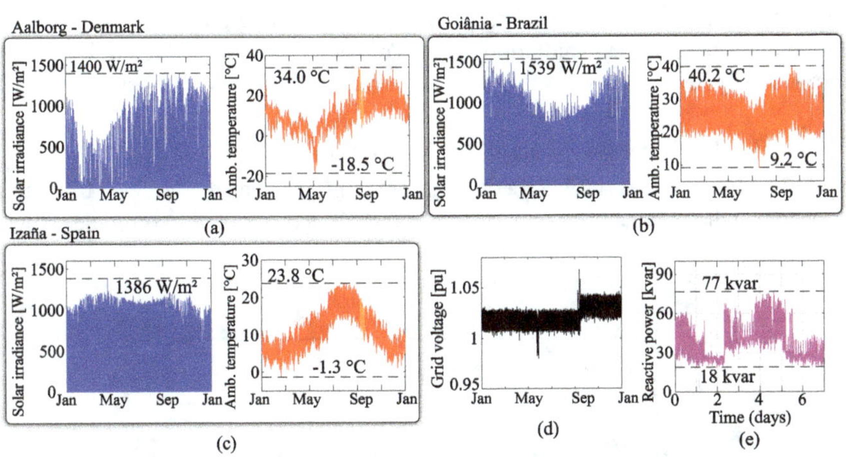

Figure 14.24 Mission profiles sampled at 1 min: solar irradiance and ambient temperature from (a) Aalborg–Denmark, (b) Goiânia–Brazil, and (c) Izaña–Spain. (d) Grid voltage, and (e) reactive power demanded by the load.

lifetime under such a mission profile is obtained by applying the appropriate stress interpretation to a dedicated lifetime model. Losses are estimated for each mission profile (MP) point by means of look-up tables, whose input parameters are: T_a, G, Q, and the feedback junction temperature/hot-spot temperature. The al-cap and power device losses are applied to a thermal model to estimate the thermal loading of PV inverter components. Then, the accumulated damage can be estimated over 1 year through the lifetime models discussed earlier and by Miner's rule. This estimated 1-year damage accumulation is fixed, which is far from reality due to deviations from model parameters and different thermal stress experienced by the PV components over the years in real-field operation. Therefore, a statistical approach based on Monte Carlo simulation is conducted according to the steps indicated in Figure 14.21. Parametric variations are introduced, modeled by a normal distribution with a 99% confidence interval. The Monte Carlo analysis is simulated for $n = 5,000$ samples, and the result is fitted using the Weibull PDF $f(t)$. The component unreliability is represented by the CDF $F(t)$, calculated from $f(t)$, while the system wear-out failure probability is obtained using the series-connected reliability model.

The multifunctional PV inverter unreliability functions for the three mission profiles are shown in Figure 14.25. Case 1 is defined by the conventional operation of the PV inverter, i.e., injecting only active power. On the other hand, Case 2 is defined by the PV inverter compensating the reactive power profile shown in Figure 14.24(e). In both cases, the inverter's apparent power limitation is always respected. In Case 1, the multifunctional inverter installed in IZA presents the lowest lifetime (10 years) due to more energy generated, followed by GOI (22.6 years) and AAL (40.8 years). Regarding Case 2, the lifetime is reduced from 10, 22.6, and 40.8 to 4.4, 6.2, and 12.4 for IZA, GOI, and AAL, respectively. These results indicate that the reactive power compensation degrades the PV inverter's lifetime regardless of the installation place. To reach a pre-defined reliability target, reference [42] investigates partial reactive power compensation in detail.

14.5.2 Control strategies for reliability improvement

As previously discussed, power electronics are configured as the weakest link in the PV system, where capacitors and power devices are the inverter components with

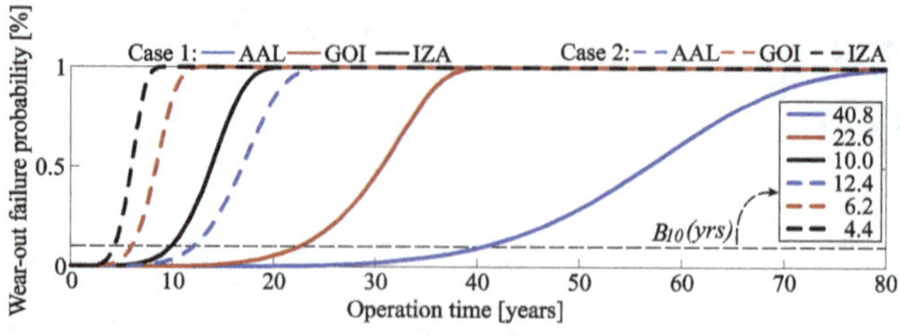

Figure 14.25 Unreliability function for both evaluated cases, considering all three mission profiles

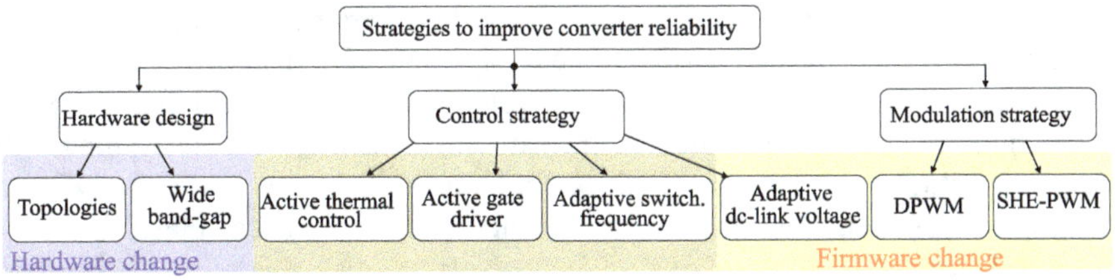

Figure 14.26 Overview of strategies proposed in the literature to improve converter reliability

the highest failure probability. Efforts to improve their reliability are goals of the PV systems industry, which has led to several strategies proposed in the literature. Figure 14.26 summarizes an overview of hardware-based, control-based, and modulation-based strategies to improve inverter reliability, featured in the following:

- **Hardware-based strategies**: the adoption of alternative converter topologies and wide-bandgap devices can considerably improve system reliability. However, large-scale production of PV inverters is undermined due to the considerable extra costs required by these solutions, in addition to the hardware changes;
- **Modulation-based strategies**: the thermal loading of the converter components can be reduced by employing discontinuous pulse width modulation (DPWM) and selective harmonic elimination PWM (SHE-PWM) techniques. Naturally, these solutions do not require inverter hardware changes;
- **Control-based strategies**: active thermal control, active gate-driver control, adaptive switching frequency control, and adaptive DC-link voltage control are the most investigated proposals in the literature for converter reliability improvement. These strategies may or may not involve retrofitting PV inverter hardware, through additional circuit instrumentation and temperature measurement.

The majority of the solutions listed above decrease the switching losses of the converter, leading to reduced thermal stress experienced by power devices and improved inverter reliability. To exemplify again reliability assessment procedures, a control-based strategy is discussed below, framed within adaptive DC-link control techniques of Figure 14.26. This method is called the minimum DC-link voltage control strategy and will be evaluated for a two-stage single-phase grid-connected PV system shown in Figure 14.23(a) [46]. It is worth mentioning that the inverter and boost converter employ power devices rated at 40 A/600 V, manufactured by Infineon, part number IKW40N60H3. Capacitor banks C_{DC} and C_{pv} are formed by 1.17 mF (1×3) and 180 μF (1×1), manufactured by TDK part number B43630. Table 14.6 summarizes all PV system parameters, including device part numbers and heat sink thermal resistance.

The minimum DC-link voltage required for the inverter to remain grid-connect depends on the adopted modulation strategy, apparent power injected, filter parameters (i.e., R_f, R_g, L_f and L_g), and the grid voltage amplitude $\hat{V}_o$. In this approach, the DC-link voltage is represented on the same basis as the converter synthesized voltage amplitude ($\hat{V}_s$), despite the different natures of the amounts (i.e., DC and AC, respectively). The minimum DC-link voltage reference is calculated by:

$$v_{DC}^{*} = k\sqrt{\left(R \cdot i_d - \omega_n \cdot L \cdot i_q + \hat{V}_o\right)^2 + \left(i_q \cdot R + \omega_n \cdot L \cdot i_d\right)^2}, \tag{14.37}$$

where $R = R_g + R_f$ and $L = L_g + L_f$. k is the factor that depends on the modulation strategy adopted. $k = 1$ for the unipolar PWM modulator at the limit of the linear region. i_d and i_q are the in-phase and quadrature components of i_o. $i_q < 0$ and $i_q > 0$ indicate inductive- and capacitive-reactive current injection, respectively. Also, the DC-link voltage reference in (14.37) depends on three signals usually estimated in the converter control algorithm, i.e., $v_{DC}^{*} = f\left(i_d, i_q, \hat{V}_o\right)$. As noted, the difference between the minimum DC-link voltage control strategy and the traditional fixed DC-link voltage technique is related to the DC-link voltage reference generation, since they are implemented with the same hardware configuration, control strategy, and controller gains. Moreover, the maximum power point tracking is not affected, due to the presence of the boost converter.

In view of the aforementioned, the inverter reliability improvement of the minimum DC-link voltage control strategy is evaluated considering the procedure shown in Figure 14.25 and the mission profiles shown in Figures 14.24(b) and (d). The comparative assessment of the inverter reliability is presented in Figure 14.27. As noted in Figure 14.27(a), converter components can be arranged according to the highest failure probability: inv. IGBTs, DC-link capacitors, boost converter capacitors, inv. diodes, and boost semiconductors. The latter is not affected by thermal short cycles, which play an important role in bond wire wear-out prediction. Thus, inverter IGBTs and DC-link capacitors are the most reliability-critical components.

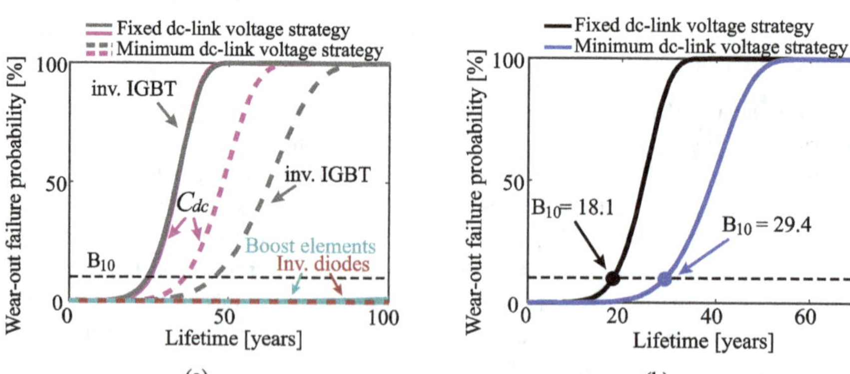

Figure 14.27 (a) Component and (b) system-level unreliability functions, considering both strategies

The system-level unreliability is presented in Figure 14.27(b) with and without the minimum DC-link voltage strategy. Through simple modification in the firmware of PV inverters, a reduction of 62.43% in the system-level wear-out failure probability is observed compared to the fixed DC-link voltage control method. In summary, the system-level B_{10} lifetime increases from 18.1 to 29.4 years, when the minimum DC-link voltage strategy is implemented.

14.6 Closure and future trends

This chapter aimed to provide insights about the reliability of power electronic systems. We introduced the basics and metrics of reliability, new modern approaches such as Physics of Failure and Design for Reliability, and also detailed the step-by-step procedure to calculate the lifetime of power semiconductors and capacitors. With this knowledge in mind, it is possible to find the system reliability, depending on the association of the components in the reliability diagram block.

Nowadays, reliability is such an important topic to be considered in the design and project of new products, as important as their performance, efficiency, power density, costs, and scalability. However, there are a lot of major challenges associated with the task of designing more reliable components, such as the increasing complexity of power electronic systems, harsher operating conditions, cost limitations, inaccuracy and uncertainties on lifetime predictions, lack of information, and so on.

Additionally, there is a major inconvenience when applying accelerated degradation tests to obtain a lifetime model. In practice, the device under test may not be submitted to the same stressors as during the tests. Therefore, it becomes very difficult to accurately predict the failure of a device under real applications. To cover that fragility, several online condition monitoring methods can be applied to predict the real-time degradation of a certain failure mechanism. However, the application of these methods also inserts complexity into the system as one of the main challenges.

Recently, artificial intelligence is a newer tool that can be applied. The main advantages are to develop automatic optimizing processes, improve monitoring methods, and support the Design for Reliability decision-making procedure. In the coming years, this approach will support the development of fast reliability evaluation tools and advanced condition monitoring methods for power converters.

Therefore, it is clear the need for more efforts from academia and industry to work together in multiphysics domains and improve the knowledge about reliability. With that, it is possible to understand better the failure mechanisms, propose new physics-based lifetime models, develop new techniques to manufacture new products with fewer parameters variance, investigate new packaging technologies, improve methods and optimization, etc.

References

[1] Chung HS, Wang H, Blaabjerg F, *et al. Reliability of Power Electronic Converter Systems*. The Institution of Engineering and Technology; 2015.

[2] Zissis G. The R3 Concept: Reliability, Robustness, and Resilience [President's Message]. *IEEE Industry Applications Magazine*. 2019;25(4):5–6.
[3] IEEE. IEEE Recommended Practice for the Design of Reliable Industrial and Commercial Power Systems. Std 493-2007 (Revision of IEEE Std 493-1997). 2007; p. 1–383.
[4] Reliability prediction of electronic equipment, MILHDBK-217A. Washington, DC; 1965.
[5] Lazzaroni M, Cristaldi L, Peretto L, *et al. Reliability Engineering Basic Concepts and Applications in ICT*. Springer; 2011.
[6] Peyghami S, Wang Z, and Blaabjerg F. A Guideline for Reliability Prediction in Power Electronic Converters. *IEEE Transactions on Power Electronics*. 2020; p. 1–1.
[7] Yang S, Bryant A, Mawby P, *et al.* An Industry-Based Survey of Reliability in Power Electronic Converters. *IEEE Transactions on Industry Applications*. 2011;47(3):1441–1451.
[8] Falck J, Felgemacher C, Rojko A, *et al.* Reliability of Power Electronic Systems: An Industry Perspective. *IEEE Industrial Electronics Magazine*. 2018;12(2):24–35.
[9] Pecht M, and Dasgupta A. Physics-of-Failure: An Approach to Reliable Product Development. In: IEEE 1995 International Integrated Reliability Workshop. Final Report; 1995. p. 1–4.
[10] Sintamarean NC, Blaabjerg F, Wang H, *et al.* Reliability Oriented Design Tool For the New Generation of Grid Connected PV-Inverters. *IEEE Transactions on Power Electronics*. 2015;30(5):2635–2644.
[11] Wang H, Liserre M, and Blaabjerg F. Toward Reliable Power Electronics: Challenges, Design Tools, and Opportunities. *IEEE Ind Electron Mag*. 2013.
[12] Abuelnaga A, Narimani M, and Bahman AS. A Review on IGBT Module Failure Modes and Lifetime Testing. *IEEE Access*. 2021;9:9643–9663.
[13] Ciappa M. Selected Failure Mechanisms of Modern Power Modules. *Microelectronics Reliability*. 2002;42(4):653–667.
[14] Schilling O, Schäfer M, Mainka K, *et al.* Power Cycling Testing and FE Modelling Focussed on Al Wire Bond Fatigue in High Power IGBT Modules. *Microelectronics Reliability*. 2012;52(9):2347–2352.
[15] Morozumi A, Yamada K, Miyasaka T, *et al.* Reliability of Power Cycling for IGBT Power Semiconductor Modules. In: Conference Record of the 2001 IEEE Industry Applications Conference. 36th IAS Annual Meeting (Cat. No. 01CH37248). vol. 3; 2001. p. 1912–1918 vol. 3.
[16] Choi U, Blaabjerg F, and Jørgensen S. Power Cycling Test Methods for Reliability Assessment of Power Device Modules in Respect to Temperature Stress. *IEEE Transactions on Power Electronics*. 2018;33(3):2531–2551.
[17] Sommer J, Licht T, Berg H, *et al.* Solder Fatigue at High-Power IGBT Modules. In: 4th International Conference on Integrated Power Systems; 2006. p. 1–6.
[18] Choi U, and Blaabjerg F. Separation of Wear-Out Failure Modes of IGBT Modules in Grid-Connected Inverter Systems. *IEEE Transactions on Power Electronics*. 2018;33(7):6217–6223.

[19] Ma K, Liserre M, Blaabjerg F, *et al.* Thermal Loading and Lifetime Estimation for Power Device Considering Mission Profiles in Wind Power Converter. *IEEE Trans on Power Electron*. 2015;30(2):590–602.

[20] Bayerer R, Herrmann T, Licht T, *et al.* Model for Power Cycling lifetime of IGBT Modules – Various Factors Influencing Lifetime. In: 5th International Conference on Integrated Power Electronics Systems; 2008. p. 1–6.

[21] Scheuermann U, Schmidt R, and Newman P. Power Cycling Testing with Different Load Pulse Durations. In: 7th IET International Conference on Power Electronics, Machines and Drives (PEMD 2014); 2014. p. 1–6.

[22] Dagdelen T, Abdel-Rahman E, and Yavuz M. Reliability Criteria for Thick Bonding Wire. *Materials*. 2018;11(4).

[23] Özkol E, Hartmann S, and Pâques G. Improving the Power Cycling Performance of the Emitter Contact of IGBT Modules: Implementation and Evaluation of Stitch Bond Layouts. *Microelectronics Reliability*. 2014;54 (12):2796–2800.

[24] Wang H, and Blaabjerg F. Reliability of Capacitors for DC-Link Applications in Power Electronic Converters – An Overview. *IEEE Trans on Ind App*. 2014;50(5):3569–3578.

[25] Alwitt R, and Hills R. The Chemistry of Failure of Aluminum Electrolytic Capacitors. *IEEE Transactions on Parts, Materials and Packaging*. 1965;1 (2):28–34.

[26] Ahmad MW, Agarwal N, Kumar PN, *et al.* Low-Frequency Impedance Monitoring and Corresponding Failure Criteria for Aluminum Electrolytic Capacitors. *IEEE Transactions on Industrial Electronics*. 2017;64(7):5657–5666.

[27] Sun Q, Tang Y, Feng J, *et al.* Reliability Assessment of Metallized Film Capacitors using Reduced Degradation Test Sample. *Quality and Reliability Engineering International*. 2013;29(2):259–265.

[28] Wang H, Nielsen DA, and Blaabjerg F. Degradation Testing and Failure Analysis of DC Film Capacitors Under High Humidity Conditions. *Microelectronics Reliability*. 2015;55(9):2007–2011. *Proceedings of the 26th European Symposium on Reliability of Electron Devices, Failure Physics and Analysis*.

[29] Levikari S, Kärkkäinen TJ, Andersson C, *et al.* Acoustic Detection of Cracks and Delamination in Multilayer Ceramic Capacitors. *IEEE Transactions on Industry Applications*. 2019;55(2):1787–1794.

[30] Wang H, Li C, Zhu G, *et al.* Model-Based Design and Optimization of Hybrid DC-Link Capacitor Banks. *IEEE Transactions on Power Electronics*. 2020;35(9):8910–8925.

[31] Yang Y, Ma K, Wang H, *et al.* Instantaneous Thermal Modeling of the DC-Link Capacitor in Photovoltaic Systems. In: 2015 IEEE Applied Power Electronics Conference and Exposition (APEC); 2015. p. 2733–2739.

[32] Menzi D, Heller M, and Kolar JW. iGSE-Cx – A New Normalized Steinmetz Model for Class II Multilayer Ceramic Capacitors. *IEEE Open Journal of Power Electronics*. 2021;2:138–144.

[33] Parler SG. *Deriving Life Multipliers for Electrolytic Capacitors*. vol. 16; 2004. p. 11–12.

[34] Wang H, Ma K, and Blaabjerg F. Design for Reliability of Power Electronic Systems. In: IECON 2012 – 38th Annual Conference on IEEE Industrial Electronics Society; 2012. p. 33–44.

[35] Fatemi A, and Yang L. Cumulative Fatigue Damage and Life Prediction Theories: A Survey of the State of the Art for Homogeneous Materials. *International Journal of Fatigue*. 1998;20(1):9–34.

[36] Zeng G, Herold C, Methfessel T, *et al.* Experimental Investigation of Linear Cumulative Damage Theory with Power Cycling Test. *IEEE Transactions on Power Electronics*. 2019;34(5):4722–4728.

[37] Wang H, Davari P, Wang H, *et al.* Lifetime Estimation of DC-Link Capacitors in Adjustable Speed Drives Under Grid Voltage Unbalances. *IEEE Transactions on Power Electronics*. 2019;34(5):4064–4078.

[38] ASTM. Standard Practices for Cycle Counting in Fatigue Analysis. E 1049-85. 2005; p. 1–10.

[39] Sousa RO, Cupertino AF, Morais LMF, *et al.* Wear-Out Failure Analysis of Modular Multilevel Converter-based STATCOM: The Role of the Modulation Strategy and IGBT Blocking Voltage. *Microelectronics Reliability*. 2022;128:114426.

[40] Reigosa PD, Wang H, Yang Y, *et al.* Prediction of Bond Wire Fatigue of IGBTs in a PV Inverter Under a Long-Term Operation. *IEEE Trans on Power Electron*. 2016;31(10):7171–7182.

[41] Brito EMS, and Cupertino AF. MATLAB Generic Codes for Reliability Analysis. GESEP/UFV; 2023. Available from: https://www.gesep.ufv.br/repositorio/.

[42] da Silveira Brito EM, Cupertino AF, Pereira HA, *et al.* Reliability-based Trade-Off Analysis of Reactive Power Capability in PV Inverters Under Different Sizing Ratio. *Int J Electr Power Energy Syst.* 2022;136:107677.

[43] Gusman LS, Pereira HA, Callegari JMS, *et al.* Design for Reliability of Multifunctional PV Inverters Used in Industrial Power Factor Regulation. *Int J Electr Power & Energy Syst.* 2020;119:105932.

[44] Xavier LS, Cupertino AF, Pereira HA, *et al.* Partial Harmonic Current Compensation for Multifunctional Photovoltaic Inverters. *IEEE Trans Power Electron*. 2019;34(12):11868–11879.

[45] IEEE. IEEE Standard for Interconnection and Interoperability of Distributed Energy Resources with Associated Electric Power Systems Interfaces. Institute of Electrical and Electronics Engineers; 2018.

[46] Callegari JMS, Cupertino AF, Ferreira VdN, *et al.* Minimum DC-Link Voltage Control for Efficiency and Reliability Improvement in PV Inverters. *IEEE Trans Power Electron*. 2021;36(5):5512–5520.

Index

www.ingramcontent.com/pod-product-compliance
Lightning Source LLC
LaVergne TN
LVHW011537200926
848484LV00001B/278